Education Division 2017

Core Programming Area at the 2017 AIChE Annual Meeting

Minneapolis, Minnesota, USA
29 October - 3 November 2017

ISBN: 978-1-5108-5793-3

Printed from e-media with permission by:

Curran Associates, Inc.
57 Morehouse Lane
Red Hook, NY 12571

CURRAN ASSOCIATES INC. proceedings .com

Some format issues inherent in the e-media version may also appear in this print version.

Printed by Curran Associates, Inc. (2018)

AIChE
120 Wall Street, FL 23
New York, NY 10005-4020

Phone: (800) 242-4363
Fax: (203) 775-5177

www.aiche.org

Additional copies of this publication are available from:

Curran Associates, Inc.
57 Morehouse Lane
Red Hook, NY 12571 USA
Phone: 845-758-0400
Fax: 845-758-2633
Email: curran@proceedings.com
Web: www.proceedings.com

TABLE OF CONTENTS

Some abstracts in this book refer to images that were not made available by the publisher

505140 Organizing Biochemical Reactions with Phase Separated Protein Droplets *in vitro* and *in vivo*

Abstract:

Research Interests:

Driven by my interest to understand the engineering principles of biological systems and to apply those principles to develop novel processes in biotechnology, I have taken training in bioengineering, soft matter, cell and molecular biology. My research focuses on the organizational behavior of soft biological matter, particularly how cells use principles of phase transitions to form protein droplets to organize biochemical reactions. With multidisciplinary tools, we showed how intramolecular interactions can tune droplet physical properties to control their function as membrane-free organelles (Zhang et al. 2015, *Molecular Cell*). In my independent research, I aim to dissect the links between droplet assembly, physical properties and functional specificity in cells, which will benefit greatly from my newly gained skills in manipulating protein interactions with optogenetic tools (Zhang, et al. 2017, *Nature Chemical Biology*). In parallel, I will apply that knowledge to design protein droplet-based bioprocesses including bioreactors in vitro and in vivo to increase reaction rates and product yields of heterologous pathways that suffer from flux imbalance because the lack of organizational systems seen in endogenous pathways.

Teaching Interests:

I have gained teaching skills by working as a teaching assistant, participating in teaching workshops and mentoring students in multiple disciplines. With my in-depth training in biochemical engineering, cell and molecular biology, I am ready to teach chemical and biological engineering courses as well as to develop interdisciplinary courses such as soft matter principles in biology. My research program will provide a great platform for training students to integrate multidisciplinary approaches and perspectives to address questions at the interface of biology and engineering.

505463 Engineering Optical Nanomaterials for Biological Sensing and Imaging

Abstract:

Research Interests:

The recent revolution in the development of nanomaterials with unique optical, chemical and material properties shows great promise for engineering sensitive and specific sensors of biological activity at critical temporal and spatial scales. My doctoral and postdoctoral training in micro and nanoscale microfluidics, engineering, and background in physics provide me with a diverse and ideal toolset for designing new infrared fluorescent nanosensors alongside building new microscopy systems for molecular imaging in vivo. My research centers on engineering nanosensors using synthetic DNA molecules, which affords precise control over modular synthesis, chemical diversity, and functionalization of optical nanomaterials, including silver nanoclusters (AgNCs) and semiconducting single walled carbon nanotubes (SWNTs). I show that DNA-functionalized AgNCs can be utilized in the design of fluorescent probes for the sensitive and specific detection of both nucleic acids and the neurotransmitter dopamine. My postdoctoral research has leveraged my knowledge in DNA-based sensing to adapting a DNA-functionalized SWNT sensor for in vivo study of dopamine neuromodulation, critical to our understanding of both brain function and psychiatric disease. Towards this goal, I developed a dual infrared imaging microscope in which two-photon excitation of DNA-SWNT nanosensors is accomplished using a 1600 nm pulsed erbium laser, pushing both excitation and emission into the near-infrared, where dopamine imaging photons are minimally scattered by brain tissue. My imaging microscope enables dopamine nanosensor imaging 2 mm deep into a brain tissue phantom with minimal scattering of light compared to single-photon excitation. My future research will merge nanosensors with this deep-tissue microscopy for non-invasive in vivo imaging of the living brain. In parallel, I will pursue techniques that expand the molecular recognition capabilities of both SWNT- and AgNCs-based fluorescent sensors to detecting other families of biomolecules.

Teaching Interests:

My diverse technical background and teaching experience has prepared me to teach a variety of course topics including mechanics, statistical mechanics, electronics and transport, with a personal emphasis on productively integrating hands-on, discovery-based laboratory curriculum with a foundational theoretical background. I intend to leverage my previous experience as a mentor both in the classroom and in the laboratory to provide the guidance needed for undergraduate and graduate students to grow into independent thinkers and scientists and prepare them to approach new challenges they may encounter in the world. Previous experience working in laboratory settings comprised of diverse technical backgrounds will aid me in growing and managing a research group that can excel at tackling interdisciplinary problems at the intersections of physics, engineering and biology.

497504 Biopolymers Produced By a Thermophile *GeobacillusÂ* sp.Â WSUCF1

Abstract:

Research Interests: Polysaccharides

Teaching Interests: Microbiology, Biochemistry, Bioengineering

Exopolysaccharides (EPSs) are high molecular weight carbohydrate biopolymers that are composed of sugar residues linked together through glycosidic bonds, and secreted by microorganisms into the surrounding environment with certain properties and functions. The EPSs are one of the adaptations developed by various kinds of extremophiles in order to survive in the environments with harsh conditions, such as extreme temperature, salinity, or pH. Extremophiles are considered as producers of EPSs with novel and unusual characteristics which may hardly be found in more traditional polymers. Recently some extremophilic EPSs have already been characterized, and the properties and functions of EPSs from related extremophiles were also partly elucidated. However cumulative knowledge of the structural and functional properties of extremophilic EPSs is still limited. In this research, a thermophilic strain, *Geobacillus* sp. (WSUCF1) which produced unique type of EPSs has been studied. WSUCF1 could utilize glucose as optimum carbon and energy source to produce significant amount of EPSs (382 mg/L). Two purified EPSs were obtained through ion-exchange and gel filtration columns, and both of them demonstrated higher degradation temperature (319.20 and 314.85 °C) than the thermophilic EPSs recently reported. Monosaccharide composition analysis showed that these two EPSs were composed of mannose and glucose with various relative molar ratios (1/0.2141 and 1/0.0270). FTIR and NMR spectrum analysis also showed several unique features of the EPSs. The EPS-based films were developed through a solvent casting technique using glycerol as plasticizer, and their characteristics, including surface morphology, thermal stability, mechanical properties and biocompatibility are currently under investigation. EPSs from the relatively unexplored WSUCF-1 strain will likely provide a valuable resource for exploitation in novel biotechnological processes and opportunities in various industrial fields including bio-based polymeric materials and antitumor-drug carrier. Meanwhile, the research on EPSs from strain WSUCF1 will also expand the scope for microorganisms isolated from hostile life environments.

505562 Cell-Free Biotechnology for a Low-Carbon Future

Abstract:

The potential of cell-free synthetic biology and bioprocessing to change the world for the better is tremendous. Previous efforts in this field to date have explored a variety of highly impactful directions, including elucidation of the biochemical basis of life, the ability to conduct highly sophisticated cell-free protein synthesis, and demonstration of complex synthetic gene circuits, to name just a few. With the modern set of biotechnological tools available, the opportunities for further exploiting this approach are continually expanding—from developing low- or negative-carbon energy and chemical systems, to accelerating agricultural engineering, to mitigating antibiotic resistance. Effective implementation of such cell-free synthetic biology tools and conversion platforms would result in significant global impact; the development of these systems is my central passion.

Leadership:

In addition to this pursuit of the future of cell-free synthetic biology, a second major professional interest is to continue developing myself as a leader. Thanks to an amazing set of experiences I had as an Army Engineer Officer and later as an entrepreneur, I've already been on the leadership development path longer than most faculty candidates. During four years in the Army, one of which was spent leading troops in Iraq, I learned a great deal about motivating others to accomplish our mission, being a highly effective team member, and my own limits, all amid trying conditions. After leaving the military and heading to graduate school, this background led to an interest in a different type of leadership – that required for technology commercialization. This entrepreneurial interest was a sentiment shared by my advisor, Percival Zhang, and over the course of the second half of my Ph.D., we co-founded two companies—Gate Fuels and Cell-Free Bioinnovations. These were fantastic learning experiences, and I remain interested in moving technology from the lab to the marketplace in the future, in addition to the upcoming opportunities I see in applying these skills to teaching and research mentorship.

My ultimate ambition is to help lead the translation of cell-free biotechnology into world-changing technologies. Recently, I had the privilege of serving in a role that I believe sets me up well to be such a pioneer in the field – as a Fellow at the Department of Energy's Advanced Research Projects Agency (ARPA-E). This role provided an incredible opportunity to lead discussions with researchers at the top of their fields around out-of-the-box, high-impact energy technologies of the future, and to carefully consider the optimal role for biology and bioengineering within the field of renewable energy. I also reviewed hundreds of concept papers, interacted extensively with researchers through pre-proposal white papers, and came to understand well the responsibilities of working on the funding decision-making side of science. My efforts at ARPA-E also dramatically bolstered my network – these contacts that have become an integral part of my research ideation, planning, and execution.

Research Interests:

For decades, the roles of proteins in living systems have been studied *in vitro*, forming the basis of our understanding of life at the smallest scales. Modern biotechnology, in harnessing this knowledge, offers the potential to improve the world in many ways, from biorenewable chemicals to agriculture to human health and beyond. Significant hurdles have limited the impact of this research enterprise in some fields however – microbes producing solvents rapidly succumb to the cytotoxic environment they create, agricultural modifications remain bound by the rate of plant growth, and the means available to efficiently debottleneck metabolic pathways are lacking. Cell-free methods—the use of cell lysate or purified enzymes to replicate complicated pathways—allow approaches with the potential to revolutionize these fields, due to their ability to operate in a wide range of conditions, serve as a rapid prototyping platform, and enable effective model-driven pathway optimization.

My previous work at Virginia Tech, and subsequently at Cell-Free Bioinnovations, focused on biological hydrogen production using a purified cell-free pathway. In this project, I combined cellulose and hemicellulose hydrolysis, novel phosphorylation steps, the pentose phosphate pathway, and a soluble hydrogenase to demonstrate the enzymatic co-utilization of C6 and C5 biomass sugars. I then improved the productivity of this pathway dramatically through the use of a kinetic model of the system—experimentally observing a 3-fold rate increase, as predicted, based on adjusting of the ratio of the enzymes in the system.

During my first months at NREL I have started applying a similar approach to a very different problem – biological lignin conversion. Once fully proven out, I anticipate model-driven pathway optimization will be widely useful for a number of metabolic engineering problems. Furthermore, this approach is highly complementary to alternative, lysate-based rapid prototyping cell-free efforts. While those approaches offer high throughput, empirical methods especially useful during discovery efforts, my approach allows rational optimization, dramatically accelerating the improvement of the rate, titer, and yield of the products of these novel pathways.

Future research objectives include the use of cell-free as a conversion platform, for the utilization of toxic substrates and/or production of value-added chemicals and fuels that can only be microbially produced at uneconomically low titers—again, a limitation that can often be mitigated through the use of purified enzymatic systems. Additional directions of interest include prototyping plant pathways, to accelerate agricultural engineering, and therapeutics-relevant cell-free applications, where complex, specific biochemical transformations are catalyzed without the introduction of self-replicating entities. I anticipate a number of continued collaborations with NREL researchers into the future as well.

Teaching Interests:

The elements of my teaching philosophy include encouraging participation and critical thinking and connecting the object of instruction with subjects about which the students are passionate. This style, sometimes called inquiry-based learning, shifts the paradigm from a one-way transfer of information, as occurs in lecture format, to encouraging an educational experience that requires active engagement. Related to this approach, I am excited to explore concepts such as research-integrated curricula as well. This approach is often used in design courses, but has significant potential in a chemical engineering laboratory or unit operations laboratory context as well. For example, exploring the concepts of reaction engineering, separations, and analytic approaches through the production of biodiesel or biogas from a waste stream, fermentative conversion of glucose into a potential nutraceutical, or pretreatment and hydrolysis of local agricultural residue. Non-standard chemical and biological engineering subjects I am interested in building courses around include metabolic engineering, enzyme engineering, cell-free systems, anaerobic digestion, process scale-up, and entrepreneurship.

I firmly believe that the example of my own passion for the subjects of biological and chemical engineering can be contagious – creatively presenting the material can make the classroom not only instructional, but inspirational. I was fortunate to witness such exemplary educational leadership multiple times at Virginia Tech. In addition to the exploration of novel research topics, these experiences included the integration of experiential learning (such as demonstrating methane release following agitation of the bottom of a pond), incorporation of specific industry-relevant examples (such as the use of polysaccharides as quick-clotting agents in modern bandages), and the use of cutting-edge web applications (for example, to assess the homology of proteins from different organisms). These experiences significantly informed my own career trajectory, research interests, and teaching philosophy, an effect I hope to replicate for my students.

At present, my teaching experience includes one semester as a teaching assistant for Virginia Tech's Unit Operations Laboratory and filling in for my advisor for four lectures of the Unit Operations classroom portion. The most rewarding teaching I conducted while a graduate student, however, was the mentorship of two REU students during the summer before I defended my Ph.D. Both contributed significantly to the success of my project, and both became co-authors on the resultant publication. While I consider these experiences a good start, over the next year I aim to significantly strengthen my teaching and undergraduate mentorship experience, by pursuing an opportunity to teach at one of the Colorado research universities associated with NREL, and by recruiting an undergraduate to work with me at NREL via the Science Undergraduate Laboratory Internships (SULI) program.

A unique role I see for myself in the teaching sphere is in my example as a professor who is a veteran and by mentoring fellow veterans. Former members of the military are quite common members of the student population, thanks to support such as the GI Bill – but faculty who have been there themselves are exceptionally rare. Student veterans are incredibly valuable – their passion, dedication, attention to detail, and, very often, ability to set the example and lead their peers, is in my opinion an underutilized resource of human capital. I hope to be lucky enough to encourage some of these veterans to build upon these skills and, if so inclined, contribute to the diversity of the research and development enterprise.

505410 Harnessing Diverse Microorganisms for Biochemical Production Using Carbon Dioxide

Abstract:

Research Interests:

Biochemical production of compounds offers advantages over conventional synthesis due to the high specificity and diversity of biological catalysts, reactions occurring under labile conditions, and the use of renewable substrates. Despite the wide array of organisms, metabolic engineers have focused on only a few as production hosts, requiring complicated modification to match the metabolic potential found in nature. My laboratory will use advances in synthetic biology, omics characterization, systems biology and bioprospecting to evaluate and harness microorganisms with unique potential in biotechnology. We aim to 1) engineer organisms that have desired properties to make specific bioproducts (i.e. alcohols, acids, alkanes, and carotenoids), 2) characterize these hosts to utilize their metabolic potential, and 3) discover alternative pathways, organisms, and communities to expand the range of chemicals to be synthesized and traits to be selected from. As work in my group progresses, we will generate design rules to unlock the full potential of microbiology in bioproduction.

A common theme in my laboratory will be the intersection of carbon dioxide with biology. As human-produced CO_2 levels continue to rise, finding creative ways to use and store CO_2 are imperative to combating climate change. Using advances made in the field of electrofuels, we plant to rapidly evaluate organisms that fix carbon dioxide without using photosynthesis as bioproduction hosts. We will explore whether uncommon CO_2 sequestration mechanisms can function in heterologous hosts as well as look for enhanced mechanisms, organisms, and communities in nature that show promise in taking up CO_2 and making products of interest. Metabolic modeling will be used to predict optimal pathways to be incorporated for energetically feasible CO_2 capture and which molecules will be best produced from these pathways. Protein homology networks will be built to discover and characterize novel carboxylase enzymes for carbon dioxide sequestration. Omics data will be collected and overlayed on existing models to further engineer the hosts and uncover new hypothetical enzymes to evaluate. In addition to using CO_2 as a source of carbon in bioproduction, we plan to use supercritical carbon dioxide ($scCO_2$) as a sustainable solvent to extract medium chain aliphatic compounds as they are made to eliminate toxicity issues. A focus of our laboratory is using systems biology to understand the complex mechanisms that endow these organisms with desired properties, and to develop clever strategies to manipulate them. This research will result in sustainable bioproduction hosts that can economically generate products from CO_2, and will teach students the principles of metabolic engineering, molecular biology, protein engineering and systems biology.

Teaching Interests:

I would like my teaching to reflect my passion for Chemical Engineering and Biology as well as project-based learning. For instance, when teaching a core Chemical Engineering course, such as Kinetics, Separations, or Control Theory, I plan on including current and related research projects as case studies or as projects the students can expand upon. These additions to core Chemical Engineering courses will add tangible examples and current relevance to each class as well as fit within my Teaching Philosophy of students learning by synthesizing their own questions, going through the scientific process, and communicating with peers. Additionally, I would like to teach elective courses such as Bioprocess Engineering, Bioproduct Design, Systems Biology, and Bioenergy due to their current relevance and fit within my research plan. For all of my classes, I plan to use reverse course design to focus content and assessment on students achieving specific learning outcomes. I will use critical reflection to evaluate progress throughout a course and after completion to ensure class goals are being met and to build strong courses year after year. I believe that teaching in higher education goes well beyond the classroom to the mentoring of students and community outreach, and I plan to be active in providing mentorship to students, service to the scientific community through peer review, and connecting my research group with local high school science programs.

505479 Streamlining Chemical Process Design with Process Systems Engineering Methods

Abstract:

Research Interests: Biorefinery, natural gas conversion, process integration, mathematical modeling, optimization

As a relatively young area focused on the understanding and development of systematic procedures for the design and operation of chemical process systems, process systems engineering (PSE) has the potential to improve the decision making processes for the chemical supply chain. This poster discusses my research vision for application of PSE methods in three paradigms.

Renewable Chemicals and Fuels

Biorefinery process development relies on PSE methods to identify primary technology drivers, prioritize research directions, and mitigate technical risk for scale-up. The objective is to synthesis detailed reaction, separation and utility systems; develop process simulation models based on experimental results; perform technoeconomic and life cycle analyses; and ultimately, suggest future research directions. We recently developed patent-pending catalytic process for production of high-value commodity chemicals (α,ω-diols, 1,5-pentanediol (1,5-PDO) and 1,6-hexanediol (1,6-HDO)) from wood and crop waste instead of petroleum feedstocks. We use process modeling, synthesis, analysis to assess economic feasibility as well as prioritize research directions based on our experimental results. We have shown that the renewable plastic precursors, serving as high value co-products of the integrated biorefinery, could lower the cost of ethanol by more than two dollars per gallon. Future work includes performing process optimization to maximize economic potential of integrated process. I envision novel generalized design methods for integrated biorefinery, where the ambiguity of prioritizing different technologies, products, and sustainability metrics is addressed.

Catalytic Natural Gas Conversion

The conversion of natural gas into chemicals and synthetic fuels could significantly upgrades the economic value of the natural gas resource and increase energy security by providing access to the fungible petrochemicals market. Yet, the direct catalytic conversion of methane to desired products such as olefins and higher hydrocarbons, remains a great challenge for the catalysis community, while it is potentially simpler in technology and more economical. We have utilized process modeling to assess the economic feasibility of a generic direct nonoxidative methane conversion strategy. We have developed a simple, flexible approach that allows for the systematic evaluation of various technology alternatives and for the identification of the key technology gaps that must be overcome in order to commercialize this technology. The results of our analysis demonstrate that an economically feasible direct methane conversion process is contingent upon fundamental research advances in the area of catalytic conversion. This generalized approach could be further extended to consider more conversion strategies and ultimately transformed to an accessible user-friendly platform for the catalyst community to assess their technologies. I hope that this study would help accelerate the realization of the methane conversion technologies on industrial scales.

Work and Heat Exchanger Network (WHEN)

We have applied advanced modeling and optimization techniques for energy recovery. Energy recovery via process integration is a long established practice in the chemical industry and scopes exist for improving the overall energy system design performance. Our work identifies several critical synthesis issues of direct practical relevance to process integration problems. Specific focus is given to non-isothermal mixing, variable stream conditions, novel superstructures and mixed-integer nonlinear programming (MINLP) models, and efficient solution methods for simultaneous work and heat exchanger network synthesis (WHENS). While current WHENs works mostly use heuristics and simplification based approaches, our work identifies, formulates and solves several important optimization issues and demonstrates significant improvement in overall cost and solution performance, which provides potentially applications for the design of a chemical process using superstructure optimization. Future research goals focus on synthesis of flexible and operable WHENs under periodical changes in the environment of a plant. Long-term goals focus on improving solution efficiency and quality for this non-convex MINLP problems, and ultimately developing global optimization algorithms.

Teaching Interests:

I am qualified to teach most of chemical engineering core courses, and especially interested in teaching process modeling, synthesis, and design. I had extensive teaching experience as an Instructor at Aspen and a TA at National University of Singapore. I prepared teaching materials, and gave lectures and tutorials. I also look forward to developing new courses related to process simulation and biorefinery.

References

Huang, K., Miller, J.B., Huber, G.W., Dumesic, J.A., Maravelias, C.T., 2017. A Generalized Approach for Evaluation of Direct Nonoxidative Methane Conversion Strategies. *In Preparation.*

Huang, K., Won, W., Barnett, K.J., Brentzel, Z.J., David Martin Alonso, Huber, G.W., Dumesic, J.A., Maravelias, C.T., 2017. Improving Economics of Lignocellulosic Biofuels: An Integrated Strategy for Coproducing 1,5-Pentanediol and Bioethanol. *Under Review.*

Huang, K., Brentzel, Z.J., Barnett, K.J., Huber, G.W., Dumesic, J.A., Maravelias, C.T., 2017. Conversion of Furfural to 1,5-Pentanediol: Process Synthesis and Analysis. *ACS Sustainable Chemistry & Engineering* 5, 4699-4706.

Huang, K., Karimi, I.A., 2016. Work-Heat Exchanger Network Synthesis (WHENS). *Energy* 113, 1006-1017.

Huang, K., Karimi, I.A., 2014. Efficient Algorithm for Simultaneous Synthesis of Heat Exchanger Networks. *Chemical Engineering Science* 105, 53–68.

Huang, K., Karimi, I.A., 2013. Simultaneous Synthesis Approaches for Cost-Effective Heat Exchanger Networks. *Chemical Engineering Science* 98, 231–245.

Huang, K., Almutairi, E., Karimi, I.A., 2012. Heat Exchanger Network Synthesis Using a Stagewise Superstructure with Non-Isothermal Mixing. *Chemical Engineering Science* 73, 30–43.

He, J., **Huang K.**, et al., 2017. New Catalytic Strategies for Alpha-Omega Diol Production from Lignocellulosic Biomass. *Faraday Discussion.*

Brentzel, Z.J., Barnett, K.J., **Huang, K.**, Maravelias, C.T., Huber, G.W., Dumesic, J.A., 2017. Chemicals from Biomass: Combining Ring-Opening Tautomerization and Hydrogenation Reactions to Produce 1,5-Pentanediol from Furfural. *ChemSusChem* 10, 1351-1355.

He, J., Liu, M., **Huang, K.**, Walker, T.W., Maravelias, C.T., Dumesic, J.A., Huber, G.W., 2017. Production of

Levoglucosenone and 5-Hydroxymethylfurfural from Cellulose in Polar Aprotic Solvent-Water Mixtures. *Green Chemistry*.

Kong, L., Avadiappan, V., **Huang, K.**, Maravelias, C.T., 2017. Simultaneous Chemical Process Synthesis and Heat Integration with Unclassified Hot/Cold Process Streams. *Computers & Chemical Engineering* 101, 210-225.

505613 Novel Biosensors for Transformative Healthcare

Abstract:

Research Interests:

Accurate, fast and low-cost detection of biomarkers is essential for effective prescreening of many chronic diseases and cancers. Conventional biosensors such as DNA microarray and real time PCR are golden standards for identification and quantification of nucleic acids. Although very sensitive, these platforms rely on (i) accurate duplication and (ii) fluorescence labeling of target molecules, which are challenging for the case of small RNAs. Ligand bias has been found in high throughput screening of microRNAs. For other types of biomarkers such as proteins, amino acids and peptides, conventional detection methods such as ELISA or Western Blot also rely on fluorescence labeling and do not achieve desirable detection limits and fast throughput. Furthermore, for novel biomarkers such as exosomes, which contain genetic information, no standard quantification methods have been developed. Therefore, novel biosensors that employ new sensing schemes are required especially for biomarkers such as microRNAs, proteins and exosomes.

Nanoplasmonics involves surface electromagnetic waves confined to a nano-scale region (hot spot) sustained by a plasmon resonator, thus allowing sensitive biomolecule sensing without the need for amplification. During my PhD studies, I developed a microRNA-screening platform that is based on a nano-cone array. A low detection limit, down to about 100 microRNAs per pixel, was achieved due to enhanced fluorescence by nanoplasmonic modes localized at the tip of the cones (Wang *et al*, Biomicrofluidics 8 021101, 2014; Wang *et al*, Optics Express 21 6609, 2013; Wang *et al* AIChE J. 59 1830, 2013). This platform is ideal for multiplexing. Detection of multiple targets in a panel can be achieved by addressing each fiber core optically and using photochemistry to selectively attach different probes. Another sensing platform that I co-developed, based on nanocapillaries, achieved similar detection limit. In this platform, plasmonic hotspots are formed by packing gold nanoparticles inside tips of a nanocapillary (Liu *et al*, Biomicrofluidics 7 061102, 2014). Hot spots in between nanoparticles can be dynamically tuned by applying external bias to achieve maximum enhancement factors. This platform overcomes the diffusion limit by electrically driving microRNAs to the detection region and achieves femto-molar concentration detection within 15 minutes.

Although providing improved sensing performance, the above-mentioned nanoplasmonic platforms still rely on labeling of target molecules with high efficiency dyes. This additional step does not only increase cost, but also affects the efficiency and yield, especially for small RNAs, proteins and amino acids. UV (Ultra-Violet) plasmonics is a young research field yet holds promise in enabling label-free biosensing. Although the native fluorescence of biomolecules resides in the UV range of the spectrum, the quantum yield of the native fluorescence is low when compared to the visible fluorescence of dyes. To improve the detection limit, I have been exploring UV plasmonics in order to modify the native fluorescence emission properties of biomolecules. In this regard, during my postdoc training at the University of Utah, I have investigated, both experimentally and numerically, novel UV plasmonic materials and geometries for the purpose of enhancing native fluorescence emission from biomolecules, in particular, amino acids. Experimentally, ultrafast fluorescence spectroscopy is used to study the change of emission rate of native fluorescence of biomolecules. Novel UV materials that I studied include aluminum (Al), magnesium (Mg), gallium (Ga) and their alloys. Geometries studied include nanoapertures, bowtie antennas and bulls eye apertures. Aluminum provides the best performance in terms of fluorescence enhancement, while Mg is a better candidate at shortening the lifetime of biomolecules (Wang *et al*, JPCC 121 11650, 2017). Moreover, Al bowtie antennas have been found to achieve the highest radiative emission rate enhancement, up to 40 times, based on numerical simulations (Lotubai *et al*, SPIE optics and Photonics, 2017). With improved emission rate, more photons are collected from each molecule, thus improving the detection limit.

Besides detection limit, multiplexing is another important functionality for biosensing platforms as multiple biomarkers might be required to identify certain pathogens. Active devices that change their responses corresponding to different biomarkers can be employed for multiplexing. For this purpose I am developing a tunable UV plasmonic device based on the interaction of pi-plasmons in graphene with surface (and localized) plasmon modes in aluminum plasmonic structures. In this regard, I recently demonstrated that by employing graphene with different doping concentrations, the plasmonic resonance of an Al hole-array can be shifted (Wang *et al*, CLEO, 2017). This is the first demonstration of graphene in a tunable UV device and paves the way for other tunable UV plasmonic devices that can be used for active biosensing.

Future research directions that I intend to undertake include:

1. **UV-plasmonics-enabled label-free nucleic-acid/protein sensor.** My research goal is to realize an integrated sensing platform that can achieve real-time detection of biomolecules based on their intrinsic fluorescence with detection limit comparable to conventional fluorescence assays. I will apply my knowledge on UV plasmonic materials and nanostructures both in experiments and simulations to design such platforms. The changes of fluorescent intensity, lifetime or diffusion coefficients upon binding of target molecules with probe molecules can be employed to detect and quantify the number of target molecules. UV plasmonic structures will be used in such platforms to achieve comparable sensing limits as conventional fluorescence assays. This platform will integrate sample pretreatment, electro-kinetic manipulation of molecules and detection of biomolecules in one chip and will be able to detect nucleic acids, proteins, or amino acids.
2. **Exosome identification and quantification for early cancer diagnosis.** Cell-derived extracellular vesicles such as exosomes are of increasing interest for understanding intercellular communication, diagnosis, and curing of cancer. One of the challenges in early cancer diagnosis is being able to detect a small amount of exosomes derived from cancer cells from a vast number of exosomes originating from healthy cells. Therefore, developing a method to rapidly identify and isolate cancer-derived exosomes from healthy exosomes is crucial for early cancer diagnosis. My goal is to develop a label-free exosome flow cytometer that can rapidly identify and sort exosomes based on their surface protein contents, which carry the signatures of their cellular origin. After sorting of exosomes, genetic analysis will be carried out to identify the type and stage of cancer using the microRNA screening platform that I developed during my early work.
3. **Investigation of novel plasmonic devices based on 2D materials in the UV and far-IR ranges.** Spectral signatures of biomolecules such as their native fluorescence lie in the UV range whereas vibrational and rotational signatures lie at the other end of the spectrum, i.e. the far-IR. In spite of this, because of its technological maturity and the easy availability of efficient materials and devices, current research is mainly performed at the visible range rather than at the UV or far-IR. From this perspective, my research interest is to investigate novel materials at these un-conventional frequency ranges and to use these new materials for biomedical applications. With this motivation in mind, I recently began exploring the plasmonic properties of structures composed of 2D materials, e.g. graphene. Graphene has been shown to have good affinity to biomolecules and also very strong UV resonances due to pi-plasmons, thus an optical sensor integrated with graphene might be able to deliver improved sensitivity. One of my current research projects, which I independently initiated while at the University of Utah, involves designing a tunable plasmonic UV biosensor using graphene. Moreover, due to its unconventional band-structure, graphene has been shown by other authors to exhibit strong plasmonic resonances at the mid and far-IR ranges. In general, studying the interaction of 2D materials with biomolecules at these "unconventional" frequency ranges, where both the native spectral features of molecules and the strongest optical response of these materials takes place is an exciting new research direction that I foresee to be highly transformative in the field.

Teaching interests:

I have taught as a teaching assistant for Biomedical Engineering Transport Phenomena' and 'Mathematical Methods for Engineers' and guest lectured for 'Electrokinetics', 'Applied Electromagnetics' and 'Nanophotonics'. I will be interested in teaching core Chemical Engineering courses such as 'Transport phenomena', 'Thermodynamics', 'Mathematical Methods for Engineers', 'Fluid mechanics', etc. and also teach specialized course such as 'Biosensing', 'Nanotechnology', 'Nanophotonics', 'Electrokinetics', 'Micro and Nano fabrication' and so on.

505459 Polymer Based Nano-Sensing Technology Platforms for Healthcare, Environmental Monitoring

Abstract:

Research Interests: Softmaterials, Nanoparticles, Microfluidics, Bio/chemical sensors, Functional sensing materials, diagnostics, medical devices, therapeutics, environmental monitoring

Teaching Interests: Microfluidics, Softmaterials, bio/chemical sensors, Interfacial science, Reaction and transport phenomena

Polymer based nanoparticle (e.g. hydrogel) sensing technologies are emerging as a nextGen screening tools for detecting various (bio)-and chemical analytes in healthcare, environment, food and agriculture sector. By presenting a case study on nucleic acid testing module using polymer nanoparticles, I would demonstrate how an interplay between interface and bulk would imparts the desirable properties and further how an interdisciplinary approach especially insights from chemical engineering would help in fine tuning the systems. Currently, nucleic acid testing (NAT) for diseases ranging from chronic to infectious conditions is exclusively performed in centralized laboratories by expensive instruments and skilled manpower. Despite several developments in the nucleic acid testing modules such as RT-PCR, microarrays, and microfluidic based PCR devices etc., they still suffer with several problems, such as longer time for hybridization/amplification, risk of sequence bias, sample preprocessing, and multiple process steps. Another major limitation is that the frequent addition of reagents externally to progress the reaction. Therefore, I would focus on "design and development of novel nucleic acid detection platforms" for diagnosis and prognosis of chronic to infectious conditions using hydrogel based polymeric materials. In my poster, I would be presenting the development of novel dual chambered particles/pads in which first half is designed to have functionalized probe molecules within the hydrogel network and another half of the polymer system provides fluid like environment to accommodate free floating target molecules, linkers, and enzymes. Further, the interface of these two chambers are separated by a stimuli responsive porous polymer layer which allows diffusion of reagents upon stimulation for subsequent reaction in a graphene-hydrogel transparent composite microwells having optical detection features. As a future faculty member, I devote on working with fluids, interfaces, and materials with broad focus on solving grave challenges in the domain of healthcare and environment. Specifically, my focus would be on fundamental understanding of reaction-transport phenomena in/on sensing surfaces, interfacial mechanisms of bio-surfaces, along with processing of soft and nano-electronic materials leading to the development of technology platforms for medical diagnostics, biological threat detection, forensics, wearables, and healthcare along with environmental monitoring. In this area, I would like to study the fundamental and applied aspects for preparing functional sensing surfaces, transport of analytes, specific recognition reaction schemes, interfacial and biological aspects, including charge transport in electronic materials along with stimuli responsive mechanisms. By utilizing the knowledge generated through fundamental aspects, I would focus on building "Field deployable nextGen technology platforms" for solving challenges in the aforementioned areas. I worked with interdisciplinary teams in various schools and industries including semiconductor Fab, and had extensive background in establishing the microfluidic sensors, polymer based research laboratories, clean rooms, and

training on range of state-of-the-art micro/nano-fabrication and characterization tools/techniques. In addition, the capabilities of establishing international collaborative projects, grant writing, and project leadership qualities along with teaching and mentoring-puts me in a unique position to pursue a career in academia for carrying out cutting edge research to solve societal challenges with technological interventions.

504524 Engineering Ligands to Control Protein Conformational Changes

Abstract:

Research Interests:

Induced conformational changes within proteins are an integral part of signal cascades and proper cellular function; however, mutations to certain cell receptors and other functional proteins that cause irregular conformations can result in a disease state. To improve our ability to understand the mechanisms by which conformational changes influence the healthy and disease state of cells, I propose a balanced approach that utilizes engineered binding proteins, high structural resolution NMR experiments, and molecular dynamic simulations. These complementary techniques provide an efficient route for the discovery of technologically useful binding ligands as well as a deeper fundamental understanding of free energy landscapes and protein dynamics.

This research vision balances laboratory experiments and computational simulations and uniquely leverages my high-throughput protein engineering expertise (PhD in ChemE, University of Minnesota, Advisor: Ben Hackel) with the structural and biophysical techniques secured during postdoctoral training with Professor Dorothee Kern (Howard Hughes Medical Institute investigator) at Brandeis University.

An essential feature of this work will include eukaryotic surface expression of small protein libraries for the discovery of binding interactions to mammalian cell-bound target receptors. Conformational changes upon binding will be detected using GFP fused signal cascade products. Positive hits will be isolated using flow cytometry, thus enabling high-throughput selection. Deep sequencing of binding populations will elucidate the most effective binding ligand attributes necessary for modulating receptor conformation. Dynamics of lead ligands will be measured by NMR spectroscopy as well as rendered using molecular dynamic simulations to provide extensive free energy landscape information for the target receptor. Importantly, the experimentally derived results will act as a feedback loop to guide the molecular dynamic system configurations for improved prediction of conformations.

This work demonstrates an innovative approach to couple high-throughput protein discovery and evolution to protein function rather than merely binding. Merging this innovation with experimental and computational structural studies allows a uniquely broad and deep perspective â€" and level of control â€" on protein conformational impacts.

Teaching Interests:

I am eager to work with undergraduate and graduate students as well as community outreach programs to promote a better understanding of science and technology. My previous teaching experiences include tutoring general and organic chemistry and teaching assistant roles for chemical engineering unit operations senior lab, diffusion and transport phenomena, and thermodynamics. In the future, I am particularly interested in leading courses that focus on biomolecular engineering and statistical mechanics.

499774 Exploiting Organization in Bacteria for Synthetic Biology

Abstract:

Research Interests:

Bacteria were once thought to be simple bags of enzymes. Today, we understand that bacteria have levels of organization that are more complex than previously thought. My research interest is in exploiting bacterial organization for useful synthetic biology applications. This includes exploring how spatial organization, temporal organization, and structural organization enhance the efficiency of cellular processes. Bacterial microcompartments are an excellent example of how bacteria utilize spatial and temporal organization. Approximately 200 nm in diameter, these structures are composed of thousands of subunits of hexameric shell proteins that encapsulate the necessary enzymes to carry out specific metabolic pathways. These pathways often produce toxic intermediates which are sequestered within the compartment, thereby protecting the rest of the cell. Using a flow-cytometry based assay for quantifying the amount of protein encapsulation, I have elucidated the design rules for an N-terminal encapsulation sequence, and have shown that a variety of heterologous proteins can be encapsulated at titratable quantities within bacterial microcompartments using variations of a short amino acid tag.

Bacterial endospores provide an example of complex structural organization. In response to nutrient starvation conditions, certain bacteria such as Bacillus subtilis can elaborate a layered protein coat comprising of approximately 80 different proteins, protecting the dormant endospore from chemical attacks, dessication, and radiation. I have worked to understand how the very first coat protein, SpoVM, localizes specifically to the developing forespore membrane within the mother cell by recognizing its convex membrane curvature. Using solid-supported lipid bilayers on silica microbeads, we reconstituted bacterial membranes in vitro and characterized the binding of SpoVM to membranes with different radii of curvature. We found that the mechanism behind the preferential binding to membranes with high degrees of curvature results from a higher kinetic adsorption rate coupled with binding cooperativity, and that this property is critically dependent on a structural kink in the protein structure that delineates the 26-amino acid peptide into an unstructured flexible N-terminus and an amphipathic alpha helical C-terminus.

Looking ahead, the groundwork I established in elucidating design principles and mechanistic understanding of bacterial organizational systems will enable my future lab to take these natural systems and exploit them for synthetic biology applications. I propose to utilize the organization of microcompartment to create customized microenvironments within the lumen that allows for synthetic pathways that are generally incompatible within the bulk cytosol. These self-contained "nanoreactors" within the bacterium can be tailored to be, for example, anaerobic, and can greatly expand the repertoire of pathways capable of being expressed in bacteria. I also propose to exploit the structural organization of bacterial endospores to create small molecule delivery systems that utilize the structural and chemical stability of the endospore coat proteins. Such a system can shield the cargo within from chemical and enzymatic attacks and increase its half-life in vivo, and also provides a platform for decorating the particle surface with additional proteins or biomolecules. Together, these proposed works to exploit the organization in bacteria will enhance the suite of biological tools available to synthetic biologists.

Teaching Interests:

My dedication to education is a major component in my decision to pursue a career in academia. I have served as a graduate student instructor for two chemical engineering courses: transport and separations, a core class taken by all undergraduates in the chemical engineering program, and biochemical engineering, an elective taken by chemical engineering majors that seek a concentration in biotechnology. In these courses, I used various strategies to actively engage students in learning, which included learning from peers, and supplementing lectures with activities designed to engage students with different learning styles. I have found great success with these strategies, as measured by positive course evaluation comments and ratings, as well as two teaching awards – the Dow Excellence in Teaching Award, awarded with a cash prize, and the campus-wide Outstanding Graduate Student Instructor award. As an instructor, I plan to keep up to date with pedagogical methods, and adapt my teaching style to fit the needs of my students, especially as new methods of learning are continually emerging.

Recent advances in *-omics* technologies allow us to rapidly collect large amounts of diverse information about biological systems. A large class of statistical models in computational biology and bioinformatics that are rapidly being developed in reponse to this data avalanche, are inherently descriptive in nature. These models need to be complemented with a mechanistic understanding of biological processes through integration of the data with physical and chemical constraints that govern biological systems.

This task demands comprehensive knowledge of biophysical phenomena from the nano- to the macro-scale. In my research, I have learnt multiple quantitative skills ranging from molecular dynamics simulation to computational population genetics. I have extensive experience in analyzing *-omics* data such as metabolomics, phosphoproteomics, and genomics. My research background place me in an ideal position to attack interesting systems-level biological problems using tools borrowed from multiple fields. My future research in computational biology will pursue two distinct directions: **(I)** Quantifying evolutionary forces in bacteria and **(II)** Gaining mechanistic insights into cell population heterogeneity in mammalian growth factor (GF) signaling networks.

Aim I: Quantifying evolutionary forces in bacteria. Bacteria play a significant role in human health as well as biogeochemical cycles. Notably, multiple evolutionary forces such as point mutations, recombination (swapping of DNA in the population), and dynamical changes in ecological niches simultaneously shape *intra species* genetic diversity in bacteria[1,2]. As a result, bacterial strains are capable of acquiring new phenotypes within months. For example, resistance to antibiotics can rapidly spread in bacterial populations through recombination of drug resistant genes.

I am interested in quantifying key evolutionary forces that dictate intra species genetic diversity in bacteria in order to better understand how environmental conditions, such as host diet, and natural selection dictate the innovation and spread of phenotypes in natural populations. There are several challenges in quantifying bacterial evolution. First, bacterial genomes are extremely variable; comprising a species-consensus 'core' genome and a variable 'auxiliary' genome[1,2]. Current practices identify the core and the auxiliary genomes on a gene-by-gene basis and are error prone as well as ultrasensitive to small perturbations in the genomes because of inaccuracies in gene annotation algorithms. Second, detailed simulations of interactions between evolutionary forces in realistic bacterial populations are numerically infeasible because of their large size ($N_e > 10^9$). To address these challenges, we have developed[1,2]: **(1)** A novel approach to identify the consensus and the variable genomes from sequence alignment. Our consensus 'basic' genome operates at the base pair level and is free from errors in gene annotation and is robust to genomic perturbations such as insertions and deletions. Moreover, the basic genome preserves the order of genes on the chromosome.

(2) A computational framework based on the coalescent theory that allows us to simulate small subpopulations within a larger bacterial population thus supplanting simulations of large populations.

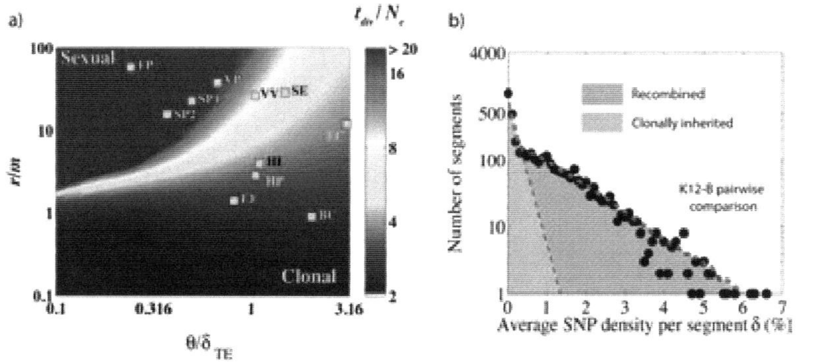

FIG. 1: **a)** Classification of bacterial species as clonal and sexual. Species names are as follows: FP: *F. psychrophilum*, VP: *V. parahaemolyticus*, SE: *S. enterica*, VV: *V. vulnificus*, SP1: *S. pneumoniae*, SP2: *S. pyogenes*, HP: *H. parasuis*, HI: *H. influenzae*, BC: *B. cereus*, EF: *E. faecium*, and EC: *E. coli*. r/m is the relative importance of recombination and mutation in generating SNPs and θ is the population diversity. **b)** A histogram of SNPs in small genomic segments quickly distinguishes clonally inherited and recombined segments in *E. coli* strains K12 and B.

In a first application of the framework[1,2], we addressed a long-standing question in bacterial evolution: can bacteria maintain a clonal species structure in the presence of recombination? We showed that depending on θ, the genetic diversity in the population, and r/m, the relative rate of recombination to point mutations, the competition between clonality and recombination results in two possible 'phases' in the evolutionary dynamics that corresponded to clonal and sexual bacteria (see **Fig. 2a**)[4]. Importantly, our analysis directly led to the development of a suite of bioinformatic metrics to quantify the relative strength of mutations and recombinations. Using these metrics, in a population genomics study of *E. coli*, we identified clonally inherited and recombined regions in wild *E. coli* strains (see **Fig. 2b**)[2]. Notably, our analysis showed that the contribution of recombination to *E. coli* genetic diversity was an order of magnitude higher than point mutations.

In my faculty tenure, I plan to further address fundamental questions about the interaction between the forces of adaptation, recombination, and ecological niche separation in dictating bacterial genetic diversity. An important problem is the rapid spread of antibiotic resistance. Due to recombination, strains can acquire antibiotic resistance genes without inventing them *de novo*. At the same time, isolation due to ecological and geographical niches suppresses genetic exchange between niches. It is thus crucial to quantify the rate of recombination required to overcome a given degree of isolation. Another example is the competition between recombination and adaptive mutations. While adaptive mutations allow strains to spread rapidly in a population, rampant recombination of these mutations will nullify the seletive benefit of advantageous mutations. At the same time, low rates of recombination result in lowered population fitness as well (a phenomena known as the

Muller's ratchet). While we suspect the existence of an optimal recombination rate as a balance between these two extremes, rigorous analysis of the competition is lacking. To address such questions, I will generalize our framework to incorporate these key evolutionary forces. In the generalized framework, the distribution of coalescent times among recombining regions will be extended to non-exponential distributions representing, for example, adaptation and ecological isolation. I will study how relative strengths of these forces affect genetic diversity by simulating the evolution of divergence in *in silico* bacterial populations via point mutations and recombination across a range of evolutionary parameters such as the distribution of fitness effects, number of ecologically separate niches, total population diversity, and rates of mutation and recombination.

This work will have broad applications in understanding bacterial evolution, for example, in the study of evolutionary dynamics of host-associated microbial communities, determinants of pathogen-host interactions, and stability of crucial phenotypes in oceanic photosynthetic bacterial populations.

Aim II: Mechanistic insights into cell population heterogeneity. Growth factors (GF) are secreted molecules that control essential cell functions in mammals. Mutations in GF stimulated signaling cascades are implicated in a variety of diseases including many cancers and diabetes. A key challenge in the mechanistic understanding of GF stimulated signaling is cell population heterogeneity of signaling dynamics. This heterogeneity has important functional consequences, for example, in emergence of chemotherapy resistant subpopulations in tumors and in cell fate decisions in the developmental embryo. I am interested in identifying biochemical network parameters that dictate heterogeneity in GF signaling and the downstream cellular phenotypes.

Recent advances in single cell measurements allow us to measure the population heterogeneity in GF signaling network components. I have developed a computational framework based on the maximum entropy (ME) principle[1,2] to infer distribution of network parameter that is consistent with experimental data. ME is an inference tool first introduced in statistical physics and is now widely used in all areas of science. Intuitively, among all possible parameter distributions that agree with imposed experimental constraints, ME predicts the distribution with the least amount of bias. Using the framework, I have studied the heterogeneity in the epidermal growth factor (EGF)/Protein kinase B (Akt) pathway.

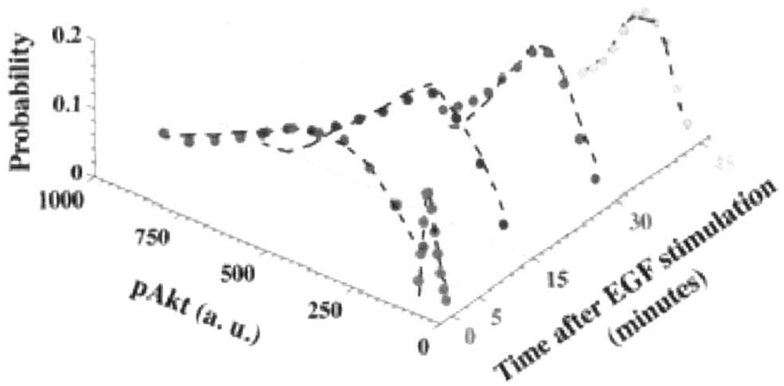

FIG. 2: Time evolution of the measured distribution of pAkt levels upon EGF stimulation (filled circles) and the corresponding predictions (dashed lines).

Aberrations in the EGF/Akt pathway are implicated in many cancers. In a combined computational and experimental study with our collaborators in Peter Sorger's lab at Harvard, we measured heterogeneity in Akt phosphorylation levels (pAkt) upon stimulation with multiple EGF doses (see **Fig. 1**). From the inferred joint parameter distribution, we could predict how key features of Akt signaling, for example, steady state pAkt and surface EGFR levels, varied from cell-to-cell.

I want to understand the mechanistic origins of drug resistance in the EGF/Akt pathway. Chemotherapy drugs such as gefitinib and lapatinib bind to EGFRs and downregulate its kinase activity. However, patients that benefit from EGFR inhibitors eventually develop drug resistance. Indeed, while a majority of 'sensitive' cells in a cultured cell population deactivate the EGF/EGFR pathway upon drug action, a small fraction of 'resistant' cells show sustained Akt phosphorylation in the presence of the drug. In order to rationally design drug regimen that target all subpopulations in a tumor, we need to understand the mechanisms behind the observed heterogeneity. Towards that end, single cell phosphoproteomics data can first be collected across multiple doses of EGF and drugs for multiple cancer cell lines. Next, using the ME framework, the joint distribution over parameters can be inferred. The inferred distribution allows us to identify *in silico* the cells with phenotypes of interest, for example, 'resistant' cells that exhibit high levels of pAkt in the presence of EGFR inhibitors. Next, we can computationally identify biochemical parameters (intracellular protein abundances and reaction rates) that are different between the 'resistant' subpopulation from the 'sensitive' subpopulation. These specific predictions can be experimentally verified. For example, if the ME framework predicts that concentration of PI3K correlates with high pAkt levels in resistant cells, we can simultaneously measure PI3K and pAkt levels in single cells to examine their correlated nature.

In the future, I plan to develop computational tools within the ME framework to analyze heterogeneity in pathways activated by other growth factors, for example, the hepatocyte growth factor cascade responsible in organ regeneration and wound healing as well as the insulin-like growth factor cascade implicated in diabetes and various cancers.

Reference: [1] **P. D. Dixit**, et al., Recombination-driven genome evolution and stability of bacterial species Genetics, 2017. [2] **P. D. Dixit** et al., Recombinant transfer in the basic genome of *Escherichia coli* PNAS, 2015. [3] **P. D. Dixit**, et al., Maximum entropy framework for inference of parameter distributions and population heterogeneity in signaling networks (*in review*, doi: 10.1101/137513).

505393 Micro-Scale Transport Processes Enables Accelerated Biochemistry, Chaotic Mixing and Inexpensive Mobile Diagnostics

Abstract:

Micro-Scale Transport Processes Enables Accelerated Biochemistry, Chaotic Mixing and Inexpensive Mobile Diagnostics

Aashish Priye, Sandia National Laboratories, Livermore, CA

Abstract Text:

Research Interests:

Recent outbreaks like Zika and Ebola highlights the challenges associated with pathogen diagnostics in the developing world. With the outbreak in Africa, and isolated cases on other continents, the need for an affordable, rapid and portable diagnostic solution has been repeatedly stressed and is one of the most critical issues confronting global health. Unfortunately, the current conventional PCR instrumentation needed to perform "gold standard" DNA-based diagnostic tests is bulky, slow, and expensive, making it unsuitable for resource limited settings in developing countries where dedicated laboratory facilities are not available. Advances in micro-fluidics, Nano-science and mobile technology have paved way for novel implementations of traditional molecular diagnostic techniques, which are extremely portable, inexpensive and user friendly.

For example, it has been recently been shown how microscale convective flow fields can be harnessed to perform the polymerase chain reaction (PCR), a very important tool in molecular biology which enables a sequence of DNA/RNA molecule to be replicated billions of time. My research led me to investigate the underlying physics of these thermally driven micro-scale Rayleigh-Bénard convective flows. This new understanding revealed how a subset of these flow fields exhibit chaotic advection that greatly accelerate temperature driven biochemical reactions such as PCR [1-4]. Having found the optimal design parameters for convective PCR, I coupled this novel isothermal setup with a versatile smartphone based detection unit and an integrated image analysis app [5-7]. This greatly simplifies the design, enabling us to integrate such biochemical analysis platforms with consumer-class quad-copter drones for rapid deployment of nucleic acid based diagnostics [8]. The ability to perform rapid in-flight assays with smartphone connectivity eliminates delays between sample collection and analysis so that test results can be delivered in minutes, suggesting new possibilities for drone-based systems to function in broader and more sophisticated roles beyond cargo transport and imaging.

The surprising interplay between reactions and micro-scale convective flows led me to consider adaptations beyond PCR. Specifically, we demonstrate that such flows, naturally established over a broad range of hydrothermally relevant pore sizes, function as highly efficient conveyors to continually shuttle molecular precursors from the bulk fluid to targeted locations on the solid boundaries, enabling greatly accelerated chemical synthesis. Insights from this study has the potential to provide a breakthrough in our understanding of the fundamental biochemical processes underlying the origin of life [9,10]. Another aspect of my research involves the hydrodynamical interaction between micron sized particles in intricate flow fields. My computational fluid dynamics models have thrown light on how the micro-fluidic flow fields can be used to tune fluid-particle interactions using secondary flow features such as curvature induced Dean flows [11,12].

At Sandia National laboratories, I have continued the development novel molecular diagnotics techniques for point of care applications, which involves an innovative approach to pathogen detection via reverse transcription loop-mediated isothermal amplification (RT-LAMP). A new kinetic model for LAMP reactions is formulated in order to shine light on the underlying reaction pathways. The new insights gained from the model are then used for efficient primer design and determining optimal reaction parameters. The LAMP technique is then coupled with an extremely simple and user-friendly smartphone enabled detection platform capable of detecting bright, unambiguous fluorescence signals produced [13]. I developed a microfluidic platform that is capable of integrating to the built-in CMOS sensors in the smartphone camera to analyze viral pathogens in bodily fluids such as blood, saliva or urine. LAMP reactions were modified to generate multiple bright fluorescent signals based on the target sample which is then mapped on a 3D chromaticity diagram to detect the presence/absence of multiple pathogens. I have also developed next generation bio surveillance platforms. These are rugged yet inexpensive robots that perform autonomous field surveillance of arboviruses by analyzing saliva from vector-borne mosquitoes with daily LAMP assays. The ability to deliver performance comparable to that of current generation systems while simultaneously providing an order of magnitude reduction in cost and turnaround time has the potential to greatly expand the use of nucleic acid based detection assays by moving them out of the laboratory and into settings where they are needed most.

Teaching Interests:

One of the most fascinating subjects I have ever come across would have to be fluid mechanics and transport processes as it is conceptually quite rich and at the same time hosts a tremendous opportunity to incorporate hands on learning experience. There is a need for innovative educational experiences that unify and reinforce fundamental principles at the interface between the physical, chemical, and life sciences. These experiences empower and excite students by helping them recognize how interdisciplinary knowledge can be applied to develop new products and technologies that benefit society. Microfluidics offers an incredibly versatile tool to address this need. In one of my educational paper [1,14], we describe our efforts to create innovative hands-on activities that introduce chemical engineering students to molecular biology by challenging them to harness microscale natural convection phenomena to perform DNA replication via the polymerase chain reaction (PCR). Experimentally, we have constructed convective PCR stations incorporating a simple design for loading and mounting cylindrical microfluidic reactors between independently controlled thermal plates. A portable motion analysis microscope enables flow patterns inside the convective reactors to be directly visualized using fluorescent bead tracers. We have also developed a hands-on computational fluid dynamics (CFD) exercise based on modeling microscale thermal convection to identify optimal geometries for DNA replication. A cognitive assessment reveals that these activities strongly impact student learning in a positive way.

1. Priye, A., Y.A. Hassan, and V.M. Ugaz, *Education: DNA replication using microscale natural convection.* Lab on a chip, 2012.

2. Priye, A., Y.A. Hassan, and V.M. Ugaz, *Microscale chaotic advection enables robust convective DNA replication.* Analytical chemistry, 2013. **85**(21): p. 10536-10541.

3. Priye, A., et al., *Selecting 3D Chaotic flow states for Accelerated DNA replication in micro-scale convective PCR.* 15th International Conference on Miniaturized Systems for Chemistry and Life Sciences, 2011. **15**.

4. Priye, A. and V.M. Ugaz, *Convective PCR Thermocycling with Smartphone-Based Detection: A Versatile Platform for Rapid, Inexpensive, and Robust Mobile Diagnostics*, in *Microfluidic Methods for Molecular Biology*. 2016, Springer International Publishing. p. 55-69.

5. Jensen, M., et al., *ANALYTICAL INSTRUMENT SYSTEMS*. 2016, US Patent 20,160,025,630.

6. Priye, A. and V. Ugaz, *PCR to Go*. 2014, Apple iOS appstore https://itunes.apple.com/us/app/pcr-to-go/id909227041?mt=8: Apple iOS appstore.

7. Priye, A. and V. Ugaz, *DNA-to-go: A portable smartphone-enabled PCR assay platform.* arXiv preprint arXiv:1606.02252, 2016.

8. Priye, A., et al., *Lab-on-a-drone: toward pinpoint deployment of smartphone-enabled nucleic acid-based diagnostics for mobile health care.* Analytical chemistry, 2016. **88**(9): p. 4651-4660.

9. Priye, A., et al. "*Synchronized chaotic targeting and acceleration of surface chemistry in prebiotic hydrothermal microenvironments.*" Proceedings of the National Academy of Sciences 114.6 (2017): 1275-1280.

10. Priye, A., Y.A. Hassan, and V.M. Ugaz, *Thermally-targeted adsorption and enrichment in microscale hydrothermal pore environments.* 17th International Conference on Miniaturized Systems for Chemistry and Life Sciences, 2013. **17**.

11. Choi B. H., Huang J. H., Priye A., Presley B., Jayaraman A. and Ugaz V. M., "A Membraneless High-Throughput Micro-Separator" 20th International Conference on Miniaturized Systems for Chemistry and Life Sciences (MicroTAS), Dublin, Ireland, October 9, 2016.

12. Priye, A. and W.H. Marlow, *Computations of Lifshitz–van der Waals interaction energies between irregular particles and surfaces at all separations for resuspension modelling.* Journal of Physics D: Applied Physics, 2013. **46**(42): p. 425306.

13. Priye A., Bird S., Light Y. K., Ball C. S., Negrete O. and Meagher, R. J., "A Smartphone Based Diagnostic Platform for Rapid Detection of Zika, Chikungunya, and Dengue Viruses" Scientific Reports (Nature Publishing Group) (2017): 7.

14. Ugaz, V.M., A. Priye, and Y.A. Hassan, *DNA to Go: A Do-it-Yourself PCR Thermocycler Lab.* American Society for Engineering Education, 2012.

505480 Complex Fluids in Complex Small Scale Geometries

Abstract:

Research Interests:

Scientific Vision

The main thrust of my research is to understand and utilize small scale flow of complex fluids in microfluidic platforms, specially as they arise at the interface of biomedical engineering. Biological complex fluids often contain clinically important biomarkers such as circulating tumor cells, bacteria, tumor derived DNA and exosomes. Microfluidic devices are multiphysics, multiplex systems and comprise various modules and components for detection and measurement of biomarkers. On one hand, microfluidics enables sensitive detection and accurate quantitative measurement of biomarkers dispersed in biological complex fluids, which is appealing from engineering perspective. Further, microfluidics provides tools to gain deeper fundamental insight into properties of complex fluids as carriers of biomarkers. By utilizing video microscopy and image analysis, microfluidics, rheology and computational methods, my engineering strategy is based upon a bottom-up design methodology: I develop innovative microfluidic platforms for bioengineering applications by gaining deeper fundamental insight into properties of complex fluids in the smaller modules comprising the system.

Previous Research Accomplishments

My background and training in fluid dynamics and rheology of complex fluids, computational methods and biomedical microfluidics make me ideally positioned to pioneer research efforts in micro scale flow of complex fluids. After conducting research on rheology of gelation process during my Master of Science studies, I joined Professor Jeffrey Morris's complex fluid research group at City College of New York. For my doctoral thesis, I focused on fluid mechanics and transport properties of inertial suspensions using computational methods. We contributed to the field of complex fluid rheology by providing in depth details of the relationship between relative particle motion, microstructure and rheological properties of inertial suspensions. We also embarked upon microfluidic flow problems and successfully provided insight into the effect of bulk inertia of the suspension on mass exchange with microfluidic vortices.

My postdoctoral research in Professor Dino Di Carlo's lab at University of California Los Angeles allowed me to add the microfluidics to my research toolbox. Most importantly, I gained valuable insight into clinical applications of miniaturized flow platforms by conducting research in four main directions: In the first project, I developed efficient microfluidic devices for separation of Circulating Tumor Cells (CTC) from a blood draw by utilizing hydrodynamic force on cancer cells. For the second project, we used microfluidics to self-assemble mammalian cells into ordered trains, followed by encapsulation in droplets. Our efforts in integration of single-phase and multi-phase microfluidics enabled us to study the instability of jets in microfluidic flow focusing geometries in great detail, which further deepened my knowledge of multiphase flow. For the forth project, I combined advanced computational geometry methods, image processing and image analysis with the precision provided by microfluidic droplet generators to gain comprehensive

engineering control of tissue generation scaffolds.

Future Research

The central focus of my lab will be on studying the properties and potential applications of complex fluids in engineering applications. In the next 3-5 years, I will focus efforts on the following research items:

1- Microfluidic measurement of cell viscoelastic properties

2- Ordered encapsulation of cells in hydrogel micro-particles

3- Extracorporeal microfluidic for detection and measurement of biomarkers

Teaching Interests:

I believe the purpose of academic education should be training critical thinkers. Solving new problems necessitates the skill of discerning relevant from irrelevant to find the best path to the solution. For being a critical thinker, students need to focus on developing a "toolbox" that includes problem solving, organization, communication, professional integrity and collaboration. This broadens student's thoughts and helps them avoid mere memorization. The main goal throughout my courses is to help students develop the skill of understanding and reducing the problems down to base principles. For graduate level courses in particular, I will pursue this goal by directing students to work on open-ended questions that reflect more advanced understand- ing of fundamentals.

During my academic career as a graduate student and a postdoctoral research scholar, I gained valuable teaching experience and I look forward to contribute to student education through developing and teaching interesting courses. I am interested and prepared to provide instruction for courses on the following topics, where I would develop a syllabus that includes integration of computational tools. *Undergraduate Level*: Transport Phenomena, Applied Computational Methods in Science and Engineering, Physics and Applications of Microfluidics and Kinetics of Chemical Reactions; *Graduate Level*: Advanced Fluid Dynamics, Rheology of Complex Fluids, Applied Microfluidics, Mesoscale Simulation Methods and Applications

505629 Design and Development of Ocular Disease Diagnostic System, and Point-of-Care Microsystem

Abstract:

Research Interests:

My research interests are below

1) Diagnosis of ocular disease by analyzing biomarkers from tear fluids

2) Diagnosis of ocular infections by influenza A virus

3) Development of point-of-care (POC) system for self-diagnosis of ocular diseases

Ocular disease influence human vision and their life quality. In the worldwide, there are 285 million people who have visual impairment, and the number of the people increases approximately 7 million in every year. Although diverse ophthalmic pharmaceuticals have been developed, people are still suffering from diverse ocular disease due to the late diagnosis of the diseases. Thus, development of early diagnostic system which can detect the ocular diseases with high speed, sensitivity, and cost effective manner are critical issues. In addition, development of point-of-care (POC) system for ocular disease detection at early stage is important to demonstrate self-diagnostic system at home or office.

1) Diagnosis of ocular disease by analyzing biomarkers from tear fluids

As the advance of the genetic and proteomic assay, tear fluid has been exploited as a source of biomarker analysis to diagnose the diseases. Tear fluid is including various biomarkers such as proteins, lipids, and oligonucleotides. Furthermore, tear is less complex body fluid than serum or plasma, and can be obtained without invasive method. Recently, proteomic research by analyzing protein biomarkers in tear fluid have been widely studied to diagnose ocular disease such as glaucoma, dry eye diseases. Since eye is a part of central nervous system, neurodegenerative diseases (i.e. Alzheimer's and Parkinson's disease) also can be identified using tear fluids. In addition, tear fluid includes high concentration of microRNA (miRNA), which have been intensely studied to identify their roles in the body, and many functions of miRNAs are not covered yet. Since miRNA can provide early, non-invasive diagnostic tools, tear fluid analysis has been getting interests as an attractive research field. In particular, it was reported that it would be possible to detect posterior ocular disease (i.e. age-related macular degeneration, diabetic retinopathy) and choroidal neovascularization at the initial stage by analyzing the miRNA in the tear fluid.

I have been working on ocular drug delivery using a microneedle. Since I have learned ocular anatomy, diseases, and pharmaceuticals, I have comprehensive knowledge to diagnose ocular diseases. In addition, since Ph.D. thesis was pathogen detection based on genetic and immune assay, I had obtained knowledge about genetic analysis technique; DNA/RNA amplification (i.e. single/multiplex PCR, isothermal amplification;

LAMP, RPA), DNA/RNA purification from human samples, DNA chemistry, DNA capillary electrophoresis, immune assay technique; immune reaction with conjugate chemistry, target detection method (i.e. ELISA), the protein structure analysis. Thus, I believe I can successfully investigate the analysis of miRNAs and proteins in the tear fluid with confidence.

2) Diagnosis of ocular infections by influenza A virus

Since ocular infection by influenza A virus have been reported, the ocular diagnosis of the influenza A virus have been issued. It was found that the influenza virus prefer eye epithelial cell because the epithelial cell has more alpha 2-3 sialic acid linkage receptors in the ocular surface which are preferred by the influenza virus for the transmission. Accordingly, the ocular infection by the influenza virus is very contagious, and the virus is rapidly transferred into the respiratory organ in the body. The multiple reassortment of the influenza virus gene have been occurred continuously by transmission of the virus from animal to animal/human. Thus, early diagnosis of the influenza A virus should be critical issues to prevent virus rage. Since eye swab is safer from RNA damage and relatively clearer than nasal swab, diagnosis using eye samples is advantageous to analyze the influenza A virus.

I have performed pathogen detection projects based on immune/genetic assay using different materials such as graphene/graphene oxide, sol-gel, and metal micro/nanoparticles. In the projects, I have learned material properties, and basic analytical chemistry (i.e. TEM, SEM, AFM, XPS, and XRD). I used the HA protein and RNA of influenza A viruses (i.e. A/H1N1, A/H3N2, and A/H5N1) in the projects to identify and subtype of the viruses. Those experiences relating to influenza A virus will be helpful to investigate the ocular infections.

3) Development of point-of-care (POC) system for self-diagnosis of ocular diseases

Diagnosis of ocular disease at early stage is important to increase drug efficacy and the fully recovery rate. Thus, development of rapid, sensitive, multiplex, and on-site detection system to diagnose ocular disease will be one of my research interest. Since Lab-on-a-chip technology can transfer extensive diagnostic technologies into the portable microchip, a POC system can be demonstrated by combination with electrical, optical, or colorimetric technology with cost-effective manner. Moreover, since the POC system enables self-diagnosis at home/office and can be developed without using expensive instrumentations, it can be used as a doctor in developing countries.

My Ph.D. thesis was developing pathogen detection system in the microfluidic platforms. I had learned different microfluidic chips. Since, I fabricated diverse microchips made by glass, silicone, quartz, polycarbonate, and polydimethylsiloxane (PDMS), I have learned various fabrication methods (i.e. photo/soft lithography, by using a CO_2 laser cutting, milling machine cutting). According to the target and detection methods, diagnostic microsystem can be optimized to maximize the detection sensitivity and specificity. Since I have experienced with various microchips, I would like to develop and launch a POC microsystem in the market.

Postdoctoral Projects

- "*Suprachoroidal Drug Delivery for Retina Disorders.*" NIH (1R01EY022097-01A1) Under supervision of Dr. Mark Prausnitz, School of Chemical and Biomolecular Engineering/Biomedical engineering, Georgia Institute of Technology.

- "*VLP Vaccine Technology, Microneedle delivery of Respiratory Syncytial Virus (RSV) virus like particle.*" NIH (5R01AI105170-02). Under supervision of Dr. Mark Prausnitz, School of Chemical and Biomolecular Engineering/Biomedical engineering, Georgia Institute of Technology.

PhD Dissertation

"*Development of a Total Integrated Rotary Genetic Analysis Microsystem for Rapid and Multiplex Pathogen Detection.*"

Under supervision of Dr. Tae Seok Seo, School of Chemical and Biomolecular Engineering, Korea Advanced Institute of Science and Technology (KAIST).

Selected Publications

B. Chiang, N. Venugopal, H.E. Grossniklaus, **J.H. Jung**, H.F. Edelhauser and M.R. Prausnitz, **2017**. "*Thickness and Closure Kinetics of the Suprachoroidal Space Following Microneedle Injection of Liquid Formulations*". Invest. Ophthalmol. Vis. Sci. 58, 555-564.

J.H. Jung, B.H. Park, S.J. Oh, G. Choi and T.S. Seo, **2015**. "*Integrated centrifugal reverse transcriptase loop-mediated isothermal amplification microdevice for influenza A virus detection*". Biosens. Bioelectron., 68, 218-224.

J.H. Jung, B.H. Park, S.J. Oh, G. Choi and T.S. Seo, **2015**. *"Integration of reverse transcriptase loop-mediated isothermal amplification with an immunochromatographic strip on a centrifugal microdevice for influenza A virus identification"*. Lab Chip, 15, 718-725.

Y.T. Kim*, **J.H. Jung***, Y.K. Choi and T.S. Seo (*co-first author), **2014**. *"A packaged paper fluidic-based microdevice for detecting gene expression of influenza A virus"*. Biosens. Bioelectron. 61, 485-490.

J.H. Jung, T.J. Park, S.Y. Lee and T.S. Seo, **2012**. *"Homogeneous Biogenic Paramagnetic Nanoparticle Synthesis Based on a Microfluidic Droplet Generator"*. Angew. Chem. Int. Ed. 51, 5634-5637.

J.H. Jung, D.S. Cheon, F. Liu, K.B. Lee and T.S. Seo, **2010**. *"A Graphene Oxide Based Immuno-Biosensor for Pathogen Detection"*, Angew. Chem. Int. Ed. 49, 5708-5711.

Teaching Interests:

Aside from my research career, I have extensive teaching experience as a TA and RA. During graduate student, I TAed Chemical and Biomolecular Engineering (CBE) Laboratory, Organic Chemistry, and DNA chemistry in School of Chemical and Biomolecular Engineering in KAIST, Korea as a part of government fellowship. All classes were performed in English. I had worked as an RA of undergraduate research program (URP) in KAIST and the research class for undergraduates in CBE during graduate school. Also, I mentored graduate students in my group. In the Georgia Institute of Technology, I have performed ocular drug delivery projects with three undergraduates for 2 year. One of students is from Thailand and she had joined as a government undergraduate research program in Chulalongkorn University, Thailand for one semester.

Since I have immersed diverse research field I believe those experiences obviously will be helpful to my teaching. My research experiences are focused in the fields including Molecular Biology, Molecular Diagnostics, Bio-conjugate Chemistry, and DNA Engineering. Thus, my teaching interests are also there in the same field. I also have interest in teaching of microfluidics, recent topics in chemical and biomolecular engineering.

In particularly, I would like to develop the class, 'Point-of-care (POC) system design and development'. In my opinion, students should have wide insights in their fields. Thus, I would like to let student develop POC system from design to demonstration in this class. Although pre-requisites, basic classes, will be required to take this class, students can widely comprehend recent technology trends in their fields by literature survey and will understand how to design POC system. Actual development will be limited, but students will learn how to use their knowledge which earned in the class. I also would like to include importance of patents in the system development.

505219 Electrokinetic Analytical Tools for Cell Characterization and Biosensing Technology

Abstract:

Research Interests:

Stem cells have great therapeutic potential because of their differentiation capacity and protective properties. Sufficiently characterizing stem cells functional behavior before use in transplant therapy is essential to the development of reliable therapeutics. Dielectrophoresis (DEP), an advancing electrokinetic technique, is used to examine cell populations and quantify their biophysical properties. One such property, membrane capacitance is an emerging biomarker that provides insight to cellular function. My research interests include using DEP technology for biosensing and stem cell characterization/processing for disease monitoring and eventual treatment.

Postdoctoral Research:

I am a postdoctoral research fellow in the Neurology Department at the University of California Irvine (UCI). I work under the direction of Dr. Lisa Flanagan and use DEP to study neural stem and progenitor cells (NSPCs). I've examined NSPCs dielectric properties from multiple sources (brain-derived, mouse-derived, and embryonic-derived) and their links to differentiation. I've quantified NSPCs membrane capacitance and harnessed this information to sort mouse-derived NSPCs as well as brain-derived NSPCs in batch and continuous systems. This is critical because NSPCs can be expanded in cell culture for post-separation processing to test in stroke and spinal cord injury models. I'm currently involved in a mouse spinal cord injury model. Embryonic-derived NSPCs were transplanted after a spinal cord injury was induced and behavioral data was collected. My specific contributions include DEP characterization of transplanted cells along with behavioral data analysis.

Postdoctoral Awards:

(1) National Science Foundation Postdoctoral Research Fellowship in Biology 2016

(2) UCI Chancellor's ADVANCE Postdoctoral Fellowship 2016

(3) Society of Women Engineers ASSIST Travel Grant 2016

(4) AES/BioMicrofluidics Art in Science Competition, 1st place Video 2016

PhD Research:

Prior to UCI I completed my M.S. and Ph.D. in Chemical Engineering at Michigan Technological University. Under the supervision of Dr. Adrienne Minerick, my PhD research focused on DEP-based cell separations for stem cell therapeutics. Specifically, I studied the polarization mechanisms of small particles such as polystyrene beads, human mesenchymal stem cells (hMSCs), and red blood cells to determine their unique dielectric properties. For hMSCs, I developed a morphology standardization procedure to accurately characterize their dielectric behavior. My major accomplishment was the quantification of hMSCs membrane capacitance and permittivity detailed in a peer-reviewed publication. Membrane capacitance may be used as a unique biomarker for hMSC subpopulations, which is essential to sorting hMSCs as a potential therapy for type 1 diabetes. Another aspect of my research was the development of a new DEP data collection and analysis technique for blood typing. I examined the use of frequency sweep rates to accurately quantify particles DEP response. I optimized the frequency sweep rate allowing a continuous DEP spectra to be obtained (nontraditional method) for polystyrene beads and red blood cells. This provides DEP researchers a fast data collection and analysis procedure, which is essential to achieve rapid blood typing (and cell separations). This frequency sweep technique has a patent No. WO2015051372-A1, April 2015.

Select PhD Awards:

(1) Doctoral Finishing Fellowship, Michigan Technological University 2013

(2) Anderson Research Scholarship, Michigan Technological University 2012

(3) Biotech Research Center Merit Award, Michigan Technological University 2012

(4) Graduate School Dean's Fellowship, Michigan Technological University 2011-2012

(5) King-Chavez-Parks Future Faculty Fellowship, Michigan Technological University 2009-2013

(6) National GEM Consortium Fellowship 2009-2010 (funded M.S. research)

Future Research Direction:

Just as cell phones have become essential and are prominent in society, portable biosensors are headed in the same direction. Chronic diseases such as heart disease, diabetes, stroke and obesity are the leading causes in preventable death and disability in the U.S. claiming more than 950,000 lives each year. It is projected that healthcare in the U.S. will reach 30% of the GDP by 2040. This necessitates the need for portable biosensors for cell characterization and processing to facilitate early disease detection, and wearable biosensors for patient monitoring as a form of preventative healthcare. My research explorations will be conducted at the cellular, organ, and exercise science levels.

Cell Characterization and Sorting: Stem cells therapeutic potential is difficult to achieve because their cultures are heterogeneous containing stem cells, partially differentiated progenitor cells, and fully differentiated cells. For example, neural stem cells contain astrocyte, neuron and oligodendrocyte progenitors, and mesenchymal stem cells contain adipocyte, myocyte, and chondrocyte progenitors. Little is known about the defining characteristics of these progenitor cells and which are best for healing. DEP uses electric fields to induce cell motion, and can be used to sort progenitor cells based on their intrinsic properties (membrane capacitance) allowing for downstream cellular analysis. DEP and other electrokinetic techniques will be implemented to develop membrane capacitance as a reliable progenitor cell biomarker. Effectively characterizing and sorting stem cells is essential to further study these cells as treatment options for chronic diseases.

Microfluidic Biosensor for Disease Monitoring: Doctors screen for stroke and heart disease risks through blood tests. DEP has been used to discriminate between healthy and diseased red blood cells (RBCs), this foundation will be used to characterize differences in healthy and unhealthy RBCs as they relate to kidney and liver function. Creatinine and albumin, protein biomarkers, are used as indicators of kidney and liver function, respectively. Creatinine is linked to the glomerular filtration rate (GFR) with high creatinine concentrations indicating low GFR. Low albumin concentrations imply diminished liver capacity, and the body is not absorbing enough essential nutrients. Creatinine and albumin are good biomarker candidates because they are used in the comprehensive metabolic panel. The goal is to develop a single microfluidic platform to quickly assess creatinine and albumin concentrations in small volumes of blood.

Exercise-related Biomarker Detection: There are chemical biomarkers related to exercise that are detectable in electrokinetic-based microfluidic devices. Consider dehydration, a person has loss high concentrations of salt affecting their muscles ability to retain water causing cramps. Salt concentrations are detectable in sweat along with other minerals, urea, and lactate. A wearable multiwell microfluidic device will be used to collect and analyze sweat for mineral content to provide health status information. Ultimately the goal is to build a profile of mineral losses during exercise to maximize recovery. This is especially important for people using exercise as an early chronic disease treatment option.

Electrokinetic techniques are promising analytical tools for stem cell characterization and sorting, chronic disease monitoring, and exercise assessments.

Teaching Interests:

My teaching interests reside within my teaching philosophy, every student needs a champion! Being a champion for students means supporting them in the classroom. My strategy is to create a comfortable, nonthreatening classroom environment where questions and feedback are encouraged. Beyond striving to ensure students learn fundamental course content my objectives are to: (a) foster critical thinking and problem solving skills; (b) develop life-long learning skills; and (c) prepare social and globally conscious engineers. As a professor, I want to inspire students and help them reach their scholarly potential.

During my Ph.D. program, I had the opportunity to gain experience as a teaching assistant (TA). I TA'd Fundamentals of Chemical Engineering II, a class with 52 students, which covered mass and energy balances, flow and piping systems, pumps, compressors, and stage-wise separations. I also TA'd Thermodynamics for Chemical Engineers a class with 60 students that covered the first and second law of thermodynamics applied to open and closed systems, energy conversions, power cycles, and entropy/enthalpy calculations. Serving as a TA was valuable because I learned how to provide students with additional classroom support and how to structure office hours. Now, I teach two chemistry labs at a local community college.

With my experience, I am prepared to teach core chemical engineering undergraduate and graduate courses. Additionally, I would like to develop content and teach courses related to Medical Microdevices, Stem Cell Engineering, and Bioengineering.

Publications

(1) Adams, T.N.G., Ro, C.R., Tiwari, S., Anderson, A.J., Cummings, B.J., and Flanagan, L.A., Characterizing human stem cell function with dielectrophoresis and flow cytometry (in preparation).

(2) Adams, T.N.G., Flanagan, L.A., and Minerick, A.R., Membrane capacitance as a live cell marker for human mesenchymal stem cell therapeutics: a review on advances in dielectrophoresis technology (in preparation).

(3) Adams, T.N.G., Jiang, A.Y.L., Vyas, P.D., Flanagan, L.A., Separation of neural stem cells by whole cell membrane capacitance, METHODS: Neural Stem Cells in Health and Disease (in review).

(4) Adams, T.N.G., Turner, P., Zhao, F., Janorkar, A., and Minerick, A.R., Characterizing the Dielectric Properties of Human Mesenchymal Stem Cells and the Effects of Charged Elastin-like Polypeptide Copolymer Treatment, Biomicrofluidics, 8, 054109, 2014.

(5) Adams, T.N.G., Leonard, K.M., and Minerick, A.R., Frequency Sweep Rate Dependence on the Dielectrophoretic Response of Polystyrene Beads and Red Blood Cells, Biomicrofluidics, 7, 064114, 2013.

(6) Adams, T., Yang, C., Gress, J., Wimmer, N., and Minerick, A.R., A Tunable Microfluidic Device for Drug Delivery, Advances in Microfluidics, InTech Open Science, 2012.

(7) Adams, T.N.G., Olson, T., King, J., and Keith, J., In-Plane Thermal Conductivity Modeling of Carbon Filled Liquid Crystal Polymer Based Resins, Journal of Polymer Composites, 32, 147-157, 2011.

505615 Engineering Devices for Diagnostics, Therapeutics and Discovery Science

Abstract:

Research Interests:

Advances in microfabrication and manufacturing techniques have opened new possibilities to build novel integrative devices that can interact with biological systems and perform controlled manipulations. My primary research interests is in building such novel devices that can address the current unmet needs in healthcare and discovery science. In the first part, I will discuss our work on developing biomimetic chips that mimic the physiological process of cell-homing to isolate immune cell populations from whole blood. Next, I will present some of our recent work on developing a microfabricated immune-isolating device that can be used to implant therapeutic cells within the body without the risk of rejection. Through a rigorous screening process, we have also developed a unique surface coating that mitigates fibrosis on these devices in mice as well as non-human primates. These biocompatible devices can maintain protein producing therapeutic cells in the body for over 6 months. Finally, I will discuss our work on developing an integrated droplet based microfluidic platform that can separate proteins and nucleic acids from single cells. We are using this platform to isolate and sequence microRNAs from single cells, and we are studying how cell to cell variability in miRNA expression drives differentiation and oncogenesis.

One of my primary interest in pursuing an academic position is also for the opportunity to mentor students and train next generation scientists. In my academic career, I have mentored over 14 undergraduates, 3 technicians and 1 graduate students. These experiences have been really impactful in my career and have helped me to become a better educator and adviser.

Teaching Interests:

I am interested and qualified for teaching multiple core classes of Chemical Engineering which includes topics in the areas of biomaterials, reaction kinetics, fluid dynamics, and thermodynamics. Additionally, I am excited to create a graduate course in the emerging fields of Molecular Engineering and Medical devices.

505389 Engineering Vascularized Organ-on-Chip Systems to Advance Biological Understanding and Therapeutic Intervention in Human Cancer and Blood Stem Cell Biology

Abstract:

Research Interests:

Traditional cell culture based on flat tissue dishes over the last few decades has greatly enhanced our understanding of cellular behaviors in 2D microenvironment. However, these models lack several characteristics of human tissue in 3D, making it challenging to fully recapitulate how cells interact and function in vivo. Organ-on-a-chip models are fantastic platforms that provide additional and controllable parameters (fluid flow, biochemical and mechanical cues) and 3D microenvironment to place cells in a closely resembling native tissue. In a way similar to the human vasculature, organ-on-chip model employs a network of microfluidic channels to establish a â€˜circulatory systemâ€™ to deliver nutrients, and gases to the cells within the culture. As we strive to better mimic cellular functions in human tissues, these artificial microfluidic networks fail to recapitulate the blood vessels in the human circulatory system largely because the endothelial cells lining the blood vessels function more than just a fluid-carrying conduits. Depending on the organs where the endothelial cells reside, they are also specialized to provide pro-survival, regenerative, and homeostatic cues to the surrounding cells within the tissues. Hence, it is crucial to incorporate organ-specific endothelial cell network in 3D model systems. To fill in the gaps, my graduate and postdoctoral work have offered an approach to engineer vascularized organ-on-a-chip systems to advance our understanding of cellular behaviors in vascularized 3D models in both physiological (angiogenesis, hematopoietic stem cells) and pathological (cancer metastasis, and immune cells) contexts.

Research experience:

My graduate research focused on engineering a biomimetic vessel. During this work, I employed lithography to fabricate a microfluidic platform where the endothelial cell channels were suspended within a 3D extracellular matrix. These endothelial cell channels serve as rudimentary blood vessels and possess properties of in vivo blood vessels. For example, they are surrounded by a thin layer of basement membrane protein and carry fluid flow. Using these rudimentary blood vessels, I further demonstrated that a gradient of chemoattractants could be incorporated into the system to trigger the endothelial cells to migrate and form new blood vessels, a process that is called angiogenesis, which often occurs during wound healing or in tumor progression. Using high resolution confocal microscopy, and imaging processing, I further characterized these newly sprouting vessels and showed that they remarkably displayed many features of in vivo angiogenesis. This biomimetic vessel on a chip also enabled me to perform screening for biochemical factors to identify molecular pathways that influence the morphogenetic processes of collective endothelial cell migration in 3D environment. On the other hand, not only endothelial cells can be seeded into the channel, various other cell types could also be seeded. For example, to model cancer cells and endothelial cell interactions, I successfully seeded pancreatic cancer cells isolated from a genetically engineered mouse pancreatic cancer model into the channel to mimic a malignant pancreatic duct. I discovered that these pancreatic ductal cancer cells invaded into the interstitial matrix towards the blood vessel. Not only did the pancreatic cancer cells invade onto the vessel, they ultimately displaced and replaced the endothelium in the blood vessel, a phenomenon that was similarly reported in human pancreatic cancer patients. Moreover, using molecular biology and genetic approaches (CRISPR/Cas9), I was able to identify the molecular signaling that was involved in vascular invasion of pancreatic ductal adenocarcinoma.

As I experienced the indispensable role of the vascular network in constructing 3D organotypic models during my graduate work, my goal in postdoctoral training is to seek to extend and develop a perfusable vascularized tissue on a chip to allow coculture of the endothelial cells with other cells from different organs such as hematopoietic cells from the bone marrow or hepatocytes from the liver. I am currently being mentored by world-renown scientists: Dr. Shahin Rafii in the field of genetics, vascular biology, hematopoietic stem cells, and Dr. Robert Schwartz in the field of liver and virology. This unique postdoctoral training opportunity has given me a chance to acquire a depth of knowledge in vascular biology, hematopoietic stem cell and organ regeneration in the liver. I am also acquiring additional knowledge and skill sets in RNA sequencing and induced pluripotent stem cells.

Future directions:

The endeavors in my graduate and postgraduate trainings enabled me to travel between different fields and provided me with unique approaches and perspectives to apply in biomedical engineering research. As an independent researcher with the strengths drawn from an engineering perspective during my PhD years and from a molecular biology perspective during my postdoctoral training, my laboratory will seek to reconstitute 3D vascularized mini-organ on a chip to understand human biology in both physiological and pathological contexts. Specifically, we will focus on the three following key areas:

1) **Pathological angiogenesis during tumor progression**. This work will initially focus on pancreatic cancer but can be extended to other angiogenic tumors such as glioblastoma or hepatocellular adenocarcinoma. We seek to elucidate the molecular mechanisms of tumor-induced angiogenesis. The ultimate goal is to identify new therapeutic targets for tumor angiogenesis as the current angiogenic therapeutics often fail due to an incomplete understanding of tumor angiogenesis.

2) **Tumor-endothelial cell communications in tumor-on-a-chip**: we are constructing a tumor microenvironment that entails tumor cells and vascular network. We will seek to elucidate the cross-talks between the blood vessel and tumor cells during metastasis.

3) **Reconstructing a bone-marrow on a chip:** This work will focus on reconstructing a vascularized tissue to coculture endothelial cell network with hematopoietic stem cells (HSCs) and understand the role of endothelial cells in maintenance and expansion of HSCs. Additional work on utilizing differentiated endothelial cells from induced human pluripotent stem cells from patients will also be included to address potential defects in maintaining stemness in hematopoietic stem cells due to genetic defects in endothelial cells.

The exciting proposed research directions above will ensure my lab contribution to the understanding of cellular interactions in human organ on a chip. Given the utility of interconnecting different microfluidic entities, I envision my lab will have an immense potential to address very exciting questions to unravel the cellular communications between different organ-on-a-chip platforms (e.g.: bone-marrow on a chip and tumor on a chip) that, thus far, have remained elusive.

Teaching Interests:

I am interested in teaching courses in the chemical engineering core curriculum such as kinetics and mass transport. With a unique background intertwined between engineering and biology, I am also particularly interested and well-prepared in teaching courses spanning the scopes of two such areas, for example, courses in tissue engineering, biotechnology, and bioengineering.

Successful Proposals:

NIH T32 Interdisciplinary Cardiovascular Training Grant

NSF Graduate Fellowship (Honorable Mentioned)

504205 Genetic Engineering of Immune Cell Recruitment to Control Inflammation

Abstract:

Research Interests:

Chronic medical conditions including heart disease, hypertension, cancer, diabetes, and chronic obstructive pulmonary disease (COPD) are responsible for 7 in 10 of the deaths per year in the United States. Furthermore, 75% of the $2.9 trillion we spend on healthcare costs per year is spent of managing chronic disease conditions. Yet despite spending nearly $2,000 dollars more per person on healthcare cost than any other nation, the United States has the second lowest life expectancy among those nations. Translational research can lead to therapeutics to combat chronic diseases and as such is of critical interest to reduce chronic disease mortality. My research focuses on the identification of critical targets of the innate immune response which can provide a possible therapeutic benefit for controlling inflammation and the development of chronic conditions. I focus on the leukocyte adhesion cascade, the sequential series of binding and activation steps in which immune cells interact with the endothelium: capture and rolling, activation, firm adhesion, migration, and eventual transmigration to sites of inflammation. My current research has placed great emphasis on manipulating the steps of this cascade as it is a prerequisite for trafficking to sites of inflammation, maintaining hemostasis, and providing immuno-surveillance.

During my Ph.D. training in the department of Chemical and Biological Engineering at the University at Buffalo under Dr. Sriram Neelamegham, I studied the interaction of carbohydrate glycans, specifically sialyl-Lewis-X, with endothelial bound selectins and impact of the glycans on neutrophil trafficking during the adhesion cascade (1). Specifically, my thesis project revolved around the genetic manipulation of leukocytes by shRNA knockdowns and CRISPR/Cas9 knockouts. I sought to disrupt the all of the members of two classes of the terminal glycosyltransferases (α1,3 fucosyltransferases (2) and α2,3 sialyltransferases (3)) thought to be critical in sLe^x formation and thus selectin-mediated adhesion. My studies were the first genetic perturbation of the critical glycosyltransferases in both human cell lines and primary human neutrophils and lead to the eventual discovery of glycolipids as E-selectin ligands in human neutrophils (4, 5). The results highlighted stark species-specific differences between humans and mice (6) and suggests caution when examining potential disease targets based solely on mouse models.

My first post-doc with Dr. Joseph Lau in the department of Molecular and Cellular Biology at Roswell Park Cancer Institute allowed me to gain experience with animal models of glycobiology and immunology and specifically experience with *in-vivo* models of inflammation. Other added benefits from this time was the ability to interact with not only the clinicians at Roswell Park which gave me a perspective of what they found useful from our research which kept the overarching goal of identifying potential therapeutics in focus. My two research aims identified the effects of how defective α(1,3) fucosylation on CXCR2 chemokine receptor increased neutrophil trafficking to the inflamed lung (7) and how increasing circulatory levels of the ST6Gal-1 enzyme can inhibit mature granulocyte production in the bone marrow (8). Both of these projects are prime examples potential checkpoints which can be leveraged for clinical benefit to control the neutrophilic response to acute inflammation.

In, my current position as a postdoc with Dr. Daniel Hammer in the Department of Chemical and Biomolecular Engineering at the University of Pennsylvania Here, I have expanded my skill repertoire to include the study of the biophysics of immune cell recruitment and force generation using engineered matrices and microfluidics. For my current research project, I am investigating the unique phenomenon where HSPCs efficiently migrate against the direction of shear flow on endothelial borne CAMs, much like a salmon swims upstream. Better understanding of this phenomenon has to potential to better deliver HSPCs to the bone marrow post-transplantation.

Future Directions:

My overall research program aspires to pursue a multidisciplinary approach melding the fields of chemical engineering, bioengineering, glycobiology, molecular biology, and immunology with an overarching goal of translating basic research on immune cell trafficking into the identification of potentially exploitable therapeutic targets against chronic inflammatory diseases. My future research will focus on the genetic engineering of potential mediators of selectin-tethering and rolling, chemokine activation, and migration along the endothelial surface which have the potential for clinical benefit.

Teaching Interests:

I am strongly dedicated to the teaching and mentoring of future generations of scientists. I have experience as a teaching assistant during graduate school, as a guest lecturer as a postdoc, and have mentored numerous students in the laboratory. These range from high school and undergraduate students doing a summer internship to the direct oversight of the thesis projects of masters and Ph.D. students during my Postdoc time. I am prepared to teach all courses of both the chemical engineering and bioengineering curriculum for both undergraduates and graduate students alike. Specific courses include fluid mechanics and transport in chemical engineering and cell engineering and biotransport in bioengineering. Also, I would be able to teach electives in molecular biology, immunology, and glycobiology owing to my experience and expertise in these areas.

Relevant Works:

1. Marathe DD, Buffone A, Chandrasekaran EV, Xue J, Locke RD, Nasirikenari M, Lau JTY, Matta KL, Neelamegham S. Fluorinated per-acetylated GalNAc metabolically alters glycan structures on leukocyte PSGL-1 and reduces cell binding to selectins. Blood. 2010;115(6):1303-12. doi: 10.1182/blood-2009-07-231480. PubMed PMID: PMC2826240.

2. Buffone A, Mondal N, Gupta R, McHugh KP, Lau JTY, Neelamegham S. Silencing α1,3-Fucosyltransferases in Human Leukocytes Reveals a Role for FUT9 Enzyme during E-selectin-mediated Cell Adhesion. Journal of Biological Chemistry. 2013;288(3):1620-33. doi: 10.1074/jbc.M112.400929.

3. Mondal N, Buffone A, Stolfa G, Antonopoulos A, Lau JTY, Haslam SM, Dell A, Neelamegham S. ST3Gal-4 is the primary sialyltransferase regulating the synthesis of E-, P-, and L-selectin ligands on human myeloid leukocytes. Blood. 2014;125(4):687-96.

4. Mondal N, Stolfa G, Antonopoulos A, Zhu Y, Wang S-S, Buffone A, Atilla-Gokcumen GE, Haslam SM, Dell A, Neelamegham S. Glycosphingolipids on Human Myeloid Cells Stabilize E-Selectin–Dependent Rolling in the Multistep Leukocyte Adhesion Cascade. Arteriosclerosis, Thrombosis, and Vascular Biology. 2016;36(4):718-27. doi: 10.1161/atvbaha.115.306748.

5. Stolfa G, Mondal N, Zhu Y, Yu X, Buffone A, Neelamegham S. Using CRISPR-Cas9 to quantify the contributions of O-glycans, N-glycans and Glycosphingolipids to human leukocyte-endothelium adhesion. Scientific Reports. 2016;6:30392. doi: 10.1038/srep30392

6. Mondal N, Buffone Jr A, Neelamegham S. Distinct glycosyltransferases synthesize E-selectin ligands in human vs. mouse leukocytes. Cell Adhesion & Migration. 2013;7(3):288-92. doi: 10.4161/cam.24714.

7. Buffone A, Nasirikenari M, Manhardt CT, Lugade A, Bogner PN, Sackstein R, Thanavala Y, Neelamegham S, Lau JTY. Leukocyte-borne α(1,3)-fucose is a negative regulator of β2-integrin-dependent recruitment in lung inflammation. Journal of Leukocyte Biology. 2017;101(2):459-70. doi: 10.1189/jlb.3A0516-215RR.

8. Dougher CWL, Buffone A, Nemeth MJ, Nasirikenari M, Irons EE, Bogner PN, Lau JTY. The blood-borne sialyltransferase ST6Gal-1 is a negative systemic regulator of granulopoiesis. Journal of Leukocyte Biology. 2017. doi: 10.1189/jlb.3A1216-538RR.

505586 Imran Rizvi, Ph.D. Assistant Professor, Department of Dermatology, Harvard Medical School; And Assistant Biomedical Engineer, Wellman Center for Photomedicine, Department of Dermatology, Massachusetts General Hospital

Abstract:

Research Interests:

I am a cancer biologist with a Ph.D. in Engineering Sciences and 20 years of experience in cancer pharmacology, biomedical optics and translational cancer research. My multidisciplinary research draws on my expertise in molecular oncology (treatment resistance pathways), photodynamic therapy (PDT) and photoimmunotherapy (PIT), 3D and in vivo tumor models, biomedical imaging and targeted combination therapies.

I trained as a postdoctoral fellow at the Wellman Center for Photomedicine at Massachusetts General Hospital (MGH) and Harvard Medical School (HMS) in the laboratory of Dr. Tayyaba Hasan, Professor of Dermatology at HMS, and Professor of the Harvard-MIT Division of Health Sciences Technology. I completed the mentored phase of my K99 (please see "Funding" section below) in the Bioengineering Division of the Department of Medicine (DOM) at Brigham and Women's Hospital (BWH), supervised by Dr. Utkan Demirci, currently Professor of Radiology at Stanford University, and Dr. Hasan.

Ph.D. Dartmouth College; Engineering Sciences (Co-Advisors: Dr. Brian Pogue and Dr. Tayyaba Hasan)

M.S. Georgetown University; Tumor Biology (Advisor: Dr. Stephen Byers)

B.A. Johns Hopkins University

Funding

1. Targeting Determinants of OvCa Metastases in Engineered 3D Microfluidic Platforms

NIH/NCI 4R00 CA175292-03

Role on Project: PI

Total Direct Costs: $436,842

Total Project Costs: $747,000

03/01/2016 - 02/28/2019

The goal of this proposal is to integrate microfluidics with heterocellular 3D ovarian cancer models to create the first system to evaluate the effects of fluid hydrodynamics on ovarian cancer progression. This platform will be used to design therapeutic strategies based on flow-induced changes in molecular target expression.

2. *Imaging-based Assay Development for Immuno-Oncology Applications*

Bristol-Myers Squibb Innovation Award

Role on Project: Co-PI (Hasan, T & Rizvi, I)

Total Direct Costs: $283,019

Total Project Costs: $450,000

01/01/2016 - 06/30/2018

This project establishes a 3D bioprinting platform to screen immuno-oncology agents, in combination with photodynamic therapy, to boost antigen specific T-cell mediated tumor killing. This platform, in conjunction with quantitative fluorescence imaging and analysis, will enable antibody drug conjugates and biologics to be screened for their capacity to enhance tumor killing by lymphocytes in a highly reproducible 3D setting that better represents the heterogeneity and complexity of the tumor microenvironment. This collaborative agreement with Bristol-Myers Squibb will form the basis for bioprinting and imaging-based 3D assay development to improve preclinical prediction of single and combination immuno-oncology agents, ultimately with goal of identifying patient-specific regimens.

3. Low-cost Enabling Technology for Image-guided Photodynamic Therapy of Oral Cancers

NIH/NCI 4UH3 CA189901-03

Role on Project: Investigator (Co-PI: Hasan, T & Celli, JP)

09/01/2016 - 08/31/2019

This proposal aims to address the problem of oral cancer by using a low-cost adaptation of PDT. My role on this project leverages my experience in sustainable engineering solutions for low and middle-income countries (LMIC) to guide the implementation and dissemination of low-cost PDT regimens and technology for the treatment of oral cancer in a mid-tier academic medical center and rural settings in India.

4. Bioluminescence-Activated Photodynamic Therapy of Breast Cancer

NIH/NCI 1R01 CA192878-01A1

Role on Project: Collaborator (PI: Yun, SH)

09/25/2015 - 08/31/2020

The goal is to develop novel photosensitizer agents that can be activated by bioluminescent molecules and to test its therapeutic potential in mouse models of breast cancer. My role on this project leverages my expertise in PDT-based combination therapies, and my research experience at the interface of oncology, bioengineering and in vivo tumor models to guide the design and evaluation of bioluminescence-based PDT regimens for early-stage invasive breast cancer.

Publications

Peer Reviewed Publications

Original Investigations: 32

Reviews: 3

Non-Peer Reviewed Publications

Book chapters: 3

Conference Proceedings: 18

h-index

Web of Science, June 2017: 25

Google Scholar, June 2017: 27

Total Number of Citations

Web of Science, June 2017: 2135

Google Scholar, June 2017: 2919

Teaching Interests:

As the beneficiary of exceptional teaching and mentorship from numerous dedicated individuals, I am highly committed to mentoring and preparing the next generation of scientists. My philosophy as a researcher and a teacher is guided by the following core principles: (i) promote critical thinking and rigorous problem-solving; (ii) emphasize integrity, reproducibility and transparency in science; and (iii) foster independent thought to identify creative and exciting questions based on an understanding of of the literature and fundamental concepts.

I regularly lecture on "Photodynamic Therapy: Basic Principles and Therapeutic Combinations" for an Introduction to Biophotonics course at the Tufts University School of Engineering (BME 51/251). I also lecture on "Photodynamic Therapy Fundamentals and Applications, and Nanotechnology-enabled Drug Delivery for Cancer Combinations" at the Harvard-MIT (HST) Summer Institute in Biomedical Optics. I am a guest lecturer for an undergraduate Cancer Biophysics course at the University of Massachusetts (UMass).

Mentoring (including Panels and Workshops): I have been a repeated Invited Panelist for a K99/R00 Pathway to Independence Panel Discussion at the Annual Meeting of the Massachusetts General Postdoc Association (MGPA) at MGH. I recently gave a well-received invited lecture on "NIH Career Development Awards" and participated in a grant writing workshop for Associate Members and junior faculty at the 38th Meeting of the ASP. I have mentored or co-mentored 1-2 undergraduate students each summer since 2011 for the HST Summer Institute in Biomedical Optics. My mentees have consistently been recognized for their productivity, and learning, through awards including the Yao Su Student Research Prize (won by my student, who is now a D.O. Candidate at the Western University of Health Sciences). I also serve as a judge, and Chair a judging subgroup, and am a reviewer for a Communications Workshop for the HST Summer Institute.

Service (Selected)

Local

2012 - MGH-HST Summer Institute for Biomedical Optics Poster Session

2012 Member, Judging Subgroup

2013 - 2015 Chair, Judging Subgroup

2016 Reviewer, Communications Workshop

2015 - Hasan Group Animal Work Compliance Committee

2016 - MGH IRB Non-ionizing Radiation Review Committee

2017 - Abstract Review Committee, MGH ORCD Poster Celebration

National and International

2011 - American Society for Photobiology (ASP)

2011 - Member, Grants and Awards Committee

2012 Session Co-Chair, 36th Annual Meeting (WE6-2) Montreal, Canada

2012 - 2014 Associate Councilor (elected position)

2013 - 2014 Member, Organizing Committee, 37th Annual Meeting

2013 - Member, Strategic Planning Committee

2013 - Member, Mentoring and Education Committee

2014 Session Co-Chair, 37th Annual Meeting, San Diego, CA

2016 - Member, ASP Council

2017- Chair, Awards Committee; Member, Organizing Committee: 38th Annual Meeting; Member, Nominations Committee

2011 - SPIE, The International Society for Optics and Photonics

2013 - 2015 Session Co-Chair, Optical Methods for Tumor Treatment and Detection: Mechanisms and Techniques in PDT; Photonics West Annual Conference

2015 - Member, Program Committee, Photonics West; BIOS Conference

2016 Session Chair, Optical Methods for Tumor Treatment and Detection: Mechanisms and Techniques in Photodynamic Therapy XXV, SPIE Photonics West Annual Conference

2013 - 2014 Society of Porphyrins and Phthalocyanines (SPP)

2013 - International Photodynamic Association (IPA)

2012 Member, Scientific Advisory Board, 14th World Congress

2012 - 2014 Member, Organizing Committee

2017 Standing Member, Awards Committee

2014 - American Society for Lasers in Surgery and Medicine (ASLMS)

2015 Session Co-Chair, Annual ASLMS Conference, PDT/PAPDT Session-PDT Science, Kissimmee, FL

2017 Session Co-Chair, Annual ASLMS Conference, PDT/PAPDT Session-PDT Science, San Diego, CA

2015 - PanAmerican Photodynamic Association (PAPDT)

2015 - Member, Board of Directors (invited position)

2016 - Secretary, Executive Committee (invited position)

External Thesis Advisor

2015 - Ph.D. Candidate: Gwendolyn Cramer (Successfully defended)

Advisor: Jonathan P. Celli, Ph.D., Department of Physics,

University of Massachusetts Boston, Boston, MA

2017 - Ph.D. Candidate: Christina Conrad

Advisor: Giuliano Scarcelli, Ph.D., Fischell Department of Bioengineering,

University of Maryland College Park, College Park, MD

2016 Society for Laboratory Automation and Screening (SLAS); Session Chair, 5th International SLAS Conference and Exhibition, Scaling Challenging Cell Models for High-Quality High Throughput Screens Session, Cellular Technologies Track

Grant Review Activities

2015 Invited reviewer, International Peer Review Committee, National Center for the Replacement, Refinement & Reduction of Animals in Research (NC3Rs); London, UK

2016 Invited reviewer, International Peer Review Committee, The Netherlands Organisation for Health Research and Development (ZonMw), More Knowledge with Fewer Animals (MKMD) Programme; The Hague, The Netherlands

Editorial Activities *(Selected)*

Ad-hoc Reviewer

ACS Applied Materials & Interfaces, Acta Biomaterialia, Advanced Materials, BBA - General Subjects, Biosensors, Biotechnology Journal, Cancers, Cancer Letters, Cancer Medicine, Cell Cycle, Gynecologic Oncology, Journal of Biomedical Optics, Journal of Biophotonics, Journal of, Photochemistry and Photobiology B: Biology, Lab on a Chip, Lasers in Surgery & Medicine, Molecular Cancer Therapeutics, Photochemical & Photobiological Sciences, Photochemistry and Photobiology, Photodiagnosis and Photodynamic Therapy, PLoS One, Radiation Research, Scientific Reports, Translational Oncology

Honors and Prizes (Selected)

2014 Ruth and William Silen, M.D. Award, New England Science Symposium

2014 Poster of Distinction, Top-tier award, 37th Meeting of the ASP; San Diego, CA

2015 Chair's Research Award, Research Performance Incentive Program, Department of Medicine, BWH.

2016 Best of Session Award, Photodynamic Therapy Session, 36th Annual Conference ASLMS-PAPDT

2016 New Investigator Award, 38th Meeting of the ASP; Tampa, FL

505404 Micro-/Nano-Fabrication and 3D-Bioprinting Technologies: An Engineering Approach Toward Translational Medicine

Abstract:

Research Interests:

Despite tremendous research and progress in surgical, chemotherapeutic, and radiological interventions for cancer treatment, the eradication of cancer tumor remains a dream as cancer recurrence and subsequently, the overall life expectancy in cancer patients is only a few months. While most studies have focused on tumor function and biology, the contribution of tumor surrounding tissues on regulation and differentiation of cancer cells, cell migration, and disease progression have been either underestimated or ignored. Therefore, the behavior of tumors is often a reflection of complex physiological interactions among cells and extracellular matrixes (ECM) in response to a treatment in a 4-dimensional domain: 3-dimentional microenvironment and time. On the other hand, the bench-to-bedside translation of novel therapeutic agents and controlled delivery systems is highly challenging due to lack of trustable pre-clinical (i.e. *in vitro* and *in vivo*) evidence as the performance of drug at the site of disease may be highly variable, making standardization and benchmarking are difficult and costly practices.

The advances that will take place over the next few years in the field of drug delivery and tissue engineering are very difficult to predict. However, based on the aforementioned problems, the time is right to construct an inexpensive, more efficient, and innovative *in vitro* screening model that can more accurately predict the *in vivo* (i.e. in human) pharmacokinetics and pharmacodynamics of new therapeutics and drug delivery systems. The recent advances in micro-/nano fabrication technologies and 3D-bioprinting have opened a new avenue for the miniaturized design and construction of organ tissue models with versatile biological functions toward personalized medicine.

As an educator and scientist, my research interests lie at the interface of drug delivery and tissue engineering, where the development and integration of advanced biomaterials with the state-of-art engineering technologies create functional tissue constructs helping to understand disease, develop therapies, and screen drugs for a range of medical applications. I intend to integrate my chemical and biomedical experience with other expertise such as biology and medicine in a multidisciplinary research team for addressing unsolved clinical problems and developing translational technologies. To achieve this goal, my research will specifically focus on the following major directions:

1) Engineering 3D-tissue construct using emerging technologies (i.e. micro-/nano fabrication and 3D-bioprinting) with controlled architectures.

2) Developing a cost-effective and reliable *in vitro* model for (a) high throughput screening of drugs, (b) testing new drug delivery systems and reducing the number of pre-clinical animal models, and (c) exploring cancer behavior via a well-designed *in vitro* model.

3) Developing novel advanced biomaterials with controlled physicochemical properties for on-demand drug delivery and tissue regenerative purposes.

In my graduate and post-doctoral research, I have mainly focused on the smart biomaterial synthesis and characterization and micro-/nano-fabrication and 3D-bioprinting as emerging technologies for drug delivery and tissue engineering. My Ph.D. research has focused on the synthesis of smart block copolymers and development of smart nanomaterials for the co-encapsulation of anticancer drugs and nucleic acids, where the new delivery system could sensibly respond to intracellular triggers and release cargoes at desired sub-cellular compartments. In addition, I developed a novel injectable, self-crosslinked, and antibacterial hydrogel/core-shell microparticle composite system for dual-agent treatment of metastasis breast cancer. The hydrogel provided an additional diffusion barrier against anticancer agents loaded into double-walled microspheres and confined their premature release. We also demonstrated that the hydrogel could be utilized as a versatile tool for retaining microparticles in a tumor resected cavity during injection after debulking surgery and preventing surgical site infection due to its inherent antibacterial properties. In another project, the core-shell microparticles (loaded with plasmid DNA nanoparticles and a protein-protein interaction inhibitor (a new small molecule drug)) were successfully fabricated and employed against liver cancer cells in a 3-D tumor construct.

In my Postdoctoral research, our multi-disciplinary team comprising of engineers, biologists, and clinicians have developed a unique biocompatible composite bioink for 3D-bioprinting elastomeric scaffolds. Specifically, we have developed a versatile 3D-bioprinted cell-laden composite hydrogel for the fabrication of a well-designed skin tissue construct. Additional efforts are exploring the use of 3D bioprinting technique for the rapid fabrication of micro-patterned 3D-vascular network tissue model for *in vitro* drug screening. My research has generated 10 manuscripts in refereed journals and more than 10 publications in prestigious international conferences.

I have also actively participated in mentoring and training 12 undergraduate students and junior research staffs and coordinated research activities in the biomedical lab over the past 5 years.

Education:

- **Postdoctoral Research Fellow:** National University of Singapore; 01/2017-present.

Postdoctoral Project: "*3D-printing of Biomimetic Skins for Skin Care Products Testing Using the Electro-hydrodynamic Jetting (e-jetting) Technique*".

- **Doctor of Philosophy (Ph.D.):** Chemical and Biomolecular Engineering; National University of Singapore, 2011-2016.

Ph.D. Dissertation: "*Fabrication of Micro-/nanoparticulate Carriers for Dual-agent (drug and gene) Cancer Therapy*".

Co-supervised by Professor Chi-Hwa Wang (National University of Singapore) and Professor M.P. Srinivasan (RMIT, Australia).

Grant writing:

- Contributed to the writing of research proposals: for instance, *National Medical Research Council, 2013, 2014, 2015, Singapore Cancer Society, 2014,* BIGHEART Grant Full Proposal, 2017.

Selected Journal Publications:

- **Pooya Davoodi**, F. Feng, Q. Xu, W.C. Yan, Y.W. Tong, M.P. Srinivasan, V. K. Sharma, C.H. Wang. Coaxial Electrohydrodynamic Atomization: Microparticles for Drug Delivery Applications. Journal of Controlled Release, 205, 2015, 70-82.
- J. Xie, J. Jiang, **Pooya Davoodi**, M.P. Srinivasan, C.H. Wang. Electrohydrodynamic atomization: Two Decades of Effort for Producing and Processing Micro-/nanoparticulate Materials. Chemical Engineering

Science, 125, 2015, 32-57.

- **Pooya Davoodi**, M.P. Srinivasan, C.H. Wang. Synthesis of Intracellular Reduction-Sensitive Amphiphilic polyethyleneimine and poly(ε-caprolactone) Graft Copolymer for On-demand Release of Doxorubicin and p53 plasmid DNA. Acta Biomaterialia 2016; 39: 79-93.
- W.C. Yan, **Pooya Davoodi**, Y.W. Tong, C.H. Wang. Computational Study of Core-shell Droplet Formation in Coaxial Electrohydrodynamic Atomization Process. AIChE Journal 2016; 62(12): 4259-4276.
- **Pooya Davoodi**, W.C. Ng, W.C. Yan, M.P. Srinivasan, C.H. Wang. Double-walled Microparticles-embedded Self-crosslinked, Injectable, and Anti-bacterial Hydrogel for Controlled and Sustained Release of Chemotherapeutic Agents. ACS Applied Materials & Interfaces 2016; 8(35): 22785-800.
- Y. Cui, Q. Xu, **Pooya Davoodi**, D. Wang, C.H. Wang. Enhanced intracellular delivery and controlled drug release of magnetic PLGA nanoparticles modified with transferrin. Acta Pharmacologica Sinica, 2017; 38(6):943-953.
- **Pooya Davoodi**, M.P. Srinivasan, C.H. Wang. Effective Co-delivery of Nutlin-3a and p53 genes via Core-shell Microparticles for Disruption of MDM2-p53 Interaction and Reactivation of p53 in Hepatocellular Carcinoma. Journal of Materials Chemistry B, 2017; DOI: 10.1039/C7TB00481H.

Teaching Interests:

My objective as a teacher is to motivate students to develop their thinking abilities, learning interests, and problem-solving capability. In particular, working as a teaching assistant, grader, and moderator of clinic sessions, I have tried to establish a learner-centered environment in the classroom where I focus on creating a dialogue with the students, asking them to raise questions and helping discover answers step-by-step by themselves. I believe a good teacher should not only prepare an interesting course plan and materials that encourage students for learning, he must consider the interconnection between student's knowledge, learning abilities, and accompanying theoretical concepts with practical examples such that students enjoy from what they learn in a class. I like to accompany my teaching materials with my own research background where students can find the applications of their knowledge in reality. Due to establishing a strong academic background in chemical engineering core modules during my undergraduate, master and PhD study, the courses that I would enjoy teaching include Heat and Mass Transfer, Unit Operation, and Separation processes at undergraduate level. Moreover, I have gained valuable experience in biomaterials preparation and processing during my PhD and would like to actively participate in developing new courses on the preparation and processing of innovative biomaterials and micro-/nanofabrication of biomedical devices for undergraduate/graduate students, where they will be introduced to the new aspects of chemical engineering science.

503667 Stochasticity, Complexity, and Multiscale Dynamics in Cancer Progression and Drug Response

Abstract:

Research Interests:

Biological systems differ from human engineered systems in three primary respects: 1) they are highly stochastic; 2) they are extraordinarily complex; and 3) they exhibit dynamics over a wide range of temporal and spatial scales. These characteristics are especially prevalent in cancer, which is a complex, adaptive disease characterized by genetic mutation, phenotypic plasticity, and cell-cell interactions among tumor cells and the tumor microenvironment. Tumors are thus highly heterogeneous and exhibit complex responses to drugs and other therapeutic interventions. Patient outcomes to therapy can vary widely, but it is not uncommon for an initial favorable response to be followed by tumor recurrence and metastasis, which is very often fatal. The mechanisms that govern drug evasion and tumor reestablishment remain poorly understood.

It is becoming increasingly clear that in order to unravel the complexities of cancer a systems engineering approach is necessary. Cancer can be seen as a reverse engineering problem, i.e., it is fundamentally a result of the breakdown of the cell's natural tumor suppressing machinery. Reestablishing or rewiring this apparatus thus requires a deep understanding of the mechanisms that underlie normal and cancer cell function. Computational models provide a means by which current understanding of biological mechanisms can be formalized mathematically. *In silico* predictions can be made, tested experimentally, and the experimental results used to update and refine the model. This iterative approach is a powerful tool for obtaining a detailed and comprehensive understanding of a system over time.

Here, I present results of combined computational and experimental investigations of *in vitro* responses of isoclonal non-small cell lung cancer (NSCLC) and melanoma cell populations to targeted drugs. Our results point to a "bet hedging" strategy employed by cancer cell populations whereby cells diversify across multiple phenotypes in the absence of drug to increase the chances of population survival to initial drug onslaught. We hypothesize that the surviving subpopulation can persist for an extended period of time and is susceptible to genetic mutation, acting as a reservoir from which tumor recurrence may emerge. However, the subpopulation may also represent a phenotypic bottleneck that may be vulnerable to a secondary treatment. Characterizing the molecular nature of the surviving subpopulation is thus of critical importance and work is underway towards this end.

Teaching Interests:

My expertise lies in chemical kinetics, thermodynamics, and numerical methods and I also have experience in biochemistry, statistics, theoretical computer science, and software engineering. During my years as a postdoctoral researcher I have had the pleasure of mentoring numerous undergraduate and graduate students and have given numerous lectures on systems biology. I enjoy teaching and make a concerted effort to engage students by providing analogies that convey complex concepts in an easy-to-understand way. My basic philosophy of teaching is that nothing is ever as complicated as it seems and that the best way to learn something is to do it yourself. I, therefore, guide students in the right direction but ultimately leave it to them to solve a problem. It is a great feeling to witness a student having a moment of true understanding. I am prepared and look forward to teaching a full semester course at either the undergraduate or graduate level in any of the chemical engineering core disciplines.

Select Publications:

L.A. Harris*, M.S. Nobile*, J.C. Pino*, A.L.R. Lubbock, D. Besozzi, G. Mauri, P. Cazzaniga and C.F. Lopez, "GPU-powered model analysis with PySB/cupSODA," *Bioinformatics* (*in press*). (*equal authors)

L.A. Harris*, P.L. Frick*, S.P. Garbett, K.N. Hardeman, B.B. Paudel, C.F. Lopez, V. Quaranta and D.R. Tyson, "An unbiased metric of antiproliferative drug effect *in vitro*," *Nat. Methods* **13**, 497-500 (2016). (*equal authors)

L.A. Harris, J.S. Hogg, J.J. Tapia, J.A.P. Sekar, S. Gupta, I. Korsunsky, A. Arora, D. Barua, R.P. Sheehan and J.R. Faeder, "BioNetGen 2.2: Advances in rule-based modeling," *Bioinformatics* **32**, 3366-3368 (2016).

J.S. Hogg*, **L.A. Harris***, L.J. Stover, N.S. Nair and J.R. Faeder, "Exact hybrid particle/population simulation of rule-based models of biochemical systems," *PLoS Comput. Biol.* **10**, e1003544 (2014). (*equal authors)

L.A. Harris*, J.S. Hogg* and J.R. Faeder, "Compartmental rule-based modeling of biochemical systems," *Proceedings of the 2009 Winter Simulation Conference*, M.D. Rossetti, R.R. Hill, B. Johansson, A. Dunkin, and R.G. Ingalls, eds., pp. 908-919 (2009). (*equal authors)

L.A. Harris and P. Clancy, "A 'partitioned leaping' approach for multiscale modeling of chemical reaction dynamics," *J. Chem. Phys.* **125**, 144107 (2006).

L.A. Harris and A.A. Quong, "Molecular chemisorption as the theoretically preferred pathway for water adsorption on ideal rutile $TiO2(110)$," *Phys. Rev. Lett.* **93**, 086105 (2004).

505451 Multiscale Multiphysics Modeling of Blood Clotting and Thrombus Biochemomechanics in the Vasculature

Abstract:

The process of clot formation and growth at the site of injury on a blood vessel wall involves a number of multiscale simultaneous processes including: multiple chemical reactions in the coagulation cascade, species transport, platelet adhesion as well as the haemodynamics and blood rheology. We model these processes at different levels of description (e.g. macro- vs. microscales). At continuum level, we solve Navier-Stokes equations for blood flow as well as the PDEs associated with advection-diffusion-reaction (ADR) of multiple species in the coagulation cascade using a spectral element solver. Depending on the required degrees of fidelity, platelets could be modeled either as a continuum concentration field or their transport and adhesion could be addressed by looking at individual platelets in a Lagrangian way. At cellular and sub-cellular levels, however, particle-based methods such as Dissipative Particle Dynamics (DPD) have proven themselves to be more effective in addressing complex biological systems. I will present the recent developments in DPD framework that extends its applicability to transport problems, and discrete models of soft biological tissues. The major drawback of modeling such fully-resolved systems at the cellular scale, however, is the long-term computation of their whole course of evolutions. There are numerous biological and materials systems that evolve at slow rates (e.g. clotting), which make the simulations prohibitively long. I propose a multifidelity technique based on co-krigging information fusion approach in statistical learning in order to facilitate acceleration in time. The multiscale framework's implications and future directions will also be discussed.

Research Interests:

- Soft Biological Systems, Multiphase and Complex Fluids (e.g., Blood)
- High-order Spectral Element and Coarse-grained Particle-based Methods for Multiphysics
- Fluid-structure Interactions and Immersed Boundary Method for Cardiovascular Mechanics
- Applied Mathematics and High-performance Scientific Computing

and Multiscale Problems

Teaching Interests:

- Fluid Mechanics (undergraduate and graduate)
- Numerical Methods and Computational Fluid Mechanics
- Heat Transfer

- Selected Topics in Multiscale Modeling
- Biomechanics

505420 Platform Technologies for Nucleic Acid-based Therapeutics

Abstract:

Research Interests:

RNA-based therapeutics, such as small-interfering (siRNAs), antisense oligonucleotides (ASOs), aptamers and synthetic mRNAs, have great potential to target a large part of the currently undruggable genes and gene products and to generate entirely new therapeutic paradigms in disease, ranging from cancer to pandemic influenza to Alzheimer's disease. However, there is still unmet need to improve their efficacy in cell, tissue and organ levels. My research interests aim to reconceptualize the way we have been delivering nucleic acids in the past decades and hence to innovate solutions to enhancing the potency of existing RNA-based therapeutics. To achieve this goal, **I plan to leverage macromolecular assembly and non-natural polypeptide synthesis to transform the delivery of siRNAs, ASOs and synthetic mRNAs**. By combining multidisciplinary approaches, including state-of-the-art protein and nucleic acid engineering (graduate work), animal models (graduate work), and polymer chemistry (postdoctoral training), our studies could potentially find utility in cancer immunotherapy, infection diseases, genetic disorders and wound healing.

Successful Proposals: My studies have led to a few successful proposals with my PhD and postdoctoral mentors including Cornell Nanobiotechnology Center Grant, DARPA, NSF and NIH.

Postdoctoral Project: Synthetic Polypeptide-Assisted Assembly of Ribonucleoproteins for Enhanced Delivery of siRNAs and mRNAs.

Under supervision of Prof. Paula Hammond, Department of Chemical Engineering, Massachusetts Institute of Technology

PhD Dissertation: Blood-Borne Cancer Metastasis: from Mechanisms to Therapeutics.

Under supervision of Prof. Michael King, Department of Biomedical Engineering, Vanderbilt University.

Research Experience:

My academic career path has been a blend of many fields of science and engineering. I pursued my PhD work with a focus on engineering novel platelet functions to neutralize circulating tumor cells through both genetic modification and biomimetic functionalization. In 2016, along with two other laboratories in the world, we demonstrated for the first time a prototype of Trojan Horse platelets for cancer prevention. In addition, my PhD work has prepared me with a variety of skillsets including genetic engineering of bone marrow stem cells (BMSCs), BMSC transplantation, and cancer mouse models. Following my graduation, I chose to conduct my postdoctoral research in Prof. Paula Hammond's lab at the Department of Chemical Engineering, MIT to learn engineering disciplines including polymer chemistry and nanotechnology.

Soon after I joined the Hammond lab, I pioneered a new strategy of co-delivering ribonucleic acids with their complementary proteins in a format of ribonucleoproteins (RNPs) by harnessing macromolecular assembly and synthetic polypeptide. Such approach has demonstrated superior efficacy in improving the delivery efficacy of siRNAs and mRNAs *in vitro* and *in vivo*. Additionally, our researches have led to collaborations with pharmaceutical companies including Moderna Therapeutics (a leading mRNA-based company) and Alnylam (a leading RNAi company).

In my future faculty position, as a follow-up study from my postdoctoral work, I am very interested in gaining a deep understanding of how synthetic polypeptides enable the assembly between RNAs and complementary proteins at an atomic level. With this knowledge in mind, I plan to further address additional challenges for each type of RNA-based therapeutics. Furthermore, I intend to apply a similar preassembly strategy spearheaded in my postdoctoral training to assemble and deliver other forms of biologics including lipids, carbohydrates and proteins. The proposed researches could potentially find utility in cancer immunotherapy, infection diseases, genetic disorders and wound healing.

Teaching Interests: One of the greatest opportunities as a mentor is to train the next generation of scientists and engineers to serve our society. Throughout my academic, I have been committed to teaching future scientists and engineers in various roles including: teaching assistant, guest lecturer, and research mentor. I have also had the privilege of mentoring five undergraduates, three master students and six PhD students in the laboratories from Cornell and MIT. Majority of them have had first-author or co-author publications under my mentorship, which is one of the most rewarding aspects of my academic career to date. I am interested and qualified in teaching both Chemical Engineering and Bioengineering courses (undergraduate and graduate level) in the areas of biomaterials, reaction kinetics, and thermodynamics. Additionally, I am excited to create a graduate course in the emerging fields of Molecular Engineering, Drug Delivery, and Synthetic Biology.

Publications:

1. **Li J**, King MR. Adhesion receptors as therapeutic targets for circulating tumor cells. Front Oncol. 2012;2:79.

2. **Li J**, Bu, P., Chen, K. and Shen, X. Spatial perturbation with synthetic protein scaffold reveals robustness of asymmetric cell division. Journal of Biomedical Science and Engineering, 2013, 6, 134-143.

3. Bu P, Chen KY, Chen JH, Wang L, Walters J, Shin YJ, Goerger JP, Sun J, Witherspoon M, Rakhilin N, **Li J**, Yang H, Milsom J, Lee S, Zipfel W, Jin MM, GÅ mÅ ÅŸ ZH, Lipkin SM, Shen X. A microRNA miR-34a-regulated bimodal switch targets notch in colon cancer stem cells. Cell Stem Cell. 2013;12:602-15.

4. **Li J**, Guillebon AD, Hsu JW, Barthel SR, Dimitroff CJ, Lee YF, King MR. Human fucosyltransferase 6 enables prostate cancer metastasis to bone. British Journal of Cancer. 2013;109:3014-22.

5. Geng Y, Chandrasekaran S, Agastin S, **Li J**, King MR. Dynamic Switch Between Two Adhesion Phenotypes in Colorectal Cancer Cells. Cellular and Molecular Bioengineering. 2014;7:35-44.

6. **Li J**, Sharkey CC, Huang D, King MR. Nanobiotechnology for the Therapeutic Targeting of Cancer Cells in Blood. Cellular and Molecular Bioengineering. 2015:1-14.

7. **Li J**, Sharkey C, King MR. Piperlongumine and immune cytokine TRAIL synergize to promote tumor death. Scientific Reports. 2015;5:9987.

8. **Li J**, Ai Y, Wang L, Bu P, Sharkey CC, Wu Q, Wun B, Roy S, Shen X, King MR. Platelet membrane-functionalized particles to target tumor cell-associated micro-thrombi. Biomaterials. 2016 Jan;76:52-65.

9. Sharkey CC, **Li J**, Roy S, Wu Q, King MR. Two-stage nanoparticle delivery of piperlongumine and tumor necrosis factor-related apoptosis-inducing ligand (TRAIL) anti-cancer therapy. Technology. 2016 Feb.

10. **Li J**, Sharkey CC, Liesveld J, King MR. Genetic engineering of platelets to neutralize circulating tumor cells. J

Control Release. 2016 Apr 28;228:38-47.

11. Chandrasekaran S, Chan MF, **Li J**, King MR. Super natural killer cells that target metastases in the tumor draining lymph nodes. Biomaterials. 2016 Jan;77:66-76.

12. **Li J**, Wang W, He Y, Li Y, Yan E, Irvine DJ, Hammond PT. Structurally programmed assembly of translation initiation nanoplex for superior mRNA delivery. ACS Nano, 2017 Mar 28;11(3):2531-2544.

13. He YP*, **Li J***, He HK, Irvine DJ, Hammond PT. Synthesis of a novel star-shaped release polymer for layer-by-layer drug films. (In review). * Equal contribution.

505415 Designing Novel Surfaces to Control the Fate of Attached Microbes

Abstract:

Designing novel surfaces to control the fate of attached microbes

Huan Gu, Ph.D.

Research Assistant Professor

Department of Biomedical and Chemical Engineering and Syracuse Biomaterials Institute, Syracuse, NY 13244, USA

Research Interests: Bacteria can be found virtually everywhere. As one of the living styles, bacteria settle onto biotic or abiotic surfaces and form communities that are called biofilms. Biofilms are the leading cause of chronic infections in humans and persistent biofouling and biocorrosion in the industry, both of which are costing the United States billions of dollars annually. Biofilms also harbor 'good' bacteria to assist human activities, for instance, wastewater treatment. A crucial part of biofilm formation is surface property (e.g., surface chemistry, topography, and charge) in which determines bacterial adhesion and the physiology/fate of attached bacterial cells. My research program exploits fundamental knowledge of the physiological response of bacteria to surface property to develop novel surfaces that either minimize the adhesion of 'bad' bacteria (e.g., multi-drug resistant pathogens) or optimize beneficial biofilms to serve our purposes better. The outcomes of my research will accelerate the control of biofilm-associated high drug resistance, hospital-acquired infections, as well as persistent biofouling and biocorrosion.

Research Experience: With a multidisciplinary background in biofilm engineering, materials and surface engineering, and biotechnology, I have investigated how to control biofilm formation and associated drug resistance by modifying surface property, and how biofilms are involved in hospital-acquired infections. During my graduate program, I established a well-defined surface system using chemical surface modification for the mechanistic understanding of biofilm formation and associated antibiotic resistance. The obtained results together with the effects of surface topography on bacterial adhesion were used to engineer 'smart' antifouling surfaces via topographic surface modification. My postdoctoral training builds upon my prior experience to control biofilm-associated hospital-acquired infections. I have gained expertise in characterizing the evolution of bacterial antibiotic resistance during hospital-acquired infections as bacteria adapt to environmental stressors (including humidity, surface topography, and immune response). Now I am focusing on developing technologies targeting the control of biofilm and associated drug resistance using magnetic force-driven engineered *Escherichia coli*and dynamic surface topography. The *E. coli* cells were engineered to respond to magnetic force differently according to their level of metabolic activity or resistance. These magnetic force-driven novel technologies, when further advanced, will have broad applications for designing innovative medical and environmental antifouling strategies or for probing and drawing out medical or environmental contaminants. Please refer to my curriculum vitae (CV) for the additional details on my research experience and relevant publications.

Teaching Interests: My goals as a teacher are to create a student-centered learning environment and develop new teaching methods to promote active and self-directed learners who will not only master the principles of chemical engineering but also be able to employ their knowledge and skills independently to solve complex problems and gain greater fluency. In the future, I envision myself teaching in both classroom and laboratory, as well as mentoring students in my research lab. Each of these settings requires a unique type of engagement with students. To address these varying needs, I have developed different teaching methods that seek to advance students' learning experience and achieve my goals as a teacher. With my training in Chemical Engineering, I am looking forward to teaching core courses in this discipline including Transport, Thermodynamics, Reaction Engineering, and Separation Processes. I am also interested in developing new courses related to nano- or micro- techniques for cell manipulation, microbes and human life, and methods for exploring cell microenvironments.

Future Directions: In my faculty position, I plan to strengthen and grow my work into a problem-driven multidisciplinary research program, with the goal of facilitating students to identify emerging problems and design new solutions from an interdisciplinary approach. Nowadays, nosocomial (hospital-acquired) infections, especially those caused by multidrug-resistant pathogens, are posing great threats to the immunocompromised individuals and public health in general due to their contributions to antibiotic resistance. Thus, the mission of my future research team will be to design more efficient antimicrobial surfaces and antibiotics for the control of nosocomial infections, based on the fundamental understanding of bacterial physiologic response to the physical and chemical stressors in both built and host environments. My previous research on the physiology and genetics of bacterial pathogens in the hospital and host environments has broadened our knowledge on how physical and chemical stressors influence the level of antibiotic tolerance. Building upon this foundation, my future translational research program will first target the design of novel molecules that will have a better accumulation in both sessile and planktonic pathogens by engineering their physiochemical property. To facilitate the development of new molecules specifically targeting the subpopulation of pathogens with high tolerance to antibiotics (also called persisters), I propose to engineer bacteria to 'sense' magnetic field. By spatially and temporally controlling gene expression using programmed magnetism regulation, I want to establish a platform that can achieve a cheaper, faster, and more robust isolation of cell groups with different levels of antibiotic tolerance for the high-throughput screening of molecules relevant to drug discovery. Finally, I aim to revolutionize the dynamic surface topography that I have established during my past research into next generation antifouling surfaces by employing magnetic force as an engine to drive the movement of polymeric surface topographies filled with iron nanoparticles. The magnetic field, with the combination of generated electric current and conductive material, can be used to design a remote technology for the detection and elimination of biofilms in both built and host environments.

I am confident that my doctoral and postdoctoral training has prepared me for establishing a successful independent research program that will make major contributions to the understanding and control of biofilms, especially, in the hospital environment. Given the interdisciplinary nature of biofilm research and close medical relevance, there are exciting opportunities to collaborate with other fields, such as biostatistics, computational modeling, microbiology, architecture science/engineering, ecology and evolution, and immunology to pursue large multidisciplinary grants. Results from these studies also have the potential to inform industry stakeholders in the development of innovative antifouling products. Beyond research studies, the biofilm-related problems are also suitable for integrating into engineering courses and for outreach with local K-12 students and teachers to promote STEM education and talent development.

505541 Enabling C1-Based Bioconversion through Metabolic Engineering

Abstract:

Research Interests:

Single-carbon (C1) substrates, such as synthesis gas and methanol, are attractive feedstocks for biochemical processes, as they are widely available, can be produced renewably, and do not compete with food supply. However, their use in industrial bioprocessing remains limited, primarily because microbes that utilize these substrates are poorly characterized biochemically, and limited tools exist for their genetic modification. This leaves the metabolic engineer with a choice: to develop genetic tools to enable engineering in the desired host, or to import the relevant catabolic pathway into a more tractable organism, such as Escherichia coli. My research program will explore both options within the context of developing strains for the conversion of C1 substrates into value-added chemicals and fuels.

Clostridium ljungdahlii is an acetogen that grows autotrophically on synthesis gas (CO, H_2, and CO_2) using the Wood-Ljungdahl pathway, and is a promising candidate for non-photosynthetic CO_2 fixation. In my PhD work, I extended its primitive genetic tools by developing a CRISPRi system for the targeted knockdown of specific genes. Constitutive downregulation of several genes with putative roles in energy conservation and carbon flux by up to 30-fold was demonstrated, and the associated phenotypes analyzed. Optimization of the promoter controlling dCas9 expression allowed for inducible knockdown, paving the way for dynamic metabolic control strategies to redirect carbon flux in engineered strains. To demonstrate this concept, several variants of a heterologous pathway for the biosynthesis of 3-hydroxybutyrate (3HB) were constructed, to probe 3HB production in the wild-type background and with various CRISPRi plasmids. Downregulation of phosphotransacetylase led to a doubling of both titer and yield of 3HB. The CRISPRi system represents a valuable contribution to the metabolic engineering field for its ability to redirect carbon flux, and is also useful to the microbiology community to probe gene function to answer open questions in the biochemistry underlying the Wood-Ljungdahl pathway.

To explore the alternative approach of importing a single-carbon catabolic pathway into a tractable host, again in my PhD work, *E. coli* was engineered to metabolize methanol. Screening various candidates of the three heterologous pathway enzymes enabled robust incorporation of 13C-labeled methanol into central carbon metabolism. To further improve methanol assimilation, a kinetic-thermodynamic modeling framework was developed and combined with novel isotopic tracing experiments to probe potential pathway limitations. Flux leakage from the cyclical ribulose monophosphate (RuMP) pathway was identified as the primary bottleneck, as this led to the build-up of the toxic intermediate formaldehyde and ablation of the thermodynamic driving force for methanol oxidation. Strategies were developed to re-wire central metabolism accordingly, which restored the driving force. Through the use of deuterated isotopic tracers and the analysis of the corresponding kinetic isotope effect, the kinetics of the first enzyme of the pathway - methanol dehydrogenase (MDH) - were identified as the next limitation. These results represent the first systematic analysis of flux limitations in E. coli engineered for methanol metabolism, and provide clear targets for further metabolic engineering to enable synthetic methylotrophy.

Through these research experiences, I developed the ability to work with non-traditional metabolic engineering hosts such as *C. ljungdahlii*, and learned advanced methodologies for analyzing heterologous metabolic pathways, such as 13C dynamic labeling. In my research program, I plan to extend the use of these techniques. For example, the CRISPRi system for *C. ljungdahlii* should allow for a high-throughput generation of libraries of knockdowns that can be assessed for production phenotype and tracked and enriched via NGS technologies. The techniques developed for the analysis of C1 assimilation in *E. coli* can be extended to substrates beyond methanol, and provide fundamental insight into thermodynamic and kinetic control and promiscuity in central carbon metabolism. Finally, I will combine the expertise I have gained in these areas, for example by applying isotopic tracer analysis to metabolic engineering of *C. lljungdahlii.*

Teaching Interests:

My chemical engineering background, combined with multidisciplinary training in chemistry, microbiology, biochemistry and molecular biology, has equipped me to teach a variety of courses. I would be particularly interested in teaching the following courses in Chemical Engineering:

1) Bioprocess Engineering - This course would introduce the core chemical engineering components of bioprocessing, including enzyme kinetics, growth kinetics, bioreactor design, and oxygen and heat transport in bioreactors, through a case-study approach designed to introduce students to the career options available in bioprocessing.

2) Metabolic Engineering - This course would introduce the theoretical and computational approaches used in modern metabolic engineering, including metabolic control analysis (MCA), metabolic flux analysis, and flux balance analysis. The course would draw examples from recent and classic literature on these topics.

505166 Engineering Metabolism for Carbon Conservation and Cellulosic Biofuel Production

Abstract:

Research Interests:

My research interest is to apply metabolic engineering for solving two major limitations in biofuel and biochemical production: (1) intrinsic carbon lost from central metabolic pathways and (2) degradation of recalcitrant biomass. To address these limitations, my research specifically focuses on rewiring central metabolic pathways for carbon conservation, and engineering an isobutanol pathway in a thermophilic cellulolytic microorganism, *Clostridium thermocellum*.

Rewiring central metabolic pathway for carbon conservation

Acetyl-coenzyme A (acetyl-CoA) is a two-carbon metabolite (C2 metabolite) and an important metabolic precursor to a variety of industrially relevant compounds including biofuels. An intrinsic limitation of acetyl-CoA derived biochemical production is the carbon loss when forming acetyl-CoA. Most organisms use some glycolytic variation, commonly the Embden-Meyerhof Pathway (EMP), to initially degrade sugar into pyruvate. Pyruvate, a C3 metabolite, is then decarboxylated to form acetyl-CoA, losing carbon to the environment. This decarboxylation limits the carbon yield to only two molecules of acetyl-CoA from one molecule of hexose, thus inhibiting the economics of any associated bioprocess. A synthetic sugar catabolism pathway, termed non-oxidative glycolysis (NOG), was recently developed to address this problem. NOG combines a phosphoketolase dependent cleavage of sugar phosphates and a carbon rearrangement cycle to directly generate C2 units per sugar in a redox neutral manner. To further expand the applications using NOG, an *Escherichia coli* strain was constructed to rely solely on NOG for sugar catabolism through rational metabolic rewiring, recombinant expression, and evolution. In this engineered and evolved bacteria, the major EMP pathway was deleted, and sugar was first converted to C2 metabolite (acetyl-CoA) through NOG prior to pyruvate formation. Under anaerobic conditions, this strain was able to convert 82% of the glucose carbon to acetate (derived from acetyl-CoA), while the maximum acetate yield from EMP is 67%. Therefore, this strain offers significant potential to be engineered for the production of a variety of acetyl-CoA derived compounds.

Consolidated bioprocessing of cellulose to isobutanol using *Clostridium thermocellum*

Biomass recalcitranceâ€"resistance to degradationâ€"currently limits the use of lignocellulose for biofuel production. Consolidated bioprocessing (CBP), in which cellulose hydrolysis and fermentation occur simultaneously in one pot without added cellulases, is a potential approach to improve lignocellulose utilization. Owing to its high cellulose deconstruction rate, *Clostridium thermocellum* is a promising thermophilic CBP host that grows at 50-60°C. The elevated temperature also promotes cellulose degradation, reduces contamination, and minimizes cooling costs. Metabolic engineering in *C. thermocellum* is usually hampered by enzyme toxicity during cloning, time-consuming pathway engineering procedures, and slow turnaround in production tests. Here, we present a streamlined approach to engineer *C. thermocellum* to produce isobutanol directly from cellulose. Essential isobutanol pathway genes under different promoters were cloned to create various plasmid constructs in *E. coli*. Then, these constructs were transformed and tested in *C. thermocellum*. Among these engineered strains, the best isobutanol producer was selected and the production conditions were optimized. Moreover, acetohydroxy acid synthase (Ahas), the first gene in the isobutanol pathway, was determined as the limiting step during the production by enzyme assay. With further Ahas overexpression, 9.7 g/L of isobutanol was produced within 120 h directly from cellulose using our engineered strain. Our study demonstrates the feasibility of CBP of lignocellulose to biofuel and biochemical compounds using *C. thermocellum*.

Teaching Interests:

My teaching interests include the all of the core classes (particularly transport phenomena and thermodynamics) of the chemical engineering discipline, as well as hands-on lab classes and research mentoring. I developed my teaching ability from multiple experiences: as a volunteer lecturer, as a teaching assistant, and as a project lead.

Teaching experience

As a graduate student at UCLA, I have provided a review of transport phenomena for students studying for the preliminary oral exam for Ph.D. candidacy. I also served as teaching assistant for thermodynamics. As a teaching assistant I delivered a one hour discussion for the whole class, conducted office hours every week, and graded homework and exams. Since thermodynamics is more abstract than other chemical engineering subjects, I always advised the students to link the problems to thermodynamic properties instead of just starting with an equation. I believe the purpose of education is to help the students develop critical thinking skills to simplify and solve problems, in addition to covering the course content.

Mentoring experience

The experience of mentoring students on research is one of my motivations to pursue a faculty career. In particular, helping students to develop their abilities, expand their competencies and achieve their potential gives me an incredible sense of fulfillment. My mentorship started at UCLA, where I established my own research team and have worked closely with 4 technicians and 9 undergraduate students in the past five years. Now, many of my former students are pursuing their advanced degree (M.D or Ph.D.) or working in related companies like DuPont and Gilead Sciences.

505531 From Integrative Metabolomics to Understanding Human Diseases and Enhancing CO_2 Fixation

Abstract:

Research Interests:

A growing concern for 21st-century humanity is the irrevocable rise of carbon dioxide (CO_2). On a more personal level, many individuals today suffer from distressing diseases like cancer. Potential solutions to these disparate problems lie in metabolism. With these applications in mind, my research goal is to advance biotechnology and medicine by integrative metabolomics, which incorporates kinetic and thermodynamic information into characterization of metabolic processes.

My postdoctoral research with Prof. Gregory Stephanopoulos has focused on enhancing CO_2 fixation by employing thermodynamic and systems-level efficiency of non-photosynthetic carbon-fixing metabolism. I developed a rapid CO_2 fixation scheme that enables precise control over the rate of CO_2 fixation (i.e., productivity) and the proportion of CO_2 incorporated into the product (i.e., yield). Using this technology, fermentation of *Moorella thermoacetica* produced acetate as a product of CO_2 fixation at high specific productivities (0.54 g per g cell dry weight (gCDW) per hour) with nearly all acetate carbons (>98%) derived from CO_2.

In establishing this technology, my graduate and undergraduate training with Profs. Joshua Rabinowitz and Bernhard Palsson helped with quantitative characterization of metabolism. Using non-equilibrium thermodynamic principles and stoichiometric metabolic models, I was able to integrate forward-to-backward reaction flux ratios with reaction quotient-to-equilibrium constant ratios via Gibbs free energy. The upshot of the integrative analysis was internally consistent and comprehensive sets of metabolite concentrations and Gibbs energies in *E. coli*, yeast, and mammalian cells. Based on these results, I found efficient enzyme usage to be a design principle across kingdoms of life.

Collaborating with other researchers, I recognized that there is ample room for new discoveries by applying the integrative metabolomic approaches in other fields. Working with Profs. Mark Brynildsen and Stanislav Shvartsman, I contributed to understanding the underlying metabolic processes in bacterial persistence and in healthy embryonic development. The resulting knowledge pointed to potential drug targets in bacteria with high antibiotic tolerance and to an essential metabolic regulation that ensures normal development.

Going forward, I would like to incorporate comprehensive enzyme cost and efficiency information into the metabolic flux and pathway analysis paradigm. Lack of efficiency in certain cells (e.g., cancer cells) could suggest potential therapeutic targets. In metabolic engineering applications, the economics of cellular proteome partitioning must be compatible with our economic goals. Optimal and feasible solutions, then, would consist of metabolic modules that simultaneously permit maximum enzyme efficiency and maximum net value created. Such integrative approaches will help disentangle complex metabolic diseases and engineering challenges, thereby facilitating better therapeutic and engineering solutions.

Teaching Interests:

Teaching and research complement each other. Without proper training of young scientists, the future of research looks bleak. Oftentimes, new unanswered questions arise from teaching and current research findings enrich the teaching component. In both settings, the importance of interdisciplinary and collaborative problem solving cannot be overstated. Many of my research projects have been fruitful collaborative efforts involving several disciplines and I can readily apply my experience to my teaching.

My teaching goals also include promoting curiosity and encouraging the endeavor to pursue new knowledge and experience. From my experience teaching in lectures, discussion and laboratory sessions as well as mentoring students, I learned that, aside from answering studentsâ€™ questions, I also have to ask them the â€˜rightâ€™ questions to engage their minds. These questions could be laid out to guide students to solve problems on their own and eventually to render them more inquisitive. I would like to incorporate my teaching principles and goals into chemical and biological engineering core courses as well as new electives that introduce students to interdisciplinary/integrative problem solving.

Selected Publications:

1. **Park JO**, Rubin SA, Xu Y, Amador-Noguez D, Fan J, Shlomi T, Rabinowitz JD. Metabolite concentrations, fluxes and free energies imply efficient enzyme usage. *Nature Chemical Biology* 2016; 12:482-489
2. Schellenberger J, **Park JO**, Conrad TM, Palsson BØ. BiGG: A Biochemical Genetic and Genomic knowledgebase of large scale metabolic reconstructions. *BMC Bioinformatics* 2010; 11:213
3. Mok WWK, **Park JO**, Rabinowitz JD, Brynildsen MP. RNA futile cycling in model persisters derived from MazF accumulation. *mBio* 2015; 6(6):e01588-15
4. Song Y, Marmion R, **Park JO**, Biswas D, Rabinowitz JD, Shvartsman SY. Dynamic control of dNTP synthesis in early embryos. *Developmental Cell* 2017
5. **Park JO**, Liu N, Holinski KM, Emerson DF, Woolston BM, Vidoudez C, Islam MA, Girguis PR, Stephanopoulos G. Enhancing CO_2 fixation by balancing reducing power and ATP. *in preparation*

505223 Selective Expansion of the Microbial Chemistry Repertoire for Metabolic and Protein Engineering

Abstract:

Post-Doctoral Research Fellow Since Sept. 2015

Research Interests:

Future Research

To better harness the potential of engineered biological systems, we must expand the repertoire of chemistry compatible with living processes. Access to additional functional groups and reaction chemistries within cells can enhance cell-based therapeutics, such as engineered probiotics, revolutionize high-throughput screens in drug development, and decrease reliance on petrochemical-based organic synthesis techniques used in manufacturing.

Two prominent examples of biochemical expansion are the introduction of non-standard amino acids (NSAAs) into the proteome and the biosynthesis of reactive aldehyde intermediates in the metabolome. NSAAs expand the structural and functional diversity of proteins, which span biocatalysts to therapeutic agents. Although systems have been engineered to enable NSAA incorporation into target proteins inside cells, these systems suffer from low selectivity for desired NSAAs. Selectivity also poses a problem for engineered biosynthesis of aldehyde intermediates inside cells, where native processes rapidly convert these molecules to alcohol byproducts and away from desired reactions.

In my research career, I have developed tools that overcome both selectivity problems, and I will use them as a faculty member to direct synthetic biology research in the following general areas:

Evolution of NSAA incorporation systems using proofreading proteins

(i) Engineering high-throughput methods to selectively discriminate amino acids from structurally similar analogs

(ii) Evolution of NSAA incorporation systems for greater selectivity

(iii) Characterization of proofreading proteins and evolved NSAA incorporating systems to inform future engineering [collaborators: Bob Sauer (MIT) and Dieter Söll (Yale)]

Biosynthesis of metabolites containing novel functional groups obtained via aldehyde intermediates

(i) Engineering of enzymes and pathways capable of converting aldehyde intermediates into further reactive bio-orthogonal chemistries, such as imines or enamines

(ii) Rational introduction of NSAAs in enzyme active sites to alter specificity towards unnatural substrates, guided by sequence alignments, crystal structures, and computational protein design

Funding strategies associated with these aims target health, energy, and defense agencies. For more information, please visit my draft future lab website: www.kunjapurlab.org.

Graduate Research

"*Microbial engineering for aldehyde synthesis*" | Dr. Kristala L. J. Prather, MIT

I received training in metabolic engineering to enable microbial synthesis of aldehydes and aldehyde-derived non-alcohol products, such as fuels, flavors, and pharmaceutical intermediates. Through combinatorial deletion of redundant genes encoding aldehyde reductases, I engineered an *E. coli* strain that could enable production of model aromatic aldehydes such as vanillin for the first time. I also tested several aliphatic aldehydes and observed widespread partial stabilization based on the same deletions. Based on these insights, my colleagues and I harnessed this strain as a platform for obtaining products derived from aldehydes. We demonstrated diverse downstream enzymatic steps that led to production of a chiral pharmaceutical intermediate through carboligation or to production of gasoline constituents through oxidative deformylation. These reactions are merely the tip of an iceberg of new potential biochemistries enabled via aldehyde retention.

Skills Acquired: Design of heterologous metabolic pathways (1–3); genome engineering strategies to achieve aldehyde retention (1, 4); identification and resolution of kinetic bottlenecks and regulatory mechanisms (5); biochemical characterization of enzymes (6).

Post-Doctoral Research

"*Engineering post-translational proofreading to discriminate non-standard amino acids*" | Dr. George M. Church, Harvard

During my post-doctoral research, my main initiative has been to expand the repertoire of amino acids that can be incorporated into proteins. I discovered how the N-end rule pathway of protein degradation applies to commonly used NSAAs, and I then rationally engineered the N-end rule pathway for tunable control over how the N-end rule selectively degrades NSAA-containing proteins. I used these insights to engineer synthetic quality control, termed "Post-Translational Proofreading" (PTP). Using my engineered PTP system, my colleagues and I can engineer cells to incorporate specific NSAAs with a high level of accuracy. This represents a dramatic change from earlier methods of NSAA incorporation, which were vulnerable to false positives that instead degrade rapidly using our system. We have used this method for evolution of more selective systems that incorporate NSAAs, and we have shown that the evolved systems improve the effectiveness of synthetic biocontainment. Thus far, this work has resulted in a first and co-corresponding author manuscript submission (7) and two provisional patent applications. I have also worked on adaptive evolution of a genomically recoded *E. coli* strain used for NSAA incorporation and am a co-first author on a submitted manuscript (8).

Skills Acquired: Site-specific *in vivo* non-standard amino acid incorporation; high-throughput screening of protein-encoding libraries; multiplexed genome engineering/editing; mass spectrometry; next-generation sequencing; PyRosetta computational protein design.

Past Proposals:

NIH NIBIB K99/R00 Resubmission (July 2017 cycle; original application score 34); DOE Genomic Sciences Center for Bioenergy 5-Year Renewal (co-author); NSF GRFP

Teaching Interests:

Teaching Experience

My interest in teaching began when I started a tutoring service that eventually served 14 clients before my high school graduation. At the University of Texas, I served as Tutor/Grader for the Transport Phenomena course for two semesters. In my final semester, I was given permission to serve as an undergraduate Teaching Assistant (TA) for the Material/Energy Balances course. At MIT, I served as a TA for two one-week courses tailored for biotechnology professionals. I also spent two summers co-teaching a seven-week "Microbial Chemical Factories" course for high-school juniors/seniors for whom a labmate and I co-designed the curriculum. Finally, I served as a TA for the graduate Kinetics and Reactor Design course at MIT, which is the most advanced course requirement for graduate chemical engineering students (my student evaluation: 6.7/7.0). My course preferences would be Kinetics, Material/Energy Balances, and Biochemical Engineering or Synthetic Biology electives.

Mentoring Experience

Parallels exist between teaching skills and research mentorship skills. In graduate school, I mentored five undergraduate research assistants, four of whom successfully co-authored publications with me (2, 5–6, 9), including one as a co-first author on a publication about the material properties of a non-ribosomally synthesized peptide (9). As a post-doc, I am mentoring two graduate students, one of whom has contributed as second author on the post-translational proofreading manuscript. Finally, having had many wonderful mentors in my own life, I support opportunities that give students access to additional mentors, and I started the SynBERC Industrial Mentorship and MIT Energy Club E-Mentors programs for this purpose.

Teaching Philosophy

My teaching philosophy for undergraduates focuses on four principles: broad relevance, participatory learning, peer-based reinforcement, and diversity.

Broad relevance: Drawing on experiences in both the petrochemical and pharmaceutical industries, and my professional network in other industries spanning consumer products and semiconductors, I will discuss scenarios and design problems that engage students with diverse career interests. Most students will not pursue academic careers, and my five undergraduate internship experiences helps me relate to those who are seeking employment after earning their bachelor's degree.

Participatory learning: Pending class size, I will encourage various degrees of interactive learning, for example by posing a question or a blank space on every board/slide that I present. When students receive questions at regular intervals rather than a one-way deluge of information, students feel more knowledgeable about material and teachers receive faster feedback on what concepts are challenging students.

Peer-based reinforcement: I will advocate for peer-based reinforcement in the form of teamwork and group discussions in every curriculum. Learning from peers is particularly valuable when groups are diverse, and those who have different backgrounds and learning styles than myself (a visually-oriented learner) will benefit tremendously from alternative explanations by their peers.

Diversity: I am committed to promoting and supporting diversity in my classroom and in my lab. This includes diversity of experience and thought in addition to demographic diversity. I have worked with exceptionally talented people from all backgrounds, including underrepresented minorities such as my first undergraduate research assistant and my graduate advisor at MIT. Additionally, I will encourage open, respectful, and fact-based sharing of information among students, and I will seek out diverse role models to support students that I mentor.

Selected Publications:

1. Kunjapur AM, Tarasova Y, Prather KLJ (2014) Synthesis and Accumulation of Aromatic Aldehydes in an Engineered Strain of Escherichia coli. *J Am Chem Soc* 136(33):11644–11654.

2. Sheppard MJ, Kunjapur AM, Wenck SJ, Prather KLJ (2014) Retro-biosynthetic screening of a modular pathway design achieves selective route for microbial synthesis of 4-methyl-pentanol. *Nat Commun* 5:5031.

3. Sheppard† MJ, Kunjapur† AM, Prather KLJ (2016) Modular and selective biosynthesis of gasoline-range alkanes. *Metab Eng* 33:28–40.

4. Kunjapur AM, Prather KLJ (2015) Microbial engineering for aldehyde synthesis. *Appl Environ Microbiol* 81(6):1892–1901.

5. Kunjapur AM, Hyun JC, Prather KLJ (2016) Deregulation of S-adenosylmethionine biosynthesis and regeneration improves methylation in the E. coli de novo vanillin biosynthesis pathway. *Microb Cell Fact* 15(1):1.

6. Kunjapur AM, Cervantes B, Prather KLJ (2016) Coupling carboxylic acid reductase to inorganic pyrophosphatase enhances cell-free in vitro aldehyde biosynthesis. *Biochem Eng J* 109:19–27.

7. Kunjapur AM, Stork DA, Kuru E, Vargas-Rodriguez O, Landon MM, Söll D, Church GM (2017) Engineering post-translational proofreading to discriminate non-standard amino acids. *BioRxiv* (preprint). DOI: 10.1101/158246.

8. Wannier† TM, Kunjapur† AM, Rice DP, McDonald MJ, Desai MM, Church GM (2017) Long-term adaptive evolution of a genomically recoded organism. *BioRxiv* (preprint). DOI: 10.1101/162834.

9. Khlystov† NA, Chan† WY, Kunjapur AM, Shi W, Prather KLJ, Olsen BD (2017) Material properties of the cyanobacterial reserve polymer multi-l-arginyl-poly-l-aspartate (cyanophycin). *Polymer* 109:238–245.

505165 Design of Synthetic C1 Carbon Assimilation Pathways

Abstract:

Research Interests:

My research interests will focus on improvement of C1 carbon assimilation, including CO_2 equivalents, formic acid, methanol and methane. I will use metabolic engineering tools to increase utilization of these C1 resources to produce useful longer chain products in common microbial platforms, or to enhance overall photosynthetic efficiency in crop systems.

C1 compounds can be produced from various sources. For example, significant amounts of methane and CO_2 are produced from anaerobic digestion of organic wastes resulting from agriculture, animal husbandry and food processing. Methanol is an important intermediate in utilization of natural gas to synthesize other feedstock chemicals. Formic acid can be produced from electrochemical reduction of CO_2, or as a byproduct in biomass pretreatment as well as some chemical synthesis. However, these C1 compounds have a common challenge in subsequent increase in carbon chain length. Chemical approaches to build C-C bonds are feasible, but require high temperature, pressure and large capital investment. Thus, biological conversion is an important step towards utilization of these C1 resources to produce useful longer chain products.

Nature has evolved several distinct pathways to assimilate various C1 compounds to form metabolites necessary for growth. Methylotrophs can grow on methane or methanol as the sole carbon and energy source through the endogenous ribulose monophosphate pathway (RuMP) or serine cycle. Some lithoautotrophs, such as *Ralstonia eutropha*, can utilize formic acid as the carbon and energy source. In addition, cyanobacteria can fix CO_2 into 3-phosphoglycerate through the Calvin-Benson-Bassham (CBB) cycle. However, engineering on these less-characterized microorganisms involves various degrees of difficulties. One solution is to engineer model organisms, such as *E. coli*, to utilize C1 compounds.

Previous studies have reported some natural C1 carbon assimilation pathways displayed several limitations when constructed into other host platforms. For example, in the most prevalent CBB cycle, the oxygenation reaction of Rubisco can cause carbon loss through photorespiration of the byproduct, glycolate. Improvement of Rubisco specificity and activity beyond the native level has proven to be ineffective. For methanol condensation via the RuMP, the toxicity of formaldehyde and the metabolite competition with the host metabolic pathways restrict its methanol assimilation efficiency.

To enhance the efficiency of C1 compounds assimilation, I designed a synthetic reverse glyoxylate shunt-glycerate (rGS-Glycerate) pathway to complement the deficiency of the CBB cycle for carbon-efficient acetyl-CoA synthesis, and also constructed a reverse glyoxylate shunt-serine (rGS-Serine) cycle, modified from the natural serine cycle in methylotrophs, for various C1 compounds assimilation.

The designed rGS-Glycerate pathway is capable of converting one C3 metabolite to two acetyl-CoA via fixation of an additional CO_2 equivalent. More important, it can also convert the CBB cycle byproduct, glycolate, into stoichiometric amount of acetyl-CoA without net carbon loss. I demonstrated the effect of the rGS-Glycerate pathway for efficient acetyl-CoA synthesis *in vitro* and in *E. coli*. Further, I implemented the pathway into a photosynthetic organism, *Synechococcus elongatus,* and showed it promoted initial growth and accumulated ketoisocaproate, an acetyl-CoA derived product, which suggests the rGS-Glycerate pathway can complement endogenous photosynthesis system for improved acetyl-CoA biosynthesis. The ultimate goal of the rGS-Glycerate pathway is to be constructed into plant systems, especially in C3 plants, in order to increase carbon fixation efficiency and further improve the production of the acetyl-CoA derived compounds in seeds.

The rGS-Serine cycle exhibits a broad capacity to assimilate various C1 compounds, which can simultaneously assimilate one inorganic carbon (CO_2 equivalents) and one reduced C1 carbon (formic acid or its upstream C1 compounds) to produce an acetyl-CoA. We demonstrated the feasibility of the cycle by complementation of growth and isotope labeling in *E. coli*. The rGS-Serine cycle opens the possibility for co-utilization of various C1 compounds as raw materials to synthesize a variety of products that *E. coli* has been engineered to produce. In the future, the rGS-Serine cycle can be developed for methane/CO_2 co-utilization to digest waste resources to produce useful products.

Teaching Interests:

My teaching interests are inseparably conjoined with my research ambitions. I strongly believe that research and teaching should always collaborate closely and that these activities complement each other in a researcherâ€™s career. An intrinsic objective of any research activity is to gain a deep and clear understanding of the underlying principles. Teaching requires the arrangement of oneâ€™s knowledge on a subject in a clear and well-organized structure. Learning how to teach a subject well is the first step in becoming well-versed in that area. Furthermore, in my experience, teaching is not a one-way process of imparting knowledge; even with a less skilled group of students, the teacher undergoes a continuous process of learning new ideas and perspectives over the same body of knowledge. I will emphasize that I am highly motivated to impart my skills and knowledge to a wider audience in order to obtain a high level of self-satisfaction and achievement.

Tong Si

600 S Mathews Ave, C-3
Urbana, IL 61801 USA

Carl R. Woese Institute for Genomic Biology
University of Illinois at Urbana-Champaign

(217) 419-1068
tongsi1@illinois.edu

EDUCATION

Doctor of Philosophy — 2014

Department of Chemical and Biomolecular Engineering
University of Illinois at Urbana-Champaign, Urbana, IL, USA

Thesis: "Genome Engineering in *Saccharomyces cerevisiae*" (advised by Professor Huimin Zhao)

Bachelor of Science — 2009

Department of Chemistry, Department of Biology
Tsinghua University, Beijing, China

Thesis: "*O*-GlcNAcylation Modulates Self-Assembly of Tau Peptides" (advised by Professor Yanmei Li)

RESEARCH EXPERIENCE

Postdoctoral Fellow — 2014-present

Biosystems Design Theme, Carl R. Woese Institute for Genomic Biology
University of Illinois at Urbana-Champaign, Urbana, IL, USA

Advisors: Profs. Jonathan V. Sweedler, Huimin Zhao, Wilfred A. van der Donk

Developed high-throughput, mass-spectrometry-based methods for profiling of microbial colonies
Devised a platform for coupled structure-activity screening of lantibiotic variants
Characterized host-microbiota interactions in rat mediated by D-amino acids
Employed synthetic biology tools for refactoring biosynthetic pathways of peptidic antibiotics
Performed correlated fluorescence and mass spectrometry imaging analyses on *Bacillus* biofilms
Profiled lipidomic changes of Parkinson's disease in a yeast model

Graduate Research Assistant — 2009-2014

Department of Chemical and Biomolecular Engineering
University of Illinois at Urbana-Champaign, Urbana, IL, USA

Advisor: Professor Huimin Zhao

Established fully automated cellular engineering methods using an integrated robotic system
Developed the RNAi-assisted genome evolution (RAGE) method in yeast
Built comprehensive, genome-wide libraries in yeast by directional cloning of cDNA libraries
Designed various high-throughput screening assays for improving diverse fermentation traits
Applied programmable nucleases (CRISPR/Cas9) for multiplex genome editing in yeast
Discovered, characterized and engineered an endogenous n-butanol pathway in yeast
Created promoter variant libraries for combinatorial optimization of metabolic pathways
Constructed a yeast strain that can grow on lignocellulose substrates as the sole carbon source

Undergraduate Research Assistant — 2007-2009

Department of Chemistry
Tsinghua University, Beijing, China

Advisor: Professor Yanmei Li

Explored the impact of *O*-GlcNAcylation on Tau protein aggregation in Alzheimer's disease
PI, National Undergraduate Innovation Training Grant (Chinese Ministry of Education)

HONORS AND AWARDS

Baxter Young Investigator Award, Baxter Healthcare Corporation, 2017

IGB Postdoctoral Fellowship, Carl R. Woese Institute for Genomic Biology, UIUC, Urbana, USA, 2014-2017

Chinese Government Award for Outstanding Self-Financed Students Abroad, Chinese Ministry of Education, 2015

Mavis Future Faculty Fellowship (2013), Drickamer Graduate Research Fellowship (2013), UIUC, Urbana, USA

Graduate College Travel Award (2013), University Block Fellowship (2009), UIUC, Urbana, USA

Invited Attendee, Global Young Scientist Summit, Singapore, Singapore, 2014

National Scholarship (2008), Academic Excellent Scholarship (2007, 2006), Social Work Excellent Scholarship (2008, 2006), Honored Freshman (2005), Tsinghua University, Beijing, China

Gold Medal, Chinese National Biology Olympiad, Xiamen, China, 2004

RESEARCH ARTICLES (*equal contribution, recent first)

16. **Tong Si***, Bin Li*, Troy J. Comi*, Yuwei Wu, Pingfan Hu, Yuying Wu, Yuhao Min, Douglas A. Mitchell, Huimin Zhao, Jonathan V. Sweedler, Optically-guided MALDI-MS profiling of microbial colonies for high-throughput engineering of multi-step enzymatic reactions, ***J. Am. Chem. Soc.***, 2017, 139(38), 12466-12473

15. **Tong Si**, Ran Chao, Yuhao Min, Yuying Wu, Wen Ren, Huimin Zhao, Automated multiplex genome-scale engineering in yeast, ***Nat. Commun.***, 2017, 15187

14. Ran Chao, Jing Liang, Ipek Tasan, **Tong Si**, L. Ju, Huimin Zhao, Fully automated one-step synthesis of single-transcript TALEN pairs using a biological foundry, ***ACS Synth. Biol.***, 2017, 6(4), 678-685

13. Bin Li, Troy J Comi, **Tong Si**, Sage JB Dunham, Jonathan V Sweedler, A one-step matrix application method for MALDI mass spectrometry imaging of bacterial colony biofilms, ***J. Mass Spectrom.***, 2016, 51(11), 1030-1035

12. **Tong Si***, Bin Li*, Ke Zhang, Yiran Xu, Huimin Zhao, Jonathan V. Sweedler, Characterization of *Bacillus subtilis* colony biofilms via mass spectrometry and fluorescence imaging, ***J. Proteome Res.***,2016, 15, 1955-1962

11. Shuobo Shi, **Tong Si**, Zihe Liu, Hongfang Zhang, Ee Lui Ang, Huimin Zhao, Metabolic engineering of a synergistic pathway for n-butanol production in *Saccharomyces cerevisiae*, ***Sci. Rep.***, 2016, 6, 25675

10. Sijin Li, **Tong Si**, Meng Wang, Huimin Zhao, Development of a synthetic malonyl-CoA sensor in *Saccharomyces cerevisiae* for intracellular metabolite monitoring and genetic screening, ***ACS Synth. Biol.***, 2015, 4, 1308-1315

9. Youyun Liang, **Tong Si**, Ee Lui Ang, Huimin Zhao, An engineered penta-functional minicellulosome for simultaneous saccharification and ethanol fermentation, ***Appl. Environ. Microbiol.***, 2014, 80, 6677-6684.

8. Zehua Bao, Han Xiao, Jing Liang, Xiong Xiong, Ning Sun, **Tong Si**, Huimin Zhao, Homology Integrated CRISPR-Cas system for one-step multi-gene disruptions in *Saccharomyces cerevisiae*, ***ACS Synth. Biol.***, 2015, 4, 585-594.

7. Jiazhang Lian, **Tong Si**, Nikhil U Nair, Huimin Zhao, Design and construction of acetyl-CoA overproducing *Saccharomyces cerevisiae* strains, ***Metab. Eng.***, 2014, 24, 139-149

6. **Tong Si**, Yunzi Luo, Zehua Bao, Huimin Zhao, RNAi-Assisted genome evolution in *Saccharomyces cerevisiae* for complex phenotype engineering, ***ACS Synth. Biol.***, 2015, 4, 283-291

5. **Tong Si**, Yunzi Luo, Han Xiao, Huimin Zhao, Utilizing an endogenous pathway for 1-butanol production in *Saccharomyces cerevisiae*, ***Metab. Eng.***, 2014, 22, 60-68

4. Mo Jiang, Min Hao Wong, Zhilong Zhu, Jieqian Zhang, Lifang Zhou, Ke Wang, Ashlee N. Ford Versypt, **Tong Si**, Lisa M. Hasenberg, Yao-En Li, Richard D. Braatz, Towards achieving a flattop crystal size distribution by continuous seeding and controlled growth, ***Chem. Eng. Sci.***, 2012, 77, 2-9

3. Jie Sun, Fei Wen, **Tong Si**, Huimin Zhao, Direct conversion of xylan to ethanol by recombinant *Saccharomyces cerevisiae* strains displaying an engineered mini-hemicellulosome, ***Appl. Environ. Microbiol.***, 2012, 78, 3837-45 (Highlighted in *Nat. Chem. Bio.*, 2012, 8, 502)

2. Jing Du, Yongbo Yuan, **Tong Si**, Jiazhang Lian, Huimin Zhao, Customized optimization of metabolic pathways by combinatorial transcriptional engineering, ***Nucleic Acid Res.***, 2012, 40, e142

1. Chunhui Yu, **Tong Si**, Yanmei Li et al., *O*-GlcNAcylation modulates the self-aggregation ability of the fourth microtubule-binding repeat of tau, ***Biochem. Biophys. Res. Commun.***, 2009, 375, 20059-62

INVITED REVIEWS (*equal contribution, recent first)

6. Ran Chao, Shekhar Mishra, **Tong Si**, Huimin Zhao, Engineering Biological systems using automated biofoundries, ***Metab. Eng.***, 2017, 42, 92-108
5. **Tong Si**, Huimin Zhao, A brief overview of synthetic biology research programs and roadmap studies in the United States, ***Synth. Sys. Biotechnol.***, 2016, 1(4), 258-264
4. **Tong Si**, Mohammad HamediRad, Huimin Zhao, Regulatory RNA-assisted genome engineering in microorganisms, ***Curr. Opin. Chem. Biol.***, 2015, 36, 85-90
3. **Tong Si***, Han Xiao*, Huimin Zhao, Rapid prototyping of microbial cell factories via genome-scale engineering, ***Biotech. Adv.***, 2015, 33, 1420-1432
2. Meng Wang*, **Tong Si***, Huimin Zhao, Biocatalyst development by directed evolution, ***Bioresour. Technol.***, 2012, 115, 117-25
1. Ryan Cobb, **Tong Si**, Huimin Zhao, Directed evolution: an evolving and enabling synthetic biology tool, ***Curr. Opin. Chem. Biol.***, 2012, 16, 285-91

BOOK CHAPTERS (*equal contribution, recent first)

2. **Tong Si***, Jiazhang Lian*, Huimin Zhao, "Strain development by whole-cell directed evolution" In *Directed Enzyme Evolution: Advances and Applications*, (M. Alcalde Ed.) pp. 173-200, Springer International Publishing AG, 2017
1. **Tong Si,** Huimin Zhao, "RNAi-assisted genome evolution (RAGE) in *Saccharomyces cerevisiae*" In *Methods in Molecular Biology*, Volume 1470, pp. 183-198: *High-Throughput RNAi Screening: Methods and Protocols*, (D. Azorsa and S Arora, Eds.), Humana Press, 2016

INVITED TALKS

1. "Post-translational Modification Modulates the Conformation and Property of the Functional Domains from Tau Protein", National Undergraduate Innovation Conference, Hunan, Changsha, China, October 27, 2008
2. "Tools and Applications for Metabolic Engineering in *Saccharomyces cerevisiae*", Beijing Institute of Technology, Beijing, China, June 14, 2013
3. "RNAi-assisted Genome Evolution", Institute of Plant Physiology & Ecology, Shanghai Institute for Biology Sciences, Chinese Academy of Sciences, Shanghai, China, July 9, 2013
4. "RNAi-assisted Genome Evolution in *Saccharomyces cerevisiae*", Metabolic Engineering Research Laboratory, Agency for Science, Technology and Research, Singapore, Singapore, January 23, 2014
5. "Automated Genome Engineering of Yeast Cells Using a Biological Foundry", 2016 Society for Industrial Microbiology and Biotechnology (SIMB) Annual Meeting and Exhibition, New Orleans, USA, July 26, 2016
6. "Accelerating Microbial Screens Using Automated Biofoundry and Mass Spectrometry", Tianjin Institute of Industrial Biotechnology, Chinese Academy of Sciences, Tianjin, China, 2018 (date TBD)

CONFERENCE PRESENTATIONS

1. **Tong Si**, Huimin Zhao, "Eukaryotic Reprogramming by Directed Genome Evolution", Biochemical and Molecular Engineering XVIII, Beijing, China, June 16-20, 2013 (**Young Investigator Poster Award**)
2. **Tong Si**, Huimin Zhao, "Eukaryotic Reprogramming by Directed Genome Evolution", "Turn on an Endogenous Pathway for 1-Butanol Production in *Saccharomyces cerevisiae*", 2013 Annual ChBE Graduate Symposium, Urbana, IL, USA, October 25, 2013 (Talk and Poster, respectively)
3. **Tong Si**, Huimin Zhao, "Fine-Tuning of Gene Expression in *Saccharomyces cerevisiae* through RNA Interference", 2013 AIChE Annual Meeting, San Francisco, CA, USA, November 3-8, 2013 (Talk)
4. **Tong Si**, Huimin Zhao, "Eukaryotic Reprogramming by RNAi-assisted Genome Evolution", 2013 AIChE Annual Meeting, San Francisco, CA, USA, November 3-8, 2013 (Talk)

5. **Tong Si**, Huimin Zhao, "Eukaryotic Reprogramming by RNAi-assisted Genome Evolution", 2014 IGB Fellows Symposium, Urbana, IL, USA, May 8, 2014 (**Best Poster Award)**
6. **Tong Si**, Huimin Zhao, "Enabling Multiplex Genome Engineering in Yeast by RNAi and CRISPR", 2014 ChBE Graduate Symposium, Urbana, IL, USA, October 24, 2014 (**Best Oral Presentation Award**)
7. **Tong Si**, Lu Zhang, Ran Chao, Huimin Zhao, "Enabling Multiplex Genome Engineering in Yeast by RNAi and CRISPR", 2014 AIChE Annual Meeting, Atlanta, GA, USA, November 16-21, 2014 (Talk)
8. **Tong Si**, Yunzi Luo, Han Xiao, Huimin Zhao, "Utilizing an Endogenous Pathway for 1-Butanol Production in *S. cerevisiae*", 2014 AIChE Annual Meeting, Atlanta, GA, USA, November 16-21, 2014 (Talk)
9. **Tong Si**, Ran Chao, Huimin Zhao, "Enabling Multiplex Genome Engineering in Yeast using RNAi and CRISPR", 2015 Synthetic Biology, Engineering, Evolution & Design Conference, Boston, MA, USA, June 10-13, 2015 (Poster)
10. **Tong Si**, Ran Chao, Huimin Zhao, "Automated Metabolic Engineering of Yeast Cells using a Biological Foundry", 2015 AIChE Annual Meeting, Salt Lake City, UT, USA, November 8-13, 2015 (Talk)
11. **Tong Si**, Bin Li, Jonathan V. Sweedler, Huimin Zhao, "Characterization of Bacillus Subtilis Colony Biofilms Via Mass Spectrometry and Fluorescence Imaging", 2016 Synthetic Biology, Engineering, Evolution & Design Conference, Chicago, IL, USA, July 18-21, 2016 (Poster).
12. **Tong Si**, Bin Li, Jonathan V. Sweedler, Huimin Zhao, "Mass Spectrometry-Based Screening for Protein Engineering" 2016 AIChE Annual Meeting, San Francisco, CA, USA, November 13-18, 2016 (Poster)
13. **Tong Si**, Wilfred A. van der Donk, Jonathan V. Sweedler, Huimin Zhao, "Accelerating Build and Test of Microbial Libraries via Integration of Synthetic Biology, Robotic Automation and Mass Spectrometry", Gordon Research Conference on Synthetic Biology, Stowe, VT, USA, July 29-Aug 4, 2017 (Poster)
14. **Tong Si**, Wilfred A. van der Donk, Jonathan V. Sweedler, Huimin Zhao, "A Versatile Synthetic Biology Platform for High-Throughput Structure and Activity Screening of Ribosomally-Synthesized and Post-Translational Modified Peptides", 2017 AIChE Annual Meeting, October 29-November 3, 2017 (Talk)
15. **Tong Si**, Bin Li, Troy J. Comi, Huimin Zhao, Jonathan V. Sweedler, "A Structure-Based High-Throughput Screening Method for Multi-Step Enzymatic Reactions Using Optically-Guided Mass Spectrometry Profiling of Microbial Colonies", 2017 AIChE Annual Meeting, October 29-November 3, 2017 (Talk)

PATENTS

1. Huimin Zhao, Byongjin Kim, Jing Du, Yongbo Yuan, Dawn Eriksen, **Tong Si**, Combinatorial Design of Highly Efficient Heterologous Pathways, US2011/054650
2. Huimin Zhao, **Tong Si**, Eukaryotic Reprogramming by RNAi-assisted Genome Evolution, United States Provisional Patent Application TF12191-PRO

TEACHING EXPERIENCE

<u>Department of Chemical and Biomolecular Engineering</u>
University of Illinois at Urbana-Champaign, Urbana, IL

Teaching Assistant, Biomolecular Engineering (ChBE473) — Spring 2011

Teaching Assistant, Techniques in Biomolecular Engineering (ChBE472) — Fall 2011

Teaching Assistant, Principles of Chemical Engineering (ChBE221) — Fall 2012

STUDENT MENTORING

<u>Graduate and Postdoctoral students</u> (6 students, PI: Professor Huimin Zhao)

Chang Sun (rotation), Zehua Bao, Dr. Han Xiao, Shekhar Mishra, Qiqi Tian, Pu Xue

Undergraduate students (11 students, PI: Professor Huimin Zhao)

Wei Yang, 2012-2013; Sherrie Lim Xuan Lu, 2013; Hejun Li, 2013-2014; Wen Ren, 2015-2016 (**Outstanding Oral Presentation**, 2016 Campus Undergraduate Research Symposium); Ke Zhang, 2015-2016; Yiran Xu, 2015; Yuying Wu, 2015-2017; Pingfan Hu, 2016-2017; Yuwei Wu, 2016-2017; Yuhao Min, 2015-date (**John E. Gieseking Scholarship for Research; Carl R. Woese Summer Research Scholar**); Linzixuan Zhao, 2017-date

SERVICE TO PROFESSIONAL ASSOCIATIONS

American Institute of Chemical Engineers (AIChE)

Co-chair, Advances in Metabolic Engineering, 2015 AIChE Annual Meeting, Salt Lake City, UT, USA

Co-chair, Advances in Metabolic Engineering and Bioinformatics for Biofuels II: Next-Generation Method Development, 2015 AIChE Annual Meeting, Salt Lake City, UT, USA

Co-chair, Advances in Metabolic Engineering of Photosynthetic/Non-Model Organisms, 2016 AIChE Annual Meeting, San Francisco, CA, USA

Co-chair, Industrial Application of Metabolic Engineering, 2017 AIChE Annual Meeting, Minneapolis, MN, USA

JOURNAL REVIEWER

Current Opinion in Chemical Biology, Metabolic Engineering, ACS Synthetic Biology, Molecular Biosystems, Journal of Industrial Microbiology & Biotechnology, Enzyme and Microbial Technology, Applied Microbiology and Biotechnology, PLOS ONE, RSC Advances, Applied Biochemistry and Biotechnology, Progress Biochemistry, Soft Matter, Bioresources and Bioprocessing, Colloid and Interface Science Communications, Molecules, Frontiers in Bioengineering and Biotechnology, Frontiers in Microbiology, World Journal of Microbiology and Biotechnology, Biotechnology and Bioprocess Engineering, Synthetic and Systems Biotechnology, Engineering in Life Sciences, Bioengineered

CAMPUS SERVICE

Department of Chemical and Biomolecular Engineering
University of Illinois at Urbana-Champaign, Urbana, IL

Search Committee for Coordinator of Research Programs — 2016

Carl R. Woese Institute for Genomic Biology
University of Illinois at Urbana-Champaign, Urbana, IL

Planning Committee for 10th Annual Biomass Conversion Symposium	2011
Search Committee for Laboratory Technician	2014
Search Committee for Laboratory Research Manager	2016
Planning Committee for IGB Fellows Symposium	2016
Planning Committee for UIUC-KU-CAS Joint Symposium	2016

OUTREACH SERVICE

"An Assembly Line of Microbes for Fuel", Public Seminar at Chambana Science Café, Urbana, IL, March 2, 2016

"Genome Day", Biosystems Design Activity Leader, Orpheum Children's Science Museum, Urbana, IL, November 14, 2015; November 12, 2016.

Research Plan: Microbiome Engineering to Study and Modulate Host Immunity

Executive Summary. The microbiota and host immunity are engaged in constant dialogs to maintain symbiotic homeostasis. Dysregulated partnerships are implicated in many severe pathologies, such as inflammatory bowel syndrome, type 1 diabetes, and cancer, which affect millions of people. Microbial regulation of the host immunity provides new therapeutic opportunities via microbiome engineering. **My research aims at applying synthetic biology principles to understand and reprogram molecular communications at the microbiota-immunity interface**. For this enormously complex symbiont, current knowledge is mainly obtained from observational sequencing studies. Mechanistic understanding is limited due to the lack of effective means for molecular profiling and genetic engineering. To overcome this grand challenge, matrix-assisted laser desorption/ionization (MALDI) mass spectrometry (MS) permits rapid, multiplex detection of microbial and immune molecules. Genetic tools will be developed to enable microbiome engineering for hypothesis testing and therapeutic modulation. Overall, **insights on how microbial genes and molecules elicit diverse immune responses will be obtained to design probes and therapeutics**. These objectives are addressed in three Projects for the next five years.

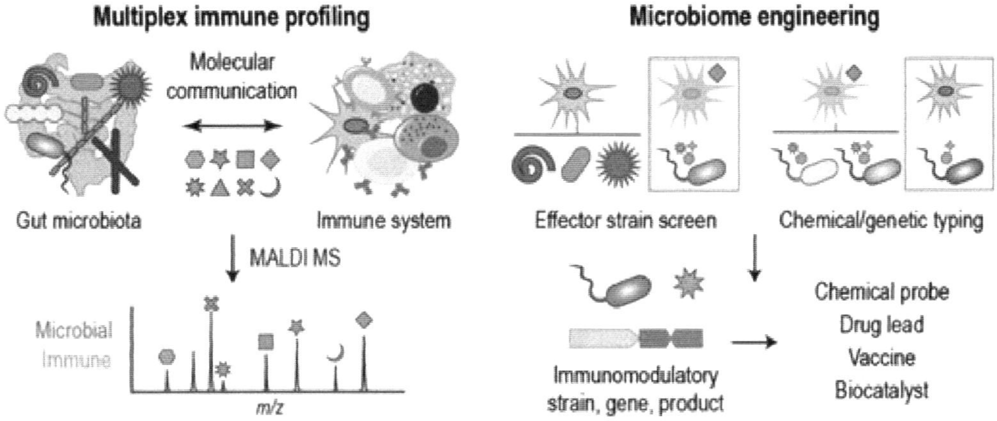

P1. Identifying immune effector strains from the gut microbiota. I plan to create clonally arrayed strain libraries from the gut microbiota to examine how individual organisms impact host immunity. Multiplex immune assays will be developed for rapid, untargeted analyses of immune factor release using MALDI MS. Cellular heterogeneity and communication of the immune network are revealed via MALDI MS imaging. Effector strains will be identified via phenotypic screens using single-type immune cells, lymphatic tissues, and animal models.

P2. Developing genetic tools for mechanistic studies of commensal-mediated immune tolerance. Conjugative plasmid libraries will be created to identify and manipulate permissive commensal recipients. For tolerogenic strains identified in P1, genetic tools will be used to identify active microbial molecules and associated genes. Tolerance-promoting molecules will be applied as chemical probes to understand and modulate regulatory immune responses.

P3. Engineering microbial vaccines for oral immunization. Microbes have been exploited as effective vectors for oral immunization, and I plan to engineer antigen presentation in microbial vaccines. Tumor-associated glycosylation will be introduced on cancer antigens to evaluate immunogenicity and specificity of engineered yeast vaccines. The unique capability of MALDI MS for both DNA sequencing and chemical phenotyping will be exploited to accelerate strain engineering via massively parallel sequence-function mapping.

Perspective. The microbiota and host immunity play central roles in health and disease, but both systems contain diverse, interacting cells that are challenging to study. I plan to use high-throughput chemical and genetic tools to dissect cellular heterogeneity and communication. New biological insights and engineered microbial agents will enable immunologic modulation to treat severe diseases. Broadly, these technologies will facilitate the grand transition from observational studies to mechanistic elucidation in microbiota research.

Past Research

My past research studied natural and engineered microbial diversity at the molecular level (Fig. 1a-c) using high-throughput analyses (Fig. 1d-f), with the goal of understanding and reprogramming microbial metabolism for industrial and biomedical applications. During my doctoral training with Dr. Huimin Zhao in the Department of Chemical and Biomolecular Engineering at University of Illinois at Urbana-Champaign (UIUC), I engineered *S. cerevisiae*, the baker's yeast, as an efficient microbial cell factory. A variety of phenotypes were targeted including inhibitor resistance,[1,2] overproduction of chemicals[3,4] and metabolic intermediates,[5,6] cell surface display of cellulase,[7,8] and substrate utilization.[9] New **genetic methods** were established to create diversities at the protein, pathway and genome levels. Notably, multiplex genome engineering in yeast was fully automated via integration of synthetic biology and robotic automation (Fig. 1a).[1,10] **High-throughput assays** were established for coupled genotyping and phenotyping, integrating various methods such as genetic biosensors,[5] next-generation sequencing (NGS, Fig. 1d),[1] and flow cytometry.[1]

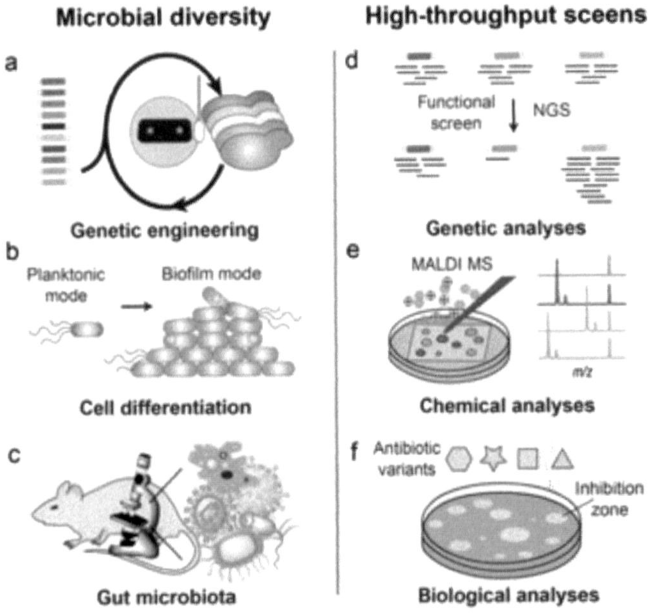

Figure 1. Schematic demonstration of my past research.

For postdoctoral research, as an IGB Fellow at the Carl R. Woese Institute for Genomic Biology (UIUC), I had the opportunities to work with Drs. Jonathan V. Sweedler, Wilfred A. van der Donk, and Huimin Zhao to develop **MALDI MS methods** for high-throughput, molecular phenotyping of microbial systems. For naturally-occurring diversities, cell differentiation and associated chemical signaling within *Bacillus* biofilms were analyzed via fluorescence and MALDI MS imaging (Fig. 1b).[11,12] Using MALDI-based strain-typing, I have been isolating clonal cultures of microbial strains from the rat gut to study **host-microbe interactions** (Fig. 1c, *in preparation*). Upon *ex vivo* stimulation with microbial metabolites such as D-amino acids, rat islets are analyzed using either MS imaging or single-cell MS profiling to monitor peptide content of pancreatic cells. For engineered diversities, optically guided MALDI MS was developed to rapidly screen molecular content of randomly distributed *E. coli* colonies (Fig. 1e).[13] In these colonies, ribosomally synthesized and post-translationally modified peptides were subjected to coupled structure-activity mapping to study and engineer antibiotic variants (Fig. 1f, *in preparation*).

Through these projects, I built an interdisciplinary foundation in synthetic biology, analytical chemistry, chemical biology, and robotic automation, which provide me with unique capabilities to tackle the grand challenges in studying and engineering complex communications between the microbiota and host immunity.

References of Past Research

(1) Si, T; Chao, R; Min, Y, et al. *Nat Commun* **2017**, *8*, 15187.
(2) Si, T; Luo, Y; Bao, Z, et al. *ACS Synth Biol* **2015**, *4*, 283.
(3) Shi, S; Si, T; Liu, Z, et al. *Sci Rep* **2016**, *6*, 25675.
(4) Si, T; Luo, Y; Xiao, H, et al. *Metab Eng* **2014**, *22*, 60.
(5) Li, S; Si, T; Wang, M, et al. *ACS Synth Biol* **2015**, *4*, 1308.
(6) Lian, J; Si, T; Nair, NU, et al. *Metab Eng* **2014**, *24*, 139.
(7) Liang, Y; Si, T; Ang, EL, et al. *Appl Environ Microbiol* **2014**, *80*, 6677.
(8) Sun, J; Wen, F; Si, T, et al. *Appl Environ Microbiol* **2012**, *78*, 3837.
(9) Du, J; Yuan, Y; Si, T, et al. *Nucleic Acids Res* **2012**, *40*.
(10) Chao, R; Liang, J; Tasan, I, et al. *ACS Synth Biol* **2017**, 6, 678
(11) Si, T; Li, B; Xu, Y, et al. *J Proteome Res* **2016**, *15*, 1955.
(12) Li, B; Comi, T; Si, T, et al. *J Mass Spectrom* **2016**, *51*, 1030.
(13) Si, T; Li, B; Comi, TJ, et al. *J Am Chem Soc* **2017**, in press

505428 Synthetic Biology for Next-Generation Plant Natural Product Discovery and Biosynthesis

Abstract:

Research Interests:

Plant natural products are chemicals of great pharmaceutical and agrochemical significance due to their structural complexity and chemical diversity. Comprehensive study of their biosynthesis lays the groundwork for studies to not only further our understanding of plant secondary metabolism, but also advance the biomanufacturing of natural product based novel bioactive compounds. Yeast has been proven as a versatile platform host for the production of various complex plant natural products, while increasingly, efforts have been made to leverage yeast as a heterologous host for plant natural product discovery. I am interested in developing synthetic biology platforms to enable the systematic discovery and production of plant natural product compounds, ultimately transforming our ability to understand how nature achieves complex compound biosynthesis. My future research will integrate plant functional genomics, synthetic biology, and metabolic engineering to develop and advance engineering approaches for putative plant biosynthetic pathway prediction and reconstruction, cellular information sensing, and metabolite production optimization. In the short term, I will develop synthetic biology technologies by discovering and engineering novel natural products derived from model plants (e.g., Arabidopsis and tomato) in yeast, which will contribute new synthetic biology toolkits, transcriptomics and metabolomics analysis methods, and reconstructed pathways that will be of great use to the broader research community. In the long term, my research agenda will expand to study and engineer less studied plant systems, thus drastically empowering plant natural product exploration. This research program will advance the development of scalable complex natural product manufacturing platforms, and will foster a variety of interdisciplinary collaborations and applications (e.g., plant biology, bioinformatics, biomass conversion, bioprocessing, and crop protection).

Research Experience:

(1) Discovery and biosynthesis of complex plant natural product in yeast utilizing synthetic biology approaches (Advisor: Professor Christina D. Smolke, Stanford University)

I have dedicated my postdoctoral research to develop cutting-edge synthetic biology technologies for the discovery and biosynthesis of complex compounds. To demonstrate the potential of yeast as a discovery platform, my research has been focused on the development of an integrated plant cluster discovery pipeline spanning multiple research fields (i.e., bioinformatics, plant biology, natural product biochemistry, and synthetic biology). The multidisciplinary approach allows us to predict novel gene clusters directly from plant genome sequences, which led to the discovery of novel tomato natural products. To leverage the versatility of yeast as a production platform, I have developed a 31-gene plant benzylisoquinoline alkaloid (BIA) biosynthetic pathway in yeast, and enhanced the titer of the anticancer BIA, noscapine, from 20 ng/L to 2 mg/L by protein engineering, metabolic engineering, and fermentation optimization.

(2) Development of synthetic biology approaches for efficient yeast engineering (Advisor: Professor Huimin Zhao, UIUC)

My graduate research focused on developing synthetic biology approaches to engineer *Saccharomyces cerevisiae* for enhanced bioethanol production using protein evolution, pathway reconstruction, and genome editing tools. To facilitate metabolite sensing and screening, I developed the first transcription factor based malonyl-CoA biosensor in yeast that enables genome-wide, high-throughput screening. I also developed a workflow for flux analysis of primary metabolism in yeast, and employed transcriptomics analysis to identify novel gene targets for further strain optimization. By leveraging these engineering and analytical methods at the protein, pathway, and genome levels, I established a highly efficient yeast strain that converts multiple carbon sources to bioethanol and enhanced the overall titer of bioethanol by three-fold, leading to direct implementation of this technology in pilot scale production.

Teaching Interests:

Teaching experience

Serving as a teaching assistant at UIUC, I taught two graduate-level courses, "Mass Transfer" and "Biomolecular Engineering". During the three years' TA experience, I have established my teaching principles: 1) inspire interests in advanced scientific research and frontier technologies; 2) enable the students to understand the concepts behind the teaching materials; 3) help students to develop critical thinking skills; 4) provide opportunities for teamwork and leadership; and 5) provide constructive feedbacks and follow-up suggestions.

Mentoring experience

I obtained extensive mentoring experience by mentoring students at both undergraduate and graduate levels. I was the formal laboratorial mentor of an undergraduate student at UIUC. I also served as the mentor of rotation students at Stanford University. My objective during mentoring is to both provide fundamental scientific and laboratorial training and to encourage academic independency. The projects were designed to provide adequate technical training and to ensure the mentees develop the skills they need to be independent researchers. I worked with the mentees to provide feedback throughout the learning process to ensure that they continue to sharpen and hone their research skills. Specifically, I insist on encouraging my students to utilize available resources and supports from experts, and to think broadly and explore their preliminary ideas, thus enabling the establishment of their own scientific vision and research goals.

Teaching philosophy

My goal of teaching and mentoring is to create a learning environment that interfaces critical thinking with solid understanding of state-of-the-art scientific knowledge. I anticipate the students to obtain the awareness of scientific and engineering principles, an interdisciplinary knowledge background, and the capability of maintaining their curiosities and interests, which will serve them regardless of their eventual career paths. To achieve this goal, I would like to motivate the students by showing the applications of the concepts taught in class, thus they are inspired to pursue the knowledge after class. I will define the scope and goals of the courses, pose questions in the beginning of the lectures, and expect the students to find the answers during the lecture. Homework will be designed to clarify the central topics of the course, normally comprise of one or two much more comprehensive problems to ensure students can learn and understand on their own. I would also emphasize written evaluations including projects and exams. In addition, students are always encouraged to discuss with peer, TA and me on any related topics to extend their vision.

Selected Publications:

1. Y. Li*, **S. Li*** (equal contribution) and C. D. Smolke, "Complete biosynthesis and optimization of the anticancer noscapine in yeast" (in preparation)
2. **S. Li***, Y. Li* (equal contribution) and C. D. Smolke, "Microbial synthesis of high-value

phytochemicals", Nature Chemistry (in review)
3. M. Wang, **S. Li** and H. Zhao, "Design and engineering of intracellular-metabolite-sensing/regulation gene circuits in eukaryotes", Biotechnology and Bioengineering, 2016, 113, 206-215
4. **S. Li**, T. Si and H. Zhao, "Development of a synthetic malonyl-CoA sensor in *Saccharomyces cerevisiae* for intracellular metabolite monitoring and genome-wide screening", ACS Synthetic Biology, 2015, 4, 1308-1315
5. **S. Li***, S. Ha*(equal contribution), H. Kim, J. M. Galazka, J. H. D. Cate, Y. Jin and H. Zhao, "Investigation of the functional role of aldose 1-epimerase in engineered cellobiose utilization", Journal of Biotechnology, 2013, 10, 1-6
6. **S. Li**, J. Du, J. Sun, J. M. Galazka, N. L. Glass, J. H. D. Cate, X. Yang and H. Zhao, "Overcoming glucose repression in mixed sugar fermentation by co-expressing a cellobiose transporter and a β-glucosidase in *Saccharomyces cerevisiae*", Molecular Biosystems, 2010, 6, 2129-2213
7. J. Du, **S. Li** and H. Zhao, "Discovery and characterization of novel D-xylose-specific transporters from *Neurospora crassa* and *Pichia stipitis*", Molecular Biosystems, 2010, 6, 2150-2156

505760 Developing Biologically Active Ionic Liquids for Therapeutic Applications

Abstract:

Research Interests:

My future research interests are directed towards the use of ionic liquids (ILs) for therapeutic applications. Ionic liquids are organic salts with melting points below 100°C, and are composed entirely by ions. They are considered materials with unique and tunable properties (e.g. melting point, miscibility, polarity, or bioavailability) which can be adjusted by the proper selection of the ionic liquid's cation and anion for a specific need. The possibility to modify the physicochemical and biopharmaceutical properties of ionic liquids is what fascinates me, and makes them very attractive candidates for a wide range of applications. In fact, they can be used to enhance human health and well-being. In particular, I'm interested in developing biologically active ionic liquids for therapeutic applications.

Prior and Current Research Experience:

My research experiences have been at the intersection of chemistry, biology, and engineering. I received my PhD in Chemical Engineering with a minor in Chemistry at the Georgia Institute of Technology in December 2013, under the supervision of Prof. Charles A. Eckert and Prof. Charles L. Liotta. My PhD research work mainly focused towards the application of green chemistry and engineering principles; looking into ways to reduce waste, improve efficiency and reduce energy consumption of a chemical process by employing green solvents. Currently, I'm a postdoctoral researcher working with Prof. Mark R. Prausnitz in the Laboratory for Drug Delivery at the Georgia Institute of Technology. My postdoctoral work involves the synthesis and characterization of ionic liquids to enhance topical delivery of active pharmaceutical ingredients (APIs). Overall, all my research experiences have provided me a comprehensive set of techniques and expertise that will enable the success of my future research interests.

Teaching Interests:

I would like to improve society as a whole by providing it with better tools, but also by educating its members. I have been privileged to have the experience of both, teaching in a classroom and mentoring students in the lab. I have been a teaching assistant in a couple of undergraduate/graduate courses including the mass and energy balance, and drug design, development, and delivery course. In addition, I have mentored both graduate and undergraduate students in the lab. My goal is to encourage their critical thinking, develop their written and oral communication skills, and prepare the students to be independent researchers. I look forward to teaching and developing courses at both the undergraduate and graduate level. My teaching interests include undergraduate and graduate level reaction engineering, thermodynamics, fluid mechanics, separations, and drug delivery courses. I'm also interested in teaching the chemical engineering laboratory courses.

Keywords: Ionic Liquids, Drug Delivery, Therapeutic

505156 Pharmaceutical System Engineering

Abstract:

Research Interests:

My main research area is process system engineering with focus on continuous pharmaceutical manufacturing. I have contributed significantly in the field of process monitoring and control, process modelling and simulation, optimization, PAT, QbD, new methodology and software tools development.

Process system engineering (PSE) is a very advanced field and has been very successful in solving complex problems of chemical based product manufacturing. However, the particulate systems are still poorly understood and there is different level of complexities involved, therefore new techniques and tools need to be developed for pharmaceutical manufacturing through extensive fundamental and applied research.

Currently, major pharmaceutical industries are shifting toward continuous manufacturing technology in which PSE can play a very significant role. Furthermore, there is a need of advanced modular continuous manufacturing platform technology for fast and robust development of solid oral products. Master formulations, process models, standardized sensing and control architecture, superior automation and novel methods and tools need to be developed to enable platform technology. Author's main research area is pharmaceutical system engineering (PSE). This area of research and education are highly encouraged and supported by pharmaceutical companies, academic institutions, regulatory authority (FDA) and national science foundation (NSF). For example, national science foundation of USA has established a 100 million dollar research center for structured organic particulate systems (C-SOPS) at Rutgers University New Jersey. Furthermore, more than 40 companies are supporting this center economically and providing scientific inputs so that the research outcomes can be directly applied to the companies.

In this work, a scientific foundation for the optimal design, optimization and control of the modular continuous process involving structured organic particulate systems has been developed. The mathematical models for continuous manufacturing of active pharmaceutical ingredients and final product tablet, has been developed for virtual experimentation and applied for plant design, optimization and control. Advanced modelling techniques such as PBM, DEM and CFD have been applied for pharmaceutical process modeling. The developed continuous tablet manufacturing plant and process model has been adapted by several pharmaceutical companies. A flexible monitoring and control system incorporating different PAT techniques and tools and control strategies for continuous pharmaceutical manufacturing process has been designed and implemented to Rutgers pilot-plant. Different control strategies such as PID, PID coupled with dead time compensator, and MPC has been compared and evaluated practically for pharmaceutical process control in terms of set point tracking and disturbances rejection ability, robustness, and stability/instability. A new hybrid MPC-PID control strategy has been developed and implemented into the pharmaceutical tablet manufacturing plant and its performance has been found better than the other control strategies. For the first time, the closed-loop operation of the tablet manufacturing plant using NIR sensor has been demonstrated, which is a significant advancement in pharmaceutical manufacturing as endorsed by the Food and Drug Administration (FDA) and other regulatory authorities as part of the Quality by Design

(QbD) paradigm. For the first time NIR has been applied for real time monitoring of powder bulk density. Feedforward control system has been integrated with feedback control system to take proactive mitigation actions on raw materials and process variability. A systematic generic framework including the methods and tools through which the cost and profit can be estimated and the Dynamic Real Time Optimization (DRTO) can be integrated with hybrid control system has been also developed. The integrated DRTO provides the optimal operational set points for the plant control system in real time optimizing an objective function. DRTO integrated with hybrid control strategy ensures the maximum possible profit irrespective of the market demand fluctuations.

Presently, I am working a research assistant professor at Department of Chemical and Biochemical Engineering of Rutgers University and currently managing and leading ten projects funded by NSF, industries and regulator with total grants exceeding 18 million USD. I was involved in proposal writing of these projects and I am PI/Co-PI of some of these projects. Prior to this, I worked in a large scale European pharmaceutical project (F^3 Factory project), which involves 25 partners from industry and Academia; where my responsibility was to manage and lead the DTU's parts of the F^3 Factory project and also to coordinate with the other partners associated with it. My doctoral research work, done at Department of Chemical and Biochemical Engineering, Technical University of Denmark, has been awarded for *prestigious "EFCE Excellence Award 2010", given in Recognition of an Outstanding PhD Thesis, from European Federation of Chemical Engineering.* I have exemplary academic records, previous teaching experiences, and an established trail of valued research accreditations'. I have published more than 48 scientific articles, wrote 10 book chapters and presented in more than 90 international conferences. I have edited one book on pharmaceutical system engineering to be published by Elsevier. I am actively serving as a conference session chair, guest editor of Journal, manuscripts reviewer and bachelor, master and PhD projects supervisor.

Teaching Interests:

My teaching experience that I acquired serving as a teaching assistant, lecturer, and research assistant professor makes me confidant to develop and teach any courses of chemical and biochemical engineering. The industrial invited crash courses that I have given provide me opportunity to extend my teaching skills to cover more applied courses and to interact with highly experienced scientist and operators from manufacturing companies. I demonstrated my skills to instruct hands-on exercise through several workshops that I conducted. Several training courses that I conducted for industrial and academic participants make me instrumented with additional capability of building highly skilled workforce. In my department, I am a lead organizer of postdoctoral training program with the aim of building highly skilled workforce for industry and academia. I developed two online course modules where I learn how to develop and give online courses. I have dual ability to teach theoretical as well as practical courses. I am open to teach any courses that your university will assign me. I am especially interested to teach courses related to process control, process modeling and simulation, process monitoring, chemical reaction engineering, process and tool integration and pharmaceuticals. If needed then I can develop new courses based on my research outcomes.

Grants: Extensive experience in funding proposals writing. Wrote several grants proposals for national science foundation (NSF), FDA, and pharmaceutical industries. 11 projects with total grant more than 18 million USD got funded.

Publications: Scientific articles (48); Book chapters (10); International conference presentations (90); Edited Book (1), Invited industrial/academic lectures (>8); Workshops (5); Thesis (3); Supervised Thesis (1 completed + 4 on going); Citations>656, h-index>15, i10-index >17.

Activities in AIChE 2017 conference

Session chair/co-chair

1. **Co-chair:** Advances in process control (35660: 10/29/2017, 3.30 - 6.00 PM, 103 E).

2. **Co-chair:** Dynamic simulation and optimization (35672: 11/01/2017, 3.15 - 5.45 PM, 103 E)

3. **Chair**: Big Data in Process Modeling, Estimation and Control (36339: 11/02/2017, 8.00-10.30 AM, 103 F).

4. **Chair:** Integrated Product and Process Design (35676: 11/02/2017, 8.00 - 10.30 AM, 103 C)

5. **Chair**: Economics and Process Control (35674: 11/02/2017, 12.30 – 3.00 PM, 103 C)

Presenter (Oral presentations)

1. **Singh, R.,** Muzzio, F. J, Ierapetritou, M., Ramachandran, R. **(2017)**. Advanced model predictive control of powder level in continuous pharmaceutical manufacturing pilot-plant. Oral presentation at AIChE annual meeting, Minneapolis, MN, USA, 29 October – 3 November. Paper 492410: 11/03/2017, 10.05-10.30 AM, 101E.

2. **Singh, R.,** Muzzio, F. J, Ierapetritou, M., Ramachandran, R. **(2017)**. Integrated control and data management system for continuous pharmaceutical manufacturing process. Oral presentation at AIChE annual meeting, Minneapolis, MN, USA, 29 October – 3 November. Paper: 493080: 10/31/2017, 5.05-5.27 PM, 203 A/B & 205 A/B.

Co-author

3. Maeda, J., Escotet-Espinoza, MS, **Singh, R.,** Ierapetritou, M. **(2017)**. Real-time monitoring and control of API concentration in a tablet press for continuous manufacturing of tablets. Oral presentation at AIChE annual meeting, Minneapolis, MN, USA, 29 October – 3 November. Paper 496960: 10/31/2017, 2:42-3:04, 204 A/B.

4. Román-Ospino, A. D., Oka, S., Mogthadernejad, S., Escotet-Espinoza, M. S., **Singh, R.,** Ramachandran, R., Ierapetritou, M., Muzzio, F. J. **(2017)**. Residence time distribution and segregation studies trough real time measurements by near infrared spectroscopy. Oral presentation at AIChE annual meeting, Minneapolis, MN, USA, 29 October – 3 November. Paper 498490: 11/01/2017, 12.35 - 12.55 PM, 201 A/B.

5. Shah, A., Ramachandran, R., **Singh, R. (2017)**. Moving horizon based real time optimization and hybrid control of continuous pharmaceutical manufacturing process. Oral presentation at AIChE annual meeting, Minneapolis, MN, USA, 29 October – 3 November. Paper 499020: 10/31/2017, 1:58-2.20 PM, 203 A/B & 205 A/B.

Selected references

1. Haas, N. T., Ierapetritou, M., **Singh, R. (2017)**. Journal of Pharmaceutical Innovation, 12(2), 110-123.

2. Ierapetritou, M., Escotet-Espinoza, M. S., **Singh, R. (2017)**. Process Simulation and control for continuous pharmaceutical manufacturing of solid drug products. Chapter 2 of book "Continuous manufacturing of pharmaceuticals" edited by Peter Kleinebudde, Johannes Khinast and Jukka Rantanen. Publisher: Wiley-VCH, pp 33-106.

3. Oka, S, Escotet-Espinoza, M. S., **Singh, R.,** Scicolone, J., Hausner, D., Ierapetritou, M., Muzzio, F. **(2017)**. Design of an integrated continuous manufacturing system. Chapter 12 of book "Continuous manufacturing of pharmaceuticals" edited by Peter Kleinebudde, Johannes Khinast and Jukka Rantanen. Publisher: Wiley-VCH, pp 405-446.

4. Román-Ospino, A. D., **Singh, R.,** Ierapetritou, M., Ramachandran, R. Méndez, R., Ortega, C., Muzzio, F. J., Romañach, R. J. **(2016)**. International Journal of Pharmaceutics, 512 (1), 61-74. http://dx.doi.org/10.1016/j.ijpharm.2016.08.029

5. Wu, S., Panikar, S. S., **Singh; R.,** Zhang, J., Donepudi, A., Glasser, B., Ramachandran, R. **(2016)**. Advanced Powder Technology, 27 (4), 1115-1127.

6. **Singh, R.,** et al. **(2015)**. International Journal of Pharmaceutics, 495, 612-625.

7. **Singh, R.,** et al. **(2015)**. PROCESSES Journal, 3, 339-356.

8. **Singh, R.,** et al. **(2015)**. Journal of Pharmaceutical Innovation, 10 (3), 233-253.

9. Simon, L. L., .. **Singh, R.,** et al. **(2015)**. Organic Process Research & Development, 19, 3-62.

10. **Singh, R.,** et al. (**2014**). International Journal of Pharmaceutics, 473, 38–54.

11. **Singh, R.,** et al. (**2014**). Computers & Chemical Engineering Journal, 66, 186-200.

12. **Singh, R.,** et al. (**2014**). Journal of Pharmaceutical Innovation, 9, 16-37.

13. Sen, M., Barrasso, D., **Singh, R.**, et al. (**2014**). Processes Journal, 2(1), 89-111.

14. Sen, M., **Singh, R.,** et al. (**2014**). Journal of Pharmaceutical Innovation, 9, 65-81.

15. Sen, M., **Singh, R.,** et al. **(2014)**. PROCESSES Journal, 2, 392-418; doi:10.3390/pr2020392.

16. Sen, M., Chaudhury, A., **Singh, R.,** et al. **(2014)**. American Journal of Modern Chemical Engineering, 1, 13-29.

17. **Singh, R.,** et al. **(2014)**. Advanced Control of continuous pharmaceutical tablet manufacturing processes. **Book** chapter 7, Publisher: Humana Press, ISBN: 978-1-4939-2995-5, 191 – 223.

18. **Singh, R.,** et al. (**2013**). European Journal of Pharmaceutics and Biopharmaceutics, 85(3), Part B, 1164-1182.

19. **Singh, R.,** et al. (**2013**). Computers & Chemical Engineering Journal, 58, 344 - 368.

20. Sen, M., Dubey, A., **Singh, R.,** et al. (**2013**). Journal of Powder Technology, http://dx.doi.org/10.1155/2013/843784.

21. Sen, M., Chaudhury, A., **Singh, R.,** et al. (**2013**). International Journal of Pharmaceutics, 445 (1-2), 29-38.

22. Sen, M., Rogers, A., **Singh, R.,** et al. (**2013**). Chemical Engineering Science, 102, 56 – 66.

23. **Singh, R.,** et al. (**2012**). International Journal of Pharmaceutics, 438 (1-2), 307-326.

24. Sen, M., **Singh, R.,** et al. (**2012**). Chemical Engineering Science, 18, 349-360.

25. Samad, N. A. F. A., **Singh, R.**, et al. **(2011)**. Computers & Chemical Engineering Journal, 35(5), 828-843.

26. **Singh, R.,** et al. **(2011)**. Process and Product Modelling: A case study approach. **Book** chapters 12.2-12.6, Publisher: Elsevier, pp 380-430.

27. **Singh, R.,** et al. **(2010)**. Computers & Chemical Engineering Journal, 34(7), 1108-1136.

28. **Singh, R.**, et al. **(2010)**. Computers & Chemical Engineering Journal, 34(7), 1137-1154.

29. **Singh, R.,** et al. **(2009)**. Computers & Chemical Engineering Journal, 33(1), 22-42.

30. **Singh, R. (2017)**. Advanced model predictive control of a novel continuous pharmaceutical manufacturing pilot-plant. Pharma, https://www.pharmafocusasia.com/. Accepted.

31. **Singh, R.,** Muzzio, F., Ierapetritou, M., Ramachandran, R. **(2016)**. Computer Aided Chemical Engineering, 38, 1473-1478.

32. **Singh, R.,** Zhang, J., Ierapetritou, M., Ramachandran, R. (2015). European Pharmaceutical Review, 20(6), 37-41.

33. Sen, M., **Singh, R.,** Ramachandran, R. **(2015)**. The Medicine Maker, February, 05, 42-45.

34. Escotet-Espinoza, M. S., **Singh, R.,** Sen, M., O'Connor, T., Lee, S., Chatterjee, S., Ramachandran, R., Ierapetritou, M., Muzzio, F. **(2015)**. Pharmaceutical Technology 39 (4), 34-42.

35. **Singh, R.,** Muzzio, F., Ierapetritou, M., Ramachandran, R. **(2015)**. Computer Aided Chemical Engineering, 37, 2183 - 2188.

36. Karry, K. M., **Singh, R.,** Muzzio, F. J. (2015). American Pharmaceutical Review, 18(4), 64 - 67.

37. **Singh, R.,** Ierapetritou, M., Ramachandran, R. (2015). European Pharmaceutical Review 20(2), 76-80.

38. **Singh, R.,** et al. (**2013**). Computer Aided Chemical Engineering, 32, 757-762.

39. Muzzio, F., **Singh, R.,** et al. (**2013**). Pharmaceutical Technology magazine, 37(6), 40-41, 77.

40. **Singh, R.,** et al. (**2013**). BioPharma magazine Asia, 2(5), 18-25.

41. **Singh, R.**, et al. (**2012**). Computer Aided Chemical Engineering, 31, 715-719.

42. **Singh, R.,** et al. (**2012**). PharmPro Magazine, Pharmaceutical Processing, 27(6), 22-25.

43. **Singh, R.,** et al. (**2012**). European Compliance Academic (ECE), http://www.gmp-compliance.org/ecanl_503_0_news_3268_7248_n.html.

44. Samad, N. A. F. A., **Singh, R.**, Sin, G., Gernaey, K. V., Gani, R. **(2011)**. IEEE, 1-6.

45. Gani, R., **Singh, R**. **(2011)**. GEN: Genetic Engineering & Biotechnology News, 31(6), 40-41.

46. Samad, N. A. F. A., **Singh, R.**, et al. **(2010)**. Computer Aided Chemical Engineering, 28, 613-618.

47. **Singh, R.**, et al. **(2010)**. Computer Aided Chemical Engineering, 29, 291-295.

48. Samad, N. A. F. A., **Singh, R.**, et al. **(2010)**. Computer Aided Chemical Engineering, 29, 86-90.

49. **Singh, R.,** et al. **(2009)**. Computer Aided Chemical Engineering, 26, 321-326.

50. **Singh, R.,** et al. **(2008)**. Computer Aided Chemical Engineering, 25, 423-428.

Contact: *Email: ravendra.singh@rutgers.edu. Website: http://ravendrasingh.wixsite.com/ppse*

NSF Engineering Research Center for Structured Organic Particulate Systems (C-SOPS), Chemical and Biochemical Engineering, Rutgers University, Piscataway, NJ, USA
Email: ravendra.singh@rutgers.edu

503630 Programmable Soft Matter for Active Reconfiguration, Biotransport and Delivery

Abstract:

Research Interests:

My research interests encompass a range of emerging, multidisciplinary topics across soft matter physics, chemistry and bioengineering. The unifying theme across these topics is the design and manipulation of materials (e.g., by acoustic, electric and magnetic fields) to direct their assembly, reconfiguration and locomotion for fundamental and applied research in biomedicine and related areas.

My research interests fall broadly into three main thrusts:

1) Development of *advanced microfluidic platforms* for processing biofluids. This entails engineering chip-based devices empowered by external fields (e.g., acoustic, magnetic and electric), that can perform multiple useful functions like drive cells into individualized compartments for encapsulation, staining and culturing.

2) Synthesis of *silicone particle scaffolds* to encapsulate, protect and controllably release drugs. Given the remarkable properties of silicones (e.g., stability, tunability and biocompatibility), I am interested in applying sol-gel techniques to synthesize particles that can biospecifically attach to substrates for field-triggered delivery.

3) Design of *reconfigurable active matter* that is capable of manipulating materials across scales. In particular, I am interested in the lithographic fabrication of particles with complex shape and patches to enable locomotion when stimulated by external fields to create the next generation of micro-robotic tools and constituents for self-healing materials, as examples.

Teaching Interests:

In college, I learned the definition of a leader is someone who sets noble objectives and pursues those objectives with such intensity, that others cannot help but follow. My perspective of teaching is the same. I believe the most important aspects to teaching are to show genuine excitement, to challenge students to identify important problems and to think critically throughout the discovery process (see my “Teaching and Mentorship” Section below).

Given my background, I am prepared to teach core courses in chemical engineering, including but not limited to, thermodynamics, kinetics, polymer science, colloidal systems and transport phenomena. Having majored in biomedical engineering, minored in materials science and graduated with high distinction, I am also prepared to teach classes in biomedical engineering and materials. Given my research and entrepreneurship experiences, I am also excited about the idea of developing and teaching new courses on advanced topics such as microfluidics, field-based manipulations and entrepreneurship.

Notable Awards:

2016 Dean's Award for Excellence in Mentoring, Duke University

2015 Exceptional Student Award, ISAC (international award to 1 Ph.D. student/yr.)

2014 NSF Graduate Research Opportunities Worldwide Award (to study in Sweden)

2013 Exceptional Student Award, ISAC

2012 NSF Graduate Research Fellowship

2011 The Raven Society, UVA

2010 Tau Beta Pi Engineering Honor Society

Postdoctoral Projects:

"*Reconfigurable assemblies of engineered microcubes and active particles powered by AC electric fields*" Under Prof. Orlin D. Velev, Department of Chemical & Biomolecular Engineering, NC State University 2016-2017

"*Mondisperse silicone gel particles for the encapsulation and controlled release of hydrophobic drugs*" Under Prof. Stefan Zauscher, Department of Mechanical Engineering & Materials Science, Duke University 2016-2017

PhD Dissertation:

"*Acoustic and magnetic techniques for the isolation and analysis of cells in microfluidic platforms*" Under Prof. Gabriel P. López, Department of Biomedical Engineering, Duke University 2011-2016

Successful Proposals:

- NSF graduate research fellowship; 2012-2015; totaling **$138,000**
- 3 startup awards for Encapsio LLC (I am the CTO); totaling **$71,000**
- 10 undergraduate research awards (written with students); totaling **$60,200**
- 7 travel awards (from NSF, MRS, ACS, ISAC, etc.); totaling **$15,450**

Proposals Under Review:

- NIH R41EB024472; I am the PI/PD; **$224,920**

Research Experiences:

During my time in graduate school, I developed expertise in designing, fabricating and operating acoustically powered microfluidic (acoustofluidic) devices. During this time, I engineered such systems to isolate rare cells from blood for templating, staining and analysis as well as assemble anisotropic particles into ordered crystallites.

During my postdoc with the NSF's Research Triangle MRSEC, I expanded this work to fabricate complex particles that, by virtue of their design, can harvest energy from external fields for active propulsion. Additionally, using assemblies of these particles, I helped develop new classes of reconfigurable materials that can bend and flex on demand in so-called "colloidal origami" that can be used to probe cell stiffness and measure the viscoelastic properties of liquid crystals.

In 2016, I co-founded a company (Encapsio LLC) based on discoveries I made at Duke University. Encapsio is a company that produces monodisperse silicone particles to encapsulate, protect and controllably release active ingredients for skin care. In the last 18 months, we raised over $70,000 to: (i) validate the core technology, (ii) develop a beta prototype of our first product (i.e., a formulation containing retinol-encapsulated microparticles), (iii) file intellectual property through Duke University, (iv) send the prototype to third-party laboratories to conduct human irritation studies and controlled release studies and (v) negotiate an exclusive licensing agreement for the technology.

Teaching and Mentorship Experiences:

As a professor, the ultimate measure of my success will be my students (as outlined below); quality research, publications and intellectual property will be enjoyable byproducts. Therefore, to hone my skills as a mentor and educator, I have tenaciously sought mentorship opportunities.

Over the last six years, I have mentored 21 undergraduate, 3 masters and 4 PhD students. Of these, 15 are female and 9 are in underrepresented or minority groups in STEM. These students are from various educational disciplines (i.e., biology, biomedical engineering, chemical engineering, chemistry, mechanical engineering and physics).

Currently 13 of my past and present mentees have peer-reviewed publications in print, and 9 others have publications that are submitted or are in preparation. Also, many of my mentees have placed in top programs after graduation (e.g., MIT Mechanical Engineering PhD Program, UPenn Medical School, UT Austin Chemical Engineering PhD Program, University of Texas Medical Branch MD/PhD Program).

Mentorship is the best part of my job, and so it is with great satisfaction that my students nominated me to win the 2016 **Dean's Award for Excellence in Mentoring**, which is an award granted by Duke to only three students annually: (https://gradschool.duke.edu/about/news/c-wyatt-shields-iv-2016-dean-s-award-winner).

Finally, I developed lectures (e.g., on acoustic and magnetic particle manipulations) and I served as a teaching assistant for two classes at Duke (Biomaterials & Biomechanics in 2013 and 2015, where I gave mini-lectures and oversaw laboratory activities) and at the University of Virginia (Biomedical Engineering: Design & Discovery in 2009 and 2011).

Future Directions:

Early on, I plan to devote my lab to three main areas that enable the study of fundamental science as well as develop engineered systems that are competitive for funding by most of the top funding agencies (e.g., NIH, NSF, DOD, ONR):

First, given the ability of acoustofluidic devices to precisely handle cells and biofluids, I would like to devote a portion of my lab to develop innovative microbioanalytical platforms, e.g., for continuous cell staining and washing, high resolution cell imaging and real-time surface potential evaluation of single cells.

Second, while the Stöber process has existed for half a century, the ability to control the properties of silicone particles for encapsulating drugs remains unexplored. By appending bioactive agents to the surfaces of the particles, it should be possible to design a set of particles that stably attach to skin for topical formulations that last days instead of hours. This advancement could revolutionize the treatment of skin disease by slowly delivering potent drugs with narrow therapeutic windows.

Third, one idea that captivates me is the marriage of active particles (i.e., particles that undergo autonomous locomotion) and biological transport. For this project, I envision is the use of engineered active particles as "cellular jetpacks", where active particles are lithographically engineered to contain small metallic patches that: (i) present antibodies via thiol chemistry for the biospecific attachment of single cells and (ii) enable particles respond to external fields for their active self-propulsion, which could enable on-chip cell separation and the bottom-up assembly of programmable tissues. This would provide a new way of creating hierarchically organized tissue constructs to enable new fundamental studies on engineered tissues.

Publications (selected from 18 articles and 2 patents):

Han, K; **Shields IV, CW**; Diwakar, NM; Bharti, B; López, GP; Velev, OD. "Sequence-encoded colloidal origami and microbot assemblies from patchy magnetic cubes," (Minor revisions at *Science Advances*).

Wang, PY;* **Shields IV, CW**;* Zhao, T; Jami, H; López, GP; Kingshott, P. "Rapid self-assembly of shaped microtiles into large, close-packed crystalline monolayers on a solid surface," *Small* 2016. 12*(2)*: 1309-1314. **Co-First Author; I.F. = 8.3*

Shields IV, CW; Sun, D; Johnson, K; Duval, K; Rodriguez, AV; Gao, L; Dayton, PA; López, GP. "Nucleation and growth synthesis of functional, monodisperse and acoustically programmable particles," *Angewandte Chemie International Edition* 2014. 53*(31)*: 8070-8073. **I.F. = 11.7*

Shields IV, CW; Johnson, LM; Gao, L; López, GP. "Elastomeric negative acoustic contrast particles for capture, acoustophoretic transport, and confinement of cells in microfluidic systems," *Langmuir* 2014. 30*(14)*: 3923-3927.

Shields IV, CW; Zhu, S; Yang, Y; Bharti, B; Liu, J; Yellen, BB; Velev, OD; López, GP. "Field-directed assembly of patchy anisotropic microparticles with defined shape," *Soft Matter* 2013. 9*(38)*: 9219-9229.

Reviews:

Shields IV, CW; Reyes, C.; López, GP. "Microfluidic cell sorting: A review of the advances in the separation of cells from debulking to rare cell isolation," *Lab on a Chip* 2015. 15*(5)*: 1230-1249. **More than 150 citations*

492843 Synthesis of Core-Shell Microparticles Containing Thermoset Resins via Suspension Polymerization

Abstract:

Thermoset powder coatings are attractive because they offer superior durability and mechanical performance without any environmental or human health risks associated from volatile organic compounds (VOCs), which are emitted from conventional solvent-based coatings. We suggest that the ideal particles to be used in powder coatings would be on the microscale and contain thermoset resins. Here, suspension polymerization was used to produce microparticles that offer many advantages: low conversion cost, mature scale-up, as well as a versatile tunability in terms of particle size and morphology. Our objective is to synthesize core-shell thermoplastic microparticles that contain a large concentration of thermosetting resins. The thermal properties, average diameter, morphology, and chemical composition of the particles were determined and tuned for the application of powder coatings that can be applied at any temperature. The expected outcome of this study is a novel sprayable thermoset powder coating that offers high performance, easy operation, significant environmental benefits, and a commercial feasibility.

Research Interests: Polymer Coatings and Adhesive, Colloids, Green Materials

Teaching Interests: Polymer Science, Thermodynamics, Colloid & Interface Science

505549 The Mesoscopic Physics of Discrete Media: Towards the Control of Dynamic Structures

Abstract:

Profile: *Process design engineer. Background in corporate R&D, management and consultancy for multinational companies. 7 years of experience in powder technology, heat transfer and water engineering. Doctorate in experimental and numerical spray drying within P&G and research in modelling multiphase flows, fouling and dynamic self-organization at Hexxcell, Imperial College and Univesity College London. Partners in consumer goods and energy sectors.*

Period	*Type*	*Job Title*	*Institution*
2016/17	Academia	Research Associate	University College London. UK
2015	Start up	Consultant	Hexxcell Ltd, Imperial College London. UK
2012/13	Academia	Visiting researcher	Imperial College London. UK
2009/12	Corporate R&D	Research Engineer	Procter & Gamble, Technical Centres. UK
2008/09	Production	Engineer	Acciona Agua, sea water desalination. Spain.
2006	Corporate R&D	R&D Manager	Procter & Gamble, Technical Centres. UK
2004	Corporate R&D	Intern	Procter & Gamble, Technical Centres. UK

Period	*Qualification/Field*	*Institution*
2014	EngD. Engineering Doctorate (Professional Doctorate)	University of Birmingham. UK.
2008	Master Environmental science and int. cooperation	School of Industrial Management. Madrid
2007	Expert International crisis prevention and management	University of Carlos III, UK.
2005	MEng. Chemical Engineer	University of Salamanca, Spain

Research Interests:

My work looks into complex dynamics in multiphase flow and the self-assembly of granular media. I am intrigued by how under the right balance between attractive forces and disruptive stresses, granular matter can be organised into well-defined spatial patterns that can propagate all the way into mesoscopic dynamic structures from relatively simple individual interactions. Examples of a rich variety of clustered structures of different sizes that are in constant formation and breakage are all around us in nature e.g. the equilibrium between the flow of sand or snow particles, flakes, clusters and static beds in a storm, a river bed, a desert or an avalanche, and industry e.g. the equilibrium of dense and dilute flow ares, cluster formation and breakage in fluidization, spray dryers, burners, combustors or mixers. Despite we can independently describe flow, growth and breakage phenomena, we do not understand well how metastable structures appear from the balance of all three. In the best cases, complex effects due to clustergins in transition regimes are avoided, in the worst, neglected. I want to explore a more organic approach to design a new generation of smart materials and processes, whereas the dynamic features of granular media are not avoided but exploited in benefit of a more efficient, more resilient and self-controlled operation. Examples include the manipulation of wall deposition, consolidation and agglomeration in spray dryers to control particle size and energy consumption, tuning heat and mass transfer rates in a fluidised bed by imposing an structured bubble flow introducing pulsed airflow, the control of reaction rates by changing the density of catalyst bed or the manipulation of the duty of shell-and-tube heat exchangers in the oil industry or the flux and selectivity of separation membranes with the dynamic control of the structure of an ubiquitous fouling layer.

Teaching Interests:

I have hands-on experience in various industries as an engineer where I had the chance to train crews, mentor student and address a diverse set of audiences, from officials to clients, management or scientists. I also have a hybrid profile as experimentalist and modeller, and from this standpoint, I believe I can produce and deliver undergraduate and postgraduate courses within transport phenomena, powder technology and core process chemical engineering, such as separation, design and heat transfer engineering or numerical modelling. I could move into a number of other specialised areas with some preparation and perhaps I will require more work to be involved in chemistry or biotechnology. My previous posts cover manufacture of granular detergents for Procter & Gamble, seawater desalinization for Acciona, and heat transfer engineering for the oil & gas sector, ExxonMobil, consulting for Hexxcell Ltd, spin off Imperial College London. The areas I have worked in include corporate R&D, management and consultancy, and institutions varying from large multinationals to universities and start-ups. I have supervised operators, researchers, technical officials and interns in addition to currently supervising PhD and Master students at UCL. I believe that I can relate my experience in the real-world challenges awaiting a young chemical engineer to my teaching responsibilities. I would be very interested in contributing to developing the student curriculum and mentor undergraduates and I am also sure that I would enjoy making them aware of the landscape and engaging into the specialization they require as function of their individual strengths and the careers they have in mind.

Research Experience:

Beyond the proprietary contributions, my public research has focused on multi-phase flow, powder technology and fluid dynamics. My doctorate dealt with droplet/particle agglomeration during spray drying of granular detergents with a strong component in fluif mechanics and turbulence. It represents a turning point in the industry because it has unravelled major deficiencies in historical practices. Developing new instrumentation methods [1] we discovered that the vortex flow employed in swirl spray dryers breaks down inside full-scale units due to wall effects [2]. This had remained undetected in smooth lab/pilot devices and so it was unnoticed for decades in academic research causing major issues in traditional models. Now we know that distinct recirculation regimes appear in response to varying levels of friction and new scale up rules are designed accordingly [3]. My research into fouling however led to a most fundamental change in the traditional view of particle formation in spray drying. We proved that most of the powder generated in a swirl unit results from the continuous formation and breakage of a clustered structure generated at its walls. It is revolutionary because it contradicts decades of research assuming spray dried granules were formed by airborne agglomeration of droplets and particles. We have refuted this hypothesis and for the first time quantified how much and for how long particles dry and agglomerate in dynamic wall structures [4]. From this new perspective, we redefined how agglomeration is envisaged in swirl spray drying [5], designed new methods to quantify how sprays interact [6] and optimized the operation and design rules of swirl dryers to improve energy efficiency based on controlling wall deposition, consolidation and resuspension [7]. On these grounds, I was relocated in 2015 to Imperial College London as visiting researcher to implement a numerical frame describing the process. In my following postdoctoral positions, I have continued to expand my numerical skills and I have explored other important fouling phenomena in the oil industry and later joining University College London. Here I deal with more fundamental granular physics involved in the control of an structured bubble flow in pulsed gas-solid fluidised beds, and I work with Prof Mark Miodownic on new avenues to develop smart materials for orthopaedics and responsive fabrics.

Selected publications:

1. Francia V, Martin L, Bayly AE, Simmons MJH. 2016. *Flow Meas. and Instr.* ***50*** *:* 216-228. [Link]

2. Francia V, Martin L, Bayly AE, Simmons MJH. 2015. *Exp. Ther. Fluid Sci.* **65**:52 - 64. [Link]

3. Francia V, Martin L, Bayly AE, Simmons MJH. 2015. *Chem. Eng. Sci.* **134** : 399-413. [Link]

4. Francia V, Martin L, Bayly AE, Simmons MJH. 2015. *AIChE J.* **61**, 6:1804â€“1821 [Link]

5. Francia V, Martin L, Bayly AE, Simmons MJH. 2016. *Powder Tech.* **301** : 1330-1343. [Link]

6. Francia V, Martin L, Bayly AE, Simmons MJH. 2016. *Powder Tech.* **301** : 1344-1358. [Link]

7. Francia V, Martin L, Bayly AE, Simmons MJH. 2017. *Chem. Eng. Sci.* **162** : 284-299. [Link]

Research Plans:

I plan to focus a first stage in my career in studying the transitional regime sitting in the interface between continuum fluid and solid mechanics. When a dilute collection of particles consolidates and moves from a so-called inertial to dense and quasi-static regime multi-particle sustained contacts become frequent, and long-range interactions, shape and anisotropy begin to have dominant effects in the system dynamics. These phenomena are extremly difficult to describe in traditional kinetic frames. So far kinetic models remain unequipped to deal with the crosscorrelation of solid velocities, an anysotropic distribution of contacts and local changes in rheology, not to mention describing systems with an intense inter-particle interaction leading to growth or breakage, where process design still heavily relies on experimentation. High-level-of-detail studies e.g. DNS, LB, LES, CFD-DEM, look often into microscale phenomena but they seldom feed new constitutive equations or rate kernels describing mesocopic physics into a macroscopic modelling frame e.g. PBE, CFD, let alone be of use to optimization engines. I want to focus precisely on this gap, connecting physics that naturally occur at a mesoscale such as clustering, breakage and consolidation with kinetic rate models, which are the only suitable alternatives to make a tangible impact in industrial optimization. This is specially crucial for sectors facing challenging granular rheologies, agglomeration and particulate or colloidal fouling.

In my first research programme, I consider developing a cross-disciplinary multi-scale modelling framework to investigate dynamic particle fouling as a combination of deposition, assembly, consolidation and removal of a particulate multi-layer. I expect to do so with the support and collaboration of many of my previous industrial and academic partners. The focus stems from my partnership with Procter & Gamble that has proven dynamic fouling crucial in manufacturing granular detergents but similar particulate and colloidal multilayers are key to many other operations e.g. combustion of biomass, heat exchangers, mixers, membranes and multiple complex fluids and biological systems. My immediate goals will include prediction of dynamic structural properties (e.g. radial distribution functions, fractal scales) and the probability and rate functions for growth, consolidation and breakage under different stimuli e.g. aerodynamic, surface or contact forces, body forces. In other words, I want to lay the experimental and numerical grounds providing transferable outputs to advanced macroscopic models of dynamic multilayer fouling in a stochastic (e.g. collision rates, adsorption or clustering frequency to use in LES/CFD) or a rate-based formulation (e.g. diffusion coefficients, agglomeration/breakage kernels to use in PBE, GPROMS). At a further stage, I will expand my research in several directions. First deepening the expertise in dynamic structures to design new processes and smart powders, then in the applications of our capability into other disciplines modifying the forces at play, the geometrical constrains and the time and spatial scales of the problem. From a theoretical standpoint, the principles and methodologies underpinning the study of self-organization are invariant and much of the expertise acquired along the way may find application in areas such as physics, robotics, data analysis or social sciences. This makes my research to offer a great opportuniy for inter-disciplinary work and a low risk profile not only in the short, but also the long term.

505606 Engineering Precision Polymers for Advanced Applications

Abstract:

Research Interests:

Synthetic polymers with precisely programmed structures and functions offer unparalleled opportunities for preparing next-generation functional materials with precisely tunable performance, processing parameters, and morphological features. While the preparation of such materials has been a grand challenge, recent scalable routes to discrete synthetic polymers opened pathways to make precision materials having specific functions that should impact myriad fundamental studies and technological applications.

To address the grand challenges of building complex adaptive materials from molecular building blocks as Nature does, and relating their structures to final properties, my research program will examine three specific areas: (1) precision polymers with tailored shapes and functions (2) structurally dynamic and asymmetric polymer networks (3) self-organization and replication of periodic polymers and hybrid materials. The materials developed in my research programs will address the challenges in the development of materials for stimuli-responsive, separation, imaging tracer/sensing and delivery applications, and serve as molecular tools for fundamental studies at the interface of synthetic and physical chemistry, chemical and materials engineering, physics, biology, and other fields.

Postdoctoral Projects:

"Synthetic Oligomers and Polymers with Discrete Structures and Hierarchical Architectures"

Department of Materials, University of California Santa Barbara

Advisor: Prof. Craig J. Hawker

PhD Dissertation:

"Functional Nanostructures from Nanoparticle Building Blocks"

Department of Polymer Science and Engineering, University of Massachusetts Amherst

Advisor: Prof. Todd Emrick

Research Experience:

In my postdoctoral research at the **Department of Materials, UCSB** I investigated synthetic strategies to prepare a series of synthetic polymers with various topologies via controlled polymerization techniques and developed a scalable method to access discrete functional oligomers/polymers from (meth)acrylates, styrenics, siloxanes, thiophenes, and other monomer families. Together with my colleagues and intern students, we investigated the effect of tailored molar mass and composition dispersity on the properties of complex polymers.

In my PhD research at the **Department of Polymer Science and Engineering, UMass Amherst I** focused on the synthesis and characterization of nanomaterials and multifunctional polymer-based ligands for making nanostructures using directed self-assembly and covalent assembly techniques. I also investigated the synthesis of zwitterionic polymer covered nanomaterials and their potential applications as nonfouling nanomaterials and photothermal therapeutic agents for in vivo studies. Before joining UMass, I was a Research Scientist at Schlumberger Research (Boston) developing optochemical and electrochemical sensing, membrane separation and microfluidic platforms for downhole technologies. I obtained my master and bachelor degree in **Chemical Engineering from the University of Tokyo, Japan** under the guidance of Prof. Takeo Yamaguchi, Prof. Steven Kraines and Prof. Hiroshi Komiyama.

Teaching Experience:

- Mentorship: 1 postdoctoral researcher, 7 graduate, 4 undergraduate and 3 high school students

- Outreach – MRSEC ASPIRE Outreach Program (UCSB 2015, UMass Amherst 2010)

- Teaching Assistant and co-leader, Polymer Synthesis Laboratory (PSE 603), University of Massachusetts Amherst, 2011–2013

- Co-leader, Polymer Science and Engineering Mentoring Program, 2010–2011

- Teaching Assistant, Life Cycle Analysis, Chemical Engineering, University of Tokyo, 2003.

Teaching Interests:

My diverse research and teaching experience has equipped me to teach a variety of courses offered in the Chemical Engineering Department. Courses of particular interest would be: Polymer Chemistry, Polymers/Material Science Laboratory, Thermodynamics, Transport Processes, Separation, Numerical/Molecular Simulation. Additionally, I am interested in developing special interdisciplinary topic courses at the graduate level on new research topics that my research group may explore.

Publications

1. **Lawrence, J.;** Goto, E..; Kim, D.S.; Ochiai Y.; Clark, P.G.; Laitar, D.; Higashihara, T.; Hawker, C. J.* A Versatile and Scalable Strategy to Discrete Conjugated Oligomers, **in revision**.

2. Oschmann, B.; **Lawrence, J**.; Schulze, M. W.; Ren, J. M.; Anastasaki, A.; Luo, Y.; Nothling, M. D.; Pester, C. W.; Delaney, K. T.; Connal, L. A.; McGrath, A. J.; Clark, P. G.; Bates, C. M.*; Hawker, C. J.*. Effects of Tailored Dispersity on the Self-Assembly of Dimethylsiloxane–Methyl Methacrylate Block Co-Oligomers. **ACS Macro Lett.** 2017, 6, 668

3. **Lawrence, J**.; Lee, S.-H.; Abdilla, A.; Nothling, M. D.; Ren, J. M.; Knight, A. S.; Fleishmann, C.; Li, Y.; Abrams, A.; Schmidt, B. V. K. J.; Hawker, M. C.; Connal, L. A.; McGrath, A. J.; Clark, P. G.; Gutekunst, W. R.*; Hawker, C. J.* A Versatile and Scalable Strategy to Discrete Oligomers, **J. Am. Chem. Soc**. 2016, 138, 6306.

4. Discekici, E.; Pester, C.; Treat, N.; **Lawrence, J.**; Mattson, K; Narupai, B.; Toumayan, E.; Luo, Y.; McGrath, A.; Clark, P.; Read d. A. J.; Hawker, C. J. Simple Bench Top Approach to Polymer Brush Nanostructures Using Visible Light Mediated Metal-Free Atom Transfer Radical Polymerization. **ACS Macro Letters**, 2016, 5, 258.

5. **Lawrence, J.***; Emrick, T. Pentafluorophenyl Ester-Functionalized Nanoparticles as a Versatile Platform for Selective and Covalent Inter-nanoparticle Coupling. **ACS Applied Materials and Interfaces**. 2016, 8, 2393.

6. Liu, Y.; Chen, Y-C.; Hutchens, S.; **Lawrence, J.**; Emrick, T.; Crosby, A. J. Directly Measuring the Complete Stress-Strain Response of Ultrathin Polymer Films. **Macromolecules**, 2015, 48, 6534.

7. **Lawrence, J.**; Yamashita, K.; Yamaguchi, T. Correlating Electronic Structure and Chemical Durability of Sulfonated Poly(arylene ether sulfone)s. **Journal of Power Sources**, 2015, 279, 48.

8. Puodziukynaite, E.; Wang, H.W.; **Lawrence, J.**; Wise, A. J.; Russel, T.P.; Barnes, M.; Emrick, T. Azulene Methacrylate Polymers: Synthesis, Electronic Properties, and Solar Cell Fabrication, **J. Am. Chem. Soc.**, 2014, 136, 11043.

9. Maiz, J.; Zhao, W.; Gu, Y.; **Lawrence, J.**; Arbe, A.; Alegria, A.; Emrick, T.; Mijangos, C.; Colmenero, J.; Russell, T.P. Dynamic Study of Polystyrene-block-poly(4-vinylpyridine) Copolymer Confined in Self-Ordered Nanoporous Alumina, **Polymer**, 2014, 55, 4057.

10. Pham, J. T.; **Lawrence, J.**; Grason, G. M.; Emrick, T.; Crosby, A. J. Stretching of Assembled Nanoparticle Helical Springs, **Phys. Chem. Chem. Phys.**, 2014, 16, 10261.

• Selected for the front cover. **Phys. Chem. Chem. Phys.**, 2014, 16, 10235

11. **Lawrence, J.**; Pham, J.T.; Lee, D. Y.; Liu, Y.J.; Crosby, A. J.; Emrick, T. Highly Conductive Ribbons from Stick-Slip Assembly of Organosoluble Gold Nanoparticles. **ACS Nano**, 2014, 8, 1173.

12. Pham, J. T.; **Lawrence, J**.; Lee, D. Y.; Grason, G. M.; Emrick, T.; Crosby, A. J. Highly Stretchable Nanoparticle Helices Through Geometric Asymmetry and Surface Forces. **Advanced Materials**, 2013, 25, 6703.

• Selected for the front cover. **Advanced Materials**, 2013, 25, 6635.

13. Lee, D. Y.; Pham, J. T.; **Lawrence, J**.; Lee, C. H.; Parkos, C.; Emrick, T.; Crosby, A. J. Macroscopic Nanoparticle Ribbons and Fabrics. **Advanced Materials**, 2013, 25, 1248.

14. Laprade, E. J.; Long, R.; Pham, J. T.; **Lawrence, J.**; Emrick, T.; Crosby, A. J.; Hui, C.-Y.; Shull, K. R. Large Deformation and Adhesive Contact Studies of Axisymmetric Membranes. **Langmuir**, 2013, 29, 1407.

15. Chen, X.#; **Lawrence, J.#**; Parelkar, S.; Emrick, T. Novel Zwitterionic Copolymers with Dihydrolipoic Acid: Synthesis and Preparation of Nonfouling Nanorods. **Macromolecules**, 2012, 46, 119. (#: equal contribution)

16. Bae, J.; **Lawrence, J**.; Miesch, C.; Ribbe, A.; Li, W.; Emrick, T.; Zhu, J.; Hayward, R. C. Multifunctional Nanoparticle-Loaded Spherical and Wormlike Micelles Formed by Interfacial Instabilities. **Advanced Materials**, 2012, 24, 2735.

17. **Lawrence, J.**; Yamaguchi, T. The Degradation Mechanism of Sulfonated Poly(arylene ether sulfone)s in an Oxidative Environment. **Journal of Membrane Science**, 2008, 325, 633.

18. **Lawrence, J.**; Robinson, K. L.; Lawrence, N. S. Electrochemical Determination of Sulfide at Various Carbon Substrates: A Comparative Study. **Analytical Sciences**, 2007, 23, 673.

Awards

• Dow Materials Institute Travel Award, University of California, Santa Barbara, 2015, 2016, 2017

• Outstanding Performance, Schlumberger Corporation, 2005, 2006, 2007, 2008

• Schlumberger Innovation Projects 2006

• Japanese Ministry of Education (Monbusho) Full Scholarship, Graduate, 2003-2005

• Japanese Ministry of Education (Monbusho) Full Scholarship, Undergraduate, 1999-2003

505455 Advanced Biologic-Synthetic Composites

Abstract:

Research Interests:

Synergistic combinations of synthetic and natural components have yielded advanced biological, thermal, and mechanical properties and intricate hierarchical assemblies that have enabled impactful technologies in medicine, engineering, and environmental remediation. Yet, there remains extensive untapped knowledge of the ways in which synthetic materials interact with living systems, and of the ways in which to harness Nature's functional, dynamic, and adaptable materials systems within the versatile, tunable frameworks of synthetic polymeric materials. Building on my graduate and postdoctoral research experiences on the synthesis, characterization, and assembly of functional polymers incorporating biological and biomimetic components, my independent research program will aim to engineer adaptive, modular composites from synthetic polymers and oligopeptides. I am interested in leveraging various polymer and peptide interactions to construct materials that disintegrate, form, or reform on-demand to capture or release active substances or to adjust mechanical properties in response to environmental cues. To this end, specific interactions between opposite-handed helical oligopeptides will be investigated as dynamic junctions within nanostructures, colloids, and hydrogels. Additionally, competition and cooperation between various interactions programmed into a given construct will be explored as a means to structure or re-structure materials in response to various inputs. For example, thermally-induced mesogen packing will be examined as a way to promote or suppress fibrillation of peptides tethered to the same scaffold. Further, I am interested to determine the role liquid-liquid and biological interfaces in the construction and assembly of biologic-synthetic composites. Interfacial mediation, acceleration, and patterning of polymer and peptide interactions is envisioned to yield complex hierarchical structures and streamline fabrication processes. Continual development and optimization of accessible, scalable methods for materials preparation represents an important goal of my proposed research program, as it will position these materials well for scale-up and translation to clinical and industrial implementation.

Research experience:

Ph.D. Dissertation Research: Functional hydrophilic polymers for solution assembly and non-viral gene therapy

Polymer Science & Engineering Department, University of Massachusetts Amherst, Thesis Advisors: Todd Emrick and Ryan C. Hayward

My graduate research involved the synthesis and assembly of functional hydrophilic polymers designed in linear and comb architectures to generate responsive emulsion-based materials and high-performance non-viral gene therapy reagents. Cationic oligopeptides facilitated complexation of nucleic acids and interaction with cellular and nuclear membranes, while zwitterionic groups imparted stimuli-responsive solution properties, reversible adhesion between emulsion droplets, and tunable polymer-nucleic acid binding strength. Reactive units enabled interfacial reactions on emulsion droplets that afforded robust capsules and triggered transitions from adhesive droplet gels to fluidic emulsions. Polymer architecture significantly impacted the performance of the cationic gene therapy reagents; comb polymers exhibited weaker polymer-nucleic acid binding strengths and afforded higher levels of gene expression compared to analogous linear systems. The high functionality of oligopeptide sequences within these polymer constructs inspired the direction of my independent research program, which will benefit from technical expertise gained during these studies.

Postdoctoral Research: Exploring impacts of stereochemistry on assembly of amphiphilic block copolypeptides

Department of Chemistry, Texas A&M University, Advisor: Karen L. Wooley

We are investigating the impact of stereochemical configuration of amphiphilic block polypeptides on solution-state and interface-mediated assembly to contribute fundamental understanding on stereochemistry-directed assembly and to address challenges in the treatment of recurrent urinary tract infections. To improve internalization of antimicrobial-loaded nanostructures into bladder cells, anisotropic particles > 100 nm in length presenting mannose-binding groups are desired to span multiple binding sites spaced approximately 20 nm apart on cell surfaces. Given that complexation of complementary stereoregular polymers (*e.g.*, syndiotactic and isotactic poly(methyl methacrylate), poly(l-lactic acid) and poly(d-lactic acid), *etc.*) impacts solubility and stabilizes micelles with stereoregular hydrophobic cores, we anticipated that programming the stereochemistry of individual blocks of amphiphilic block polypeptides would allow access to non-spherical morphologies. Further, dramatic changes in thermal and mechanical properties reported upon complexation of stereoregular polymers suggests the potential for imparting advantageous materials properties, such as controlled release of antimicrobials. Well-defined block polypeptides, in which the stereochemistry of each block was dictated by that of the corresponding monomer, were synthesized and assembled into spherical and cylindrical nanostructures. Preliminary solid- and solution-state atomic force microscopy revealed distinct differences in the dispersion of polypeptide nanostructures with left- *vs.* right-handed helical hydrophilic shells connected to left-handed helical hydrophobic cores. Motivated by the power of stereocomplexation in bulk polymers and nanoscale assemblies, I wish to further pursue these interactions in my independent research program towards reformable materials with dynamic junctions composed of stereocomplexed oligopeptides. In this context, the rich functional and compositional variety of natural and non-canonical amino acids offers an exciting feedstock from which to pioneer the development of tailorable, adaptive materials systems poised to meet central challenges in medicine and engineering.

Teaching Interests:

Through classes, my research group, and outreach and service activities, I aspire to guide students and mentees toward fundamental scientific and engineering principles, quality, safe, and ethical scientific practices, and proficiency in oral and written communication. Future technological advances rely on the continual mentorship of the next generation of scientists and engineers, and having worked with outstanding mentors, I recognize the importance of endowing young scientists with technical, leadership, and communication skills within an empowering, collaborative, and interdisciplinary environment. Therefore, my classes will integrate recent discoveries in fundamental and applied research at universities, national laboratories, and companies to inspire students toward creative problem-solving in a variety of contexts. Quality, safe, and ethical practices will be discussed as they relate to concepts taught in the classroom, encountered in the laboratory and in research meetings, and reported in the literature and general media. Collaborative work has been an essential component of my graduate and postdoctoral research experience, and will be encouraged through classroom-based group problem-solving activities, as well as in my research program through intra- and inter-group collaborative research to promote innovation and develop communication skills. Written assignments and oral presentations in the classroom, preparation of manuscripts and research presentations, and involvement of my research group in the proposal drafting process will assist in developing essential communication skills. Finally, I am committed to being an active leader in service, outreach, and in the promotion of diversity, and wish to both participate in and lead activities that encourage engagement in science and engineering by future scientists and engineers, and the general public. I will recruit volunteers from my classes, research group, and department as further opportunities to improve communication skills and to enkindle interest in pursuing education, service, and outreach in their own careers.

Given my research and education background spanning chemical engineering, chemistry, and polymer-based materials science disciplines, I very much look forward to drawing from my polymer background to apply elements of polymer science to teaching chemical engineering core classes. Among the chemical engineering core classes, I am most interested in teaching introductory chemical engineering, thermodynamics, reaction kinetics, separations, and transport phenomena. Non-Newtonian flow behaviors exhibited by polymers, as well as processing techniques, such as spin-coating, electrospinning, and extrusion, will provide stimulating, industrially-relevant examples with which to explore fluid dynamics. Similarly, size exclusion chromatography used routinely in industrial and research environments to characterize polymer molecular weight can be included in separations classes, while polymer purification by dialysis represents a tangible example with which to teach equilibration across a semi-permeable membrane. Apart from core classes, I am keen to design a course on scientific written and oral communication and interdisciplinary courses related to polymer synthesis and characterization.

Selected service, leadership, and outreach activities:

Vice President, Postdoctoral Association of Chemistry, February 2017-present: Co-organized networking and professional development events for postdoctoral associates

Science Night, Lead organizer, April 2017: Coordinated the organization of a science night at a local elementary school, in which approximately 300 students engaged in 13 activities contributed by S.T.E.M. groups at Texas A&M University and two local companies

NSF East Asia Pacific Summer Institutes (EAPSI) Singapore Alumni Resource Advisor, 2015 and 2017: Prepared a presentation and documents to assist new fellows in making the most of their international research experience

University of Massachusetts Amherst Polymer Science & Engineering ASPIRE (A Student-led Program in Research and Education) program, 2013-2015: Designed a polymer hydrogel module implemented in the 2014 and 2015 programs

Selected Publications (10 total, 4 first author, 22 citations):

Letteri RA*, Santa Chalarca CF*, Bai Y, Hayward RC, Emrick T. "Forming sticky droplets from slippery polymer zwitterions." *Advanced Materials,* in press. *Equal contributors

Zigmond, JS, **Letteri RA**, Wooley KL. "Amphiphilic Cross-Linked Liquid Crystalline Fluoropolymer-Poly(ethylene glycol) Coatings for Application in Challenging Conditions: Comparative study between Different Liquid Crystalline Comonomers and Polymer Architectures." *ACS Appl. Mater. Interfaces* **2016**, *8*, 33386-33393

Ghobadi AF*, **Letteri R***, Parelkar SS, Zhao Y, Chan-Seng D, Emrick T, Jayaraman A. "Dispersing Zwitterions into Comb Polymers for Nonviral Transfection: Experiments and Molecular Simulation." *Biomacromolecules* **2016**, *17*, 546-557. *Equal contributors

Chang CC*, **Letteri R***, Hayward RC, Emrick T. "Functional Sulfobetaine Polymers: Synthesis and Salt-Responsive Stabilization of Oil-in-Water Droplets." *Macromolecules* **2015**, *48*, 7843-7850. *Equal contributors

Parelkar S*, **Letteri R***, Chan-Seng D*, Zolochevska O, Ellis J, Figueiredo M, Emrick T. "Effect of oligopeptide orientation on polymer-based DNA delivery." *Biomacromolecules* **2014**, *15*, 1328-1336. *Equal contributors

505490 Building New Materials and Electronics within Intact, Living Biological Systems: from Nanoelectronics through Polymeric Device to Genetically-Targeted Electronics

Abstract:

Research Interests:

Rapidly advancing material science and nanoelectronics has been blurring the distinction between man-made and biological systems, providing tremendous opportunities in both fundamental research and biomedical applications. Exploring and exploiting cutting-edge, unconventional materials and electronics to solve specific biological questions and develop biomedical tools will form the core of my future research. In particular, my career objectives will focus on three areas: 1) Investigating nanomaterials and polymers for high-performance soft bioelectronics, and seeking their advanced human-machine interface; 2) developing optoelectronics tools and strategies for the intact brain activity mapping with single cell and single-spike resolution in behaving animals and 3) building cell-specific and genetically-targeted neuromodulation systems without genome editing.

Postdoctoral Research Summary: "Building functional materials and electronics within intact, living brain and cardiac systems"

Advisor: **Zhenan Bao** (Chemical Engineering), in a close collaboration with **Karl Deisseroth** (Bioengineering & Psychiatry and Behavioral Sciences), **Sergiu P. Pasca** (Psychiatry and Behavioral Sciences) and **Anson Lee** (Cardiothoracic Surgery), Stanford University

PhD Dissertation: "Biomimetics through nanoelectronics"

Advisor: **Charles M. Lieber**, Department of Chemistry and Chemical Biology, Harvard University

Research Experience

My academic training and research experience have provided me with a background in highly interdisciplinary area combing scientific disciplines including nanomaterial & nanoelectronics, polymer chemistry, tissue engineering, neuroengineering and medical engineering.

As a graduate student with Prof. Charles Lieber at Harvard University, I investigated semiconductor field-effect transistor as nanoscale sensors for the biological signal recording and then later develop the nanoelectronics into a three-dimensional (3D) flexible structure for sensing applications. I designed the nanoelectronic network to tissue scaffold as *nanoelectronic scaffold (nanoES)*. Through the collaboration with Prof. Daniel Kohane and Prof. Robert Langer, within this nanoES, I cultured synthetic neural tissue and cardiac patch, representing the first example of "cyborg tissue". Furthermore, I showed the response of these synthetic tissues to the 3D pharmacological stimulation and electrical stimulation as cardiac disease models, further proving the embedding of electronics with tissue in a seamless manner. I extended such nanoelectronics into the brain related research through inventing a "*syringe-injectable mesh electronics*" system. I precisely delivered such flexible and open-mesh nanoelectronic network into the rodent brain with targeted location, controlled unfolding and tissue penetration. Significantly, I demonstrated that this ultra-soft and open electronic network could form a long-term stable neural interface with vanishingly small immunoresponse from the neural tissue and thus enable a stable long-term neural activity recording at single-cell and single-spiking resolution.

My postdoctoral research with Prof. Zhenan Bao at Stanford University allows me to get further training in polymer science with a focus on polymeric electronics, as well as training in genome engineering, neuroengineering and cardiology. I have developed a *genetically-targeted conductive polymer synthesis for connecting neuron* in the intact, living brain tissue, and further demonstrated its biocompatibility and connection to external hardware with Prof. Karl Deisseroth. I applied such technology to the human brain organoid derived from neuropsychiatric disease modelled stem cells to study the tissue-wide activity modulation with Prof. Sergiu P. Pasca as a potential electronic therapeutic strategy. In addition, I have discovered several new materials systems including a conductive hydrogel with significantly reduced impedance for electrical recording, a photopatternable perfluorinated polymer system with low-k and ultrastability in solutions and a stretchable semiconductor that can be patterned through inkjet-printing. Accordingly, I have designed strategies to enable the micropatterning of those materials by the conventional photolithography to enable 1) a fully stretchable transistor array to drive an intrinsically stretchable organic light emitting electrochemical cell array and 2) a *high-density electrode array for the cardiac arrhythmogenic activity mapping in living animal* with Prof. Anson Lee.

Teaching Interests:

Aside from my research career, I also have extensive experience teaching Harvard and Stanford undergraduate students and graduate students. At Harvard University, I have TAed undergraduate-level classes for the pre-medical student in Physical Science. At Stanford University, I have guest-lectured graduate-level chemical engineering classes in Micro & Nanoscale Fabrication Engineering. I have received good reviews from both teaching experiences. I have been working in Chemical Engineering Department at Stanford and taken some Chemical Engineering core classes in both undergraduate-level and graduate-level thus I am willing to study and teach any existing classes, as well as design new classes. Last but not least, I mentored 1 undergraduate student in Pre-med and 1 graduate student in Chemistry at Harvard, and 3 graduate students in Chemical Engineering and Bioengineering at Stanford.

Selective Publication List (22 articles in total, citation: 2104)

(# indicates co-first author and * indicates co-corresponding author)

1. **Jia Liu#**, T. Fu#, Z. Cheng#, G. Hong, T. Zhou, L. Jin, M. Duvvuri, Z. Jiang, P. Kruskal, C. Xie, Z. Suo, Y. Fang* and C. M. Lieber*, "Syringe-injectable electronics," *Nature Nanotechnology* 10, 629-636 (2015). **Cover Article**
   - **10 World Changing Ideas, 2015** awarded by *Scientific American*
   - **Top Research of 2015** awarded by *Chemical and Engineering News*
2. **Jia Liu#**, C. Xie#, X. Dai#, L. Jin, W. Zhou and C. M. Lieber, "Multifunctional three-dimensional macroporous nanoelectronic networks for smart materials," *Proceedings of the National Academy of Sciences USA* 110, 6694-6699 (2013).
3. **Jia Liu**, Z. Sun, Y. Deng, Y*. Zou, C. Li, X. Guo, L. Xiong, Y. Gao, F. Li and D. Zhao*, "Highly water-Dispersible biocompatible magnetite particles with low cytotoxicity Stabilized by citrate groups,"

Angewandte Chemie International Edition 121, 5989-5993 (2009).
4. C. Xie#, **Jia Liu#**, T. Fu#, X. Dai, W. Zhou and C. M. Lieber, "Three-dimensional macroporous nanoelectronic networks as minimally-invasive brain probes," *Nature Materials* 14, 1286-1292 (2015).
5. B. Tian#, **Jia Liu#**, T. Dvir#, L. Jin, J. H. Tsui, Q. Qing, Z. Suo, R. Langer, D. S. Kohane* and C. M. Lieber*, "Macroporous nanowire nanoelectronic scaffolds for synthetic tissues," *Nature Materials* 11, 986-994 (2012).
    - **Special issued in** *Science*: "The cyborg era begins". (*Science*, 340, 1162-1165 (2013).)
6. A. Chortos#, **Jia Liu#** and Z. Bao, "Pursuing prosthetic electronic skin," *Nature Materials* 15, 937-950 (2016).

505638 Deep Learning in Chemical Engineering

Abstract:

Research Interests:

My reasearch interests are in using deep learning for discovering new chemical molecules for variety of applications. I'm using deep learning for high throughput material and molecular discovery. The technology (atomic convolutional neural network (ACNN)) that we have developed at our lab during my post-doc is capable of predicting the molecular descriptors with high accuracy and efficiency. I will show two examples of success stories of ACNN for high through put screening of polymers for organic photovoltaic and G-Protein Coupled Receptors (GPCRs) drug discovery.

Teaching Interests:

Transport phenomena, polymer physics, interfacial physical chemistry, fluid mechanics

504796 From Soft Materials to Soft Circuits

Abstract:

Conjugated polymers possess outstanding properties including high electrical conductivity (in the doped state), tunable band gap, and the ability to emit and absorb a wide spectrum of light. Hence, they are widely used for applications including artificial skin, organic photovoltaics (OPVs), organic light emitting diodes (OLEDs), energy storage devices, and flexible sensors. Recent investigations also show that conjugated polymers display great potential to exhibit high thermal conductivity, and thus are promising candidates for next-generation soft heat transfer materials. In this poster, I will present my PhD research focusing on three branches of conjugated polymer study. First is the synthesis of poly(3,4-ethylenedioxythiophene) (PEDOT) thin films possessing ultrahigh conductivity and mobility, complemented by theoretical modeling based on Boltzmann transport and wafer scale device fabrication of radio frequency AC to DC rectifier. Second is achieving record high cross-plane thermal conductivity (>10x common polymers) using undoped poly(3-hexylthiophene) (P3HT) thin films along with theoretical modeling. Third, is the fabrication of chemical sensing devices, using various nanostructured PEDOT and related copolymers, and the integration of the sensing devices in printed circuit boards (PCBs). Oxidative chemical vapor deposition (oCVD) is used as the synthesis method for the aforementioned polymers. As a solvent free method, oCVD circumvented the problem of low solubility of conjugated polymers and compatible issue between substrate and the solvent. In addition, the unique growth mechanism of oCVD P3HT synthesis essentially enhances the thermal conductivity. oCVD also makes it easy to pattern the polymer film during device fabrication, and provides conformal coating on nanostructured scaffolds. In addition to my three primary focus areas, side projects during my doctoral research include collaborative work with Eni S.p.A on large-scale flexible photovoltaic fabrication; collaborative work with the Department of Electrical Engineering and Computer Science at MIT on flexible energy harvesting device fabrication; DFT calculations and machine learning assisted monomer selection for light emitting polymers.

Research Interests:

1. *Material development:*

Material synthesis and study for conjugated polymers and other soft semiconductors, with special focus on their thermal,electrical, and optical properties, and the related theoretical modeling.

2. *Flexible circuit development:*

Fabrication of flexible electronics and circuits using newly discovered high performance soft materials. The targeted applications include energy harvesting, photovoltaic, energy storage, flexible sensor, and soft robotics.

3. *Machine learning assisted material design:*

The main goal is using machine learning (ML) as a powerful tool to facilitate the process of conjugated polymer discovery. There are two stages of the project: 1. predicting interested properties of new conjugated polymers using known experimental data, and additional knowledge; 2. generating new molecular structures from the ML algorithms. The results of ML can guide the experiments. And the experimental results, in turn, will improve the development of the ML algorithms.

Teaching Interests:

I would be happy to teach chemical engineering core classes, including transport phenomena, thermodynamics, reaction engineering and numerical methods. Other classes such as polymer chemistry and polymer physics also fit my background. I had experience working as a teaching assistant with Professor Deen and Professor Braatz teaching graduate level analysis of transport phenomena at MIT.

I would also be happy to design and teach new classes, such as metallic/semiconducting polymers, chemical and biological sensors, solid state physics for chemical engineers, or applications of machine learning in chemical engineering.

Selected Publications:

[1] Karen K. Gleason and **Xiaoxue Wang** "Oxidative Chemical Vapor Deposition for Conjugated Polymers: Theory and Applications", invited chapter for the fourth edition of the Handbook of Conducting Polymers, CRC Press/Taylor & Francis, to be submitted in December 2017, anticipated publication in late 2018

[2] **Xiaoxue Wang**, Xu Zhang, Lei Sun, Dongwook Lee, Sunghwan Lee, Yang Shao-Horn, Mircea Dinca, Tomas Palacios and Karen K. Gleason "Ultrahigh Electrical Conductivity of oCVD PEDOT Thin Films and the Wafer Scale Fabrication of the 13.6MHz Rectifiers based on the PEDOT-Si Diode", to be submitted

[3] Yanfei Xu†, **Xiaoxue Wang†**, Jiawei Zhou, Elizabeth Lee, Karen K. Gleason and Gang Chen "High Thermal Conductivity of Amorphous oCVD Grown P3HT" († authors contributed equally), to be submitted

[4] **Xiaoxue Wang**, Sema Ermez, Hilal Goktas, Silvija Gradečak and Karen Gleason "Room Temperature Sensing Achieved by GaAs Nanowires and oCVD Polymer Coating", *Macromolecular Rapid Communications* (2017)

[5] Minghui Wang†, **Xiaoxue Wang†**, Priya Moni†, Andong Liu, Do Han Kim, Won Jun Jo, Hossein Sojoudi and Karen K Gleason "CVD Polymers for Devices and Device Fabrication"(†contributed equally), *Advanced Materials*, 29.11 (2017): 1604606

[6] **Xiaoxue Wang**, Asli Ugur, Hilal Goktas, Nan Chen, Minghui Wang, Noa Lachman, Estelle Kalfon-Cohen, Wenjing Fang, Brian L Wardle and Karen K Gleason "Room Temperature Resistive Volatile Organic Compound Sensing Material based on a Hybrid Structure of Vertically Aligned Carbon Nanotubes and Conformal oCVD/iCVD Polymer Coatings" *ACS Sensors*, 1.4 (2016): 374-383.

[7] **Xiaoxue Wang**, Sichao Hou, Hilal Goktas, Peter Kovacik, Frank Yaul, Arun Paidimarri, Nathan Ickes, Anantha Chandrakasan, Karen Gleason "Small-area, Resistive Volatile Organic Compound (VOC) Sensors using Metal-polymer Hybrid Film based on Oxidative Chemical Vapor Deposition (oCVD)." *ACS Applied Materials and Interfaces*, 7.30(2015): 16213–16222

Proposal Writing Experience:

[1] "Improving oCVD PEDOT Properties: Transparent, Conducting, and Flexible Electrodes for Photovoltaics" full proposal for Eni S.p.A under the supervision of Professor Karen Gleason, 2016. 2-year financial support awarded beginning in September, 2017

[2] "oCVD/iCVD Structural Electronics Gas Sensor" full proposal for Analog Devices under the supervision of Professor Karen Gleason and Professor Brian Wardle, 2016. 1-year financial support awarded beginning in January, 2018

[3] "oCVD Polymers for Next Generation Flexible Photovoltaic" pre-proposal for Eni S.p.A under the

supervision of Professor Karen Gleason, 2016

[4] “Surface Modification to Reduce Nanoparticle Diffusion Through Flexible Polymers” white paper for Semiconductor Research Corporation under the supervision of Professor Karen Gleason, 2014

Invited Talk:

Wang, Xiaoxue et al“ Novel Resistive Volatile Organic Compound (VOC) Sensors” , invited talk in Nano-Engineered Composite aerospace Structures (NECST) Consortium all-day industry meeting, June 2016

505579 Intrinsically Stretchable Skin Electronics for Wearable Biomedical Applications

Abstract:

Research Interests:

The next generation of consumable electronics and biomedical devices feature more intimate integration with human bodies and tissues, which will be subject to the requirements of the same mechanical properties with epidermis including being soft and stretchable. Using skin-like intrinsically stretchable polymer materials to build up electronics can effectively blur the interface between human body and electronics, both chemically and mechanically. The development of this emerging generation of electronics is still at its premature stage, which requires both materials platforms and technology platforms. For the materials platform, stretchable semiconductors are indispensible, but existing ones typically sacrifice charge transport mobility to achieve stretchability. I, and my colleague, developed a concept based on the nanoconfinement of polymers to substantially improve the stretchability of polymer semiconductors, without affecting charge transport mobility. The increased polymer chain dynamics under nanoconfinement significantly reduces the modulus of the conjugated polymer and largely delays the onset of crack formation under strain. As a result, our fabricated semiconducting film can be stretched up to 100% strain without affecting mobility, retaining values comparable to that of amorphous silicon. The fully stretchable transistors exhibit high biaxial stretchability with minimal change in on current even when poked with a sharp object. This methodology has been shown to be broadly applicable to various polymer semiconductors. In the next major step, the technology platform needs to be developed for fabricating multiple devices and connecting them together to form skin-like signal processing module. I have developed the first fabrication platform for intrinsically stretchable transistor array, which has material universal applicability, ultrahigh yield and uniformity for performance. The fabricated transistor array has the mobility comparable to amorphous silicon. High stretchability of all the material components concurrently gives rise to supreme stretchability with stable performance up to 100% strain, robustness up to 1000 cycles, and an unprecedentedly high device density. Using this platform, I have further built up the major circuit foundations for intrinsically stretchable skin electronics. Bridging materials to electronic development, this technological platform takes an important step forward towards the realization and wide application of intrinsically stretchable skin electronics. My future research interest involves the development of intrinsically stretchable high-performance polymer electronic materials, intrinsically stretchable transistors, sensors, actuators and human-interactive integrated system with unprecedented functionality and skin-like nature.

The skin-attached biomedical devices are desired to operate independently and maintenance-free, which requires power sources scavenging energy from human body. Mechanical energy is one of the most universally-existing, diversely-presenting, but usually-wasted energies in the natural environment. The triboelectric effect is a universal phenomenon that can generate electrostatic charges from mechanical contact. During my PhD research, I developed a new type of technology—triboelectric nanogenerators (TENGs), which can efficiently convert mechanical energy into electricity based on the coupling of triboelectrification and electrostatic induction. My research achievements expand the establishment of three basic modes/mechanisms of TENGs that serve as the basis for most of the TENG structures, the optimization of materials and device structure for performance enhancement and different applications, and the development of self-charging power unit for electronic devices by hybridizing a TENG with a flexible Li-ion-battery. This technology has shown the potential to utilize (bio)mechanical energy to realized self-powered systems. My future research interest would involve the development of bio-compatible mechanical energy harvesting materials and technology and the stretchable self-powered bio-medical and wearable systems.

Teaching Interests:

After an extensive training in interdisciplinary fields including, materials science and engineering, chemical engineering, electrical engineering, nanotechnology and electrochemistry, I am more than ready to share my knowledge with new generations of students. I am interested in and capable of teaching most of the major courses in Chemical Engieering, including *Thermodynamics*, *Kinetics, Crystal structures and defects, Materials Characterization*, *Electrochemistry*, *Microfabrications*, *Nanoscience and Nanotechnology*, *Organic electronics, Green Energy Technology*, *Semiconductor Physics, etc.*

I have served as TA for multiple courses during my graduate study. Also, I have the experiences of guest lecturer both in Stanford University and in Georgia Tech.

Selected Publications:

J. Xu*, **S. Wang***, G.-J. N. Wang, C. Zhu, S. Luo, L. Jin, X. Gu, S. Chen, V. R. Feig, J.W.F. To, S. Rondeau-Gagné, J. Park, B. C. Schroeder, C. Lu, J. Y. Oh, Y. Wang, Y.-H. Kim, H. Yan, R. Sinclair, D. Zhou, G. Xue, B. Murmann, C. Linder, W Cai, J. B.-H. Tok, J. W. Chung, Z. Bao "Highly stretchable polymer semiconductor films through the nanoconfinement effect" *Science*, 355 (2017) 59.

S. Wang, Y. Zi, Y.S. Zhou, S.M. Li, F.R. Fan, L. Lin, Z.L. Wang "Molecular surface functionalization to enhance power output of triboelectric nanogenerators" *Journal of Materials Chemistry A*, 4 (2016) 3728.

Y.L. Zi*, J. Wang*, **S. Wang***, S.M. Li, Z. Wen, H.Y. Guo, Z.L. Wang "Effective energy storage from triboelectric nanogenerators" *Nature Communications*, 7 (2016) 10987.

S. Wang, Y.N. Xie, S.M. Niu, L. Lin, C. Liu, Y.S. Zhou, Z.L. Wang "Maximum surface charge density for triboelectric nanogenerators achieved by ionized-air injection: methodology and understanding" *Advanced Materials,* 26 (2014) 6720.

S. Wang, S.M. Niu, J. Yang, L. Lin, Z.L. Wang "Quantitative measurements of vibration amplitude using a contact-mode freestanding triboelectric nanogenerator" *ACS Nano,* 8 (2014) 12004.

S. Wang, Y.N. Xie, S.M. Niu, L. Lin, Z.L. Wang "Freestanding-triboelectric-layer based nanogenerators for harvesting energy from a moving object or human motion in contact and non-contact modes" *Advanced Materials*, 26 (2014) 2818-2824.

S. Wang, L. Lin, Z.L. Wang "Triboelectric nanogenerators as self-powered active sensors" *Nano Energy*, 11 (2014) 436.

Y.N. Xie*, **S. Wang***, S.M. Niu*, L. Lin, Q.S. Jing, J. Yang, Z.Y. Wu, Z.L. Wang "Grating-Structured Freestanding Triboelectric-layer Nanogenerator for Harvesting Mechanical Energy at 85% Total Conversion Efficiency" *Advanced Materials*, 26 (2014) 6599.

S. Wang, L. Lin, Y.N. Xie, Q.S. Jing, S.M. Niu, Z.L. Wang "Sliding-triboelectric nanogenerator based on in-plane charge-separation mechanism" *Nano Letters*, 13 (2013) 2226.

S. Wang, Z.–H. Lin, S.M. Niu, L. Lin, Y.N. Xie, K.C. Pradel, Z.L. Wang "Motion charged battery as sustainable flexible-power-unit" *ACS Nano,* 7 (2013) 11263.

S.M. Niu*, **S. Wang***, L. Lin, Y. Liu, Y.S. Zhou, Y.F. Hu, Z.L. Wang "Theoretical Study of the Contact-Mode

Triboelectric Nanogenerators as Effective Power Source" *Energy & Environmental Science*, 6 (2013) 3576.

S. Wang, L. Lin, Z.L. Wang "Nanoscale-triboelectric-effect enabled energy conversion for sustainably powering of portable electronics" *Nano Letters*, 12 (2012) 6339.

Molecular Simulation of Gas Transport in Polymeric Membranes

Computational Soft Matter Materials

Kai Zhang

B.S. 07' Tsinghua University; Ph.D. 12' Duke University

Department of Chemical Engineering, Columbia University, New York, NY

17AIChE Annual Meeting, Minneapolis, MN

DMR-1507030

The Upper Bound of the Robeson Plot

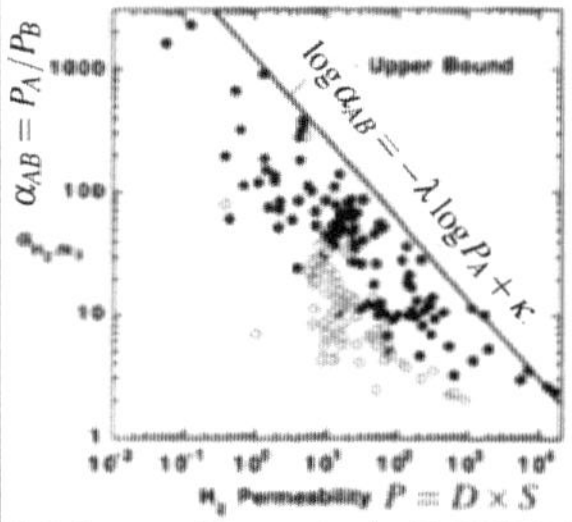

$$\log D = -ad^n + b$$

$$\lambda_n \equiv (d_B/d_A)^n - 1$$

$$\kappa = \lambda_n b + \lambda_n \log S_A + \log \alpha^S_{AB}$$

Gas diffusion in polymers is an activated process $D \sim e^{-E_a/k_BT}$

The activation energy scales with gas diameter as $E_a \sim d^2$

Freeman's theory for the upper bound slope

$$\lambda = (d_B/d_A)^2 - 1$$

B. D. Freeman, *Macromolecules* **32**, 375 (1999)

Coarse-grained Molecular Dynamics Simulation of Penetrant Transport in Polymers

Three polymer models

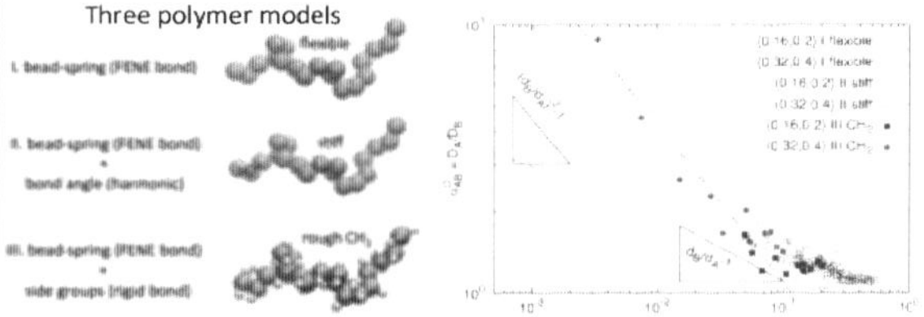

We measure the diffusivity (D) of gas pairs with sizes (0.16,0.2) and (0.32,0.4) in polymers at various monomer densities and construct the diffusion Robeson plot. For all three polymer models, the data from the small gas pair (0.16,0.2) has a slope of λ~0.2, close to d_A/d_B-1=0.25. For the large gas pair (0.32,0.4), the slope rises to λ~0.58, close to the Freeman value $(d_A/d_B)^2$-1=0.5625. Stiff polymers are more selective. This occurs when segmental dynamics is caged and the penetrant size d>d_c, such that the penetrant motion is hopping-like.

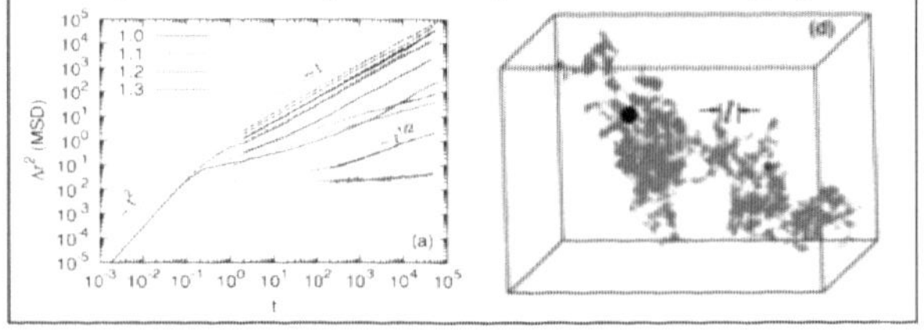

Three Dynamic Regimes of Penetrant Diffusion

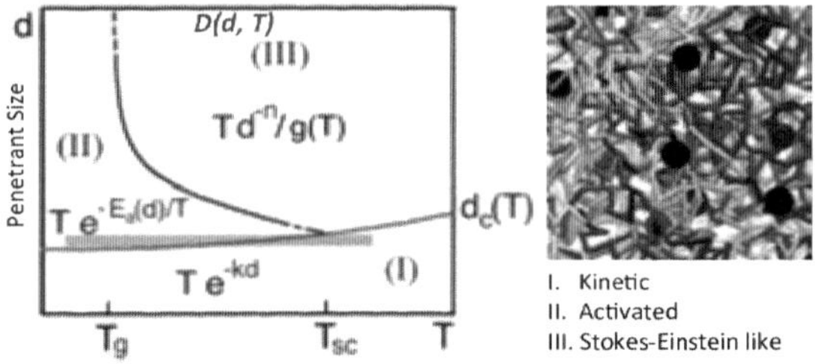

I. Kinetic
II. Activated
III. Stokes-Einstein like

K. Zhang and S. K. Kumar, *ACS Macro Letters* **6**, 864-868 (2017)

Glass-forming Ability of Metallic Glasses

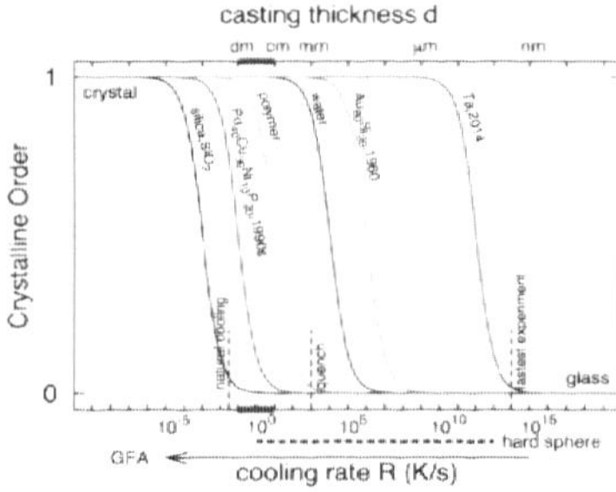

The final state of a liquid under cooling, being crystalline or amorphous, depends on the cooling rate R. The **critical cooling rate** R_c and the casting thickness d can be used to characterize the glass-forming ability (GFA) of the material. Metals and alloys are typically poor glass formers.

R_c of binary hard spheres as a function of size ratio and composition compared with typical binary bulk metallic glasses

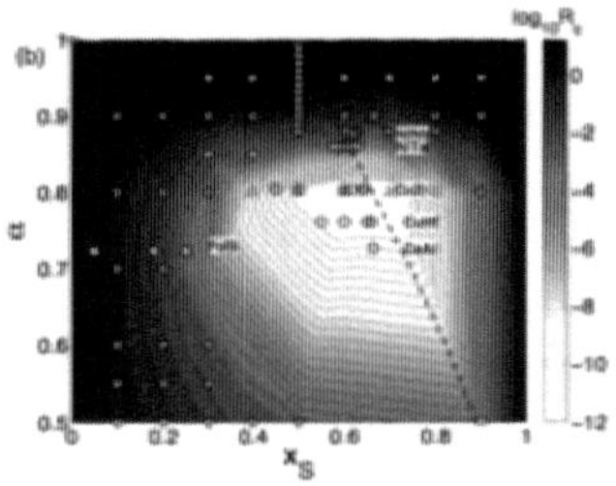

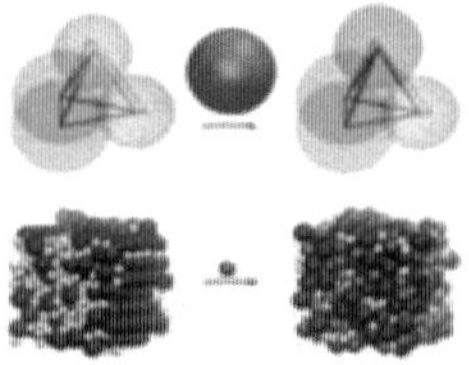

Packing of hard spheres also explains why ternary metallic alloys are better glass formers.

K. Zhang, W. W. Smith, M. Wang, Y. Liu, J. Schroers, M. D. Shattuck, C. S. O'Hern, *Phys. Rev. E* **90**, 032311 (2014)

Free-energy based method for solving the phase diagram of modulated phases

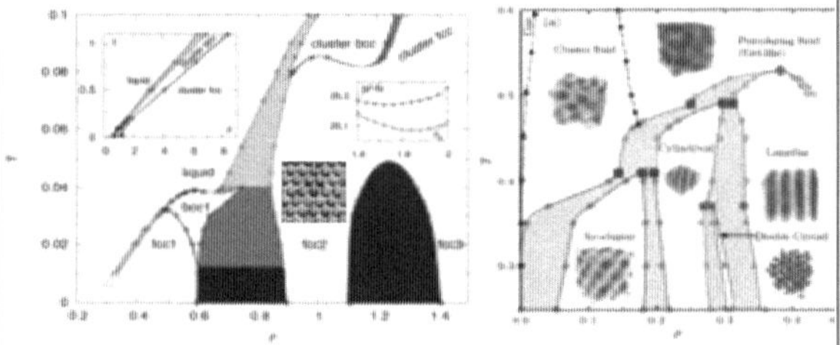

We develop a free-energy based simulation method to calculate the equilibrium modulated phases and solve the phase diagram.

K. Zhang and P. Charbonneau, *Phys. Rev. Lett.* **105**, 245701 (2010)
Y. Zhuang, K. Zhang and P. Charbonneau, *Phys. Rev. Lett.* **116**, 098301(2016)

Future Research Interests

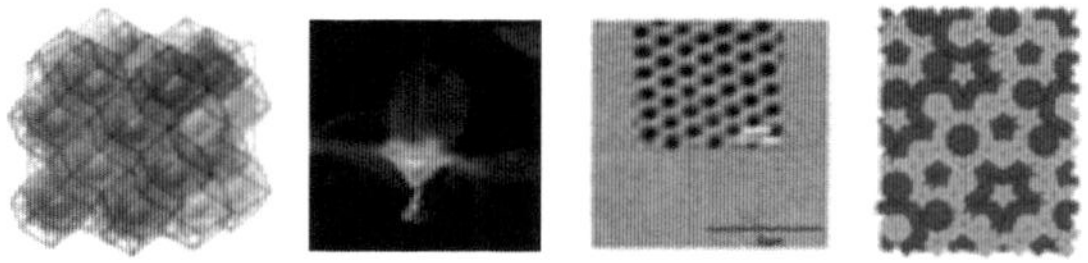

Z. Wang *et al.*, *J. Cell Biol.* **206**, 619 (2014); H. B. Park *et al.*, *Science* **356**, 1137 (2017)

In the future, I intend to make a marked impact on soft matter by solving problems key to the grand challenges of energy and biomedical technology, such as nanomaterials stability, mechanisms for nanofiltration membranes, cell polarity and membrane invasion, as well as to pursue more fundamental materials questions, such as quasicrystal formation. By harnessing the ever-increasing capabilities of computers, I will strive to optimize the research strategy for materials discovery and to create tools that will transform soft matter research.

Teaching Interests

Thermodynamics, Transport, Physical Chemistry, Materials Science, Polymer Physics, Soft Condensed Matter, Statistical Mechanics, Quantum Chemistry, Numerical Methods, Molecular Simulation

Morphology, Electrical Conductivity and Electromagnetic Interference Shielding of Fe-Catalyst-Grown Carbon Nanotube/Polyvinylidene Fluoride Nanocomposites: Impact of Synthesis Temperature

Mohammad Arjmand, Uttandaraman Sundararaj

Department of Chemical and Petroleum Engineering, University of Calgary, Canada
arjmand64@yahoo.com
u.sundararaj@ucalgary.ca

INTRODUCTION

Nowadays, ease of use and availability of cellphones provide a fertile land for electro-communication industries. It is estimated that the number of cellphones reaches 5.07 billion in 2019. This technological breakthrough might shorten the communication gap in the globe, but it also manifests a prominent threat to the health of people. Most wireless devices like cellphones employ the electromagnetic waves to convey messages and voice signals. Numerous studies have shown that long-term exposure of human being to electromagnetic fields would be associated with health problems. Owing to this, international organizations have formulated guidelines establishing limits for occupational and residential electromagnetic exposure. One of the strategies alleviating the harmful effect of electromagnetic wave emission is electromagnetic interference (EMI) shielding. As its name suggests, EMI shielding refers to mechanisms in which a compound acts like a shield to diminish the penetration of electromagnetic radiation.

Polymer–based composites containing conductive fillers are attractive for shielding owing to their tunable electrical conductivity, ease of processability (e.g., moldability), low density, low cost and corrosion resistance [1-3]. High conductivity and connectivity within the polymer matrix are the two key characteristics for ideal fillers employed for shielding applications [2]. In this regard, carbon nanotubes (CNTs) are promising options owing to their very high aspect ratio and outstanding electrical, mechanical, and thermal properties.

To the best of our knowledge, there are a rare number of papers correlating the effect of CNT synthesis conditions with the physical characteristics of CNTs, quality of CNT-polymer interaction, and EMI shielding of CNT/polymer nanocomposites (CPNs) [4-6]. Accordingly, this study strives to cover the whole hierarchy of synthesis-processing-structure-property of CPNs. The first part of this article is devoted to a detailed characterization of CNTs synthesized at different temperatures using a chemical vapor deposition (CVD) setup. After that, effects of synthesized carbonaceous fillers on the morphology and electrical properties of the developed CPNs, i.e. electrical conductivity and EMI shielding, are studied. Huge discrepancies between the physical properties of the generated nanofillers, and therefore properties of the developed CPNs were observed. This provided an in-depth discernment on correlation between the physical properties of the nanofillers and final properties of the nanocomposites. Moreover, surprisingly, this study revealed an optimum synthesis temperature with electrical properties far superior to the other synthesis temperatures.

METHODOLOGY

CNTs were synthesized via CVD method, wherein source and carrier gases were conveyed over an alumina-supported iron catalyst in a quartz tubular reactor with an inner diameter of 4.5cm. The catalyst was prepared by incipient wetness impregnation of iron (iron (III)) nitrate nanohydrate, Baker Analyzed® ACS Grade) dissolved in water on aluminum oxide support (Sasol Catalox Sba-200), followed by drying, calcination and reduction. CNT samples were synthesized at five different temperatures (550°C, 650°C, 750°C, 850°C and 950°C) using a mixture of ethane, hydrogen and argon. The synthesis time was 2h for all samples.

Synthesized CNTs were mixed with a PVDF matrix (11008/0001, 3M Canada) using a miniature melt mixer at 240°C. The samples were prepared at different CNT concentrations, i.e., 0.3, 0.5, 1.0, 2.0, 2.7 and 3.5wt%. A Carver compression molder (Carver Inc., Wabash, IN) was used to make the samples with the dimensions of $24\times11\times1.1mm^3$ at 220°C under 38MPa pressure for 10min.

The carbon purity and thermal stability of synthesized CNTs were evaluated using a Thermogravimetric Analyzer (TGA), TA instruments Q500, under air atmosphere (Praxair AI INDK) from room temperature to 900°C at a ramp rate of 10°C/min. The electrical conductivity of CNT powders was measured using an especially designed instrument. The powder of 40–50mg was filled into a cylinder and compressed with a piston, and then the electrical conductivity of the powder was measured using a Keithley 2001 electrometer at various piston pressures, i.e. 5, 10, 20, 25, and 30MPa. Further details about this procedure are provided elsewhere [7]. In order to investigate the morphology of synthesized CNTs, a high-resolution Philips XL30 was used to obtain scanning electron microscopy (SEM) images. A transmission electron microscope (Tecnai TF20 G2 FEG-TEM (FEI, Hillsboro, Oregon, USA)) was used to take the images to measure the length and diameter of CNTs (100 individual CNTs) for each sample using the MeasureIT software (Olympus Soft Imaging Solutions GmbH).

The microdispersion state of CNTs within the PVDF matrix was quantified using light transmission microscopy (LM) on thin cuts ($5\mu m$ thickness) of the compression-molded samples, prepared with a Leica microtome RM2265 (Leica Microsystems GmbH, Wetzlar, Germany). Two conductivity meters with 90V as the applied voltage were employed to measure the DC electrical conductivity of the generated materials. For the nanocomposites with an electrical conductivity higher than $10^{-2} S\cdot m^{-1}$, the measurements were conducted according to the ASTM 257-75 standards using a Loresta GP resistivity meter (MCP-T610 model, Mitsubishi Chemical Co., Japan) connected with an ESP probe. For electrical conductivities less than $10^{-2} S\cdot m^{-1}$, the measurements were performed with a Keithley 6517A electrometer connected to a Keithley 8009 test fixture (Keithley instruments, USA). EMI shielding measurements in the X-band frequency range (8.2-12.4GHz) were performed using an E5071C network analyzer (ENA series 300KHz - 20GHz). In order to perform the measurements, the samples under the test were sandwiched between two waveguides of the network analyzer and then the scattering parameters (S-parameters) of each sample were recorded and used to measure EMI shielding effectiveness

(SE). SE is expressed in dB, and is the logarithm of the ratio of the incident power to the transmitted power.

RESULTS AND DISCUSSION

Figure 1 depicts TG, DTG and powder conductivity graphs of CNTs synthesized at various temperatures. These three curves provide invaluable information about the thermal stability, carbon purity and quality of the synthesized nanofillers. The catalyst particles comprised 80wt% alumina and 20wt% metallic particles. Alumina is insulative and metallic particles have much lower surface area than synthesized CNTs, and their surface area reduced further due to sintering phenomenon at the synthesis temperatures. This explains why low carbon purity can drastically deteriorate the electrical properties, and confirms the significance of carbon purity on the final electrical properties. As shown in **Figure 1(a)** and tabulated in **Table 1**, $CNT_{550°C}$ showed by far the lowest carbon purity about 43% and $CNT_{650°C}$ presented the highest carbon purity around 85%. Carbonaceous fillers synthesized at the other three temperatures presented carbon purities around 70%. In this regard, $CNT_{650°C}$ stands out in terms of the carbon purity.

Figure 1(b) shows that $CNT_{550°C}$ depicts a much broader DTG curve than CNTs synthesized at the other temperatures. Moreover, the onset degradation point for $CNT_{550°C}$ is much lower than the other types of CNTs. This can be ascribed to low synthesis temperature, facilitating the formation of amorphous carbon. $CNT_{650°C}$ and $CNT_{750°C}$ showed narrow DTG curves, denoting superior quality of CNTs. Nonetheless, it was surprisingly observed that $CNT_{850°C}$ and $CNT_{950°C}$ indicated a very different shape of DTG curve compared to the other three temperatures. The yellow box presents the portion of carbonaceous fillers synthesized at 850°C and 950°C made of CNTs. The left side of the yellow box coincides with the onset degradation of $CNT_{850°C}$ and $CNT_{950°C}$, which is far higher than the other synthesis temperatures. This shows that no amorphous carbon formed at these two synthesis temperatures, as confirmed by high inflection points defined in **Table 1**. The right side of the yellow box signifies the temperature in which CNTs were fully decomposed, matching the plateau in the TG curves of $CNT_{550°C}$, $CNT_{650°C}$, and $CNT_{750°C}$. This shows that the final material synthesized at 850°C and 950°C consisted of a thermal-resistant carbonaceous filler, CNT and catalyst. We believe that the thermal-resistant carbonaceous filler synthesized at high temperatures arose from sintering and coalescence of catalyst particles, leading to the coverage of large catalyst particles by the carbonaceous filler. **Figure 1(c)** shows the powder conductivity of the synthesized carbonaceous fillers as a function of piston pressure. It is vivid that the powder conductivity had an ascending trend with the synthesis temperature. Given the high conductivity of the materials synthesized at 850°C and 950°C, it can be asserted that carbonaceous fillers formed on the surface of large catalyst particles had graphitic sp^2 structure.

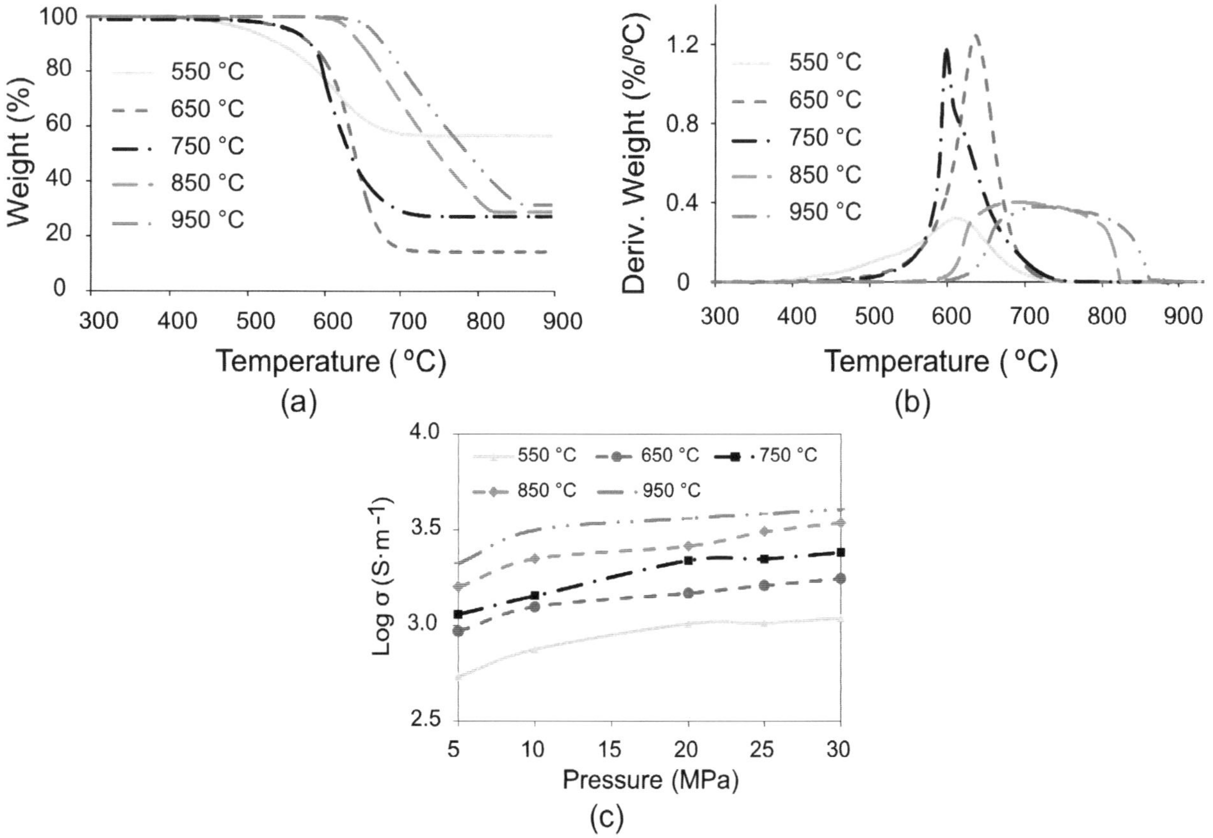

Figure 1. (a) TG, (b) DTG and (c) powder conductivity of CNTs synthesized at various temperatures.

TABLE 1. Physical characteristics of CNTs synthesized at different temperatures and their polymeric nanocomposites.

	550°C	650°C	750°C	850°C	950°C
Inflection Point (°C)	612	636	599	696	711
Carbon Purity %	43.1	85.4	72.7	71.1	68.7
Length (nm)	853	629	1696	-	-
Diameter (nm)	22	15	29	-	-
Agglomerate Area Ratio % (Nanocomposite)	3.2	0.8	4.7	2.7	1.9
Relative Transparency % (Nanocomposite)	45.0	23.3	46.2	86.9	90.6

Figure 2 depicts the SEM micrographs of CNT powders and LM micrographs of CNT/PVDF nanocomposites. SEM reveals that distinctive morphological features of our CNT samples stem from manipulation of synthesis temperature. As shown in **Figure 2**, micrograph of $CNT_{550°C}$ shows highest density of catalyst particles among all samples, covered by small-diameter CNTs. This synthesis temperature is also in favor of generation of poor thermally stable amorphous carbon as validated by TGA analysis. The micrographs for $CNT_{650°C}$ and $CNT_{750°C}$ reveal the rich formation of entangled low-diameter CNTs over catalyst particles. Lower CNT diameters as well as higher carbon purity are associated with $CNT_{650°C}$ samples. At higher synthesis temperatures (850°C and 950°C), the formation of other carbonaceous materials like graphite and

nanoparticles are more dominant over CNT synthesis. In addition, much higher diameters of CNTs were observed than the lower synthesis temperatures, which is attributable to sintering of catalyst particles at higher temperatures. The possibility of higher sintering of catalyst particles at 850°C and 950°C was also validated by low BET surface of catalyst samples, denoting iron catalyst particles agglomeration, i.e. large iron agglomerates block alumina pores. BET surface areas of the studied samples are tabulated in **Table 2**.

Measurement of the length and diameter of CNTs using the TEM images indicated that $CNT_{750°C}$ had the largest length, whereas $CNT_{650°C}$ presented the lowest diameter (largest surface area). Nonetheless, we could not find sufficient number of CNTs at 850°C and 950°C to obtain the average length and diameter.

State of dispersion of nanofiller in polymer matrix is a key parameter to justify the electrical properties of the resultant nanocomposites. The better dispersion, the more enhanced the network formation and the superior electrical properties. LM provides a powerful tool to assess the microdispersion state of nanotubes that are not dispersed and appear as agglomerates. This non-dispersed part was quantified using the agglomerate area ratio. Moreover, the concept of relative transparency ratio (transparency of sample over glass slide/Aquatex/cover glass assembly) was implemented to study the dispersion of small agglomerates with sizes equal to or slightly larger than the wavelength of visible light, ca. 400-700nm. Referring to **Figure 2** and **Table 1**, the lowest transparency was observed for the nanocomposites with $CNT_{650°C}$, implying decent dispersion of carbon nanotubes. On the other hand, samples prepared from $CNT_{850°C}$ and $CNT_{950°C}$ had the highest transparency far beyond the other synthesis temperatures. Moreover, the agglomerate area ratio was the lowest and highest for 650°C and 750°C, respectively. These results clearly reveal that 650°C was the best temperature for the CNT dispersion, while 850°C and 950°C had the worst state of dispersion.

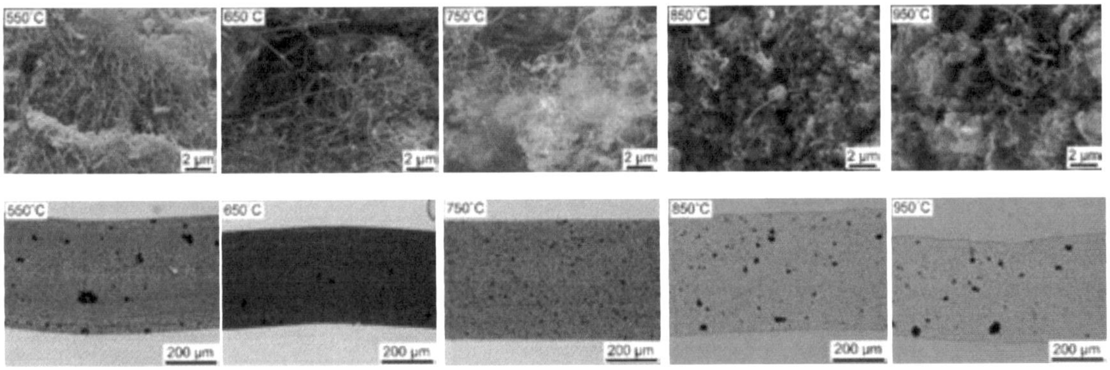

Figure 2: Row 1: SEM micrographs of CNT powder and Row 2: LM micrographs of CNT/PVDF nanocomposites. CNTs were synthesized at different temperatures.

TABLE 2. BET surfaces of catalysts at different synthesis temperatures.

Sample	BET Surface [m^2/g]
Catalyst Before Reduction	125.17
Catalyst After Reduction at 550°C	140.55
Catalyst After Reduction at 650°C	135.73
Catalyst After Reduction at 750°C	123.17
Catalyst After Reduction at 850°C	108.46
Catalyst After Reduction at 950°C	92.32

Figure 3 shows the percolation curves of CNT/PVDF nanocomposites fabricated from different synthesis temperatures. The values of percolation threshold derived from statistical percolation theory [7] for $CNT_{550°C}$, $CNT_{650°C}$ and $CNT_{750°C}$ are 0.4wt%, 1.0wt% and 2.3wt%, respectively. No conductivity was observed for both $CNT_{850°C}$ and $CNT_{950°C}$ samples in a whole concentration range. The insulative nature of $CNT_{850°C}$ and $CNT_{950°C}$ nanocomposites was due to lack of formation of CNTs in these two temperatures. This led to lack of formation of conductive network within the PVDF matrix for these two temperatures.

$CNT_{650°C}$ nanocomposites showed electrical conductivity and percolation threshold far superior to the other temperatures. The percolation threshold obtained for $CNT_{650°C}$ nanocomposites is comparable or lower than the results presented for CNT/PVDF nanocomposites in the literature [8, 9]. It was also observed that the maximum conductivity obtained for $CNT_{650°C}$ nanocomposites is much higher than the other temperatures.

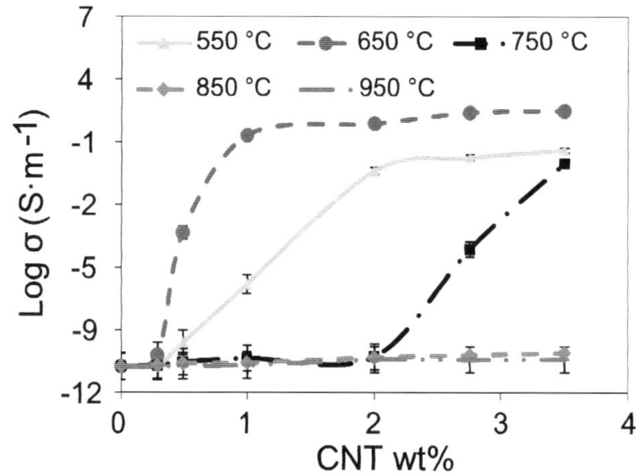

Figure 3: Electrical conductivity of CNT/PVDF nanocomposites made at different synthesis temperatures as a function of CNT content.

Figure 4 depicts EMI shielding of the generated nanocomposites as a function of CNT content. Similar to electrical conductivity, both $CNT_{850°C}$ and $CNT_{950°C}$ nanocomposite samples presented very poor SE results. Among the other three synthesis temperatures, $CNT_{650°C}$

nanocomposites showed SE far away superior to the other synthesis temperatures. Preparation of nanocomposite with 1.0 and 3.5wt% of $CNT_{650°C}$ increased SE of pure PVDF from 0.3dB to 8.9 and 20.2dB, respectively. However, for instance, SE of $CNT_{550°C}$ nanocomposite was 1.0 and 4.2dB for 1.0 and 3.5wt% CNT content, respectively.

Differences in length and diameter of CNTs, carbon purity, crystallinity of CNTs, and state of dispersion within the PVDF matrix are the key parameters responsible for different behavior of percolation curves for $CNT_{550°C}$, $CNT_{650°C}$, and $CNT_{750°C}$. It has been established well, both theoretically and experimentally, that well-dispersed CNTs with higher aspect ratio (higher length and lower diameter) demonstrate lower percolation threshold, and loftier electrical conductivity and EMI shielding [10]. As a matter of fact, lower diameter signifies higher interacting surface area per mass of CNTs, and CNTs with higher length have more likelihood to be in close contiguity or contact each other. In addition, CPNs with better state of CNT dispersion present enhanced conductive network [11-13]. Our results show that decent state of dispersion, high carbon purity, high quality, and low diameter of $CNT_{650°C}$ were the contributing factors to its superior electrical behavior.

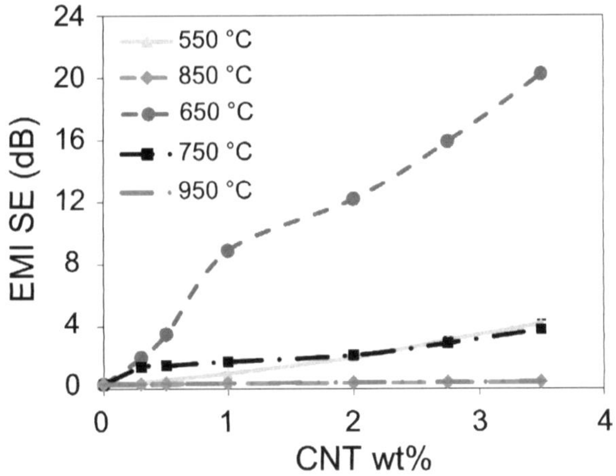

Figure 4: (a) Overall EMI SE of CNT/PVDF nanocomposites made at different synthesis temperatures as a function of CNT content.

AKNOWLEDGMENT

Financial support from the Natural Sciences and Engineering Research Council (NSERC) of Canada is highly appreciated. Dr. Mohammad Arjmand thanks IPF Dresden for granting a research stay.

References

1. W. Bauhofer, J.Z. Kovacs. A review and analysis of electrical percolation in carbon nanotube polymer composites. Composites Science and Technology. 69 (2009) 1486–1498.
2. F. Sharif, M. Arjmand, A. Abbasi Moud, U. Sundararaj, E.P.L. Roberts. Segregated hybrid graphene/magnetite/PMMA polymer nanocomposites towards electromagnetic interference shielding. ACS Applied Materials and Interfaces. 9 (2017) 14171-14179.

3. K. Chizari, M. Arjmand, L. Zhe, U. Sundararaj, D. Therriault. Three-dimensional printing of highly conductive polymer nanocomposites for EMI shielding applications. Materials Today Communications. 11 (2017) 112-118.

4. M. Arjmand, U. Sundararaj. Electromagnetic interference shielding of nitrogen doped and undoped carbon nanotube/polyvinylidene fluoride nanocomposites: a comparative study. Composites Science and Technology. 118 (2015) 257–263.

5. M. Arjmand, K. Chizari, B. Krause, P. Pötschke, U. Sundararaj. Effect of synthesis catalyst on structure of nitrogen-doped carbon nanotubes and electrical conductivity and electromagnetic interference shielding of their polymeric nanocomposites. Carbon 98 (2016) 358–372.

6. M. Arjmand, A. Ameli, U. Sundararaj. Employing nitrogen doping as innovative technique to improve broadband dielectric properties of carbon nanotube/polymer nanocomposites. Macromolecular Materials and Engineering. 301 (2016) 555-565.

7. M. Weber, M.R. Kamal. Estimation of the volume resistivity of electrically conductive composites. Polymer Composites. 18 (1997) 711–725.

8. C.-C.M. Ma, Y.-L. Huang, H.-C. Kuan, Y.-S. Chiu. Preparation and electromagnetic interference shielding characteristics of novel carbon-nanotube/siloxane/poly- (urea urethane) nanocomposites, Journal of Polymer Science B: Polymer Physics. 43 (2005) 345–358.

9. R. Ram, M. Rahaman, D. Khastgir. Electrical properties of polyvinylidene fluoride (PVDF)/multi-walled carbon nanotube (MWCNT) semi-transparent composites: modelling of DC conductivity. Composites Part A: Applied Science Manufacturing. 69 (2015) 30–39.

10. J. Li, P.C. Ma, W.S. Chow, C.K. To, B.Z. Tang, J.K. Kim, Correlations between percolation threshold, dispersion state, and aspect ratio of carbon nanotubes. Advanced Functional Materials. 17 (2007) 3207–3215.

11. M. Arjmand, A. Abbasi Moud, Y. Li, U. Sundararaj. Outstanding electromagnetic interference shielding of silver nanowire: comparison with carbon nanotube. RSC Advances. 5 (2015) 56590-56598.

12. M. Arjmand, S. Sadeghi, M. Khajehpour, U. Sundararaj. Carbon nanotube/graphene nanoribbon/polyvinylidene fluoride hybrid nanocomposites: rheological and dielectric properties. Journal of Physical Chemistry C. 121 (2017) 169-181.

13. Y. Zamani Keteklahijani, M. Arjmand, U. Sundararaj. Cobalt Catalyst Grown Carbon Nanotube/Poly(Vinylidene Fluoride) Nanocomposites: Effect of Synthesis Temperature on Morphology, Electrical Conductivity and Electromagnetic Interference Shielding. ChemistrySelect. 2 (2017) 10271-10284.

Nanorheology of Entangled Polymer Melts

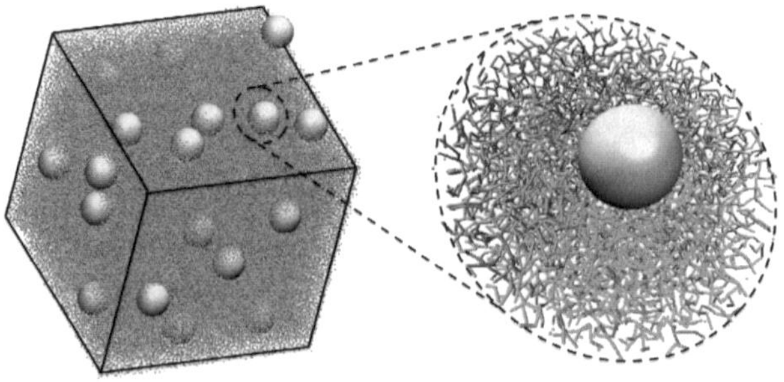

explore the extension of microrheology to nanorheology

Ting Ge and Michael Rubinstein
University of North Carolina at Chapel Hill
Gary S. Grest
Sandia National Laboratories, New Mexico

2017 AIChE Annual Meeting
Oct. 29 – Nov. 3, 2017, Minneapolis, MN

1

Model and Methodology

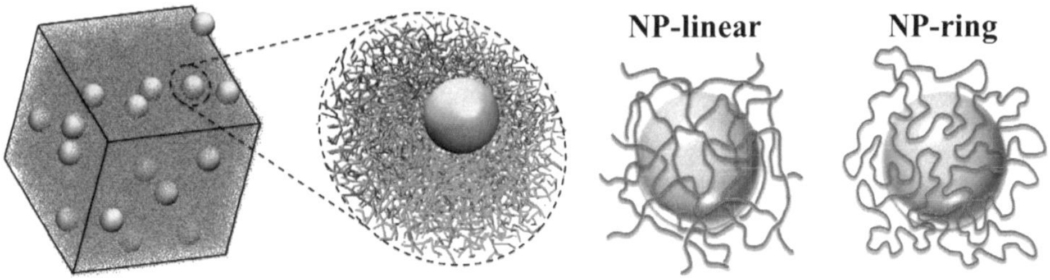

Molecular Dynamics Simulations

- Weakly interacting mixtures of NPs and ***entangled*** linear / non-concatenated ring polymers chain length ≈ *30* entanglement length
- NPs of diameter ***d*** are well dispersed at volume fraction ~ *10%*
- NVT simulation of NPs in polymer melts

2

Stress Relaxation Modulus

Green-Kubo formula

$$G_{GK}^{ij}(t) = \frac{V}{k_B T}\left\langle \overline{\sigma^{ij}}(t)\overline{\sigma^{ij}}(0)\right\rangle \quad G_{GK}(t) = \frac{1}{3}\left[G_{GK}^{xy}(t) + G_{GK}^{xz}(t) + G_{GK}^{yz}(t)\right]$$

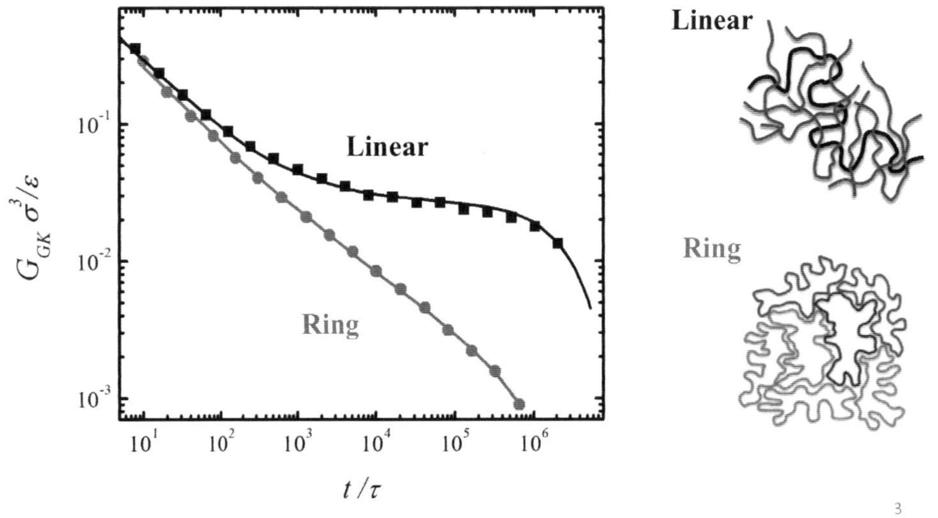

3

Generalized Stokes-Einstein Relation

$$\widetilde{G}(s) = \frac{6k_B T}{f\pi ds\left\langle \widetilde{\Delta r^2}(s)\right\rangle}$$

Stick Boundary --- $f = 3$

Slip Boundary --- $f = 2$

Estimate the transform by describing MSD $\langle \Delta r^2(t)\rangle$ as a local power law

T. G. Mason, *Rheol. Acta*, 2000, 39, 371

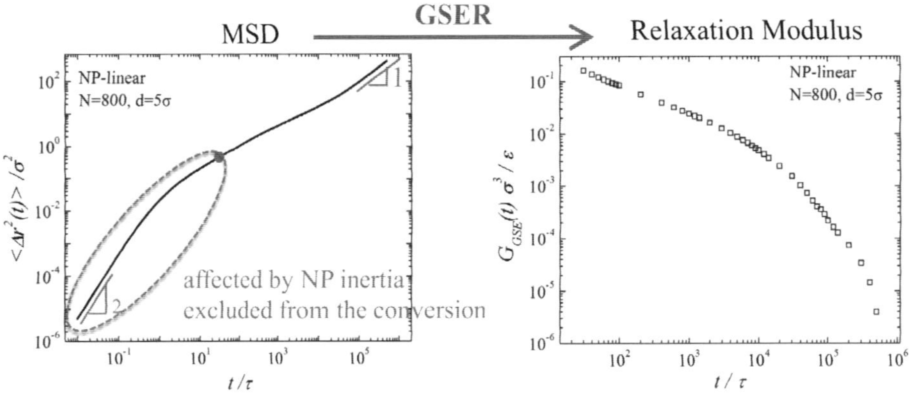

4

NPs in Entangled Linear Polymer Melt

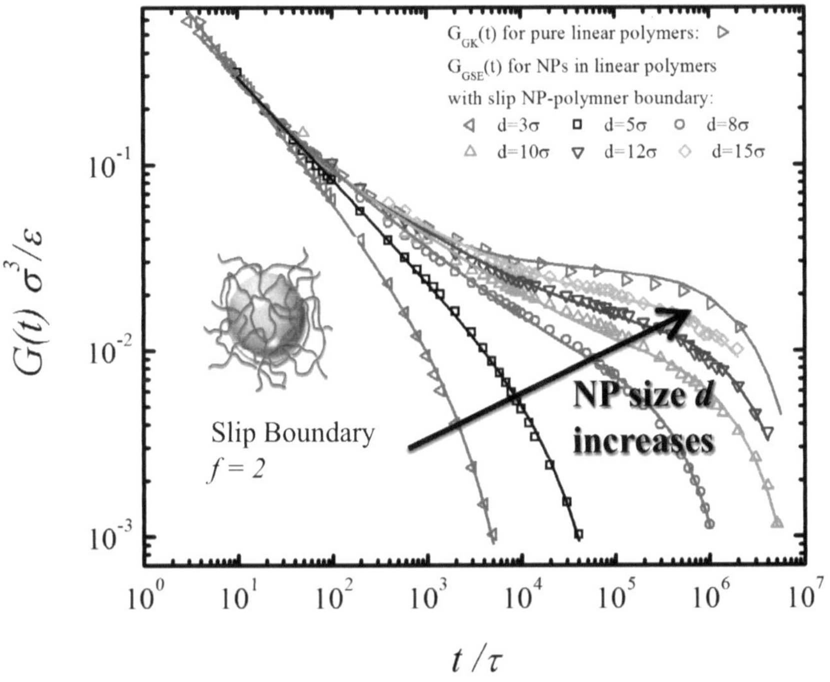

5

NPs in Entangled Linear Polymer Melt

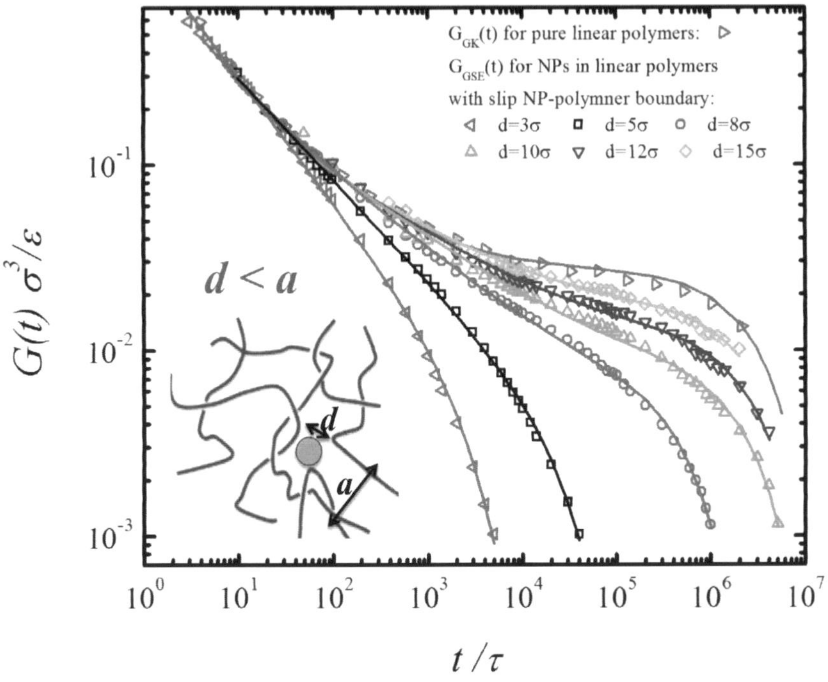

6

NPs in Entangled Linear Polymer Melt

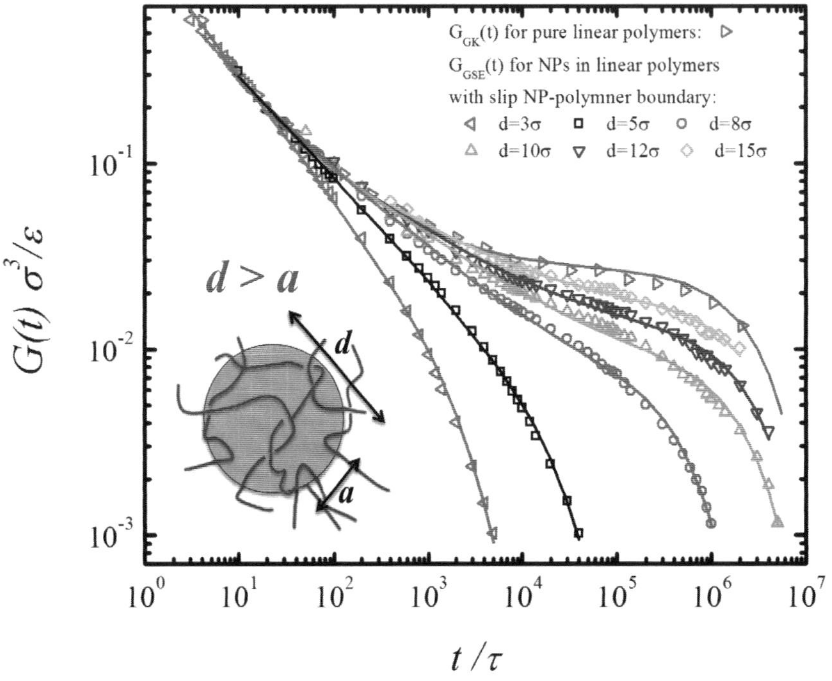

7

NPs in Entangled Ring Polymer Melt

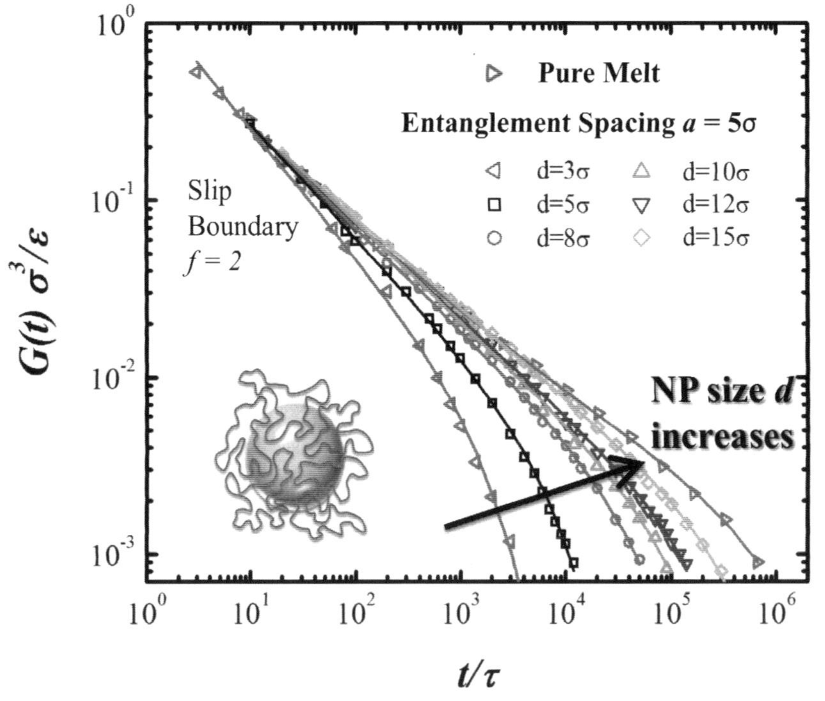

8

Terminal Diffusion of NPs

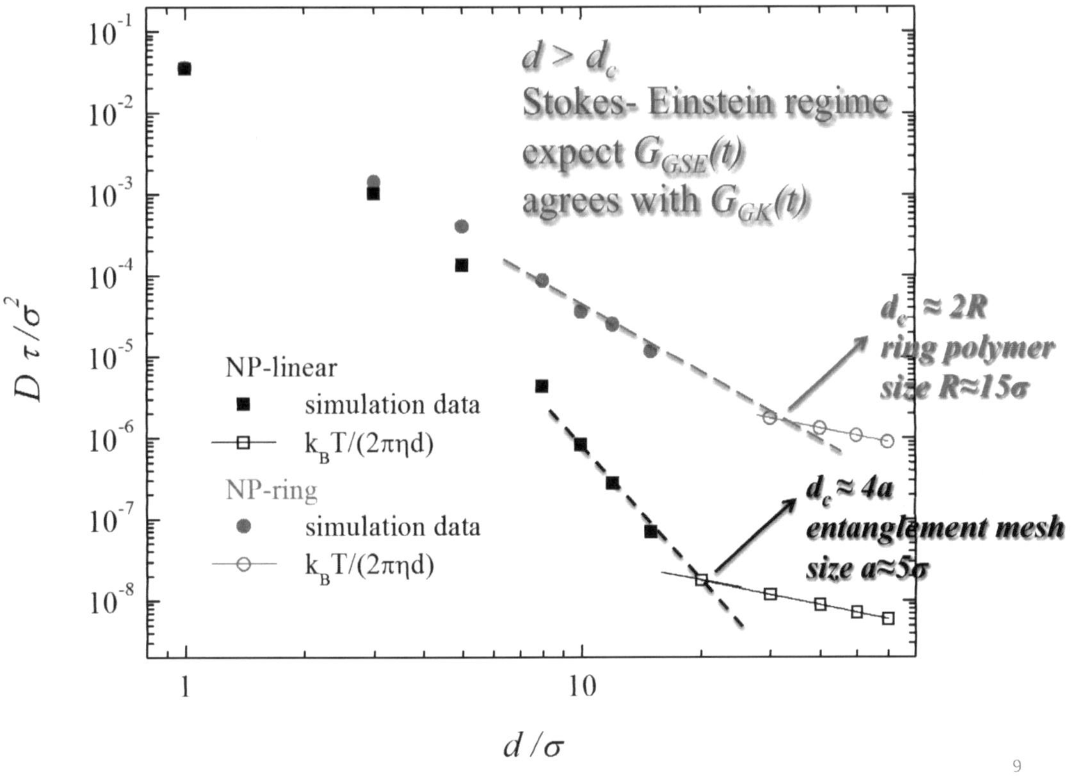

9

505436 Polymer Process Design and Modelling to Fabricate and Understand Unique Composite Architectures

Abstract:

Research Interests: Soft materials, such as polymers, continue to play an increasingly important role in virtually all engineering disciplines. Understanding the complex interplay between synthesis, processing, structure, and properties of polymers is critical to make any substantial advances in fields where polymers play a dominant role, such as packaging, filtration, bio-scaffolding, and energy storage to name just a few. As a graduate student at Case Western Reserve University studying under Prof. LaShanda Korley I gained expertise in applying polymer processing technologies to achieve unique structures. While studying deformation mechanics in multilayer films, I wrote a successful Department of Energy experimental grant for synchrotron time at Brookhaven National Laboratory that led to multiple publications for not only myself, but many others in our department. One of these manuscripts highlighted the role of structural evolution during post-extrusion processing of poly(Îµ-caprolactone) (PCL) nano-fibers. We not only correlated this structural evolution to mechanical properties of the PCL fiber mats, but also the reduced efficacy of peptide modification of the fiber surfaces which substantially altered their use as tissue engineering scaffolds. I was also able to use existing multilayer fiber/matrix coextrusion technology to develop a one-step method to fabricate well-distributed, randomly aligned, sub-micron scale polyester fiber-reinforced poly(ethylene oxide) (PEO) hydrogels. By varying fiber loading, post-extrusion take-up, and fiber material type, I was able to substantially improve the mechanical properties of these hydrogels, reaching similar stiffness to articular cartilage. By manipulating the mechanical environment of the hydrogels, we also manipulated cellular response and differentiation when using the hydrogels as bio-scaffolds. This work highlighted the ability to use polymer processing technology and commodity polymers to address a substantial challenge in the biomaterials community. In addition to existing publications on this technology, we have also filed a US Patent application and I assisted with writing a National Institute of Health grant to continue this work after moving on to my post-doctoral fellowship position.

As a post-doctoral fellow at the University of Minnesota working with Prof. Chris Macosko and Prof. Frank Bates, I have developed a substantial knowledge base in polymer processing fundamentals and rheology. My current research is focused on understanding polymer-polymer interfaces. I have been able to utilize rheological and thermal transport modelling to better understand the relationship between processing parameters, molecular structure, and catalyst selection on adhesion between polyolefins. The observed interfacial weld strengths between polyethylene (PE) and polypropylene (PP) vary over ~2 orders of magnitude, based on these variables. Each interface was also characterized with a suite of microscopy techniques. Further, we have been able to connect these interfaces and morphology derived from various processing techniques (extrusion and molding) with the final mechanical properties of polyolefin blends. This work demonstrates the role of not only molecular structure but also process modelling to maximize adhesion between immiscible polymer pairs to achieve tailored interfaces with specific mechanical properties, and forms the basis for two manuscripts currently in preparation. As I continue on as an independent faculty member, I would leverage my lifelong passion and ability to design new systems to better understand fundamental concepts at the intersection of chemical, mechanical, and materials engineering.

Selected Publications (13 total, 5 lead author, 98 citations via Google Scholar)

1. Jordan, A. M.; Kim, S.-E.; Van de Voorde, K.; Pokorski, J. K.; Korley, L. T. J., In Situ Fabrication of Fiber Reinforced Three-Dimensional Hydrogel Tissue Engineering Scaffolds, *ACS Biomater. Sci. Eng.*, **2017**, ***DOI:*** *10.1021/acsbiomaterials.7b00229.*

2. Jordan, A. M.; Viswanath, V.; Kim, S-E.; Pokorski, J. K.; Korley, L. T. J., Processing and surface modification of polymer nanofibers for biological scaffolds: a review, *J. Mater. Chem. B*, **2016**, *4 (36)*, 5958-5974, ***DOI:*** *10.1039/C6TB01303A.*

3. Jordan, A. M.; Marotta, T.; Korley, L. T. J., Reducing Environmental Impact: Solvent and PEO Reclamation During Production of Melt-extruded PCL Nanofibers, *ACS Sustainable Chem. & Eng.*, **2015**, *3* (*11*), 2994-3003, ***DOI:*** *10.1021/acssuschemeng.5b01019.*

4. Jordan, A. M.; Korley, L. T. J., Toward a Tunable Fibrous Scaffold: Structural Development During Uniaxial Drawing of Coextruded Poly(Îµ-caprolactone) Fibers, *Macromolecules,* **2015**, *48* (*8*), 2614-2627, ***DOI:*** *10.1021/acs.macromol.5b00370.*

Teaching Interests: I believe that teaching in a university setting can be divided into three separate but related roles: individual mentoring, classroom instruction, and community interaction. The role of faculty members as individual mentors cannot be understated. As a graduate student and post-doctoral fellow, I have personally mentored 11 undergraduate students over multiple academic terms. Each of these students has contributed to the progress outlined above and have gone on to successfully fill roles ranging from technical sales positions to graduate students at academic institutions. In addition to working with undergraduate students, I have also had the opportunity to work with high school students from the Cleveland Metropolitan School District (CMSD). Working closely (5 hours per week over a 2 year period) with two different high school students in a laboratory setting provided them with hands-on experience working in a scientific role and allowed me to adjust my mentoring style to reach students at different points in their lives and educational experience. Both of these students are currently enrolled in 4-year university programs studying forensic science and biology. These experiences highlight my ability to effectively lead a research team while simultaneously working on an individual basis to help students find their own path in science, technology, engineering, and medical careers.

In addition to their role as mentors, university faculty must be excellent classroom instructors. During my academic career I have served as an instructor for laboratory classes emphasizing polymer characterization techniques for both undergraduate and graduate students. I was also given the opportunity to lead the introduction to polymer mechanics class during the summer offered to CMSD high school students participating in our Polymer Envoys program. As a classically trained chemical engineer, I would feel comfortable teaching core courses ranging from introductory level to transport phenomena. Having gained substantial expertise in soft materials processing and mechanics throughout my graduate career, I would also feel comfortable teaching any advanced courses in polymer rheology, physics, processing, and mechanics. I would also welcome the opportunity to combine these concepts to develop a special topics course with a focus on innovative processing strategies for large-scale production of functional soft materials. While teaching courses, university faculty are often called to serve the needs of the department as a whole. As a graduate student, I was responsible for training users, obtaining and interpreting data for difficult experiments, and repairing departmental equipment including tensile testing equipment (4 years), small/wide-angle X-ray scattering (3 years), atomic force microscope (3 years), and equipment in the thermal analysis laboratory (2 years); currently as a post-doctoral fellow, I am responsible for supervising the polymer processing laboratory.

Finally, as community support of scientific funding is increasingly crucial, I feel that it is our role to interact with our local communities and stress the importance of not only scientific research, but also scientific literacy. This can be accomplished through local coffee shop chats, where members of the community can ask questions of university personnel in a relaxed setting to better understand our day-to-day job functions and added value. We can also achieve this through outreach events at local schools with 1-2 events per semester. I have also led these efforts both as a member of the Korley research group as we travelled to K-8 schools in the Cleveland area and also as the president of the graduate student organization where I worked with the CWRU community to organize and host groups for outreach days such as â€œIntroduce a Girl to Engineering Dayâ€. While not the event organizer as a post-doctoral fellow, I have continued this commitment to community involvement as a participant in numerous outreach events at the University of Minnesota. This multi-faceted approach to education and community engagement along with a proven track record of contributing to the academic community in both research, teaching, and support roles makes me an ideal candidate for a faculty position.

510215 Biosensor mediated evolution of biosynthetic pathways for biomanufacturing

Abstract:

Research Interests: Depleting petroleum reserves, energy security concerns and global climatic changes have driven our search for alternate technologies in order to shift to a more bio-based sustainable society. Biosynthetic pathways provide an inexpensive route from renewable materials for the production of molecules that are currently sourced from petroleum. the success of a biosynthetic pathway is highly context dependent - microbial host, pathway genes and their expression levels, promoters, enzyme activities, product and intermediate metabolite titers, cofactor availability, and often their delicate balance is offset owing to the complexities of biology. Advances in synthetic biology methods has enabled the construction and rapid generation of billions of cells with varying metabolic design. However, the evaluation of such designs by traditional metabolite measurement techniques currently available, lag by several orders of magnitude. Genetically encoded biosensors that act as screens for productive cells can solve these problems, identify optimal combination of genetic components, and enhance product titers and yields.

My research interest is in the development of biosensors and their application in the engineering of microbial metabolic pathways for human needs; including polymers, fuels and pharmaceuticals.

Research Experience: My research background has provided me with a wide range of experience in synthetic biology, metabolic engineering, protein engineering, recombineering, molecular biology, biochemistry, fermentation technology, downstream processing, and analytical methods, some are described below:

Postdoctoral Study:

1. "Genetic biosensors for small molecules to engineer biosynthetic pathways in *Pseudomonas putida*"
   Advisors: Dr. Ramesh Jha and Dr. Taraka Dale, Los Alamos National Laboratory

This project focusses on using transcription factor based whole cell biosensors with fluorescence readouts for enhancing product titers and yields, evolving their promoters for transition to alternate production hosts, enzyme evolution, strain performance and selection of best producers, adaptive evolution for better growth of engineered host strains.

2. Biosynthesis of Isopentenol in *Escherichia coli* via the Methyl Erythritol Phosphate Pathway

Advisor: Prof. Greg Stephanopoulos, Massachusetts Institute of Technology

The project was focused on improving isopentenol titers produced by the MEP pathway by soluble and functional expression of dephosphorylase and the pathway enzymes, maturation of the Fe-S cluster proteins, investigation of possible iron donors, NADP regeneration routes for cofactor supply, development of product and pathway metabolite analysis methods, culture media and isopentenol extraction procedures.

3. Pathways for synthesis of Advanced biofuels

Advisor: Prof. Greg Stephanopoulos, Massachusetts Institute of Technology

This project was focused on isovaleryl coA pathway for synthesis of C5 compounds, identification, cloning, characterization and variants of putative pivalyl coA mutases, investigation of prenol synthesis from isovaleryl coA pathway, development of analytical methods for pivalic acid and isovaleric acid determination.

PhD Dissertation: "Molecular and genetic strategies to enhance the functional expression of recombinant proteins in *Escherichia coli*"

Advisors: Prof. C. Perry Chou and Prof. Bill Anderson, University of Waterloo

The research work during graduate study was focused on the exploration of 'strategies for functional expression of recombinant proteins in *E. coli* using model proteins (lipase, penicillin acylase, yellow fluorescent protein, β-galactosidase, hCD83ext). They included posttranslational modifications, cytoplasmic and periplasmic expression, surface display, extracellular secretion, molecular chaperones, site directed and error-prone mutagenesis, stress regulatory pathways, host strains, and optimization of culture conditions.

Industrial collaboration project with Argos Therapeutics at University of Waterloo

"Development of 'bioprocess for production of recombinant human therapeutic protein hCD83ext"

Advisor: Prof. Murray Moo-Young and Prof. C. Perry Chou

The project included optimization of culture media and bioreactor conditions for high level expression of hCD83ext, Development and optimization of purification protocol for hCD83ext: GST affinity column, thrombin cleavage, ion-exchange chromatography, ultrafiltration, analytical methods for GST assay and endotoxin assay, construction of hCD83 mutants for improved protein stability

Funding/Awards:

Postdoctoral Fellowship by the Natural Sciences and Engineering Research Council of Canada

Murray Moo-Young Biotechnology Scholarship

NSERC Graduate Studentship and International Doctoral Student

Teaching Interests: My background in Chemical Engineering, coupled with my research training in Biological Engineering has equipped me to teach a variety of courses in Chemical Engineering, Biotechnology, Bioengineering, Biochemistry, Protein Engineering, Metabolic Engineering and Synthetic Biology. However, I would be particularly interested in teaching 1. Biochemical Engineering and 2. Synthetic Biology. The coursework would be designed to encompass theoretical and computational approaches, practical laboratory classes, experimental model systems, and current research & classic literature.

Teaching Experience: I have had the opportunity to be a teaching assistant for various courses at the University of Waterloo and at Indian Institute of Technology Delhi. Some of them include Bioprocess Engineering, Advanced Biochemical Engineering, Engineering Biology, Introduction to Biotechnology, Chemical Engineering Concepts, and Physical Chemistry. This has given me teaching experience for 5-6 years at undergraduate level and around 8 months at the graduate level. Teaching helped me to hone my technical knowledge in the curriculum subjects and develop my interpersonal and communication skills and I was able to enjoy a good rapport with my students and colleagues.

Supervisory experience: I have had the chance to mentor and work with 6 graduate students, 4 visiting scholars, 7 undergraduate co-op students and 3 technical staff during the course of my research career of graduate studies at UWaterloo, the Industrial collaboration project for therapeutic protein production at

UWaterloo, and postdoctoral study at MIT. Most students who started without any background in biotechnology and bioengineering, developed the essential skills within a short period of time through research training. This is also demonstrated by their co-authorship on our publications.

Teaching Certificate Program: I have completed a teaching certificate program at MIT which was based on seven workshops aimed at development of teaching skills, exposure to research on teaching and learning, and structure of future learning. The workshop was followed by two microteaching sessions that are videotaped, evaluated and feedback provided.

Manuscripts in preparation

1. Jha, R. K., **Narayanan, N.**, and Dale, T. Whole cell biosensor mediated enzyme evolution for biomanufacturing in *Pseudomonas putida.*
2. **Narayanan, N**., Edgar, S., Woolston, B., and Stephanopoulos, G. Maturation of Fe/S clusters of Methyl erythritol phosphate pathway enzymes in *Escherichia coli.*
3. **Narayanan, N**., Woolston, B., and Stephanopoulos, G. NADPH regeneration in *Escherichia coli* improves the biosynthesis of isopentenol via Methyl erythritol phosphate pathway.
4. **Narayanan, N**., Duggar, D., Pereira, B., and Stephanopoulos, G. Identifying and Engineering of putative pivalyl CoA mutase for the biosynthesis of C5 branched chain compounds.

Publications in refereed journals

1. **Narayanan, N.**, Khan, M., and Chou, C. P. **(**2011**)** Enhancing functional expression of heterologous *Burkholderia* lipase in *Escherichia coli*. Molecular Biotechnology. 47: 130-143
2. Srivastava, P., Singh, P., **Narayanan, N.** and Deb, J.K. **(**2011**)** Physiological and biochemical consequences of host–plasmid interaction – A case study with *Corynebacterium renale*, a multiple cryptic plasmid containing strain. Plasmid. 65: 110-117
3. Zhang, L., **Narayanan, N.**, Brand, S. R., Nicolette, C. A., Baroja, M., Arp, J., Wang, H., Moo- Young, M., and Chou, C. P. **(**2010**)** Structural identification of recombinant human CD83 mutant variant as a potent therapeutic protein. Protein Expression and Purification. 73: 140-146
4. **Narayanan, N.**, Khan, M. and Chou, C. P. **(**2010**)** Enhancing functional expression of heterologous lipase B in *Escherichia coli* by extracellular secretion. Journal of Industrial Microbiology and Biotechnology. 37: 349-361
5. **Narayanan, N.**, and Chou, C. P. (2009) Alleviation of proteolytic sensitivity to enhance recombinant lipase production in *Escherichia coli*. Applied and Environmental Microbiology.75: 5424-5427
6. **Narayanan, N.**, and Chou, C. P. (2008) Physiological improvement to enhance *Escherichia coli* cell-surface display via reducing extracytoplasmic stress. Biotechnology Progress 24: 293-301
7. **Narayanan, N.**, and Chou, C. P. (2008**)** Periplasmic chaperone FkpA reduces extracytoplasmic stress response and improves cell-surface display on *Escherichia coli*. Enzyme and Microbial Technology. 42:506-513
8. **Narayanan, N.**, Follonier, S., and Chou, C. P. **(**2008**)** In vivo monitoring and alleviation of extracytoplasmic stress to recombinant protein overproduction in the periplasm of *Escherichia coli*. Biochemical Engineering Journal. 42: 13-19
9. **Narayanan, N.**, Xu, Y., and Chou, C. P. **(**2006**)** High-level gene expression for recombinant penicillin acylase production using the *ara*B promoter system in *Escherichia coli*. Biotechnology Progress. 22: 1518-1523
10. **Narayanan, N.**, Hsieh, M. Y., Xu, Y., and Chou, C. P. (2006**)** Arabinose-induction of *lac*-derived promoter systems for penicillin acylase production in *Escherichia coli*. Biotechnology Progress. 22: 617-625
11. Xu, Y., Hsieha, M. Y., **Narayanan, N.**, Anderson, W. A., Scharer, J. M., Moo-Young, M., and Chou, C. P. **(**2006**)** Cytoplasmic overexpression, folding, and processing of penicillin acylase precursor in *Escherichia coli*. Biotechnology Progress. 21: 1357-1365
12. Xu, Y., Weng, C. L., Hsieh, M. Y., **Narayanan, N.**, Anderson, W. A., Scharer, J. M., Moo-Young, M., and Chou, C. P. (2005) Chaperone-mediated folding and maturation of the penicillin acylase precursor in the cytoplasm of *Escherichia coli*. Applied and Environmental Microbiology. 71: 6247- 6253
13. **Narayanan, N.**, Roychoudhury, P. K., and Srivastava, A. **(**2004**)** Isolation of *adh* mutant of *Lactobacillus rhamnosus*.sp for production of L(+) lactic acid. Electronic Journal of Biotechnology. 7:

72-84
14. **Narayanan, N.**, Roychoudhury, P. K. and Srivastava, A. **(**2004**)** L (+) lactic acid fermentation and its product polymerization. Electronic Journal of Biotechnology.7: 167-179

505227 Programmable Assembly and Deformation of Polymers and Networks

Abstract:

Research Interests:

The main theme of my research was to understand the role that surface tension plays on sub-millimeter scale objects, especially on polymeric material systems, and how to utilize this phenomenon to assemble and deform such objects. My research addressed several phenomena of nano-and micro-sized polymeric soft objects at fluid interfaces. On the nano-scale, amphiphilic block copolymers were used to explore interfacial behaviors due to their enhanced stability, mechanical properties, and tunability compared to other interfacially active materials such as small molecule surfactants or lipids. I investigated the tailoring of amphiphilic block copolymer assemblies through deformation at the oil/water interface by inducing interfacial instabilities to incorporate inorganic nanoparticles into micelles, and by controlling osmotic stresses to prepare multi-compartment emulsions and capsules. Next, thin temperature responsive polymer networks (i.e., hydrogel sheets) were used to probe elastic properties, buckling instabilities, and capillary attraction at the micro-scale. I exploited the competition between surface energy and elastic bending energy to quantitatively determine key elastic properties of thin films that are otherwise challenging to measure. Additionally, I found significant edge imperfections due to the finite resolution of photolithography by observing interfacial deformations. The edge imperfections were then employed to drive the buckling of narrow, photo-crosslinked hydrogel ribbons. Further, I introduced a new concept of capillary assembly of soft hydrogel particles with programmed three-dimensional geometry. This offered opportunities to study correlations between elastic properties and surface tension, as well as to enabled unprecedented levels of control over inter-particle interactions, and therefore, higher-order assemblies.

The overall goal of my independent research program is to achieve programmable assembly and deformation of polymeric materials systems for applications ranging from biomedical devices to soft actuators by probing unexplored fundamental questions. The early stage of my independent research will involve 1) development of a new 4D microfabrication technique for programming internal heterogeneity in nanocomposite polymer networks, 2) exploration of a photo-switchable capillary assembly of shape programmed soft microparticles; and 3) analysis of bioenergetics at the single cell level by integrating a calorimetric sensor into the microfabrication-chip.

Teaching Interests:

I am excited to teach a wide range of existing courses in Chemical Engineering at the undergraduate and graduate levels. Moreover, I would be pleased to draw from my multi-disciplinary academic background to develop new courses related to polymer physics, polymeric material characterization, capillarity and wetting, and nano/micro-fabrication.

505565 Structureâ€"Property Relationships in Polymer-Based Transistors

Abstract:

Research Interests:

Polymer electronic materials have been introduced as potential solutions in a wide range of electronic products including light-weight, flexible organic field-effect transistors (OFETs), organic photovoltaics (OPVs), and organic light-emitting diodes (OLEDs). The primary aim of my research is to address the fundamental structure-property relationships for improved performance and stability of flexible organic devices based on multi-component electronically-active materials (e.g., semiconducting polymer-polymer blends) when they are cast into thin films. For instance, OPV devices based on electron donor and electron acceptor (D-A) bulk heterojunctions show a critical relationship between the phase-separated structures and device efficiency. In addition, organic transistors based on blended films are one of the emerging device classes that exhibit excellent performance and stability relative to common transistors based on a single semiconducting film. The OFETs have attracted a significant amount of academic and industrial research attention over the past three decades due to their promise of delivering mechanically-robust (e.g., flexible and stretchable materials), low-cost, and easily processed electronic devices. Moreover, the study of multi-component polymer films has been another important arena of research as properly combining materials allows for the creation of attractive features of the constituent components that goes beyond the sum of the individual parts.

In the well-known realm of bulk heterojunction OPVs, the two components of blends are strongly involved in the charge exchange and transport in the active layers. In the less-studied realm of organic transistor composite active layers, the blends of a polymer and small molecule provide a synergistic effect that combines the high electrical performance from the small molecule and the easy film processing from the polymer, leading to high-performance organic transistors (i.e., high charge carrier mobility and high ON/OFF current ratio). Furthermore, the polymer blends based on an organic semiconductor and an insulator allow for superior performance and environmental stability of organic transistors, due to the enhanced doping effect by an insulating matrix. Critically, the performance and stability of these transistors based on blended films significantly depends on the active materials and their interfacial morphologies. However, elucidating the underlying mechanism and impact on the device performance of the interpenetrating structures is still an open opportunity in the literature. Also, the scope of application is currently limited. In order to meet these needs, we propose three specific research efforts. We will: (1) elucidate the fundamental mechanism of phase-separated structures by a range of analysis techniques and systematically investigating the structures by the control of the active structure through lithographic techniques; (2) fabricate optimized organic transistors based upon multi-component polymer blends in order to clarify the impact of the nanostructure on device performance; and (3) synthesize new linear, branched, star, and block polymers consisting of at least two components in order to develop the critical structure-property-performance relationships that are required for the success of multi-component organic transistors. I believe there are great challenges and opportunities in this multi-component system for the development of flexible organic transistors.

Research Experience:

My research at both Seoul National University (Mentor: Professor Kookheon Char) and Purdue University (Mentor: Professor Bryan W. Boudouris) has focused on the design of soft materials and their integration into next-generation flexible electronic devices. In brief, I have focused on: (1) determining the impact of nanostructure in phase-separated organic memory diodes in order to advance flexible logic circuits; (2) utilizing interfacial engineering with radical polymers in organic field-effect transistors for enhanced device performance; (3) and elucidating the charge transport mechanism of radical polymers in order to generate ambipolar, transparent thin film transistors. As a graduate student, I have a significant amount of experience with respect to the design, fabrication, and optimization of organic electronic devices. Moreover, I have been able to utilize these devices to probe the fundamental physics of charge transport in a unique and emerging conducting polymer system (i.e., open-shell macromolecules).

In addition, I have intentionally switched from more synthetic and device-focused work during my graduate career to a postdoctoral research position that is centered on polymer physics during my upcoming time at the University of Delaware (Mentor: Professor Thomas H. Epps III). This will allow my future research team to understand and improve the self-assembly and implementation of homopolymer and block polymer materials for energy storage devices. This work mainly involves the design, synthesis, and nanoscale characterization of macromolecules for energy applications (i.e., organic batteries). With well-established charge transport and synthetic optimization of energy materials, my future research will pursue materials development for advanced energy applications. Thus, my work to date has focused on linking nanoscale scientific phenomena to device engineering at the macroscale, and I anticipate utilizing this previous experience to launch my own research group in order to impact the realms of flexible and stretchable electronics, energy conversion devices, and energy storage devices in the near future.

Teaching Interests:

I was born and raised in South Korea, and I experienced military service for 2 years while I was an adult in South Korea. During this military service period, I learned the initiative and drive arising from positive attitudes play a crucial role in military operations. In a similar manner, a positive and supporting attitude from an instructor can allow students to overcome many difficulties in classroom as well as in the laboratory. Moreover, I have considerable formal teaching experience, and I have previously served as a teaching assistant for undergraduate courses on Engineering Mathematics, Transport Phenomena, and Design of Staged Separation Processes. In addition, this previous experience has allowed me to instruct many of the core, fundamental courses in chemical engineering at my previous institutions of Seoul National University and Purdue University. Thus, I was able to gain a deep insight into how to instruct undergraduate students in the fundamentals of chemical engineering from some of the most respected instructors in the field (e.g., Professor Phillip Wankat). Specifically, I delivered lectures for students, had regular office hours, and developed course materials. Thus, I can contribute to lead any course in the core chemical engineering curricula and more specific electives, such as polymer chemistry and organic electronics. In addition, I participated in the mentorship of undergraduate students as a part of their undergraduate thesis at Seoul National and Purdue. Based on my previous experiences, I can lead the next generation to the high level of learning and training.

503355 Three-Dimensional Responsive Soft Micro/Nano-Structures for Biomedical and Electronic Applications

Abstract:

Three-dimensional responsive soft micro/nano-structures for biomedical and electronic applications
Weinan Xu, David H. Gracias
Department of Chemical and Biomolecular Engineering, Johns Hopkins University, Baltimore, Maryland 21218, USA

Responsive soft micro/nanostructures have attracted much attention in recent years due to their abilities to adapt and respond to external stimuli, and promising applications in biosensing, drug delivery, self-healing materials and flexible electronics. To further improve the performance of the 3D responsive structures, one promising way is to use novel and smart materials as the building blocks. In this work, novel polymers with branched architecture and multi-responsive properties were used to fabricate hierarchical microcapsules, which have the ability to simultaneously encapsulate multiple types of cargo molecules, and release them in a programmable manner triggered by external stimuli. On the other hand, for the fabrication of 3D bioelectronics, 2D nanomaterials including graphene and transition metal dichalcogenides, were utilized as the major components. The 2D materials were folded into well-defined 3D geometries by combining surface functionalization and top-down fabrication. Such 3D graphene microstructures have the ability to encapsulate biological samples such as live cells, which enables highly sensitive 3D analysis, mapping and sensing. To sum up, responsive 3D soft micro/nano-structures were built with novel and smart components, which leads to hierarchical internal structures and superior performance compared with traditional ones, such ultra-thin, flexible, and biocompatible 3D structures provide a new platform for bioelectronics, biosensing and drug delivery.

Related publications:

[18] Xu, W.; Qin, Z.; Chen, C.; Kwag, H. R.; Ma, Q.; Sarkar, A.; Buehler, M. J.; Gracias, D. H. Ultra-thin thermo-responsive self-folding 3D graphene. Science Advances 2017, In revision.
[17] Erwin, A. J.; Xu, W.; He, H.; Matyjaszewski, K.; Tsukruk, V. V. Linear and Star Poly(ionic liquid) Assemblies: Surface Monolayers and Multilayers. Langmuir 2017, 33, 3187-3199.
[16] Korolovych, F. V.; Ledin, P. A.; Stryutsky, A.; Shevchenko, V. V.; Sobko, O.; Xu, W.; Bulavin, L. A.; Tsukruk, V. V. Assembly of Amphiphilic Hyperbranched Polymeric Ionic Liquids in Aqueous Media at Different pH and Ionic Strength. Macromolecules 2016, 49, 8697–8710.
[15] Xu, W.; Ledin, P. A.; Iatridi, Z.; Tsitsilianis, C.; Tsukruk, V. V. Multicompartmental Microcapsules with Orthogonal Programmable Two-way Sequencing of Hydrophobic and Hydrophilic Cargo Release. Angew. Chem. 2016, 55, 4908-4913.
[14] Xu, W.; Steinschulte, A. A.; Plamper, F. A.; Korolovych, V. F.; Tsukruk, V. V. Hierarchical Assembly of Star Polymer Polymersomes into Responsive Multicompartmental Microcapsules. Chem. Mater. 2016, 28, 975–985.

[13] Xiong, R.; Hu, K.; Grant, A. M.; Ma, R.; Xu, W.; Lu, C.; Zhang, X.; Tsukruk, V. V. Ultrarobust Transparent Cellulose Nanocrystal-Graphene Membranes with High Electrical Conductivity. Adv. Mater. 2016, 28, 1501-1509.
[12] Ledin, P. A.; Xu, W.; Friscourt, F.; Boons, G.; Tsukruk, V. V. Branched Polyhedral Oligomeric Silsesquioxane Nanoparticles Prepared via Strain-Promoted 1,3-Dipolar Cycloadditions. Langmuir. 2015, 31, 8146-8155.
[11] Xu, W.; Ledin, P. A.; Shevchenko, V. V.; Tsukruk, V. V. Architecture, Assembly, and Emerging Applications of Branched Functional Polyelectrolytes and Poly(ionic liquids). ACS Appl. Mater. Interfaces 2015, 7, 12570-12596.
[10] Xu, W.; Ledin, P. A.; Iatridi, Z.; Tsitsilianis, C.; Tsukruk, V. V. Multi-Responsive Star-Graft Quarterpolymer Monolayers. Macromolecules 2015, 48, 3344-3353.
[9] Steinschulte, A. A.; Xu, W.; Draber, F.; Hebbeker, P.; Jung, A.; Bogdanovski, D.; Schneider, S.; Tsukruk, V. V.; Plamper, F. A. Interface-Enforced Complexation between Copolymer Blocks. Soft Matter 2015, 11, 3559-3565.

Microcapsules Based on Multilayer Assembly of Star Polyelectrolytes. Macromolecules 2014, 47, 7858-7868.
[7] Ledin, P. A.; Tkachenko, I. M.; Xu, W.; Choi, I.; Shevchenko, V. V.; Tsukruk, V. V. Star-Shaped Molecules with Polyhedral Oligomeric Silsesquioxane Core and Azobenzene Dye Arms. Langmuir 2014, 30, 8856-8865.

Induced Limited Aggregation of Responsive Star Polyelectrolytes. Macromolecules 2014, 47, 2112-2121.
[5] Ren, H.; Kulkarni, D. D.; Kodiyath, R.; Xu, W.; Choi, I.; Tsukruk, V. V. Competitive Adsorption of Dopamine and Rhodamine 6G on the Surface of Graphene Oxide. ACS Appl. Mater. Interfaces 2014, 6, 2459–2470.
[4] Choi, I.; Kulkarni, D. D.; Xu, W.; Tsitsilianis, C.; Tsukruk, V. V. Star Polymer Unimicelles on Graphene Oxide Flakes. Langmuir 2013, 29, 9761-9769.

Tuning of Layer -by-Layer Microcapsule Permeability. ACS Nano 2013, 7, 598-613.
[2] Choi, I.; Malak, S. T.; Xu, W.; Heller, W. T.; Tsitsilianis, C.; Tsukruk, V. V. Multicompartmental Microcapsules from Star Copolymer Micelles. Macromolecules 2013, 46, 1425-1436.
[1] Xu, W.; Qin, Z.; Yu, H.; Liu, Y.; Liu, N.; Zhou, Z.; Chen, L. Cellulose nanocrystals as organic nanofillers for transparent polycarbonate films. J. Nanopart. Res. 2013, 15, 1-8.

Research Interests:

Stimuli-responsive polymers, 2D nanomaterials, 3D micro- and nano-fabrication, Additive manufacturing, Drug delivery

Postdoctoral Fellow, Department of Chemical & Biomolecular Engineering, Johns Hopkins University. Jan. 2016-Present

Ph.D. in Materials Science and Engineering, Aug. 2011 – Dec. 2015, Georgia Institute of Technology, Atlanta, GA, USA

B.S. in Polymer Science and Engineering, Sept. 2007 – Jul. 2011, Donghua University, Shanghai, China

My PhD research is focused on stimuli-responsive polymers, and their assembly into well-defined micro- and nano-structures, for smart coating, self-healing materials, and drug delivery applications. My postdoc work is focused on the fabrication of 3D micro- and nano-structures from atomically thin 2D materials, including graphene, BN, and TMDs, with applications in miniaturized 3D electronic devices, biosensing and bioelectronics.

For my future work as an independent researcher, I will focus on combining the structural flexibility and easy processability of polymeric materials, with the superior electrical, optical and catalytic properties of 2D nanomaterials. And it is distinct from traditional polymer nanocomposite research which simply mixing polymers with nanomaterials. In the first place, functional and responsive polymers will be attached to 2D nanomaterials mostly through non-covalent interaction, including electrostatic attraction, hydrophobic interaction, π - π interaction and host-guest interaction, which enables the precise control of the structures in a molecular and nanosale level. After the functionalization with responsive polymers, the 2D nanomaterials will be used to fabricate smart and intelligent devices, such as transistors or sensors. The changes in external conditions induce the structural changes of the functionalized nanomaterials, which will lead to the actuation and detection of the devices. On the other hand, engineering the interface between responsive polymers and 2D nanomaterials in a layer-by-layer way will lead to high-performance membranes, which have promising applications in ion and gas separation, water purification and ion conducting membranes for batteries.

Teaching Interests:

Introduction to Materials Science

Introduction to Polymer Materials

Polymer Physics

Polymer Chemistry

Polymer Characterization

Micro- and Nano-fabrication

I believe teaching is as important, if not more important than research in any university including research universities, because the ultimate goal of college and graduate education is to transform the students into all-round developed talents with expertise in certain fields, so that they can contribute to the advancement of our society in various ways. To this purpose, teaching is the most direct and effective approach. The materials science and engineering major is one which closely related to our real life, and can have significant influence in our way of thinking and view of the world, that's the main reason that I am very enthusiastic about teaching class in the materials science and engineering field.

My main objective of teaching is to provide the students with a more scientific and deeper view of the objects around us and the world, during the same time, the students develop strong ability of critical thinking and self-learning. Moreover, the students can also use the knowledge from materials science to make inventions for the future, for example, the concept of smart clothes, which have large number of miniaturized flexible electronic devices embedded in them, and can communicate with our body, is an excellent showcase of advanced materials science. I will adopt active teaching and teaching as scholarship methods in my class, such as small-group discussions, group presentations and frequent assessments, to make my teaching more effective and inclusive.

505692 Functional Materials Interfacing Chemistry and Biology

Abstract:

Research Interests:

My research interests focus on fabricating various functional materials and exploring their applications in different fields. Primarily, I am interested in rationally designing and synthesizing molecules with different functionalities. Using chemical methods and microfluidic technology, I engineer these as-synthesized molecules to construct nanoscale, microscale and bulk materials, including carbon nanomaterials, silicon nanowires, stimuli-responsive microcapsules, porous materials and photonic crystals. Furthermore, I am also interested in applying these as-prepared materials to address both fundamental and practical issues in chemistry, biology and energy fields.

Research Experience:

My research has involved many fields of science and engineering. My formal research training started in Chemistry Departments at both Tsinghua University and Purdue University, in which I have studied organic synthesis, self-assembly, nanomaterials synthesis and biomedical applications of nanomaterials. After completing my PhD, I joined the School of Engineering at Harvard University working on microfluidics for producing functional materials, such as stimuli-responsive microcapsules for controlled release and bio-sensing applications.

Synthesis of Nanomaterials: With my chemistry background for molecular design and organic synthesis, I fabricated nanomaterials through rationally designing and synthesizing amphiphilic molecules with different functional groups, followed by investigating their self-assembly behavior, and finally using these self-assembled structures as templates and/or precursors to controllably prepare highly ordered mesoporous materials, ultrathin carbon nanofiber and pure single layer graphene. Besides nanomaterials fabrication, I also synthesized ionic liquids and incorporated them into 3D photonic crystal structure to develop chemical sensors with capability for naked-eye detection.

Bio-applications of Nanomaterials: My PhD research mainly focused on biomedical applications of nanomaterials. I used chemical vapor deposition methods to controllably prepare silicon nanowires (SiNWs) and graphene. By visualizing SiNWs with their intrinsic non-linear optical (NLO) signals, I investigated the impact of size and surface functionalities of SiNWs on their cellular binding and internalization. In addition, I demonstrated that graphene protects metals from corrosion in biological environment and reduces immune responses toward metallic devices. Moreover, I worked with a collaborator to develop a new NLO imaging method based on highly sensitive transient absorption to characterize graphene and graphene oxide in bulk and in biological systems.

Microfluidics for Functional Materials Production: As a postdoc in Prof. David Weitz's group at Harvard University, I am engaged to produce novel functional materials using microfluidic processes. By using glass capillary microfluidic devices, I efficiently encapsulated a variety of enzymes and antibodies within microcapsules with different shell structures. These encapsulated bio-actives can be well maintained for long-term storage without losing their biological activities and can be controllably released upon applying osmotic pressure changes. This encapsulation technique shows great potential in protection and delivery of enzymes and antibodies. Additionally, I prepared biosensors by encapsulating quantum dots within microcapsules for glucose detection. This study opened the possibilities of developing implantable biosensors with high sensitivity. Besides microfluidics, I also developed a versatile strategy to fabricate 3D porous conductive collectors using various inorganic, organic and metallic substrates, as lithium battery anodes. I demonstrate the performance, such as battery life and efficiency, of lithium batteries with these anodes is as good as the best reported results in the literature.

Future Direction:

As a research faculty member, I would like to continue the philosophy of designing and fabricating novel functional materials based on chemistry and microfluidics to explore and solve challenges in chemistry and biology. I will contribute my efforts to prepare functional materials with hierarchical structures and novel properties to meet requirements for studying both fundamental and practical problems.

One interesting direction is to further explore biomedical applications of nanomaterials, particularly, graphene. Due to its biocompatibility, excellent mechanic and electric properties, graphene is an ideal candidate to study cellular biomechanics and differentiation behavior of neuron stem cells under an external electrical field.

I am also anticipated to fabricate "smart" microcapsules for specific release. In comparison to the release triggered by osmotic pressure, the release of encapsulated actives triggered by specific chemical or biological species can be precisely controlled. The specific release can be achieve by introducing functional groups into shell materials, which can degrade by adding specific additives. This precisely controlled specific release will have broad applications in both industry and biomedical fields.

Additionally, I am interested in fabricating novel chemical and biological sensors. Based on photonic crystals, I want to prepare sensitive sensors for detecting chemicals without using sophisticate instruments, which will have wide applications in chemistry and environmental detection. I also like to develop implantable biosensors using microfluidic approach.

In addition to the above directions, due to the interdisciplinary nature of my research work, I look forward to establishing a strong and active collaboration with fellow faculties in science and engineering to explore new opportunities at the interface of chemistry, biology and materials science.

Teaching Interests:

Teaching is a very important part in academia, since teaching is not only transferring knowledge, but also a process of learning and inspiring. I am always eager to involve myself in teaching. My teaching experience came from teaching in classes as a teaching-assistant, and mentoring undergraduate and graduate students in the lab. During my PhD study at Purdue University, I was repeatedly chose by my department to teach several courses including both lectures and laboratory for students from Chemistry and Chemical Engineering Departments. My duties included lecturing undergrads, experiment demo, holding office hours, preparing course materials and exams. In addition, I mentored undergraduate students as a part of their undergraduate thesis, and one of these undergraduates received the undergraduate research award and graduated in Purdue Chemistry Honors Program with the Highest Distinction. I have also been mentoring about 10 undergraduate and graduate students to conduct research programs in my labs at both Purdue University and Harvard University.

With my education and research background, I am well prepared to teach a broad range of courses in chemistry, chemical engineering and materials science at both undergraduate and graduate. I also envision instructing and encouraging students and postdocs to deliberate about questions and to conduct independent research studies in my lab.

Selected Peer-Reviewed Publications:

-**24** peer-reviewed journal articles (**12** first-authored or corresponding-authored papers), **2** patents, **1** book chapter († co-first author, * Corresponding author).

W. Zhang†, L. Qu†, H. Pei, Z. Wu, J. Didier, D. Ingber, D. Weitz, Asymmetric Swelling and Rupture of Inhomogeneous Microcapsules for Biologics Release. (In Preparation).

H. Zhao, R. Ding, X. Zhao, Y. Li, L. Qu, H. Pei, L. Yildirimer, Z. Wu, **W. Zhang***, Graphene-based Nanomaterials for Drug/Gene Delivery, Bioimaging and Tissue Engineering. *Drug Discov. Today*, Accepted, (Invited Review).

W. Zhang†, A. Abbaspourrad†, D. Chen, E. Campbell, H. Zhao, Y. Li, Q. Li, D. Weitz, Osmotic Pressure Triggered Rapid Release of Encapsulated Enzymes with Enhanced Activity. *Adv. Funct. Mater.*, In print, (Featured as inside front cover).

Y. Li†, **W. Zhang†**, et al., Rapid Assembly of Large Scale Transparent Circuit Arrays Using PDMS Nanofilm Shaped Coffee Ring. *Adv. Funct. Mater.*, 2017, 27, 1606045.

X. Xie†, **W. Zhang†**, A. Abbaspourrad†, et al., Microfluidic Fabrication of Colloidal Nanomaterials-Encapsulated Microcapsules for Biomolecular sensing. *Nano Lett.*, 2017, 17, 2015-2020.

J. Li†, **W. Zhang†**, et al., Highly sensitive transient absorption imaging of graphene and graphene oxide in living cells and circulating blood. *Sci. Rep.*, 2015, 5, 12394.

W. Zhang, S. Lee, et al., Use of graphene as protection film in Biological Environments. *Sci. Rep.*, 2014, 4, 4097.

W. Zhang, L. Tong, C. Yang, Cellular Binding and Internalization of Functionalized Silicon Nanowires. *Nano Lett.*, 2012, 12, 1002-1006.

W. Zhang, J. Cui, et al., A strategy of Producing Pure-Layer Graphene Sheets Based on a Confined Self-Assembly Approach. *Angew. Chem.*, 2009, 121, 5978-5982.

- Selected as **Hot Paper** by *Angewandte Chemie*
- Reported by *C&EN*, Latest News, “Graphene Via Self-Assembly”. July 8 (2009)
- Reported by *Nature China*, Research Highlight, “Material chemistry: Casting graphene”, DOI:10.1038/nchina.2009.147, July 22 (2009)
- Reported by *NPG Asia Materials*, Research Highlight, “Graphene: Confined approach”, DOI:10.1038/asiamat.2009.22, October 27 (2009)

W. Zhang, J. Cui, et al., Pyrrole containing ionic liquid as tecton for construction of ordered mesoporous silica with aligned polypyrrole nanowires in channels. *J. Mater. Chem.*, 2009, 19, 3962-3970.

W. Zhang, J. Cui, et al., Confined Self-Assembly Approach to Produce Ultrathin Carbon Nanofibers. *Langmuir*, 2009, 25, 8235-8239.

W. Zhang, Y. Li, et al., Electrochemical Polymerization of Imidazolum-Ionic Liquids Bearing a Pyrrole Moiety. *J. Polym. Sci. A: Polym. Chem.*, 2008, 46, 4151-4161.

505009 Plasmonic Perovskites Nanolasers in Accelerating Emission Dynamics

Abstract:

Research Interests: Self-assembled Nanophotonics

Teaching Interests: With my unique background (Chemical Engineering, B.S, Chemistry, M.S and Applied Physics, Ph.D), My teaching interests will be broad and cover mutiple dimension of science and engineering.

Abstract body:

Halide perovskites are an emerging class of semiconductors with their long exciton lifetimes that drives perovskites' outstanding photovoltaic performance. However, such slow exciton recombination brings a fundamental limitation for nanophotonic and optoelectronic devices, eg. lasers, in applications such as ultrafast optical information processing and communications. Here we demonstrate an ultrafast perovskite plasmonic nanolaser with high emission enhancement at room temperature by a scalable solution process. The accelerated emission dynamics has been enabled by the strong plasmonic confinement of laser modes via perovskite emission hybridized with surface plasmons. It not only downscales the device but also facilitates its emission efficiency dramatically exhibiting a clear single mode lasing behavior at room temperature. Such accelerated emission dynamics also significantly facilitates laser gain compensation through Einstein equation. The capability to combine plasmonic confinement with ultrafast amplification through perovskites not only promises fundamental physics for accelerating light–matter interactions but also opens new avenues for applications from ultrafast integrated optical communications to implantable computing microchips.

505612 First-Principles Study for Detailed Understanding of Nanoporous Materials

Abstract:

Research Interests:

1. Detailed understanding of adsorption phenomena within nanoporous materials

Nanoporous materials such as metal-organic frameworks have drawn interest for potential applications in areas such as gas separations, catalysis, and chemical sensing for their high crystallinity, large adsorption surface areas, and chemical customizability. These materials can possess highly specialized chemistry, and the variety of materials and applications creates a need for deep understanding of the interactions of adsorbates with these materials in order to identify materials suited to these applications. This has been an area of substantial interest, but the wide variety of potential applications and the constant discovery of new materials leaves much work remaining to be done in this area. I propose to use electronic structure methods including density functional theory to characterize the interactions of adsorbates with new materials to both establish fundamental understanding and identify materials for applications of interest. Within this area, there is great potential for work with experimental collaborators in identifying, characterizing, and understanding emergent materials of interest.

2. Unique adsorption phenomena at sites of reduced symmetry in complex porous materials

Much of the focus on nanoporous materials to date has focused on adsorption to the internal surfaces of these materials, yet the small particle size of many materials results in the external surfaces contributing a non-negligible percentage of the total adsorption surface area. Additionally, defects within these crystals will exist, interrupting the perfect crystallinity on which most theory efforts have focused. Both of these traits break the symmetry of the bulk crystal, creating unique sites for adsorption interactions not present in the perfect crystalline bulk. In addition to understanding adsorption in the ideal bulk, it is important to understand unique adsorption phenomena at or near these features with reduced symmetry. I propose to study the character and interactions present at these reduced symmetry sites for understanding both how they affect the overall stability of these materials as well as the potential to exploit special electronic properties at these sites for potential applications.

3. Increasing stability of metal-organic frameworks and maintaining chemistry under application conditions

A key challenge that exists for metal-organic frameworks in many applications is the potential for these materials to degrade. Often applications involve exposure to elevated temperatures, moisture, and trace impurities, all of which can cause degradation in metal-organic frameworks. Surfaces and defects have been identified as sites key to the stability or degradation of these materials. I propose detailed study of protection and modification of vulnerable surface and defect sites in order to improve their process stability, maintain relevant chemical properties for applications of interest, and improve their potential to interface with other materials.

Teaching Interests:

Courses: Statistical Mechanics/Thermodynamics, Transport Phenomena, Classical Thermodynamics, Reaction Engineering/Kinetics, Mass Balances, Molecular Simulations, Solid State Physics.

General: I have particular interest in the incorporation of active learning and promotion of metacognitive processes in students. I plan to work on integration of these concepts within traditional chemical engineering pedagogical approaches as well as in the restructuring of education to strongly promote deep, critical thinking and learning.

505228 Colloidal Assemblies for Mesoscale Materials

Abstract:

Research Interests:

Many physical effects are enhanced or controlled by structuration on the mesoscale between 100 nm and 10 μm, such as light scattering, capillary forces, and mass transport. These phenomena are critical for a variety of fields including optics, wetting, and electrochemistry, with applications in energy, defense, and healthcare. I plan to develop materials that can help us understand and control the phenomena that occur on these length scales, as well as develop synthetic processes that will allow the production of these materials to be scaled up.

My research focuses on engineering particles and surfaces to control their chemical and physical properties. By understanding and controlling these structure-property relationships, materials can be designed from the bottom up. My ultimate goal is to develop materials that can be used to solve global problems, such as those related to energy and the environment. While nanoscale energy materials and phenomena have been widely studied, mesoscale materials provide unique opportunities to take advantage of length scale-dependent properties.

My doctoral research at Harvard probed the self-assembly of colloids into mesoscopically ordered, porous structures known as inverse opals. Using chemistry, I was able to control the hierarchical structure of the inverse opal, which allowed for control over the resulting wetting and optical properties; I was particularly struck when I achieved a macroscopic color change of an inverse opal by using temperature to control the degree of condensation of the silica matrix material. As a postdoc at MIT, I have gained expertise in electrochemistry, working on several topics related to electrochemical CO_2 capture and conversion. These include a project developing novel electrocatalysts for CO_2 conversion, a collaboration to engineer CO_2 bubble delivery to catalyst surfaces, and more recently a focus on electrochemical carbon capture. As an independent scientist, I plan to work at the interface of these areas by developing methodologies to make novel hierarchical nano/mesomaterials, with an initial focus on carbon capture and conversion applications. This advanced understanding will open the door to future applications beyond carbon capture, such as sensing or drug delivery.

Teaching Interests:

I have experience teaching a wide range of courses; I was a teaching fellow (~TA) for an introductory chemistry course, an introductory food science/soft matter course, a quantum chemistry course, a graduate teaching course, and an advanced course on the chemistry of materials. These represent introductory and advanced undergraduate and graduate classes in chemistry and engineering. I would be interested in teaching similar courses in the future, as well as classes related to materials synthesis, nanotechnology, and colloid chemistry. I also have extensive undergraduate advising experience.

505273 Colloidal Fluids As Electrical Current Collectors

Abstract:

Research Interests: Advances in synthetic techniques have enabled a revolution in the design of colloidal dispersions containing particles with a great diversity of sizes, shapes and compositions. This has provided not only for a sophisticated understanding of the intimate link between a dispersion's microstructure and its macroscopic properties, but also the basis for a powerful design paradigm for the development of new materials to address a wide range of engineering problems. To realize this future, there remain significant scientific challenges both in terms of formulation science and the continued development of a detailed understanding of the relationships between a dispersion's microstructure and its macroscopic behavior. In situ measurements of colloidal microstructure are an increasingly important component to inform this goal as they provide information both about the quiescent properties of colloidal dispersions but also their properties under simple shear where time-dependent structural evolution proves important to understand their behavior.

Current Research: I will highlight the progress I have made toward understanding the relationship between the electrical conductivity and the rheological properties of suspensions of electrically conductive nanoparticles in the quiescent state and under steady shear. Such suspensions find use as the conducting current collectors in electrochemical flow applications. Key to their performance in this application is a trade-off that arises between maximizing electrical conductivity and minimizing viscosity. To study the origin of this trade-off, suspensions of high structured carbon blacks are formulated in neat propylene carbonate and are characterized, using small amplitude oscillatory shear and impedance spectroscopy, to elucidate their electrorheological behavior spanning their fluid-gel transition. Using these methods, I identify the electrical and mechanical percolation transitions and couple these measurements with small angle neutron scattering experiments to probe their microstructural origin. The comprehensive picture provided by the combination of these three measurements harmonizes several emerging experimental results and guides the development of current collectors with improved performance. Future directions for this research involve looking at the field dependence mobility of electrons through suspensions of conducting particles and its link to particle dynamics.

Education:

2015-Present NRC Postdoctoral Researcher – National Institute of Standards and Technology – NCNR Advisor: Paul D. Butler and Norman J. Wagner

2009-2014 Doctor of Philosophy - University of Washington - "Structural Characterization of Composite Thin-Film and Dispersed Phase Conjugated Polymer/Fullerene Composites": Advisor: Lilo D. Pozzo

Teaching Interests: I know that the success of my career as a professor in chemical engineering will be evaluated not only by my research but also based on my efficacy as an instructor, a mentor, and a research advisor. True mastery of each of these roles is a lifetime endeavor, but I have developed a strong foundation on which to build a successful teaching career. The training I received as I acquired my B.S. and PhD in chemical engineering as well as my research interests has natural connections with undergraduate Heat and Mass Transport and Thermodynamics coursework. I also have extensive experience in instructing students in the field of colloids and interfacial sciences both at the undergraduate and graduate level and materials characterization.

505363 Complex Fluids and Anisotropic Liquids for Molecular Engineering and Rational Material Design

Abstract:

I received my PhD degree in Polymer Physics at the *ETH*, in Zürich, under the supervision of Prof. Hans Christian Oettinger. As a PhD student, I worked on various aspects of complex-fluid rheology, involving the interplay between mesoscale structure and dynamics. I designed and built a novel experiment, in which a lubricated cross-slot channel was employed to deform polymeric molecules in a complex flow geometry, thereby allowing me to generate some of the most precise data available for the rheology of polymer melts in a mixed shear and extensional flow. In addition, I developed a finite-element based data analysis technique to reconstruct accurate field kinematics from experimental data, with which I could examine the performance of various rheological models using numerical simulations. I worked closely with theorists, computational scientists, and other experimentalists to address what is widely viewed as a particularly challenging problem.

As a postdoctoral fellow in Prof. Fredberg's lab at *Harvard University*, building on my background in polymer rheology, I developed an assay to study the dynamics of living cells under geometrical confinement, and I designed a new device to explore intercellular shear deformation. My work helped discover new mechanism for the response of the epithelial cells to large intercellular shear deformations. In my current postdoctoral position at the *University of Chicago*, I have worked on the coupling of structure and dynamics of structured fluids, including liquid crystals, under the supervision of Prof. Juan de Pablo. We have shown that liquid crystal-aqueous interfaces are capable of reporting the aggregation of polypeptides in its early stages, and have shown that reducing water activity by adding salts to an aqueous phase can initiate transport of dissolved water across liquid crystal membranes. These results have provided fundamental new principles for the design of biological sensors, drug delivery systems, and molecular machines. Working with scientists at the Argonne National Laboratory, we have also developed a detailed structural view of liquid crystal molecules at interfaces, using synchrotron X-ray reflectivity and Grazing-incidence small-angle scattering measurements. On a different project, I have developed a new platform for deformation of chiral liquid crystals, which has allowed me to discover new phases of these systems, and to propose exciting new applications in the area of mechano-optical metamaterials.

Selected Publications (15 published and 4 under preparation)

1- M. Sadati, H. Ramezani, W. Bu, E. Sevgen, Z. Liang, C. Erol, M. Rahimi, N. Taheri Qazvini, B. Lin, N. L. Abbott, B. Roux, M. L. Schlossman and J. J. de Pablo, "Molecular Structure of Canonical Liquid Crystal Interfaces", *Journal of the American Chemical Society* (JACS), 139, 3841 (2017)

2- H. Ramezani, M. Sadati, M. Rahimi, B., Roux and J. J. de Pablo, "Understanding Atomic-Scale Behavior of Liquid Crystals at Aqueous Interfaces", *Journal of Chemical Theory and Computation*, 13, 237 (2017)

3- J. Martinez-Gonzalez, X. Li, M. Sadati, Y. Zhou, R. Zhang, P. F. Nealey and J. J. dePablo, "Directed Self-assembly of Liquid Crystalline Blue-phases into Ideal Single-crystals", *Nature Communications* (2017), doi:10.1038/ncomms15854

4- M. Sadati, A. Izmitli Apik, Julio C. Armas-Perez, Jose Martinez-Gonzalez, Juan P. Hernandez-Ortiz, N. L. Abbott and J. J. de Pablo," Liquid Crystal Enabled Early Stage Detection of Beta Amyloid Formation on Lipid Monolayers", *Advanced Functional Materials*, 25, 6050 (2015), (Featured on the Journal's Cover).

5- M. Sadati, N. Taheri Qazvini, R. Krishnan, C. Y. Park and J. J. Fredberg, "Collective Migration and Cell Jamming", *Differentiation*, 86, 121 (2013).

6- M. Sadati, C. Luap, M. Kröger, B. Lüthi, and H. C. Öttinger, "Application of Full Flow Field Reconstruction to a Viscoelastic Liquid in a 2D Cross-slot Channel", *Journal of Non -Newtonian Fluid Mechanics*, 192, 10 (2013).

Research Interests: As a new faculty member, my future research will include exploring structure rheology-property relationships of polymeric materials and lyotropic and biological liquid crystalline mesophases, with a focus on linking molecular information to micron-scale phenomena, and on finding new routes to engineer innovative functional materials with tunable properties for drug delivery, biomedical and photonics applications. Building on my background in polymer physics, I also intend to pursue my research goals through a combination of both experimental and theoretical means and to create a rational framework with which to develop the necessary rheological and processing aspects that arise in additive manufacturing of soft materials.

Teaching Interests: I have been a teaching assistant for Polymer Physics and Finite-Elements at the Materials Department Institute at the ETH Zurich. I have also taught undergraduate students in three rheology laboratory courses. In addition, I have supervised eight undergraduates and two graduate students at Harvard University and at the University of Chicago on various research projects. With my background in Polymer Physics and Engineering, I have a strong foundation in materials science and chemical engineering, with an emphasis in polymer science related courses such as polymer physics, polymer chemistry, crystallization, rheology and polymer characterization.

505626 Contorted Molecular Semiconductors for Organic Electronics

Abstract:

Organic semiconductors are promising for light-weight, flexible and bio-compatible electronic devices. The implementation of diverse functions in organic electronics requires chemical engineering to create molecules that go beyond the capability of traditional materials. While conventional wisdom is that planar molecules with strong intermolecular coupling in the solid state are ideal for charge transport, molecules with contorted structures sometimes have superior properties over planar molecules. In this poster, I will present two types of contorted molecules with extraordinary characteristics and applications. First, I will show the property and application of a conjugated macrocycle that consists of diphenyl perylene diimides. This molecule enables us to make ultra-sensitive organic photodetectors with unparalleled detectivity. We elucidate the importance of the cyclic molecular conformation to the high device performance. Second, I will introduce a series of helical, molecular ribbons that act as excellent acceptors in organic solar cells. They are among the best materials for non-fullerene acceptors. Moreover, they are ideal optoelectronic materials for narrowband photodetectors that have narrow spectral response with fullwidth at half maxima of < 20 nm. This figure-of-merit is comparable to the state-of-the-art narrowband photodetectors made from methylammonium lead trihalide perovskite single crystals. These molecules represent new design motifs of functional organic materials with broad and promising applications. The physical and chemical principles underlying the molecule and device design will be discussed. I will show what we learn from the structure-property relationships and how we make a connection between properties and applications from the molecular aspect.

Research Interests: Organic electronics, Organic-inorganic hybrid nanomaterials, Energy conversion, Ion and charge transport.

Teaching Interests: Materials Chemistry, Nanoscience, Physical chemistry, Electronics.

505190 Controlling the Dynamics of Soft Materials at Interfaces

Abstract:

Research Interests:

My research program is focused on controlling and engineering the dynamics, transport and mechanics of soft materials (i,e., complex fluids, bacterial biofilms, swarming bacteria, polymer hydrogels, colloidal systems, etc.) at surfaces and interfaces. Examples of questions that my research program addresses include — How can we measure *in-situ* the collective spatio-temporal patterns of cell differentiation within microbial communities and biofilms? How can we understand and mathematically model the resulting emergent collective mechanical properties (such as stiffness, surface tension, osmotic pressure, fluid transport, jamming and clogging) in both these complex systems, and simplier model experimental analogs? How should we engineer interfaces to control and direct the growth of living matter? By applying surface patterning, wrinkling and chemical functionalization techniques, fluorescence microscopy visualization, reporter labeling and genetic techniques, image analysis and mathematical modeling, I am interested in studying the out-of-equilibrium properties and dynamic transitions in diverse surface-associated biological and physical processes across many length and time scales.

My broad research interests are in structure-property relationships in physical and biological processes, wherein a quantitative understanding of the physicochemical hydrodynamics at the micron scale can illuminate complex emergent behavior at larger scales. My research program aims to to explore both practical and fundamental questions in interfacial fluid mechanics, polymeric materials, surface engineering, and biologically active matter by applying experimental imaging methods, combined with analytical and modeling techniques. In synergy with my skills in analysis and modeling, the data sets generated by my experimental program will allow my group to quantitatively develop useful predictive and descriptive theories and models. Additionally, I envision strong collaborations with other experimental groups for this purpose. In the long-term, I envision building a collaborative research group that is at the forefront of discovering fundamental physical and biological mechanisms in complex interfacial systems leading to innovative engineering solutions to control the growth of living matter at interfaces.

Teaching Interests:

My background and mathematical training allows me to teach a broad set of courses that encompasses the core chemical engineering syllabus including fluid dynamics, transport phenomena, statistical thermodynamics, reaction and kinetic engineering and colloid and interfacial phenomena at both the graduate and the undergraduate level. My immediate preferences would be in teaching fluid dynamics and transport to graduate and undergraduate students, a subject that I am especially passionate about and is related to my research interests. In the long term, I intend to develop an advanced graduate level class on Biological Soft Matter.

505202 Design of Advance Materials by Using a*b initio* Structural Search

Abstract:

Research Interests: Crystal structure prediction, X-ion batteries (X=Li, Na, K, Mg,), Lightweight materials, super hard materials, molecular dynamics, transport properties

Teaching Interests: Inorganic chemistry, Physics, Thermodynamics, Computational physics

In solid-state physics, it is well known that the arrangement of atoms in a crystal structure determines almost all physical and chemical properties of a given material. In recent years, the discovery and computational design of novel materials has been performed by exploring similar compounds and structures to the one of interest, experimentally as well as theoretically. However, an open question remains as how we can proceed if we do not know both the structure and composition? Can we predict stable structures based on the knowledge of the chemical composition alone? In recent years, progress in computational materials science has helped the development of structural search algorithms that have tried to solve these questions.

Currently, crystal structure prediction is one of the most important fields of study and an open challenge. This is because the only input needed is the chemical composition and external conditions. Furthermore, novel materials with specific chemical and/or physical properties can be predicted (i. e. hardness, band gap, density, dielectric constant, etc.). Also, we can study systems under extreme conditions such as high pressure and high temperature. With the help of these methodologies the cost of production and design of new materials can be reduced, as well as the number of potential experiments. Another advantage of crystal structure prediction is that multiple compositions and conditions can be theoretically studied. Additionally, their properties can be studied before they are synthesized. Very recently, novel methodologies to predict new crystal structures have been developed. In some of these methodologies, the search is based on experimental experience where the information is stored in databases. Then, novel materials can be predicted by using this information along with the help of methods developed in data mining. Other methods are based on exhaustive searches of stable structures over the potential energy surface. Here, the main goal is to find the global minimum, which corresponds to the most stable structure (ground state). Some of these methods are random search, particle swarm, evolutionary algorithms, basin hopping, and minima hoping.

In my future work as faculty, I would like to continue working in the crystal structure prediction field. My research will be focus in tree different topics: 1) lightweight materials, 2) superhard materials, and 3) X-ion batteries (X=Li, Na, K, and Mg).

491106 Engineered Porous Materials for Advanced Chemical Conversions: Understanding Structure-Property-Acitivty Relationship

Abstract:

Engineered Porous Materials for Advanced Chemical Conversions: Understanding Structure-Property-Activity Relationship

Satish K. Nune

Senior Research Scientist, Applied Functional Materials

Pacific Northwest National Laboratory (PNNL), Richland, WA

Engineering, materials and catalysis play a vital role in achieving energy security for sustainable future and growth. Accelerated development of new advanced materials with well-defined active sites for achieving unique, reproducible and reliable properties is a key in achieving global competitiveness in 21st century. Although recent advances in porous materials have provided significant progress, there is an explicit and growing need for the development of new advanced materials with improved properties. If one could successfully engineer porous materials that are robust, structurally tunable and inexpensive but utilize all active sites, porous materials could potentially offer a clear advantage over other porous materials. To address this, Dr. Nune has developed various nanoscale porous materials with varied functionalities and hydrophobicity by taking advantage of metal-ligand cooperation effect for sustainable, economical and energetically favored chemical transformations. Wide variations in metal ions and linkers allowed structural tailorability to precisely control pore size, surface area, and Lewis acidity. He also developed approaches to further synthesize and functionalize porous nanomaterials with well-defined active sites and activity that cannot be synthesized otherwise. Larger pore sizes translated to smaller mass transport limitations compared to other inorganic materials and greater accessibility to active sites in the framework that are critical for inducing selectivity. Dr. Nune will discuss on Engineered Porous Materials for Advanced Chemical Conversions: Understanding Structure-Property-Activity Relationship.

Research Interests:

Hybrid Porous Materials, Catalyst Design, Doped-Porous Carbons, Gas Separation, Advanced Chemical Conversions, Energy Storage, Critical Metal Extraction, Nanomaterials-Synthesis and Functionalization.

Teaching Interests:

1. Introduction to Materials Science and Engineering

2. Principles of Chemical Engineering

3. Kinetics and Reactor Design

4. Advanced Composites

505567 Engineering Materials and Devices for Energy, Environment and Human Health: From Capillary Foams to Wearable Sensors and Implantable Neural Probes

Abstract:

Research Interests:

My primary research interest lies at applying the knowledge of materials, chemistry, and micro- and nano-manufacturing to engineer materials and devices for energy, environment, and human health. My doctoral research focuses on the design and development of a new class of colloidal materials, capillary foams, that can be applied and greatly enhance the efficiency in oil recovery, oil spill cleanup and chemical separations. For my postdoctoral research, I am currently working on wearable sensors and implantable neural probes.

Here, I will provide a brief overview of my research from both a fundamental point of view and their emerging applications. In particular, this overview includes: (i) Development of capillary foams for surfactant-free and energy-efficient separations; (ii) Design of reconfigurable wetting systems, applicable to new methods to collect and purify water and recover oil from reservoirs and spills; (iv) Engineering of soft, skin-mounted sensors for in-situ perspiration analysis; (iv) Development of implantable neural probes for programmable drug delivery and opotogenetics.

Postdoctoral Advisor: Professor John A. Rogers, Northwestern University

Ph.D. Advisors: Professor Sven H. Behrens and Professor J. Carson Meredith, Georgia Tech

Selected Publications:

1. **Y. Zhang**, S. Wang, J. Zhou, R. Zhao, G. Benz, S. Tcheimou, J. C. Meredith, S. H. Behrens, "Interfacial activity of non-amphiphilic particles in fluid-fluid interfaces," *Langmuir,* 2017, 33 (18):4511–4519.

2. **Y. Zhang**, A. Shitta, J. C. Meredith, S. H. Behrens, "Bubble meets droplet: particle-assisted reconfiguration of wetting morphologies in colloidal multiphase systems," *Small*, 2016, 12(24):3309-3319.

3. **Y. Zhang**, M. C. Allen, R. Zhao, D. D. Deheyn, S. H. Behrens, J. C. Meredith, "Capillary foams: stabilization and functionalization of porous liquids and solids," *Langmuir*, 2015, 31(9): 2669-2676.

4. **Y. Zhang**, J. Wu, H. Wang, J. C. Meredith, S. H. Behrens, "Stabilization of liquid foams through the synergistic action of particles and an immiscible liquid," *Angewandte Chemie*, 2014, 126(49): 13603-13607.

Teaching Interests:

My teaching philosophy is stimulating the curiosity of my students and equipping them with an effective way of learning that they can benefit from for their lifetime. In this rapidly changing world, the knowledge in science and engineering is expanding exponentially. Therefore, instead of solely subject-oriented teaching, it is critical to stimulate students' curiosity to explore different fields and teach them learning skills that they can apply to various learning processes.

At Georgia Tech, I have assisted teaching both undergraduate and graduate courses. During such processes, I had opportunities to practice and improve my teaching skills through leading pre-lab meetings, supervising experiments and data analysis, grading laboratory reports, and holding office hours. In addition, I have advised five undergraduate students for two years and resulted in three publications, five posters, and two awards. Throughout these mentoring processes, I have learned how to motivate students' research interest, manage a project, and communicate with students in a clear, concise and effective manner.

I am prepared to teach both undergraduate- and graduate-level courses in chemical engineering and material science and engineering. I feel confident to teach a comprehensive list of courses, including chemical engineering thermodynamics, transport phenomena, kinetics and reactor design, unit operations laboratory, and polymer science and engineering. I will also develop new courses on cutting-edge research topics, such as nanomedicine and bio-integrated electronics.

505489 Engineering Molecular Interactions in Biological and Electrochemical Interfaces

Abstract:

Research Interests:

The structure and dynamics of ions and biomolecules in confined interfaces critically impact the behavior of many complex systems, ranging from energy storage devices to the human brain. However, it remains extremely difficult to rationally tune the subtle balance of strong and weak intermolecular forces that determine molecular assembly in active interfaces. A major origin of this knowledge gap is the dearth of experimental techniques that can directly measure intermolecular forces while simultaneously observing nanoscale electric and magnetic field fluctuations in confined interfaces.

I will draw on my background in molecular force spectroscopy, materials synthesis, and advanced microscopy to develop new approaches to observe and influence ion and biomolecule assembly in active interfaces. One of my primary aims will be to develop a platform for optically measuring electric and magnetic field fluctuations with high spatial and temporal fidelity. I will combine these electromagnetic sensing capabilities with molecular force spectroscopy to tackle fundamental problems in materials science and chemical engineering and to explore dynamic electromagnetic phenomena in complex systems, like neuronal networks.

My research program will initially focus on three questions:

What is the mechanism of ion conductivity in ionic liquids and energy storage electrolytes?

How can ion and solvent dynamics be tuned to influence protein aggregation and assembly?

How can magnetic soft materials be used to measure and influence biological processes?

My vision bridges and expands upon my prior research experiences. During my Ph.D. at UC Santa Barbara, I worked with Jacob Israelachvili, using molecular force spectroscopy to study ionic liquids and peptides in confined interfaces. One of my major scientific breakthroughs was the astonishing discovery that less than 0.1 % of the ions that compose ionic liquids contribute to electrostatic screening in electrochemical interfaces, resulting in a new perspective on ionic liquids and energy storage electrolytes. Further, my work raised many important additional questions, and the ion conductivity mechanism in ionic liquids and high concentration electrolytes remains controversial.

While working in the colloid and interfacial science field, I realized that combining the measurement of nanoscale electromagnetic field fluctuations with molecular force spectroscopy would provide revolutionary opportunities to address many important scientific questions. As a result, I moved to Stanford University, where I was awarded the highly competitive 2016-2018 GLAM Postdoctoral Fellowship to work with Nicholas Melosh to develop diamond nanomaterials for advanced bio-imaging and electric field sensing. I now have experience in the synthesis and processing of diamond nanomaterials and using diamond quantum sensing to optically measure electromagnetic fields.

Consequently, I am uniquely poised to combine fluorescence-based electric and magnetic field sensing with molecular force spectroscopy to develop a research program that addresses critical scientific challenges, from developing novel energy storage materials to fingerprinting and influencing neuronal misfiring in neurological disorders.

Selected Publications:

Gebbie MA, Valtiner M, Banquy X, Fox ET, Henderson WA, Israelachvili JN. Ionic liquids behave as dilute electrolyte solutions. *Proc. Natl Acad. Sci. USA* 110, 9674–9679 (2013).

- **DOE Basic Energy Sciences Research Highlight,** June 2013

Gebbie MA,** Smith AM, Dobbs HA, Lee AA, Warr GG, Banquy X, Valtiner, M, Rutland** MW, Israelachvili** JN, Perkin** S, Atkin** R. Long range electrostatic forces in ionic liquids. *Chem. Commun.* 53, 1214–1224 (2017).

- **** - Co-corresponding authorship**
- **Invited Feature Review**

Gebbie* MA, Wei* W, Schrader AM, Cristiani T, Waite JH, Israelachvili JN. Tuning underwater adhesion with cation-π interactions. *Nature Chemistry* 9, 473–479 (2017).

- * - Equal contribution
- **Highlighted in Chemical & Engineering News,** February 2017

Teaching Interests:

I have diverse teaching interests and would be excited to teach courses in statistical mechanics, general thermodynamics, chemical engineering unit operations, surface and interfacial science, and introductory materials science.

505435 Engineering Precision Polymers for Advanced Materials Applications

Abstract:

Research Interests: A long-standing challenge in materials science is the development of functional materials with predictable emergent bulk properties. Soft materials, such as polymeric supramolecular assemblies, have unique properties including flexibility, self-organization, hierarchical structure, self-healing, and stimuli-responsiveness that make them ideal candidates for health, energy, and infrastructure related applications. In such systems, the polymer microstructure and architecture strongly affect material functionality. However, **studying the static and dynamic properties of these complex macromolecules in bulk is challenging due to their inherent polydispersity**. My research aims to directly address this challenge by developing new approaches to **synthesize polymers with precise control over dispersity, sequence, and topology (*e.g.*, star, comb, brush polymers) to elucidate the relationship between molecular-scale properties and macroscopic material response**, with the overarching goal of enabling better predictions of structure and properties of polymeric materials.

Successful Proposals: NIH T32 Postdoctoral Fellowship – Trauma, Burn, Wound Healing; Oak Ridge National Laboratory, Spallation Neutron Source, SANS (2017); European Synchrotron Radiation Facility, Time-resolved SAXS (2017).

Postdoctoral Project: *Polyelectrolyte Driven Complexation of Sequence Specific Polypeptides,* Institute for Molecular Engineering, The University of Chicago (Advisor: Professor Matthew V. Tirrell).

PhD Dissertation: *Synthesis and Supramolecular Assembly of Biomimetic Polymers*, The Center for Biophysics and Quantitative Biology, University of Illinois at Urbana-Champaign (Advisor: Associate Professor Charles M. Schroeder)

Research Experience: My research focuses on the development of new classes of soft materials by studying and engineering the effects of polymer sequence, composition, and topology on polymer structure and properties. In my PhD research, I studied the dynamic and elastic properties of branched polymers. Specifically, I **developed a hybrid synthetic strategy to produce branched polymer architectures based on chemically modified DNA that enables precise control of backbone length and flexibility, as well as branch grafting density and chemical identity**. Using this approach, I synthesized a variety of polymer architectures, including three-arm stars, H-polymers, graft block copolymers, and brush polymers and characterized their material properties using single molecule studies. [1-3] These studies have increased our understanding of the dynamic and elastic properties of industrially relevant branched polymer topologies.

In my postdoctoral research, I developed synthetic polypeptide sequences using solid-phase synthesis to precisely control polypeptide length and sequence to create alternating, block, and gradient polypeptide copolymers comprising charged, polar and aromatic amino acids. Using this approach, I studied **the effects of sequence patterning on phase separation** using a variety of microscopic and spectroscopic techniques, including optical microscopy, fluorescence spectroscopy, and nuclear magnetic resonance spectroscopy. In addition, I investigated their chain conformations in polyelectrolyte complexes using SAXS and SANS. This work holds the potential to develop fundamental understanding of complex biopolymer phase separation, impacting the design of novel functional biomimetic materials for a variety of applications in drug delivery, coatings, and reaction engineering in controlled compartments.

Future Research Directions: As an independent investigator, I will study the impact polymer microstructure has on bulk materials properties. Initial investigations in the Marciel laboratory will consist of three aims (1) synthesis and investigation of dynamic properties of sequence specific polyampholytes, (2) correlation of branch polymer microstructure and mechanical and rheological properties, and (3) microfluidic-directed assembly of charged polymers. This work will require a multidisciplinary approach, combining my PhD work in synthesis and directed-assembly with my postdoctoral work in characterization of bulk polyelectrolyte complexes.

Teaching Experience: During my Ph.D., I served two semesters as a teaching assistant (TA) for Molecular Genetics and was awarded the **2010 and 2011 List of Teachers Ranked as Excellent** by the University of Illinois at Urbana-Champaign (UIUC). In my second semester, I was selected to lecture the honors section, which afforded me the opportunity to lecture weekly and prepare problem sets. Furthermore, I served as a guest lecturer for the course **Polymer Science and Engineering** (**ChBE 456**). In addition to my teaching experiences, I have made an ongoing effort towards the advancement of underrepresented groups in STEM fields. Throughout my graduate and postdoctoral career, I have participated in a broad variety of outreach programs ranging from hands-on experiments with primary school students to graduate level recruitment in engineering with the Multicultural Engineering Recruitment for Graduate Education (MERGE) program. In addition, I had the privilege to mentor several outstanding female undergraduate and graduate students, who are currently pursuing engineering careers at ExxonMobil and Intel and postdoctoral research in chemical engineering at MIT.

Teaching Interests: My teaching interests include the central courses in polymers – Polymer Science and Engineering and Polymer Physics. In addition, I am also interested in teaching Thermodynamics, Reaction Kinetics, Biomolecular Engineering and courses on materials characterization techniques.

Selected Publications:

1. **A.B. Marciel**, D.J. Mai, C.M. Schroeder, "Template-directed synthesis of structurally-defined branched polymer architectures", *Macromolecules*, 48, 1296-1303 (2015).
2. D.J. Mai, **A.B. Marciel**, C.E. Sing, C.M. Schroeder, "Topology-controlled relaxation dynamics of single comb polymers", *ACS Macro Letters*, 4, 446-452 (2015).
3. J. P. Berezney*, **A.B. Marciel***, C.M. Schroeder, O.A. Saleh, "Scale-dependent stiffness and internal tension of a model brush polymer", *submitted,* (2017). (*co-first)
4. **A.B. Marciel**, M. Tanyeri, B.D. Wall, J.D. Tovar, C.M. Schroeder, W.L. Wilson, "Fluidic-directed assembly of aligned p-conjugated oligopeptide nanofibers", *Advanced Materials,* 25, 6398-6404 (2013).
5. **A.B. Marciel** and C.M. Schroeder, "New Directions in Single Molecular Polymer Physics" *Journal of Polymer Science, Part B: Polymer Physics*, 51, 556-566 (2013).
6. K.T Nam, S.A. Shelby, P.H. Choi, **A.B. Marciel**, R. Chen, L. Tan, T.K. Chu, R.A. Mesch, B.C. Lee, M.D. Connolly, C. Kisielowski, R.N. Zuckermann, "Free-floating ultra-thin two-dimensional crystals from sequence-specific peptoid polymers", *Nature Materials*, 9, 454-460 (2010).
7. **A.B. Marciel**, E.-J. Chung, B.K. Brettmann, L. Leon, "Bulk and nanoscale polypeptide based polyelectrolyte complexes" *Advances in Colloid and Interface Science*, 239, 187-198 (2016).
8. J. Vieregg*, M. Lueckheide*, **A.B. Marciel**, A. Bologna, J. Reyes, L. Leon, M. Tirrell, "Oligonucleotide-peptide complexes: phase control by DNA hybridization" *in revision*, (2017). (*co-first)

495518 Metallurgy-Mimic Thermal Processing and Morphology of Particle-Forming Diblock Copolymers

Abstract:

Low-symmetry phases including the dodecagonal quasicrystal and tetrahedral close-packed Frank-Kasper Ïƒ phase recently have been identified in various forms of soft materials including dendrimers, surfactants, and block polymers. In block polymers these complex phases emerge from the supercooled fluctuating disordered state, comprised of well-formed micelles. Rapid cooling to sufficiently low temperatures below the order-disorder transition extinguishes molecular exchange resulting in non-ergodic â€œliquid-like packingâ€. Here we present yet another family of the Frank-Kasper phases obtained as a function of temperature and time from polyisoprene-*b*-polylactide (PI-*b*-PLA) diblock copolymers containing 15 to 20% PLA, utilizing small-angle X-ray scattering as a probe. The development of distinct periodic or aperiodic states depends on how specimens are cooled from the disordered state to temperatures below the order-disorder transition temperature. Whereas direct cooling leads to the formation of previously documented morphologies, rapidly quenched samples that are then heated from low temperature form the hexagonal C14 and cubic C15 Laves phases commonly found in metal alloys. Self-consistent mean-field theory calculations show that these, and other associated Frank-Kasper phases, have nearly degenerate free energies, suggesting that processing history drives the material into long-lived metastable states defined by self-assembled particles with discrete populations of volumes and polyhedral shapes.

Research Interests: Structure-property relationship of block copolymers, polyolefins, and functional polymer materials

I earned a Ph.D. in Polymer Chemistry at Pohang University of Science and Technology (POSTECH), South Korea. My Ph.D. work was mainly evaluating structure-property relationship of thin-film polymer electronic materials. The study has familiarized me with (i) polymer chemistry and physics, (ii) characterization of thermal, mechanical, and electrical properties of functional polymers, and (iii) synchrotron X-ray techniques such as X-ray scattering, reflectivity, and spectroscopy, along with fundamental theories governing such techniques. My current research as a postdoc involves two very different projects. One aims to reveal the underlying principles of microphase separation in sphere-forming block polymers at equilibrium and non-equilibrium conditions (which will be presented here). The other focuses on interfacial properties of commercial polyolefin blends and multilayer films. These projects rely on the integration of polymer synthesis and molecular characterization, synchrotron X-ray scattering, rheological characterization, and imaging techniques such as scanning electron microscopy (SEM) or transmission electron microscopy (TEM).

I enjoy learning any theories or techniques that help further my understanding of nature, and proving their validity. Also, I am willing to find fundamental ideas that govern complex natural phenomena. Thus, I would like to pursue my next career as a faculty member in research institution. My future research interst lies on several aspects. The most insteresting theme would be the study of complex phase formation in simple block polymer systems. My synthesis and characterization skills for the model block polymers will enable extensive research on particle-forming phase portrait of the system and investigation to find governing principles of such complex phase formation. Diblock copolymers are an obvious candidate, but the research can be extended to relevant systems (e.g., multiblock polymers). My research interest also includes interfacial phenomena occurring in polyolefins and polymers having functional groups that have potential application for electronic thin film devices.

Teaching Interests: With my expertise in X-ray scattering technique, I can teach courses such as Structure of Materials or X-ray Scattering/Diffraction/Crystallography. My solid knowledge of basic polymer science enables me to teach Polymer Chemistry, Polymer Characterization or Polymer Physics. Furthermore, thanks to my educational background as a chemist, I can teach chemistry-related core undergraduate courses for materials science or chemical engineering students, such as Physical Chemistry or Instrumental Analysis.

Relevant Work: Kyungtae Kim, Morgan W. Schulze, Akash Arora, Ronald M. Lewis III, Marc A. Hillmyer, Kevin D. Dorfman, Frank S. Bates, Thermal processing of diblock copolymer melts mimics metallurgy, Science 2017, 356, 520-523. DOI:10.1126/science.aam7212 (K.K., M.W.S., and A.A. contributed equally)

505542 Nuclear Spin Hyperpolarization for Characterization of Materials, Surfaces, and Interfaces

Abstract:

Research Interests:

With unmatched chemical specificity and the ability to probe materials across length- and timescales, nuclear magnetic resonance spectroscopy (NMR) and imaging (MRI) are two of the most powerful analytical techniques available to engineers, scientists, and medical professionals. Despite these strengths, sensitivity is limited by the degree of equilibrium nuclear spin polarization, typically at parts per million levels. In order to probe to surfaces, interfaces, and low-concentration solutions, methods for nuclear spin "hyperpolarization" are needed. My laboratory will pursue the development of methodology for targeted, high sensitivity spectroscopy for chemical engineering applications such as structure and chemistry at surfaces and interfaces.

I completed my PhD in Chemical Engineering with Prof. Jeffrey Reimer at UC Berkeley where I investigated optical methods of nuclear spin hyperpolarization in semiconductors, culminating in the discovery of a method to spatially control hyperpolarization and directly image depletion layers at the surface of bulk gallium arsenide[1]. This work is an excellent example of how controlling spin hyperpolarization provides both the sensitivity and selectivity to probe surfaces and interfaces.

During my postdoctoral research with Prof. Alexander Pines I have worked on a wide range of topics ranging from fundamental spin dynamics to geochemical exploration with novel magnetic resonance detection in boreholes. My work in combining dynamic nuclear polarization with photophysics in diamonds has resulted in one of the highest levels of room-temperature hyperpolarization ever observed[2]. As a faculty member, I will leverage my fundamental understanding of spin interactions for chemical engineering applications. This project will be generally organized as follows:

1. **Understanding the transport and equilibrium properties of interacting spins from microscopic to macroscopic scales.**
2. **Targeting and localization of hyperpolarization as a selective chemical and structural probe at surfaces and interfaces.**
3. **Chemical engineering applications to interfaces, surface science, and catalysis.**

I also seek to extend NMR beyond the traditional model requiring a high-field superconducting magnet in a dedicated facility. Low-field NMR allows the use of cheap, portable magnets in benchtop and truly portable systems. Zero-field NMR allows the observation of spin-interactions that are invisible in the presence of a magnetic field, providing additional information about material structure, geometry, and dynamics[3]. Furthermore, zero-field experiments are free from the deleterious effects of inhomogeneous magnetic susceptibility, enabling high-resolution study of heterogeneous materials.

1. **King, J. P.** *et al.* "Optically rewritable patterns of nuclear magnetization in gallium arsenide." *Nature Communications* 3:918 doi: 10.1038/ncomms1918 (2012).
2. **King, J. P.** *et al.* "Room-temperature *in situ* nuclear spin hyperpolarization from optically pumped nitrogen vacancy centres in diamond." *Nature Communications* 6:8965 doi: 10.1038/ncomms9965 (2015).
3. J. W. Blanchard, T. F. Sjolander, **J. P. King,** M. P. Ledbetter, E. H. Levine, V. S. Bajaj, D. Budker, and A. Pines. "Measurement of untruncated nuclear spin interactions via zero- to ultralow-field nuclear magnetic resonance." Phys. Rev. B 92, 220202(R) (2015).

Teaching Interests:

My teaching philosophy can be stated concisely: effective teaching is the facilitation of effective learning. While this statement is simple, in practice it presents these challenges: How do we define effective learning? How do we assess if learning has occurred? What steps do we take based on this assessment? I will present my plan to address these challenges in the chemical engineering curriculum. I am interested in teaching all aspects of the core curriculum, but with my background would be especially interested in teaching thermodynamics, fluid dynamics, and transport phenomena at the undergraduate and graduate levels. I would also develop special classes in the theory and application of spectroscopic techniques.

505194 Porous Materials Chemistry for Catalysis and Separations

Abstract:

Research Interests:

My research career has spanned areas from surface science and catalysis of complex multifunctional biomass-derived molecules to synthesis of nanoporous materials for adsorption, separations and understanding of materials stability. I believe my combination of skills will allow for design and synthesis of uniquely functionalized porous materials and evaluation of these materials in catalytic and separations processes, with complementary spectroscopic and structural characterization.

My Ph.D. work at the University of Colorado Boulder under Professor Will Medlin focused on elucidating the fundamental surface reaction mechanisms of biomass-derived molecules, such as furfural, on transition metal surfaces and utilizing those insights to design catalysts for atom-efficient selective transformations. Using surface science techniques, I proposed different surface intermediates that led to either the desired hydrodeoxygenation or the atom-inefficient decarbonylation reactions. This fundamental insight led to the design of surface-modified and bimetallic oxide-supported catalysts with superior hydrodeoxygenation selectivity and a spectroscopic look at the mechanisms by which the catalytic selectivity was improved. For my dissertation "Improving Catalytic Selectivity through Control of Adsorption Orientation," I received the Outstanding Dissertation Award from CU Boulder's College of Engineering.

During my tenure as a postdoc at the Georgia Institute of Technology under Professors Chris Jones and Ryan Lively, I have broadened my skills in the areas of porous materials chemistry, adsorption, and separations, working with a variety of nanoporous materials from metal-organic frameworks to structured porous oxides. The majority of my work has focused on designing and characterizing materials systems for CO_2 adsorption from dilute sources such as ambient air and investigating the how nanoporous materials interact with and are transformed by other acid gases. These projects have required engineering at length scales from the atomic-level design of adsorbents and surfaces to macroscopic contactors such as honeycomb-shaped monoliths and hollow fibers.

Research in my independent career will combine these diverse skill sets to develop advanced nanostructured porous materials platforms for a variety of applications in catalysis and separations. My research program will span three areas in porous materials chemistry and engineering: (1) enzyme-mimetic nanoporous systems for selective catalysis and adsorption, (2) crystal morphology engineering of soft nanoporous materials, and (3) highly-dispersed heterogeneous catalysts derived from mixed-metal-organic frameworks. My research will be motivated by challenges in asymmetric catalysis, molecular separations, biomass valorization, and partial hydrogenation and oxidation reactions.

Selected Publications (10 first author, 20 total):

1. S.H. Pang, L.-C. Lee, M.A. Sakwa-Novak, R.P. Lively, C.W. Jones "Design of Aminopolymer Structure to Enhance Performance and Stability of CO_2 Sorbents: Poly(propylenimine) vs. Poly(ethylenimine)," *J. Am. Chem. Soc.* **2017**, *139*, 3627-3630.
2. S.H. Pang, C. Han, D.S. Sholl, C.W. Jones, R.P. Lively "Facet-Specific Stability of ZIF-8 in the Presence of Acid Gases Dissolved in Aqueous Solutions," *Chem. Mater.* **2016**, *28*, 6960-6967.
3. S.H. Pang, N.E. Love, J.W. Medlin "Synergistic Effects of Alloying and Thiolate Modification in Furfural Hydrogenation over Cu-Based Catalysts," *J. Phys. Chem. Lett.* **2014**, 5, 4110-4114.
4. S.H. Pang, C.A. Schoenbaum, D.K. Schwartz, J.W. Medlin "Directing Reaction Pathways by Catalyst Active-Site Selection using Self-Assembled Monolayers," *Nature Commun.* **2013**, 4:2448.
5. S.H. Pang, J.W. Medlin "Adsorption and Reaction of Furfural and Furfuryl Alcohol on Pd(111): Unique Reaction Pathways for Multifunctional Reagents," *ACS Catal.* **2011**, 1, 1272-1283.

Teaching Interests:

As an undergraduate student, I developed and led recitation sessions for classes in Thermodynamics and Separations. As a graduate student teaching assistant for Process Controls, I taught the lab portion of the class and earned the department's "Outstanding Graduate Student Teaching" award in 2012. I also served as the Lead Graduate Teacher for the department, educating fellow graduate students in engineering pedagogy to help them become more effective teachers. In 2013, I was given the opportunity to serve as the department's Teaching Fellow, developing course materials and teaching General Chemistry for Engineers for half a semester, gaining valuable experience in active learning techniques, while continuing my graduate research.

My teaching philosophy is centered on two major goals for student learning: (1) the students develop a fundamental understanding of the theories and concepts central to Chemical Engineering, and (2) they become critical thinkers with the problem solving and teamwork skills that will be necessary for them to be successful in the future. The achievement of these two goals will be accomplished via a combination of active learning activities designed to engage students and allow for more teacher-student interaction.

Because of my emphasis on conceptual learning and problem solving, I am primarily interested in teaching chemical engineering courses that form the foundation onto which the other courses build, such as Material and Energy Balances and Thermodynamics, though would be comfortable teaching any core Chemical Engineering course. I am also interested in developing upper-level elective courses in Chemical Kinetics and Catalysis, Nanoporous Materials and Nanotechnology, and Spectroscopic Methods.

I have also served as a research mentor for a dozen undergraduate students throughout my career, five of which are from underrepresented groups in engineering, and am especially proud to have published papers with two of these students (one female, one Hispanic). I look forward to continuing my role as an educator and developing problem solving skills in the next generation of engineers and researchers. In the long term, I am interested in applying the latest concepts from STEM education research to advancing the methodology of teaching chemical engineering to make it accessible to a larger population of students.

490443 Self-Aligned Strategies for Printed Electronics

Abstract:

Research Interests:

My research focuses on nano/micro manufacturing with printing techniques and advanced nanomaterials for flexible electronics. At University of Minnesota, I have been developing novel printing processes compatible with roll-to-roll systems for continuous manufacturing of flexible electronics. These include self-aligned inkjet printing based on capillary action of liquid inks in microfluidic channels, high-resolution screen printing with silicon stencils, and rapid photonic annealing with intensive pulsed light. In addition to developing printing processes, I have been studying electrical and mechanical properties of nanomaterials (*e.g.*, graphene, silver) printed on flexible substrates, and demonstrating fully-printed flexible electronics using the processes and nanomaterials. This research has established practical printing strategies for manufacturing flexible electronics in a low-cost and high-through manner.

I will pursue my research interests in printed and flexible electronics for my future research group. My group will lead the research field by developing novel processes, using advanced materials, designing high-performance devices, and pioneering new application areas. The research projects that my group will first work on include, but will not be limited to, microfluidics for self-aligned printing and novel device architectures of electronic devices, nano/micro 3D structuring of electronic materials by 2D printing, and printed electronic sensors for healthcare application. This future research will provide many opportunities to obtain research grants and induce extensive collaborations within the university and across the country.

Teaching Interests:

Along with research, teaching is one of the most important roles as a faculty member in academia to transfer the knowledge to others. One of my main motivations for applying to a position in academia, rather than industry, is the opportunities to educate, advise, and inspire students. To be a good teacher, I have made strong efforts to build diverse teaching experiences in classes and laboratories during my tenure as a graduate student and postdoctoral scholar. These experiences have helped me create my own teaching philosophy and built my confidence and interest in delivering lectures, developing new courses, and supervising students.

I am prepared to teach a variety of courses for both the undergraduate and graduate programs in Chemical Engineering and Materials Science. In particular, I would be interested in teaching courses on fluid mechanics, transport phenomena, thermodynamics, colloid science, materials processing, and electrical/optical/mechanical properties of materials. Moreover, I would also welcome opportunities to teach other courses in Chemical Engineering and Materials Science, and be delighted to develop new courses for both the undergraduate and graduate programs.

Selected Publications (21 total, 15 first author):

W. J. Hyun, E. B. Secor, C.-H. Kim, M. C. Hersam, L. F. Francis, C. D. Frisbie, Scalable, Self-Aligned Printing of Flexible Graphene Microsupercapacitors, *Adv. Energy Mater.* 2017, In press (DOI: 10.1002/aenm.201700285).

W. J. Hyun, F. Z. Bidoky, S. B. Walker, J. A. Lewis, L. F. Francis, C. D. Frisbie, Printed, Self-Aligned Side-Gate Organic Transistors with a Sub-5 μm Gate-Channel Distance on Imprinted Plastic Substrates, *Adv. Electron. Mater.* 2016, *2,* 1600293.

W. J. Hyun, E. B. Secor, G. A. Rojas, M. C. Hersam, L. F. Francis, C. D. Frisbie, All-Printed, Foldable Organic Thin-Film Transistors on Glassine Paper, *Adv. Mater.* 2015, *27,* 7058-7064.

W. J. Hyun, E. B. Secor, M. C. Hersam, C. D. Frisbie, L. F. Francis, High-Resolution Patterning of Graphene by Screen Printing with a Silicon Stencil for Highly Flexible Printed Electronics, *Adv. Mater.* 2015, *27,* 109-115.

W. J. Hyun, S. Lim, B. Y. Ahn, J. A. Lewis, C. D. Frisbie, L. F. Francis, Screen Printing of Highly Loaded Silver Inks on Plastic Substrates Using Silicon Stencils, *ACS Appl. Mater. Interfaces* 2015, *7,* 12619-12624.

497183 Synthesis of Crumpled Graphene-Based Materials Using Aerosol Techniques and Their Application to CO_2 Photoreduction

Abstract:

Research Interests: Synthesis of novel materials and the applications to energy storage.

Synthesis of crumpled graphene-based materials using aerosol techniques and their application to CO_2photoreductionÂ

With the rapid development of the worldâ€™s economy, increasing amounts of fossil fuels are being consumed to meet the growing energy demand. The burning of fossil fuels is widely considered to be the main anthropogenic source of atmospheric carbon dioxide (CO_2) emissions, which have been causing the global climate change. Therefore strategies to urgently mitigate CO_2 have attracted global attention. CO_2 capture and conversion (CCC) can not only capture CO_2, but also convert it to useable products such as hydrocarbon fuels. A CCC technique, photocatalytic reduction is an attractive option that directly harnesses inexpensive and abundant solar energy.

Graphene nanosheets, with their excellent electronic conductivity and extremely large specific surface area, are a promising material for use as a CO_2photoreduction catalyst. However, the application of these flat-structured materials can be significantly limited by restacking of the two-dimensional (2D) sheets. Restacking of these sheets, which results from strong Ï€-Ï€ interactions, reduces the accessible surface area and thus affects the overall material efficacy. Converting the flat 2D sheet to a 3D crumpled structure by an aerosol technique is a feasible approach to overcome the above hurdles without compromising its electrical properties.

A thorough understanding of the physical properties (including mobility and charging characteristics) of crumpled graphene oxide (CGO) nanoparticles is needed for better application of these materials. Our study has demonstrated the enhanced CO_2 photoreduction efficiency of titanium dioxide (TiO_2) with modification of CGO nanoparticles. Moreover, the morphology (flat, wrinkled, or crumpled) of GO nanoparticles was also shown to affect the mobility and charging characteristics of GO-based materials. However, to further understand the effects of size and morphology of CGO nanoparticles on the CO_2 photoreduction efficiency, we are now combining both experimental results and theoretical modeling work.

Teaching Interests: Aerosol science and technology, reaction engineering, transports, et at.

505197 Kinetic of Biomass Fast Pyrolysis

Abstract:

Research Interests:

My research over the last nine years has focused on the energy regarding relation between surface and interface of fluids for the purpose of enhanced oil recovery and asphaltene behavior in petroleum reservoir under different condition. My Ph.D. thesis focus on Kinetics of Biomass Fast Pyrolysis, which is funded by National Science Foundation (NSF).I designed a new microreactor micro spherical biomass particle technique for the study of kinetic of biomass fast pyrolysis which is the best method for production of biofuel with less pollution.

In the beginning, my research was about the effect of temperature and pressure on kinetic of asphaltene deposition during of capturing CO2 as the global gas environment. One of my papers was published in American Chemical Society (ACS) publication which its detail is on my C. V. Then I started my thesis on investigating the effect of temperatures, pressures and composition on CO2 solubility and miscibility by measuring interfacial tension during CO2 caption. My first paper of thesis was published in chemical engineering data journal (American Chemical Society (ACS Publication)) on the first page of the advertising board. This paper was in the title of "Temperature and Composition Effect on CO2 Miscibility by Interfacial Tension Measurement ". I completed my second paper of my thesis in the title of " Investigating the Effects of Temperature, Pressure, and Paraffin Groups on the N2 Miscibility in Hydrocarbon Liquids using the Interfacial Tension Measurement Method" by the march of 2013 which is published in I&EC journal (American Chemical Society (ACS Publication)).

After finalizing my master thesis, I became a member of a group that is conducting research on modifying the parachor model, neural network, molecular dynamics simulation by interfacial tension technique to measure the solubility of the global gas environment (such as CO2) at different pressure and temperatures.

My main specialties are in the investigation of kinetic, thermodynamic and heat transferring on the solubility of different materials, also relation to surface properties (such as wettability alteration and interfacial tension).

Since I am so much interested in research and development, also the today's problem with population and global warming I was searching for a project to help the environment and ultimately beneficial to humankind. I was searching the internet and interesting enough I found the project in the title of “A multi-scale environmental and kinetic study on the pyrolysis of sustainable biomass feedstock” at Tennessee Tech University. During my Ph.D. research, a novel laboratory-scale fast pyrolysis technique is demonstrated using biomass microspheres and a unique single-particle micro-reactor technology. The biomass microspheres technology provides a controlled geometry feedstock while the novel reactor enables the control of residence time factors. With this new technique, the amount and rate of hydrocarbon gas production are measured by flame ionization detector. The effects of biomass particle size on the fast pyrolysis within three different types (crystal cellulose, switchgrass, and tall fescue) is investigated to determine the char yield and reaction dynamics. The Calculated kinetic parameters were used to predict the dynamic temperature of spherical biomass.

Summary of Top Achievements:

1- New technology (microreactor microsphere) for the study of kinetic of biomass fast pyrolysis to produce biofuel.

2- Relation between temperature and pressure on CO_2 –N_2 injection process in petroleum reservoir by Interfacial tension properties, and one of my papers which have been accepted in Journal of Chemical & Engineering Data (has selected on advertising board on ACS Publications).

3- Investigation of CO_2 injection in asphaltene deposition at reservoir condition by new visual apparatus.

4- Effect of native and non-native resin in asphaltene deposition at reservoir condition.

5- Develop one semi-empirical equation for measuring of interfacial tension of hydrocarbon components with CO_2.

6- Investigation of mass transfer and molecular behavior of components in surface of CO_2-N_2 and paraffin component

My future research plans include completing my current research on kinetic of biomass fast pyrolysis and also using my previous experiences on Asphaltene deposition, Surface and interface properties, Molecular modeling simulation and Enhanced oil recovery methods to write a proposal to be qualified to receive a grant.

RESEARCH INTERESTS

Kinetics of Biomass Fast Pyrolysis; Synthesis and Interfacial Tension Measurement of Green Solvents (Ionic Liquids); Interface and Surface properties; Wettability Alteration using Synthesis Nano Material; Material Characterization Micro-environmental systems

Asphaltene Deposition during Miscible Gas Injection; Experimental Determination of Optimum Gas Injection Conditions in EOR Process by VIT technique.

Impact Factor (Google Scholar)

Citation indices	All	Since 2010
Citations	204	203
h-index	6	6
i10-index	5	5

https://scholar.google.com/citations?user=5atMn1EAAAAJ&hl=en

Teaching Interests:

I believe in the real and complex engineering and science environment, the principle key for solving sophisticated problems and engineering obstacles is creativity and how to learn the problem-solving process. During my first mathematics and chemical engineering design courses in university, I made a great effort to improve my creativity skills along with learning the problem-solving process. That became a fundamental motivation for me during graduate school. Therefore, as a teacher, I will strive to teach my students with these two essential factors to motivate them and provide them with extensive understanding. I believe it will encourage students learning challenging topics, guide them to recognize and completely understand the fundamental concepts in chemical engineering and build confidence in their abilities. I would like to make my students capable of analyzing and evaluating on their own to become an independent learner.

Chemical engineering is one of a fascinating major among all engineering disciplines with distinct theoretical and experimental aspects. From the theoretical point of view, strong knowledge and deep understanding of applied mathematics are ineluctable specifically at senior and graduate levels. Applied mathematics brings about great insight for the student to trust their intuition and problem-solving skills to assess and investigate widespread problems. In my Thermodynamic course, I encouraged students to focus on conceptual aspects of each problem by solving selected practical problems in class and providing alternative approaches to express different perspectives at each problem. This strategy of teaching will give rise to more motivation, satisfaction, and self-confidence.

Additionally, my experience as a teaching assistant in Thermodynamic course enabled me to realize the importance of practical and experimental aspects of chemical engineering by suggesting different methodologies for involving students in practical perspectives of materials and problems. First, I will present the theories for each section concisely and will spend more time on technical and practical examples. As an example, in Fluid mechanics course, I provided lab demonstration and placed students in groups to work with experimental setups and software to build their practical skills. Second, I will take advantage of visual demonstrations and graphs. Finally, I will involve my students in practical discussions and by offering them the opportunity to experience challenging concepts. During my teaching experience, I recognized that a proper evaluation method was to assign students homework problems which required comprehensive work throughout the semester. Homework assignments and examinations helped students demonstrate their knowledge and problem-solving skills and analyze solutions properly. In most chemical engineering courses, results can be reached by more than one method. I require my students to verify their analysis by computer simulation, as well. Moreover, individual and group assignments or final projects can help students acquire deeper and conceptual understanding. Projects should be flexible enough to serve different student's interests and to evaluate each student's skill and expertise. I take advantage of various types of exams and projects, such as problem-solving and take home exams. Additionally, I welcome feedback from students which allow me to evaluate my performance during a semester by conducting teacher evaluation survey in my class to assure what my students expect of me to improve my teaching skills and methods.

TEACHING Interests:

Instructor

University of Applied Sciences and technology

- **Industrial security and safety**
- **Thermodynamic**
- **Fluid mechanics**

Assistant Instructor

Shiraz University, Shiraz, Iran

- Fluid and rock properties Lab:spring 2011- Fall 2014.

The crystal quality and structure of AM-6

Rumeysa Tekin, Juliusz Warzywoda, Albert Sacco Jr.*

Materials Characterization Center, Whitacre College of Engineering, Texas Tech University, Lubbock, TX 79409, USA

Abstract

Microporous vanadosilicate AM-6, isostructural with titanosilicate ETS-10, is a novel promising material in the traditional zeolite applications such as catalysis, photocatalysis, gas separation, sorption and the monatomic –O-V-O-V-O chains have shown potential for use in quantum wire application. High purity vanadosilicate AM-6 products with different crystal quality (disorder, or average length of the V-O-V chains in the AM-6 framework) and sizes in the range of ~50 nm to ~6 μm were hydrothermally synthesized at 448-503 K using tetramethylammonium (TMA^+) ions. The combined SEM, EDX, XRD, Raman, TGA, XPS, and nitrogen adsorption isotherm data showed that by treating these as-synthesized materials in gaseous ammonia at 673 K it is possible to completely remove TMA^+ ions from the AM-6 pores without adversely affecting product crystallinity and the micropore structure of AM-6, and with only minimal degradation of crystal quality of the resulting products. However, the success in a complete removal of TMA^+ ions without affecting other crystalochemical characteristics of the product depended on its initial crystal quality. TMA^+ ions could not be completely removed from the lowest crystal quality product (FWHM=20.3 cm^{-1}) without completely decomposing the V-O-V chains. The intermediate crystal quality product (FWHM=15.0 cm^{-1}) could be successfully ammonia-treated only at 673 K to remove TMA^+ ions, but higher temperatures resulted in a complete decomposition of the V-O-V chains. The highest crystal quality product (FWHM=12.7 cm^{-1}) retained the integrity of V-O-V chains as well as the long range order and the micropore structure when ammonia-treated up to 748 K.

Keywords: AM-6; Crystal quality; Gaseous ammonia treatment; Hydrothermal synthesis; Vanadosilicates

Research Interests:

- Synthesis of zeolites and related microporous (zeotype) and mesoporous materials
- Growth/preparation of nanoporous thin films.
- Photocatalysis using novel molecularly-selective titanosilicate and vanadosilicate materials
- Zeolite/zeotype nucleation and crystal growth

Molecularly selective photocatalysis is a potential method for using solar energy to destroy environmental pollutants and simultaneously recover useful products. However, commonly investigated photocatalysts are activated by ultraviolet (UV) irradiation because of their wide bandgaps, and UV light accounts only for a small fraction (~4-5%) of solar energy compared to visible light (~45%). In addition, common photocatalysts like TiO_2, tend to completely oxidize organic pollutants, and thus preclude the recovery of desirable intermediate products. A novel nanostructured microporous vanadosilicate material, AM-6, with its 3-D pore structure and its ability to be photocatalytically active in the visible range provides a unique opportunity to investigate rate improvements through improved visible light activity, and selectivity through structure and electronic properties of the material for a potentially new and novel photocatalyst.

Postdoctoral Project: Synthesis and characterization of vanadosilicate AM-6 for photocatalytic applications.
During my tenure as a postdoc at the Texas Tech University under Albert Sacco Jr. and Juliuzs Warzywoda, I have broadened my skill set in the areas of materials synthesis and characterization working with a variety of inorganic porous materials. The majority of my work has focused on the quality of AM-6 crystals as well as the development of new synthesis methods and improvement of the crystal quality for photocatalysis applications. Utilization of zeolite and zeo-type materials as thin films has attracted significant relevance as an alternative way in many device applications such as membranes and sensors. In parallel with this purpose, I have worked on the growth of AM-6 thin films for advanced applications. Overall, my postdoctoral work has resulted in the filing of an article published in Microporous and Mesoporous Materials Journal.
I have also collaborated on projects including the investigation of catalytically active zeolites, ZSM-5 and MCM-22 to improve the yield to the desired chemical products.

PhD Dissertation: Encapsulation of fragrance molecules in microcapsules and nanoporous zeolites.
Nanotechnological approach has recently been introduced into fragrance applications. One of these processes involves microencapsulation of fragrances in polymeric microcapsules, and the other is encapsulation of fragrances in nanoporous zeolites to tailor products with controlled release properties.
My Ph.D. work at the Yeditepe University under the supervision of Nurcan Bac, focused on the encapsulation of fragrance using both polymeric microcapsules and zeolites as fragrance carriers. I produced microcapsules, containing fragrance for textile applications. This study was an industry funded project conducted in collaboration with EPS Fragrances Co. I also worked as a Quality Control Engineer in Yeditepe University Cosmetic Research and Production Facility, which provided me an opportunity to develop and transfer into practice various applications of nanotechnology to cosmetics.
In the 5th semester of my Ph.D., I was accepted to the Research Fellowship Program at Texas Tech University and worked under Albert Sacco Jr. and Juliuzs Warzywoda. During my visit, I focused on synthesis and characterization of zeolite X crystals for adsorption and desorption applications and improved my instrumental skills by learning XRD, SEM, EDX, and TGA/DSC. In addition to adsorptive properties of zeolites, I studied on the ion-exchanged zeolites in collaboration with the Genetics and Bioengineering Department to understand the antimicrobial behavior of zeolites. My Ph.D. work resulted in publications (5 first-authored papers and 1 book chapter).

<u>*Publications*</u>

• **Tekin, R.**, Warzywoda, J., and Sacco, A. (2017). The effect of gaseous ammonia treatment of AM-6 crystallized in the presence of tetramethylammonium cations on their removal from crystal pores, and the crystal quality and structure of AM-6, 251, 94-104.
• **Tekin, R.**, Bac, N., and Erdogmus, H. (2017). Assessment of Zeolites as Antimicrobial Fragrance Carriers, Antimicrobial Research: Novel bioknowledge and educational programs, 379-384.
• **Tekin, R.** and Bac, N. (2016). Antimicrobial behavior of ion-exchanged zeolite X containing fragrance. Microporous and Mesoporous Materials, 234, 55-60.
• **Tekin, R.**, Bac, N., Warzywoda, J., and Sacco, A. (2015). Encapsulation of a fragrance molecule in zeolite X. Microporous and Mesoporous Materials, 215, 51-57.
• **Tekin, R.**, Bac, N., Warzywoda, J., and Sacco, A. (2015). Effect of reaction mixture composition and silica source on size distribution of zeolite X crystals. Journal of Crystal Growth, 411, 45-48.
• **Tekin, R.**, Bac, N., and Erdogmus, H. (2013). Microencapsulation of Fragrance and Natural Volatile Oils for Application in Cosmetics, and Household Cleaning Products. In Macromolecular Symposia (Vol. 333, No. 1, pp. 35-40).

Teaching Interests:

I have considerable teaching experience and previously served as a teaching assistant for Design of Chemical and Biological Processing Units and Experimental Chemical and Bioprocess Engineering courses in the Chemical Engineering Department at Yeditepe University, Istanbul. I delivered several lectures and led tutorial and review sessions. I was praised by the students for the clarity of my presentation and conveyance of learning objectives, and my empathy for student advancement. I believe, I have the credentials to make significant contributions to teaching in the Chemical Engineering or Material Science Department. My research and teaching expertise has prepared me to teach Design of Chemical Processing Units as well as Mass Transfer. I can also teach any course in the core chemical engineering curricula and technical electives such as Materials Science and Instrumental Analysis. Students will be given the opportunity to work in project teams to investigate a topic of their choosing. Over the course of the semester, they will teach their classmates, using the educational tools, and give a presentation on their chosen topic. In addition to enhancing their understanding through peer instruction, they will gain valuable practice giving presentations. My goal is to help students identify, and formulate important problems and to reach at intelligent solutions by teaching critical thinking skills. I will also motivate them to engage in lifelong learning.

I am currently a Research Associate in the Materials Characterization Center at College of Engineering, Texas Tech University and conduct comprehensive hands on training for the users of X-ray powder diffractometers (XRD); transmission electron microscope (TEM); field emission scanning electron microscope (FE-SEM); TGA/DSC; XPS, FTIR, Raman, UV-Vis, EDX and GC.MS spectrometers; confocal microscope; particle size analyzers; zeta potential analyzer; gas sorption analyzer for characterization of surface area, pore volume, and pore size distribution of microporous and mesoporous solids. I have mentored several undergraduate and graduate students in Texas Tech and Yeditepe University. I found each of these experiences to be very rewarding and believe they have helped prepare me for future mentorship and teaching roles. I have obtained the Global Scholarship for the 21st Century Award in recognition of the distinguished academic achievements during my tenure as a visiting international scholar at Texas Tech University.

505625 Theoretical and Computational Study of Soft Matter Systems: From Classical Challenges to Rational Design of New Materials

Abstract:

Research Interests:

Modern chemical engineering becomes increasingly molecular-based. The development of theoretical and computational tools that capture the essential physics, address the molecular feature and quantify the systems across multiple time and length scales is vitally important. Soft materials are widely used from commodity to biomedical materials. Many phenomena in soft matter systems, ranging from classical experiments to emerging interest, cannot be satisfactorily explained, which necessitates theories that can provide fundamental insights to these seemingly complex problems. The large design space due to the rapid development of new synthetic and processing methods also requires predictive tools to accelerate the design process of new materials. My research focuses on using statistical mechanics, which bridges the molecular information and macroscopic phase/interfacial properties, to study a variety of soft matter systems, including electrolyte solutions, liquid crystals, membranes, colloidal dispersions, polymers, polyelectrolytes, polymer-protein conjugates and polymer networks/gels.

I did my PhD work at Caltech with Prof. Zhen-Gang Wang, where we developed renormalized Gaussian fluctuation theory to study the effect of self energy of ions on the double layer structure and interfacial properties. Our theory incorporates different components of self energy in a unified framework and involves a minimum set of intrinsic parameters. We show the consequence of the self energy by revealing three essential effects: the image charge effect, inhomogeneous screening effect and specific ion effect. We are able to quantitatively explain a couple of long-standing puzzles in physical chemistry and soft matter physics, such as the salt concentration dependence of surface tension, like-charge attraction and charge inversion, Hofmeister series in the interfacial affinity and strong interfacial adsorption of hydrophobic ions.

My current postdoctoral research in MIT with Professors Bradley Olsen, Alfredo Alexander-Katz and Jeremiah Johnson is to develop theoretical and computational tools to quantify the topology, gelation and elasticity of polymer networks. I developed a kinetic graph theory which demonstrates a universal dependence of loop defects on the preparation condition and reveals the intrinsic relation between different cyclic topologies. I developed a kinetic Monte Carlo simulation which quantifies the loop effects on the gel-point suppression and the change of critical exponents. I also developed a real network theory which bridges topological defects to gel elasticity. These theories and simulations show good agreement with experimental measurements, providing, for the first time, a quantitative understanding of gel elasticity based on molecular details. Using this platform, we are now seeking to optimize the mechanical properties through controlling the molecular connectivity of the network.

In the future, I would like to carry theoretical and computational research to solve fundamental challenges in soft matter systems and facilitates the design of new materials that meet the emerging requests in environment, energy and biology. I will develop simple models that capture the most essential features behind the complexity of the problems at hand. With the fundamental insights, I will develop more detailed theoretical and computational tools as low-cost, high-performance platform to aid the rational design of novel materials. Meanwhile, I will seek for extensive collaboration with experimentalists and data scientists to accelerate the material design process via coupling the predictive physics-based models, high-throughput experimentations and machine-learning techniques. I will initially focus on the following topics:

1) *Complex interfacial phenomena of salt ions and polyelectrolytes driven by electrostatics*, with fundamental challenges including charge inversion and like-charge attraction, and applications such as layer-by-layer deposition techniques and membrane fouling in water purification.

2) *Phase behavior of polyelectrolyte hydrogels*, with fundamental challenges including swelling-collapse transition and nanophase patterns, and applications such as stimuli-responsive materials and micro/nano soft particles used as actuators, sensors and microreactors.

3) *Interfacial topology and properties of polymer networks*, with fundamental challenges including erosion, friction, fracture and adhesion of gels, and applications such as self-healing materials and cell-matrix interactions.

Teaching Interests:

Teaching is extremely important for the professional development of my future students, and also benefits my own scientific research. Based on my teaching experience in Caltech and MIT, I feel that the traditional design of the chemical engineering courses cannot fully satisfy the emerging requests in both the scientific research and industrial employment. I suggest two modifications. 1. Besides the traditional courses that focus on the reactor scale, knowledge at molecular level such as synthetic chemistry, statistical (or even quantum) mechanics and molecular biology are needed. 2. Interdisciplinary courses referring to material science, environment, energy and biology need to be involved.

I would like the students in my class to be actively engaged in the class and get passion in learning and doing science, obtain clear basic concept, think critically and keep challenge themselves, and gain the ability of solving real problems. To achieve this goal, I will organize the course materials and develop the teaching methods by addressing the following aspects: 1) encourage the students to think about one problem from different perspectives; 2) guide the students to find the underlying physics that explains different phenomena; 3) design problem sets and term projects from recent research progresses.

As a graduate student at Caltech, I had the opportunity to work as a teaching assistant in graduate level thermodynamics, statistical mechanics and polymer physics. The evaluation from the students shows that I was one of the best teaching assistant in recent years. Besides these courses, my education and research experiences also make me well quantified to teach kinetics, transport process, interfacial phenomena, mathematics in chemical engineering and computer modeling and simulation. Currently as a postdoc in MIT, I had the opportunity to give guest lectures in the course of synthetic polymers. I am also training for mentoring undergraduate and graduate students, which is extremely helpful for my teaching skills. In the future, I plan to develop an in-depth course for soft matter physics that can meet the rapid development of this field. This interdisciplinary course will cover both the fundamental concepts and broad applications, which will benefit students from different background.

505588 Vapor-Phase Deposition for Functional Metal-Organic Framework (MOF) and Polymer Thin Films

Abstract:

Research Interests:

Recent chemical attacks in Syria have caused many civilian casualties. The nerve agent sarin used in these incidents is a kind of highly toxic organophosphorus compounds that can paralyze human central nerve systems and rapidly lead to severe injuries and deaths. The threat of these chemical warfare agents (CWAs) motivates the development of new materials for protecting soldiers, first-responders and the public. Current state-of-the-art filters are based on impregnated activated carbon and do not have catalytic activity for destructing CWAs. In contrast, metal-organic frameworks (MOFs) have been reported with exceptional catalytic performance and are therefore promising candidates for decontamination applications. Integrating thin films of MOFs with textiles could enable protective garments and masks.

It is nontrivial to form conformal and continuous MOF thin films on polymer fiber surfaces, since conventional synthesis methods often lead to sparse crystals on fibers. My Ph.D. research was mainly focused on developing new approaches to enhance MOF thin film growth on fibers,[1–4] investigating MOF-functionalized fibers for defense-related applications,[1,2,4] and searching for rapid reaction routes to synthesize MOF and MOF-based composites at room temperature for future manufacturing.[3] We have demonstrated that atomic layer deposited (ALD) metal oxide layers enhance the heterogeneous nucleation of MOFs on fiber surfaces, enable the formation of uniform, continuous and high quality MOF thin films, and significantly improves the mass loading and adhesion of MOFs to the fiber substrates.[1,2,4] Our MOF-functionalized nonwoven fiber mats and electrospun nanofibers exhibit large adsorption capacity for hazardous gases such as NH_3 and H_2S,[1–3] and high catalytic activity for degrading CWA simulants and nerve agents in minutes.[4] These results all indicate that our MOF-coated textiles are very promising for the next generation of protective masks and suits. To push forward the industrial applications and commercialization of MOF-based materials, we developed a new synthesis method using hydroxy double salts (HDS) as key intermediates for rapid formation of MOFs and MOF thin films in just 1 minute. The ultra-high space-time-yield (up to 2.9×10^4 $kg\cdot m^{-3}\cdot d^{-1}$ for Cu-BTC powder) shows the great potential of our process for scaling up MOF manufacturing.[3]

In addition to the research achievements highlighted above, I have also investigated CO_2 diffusion in mm-scale Cu-BTC crystals (collaboration with Vanderbilt University),[5] studied the growth mechanism and kinetics of MOF thin films using *in situ* ATR-FTIR with a flow cell (collaboration with NIST), and applied ALD for synthesis and modification of nanostructures. Currently, I am investigated initiated chemical vapor deposited (iCVD) polymer thin films for desalination-related applications at MIT.[6] In the future, I will actively explore functional thin film materials for applications in energy, environment and artificial intelligence.

Teaching Interests:

My teaching interest falls in reaction kinetics and chemical reactor design courses for both undergraduate and graduate students, which will be well supported by my past experience as a teaching assistant for undergraduate chemical reaction engineering classes. I am also interested in designing a course to introduce chemical processes involved in advanced nano-manufacturing, based on my research and work experience in semiconductor industry. My past experience in mentoring undergraduate and graduate students will also be beneficial for teaching.

(1) Zhao, J.; Losego, M. D.; Lemaire, P. C.; Williams, P. S.; Gong, B.; Atanasov, S. E.; Blevins, T. M.; Oldham, C. J.; Walls, H. J.; Shepherd, S. D.; Browe, M. A.; Peterson, G. W.; Parsons, G. N. Highly Adsorptive, MOF-Functionalized Nonwoven Fiber Mats for Hazardous Gas Capture Enabled by Atomic Layer Deposition. *Adv. Mater. Interfaces* **2014**, *1* (4), 1400040.

(2) Zhao, J.; Gong, B.; Nunn, W. T.; Lemaire, P. C.; Stevens, E. C.; Sidi, F. I.; Williams, P. S.; Oldham, C. J.; Walls, H. J.; Shepherd, S. D.; Browe, M. A.; Peterson, G. W.; Losego, M. D.; Parsons, G. N. Conformal and highly adsorptive metal–organic framework thin films via layer-by-layer growth on ALD-coated fiber mats. *J. Mater. Chem. A* **2015**, *3* (4), 1458–1464.

(3) Zhao, J.; Nunn, W. T.; Lemaire, P. C.; Lin, Y.; Dickey, M. D.; Oldham, C. J.; Walls, H. J.; Peterson, G. W.; Losego, M. D.; Parsons, G. N. Facile Conversion of Hydroxy Double Salts to Metal-Organic Frameworks Using Metal Oxide Particles and Atomic Layer Deposition Thin-Film Templates. *J. Am. Chem. Soc.* **2015**, *137* (43), 13756–13759.

(4) Zhao, J.; Lee, D. T.; Yaga, R. W.; Hall, M. G.; Barton, H. F.; Woodward, I. R.; Oldham, C. J.; Walls, H. J.; Peterson, G. W.; Parsons, G. N. Ultra-Fast Degradation of Chemical Warfare Agents Using MOF–Nanofiber Kebabs. *Angew. Chem. Int. Ed.* **2016**, *55* (42), 13224–13228.

(5) Tovar, T. M.[#]; Zhao, J.[#]; Nunn, W. T.; Barton, H. F.; Peterson, G. W.; Parsons, G. N.; LeVan, M. D. Diffusion of CO_2 in Large Crystals of Cu-BTC MOF. *J. Am. Chem. Soc.* **2016**, *138* (36), 11449–11452.
([#]Equal Contribution)

(6) Zhao, J.; Wang, M.; Gleason, K. K. Stabilizing the Wettability of Initiated Chemical Vapor Deposited (iCVD) Polydivinylbenzene Thin Films by Thermal Annealing. *Adv. Mater. Interfaces* **2017**, DOI: 10.1002/admi.201700270.

505922 Computational Design of Surfaces and Nanostructures for Energy Applications

Abstract:

Research Interests:

-Efficient materials screening and design, particularly for catalysts:

Adsorption energies are good predictors of catalytic performance, but calculating adsorption energies on a large number of potential catalytic surfaces remains a challenge. I have developed methods for efficiently and accurately predicting adsorption energies of important intermediates on transition metal surfaces. These methods are quite general, applying to any transition metal alloy surface, and linking any adsorbate to electronic structure or to any other adsorbate. They are also efficient and physically transparent, allowing high-throughput screening or rational design. Future work will involve applying these methods to design novel catalysts (particularly those that break traditional scaling relations), and further improving their accuracy, efficiency and applicability.

More generally, machine-learning methods will be applied to develop improved molecular dynamics methods, based on an ensemble of simple force-fields. This will allow would allow accurate, efficient studies of long-time scales and rare events, based on relatively small datasets.

-Excitations in metals and metal interfaces, for catalysis and solar cells:

Since metals have no bandgap, reactions on metal surface may induce electronic excitations, which could affect surface chemistry. However, most studies of processes on metal surfaces assume that the system remains in the ground state. Using an efficient nonadiabatic dynamics code employing density functional theory and Ehrenfest dynamics, I have tested this assumption for nitrogen dissociation on Ru. These calculations show that even in thermal catalysis, surface processes can excite electrons in metal surfaces, and these excitations can affect surface chemistry. Future work will focus on light-induced excitations (e.g., plasmons) in metal nanoparticles and hybrid systems involving metals and supports or molecules. This will bring improved understanding and design principles for plasmonic solar cells and plasmonic catalysis.

Teaching Interests:

-Mathematical methods

-Materials chemistry and structure

-Thermodynamics

-Fluid dynamics and transport

-Quantum and computational chemistry and materials science

505447 A Marriage of Convenience: Uniting Polymer Chemistry and Polymer Physics to Craft Advanced, Functional Materials

Abstract:

A Marriage of Convenience: Uniting Polymer Chemistry and Polymer Physics to Craft Advanced, Functional Materials

Research Interests:

Research Aims

i. Development and application of facile polymer synthesis methods

ii. Fundamental studies of self-assembly of functional nanoparticles and polymers

iii. Production and characterization of functional materials for energy and device applications

Motivation

Polymer nanocomposites (PNCs) have been successfully implemented in a variety of disparate areas such as aerospace, batteries, and biosensing. However, while there is a wide diversity of available polymers and nano-sized filler materials, PNCs are under-utilized, largely due to cost considerations. Developing PNCs into the *commercial* materials of the future necessitates the design, synthesis, and characterization of new, scalable combinations of polymers and nanosized fillers, *which requires a fundamental understanding of the underlying physics that dictates matrix-filler interactions and the ability to modify the chemistry that can tune these interactions.* Development and application of a versatile and facile polymer synthesis platform will allow for arbitrary control of polymer material properties and chemical functionality, which will allow us to tune matrix-filler interactions to control PNC properties and create novel, *useful* materials in a rational way.

Future Directions

As faculty, I want to unite polymer chemistry and polymer physics to create materials that solve real world problems. Specifically, I want to marry the polymer chemistry tools I have developed in my postdoctoral work with the soft matter physics training I received during my Ph.D. and apply this combination to rationally design, synthesize, characterize, and develop advanced materials with commercially relevant properties. My research goals are threefold: (1) Develop and utilize facile polymerization methods that allow for the synthesis of functional polymers with a wide range of properties, (2) understand the self-assembly properties of polymers and/or polymer grafted nanoparticles, and (3) combine (1) and (2) to tune any aspect of the PNCs, allowing for complete control over properties (*e.g.*, dielectric constant, electrical conductivity) for a variety of energy applications.

In the short term, I envision utilizing the polymer synthesis technique I developed in my postdoctoral work, which allows us to easily create polyethers with a diverse range of functionalities and architectures. I will apply these polymers to develop an understanding of how different brush structures (*e.g.,* block, graft, hyperbranched) affect the dispersion of nanoparticles in a polymer matrix, which affects the overall nanocomposite properties. I would also like to use polyether-grafted nanoparticles in polymer electrolytes for battery applications. In this case, polyethers are a great choice as their dielectric properties can be tuned through the pendant group and grafting high dielectric polyether copolymers to the nanoparticles should further enhance the ionic conductivity.

Longer term, I am interested in the self-assembly of polymers of different architectures for lithography. Currently, linear block-co-polymers which create lamellar morphologies are the industry standard, but they are reaching their limit in terms of feature size. By using the polymerization platform I developed in my postdoctoral work, we could access different morphologies (*e.g.*, gyroidal), architectures (*e.g.*, bottle brush), and combinations of monomer units (*i.e.*, for high χ), which would allow us to achieve industrially coveted domain spacing (*i.e.* sub 10 nm). I am also interested in stimuli-responsive and hybrid materials for energy applications. Creating materials for advanced applications necessitates individual components to act synergistically to achieve the target property set. Combining conducting polymers with metal and/or semiconducting nanoparticles in a rational way would allow us to make improved materials for energy harvesting applications (*e.g.*, solar cells, thermoelectric generators).

Research Background:

Research Experience

Throughout my academic career I have worked toward understanding and controlling the interactions between polymers and nanoparticles. As an undergraduate and master's student in the Physics and Materials Science and Engineering departments at **Drexel University**, I utilized functional polymer crystals to create Janus magnetic nanoparticles that could be co-assembled with other nanoparticle types to create hybrid nanomaterials. As a Ph.D. student in the Chemical and Biomolecular Engineering department at the **University of Pennsylvania**, I began working with polymer-grafted gold nanorods, which have unique optical properties due to their anisotropic shape. I designed methods to assemble gold nanorods permanently or reversibly through chemical (*e.g.,* using peptides) or temperature (*e.g.*, using DNA) responsive triggers, which allowed me to control optical properties of polymer films. I wanted to apply these ideas to other anisotropic particles and other matrices, which led me to work with both experimentalists and theorists in other fields. In this way, I have had many fruitful collaborations across campus (materials science, physics, chemistry) and across the world (**Kyoto University** (Japan), **CNRS** (France)). The collaborations led me to want to explore more diverse material systems; to do this, I felt I needed to increase my synthetic toolkit. As such, I joined a polymer chemistry laboratory in the Chemical Engineering department at the **University of Texas at Austin**. Here, I have learned various polymerization techniques such as anionic polymerization and radical polymerization. I have also developed a new facile polymerization method that has allowed us to synthesize unique, functional polymers at controlled molecular weights. I remain interested in polymer nanocomposites and I am currently using the polymerization platform I developed to create new nanocomposite materials.

Past Research Projects

Postdoctoral Project: "*A New, Facile Approach for Epoxide Polymerizations Utilizing Organoaluminum Initiators*" Supervised by Prof. Nathaniel A. Lynd in the Chemical Engineering Department at the *University of Texas at Austin*, 2016-Present

International Research Fellow Project: "*Probing the Interaction of Gold Nanorods and Organic Semiconductors*" Supervised by Drs. Patrice Rannou, Brigitte Pépin-Donat, and Didier Gasparutto in INAC at *CNRS in Grenoble, France*, Summer-Fall 2013

NSF-EAPSI Fellow Project: "*ATRP Initiated from Nanorod Surfaces*" Supervised by Prof. Kohji Ohno in the Institute for Chemical Research at *Kyoto University*, Japan, Summer 2012

Ph.D. Thesis: "*Surface Chemistry Mediated Assembly of Polymer-Grafted Nanorods in Solution and Polymer Matrices*" Supervised by Prof. Russell J. Composto in the Chemical and Biomolecular Engineering Department at the *University of Pennsylvania*, 2009-2015

M.S. Thesis: "*Polymer Templated Janus Magnetic Nanoparticles*" Supervised by Prof. Christopher Y. Li in the Materials Science and Engineering Department at *Drexel University*, 2007-2009

Funded Proposals

NSF EAPSI Fellowship

Labex-ARCANE Fellowship

NBIC International Research Fellowship

Selected Publications

Robert C. Ferrier, Jr.†, Jennifer Imbrogno†, Christina G. Rodriguez†, Malgorzata Chwatko, Paul W. Meyer, Nathaniel A. Lynd, Four-fold Increase in Epoxide Polymerization Rate with Change of Alkyl-substitution on mono-μ-oxo-dialuminum Initiators, *Polymer Chemistry,* **2017** (*in press*). † equal contribution

Christina G. Rodriguez†, **Robert C. Ferrier, Jr.**†, Alysha Helenic, Nathaniel A. Lynd, Ring Opening Polymerization of Epoxides: Facile Pathway to Functional Polyethers via a Versatile Organoaluminum Initiator, *Macromolecules*, **2017**, *50* (8), pp 3121 – 3130. † equal contribution

Robert C. Ferrier, Jr. †, Jason Koski†, Robert A. Riggleman, Russell J. Composto, Engineering the Assembly of Gold Nanorods in Polymer Matrices, *Macromolecules*, **2016**, *49* (3), pp 1002-1015 † equal contribution

Elaine Lee†, Yu Xia†, **Robert C. Ferrier, Jr.** †, Hye-Na Kim, Mohamed A. Gharbi, Kathleen J. Stebe, Randall D. Kamien, Russell J. Composto, Shu Yang. Fine Golden Rings: A Tunable Surface Plasmon Resonance from Liquid Crystal Assembled Nanorods. *Advanced Materials*, **2016**, *28*, pp 2731-2736 †equal contribution

Robert C. Ferrier, Jr., Yun Huang, Kohji Ohno, Russell J. Composto. Dispersion of Polymer Grafted, Mesoscopic, Iron-oxide Rods in Polymer Matrices. *Soft Matter*, **2016**, *12*, pp 2550-2556

Robert C. Ferrier, Jr., Guillaume Gines, Didier Gasparutto, Brigitte Pépin-Donat, Patrice Rannou, Russell J. Composto. Tuning Optical Properties of Functionalized Gold Nanorods through Controlled Interactions with Organic Semiconductors. *Journal of Physical Chemistry C*, **2015**, *119* (31), pp 17899-17909

Robert C. Ferrier, Jr., Hyun-Su Lee, Michael J. A. Hore, Matthew Caporizzo, David M. Eckmann, Russell J. Composto. Gold Nanorod Linking to Control Plasmonic Properties in Solution and Polymer Nanocomposites. *Langmuir*, **2014**, *30* (7), pp 1906–1914

Teaching Interests:

Teaching and Mentoring Experience

For two years at the University of Pennsylvania, I was a teaching assistant (TA) and guest lecturer for the undergraduate class Soft Matter in the Materials Science department. As a TA, I prepared exam and homework questions, graded exams and homework, and helped determine the final curve and final grades. I participated in several teaching workshops that helped expose me to a variety of different teaching styles. I also had one of my lectures evaluated to help me understand my strengths and weaknesses as a lecturer. Finally, mentorship is extremely important to me and I take it very seriously. I have been extremely lucky to have been able to nurture the scientific minds of a number of REU students, undergraduate senior design students, and graduate students during my Ph.D. and postdoctoral careers. Continuing to be a good mentor is something I look forward to when I become faculty.

Teaching Interests

I have an academic background that spans four different disciplines (chemistry, physics, materials science, and chemical engineering) so I would be suited to teach a wide variety of courses. Courses of particular interest would be: polymer physics, soft matter, polymer chemistry, and thermodynamics.

505498 Beyond Graphene: Two-Dimensional Transition Metal Carbides and Nitrides (MXenes)

Abstract:

The research community in the field of materials science has been steadily expanding the family of few-atom-thick crystals beyond graphene, discovering new 2D materials or producing known materials in a 2D state and demonstrating their unique properties. Transition metal carbides (TMCs), which are prepared by integrating carbon atoms into the interstitial sites of their parent metals, combine the properties of ceramics and metals and are emerging as new candidates for applications in sensors, electronics, catalysis, and energy storage and conversion (ESC) devices due to controllable band gap, superior reactivity, and strong mechanical and chemical stability. Recently, a breakthrough was carried out by our group (the Gogotsi Group), by extending the 2D concept to TMC family. Benefited from the top-down nanostructuring, the 2D TMCs (so-called MXenes) show significantly improved electrochemical performance, especially in high-volume energy densities (**Science**, 341 (6153), 1502-1505; **Nature**, 516 (7529), 78-81). Besides, they can be either metallic materials for semiconductors depends on their compositions and surface functionality. However, the lack of large-area and high-quality crystals hindered investigations of their intrinsic physics and properties for sensing applications. Besides, it is still a great challenge to synthesis 2D TMCs or MXenes in large quantity and high stability due to the critical synthesis process, which hinders their large scale applications.

Research Interests:

To solve these issues, my research will focus on the synthesis, characterizations, and nanofabrication of 2D TMCs and their heterostructures for sensors and ESC device applications. In particular, initial interests will lie in chemical vapor deposition of large-area, high-quality 2D transition metal carbides (TMCs) for field emission transistor-based sensing applications. Besides, I would also propose the research on template-synthesis of large quantity 2D TMCs for energy storage. Efforts will also be devoted to the fabrication of 2D TMC-containing heterostructures, towards the performance improvement in sensor and energy storage applications.

Teaching Interests:

In the curriculum of the Chemical Engineering, there are a number of courses which I would be particularly interested to teach, including Intro to Physical Chemistry, Reaction Engineering, Principle of Transfer Processes, and others. There are 2 courses that I am eager to develop (either in collaboration with other faculty members or by myself). The first course would be "Emerging Nanomaterials for Electrochemical Energy Storage"; it would familiarize students with different nanomaterial synthesis techniques, energy storage technologies, and discuss the physicochemical properties of several typical nanomaterials and their applications in batteries and supercapacitors. The second course I plan to develop would be "Electron Microscopy Characterizations in Catalysis". This class would introduce students to different electron microscopy techniques that are commonly used to characterize various catalysts and will describe important structure-activity relationships that exist in catalysis.

505324 Biomolecular Sensing Using Fluorescent Single Wall Carbon Nanotubes

Abstract:

Research Interests:

My research has been focusing on using nanomaterials for controlled drug delivery and for optical biomolecular sensing. For my Ph. D work, I applied surface modified mesoporous silica nanoparticles for anti-cancer drug delivery, and studied the particle interior magnetic heating effect using temperature-dependent luminescence of nanocrystals. The knowledge on the physical and chemical properties of silica nanoparticles facilitated their biomedical use for in vitro and in vivo studies. During my postdoctoral training, I investigated on using near IR fluorescent carbon nanotubes to detect protein content as well as to examine their binding kinetics. I have developed a miniature microarray platform based on optical signals for label-free multiplexed protein molecule sensing. In addition, I have also examined cancer biomarker expressions in extracellular vehicle exosomes, in order to distinguish exosomes of different cell origins to perform cancer early diagnosis.

Postdoctoral Project: "*Multiplexed protein detection using single wall carbon nanotube microarray*"

PI: Michael S. Strano, Department of Chemical Engineering, MIT

PhD Dissertation: "Physical Properties of Mesoporous Silica Nanoparticles for Stimuli-Responsive *Drug Delivery*"

PI: Jeffrey I. Zink, Department of Chemistry and Biochemistry, UCLA

Research Experience:

- Label-free Single Wall Carbon Nanotube Microarray for Protein Profiling
  - Surface modification of single wall carbon nanotubes with antibody and ssDNA
  - Protein binding kinetic analysis and multiplexed detection
  - Carbon nanotube near IR fluorescence spectroscopy and microscopic imaging
  - Near IR microarray detector construction
  - *In Situ*Optical Measurement of Nanoparticle Interior Temperature
    - Synthesis of thermally sensitive upconversion lanthanide nanocrystals and dual core nanoparticle
    - *In situ* interior temperature study of magnetically heated silica nanoparticles
    - Mechanism investment of nanomachine operation in response to continuous near-IR irradiation
    - Simultaneous spectral detection of nanoparticle interior temperatures and the corresponding cargo release

    Mesoporous Silica Nanoparticles for Controlled Drug Delivery
      - Synthesis of polymer coated silica particles with nanovalves for enhanced biomedical properties
      - Construction of Gd incorporated nanomachines for a multi-functional delivery vehicle
      - Synthesis of functionalized mesoporous silica nanoparticle for *in vivo* and *in vitro* drug delivery studies

Teaching Experience:

- Teaching certificate, MIT Kaufman Teaching Certificate Program
- Chemistry content expert, MIT BLOSSOMS (outreach program)
- Teaching assistance for general chemistry discussion sessions, UCLA
- Teaching assistance for chemical laboratory sessions, UCLA
- Mentor for two graduate students and four undergraduate students

Teaching Interests:

I believe the goal of a college education is not only to teach the knowledge and skill sets that are needed for work but also to prepare students with the curiosity and methods to continue the learning throughout their lives, for career and personal development. To meet this goal, I will try to improve the learning efficiency by integrating more critical thinking, active learning and connection-building components into our teaching process. Especially for upper level classes, I will introduce advanced projects, in which students are asked to pick research papers that they are interested in, orally present the work to the class and articulate their own opinions about it in the format of a short study report.

Future Direction:

I aim at developing novel molecular recognition sites using single wall carbon nanotube, which are readily applicable to a variety of biomedical fields including cancer diagnosis, treatment and biopharmaceutical development.

Selected Publications:

- **J. Dong** and J. I. Zink. "Simultaneous Spectroscopic Measurements of the Interior Temperature and Induced Cargo Release from Pore-restricted Mesoporous Silica Nanoparticles." *Nanoscale* **8**, 10558-10563 (2016).
- **J. Dong** and J. I. Zink. "Light or Heat? The Origin of Cargo Release from Nanoimpeller Particles Containing Upconversion Nanocrystals under IR Irradiation." *Small 11*, 4165-4172 (2015).
- **J. Dong** and J. I. Zink. "Taking the Temperature of the Interiors of Magnetically Heated Nanoparticles." *ACS Nano 8*, 5199-5207 (2014). **Highlighted** in *Nature Physics*.
- **J. Dong**, M. Xue and J. I. Zink. "Functioning of Nanovalves on Polymer Coated Mesoporous Silica Nanoparticles." *Nanoscale 5*, 10300-10306 (2013).

505350 Interaction of Nanostructures Leads to Macroscopic Behaviors: Towards Designing Multiple-Component Nanostructures with Functionalities for Energy-Related Applications

Abstract:

Postdoctoral Projects:

Lawrence Berkeley National Laboratory, Berkeley, CA, USA March 2015-Current

- Designing Metal Organic Framework/Polymer Films and Graphene Derivative Systems as Scalable Barrier Layers for Enhanced Protection on Photovoltaics
- Hybrid Nanocomposites/Superlattices for Solar Energy Conversion
- Designing Iron Oxide-Metal Organic Framework Superstructures by Ligand-mediated Self-assembly
- Pt-Iron Oxide Hetero-nanostructures with Enhanced Electron Transfer for Photocatalysis

PhD Dissertation: "Semiconductor nanocrystals for photocatalytic hydrogen generation" 2015

Advisor: Prof. Todd D. Krauss (University of Rochester)

Background and Research Interests:

Designing nanostructures with desired functionalities for energy-related applications has always been a central aim of research in nanotechnology. While facing the challenges of finding a catalytic system with good activity, selectivity, and durability, especially under visible light, researchers have put tremendous effort into designing suitable materials or systems. My research aims to solve this problem via developing multiple-component nanostructure systems and understanding key chemical/physical properties as well as inter-component coupling interactions. Through this approach, it is possible to enable rational optimization and practical application of these materials.

Homogeneous systems for light-driven reduction of protons to H_2 typically suffer from short lifetimes because of decomposition of the light-absorbing molecule. A complementary approach would be to incorporate semiconductor nanocrystals, into the system. We recently found a robust and highly active system for photocatalytic hydrogen generation in water that uses CdSe nanocrystals capped with dihydrolipoic acid (DHLA) as the solar energy harvester and a soluble Ni^{2+}-DHLA catalyst for proton reduction. In order to understand the mechanism of photo-induced electron transfer in the system, I compared photocatalytic H_2 production for core CdSe nanocrystals to other more highly engineered nanoparticle systems such as core/shell CdSe/CdS nanocrystals and CdSe/CdS dot-in-rod nanoparticles, achieving a result suggesting that surface charge density plays an important role in charge transfer and ultimately hydrogen production activity. The ultrafast transient absorption spectroscopy studies of electron transfer from the CdSe nanocrystals to the Ni^{2+}-DHLA catalysts reveal extremely fast electron transfer, providing a fundamental explanation for the previous experimental results of photocatalytic H_2 production.

The above is a summary of work from my PhD on semiconductor nanocrystals-based homogenous photocatalysis. During my post-doctoral study, conducted at the Molecular Foundry at Lawrence Berkeley National Laboratory (advisor Dr. Jeffrey J. Urban), I have enlarged my research scope to cover an extensive variety of materials including inorganic nanoparticles, graphene derivatives, MOFs, and polymers. Using all of these types of materials, I fabricated multiple-component nanostructures/superstructures through self-assembly and atomic layers deposition.

In the future, my efforts will be focused on crossing the traditional boundaries of materials science, chemical engineering, chemistry and nanotechnology. With broad experience in synthesis, fundamental understanding of charge transport and photocatalysis, my future research will focus on three different categories: (1) Synthesis and self-assembly of multifunctional nanostructures. (2) Multiple-component nanostructures for solar energy conversion. (3) New materials and devices for energy storage.

Teaching Interests: With a background in physics, chemistry, and nanomaterials, I would be interested in teaching materials related courses, thermodynamics and nanomaterials engineering. Further, I would like to develop some courses on the interface between science and engineering.

Selected Publications:

1. Zhiji Han*, **Fen Qiu***, Richard Eisenberg, Patrick L. Holland, Todd D. Krauss. "Robust Photogeneration of H_2 in Water Using Semiconductor Nanocrystals and a Nickel Catalyst" *Science* **2012, 338, 6112, 1321-1324 (* equally contributed)**

2. **Fen Qiu**, Zhiji Han, Jeffrey J. Peterson, Michael Y. Odoi, Kelly L. Sower, Todd D. Krauss. "Photocatalytic Hydrogen Generation by CdSe/CdS Nanoparticles" *Nano Lett.* **2016**, 9, 5347-5352

3. Cunming Liu, **Fen Qiu**, Jeffrey Peterson, Todd D. Krauss. "Aqueous Photogeneration of H_2 with CdSe Nanocrystals and Nickel Catalysts: Electron Transfer Dynamics" *J. Phys. Chem. B* **2015**, 119, 7349-7357

4. Eun Seon Cho, **Fen Qiu**, Jeffrey J. Urban "Polymer Size and Solvent Effects on Mg Nanocrystal Growth in the Composite" *Small* **2017**, 13, 1602572

5. **Fen Qiu**, Jun Feng, Dequn Wu, Xianzheng Zhang, Renxi Zhuo. "Nanosized Micelles Self-Assembled from amphiphilic dextran-graft-methoxypolyethylene glycol/poly(ε -caprolactone) copolymer" *Eur. Polym. J.* **2009**, 45, 4, 1024-1031

6. Gregory Pilgrim, Joanne Leadbetter, **Fen Qiu**, Anni Siitonen, Steven Pilgrim, Todd D. Krauss. "Electron Conductive and Proton Permeable Vertically Aligned Carbon Nanotube Membranes" *Nano Lett.* **2014**, 14, 4, 1728-1733

7. Youn Jue Bae, Eun Seon Cho, **Fen Qiu**, Daniel T. Sun, Teresa E. Williams, Jeffrey J. Urban, Wendy L. Queen "Transparent Metal-Organic Framework/Polymer Films as Water Vapor Barriers" *ACS Appl. Mater. Interfaces* **2016**, 8, 10098-10103

505379 Light and Heat-Managing Nanomaterial for Energy Efficiency and Human Health

Abstract:

Research Interests:

Light is one of the most useful high-grade energy sources. Heat, on the other hand, is extremely ubiquitous but less straightforward to manipulate. Controlling light and heat smartly and efficiently would bring huge benefits for fields such as sustainable energy, power electronics and human health. Indoor temperature control contributes to 12% of the total energy consumption in the USA, and energy demand for cooling is very likely to increase due to global warming. To tackle this problem, I focused onto the most universal way of heat transfer from the human body, that is, thermal radiation. When indoors, sweat evaporation and wind convection are minimal, and radiation contributes 50% of the total heat flux. However, traditional textiles are all opaque in mid-infrared and thus are radiatively resistive. By using the principle of polymer spectroscopy and size dependence of electromagnetic wave scattering, I found a material that can be the "coolest" material for textile – nanoporous polyethylene (nanoPE). NanoPE is transparent to human body radiation but appears opaque to human eyes. Its interconnected nanopores also provide great vapor permeability for humid air. Experimentally, nanoPE can be 2°C cooler than cotton, which can save 20% of energy. Based on this new class of material, the radiative thermal property of textile can finally be explored and benefit the next generation smart cloth and wearable technology. For example, following up the first finding of nanoPE, I developed the textile which can perform both cooling and heating, without any additional energy input. My colleague and I also expand the cooling textile to outdoors, again using the light management concept. My future research interest would like to include other heat transfer routes, e.g., moisture control, convection control, onto the radiative-smart textile. Also, I am interested in applying my knowledge in nanomaterials, heat pumps, and microfluidics onto human body thermal management, which can greatly expand the applicability from ordinary to extreme activities.

Besides textiles worn by humans, the building solar heat gain is also significantly affecting the cooling efficiency. Smart windows, which can electronically control their transparency, can reduce the amount of sunlight and therefore reducing the cooling load. However, transparent conducting film with ultra-low resistance is the bottleneck for large-scale and fast transmittance switching. Traditional transparent conducting oxides have reached the material limit and become difficult to break the trade-off between transmittance and resistivity. I used metal nanowires and graphene film to realize transparent conducting films with <1 Ω/sq sheet resistance, >95% transmittance, and >1000 cycle life. The combination of these research for light and heat management bring about exciting opportunities for energy efficiency and human health applications.

Teaching Interests:

I was the TA for a graduate level class of 100 students. The class title was nanomaterials and technology. I also mentored 15 students for their final project in the heat transfer class.

I enjoys teaching materials science, thermodynamics, materials kinetics, transport phenomena, nanoscience, semiconductor physics, electrochemistry, and other major materials science and engineering courses.

Selected publications:

1. **P.-C. Hsu**, A. Y. Song, P. B. Catrysse, C. Liu, Y. Peng, J. Xie, S. Fan, Y. Cui. "Radiative human body cooling by nanoporous polyethylene textile" **Science**, 2016, 353 (6303), 1019-1023
2. **P.-C. Hsu**, X. Liu, C. Liu, X. Xie, H.R. Lee, A.J. Welch, T. Zhao, Y. Cui. "Personal thermal management by metallic nanowire-coated textile" *Nano Letters*, 2015, 15 (1), 365-371
3. **P.-C. Hsu**, D. Kong, S. Wang, H. Wang, A.J. Welch, H. Wu, Y. Cui. "Electrolessly deposited electrospun metal nanowire transparent electrodes" *JACS,* 2014, 136 (30), 10593-10596
4. **P.-C. Hsu***, S. Wang*, H. Wu, V.K. Narasimhan, H.Y. Lee, Y. Cui. "Performance enhancement of metal nanowire transparent conducting electrodes by mesoscale metal wires" *Nature Communications*, 2013, 4:2522
5. **P.-C. Hsu**, H. Wu, T.J. Carney, M.T. McDowell, Y. Yang, E.C. Garnett, M. Li, L. Hu, Y. Cui. "Passivation coating on electrospun copper nanofibers for stable transparent electrodes" *ACS Nano*, 2012, 6 (6), 5150-5156
6. B. Deng*, **P.-C. Hsu***, G. Chen, B.N. C, L. Liao, Z. Ayitimuda, J. Wu, Y. Guo, L. Lin, Y. Zhou, M. Aisijiang, Q. Xie, Y. Cui, Z. Liu, H. Peng. "Roll-to-roll encapsulation of metal nanowires in graphene and plastic substrate for high-performance flexible transparent electrodes" *Nano Letters*, 2015, 15 (6), 4206-4213

505139 Multiscale Design of Heterogeneous Nanomaterials for Energy Applications: Solution Synthesis, Structures, and Properties

Abstract:

Research Interests:

Research experience. Developing superior materials for energy production, conversion, and storage is crucial for meeting the global energy demand. Bottom-up design of heterogeneous nanomaterials offers abundant opportunities for tuning the electronic, optical, thermal and catalytic properties of these novel materials to address the various challenges we face in energy applications. My research centers around solution phase synthesis of complex nanomaterials, bottom-up fabrication of functional nanocomposites, and understanding the structure-property relationship from atomic scale to macro scale.

My graduate research at Purdue University focused on the solution phase synthesized metal chalcogenide nanomaterials for thermoelectric applications. I developed or optimized the synthesis of copper zinc tin sulfide (CZTS) nanocrystals, bismuth telluride, lead telluride, and silver telluride nanowires and/or heterostructures. Upon purification, ligand removal and consolidation of the solution-synthesized nanomaterials, we developed functional thermoelectric composites with nano-sized grains and designable nanoscale heterostructures, which were demonstrated to exhibit superior thermoelectric properties.

My postdoc work at University of Pennsylvania emphasized the charge transport behavior in semiconducting nanocrystal (NC) solids, in which the charge transport properties are determined not only by the nature of individual NCs but also by the interface and organization of the NCs. I showed that doping semiconductor NC solids with metal nanoparticles, in analogy to atomic doping, modifies the macroscopic carrier density and the charge transport dynamics of the NC solids. The doped NC solids represent a novel family of artificial materials which can be designed in nanometer scales and are of great interest for many emerging nanocrystal-based technologies including electronics, optoelectronics, and thermoelectrics.

Recently, I moved to Lawrence Berkeley National Laboratory, where I extended my research area into nanocrystal growth kinetics and high-throughput synthesis of core-shell CdS-CdS_xSe_{1-x} nanocrystals based on kinetic designed reactions. By using a mixture of S and Se precursors and by controlling the relative decomposition rates of the two precursors, we could theoretically create desired heterostructures from the one-pot synthesis, which greatly simplifies the production of heterogeneous nanocrystals compared with the existing layer-by-layer method. To validate the hypothesis, we perform the high-throughput robotic synthesis to explore the large parameter space of the reaction.

Future directions. I am fascinated by the ability to design novel materials from nanoscale through scalable, wet-chemical, bottom-up approach. As many physical events, such as electrical and thermal transport and light absorption and emission, have characteristic lengths in the nanometer range, the ability to design materials on nanoscale will enable more effective manipulation of electrons, heat, and light. This requires the capability to create nanostructures with atomic precision and to integrate these nanostructures into working devices. I envision this as a multi-scale problem which needs to be tackled from the following three scales/aspects: (1) on the atomic scale, understand the kinetics of the synthesis, especially for complex materials or structures, to enable the atomic-precision control of the nanomaterials; (2) on the mesoscale, understand the interfacial properties which dominate the assembly of the nanomaterials and therefore affect the device integration; (3) on the device scale, understand the collective properties of the assembled nanomaterials and their surroundings. Finally, energy applications will be built upon the multi-scale design of nanomaterials, including but not limited by waste heat recovery, solid state lighting and display, etc.

Teaching Interests:

I have developed my interest in teaching the fundamental courses in chemical engineering, including thermodynamics, kinetics, reaction engineering, and mass and energy balances since my service as a teaching assistant at Purdue University. I enjoy leading the students to deepen their understanding on the concepts in these courses and to develop their problem-solving skills utilizing their new knowledge. As a future instructor, my goal is not only to deliver the knowledge, but also to connect the knowledge to practical problems and to inspire the students to think critically using the knowledge.

In addition to the core courses, I am also interested in developing and teaching elective courses to the senior undergraduate students and the graduate students based on my own research. During my postdoc training, I have obtained valuable experience in designing and teaching such courses as a guest lecturer both at University of Pennsylvania and University of Virginia. The course will focus on the synthesis, properties, and applications of nanomaterials, and my goal is to introduce both the fundamentals and the cutting-edge research progress of nanomaterials to the interested students.

Rational Materials Design for Energy and Heterogeneous Catalysis Applications: Noble Metal Single Atom Catalysts and 1D Nano-array Support Materials.

Son Hoang

Department of Materials Science and Engineering & Institute of Materials Science, University of Connecticut, Storrs, CT 06269-3136, USA
E-mail: son.hoang@uconn.edu

Research Interests

Many industrial chemical production, petroleum refining, environmental remediation, and energy conversion processes rely on noble metal catalysts dispersed on high-area porous supports. For examples, in 2014, 98 tonnes of platinum were used for vehicle emission control devices and 20 tonnes for chemical production and petroleum refining, constituting 45% and 9% of global platinum use, respectively. Noble metals are also effective co-catalysts for boosting the efficiency of photocatalysis, in which the semiconductor supports act as both light absorbers and catalysts. However, noble metals are expensive and of limited supply, thus it is critical to improve their material utilization efficiency. Rational material design for both noble metal catalysts and supports has been proved effective for promoting activity, selectivity, and stability while reducing material usage. Downsizing catalyst particle/cluster size, ultimately to single atom dispersion, increase the surface metal atoms, on which most catalysis occurs. Additionally, single atom catalysts (SACs) often possess unique properties that cannot be achieved by the nanoscale counterpart since the catalytic behavior of metals can be affected by size reduction via several channels including increased unsaturated coordination, quantum confinement effects, metal–support interactions, and cluster configurations. However, SACs are often unstable in the reaction conditions, hindering their practical applications. Anchoring and localizing single metal atoms onto support surfaces are essential to designing and developing efficient and durable supported metal SACs. One-dimension nano-array structured support materials such as nanowire-, nanorod- and nanotube-arrays have demonstrated as an emerging class of catalysts and supports. The combination of 1D confinement effects and collective properties of long-range ordered structure often leads to interesting chemical, electrical, optical, and thermal properties. Recent advances in synthesis and fabrication methods allow engineering of 1D nano-array materials with well-defined exposed crystal facets and defects, thus enabling tunable metal-support interaction, which in turn can be exploited to stabilize supported single atom catalysts.

As faculty, I plan to apply my interdisciplinary expertise in nanomaterials, photoelectrochemistry, surface science, heterogeneous catalysis, and micro/nanofabrication to establish material design principles for efficient and stable noble metal SACs supported on 1D nano-array materials. My initial focus will be directed on catalytic and photocatalytic properties of the materials for chemical production, environmental remediation and solar energy conversion applications, including propane/ethane dehydrogenation, ammonia synthesis, selective hydrogenation, photocatalytic water splitting, thermal/photoelectrochemical CO_2 reduction, low-temperature oxidation of volatile organic compounds (VOCs).

Proposal Experience. Assisting PIs in drafting several proposals for funding opportunities, including three most recently below:

1. DOE-BES: Sustainable Ammonia Synthesis (FOA #: DE-FOA-0001569). Program manager: Raul Miranda.

2. NSF-Directorate for Engineering: PFI: AIR-TT Program (solicitation #: NSF 16-583). Program director: Barbara H. Kenny.

3. DOE-VTO: Fiscal Year 2017 Vehicle Technologies Program Wide Funding Opportunity (FOA #: DE-FOA-0001629). Contract Specialist: Raelynn Honkus.

Research Experience.

My research career path has been a blend of heterogeneous catalysis, nanomaterials and nanotechnology, (photo)electrochemistry, and high-vacuum instrument design. In my dissertation work, I set about to tackle one of the most challenging problems in the field of photocatalysis, which is finding efficient photocatalyst materials that satisfy stringent conditions for photoelectrochemical water splitting. Our highly collaborating work on this topic resulted in several publications in peer-reviewed journals such as Nano Letters, Journal of American Chemical Society, ACS Nano, some of which are among the top 1% highly cited by Web of Science. For examples, my works on synthesis and characterization of visible light active TiO_2 nanowire arrays have set a benchmark for the performance of N-doped TiO_2 materials and helped to elucidate the role of incorporated N to the visible light activity, which has been a subject of debate for decades. Furthermore, I discovered a new method to reduce the band gap of TiO_2 using the interaction of N and Ti^{3+}. In one of my postdoc research work, we applied excitonic energy transfer schemes to enhance light harvesting in ultrathin crystalline Si solar cells. We developed a hybrid solar cell platform, which utilizes a proximity quantum dot layer as energy transfer donor to sensitize the Si active layer. This study offers an approach for the rational design and integration of energy transfer and optical coupling structures for the improvement of light harvesting in optoelectronic devices. Currently, I am leading an effort in the development a scalable method to integrate single atom Pt catalysts/TiO_2 nano-arrays for diesel oxidation catalysts, which involves the collaboration between four universities, three DOE national labs, and one industry manufacturer.

Teaching Interests

In addition to my passion for doing research, the desire to teach/mentor is one of the primary reasons I want to pursue a career in academia. I really enjoyed the experience as a teaching assistant for several chemical engineering laboratory courses. I also have extensive experience in mentoring undergrad, grad, and postdoc researchers from diverse backgrounds. I take pleasure in encouraging students' critical thinking, facilitating a team-working environment, and improving students' organizing and goal-orientating skills, as well as creating lesson plans so that students can apply their knowledge to solve practical problems.

Due to my diverse education and research background, I am confident teaching courses in different topics, including most general subjects within the chemical engineering curriculum for both undergraduate and graduate students. I would be most effective in teaching kinetics and reactor design, electrochemical engineering, materials science, transport phenomena, and fundamental laboratory classes.

Selected Publications.

1. Hoang, S; Gao, P., *"Nanowire Array Structures for Photocatalytic Energy Conversion and Utilization: A Review of Design Concepts, Assembly and Integration, and Function Enabling,"* **Advanced Energy Materials,** http://dx.doi.org/10.1002/aenm.201600683

2. **Hoang, S.**; Ashraf, A.; Eisaman, M. D.; Nykypanchuk, D.; Nam, C.-Y., "*Light Harvesting and Energy Transfer in Ultrathin Inorganic Solar Cells,*" **Nanoscale** 2016. http://dx.doi.org/10.1039/C5NR07932B

3. **Hoang, S.**; Guo, S. W.; Hahn, N. T.; Bard, A. J.; Mullins, C. B., "*Visible Light Driven Photoelectrochemical Water Oxidation on Nitrogen-Modified TiO2 Nanowires,*" **Nano Letters**, 2012, *12* (1), 26-32. http://dx.doi.org/10.1021/nl2028188

4. **Hoang, S.**; Berglund, S. P.; Hahn, N. T.; Bard, A. J.; Mullins, C. B., "*Enhancing Visible Light Photo-oxidation of Water with TiO_2 Nanowire Arrays via Cotreatment with H_2 and NH_3: Synergistic Effects between Ti^{3+} and N*", **Journal of the American Chemical Society**, 2012, *134* (8), 3659-3662. http://dx.doi.org/10.1021/ja211369s

505527 Smart Magnetic Nanomaterials for Sustainable Applications in Biomedicine and Catalysis

Abstract:

Research Interests:

Magnetic nanomaterials are gaining steadfast attention from scientists, as quintessential components in a vast array of fields from heterogeneous catalysis to medicine. In particular, biomedical applications such as drug delivery, microfluidics, and tissue engineering that incorporate magnetic nanoparticles (MNPs) have become prominent, in recent years. Hybrid composite materials that combine organic and inorganic materials, together with magnetic components, offer enormous possibilities in developing new and enhanced biomedical practices and devices.

Magnetic scaffolds consisting of composites of hydrogels incorporated with magnetic nanoparticles, are currently gaining prominence in tissue engineering. This is due to the enhanced cell-compatibility properties introduced by the magnetic component such as increased cell adhesion, penetration and viability, as well as unique chemical, physical and mechanical properties. Great strides in this area have been made in bone regeneration, as magnetic composite scaffolds seem to significantly increase osteoblast integration and proliferation compared to the non-composite scaffolds. Another significant factor is that such scaffolds can either be actuated or stimulated using an external magnetic field, which can be greatly beneficial for targeted/triggered tissue engineering or drug delivery. However, due to the limitations in bio-stability and cost-effectiveness, such magnetic-polymer actuators have not yet been transformed into real-world applications.

Building on the above premise, a novel technique for fabrication of magnetic hydrogel actuators have been developed. These materials comprise of an organic polymer and Fe_3O_4 MNPs, and form a biocompatible platform, targeting multiple biomedical applications. The magnetic hydrogel has customizable physical and mechanical properties achieved *via* variation of polymer and MNP concentrations. Due its biocompatibility, biodegradability, facile synthesis, cost-effectiveness, scalability and self-healing properties, this material has the potential to be used in a vast range of biomedical applications, including implants, support materials, microfluidics, self-healing scaffolds and adaptive artificial muscles.

These magnetic polymeric composites have led to the development of a platform technology, and have generated multiple spin-off projects, including the fabrication of magnetic fibers, foams and 3D printed materials, all with actuation ability, coupled with biocompatibility. They are shown to be excellent vehicles for controlled drug release, due to their ability to respond to external magnetic fields. This technology is patent pending, and will continue to create multiple research opportunities as a result of their versatile nature and the ability to be synthesized in a facile, energy-efficient manner.

Another field that benefits significantly from the unique properties of MNPs is heterogeneous catalysis. Iron/iron(III) oxide core-shell nanoparticles are highly valuable catalysts, with industrial appeal. Sulfonic acid catalyzed core-shell nanoparticles were used as a highly selective catalyst to hydrolyze cellulose into glucose. A statistical technique was used to systematically optimize the reaction conditions of the above. The statistically-guided optimization led to high selectivity toward glucose, while production of 5-HMF and other monosaccharides were decreased. Reduction of 5-HMF is particularly significant, as it is toxic to yeast, which are used industrially, to convert cellulose degradation products into ethanol. The catalyst was found to be highly stable and reusable, due to its magnetic property, thus, leading to a potential sustainable, technology for conversion of plant matter into ethanol.

The above is a selected summary of work from my post-doctoral research, conducted at University College London (advisor Prof Marc-Olivier Coppens) and my PhD at Kansas State University (advisor Prof Stefan H. Bossmann). My future research plans include development of novel nanomaterials *via* sustainable routes for biomedical applications, and heterogeneous catalysis.

Teaching Interests:

My teaching experience includes conducting lectures, at graduate and undergraduate levels, and lab classes, in areas of physical organic chemistry, synthetic organic chemistry, organometallic chemistry and general chemistry. I am interested in further expanding my scope of teaching into thermodynamics, process engineering and nanomaterials engineering. My teaching philosophy is based on promoting the following principles: 1) generating a true enthusiasm for science, 2) effective communication between student and teacher, 3) bridging the gap between lecture and lab classes and 4) real world applications. I also strongly believe that undergraduate and graduate teaching must go hand in hand with public engagement and outreach. I have been a volunteer for multiple outreach programs that aim to promote interest in STEM disciplines at the middle and school levels. I have also been an invited speaker at multiple science festivals and events in London, aimed at educating the general public on current research.

505494 Solution Processable Multicomponent Nanomaterial for Next Generation Transparent Electronic/Optoelectronic Devices

Abstract:

Research Interests: Material Science/Nanoscale Science & Engineering

Multicomponent nanomaterials combine several chemical compositions, crystal structures, and/or morphologies into a single hybrid nanocomposite structure. The "hybrid" nature of such systems extends beyond the physical description of the material components to their functionality. In such architectures, it is possible to design a single structure for multiple functionalities (optical, plasmonic, magnetic), as well as to design for new amplified functionality, where one component modifies the behavior of the other. Such systems bring many potential opportunities to the field of nanostructured materials. Their incorporation into functional devices can present challenges. In particular, interface engineering (structure and quality) in these multicomponent nanomaterials often holds the key to device performance. My vision for this research direction is to develop a novel combinatorial approach to solution-processable inorganic multicomponent nanomaterials as a new route to the complex physical properties required for efficient transparent electronic/optoelectronic devices. In particular, I plan to experimentally observe correlations between material synthesis (doping crystal structure, morphology) and structural (defects and interface) properties/device performance. The proposed research ideas will not only result in the development of multicomponent/nanocomposite materials that can address varying problems from electrode materials to opto-electronic devices, but will also provide fundamental understanding towards structure, morphology, dopants and the role of interfaces in surface functionality.

My PhD thesis and post-doctoral work focused on the development of novel synthetic routes and macroscale assembly for semiconductor nanocrystals (NC). A versatile colloidal approach was developed to synthesize both simple binary semiconductors NCs and more complex compound semiconductor NCs [Cu_2ZnSnS_4, $CuInGaS_2$, CdSe/CdS/ZnO, Sn/F-In_2O_3, etc.], with tight control of shape, size and crystal phase. Specifically, I developed a synthesis in which the shape (spherical, rod, cube, octahedral etc.), doping, and crystal structure (stable and metastable phases) of the NCs are precisely controlled by reaction conditions – the culmination of this work was the synthetic achievement of the hexagonal wurtzite phase in alloyed copper chalcogenide (CIGS and CZTS), which is a metastable structure for these materials. Annealing of these metastable NC precursor films not only enabled growth of micron size grains, but did so at much lower temperatures (325-400° C) than traditional processing. This work uncovered a simple process to make large grain high quality thin films of compound semiconductors from nanocrystals using solution processing. This synthesis path not only reduces the cost and temperature requirement from an industrial prospective, but also contributes new understanding to the semiconductor thin film community.

In the current project, I am trying to elucidate, at the atomic scale, the nature of the core-shell interface that is buried within the nanostructure of core-shell nanocrystal. These observations will provide unparalleled insight into structural and interface properties of core/shell nanocrystals. The findings will also be highly useful for the application of these materials to the solid-state lighting industry. Furthermore, I am interested in developing a new type of optical nanomaterial, namely, novel plasmonic metal oxides. I plan to explore their potential to dramatically enhance the light-emitting properties of traditional semiconducting nanomaterials used in high-efficiency solid-state lighting and as single-photon sources for quantum communication/computing devices. The plasmonic metal oxides could also serve as a new material for transparent conducting electrodes. My poster will highlight the key aspects of my research along with an outline of future projects.

Teaching Interests: Materials, Electronics

505595 Ubiquitous Energy Harvesting through Chemically Engineered 2D Materials

Abstract:

The *Internet of Everything (IoE)* promises to build a "nervous system" for our society and fuse the digital and physical worlds, by bringing pervasive and ambient intelligence to daily objects. Especially, IoE provides the hardware foundation for smart control and green manufacturing for the next-generation "intelligent chemical plant". In order to make this vision a reality, a flexible, efficient and always-on energy harvester that can be seamlessly integrated with arbitrarily-shaped objects is a must, but still missing. Among the various possible energy sources, Wi-Fi wireless charging seems to be an ideal choice thanks to the ever-increasing Wi-Fi coverage. However, the high frequencies used for Wi-Fi communications (2.4 GHz and 5.9 GHz) have remain elusive to all the RF harvesters (*i.e.* rectennas) made of flexible semiconductors. In this regard, two-dimensional (2D) materials open up intriguing possibilities owing to their unique electronic properties, excellent mechanical flexibility and robustness. Moreover, chemically engineered 2D materials provides a new degree-of-freedom to further tailor the properties of 2D materials for energy applications. During my PhD, I have successfully demonstrated a unique MoS_2 metallic-semiconducting phase-junction through a self-aligned chemical approach. Such a chemical phase-junction of MoS_2 enables RF energy harvesting up to 10 GHz: one order of magnitude improvement in speed compared to the state-of-the-art flexible rectifiers. It is the first flexible rectifier operating up to the X-band and it fully covers the Wi-Fi channels, as well as the global satellite positioning (GPS), the cellular communication 4G band, the Bluetooth, and even the next-generation 5G radio system. By integrating with an antenna, our MoS_2-enabled rectenna (rectifying antenna) successfully demonstrates direct energy harvesting of electromagnetic radiation in the Wi-Fi band to power a commercial LED. Importantly, the operation of our device requires zero external bias and therefore eliminates the need of a battery. This work opens up exciting opportunities for a new generation of large-area distributed electronics based on 2D materials and paves the way towards using the existing Wi-Fi infrastructure as an energy hotspot for wireless charging. Moreover, our MoS_2rectifier also realizes successful frequency conversion as a flexible mixer (beyond 10 GHz) which is a critical component for any wireless communication systems.

In addition to smart chemical manufacturing, I'm also dedicated to the chemical engineering of 2D material itself. We developed a highly nonintrusive doping technology for graphene based on chlorination, which can introduce strong hole doping and maintain the 2D nature and long-range periodicity of the electronic states of graphene intact. The all-surface nature of 2D materials provides us with a unique tool to tailor the electronic properties through chemical functionalization. We have used a suite of X-ray techniques to investigate the effect of surface functionalizing dopants on the electronic and chemical states of 2D materials, with special focuses on their surface binding energy, bonding configuration and work function shift.

Research Interests:

1. System-level Integration of Self-powered Sensor Network for Intelligent Chemical Plants and Healthcare

The future Internet will not only connect people but also almost every object and machine in the world. It is predicted that the number of sensors will grow from billions in 2012 to trillions by 2025. It becomes extremely challenging to provide power and connectivity to sensor nodes as their number grows exponentially. We believe that 2D materials can play a significant role in this perspective. Our previous work on chemically engineered MoS_2 RF energy harvester and mixer provide key enabling technologies for this application. By integrating them with sensors, we aim to demonstrate a system-level largely distributed ultralight sensor network for ubiquitous sensing in next generation of intelligent chemical plants and healthcare. In particular, such a system will allow (1) ubiquitous and wireless energy harvesting and (2) flexible radios that can be wirelessly read out by a cell phone or a Wi-Fi station for real-time data-collection and control.

2. “Smart Skin” for Soft-Robotics and Bio-medical Applications

The ultralight, flexible and mechanically robust 2D materials represent an ideal platform to build a “Smart Skin” that can be integrated with robots as an interactive surface with perceptions of environmental stimuli. By putting together the building blocks of energy harvester, sensing, communication and logic circuits that I have previously worked upon separately, it is exciting to build a self-powered and transparent “smart skin” for applications such as soft-robots, human-machine interfaces and assistive medical devices, etc.

3. Adaptive "Synthetic Cells and Neuro Synapses"

The biological cell represents one of the most amazing micro-machines with unprecedented performance-to-volume ratio (i.e. performance density) and adaptability. By using the unique properties of 2D materials, especially their outstanding mechanical flexibility, and electrical and chemical properties, we will develop a foldable and adaptable "synthetic cell" with autonomous behaviors. This unique family of van-der-Waals materials can enable foldings into various geometric surfaces of specific funtionalities such as antennas, sensors, transducers and energy harvesters, and eventually into a super-structure that mimic the behaviors of a biological cell. Moreover, The chemically engineered 2D materials provide a full spectrum of electronic materials from metal, semiconducters and insulators. The stacking of different 2D van-der-Waals layers provides novel voltage-tunable junctions to mimic the different states of the synaptic connections and achieve the dynamic reconfigurability between different post-synaptic effects. In our biological neuro cells, both excitatory and inhibitory neurotransmitters can be co-released from axon terminals which are fundametally important to learning and memory processes. In our 2D materials-based synthetic cells and synapses, the highly tunable charge transfer mechnism and junction formation will be exploited to resemble the various biological synaptic activities. The 2D materials-based artificial neural network would be an ideal architecture which is highly parallel, extremely compact and power-efficient.

Teaching Interests:

I have worked as a teaching assistant for MIT graduate level course “Applied Quantum and Statistical Physics” (Prof. Peter Hagelstein, 2015), and have been highly rated by the students (6.9/7.0, the highest evaluation score in the last five years). In 2016, I was also invited as a guest lecturer for the same class for one week. During my PhD, I have also mentored 1 PhD student, 1 master student, and 2 undergraduate students.

I would like to teach various physics-related and device-related chemical engineering classes, such as Thermodynamics and Statistical Mechanics, Transport Phenomena, Micro/Nano Processing Technology and NanoElectronics. With a bachelor degree in condensed matter physics and a PhD degree in electrical engineering, these classes are also core courses in my departments. I would be very excited to bring my physics insights and device perspectives into these traditional chemical engineering classes. I would also be happy to teach new classes, such as Fundamentals of Energy Harvesting, Physics of Low-dimensional Materials, Nanoscale Fabrication, etc.

Grant Proposal Writings:

[1] "Ubiquitous 2-Dimensional Electronics (U2DE) Initiative", pre-proposal for NSF program under the supervision of Professor Tomas Palacios, 2013

[2] "Foldable and Adaptive Two-Dimensional Electronics", full proposal for MURI Program (AFOSR) under the supervision of Professor Tomas Palacios. 3+2 year financial support awarded beginning in September 2015.

[3] "Graphene-Based High Efficiency CMOS-integrated Long Wavelength Infrared Detection", full proposal for DAPAR program under the supervision of Professor Tomas Palacios, 2015

[4] "Providing Power and Connectivity for Next Generation of Ubiquitous Sensing Systems", white paper for MIT.nano Center of Excellence, under the supervision of Professor Tomas Palacios, 2017 (pending)

Selected Publications:

[1] **Zhang, X.**; Schiros, T.; Nordlund, D.; Shin, Y. C.; Kong, J.; Dresselhaus, M.; Palacios, T. X-Ray Spectroscopic Investigation of Chlorinated Graphene: Surface Structure and Electronic Effects. **Adv. Funct. Mater.** 2015, 25, 4163–4169.

[2] **Zhang, X.**; Hsu, A.; Wang, H.; Song, Y.; Kong, J.; Dresselhaus, M. S.; Palacios, T. Impact of Chlorine Functionalization on High-Mobility Chemical Vapor Deposition Grown Graphene. **ACS Nano** 2013, 7, 7262–7270.

[3] **Zhang, X.**; Grajal, J.; Ujwal, R.; Wang, X.; Chern, W.; Zhou, L.; Lin, Y.; Shen, P.; Ji, X.; Ling, X.; Zubair, A.; Zhang, Y.; Wang, H.; Dubey, M.; Kong, J.; Dresselhaus, M. S.; Palacios, T. Two-dimensional MoS_2-enabled Rectenna for Ubiquitous Energy Harvesting in the Wi-Fi Band. *(In submission)*

[4] **Zhang, X.**; Grajal, J. ; Wang, X.; Ujwal, R.; Xiang, J.; Shen, P.; Zhang, Y.; Kong, J.; Dresselhaus, M. S.; Palacios, T. Large-scale Chemical Vapor Deposition Grown MoS_2 Schottky diode for Frequency mixing and Wireless Charging. *(In submission)*

[5] Wang, X.; **Zhang, X.**; Sun, L.; Lee, D.; Lee, S; Yang, S.; Dinca, M.; Palacios, T.; Gleason, K.; Ultrahigh Electrical Conductivity of oCVD PEDOT Thin Films and the Wafer Scale Fabrication of the 13.6MHz Rectifiers based on the PEDOT-Si Diode. *(In submission)*

505279 Understanding and Controlling Interfaces of Nanomaterials Via Electrochemistry

Abstract:

Research Interests:

My research is at the interface of chemistry, materials engineering, and nanotechnology. So far, I have focused on building new synthetic tools for applications in energy conversion and generation platforms (photovoltaic/white light source) along with investigating fundamental light-matter (plasmon/exciton) interactions and catalytic processes at the nanoscale.

During my doctoral studies, I focused on the electrochemical synthesis of multi-segmented nanorods. Prior to my work, electrochemical synthesis was limited to one-dimensional modifications. I co-invented a technique, termed coaxial lithography (COAL), which allows for the synthesis of coaxial nanowires in a parallel fashion with sub-10 nanometer control in both axial and radial dimensions (US Patent Publication Number: 20160013340). My work has dramatically expanded current synthetic capabilities with respect to materials generality and the ability to tailor two-dimensional growth in the formation of core-shell structures. I believe that COAL will be extremely useful for prototyping device architectures that demand components and features not readily made by any existing technique, and as such should become a valuable research tool in the nanophotonics, energy harvesting, and nanotechnology fields for studying fundamental light-matter interactions.

In my postdoctoral research, I have integrated electrodeposition methodology with Si nanowires to evolve novel functional structures for selective catalysis and on-demand drug delivery applications. The work takes advantage of the scalability and low cost of electrochemical methods while working in the nanoscale regime with high fidelity and materials compatibility at room temperature. Thus, relatively sophisticated nanowire structures may be produced simply. The compatibility with organic molecules hold promise for the integration of this approach in the well-established silicon-based semiconductor industry as a potential route for fabrication of an array of single nanowire devices in parallel. I expect that the versatility and materials generality of this innovative approach will find diverse applications in chemistry, physics, and medicine. In particular, I am interested in exploring immediate applications of this work in the fields of energy conversion and storage, sensing, and bioelectronics. Additionally, I am currently working on the development of nanostructured crystalline catalysts for sustainable solar to chemical energy conversion by utilizing simple electrochemical techniques. Another exciting project that I am working on lies at the bio-nano interface for which I am modifying the tips of nanowire probes for the on-demand release of drugs and molecules by external stimuli.

In my independent career, my research goal is to contribute significantly to our fundamental understanding of how nanomaterials can be designed to form new pathways for crucial chemical reactions, and is potentially transformative for our worldwide energy infrastructures. I will focus on the design, characterization, and applications of nanoscale materials for solar energy conversion, energy storage, and biomedical applications. I will engineer nanostructured surfaces along multiple directions, utilizing the combination of radial and longitudinal degrees of compositional freedom provided by the electrochemical approaches that I developed so far. One of the key problems that I will tackle is controlling the efficiency and product selectivity of chemical processes by using light as a driving force. Understanding the difference between hot electron transfer versus local heating effects in a plasmon-mediated photocatalytic process is fundamentally important. The development of nanostructures showing such properties is vital as they can be used as foundational components in next-generation photocatalytic and photovoltaic systems.

Selected publications (6 out of 27)

1- **T. Ozel**, B. A. Zhang, R. Gao, R. W. Day, C. M. Lieber, and D. G. Nocera. Electrochemical deposition of conformal and functional layers on high aspect ratio silicon micro/nanowires, ***Nano Letters,*** DOI: 10.1021/acs.nanolett.7b01950 (2017).

2- **T. Ozel**, G. R. Bourret, C. A. Mirkin. Coaxial lithography, ***Nature Nanotechnology*** 10, 319 (2015).

3- **T. Ozel**, M. J. Ashley, G. R. Bourret, M. B. Ross, G. C. Schatz, C. A. Mirkin. Solution-dispersible metal nanorings with deliberately controllable compositions and architectural parameters for tunable plasmonic response, ***Nano Letters*** 15 (8), 5273 (2015).

4- **T. Ozel**, G. R. Bourret, A. L. Schmucker, K. A. Brown, C. A. Mirkin, Hybrid semiconductor core-shell nanowires with tunable plasmonic nanoantennas, ***Advanced Materials*** 25, 4515 (2013).

5- **T. Ozel**, P. L. Hernandez-Martinez, E. Mutlugun, O. Akin, S. Nizamoglu, I. O. Ozel, Q. Zhang, Q. Xiong, H. V. Demir. Observation of selective plasmon-exciton coupling in nonradiative energy transfer: Selective plexcitons, ***Nano Letters*** 13 (7), 3065 (2013).

6- **T. Ozel**, S. Nizamoglu, M. A. Sefunc, O. Samarskaya, I. O. Ozel, E. Mutlugun, V. Lesnyak, N. Gaponik, A. Eychmuller, S. V. Gaponenko, H. V. Demir. Anisotropic emission from multilayered plasmon resonator nanocomposites of isotropic semiconductor quantum dots, ***ACS Nano*** 5 (2), 1328 (2011).

Teaching Interests:

Based on my previous experience, I am interested in teaching both graduate and undergraduate level courses in both lecture and laboratory setting that fall under the following broad topics and areas in Chemical Engineering, Materials Science, and Mechanical Engineering:

- Materials Engineering

- Physical Chemistry

- Solar Energy Conversion

- Electron Microscopy

- Electrochemistry

- Quantum Physics of Materials

- Semiconductor Synthesis, Nanofabrication, and Device Processing

- Principles of Optoelectronic Devices (Photovoltaics, Sensors, and Light Emitters)

- Electricity and Magnetism

- Mechanics of Materials

- Fundamentals and Applications of 0D, 1D, and 2D Materials
- Nanoscience and Nanotechnology
- Nanophotonics (Plasmonics and Quantum Confined Materials)

504633 Directed Self-Assembly of Blue Phases Single Crystal By Chemically Patterned Surfaces

Abstract:

Research Interests: My general research interests are the nanostructure of the soft matters (polymer and liquid crystals) and its applications, self-assembly behaviors, particle trapping, optical properties, and multifunctional nanocomposite materials. These include e-beam lithography to produce chemical patterned surface, assembly of liquid crystals at surface with designed alignment, 3D structures and optical property, characterization of nanostructures with various microscopes and spectroscopes, and explore their potential applications in biological sensors, photonic materials, liquid-crystal displays and energy related devices.

Teaching Interests: Through my teaching experiences, I have learned how to deliver a complicated idea in a simple and concise way. In addition, I have refined my way of communicating with students, simulating their interest in the courses, and helping to formulate their own ideas. I could readily offer courses in the following ares: Polymer Physics, Material Science, Polymer Thin Film, Physical Properties of Liquid Crystal, Nano Structure and Fabrication, Advanced Instrumental Analysis.

505317 Multifunctional Soft-Nano Interfaces for Energy, Environment, and Healthcare

Abstract:

Research Interests:

My research interests are in micro/nano fabrication of functional materials, colloids and interfaces of soft nanostructures, self and directed assembly, nano/micro-electronics, microfluidics, 3D printing, photovoltaics, polymer thin-films, carbon nanomaterials, carbon composites, soft elastomers, liquid metals, flexible electronics, stretchable transistors, carbon MEMS/NEMS in security, health, energy and environmental applications.

Teaching Interests:

In addition to research, I would also like to teach Chemical Engineering, Materials Engineering and Physics based on my interdisciplinary academic background and experiences.

Multifunctional Soft-nano Interfaces for Energy, Environment, and Healthcare

Kunal Mondal[1,2], Michael D Dickey[1], Ashutosh Sharma[2], and Jan Genzer[1]

[1]Department of Chemical and Biomolecular Engineering, North Carolina State University, 911 Partner's Way, Raleigh, NC-27695, USA

[2]Department of Chemical Engineering, Indian Institute of Technology Kanpur, Uttar Pradesh-208016, India

Abstract

I would like to help with the country's emerging problems in healthcare, agriculture and security sectors to benefit a broad spectrum of society while using minimal resources. Multifunctional nanointerfaces can be used to develop sensor and detection devices such as biosensors, explosives trace detectors, mechanical-stress sensors, wastewater management and energy storage owing to their nanoscopic surface properties. During my doctoral study in Chemical Engineering at Indian Institute of Technology Kanpur, I have performed research on synthesis of various nanostructured materials and their application towards electrochemical biosensing, waste water treatment, Li-ion batteries and supercapacitors. I have used electrospun carbon nanofibers and nanocomposites with metal and metal-oxide nanophase to fabricate sensors for enzymatic detection of cholesterol, breast cancer, food toxin and soil nutrient. Catalytic and photocatalytic properties of those nanostructures were used to degrade industrial and agricultural wastes. The pseudo capacitance of carbon nanomaterials was integrated with metal oxides to store electrochemical energy in supercapacitors and Li-ion batteries. At present in my postdoctoral research here at NC State University, I am focusing on flexible electronics such as soft stretchable transistors, strain and pressure sensors. I have used microfluidic and 3D printing approaches to pattern liquid metal alloy on soft elastomers to create stretchable conducting electrodes. I am also working with Eastman Chemical Company, USA on automotive coating formation and impact resistive coating on glass windshields.

Research Interest: My research interests are in micro/nano fabrication of functional materials, colloids and interfaces of soft nanostructures, self and directed assembly, nano/micro-electronics, microfluidics, 3D printing, photovoltaics, polymer thin-films, carbon nanomaterials, carbon composites, soft elastomers, liquid metals, flexible electronics, stretchable transistors, carbon MEMS/NEMS in security, health, energy and environmental applications.

Teaching Interest: In addition to research, I would also like to teach Chemical Engineering, Materials Engineering and Physics based on my interdisciplinary academic background and experiences.

Keywords

Soft-nano fabrication; Nao-interface; Chemical engineering' Materials Engineering; Materials Physics; Microfluidics; Flexible electronics

Contact Email: kmondal@ncsu.edu; mddickey@ncsu.edu; ashutos@iitk.ac.in; jgenzer@ncsu.edu

References:

1. Mondal, K. *et al.* Highly sensitive porous carbon and metal/carbon conducting nanofiber based enzymatic biosensors for triglyceride detection. *Sens. Actuators B Chem.* **246,** 202–214 (2017).

2. Ali, M. A., Mondal, K. *et al.* In situ integration of graphene foam–titanium nitride based bio-scaffolds and microfluidic structures for soil nutrient sensors. *Lab Chip* **17,** 274–285 (2017).

3. Ali, M. A. Mondal, K. *et al.* Microfluidic immuno-biochip for detection of breast cancer biomarkers using hierarchical composite of porous graphene and titanium dioxide nanofibers. *ACS Appl. Mater. Interfaces* **8,** 20570–20582 (2016).

4. 11. Mondal, K. et al. Metal-oxide decorated multilayered three-dimensional (3D) porous carbon thin films for supercapacitor electrodes. *Ind. Eng. Chem. Res.* **55,** 12569–12581 (2016).

5. Ali, M. A. Mondal, K. *et al.* Mesoporous few-layer graphene platform for affinity biosensing application. *ACS Appl. Mater. Interfaces* **8,** 7646–7656 (2016).

6. Mondal, K., Ali, M. A et al. Electrospun functional micro/nanochannels embedded in porous carbon electrodes for microfluidic biosensing. *Sens. Actuators B Chem.* **229,** 82–91 (2016).

7. Ali, M. A., Mondal, K., et al. Anti-epidermal growth factor receptor conjugated mesoporous zinc oxide nanofibers for breast cancer diagnostics. *Nanoscale* **7,** 7234–7245 (2015).

8. Khosla, R. Mondal, K. et al. Effect of electrical stress on Au/Pb ($Zr_{0.52}$ $Ti_{0.48}$) O_3 /TiO_xN_y /Si gate stack for reliability analysis of ferroelectric field effect transistors. *Appl. Phys. Lett.* **105,** 152907 (2014).

9. Ali, M. A., Mondal, K. *et al.* A surface functionalized nanoporous titania integrated microfluidic biochip. *Nanoscale* **6,** 13958–13969 (2014).

10. Mondal, K., et al. Highly sensitive biofunctionalized mesoporous electrospun TiO_2 nanofiber based interface for biosensing. *ACS Appl. Mater. Interfaces* **6,** 2516–2527 (2014).

11. Singh, P., Mondal, K. & Sharma, A. Reusable electrospun mesoporous ZnO nanofiber mats for photocatalytic degradation of polycyclic aromatic hydrocarbon dyes in wastewater. *J. Colloid Interface Sci.* **394,** 208–215 (2013).

505392 Advanced Materials and Nanotechnologies for Water-Energy Applications

Abstract:

Research Interests:

My research interest is to develop advanced materials and nanotechnologies for energy and environment including resource mining from seawater, desalination, wastewater resource recovery, water/wastewater treatment, and fuel production from wastewater. Earth water system including the ocean, surface water, and groundwater provide tremendous resources including fresh water, food, minerals and so on. To preserve the water system and at the same time withdraw valuable resources and energy in an eco-friendly manner is extremely important and challenging.

During my postdoc, I have developed several technologies for resource mining/recovery from seawater and wastewater. One of such technologies is uranium extraction from seawater. In total there is hundreds of times more uranium in sea water than on land, but extracting it for use in nuclear power generation is challenging due to its low concentration and the high salinity background. Current approaches based on sorbent materials are limited due to their surface-based physicochemical adsorption nature. I have developed a half-wave rectified alternating current electrochemical (HW-ACE) method for uranium extraction from seawater based on an amidoxime-functionalized carbon electrode. The amidoxime functionalization enables surface specific binding to uranyl ions, while the electric field can migrate the ions to the electrode and induce electrodeposition of uranium compounds, forming charge-neutral species. Extraction is not limited by the electrode surface area, and the alternating manner of the applied voltage prevents unwanted cations from blocking the active sites and avoids water splitting. The HW-ACE method achieved the highest reported capacity for uranium extraction which is nine-fold higher than conventional physicochemical methods. The HW-ACE method also shows much faster kinetics with great selectivity and reusability.

Besides, I have also worked on water/wastewater treatment during my Ph.D. Point-of-use water disinfection is extremely important for public health, especially for developing areas with limited energy supply and insufficient infrastructures. I have developed two types of point-of-use water disinfection technologies for rapid and effective pathogen inactivation. In the first technology, I have developed a new photocatalyst, few layered vertically aligned MoS_2 nanofilms (FLV-MoS_2), to harvest not only UV light but also visible light for photocatalytic water disinfection. By decreasing the domain size of MoS_2 to few layers, its bandgap was increased from 1.3 to 1.55 eV which allowed the FLV-MoS_2 to harvest both UV and visible light to generate reactive oxygen species for bacterial inactivation. FLV-MoS_2 showed much better photocatalyst water disinfection performance comparing to the most adopted photocatalyst TiO_2. In the second technology, I have invented a new technology, one-dimensional nanowire-assisted electroporation (1D-NE), for pathogen inactivation. High electric field ($>10^6$ V/m) was built by introducing 1D nanomaterial electrode under a small and safe voltage of less than 10V. Bacteria and viruses were electroporated under high electric field. Since electroporation occurs on the scale of nanosecond to microsecond, the 1D-NE based filtration achieved a speed that was several orders of magnitude faster than traditional filtration technologies with comparable disinfection efficiency.

In the future research, I will take the advantages of nanotechnologies and electrochemical methods to bring in a new perspective and better solutions for water research that can contribute to the water-energy-food security nexus.

Teaching Interests:

I am interested in teaching most of the major courses for Chemical Engineering including thermodynamics, spectroscopy, catalysis, separation process, nanoscale fabrication, and electrochemistry. I have experience as guest lecturers for both graduate and undergraduate level courses related to nanotechnologies at Stanford.

Selected Publications:

C. Liu, P.-C. Hsu, J. Xie, J. Zhao, T. Wu, H. Wang, W. Liu, J. Zhang, S. Chu, Y. Cui. A Half-Wave Rectified Alternating Current Electrochemical Method for Uranium Extraction from Sea Water. ***Nature Energy***, 2017, 2, 17007.

C. Liu, D. Kong, P.-C. Hsu, H. Yuan, H. W. Lee, Y. Liu, H. Wang, S. Wang, K. Yan, D. Lin, P. Maraccini, K. Parker, A. B. Boehm, Y. Cui. Rapid water disinfection by vertically aligned MoS_2 nanofilms and visible light. ***Nature Nanotechnology***, 2016, 11, 1098â€"1104.

C. Liu*, P.-C. Hsu*, H. W. Lee, M. Ye, G. Zheng, N. Liu, W. Li, Y. Cui. Transparent air filter for high-efficiency $PM_{2.5}$ capture. ***Nature Communications***, 2015, 6, 6205.

C. Liu, X. Xie, W. Zhao, J. Yao, D. Kong, A. B. Boehm, and Y. Cui. Static electricity powered copper oxide nanowire microbicidal electroporation for water disinfection. ***Nano Letters***, 2014, 14, 5603â€"5608.

C. Liu, X. Xie, W. Zhao, N. Liu, P. A. Maraccini, L. M. Sassoubre, A. B. Boehm, and Y. Cui. Conducting nano-sponge electroporation for affordable and high-efficiency disinfection of bacteria and viruses in water. ***Nano Letters***, 2013, 13, 4288â€"4293.

505151 Multiscale Design of Aerosol Synthesis of Nanomaterials

Abstract:

Nanoparticles have attracted strong interest due to their novel properties and have been explored widely in photonics, electronics, catalysis, sensors and information storage. Gas-phase methods, such as flame reactors, are used routinely for commercial synthesis of nanostructured particles (e.g. fumed silica, pigmentary titania, carbon black, Ni) as well as for advanced materials (e.g. photocatalysts, nanofluids and biomaterials). Such nanostructured commodities consist of clusters of primary particles (PPs) that are formed by chemical reactions, condensation/evaporation or surface growth and grow further by sintering and coagulation. Depending on process conditions, gas-phase synthesis typically results in aggregates (highly-sintered particles) which are attractive in catalysis, lightguide preforms and electronics and, downstream in the process in agglomerates (PPs held together by rather weak, physical forces) that are attractive in nanocomposites, pigments and liquid suspensions.

The dynamics of aerosol reactors and product nanoparticle characteristics span 10 and 15 orders of magnitude in length and time requiring different models for each scale (Goudeli and Pratsinis, 2016a). More specifically, Molecular Dynamics (MD) simulations are used for the accurate quantification of small (below 10 nm) nanoparticle sintering rates and crystallinity dynamics (Goudeli and Pratsinis, 2016b). Even though progress in particle synthesis has allowed close control of collision and sintering rates, crystallinity is controlled rather empirically in practice (e.g. by the use of dopants). Crystalline structure plays crucial role in electronics as it affects transport properties and in catalysis where high-index facets enhance catalytic activity. Furthermore, reactive MD simulations allow the determination of binding coefficient among small clusters (such as protons and polycyclic aromatic hydrocarbons) required for the accurate calculation of cluster collision rates employed in mesoscale simulations of nascent soot formation (Kelesidis et al., 2017).

Mesoscale models provide the transport properties (diffusion, settling rates, etc.) and interaction (coagulation rate) of multi-particle structures. Discrete Element Modeling (DEM) is used here to track the detailed structure and size distribution of fractal-like agglomerates. Easy-to-use relations are derived (Goudeli et al., 2015a; 2016a) and validated with experiments (Goudeli et al., 2016b) that can be readily interfaced with population balance equations models (Goudeli et al., 2015b), climate dynamics, meteorological models or computational fluid dynamics describing the reactor operation and particle production.

Interfacing the above models can facilitate the understanding and design of aerosol reactors for synthesis of nanoparticles whose properties can be closely controlled during scale-up from laboratory scale to commercial products. This systematic approach to study particle formation can offer significant insight into fundamental physical principles and mechanisms that may be exploited by chemical industry and nanotechnology.

Goudeli, E., Eggersdorfer, M.L., and Pratsinis, S.E. (2015a) *Langmuir* **31**, 1320-1327.

Goudeli, E.,Eggersdorfer, M.L. and Pratsinis, S.E. (2015b) *J. Aerosol Sci.* **89**, 58-68.

Goudeli, E., Eggersdorfer, M.L., and Pratsinis, S.E. (2016a) *Langmuir* **32**, 9276-9285.

Goudeli, E.,GrÃ¶hn, A.J. and Pratsinis, S.E.(2016b) *Aerosol Sci. Technol.* **50**, 591-604.

Goudeli, E. and Pratsinis, S.E. (2016a) *Particul. Sci. Technol.* **34**, 483-493.

Goudeli, E. and Pratsinis, S.E. (2016b) *AIChE J.* **62**, 589-598.

Kelesidis, G.A., Goudeli, E. and Pratsinis, S.E. (2017) *Proc. Comb. Inst.* **36**, 29-50.

Research Interests:

- Multiscale modeling and simulation of aerosol reactor design for the synthesis of nanostructured materials (discrete models â€“ Molecular Dynamics & Brownian Dynamics, continuum models â€“ balance equations), aerosol particle dynamics and growth, chemical reactions
- Theory and simulation of polymer nanocomposites, modeling of their rheological and mechanical properties (effect of polymer architecture, molecular weight, particle size, polydispersity, temperature), nanoparticle interactions with biological interfaces
- Molecular Simulations, Statistical Mechanics, reactive Molecular Dynamics

Teaching Interests:

- Physical Chemistry, Chemical Engineering Thermodynamics, Statistical Thermodynamics
- Reaction Engineering and Surface Chemistry
- Transport Phenomena
- Molecular Simulations and Statistical Mechanics

505381 Nano Material Based Protein Sensor Design for Complex Cellular Environments By a Fast Integrated Simulation System.Â

Abstract:

Research Interests:

1) Nanoparticle based protein-sensor design.

Nanoparticle (2D or 3D) based materials are emerging as key components in many techniques in biological applications, whereas their effects on surrounding proteins are not accurately predictable. I am interested in understanding and predicting protein-NP interactions by developing an integrated molecular dynamics simulation system, which will function in predicting protein structural conformation, thermal or denaturant stability, protein-NP binding affinity, and protein orientations with a variety of novel NP biosensor materials.

2) Protein folding *in vivo*.

How a protein would fold/misfold/aggregate in the cellular environments compared to the traditional in vitro experimental solvent condition recently attracts attentions from researchers. Both experimental methods and simulation methods were recently developed to mimic such complexity. I am particularly interested in the method development that could accurately simulate protein behavior in any mixture of co-solvents and thus have potential scalability leading to the â€œwhole-cellâ€ simulation.

3) Transmembrane proteins: folding, insertion, and signaling.

Transmembrane proteins, such as the G protein-coupled receptors (GPCRs), are a major group of drug-targeting proteins. They could regulate cell responses to its environments by sensing molecules (e.g. drugs) outside of the cell membrane. Therefore, more and more studies have been performed to understand the folding behavior of transmembrane proteins near or in cell membranes both experimentally and computationally. To disentangle the complex effects of the cell membrane environments on protein folding, an important expansion of the co-solvent method leads to the development of a very fast simulation tool for studying protein folding in membranes. I will focus on the computationally understanding of the biophysics of GPCRs such as folding, transportation, and allosteric regulations in cell membranes, which facilitate the drug discovery targeting GPCRs.

Teaching Interests:

I would like to teach general engineering classes that are currently given by the department on undergraduate or graduate levels, such as Engineering Mathematics, Thermodynamics, Kinetics, and Transport Phenomena. Also, I will contribute classes such as Bioinformatics, Computational Biology, Molecular Simulation and Statistical Mechanics for chemical engineers, which are directly related to my research experiences and interests.

505408 Sustainability through Nanoscience: Green, Smart, and Controllable Synthesis and Characterization of One-Dimensional Metal Nanostructures

Abstract:

Research Interests:

I strongly believe there is no firm border between the various fields of engineering and science. This motivates me to pursue my future career in the multidisciplinary field of nanomaterials. I want to contribute to the creation and application of the next generation of green functional nanomaterials for electronic devices, solar cells and photovoltaic. I envision a lab capable of providing these functional nanomaterials for application to many fields such as biosensing, energy, drug delivery, tissue engineering, and food science.

In recent years much attention has been paid to the importance of metal nanoparticles due to their unique physicochemical characteristics, nano-scale size, and surface Plasmon behavior. Among metal nanoparticles, silver nanoparticles (AgNPs) have unique and outstanding shape and size dependent electrical, thermal, optical, catalytical, optoelectronic, anticancer, biosensing, medicinal, antiviral, and biological characteristics which make them exceptional for applications in a variety of fields and disciplines. Palladium (Pd) is well known for its high capacity in hydrogen affinity and corresponding adsorption or absorption that make its nanostructure broadly applicable as a primary catalyst for hydrogenation reactions, petroleum cracking, reduction of automobile pollutants, hydrogen purification, and many electrochemical applications. One dimensional (1D) metal nanostructures are important in several applications where nano spheres are less effective. Green noble nanoparticle syntheses utilizing natural precursors are of great interest because of cost effectiveness, facile reactions, less harmful feedstocks and procedures, and sustainability. However, these new techniques should be optimized not only in terms of scale-up capability, but also with respect to product quality and performance. These green and smart technologies can be achieved by either application of green reagents including the reducing agent, solvent, and capping agent or application of green templates such as plant viruses as biotemplate for their synthesis.

My graduate studies at University of New Hampshire (UNH), and my post-doctoral experiences at Purdue University in chemical engineering have uniquely prepared me to be able to synthesize 1D metal nanostructures with controllable size and morphology for specific practical applications. In my graduate studies I synthesized high aspect ratio silver nanowires (AgNWs) with high yield in both large-scale batch and continuous millifluidic reactors. I investigated both the reaction mechanism and the flow behavior of the reaction mix. The synthesized AgNWs were used to formulate conductive inks adaptable for screen-printing to create conductive patterns. In my postdoc studies I have utilized several methods for green synthesis of 1D Ag and Pd nanostructures. Moreover, my duties as a postdoc include mentoring and supervising the research of undergraduates. We are using wire-shaped plant viruses such as the tobacco mosaic virus (TMV) and barely strip mosaic virus (BSMV) expressed in e-coli in collaboration with synthetic biology and plant pathology laboratories as biotemplates for the synthesis of Pd and Ag nanorods. We are using several green reducing agents such as sugar substitutes and artificially sweeteners with the assistance of various capping agents for synthesis of 1D Pd and Ag nanostructures. We are also utilizing several polymers and compounds containing OH-groups to assess the utility of the OH- group in the reduction of Pd and Ag ions and formation and stabilization of the corresponding nanoparticles.

As an independent scientist and new faculty member I will utilize different millifluidic methods for synthesis of 1D metal nanostructures with the goal of reducing their final cost by an order of magnitude through the application of less expensive water based reducing agent at lower reaction temperature in a continuous manner. These methods will be tailored to control the morphology of the nanostructures for specific practical applications as their characteristics strongly depend on their size and morphology. The reaction mechanisms will be investigated with the assistance of several characterization methods such as mass spectroscopy, X-ray absorption spectroscopy, scanning electron microscopy, transmission electron microscopy, Fourier transfer spectroscopy, and UV-Visible spectroscopy. These detailed fundamental investigations and characterizations will make us able to assess which functional groups in each green reducing agent are responsible for the reduction of specific metal ions, and consequently selection of other green reducing agents offered by nature suitable for green synthesis of a specific metal nanostructures. My lab will be able to provide metal nanostructures with specific morphology for use by other labs and groups for further practical applications. As the commercially available 1D metal nanostructures are expensive, the finding of my research will provide a tool to control the morphology of 1D metal nanostructures in an inexpensive way, and consequently synthesize them with specific dimensions. The result will be increased in efficiency and lower cost of various 1D metal nanostructure based devices including but not limited to sensors, biosensors, nanoelectronics, solar cells, and touch screens. This will affect several research areas such as health care, energy, and food science both technically and scientifically.

I will also address specific applications of metal nanostructures to formulate conductive inks, and more generally film deposition processes including screen printing, doctor blading, spin coating, inkjet printing, and spray coating to create conductive patterns for solar cells and photovoltaic applications. Finally, I will be able to fully characterize these new devices using several electrical characterization techniques.

Teaching Interests:

My teaching philosophy has developed based on my experience in the classroom. As a new faculty member, I will actively try to work closely with students to be sure they grasp not only the details of technical procedures but broader theoretical concepts as well. I also have a strong belief in peer learning, having found that students learn better through interactive participation rather than passive listening. All in all, I believe that my experience and my beliefs have prepared me to implement successful pedagogy by encouraging student curiosity in an open and collaborative environment. I enjoy teaching courses on Introduction to Chemical Engineering, Transport Phenomena, Reaction Kinetics, and Rector Design. I would also be interested in complementing current courses or creating new course material related to topics from my own research, including nanomaterial characterization.

505148 Wearable/Implantable Ultrathin Electronic/Optoelectronic Devices with Engineered Semiconductor Nanocrystals

Abstract:

Research Interests:

The pervasiveness of electronic devices in our everyday life is increasing, heading towards a ubiquitous environment involving cutting-edge technology in fields of artificial intelligence, the "Internet of Things," autonomous vehicles, etc. In particular, wireless communication between the human body and objects is an emerging multi-disciplinary field that finds numerous applications from many diverse research areas. Merging electronic devices with our human body can be realized through the development of wearable/implantable ultrathin electronic devices which conform to the soft human tissues. It is expected that the wearable/implantable technology will lead to electronic skin displays, real-time health monitoring systems, brain-computer interfaces, and many others. Colloidal semiconductor nanocrystals have drawn considerable interest as attractive building blocks for flexible and printable electronic/optoelectronic devices. A critical barrier to commercialization, however, is the toxicity associated with usage of heavy metals (*e.g.* Pb or Cd), which makes RoHS compliant nanocrystals (*e.g.* CISeS, CZTS, Si, Ge etc.) particularly attractive for potentially biocompatible electronic devices. Hence, understanding the charge carrier transport characteristics of non-toxic semiconductor nanocrystals thin films and their practical application are crucial to their implementation in wearable/implantable electronic/optoelectronic technologies.

I have previously demonstrated the first switchable-logic devices (CMOS inverter, and NAND or NOR gate devices) using engineered *non-toxic* colloidal semiconductor nanocrystals. To expand on this, starting from the functionalized nanocrystals, I seek to develop ultrathin printable logic-gate and optoelectronic devices on flexible substrate for achieving wearable/implantable electronic devices. This research will involve: 1) synthesis of solution-processed non-toxic semiconductor nanocrystals; 2) exploration of suitable surface ligands for biocompatible electronic devices; and 3) design and fabrication of ultrathin logic-gate and optoelectronic device architectures. In particular, engineering the surface ligands of tailored nanocrystals strongly influences the carrier transport polarity (unipolar *p*- or *n*-type, or ambipolar) of electronic devices, which are basic characteristic for fabricating efficient logic-gate devices and other devices such as light-emitting field-effect transistors (FETs). These devices are highly applicable to wearable/implantable ultrathin electronic gadgets. Therefore, understanding the ligand chemistry for non-toxic semiconductor nanocrystals will be crucial for fabricating high-performance electronic/optoelectronic devices. Recently, inorganic ligands such as halide, chalcogenide and metal-chalcogenide complex ions have been extensively used to produce strong electronic coupling between nanocrystals, and tune the transport polarity. My research group will employ various classes of atomic inorganic ligands to non-toxic semiconductor nanocrystals for achieving high performance logic-gate and opto-electronic devices on the ultrathin substrates. The versatile tunability of transport polarity of nanocrystal film with applying various inorganic ligands is expected to enable inexpensive and high-throughput solution-based processing of ultrathin electronic/optoelectronic devices in ambient environment. In particular, this significant simplification of integration processes implies the feasibility of high profile fabrication of various device geometries.

In order to pursue such ambitious lines of research, significant financial resources will be needed. All funding opportunities available to me will be pursued, starting with National Science Foundation (NSF) fellowship applications for all grad students with a reasonable chance of success. In the beginning phases, I will also apply for early career grants such as the ACS PRF Doctoral New Investigator Grant and the NSF Early Career Award. Furthermore, national research foundation of South Korea (my home country) opens the chance to make international collaboration with exceptional research groups. I will tackle the main funding agencies with my previous collaborator in Seoul National University, KAIST, Samsung, and so on.

Research Experiences:

My research focus has been on the synthesis of semiconductor nanomaterials and characterization and control of carrier interfacial transfer and mesoscale transport behavior. My PhD studies at Seoul National University involved the development of semiconductor nanomaterials to enhance both the photo-generation of electrons and the charge transfer kinetics for solar energy conversion. I developed various impurity-doped TiO_2 photocatalysts and designed a biomimetic Z-scheme photocatalyst by combining two visible light responsive photocatalysts. At University of Pennsylvania, I focused on designing and demonstrating new variants on CdSe nanocrystal quantum-dot-sensitized liquid junction solar cells. I first applied atomic scale inorganic ligand exchange techniques to fabricate a high performance photoanode to enhance the rate of charge transfer. I am currently conducting research on the characterization of carrier transport in quantum dot thin films in Los Alamos National Laboratory. Specifically, I fabricate field-effect transistors (FETs) and photoresponsive switches for investigating the carrier dynamics of various semiconductor nanocrystals. Very recently, I successfully fabricated the first environmentally-friendly CMOS inverter, NAND and NOR gate logic devices by combining p- and n-channel non-toxic quantum dot FETs, whose polarities are well-tuned by controlling their intrinsic defect states.

Teaching Interests:

As a teaching assistant, I helped students at undergraduate courses of Chemical Engineering Experiments and Transport Phenomena in both Ajou University and Seoul National University. In addition, I mentored undergraduate and graduate students in research when I was senior PhD graduate students and post-doctoral associate. I was also invited to give research talks several universities and high school. Based on my research career, I can give lectures related semiconductor science and technology, inorganic chemistry, renewable energy, surface science and engineering, and so on.

I have three important key words as mentoring philosophy. 1) Socratic elenchus. As mentor, I lead a mentee to find their own answer by themselves. For this great endeavor, I asked and got answers to stimulate critical thinking. 2) Open mind. In research society, somebody was really unwilling to show their results because of a fear of being scooped. But I emphasized we are studying to make better world, not to get a fund or good papers. Open mind always makes scientific revolution in various research fields. 3) Beyond the ivory tower. As an engineer, I lead mentee to link our knowledge from text book and our real life. This kind of linkage made the learner feel fun of studying and motivate their creativity.

Postdoctoral Project:

"Transport Behavior of Non-Toxic $CuInSe_xS_{2-x}$ *Quantum Dot Films"*

Under the supervision of Victor I. Klimov, Division of Chemistry, Physical Chemistry and Applied Spectroscopy

"Effective Conversion of Solar Light Using A Variety of Semiconductor and Metal Nanocrystals"

Under the supervision of Christopher B. Murray, Department of Chemistry, University of Pennsylvania

PhD Dissertation: *"Preparation and Characterization of* TiO_2*-Based Photocatalytic Materials for the Solar*

Light Utilization"

Under the supervision of Jongheop Yi, School of Chemical and Biological Engineering, Seoul National University

Successful Proposals: Postdoctoral fellowship from the Basic Science Research Program through the National Research Foundation of Korea (NRF) funded by the Ministry of Education

Selected Publications:

- **H. J. Yun,** J. Lim, S. Keene, A. Fuhr, M. Law, J. M. Pietryga, and V. I. Klimov, "Defect Control in $CuInSe_xS_{2-x}$ Quantum Dot Films for Tuning Their Carrier Transport Behavior", *In Preparation*
- Q. Lin†, **H. J. Yun†,** W. Liu, H.-J. Song, N. Makarov, O. Isaienko, T. Nakotte, G. Chen, H. Luo, V. I. Klimov, and J. Pietryga, "Phase-Transfer Ligand Exchange of Lead Chalcogenide Quantum Dots for Direct Deposition of Thick, Highly Conductive Films", *J. Am. Chem. Soc.* **2017,** 139, 6644. (†Qianglu Lin and Hyeong Jin Yun contributed equally to this work.)
- **H. J. Yun,** T. Paik, M. Edley, J. Baxter, and C. B. Murray, "Size-Dependent Efficiency of Quantum Dot Sensitized Solar Cell in Strongly-Coupled CdSe Nanocrystals/TiO_2 System", *ACS Appl. Mater. Interfaces* **2016,** 8, 14692.
- **H. J. Yun,** T. Paik, M. Edley, J. Baxter, and C. B. Murray, "Enhanced Charge Transfer Kinetics of CdSe Quantum Dot Sensitized Solar Cell by Inorganic Ligand Exchange Treatments", *ACS Appl. Mater. Interfaces* **2014**, 6, 3721.
- **H. J. Yun**, H. Lee, N. D. Kim, D. M. Lee, S. Yu, and J. Yi, "A Combination of Two Visible-Light Responsive Photocatalysts for Achieving the Z-Scheme in the Solid State", *ACS Nano* **2011,** 5, 4084.
- **H. J. Yun**, H. Lee, J. B. Joo, N. D. Kim, M. Y. Kang, and J. Yi, "Facile Preparation of High Performance Visible Light Sensitive Photo-Catalysts", *Appl. Catal. B* **2010,** 94(3-4), 241.

505399 Microbiome Engineering for Human Health and Agricultural Productivity

Abstract:

Research Interests:

Microbial communities are ubiquitous in nature and contribute to human health, agricultural productivity, and industrial and environmental processes. Organisms within these communities exchange metabolites, actively inhibit or enhance growth of each other, and have direct and indirect interactions with their environment. Despite the impact these communities can have on our lives, we do not yet know how to control their behavior. To enable the engineering of microbial community function, my research program will use experimental data from natural and model microbiomes to develop predictive models to engineer complex emergent behavior of microbial communities.

Human Microbiome: My current research is focused on microbes that colonize the human skin, and how they impact wound healing, attraction/repulsion of biting insects, and interaction with the immediate environments. Microbiome representatives from the human skin have diverse activities including secretion of host-responsive metabolites, UV resistance, production of volatile organic compounds, and secretion of pigments such as melanin. In this work we focus on developing systems to study model communities of skin microbiota and determine how their diverse functions are combined to impact human health. Our overarching goals it to modulate the microbiome *in situ* to promote positive outcomes in wound healing and interaction with the environment.

Microbiomes in Agriculture: In agricultural systems, plant-associated microbes contribute to disease resistance, drought tolerance, and nutrient uptake. Plant systems are amenable to microbiome engineering research, as germ-free plants are easily generated and induced phenotypes are easily measured. My research group will interface with agricultural sciences to solve current and emerging problems in agricultural productivity, including the rapidly approaching peak phosphorous, inefficient fertilizer uptake, and the spread of emerging plant pathogens.

My research group will build on the above research by using engineering principles to design communities of genome-sequenced bacterial isolates using individual phenotypes, phylogenetic relationships, and predicted genomic content and test the ability of the community to mimic the natural microbiome response to stress. Models of bacterial growth kinetics, metabolism, and interactions in culture conditions will be used to predict behavior in a community. Integrating quantitative experimental data with predictive models will help us understand how the microbiome assembles in nature, ultimately leading to the ability to engineer constructed communities of bacteria with specific and predictable functions.

PhD Dissertation: Kinetics of vesicular stomatitis virus mRNA and genomes during infection

Mentor: John Yin, University of Wisconsin-Madison, Department of Chemical and Biological Engineering

Research Career

My research career began as an undergraduate where I studied gas hydrate formation at the Colorado School of Mines. I helped develop software for predicting hydrate formation in gas pipelines and tested predictions using a novel reactor in the laboratory. This valuable experience familiarized me with quantitative experiments and laboratory techniques, but more importantly convinced me to go to graduate school to continue my academic career. In graduate school I changed my focus to biological engineering, where I learned how to design experiments and control biological systems to make quantitative measurements of viral replication. I used the data I collected to build kinetic models of viral infection, and showed that viral mRNA production is independent from host environment and only depends on viral genome number. This suggests that transcription is robust and insensitive and helps to identify targets for anti-viral strategies for RNA viruses. While studying interactions between two organisms (viruses and cells), I became interested in complex ecological systems with multi-member interactions. I was fortunate to find a post-doctoral position as a part of the Plant-Microbe Interfaces project at Oak Ridge National Laboratory studying the microbiome of *Populus* trees. At ORNL I learned about the contribution of the plant microbiome to host growth and productivity. I am currently at the Johns Hopkins University Applied Physics Laboratory, where our goal is to engineer biological systems to meet the needs of the Department of Defense. At APL I am studying the microbiome of the human skin and how it contributes to wound healing and interaction with the immediate environment of the host. In my postdoctoral studies, I have focused specifically on understanding bacterial genetic and phenotypic diversity in the system to help determine how interactions between microbiome members and the host lead to overall system function.

Teaching Interests:

As I have moved through my academic career I have become more active in my role as a teacher and mentor. During my undergraduate studies I worked closely with many of my peers, and began to appreciate the different learning and teaching styles present in the group. I had the opportunity to informally tutor younger students in heat and mass transfer classes; in both cases their grades improved after we began working together. In graduate school I got my first opportunity to formally teach classes to students. As a teaching assistant for the introductory transport course, I enjoyed spending time developing lectures and example problems for students, and focused on how I could engage the entire class in a single topic. I also worked as a TA for the process control laboratory where I helped students run experiments and taught them to present their results in concise reports. As a post-doc I have worked with undergraduate summer students and new graduate students from chemistry, biochemistry, biological systems engineering and chemical engineering. Their unique backgrounds have helped me generalize my teaching methods. I have helped them learn how to work in the laboratory, analyze data, and present results. The students helped identify metabolic pathways, isolated bacteria that grow on *Populus* metabolites, identified genes that affect plant growth, and developed methods for studying root growth in microfluidic devices. Watching students succeed with my input has been extremely important to my development and has helped me appreciate the commitment of my teachers and mentors throughout my academic career.

505491 A Holisitic Design Approach for Zeolite Catalysts

Abstract:

Research Interests:

With the steady increase in computational power, the targeted, computational design of materials for specific applications becomes ever more important. In some cases the performance of materials can be linked to one specific parameter, but how would a similar approach look like for materials as complex as zeolites? Or even more importantly: How is the actual performance of a zeolite linked to the synthesis process, the nature of the active sites, the preparation of the material and the operating conditions?

The first step in the design of zeolite catalysts is to understand the synthesis process and its impact on the distribution of the anchoring points for possible active centers in dependence on the synthesis conditions (i.e. the present template and slurry composition). In a subsequent step it is then possible to use thermodynamic modeling to predict the condition dependent distribution (type of active site, temperature, gases and their relative pressures) of active centers in the material. Finally the macroscopic activity of the catalyst can be predicted by identifying the contribution of each active site and the application of microkinetic modeling. Clearly the variations of each single parameter will lead to significant differences in the performance of the material and therefore a holistic design approach, taking into account the synthesis, distribution of active centers and the contribution of each one of them, is necessary for accurate predictions.

In my future research, I will develop and apply a holistic design approach for zeolite catalysts. This includes to use electronic structure calculations based multi-scale modeling of the synthesis process, the subsequent prediction of the distribution of active sites and the prediction of the contribution of each site to the reactivity.

Teaching Interests:

Besides my achievements in science, I have always had a passion for learning and teaching, which has been reflected in mentoring graduate and undergraduate students as well as teaching selected classes during my current post-doc. In my time as a faculty member I plan to develop a course describing how state of the art electronic structure methods can be applied to gain insights into heterogeneous catalysis, either by understanding reaction pathways or by predicting observed spectra. It will provide the necessary theoretical background as well as practical examples and hands-on experience for the students. Among the undergraduate core classes I am ready to teach thermodynamics and, depending on the departmental needs, am looking forward to develop classes close to my core competencies in fundamental physics, quantum mechanics, catalysis or materials science.

505233 Catalysis for Energy: Catalyst Design Based on Spectroscopy and Fundamental Structure-Function Relationships

Abstract:

Research Interests:

My research is primarily focused on addressing the most important research question of our time: anthropogenic climate change. In this context, catalysis has a very important role to play, by opening up pathways for the sustainable production of chemicals and fuels. Both my doctoral and postdoctoral research focused on the catalytic upgrading of biomass, through C-C bond formation, as well as hydrodeoxygenation processes. Going forward, I envision using kinetics and spectroscopy to design catalysts and processes for sustainable chemistry.

In my PhD research, with Prof. Dean Toste at UC Berkeley, I developed a selective, active and stable catalyst for the upgrading of fermentation products to diesel fuel precursors. This was achieved by combining theoretical calculations with kinetic measurements and advanced characterization techniques, including X-ray absorption and tomography. In doing this, we correlated the decarbonylation and hydrogenation selectivity of alloy catalysts with the extent of surface enrichment and the strength of binding of reactive species, and developed a kinetic model for reactions of oxygenates over monometallic and bimetallic catalysts. During my postroctoral research, with Prof. Dion Vlachos at the University of Delaware, I discovered structureâ€"function relationships that determine the activity and stability of oxides in hydrodeoxygenation reactions. We combined insights from DFT calculations, experimental kinetics, and spectroscopy to understand the active site requirements for HDO over oxides and to prove the generality of C-O bond scission reaction mechanisms over oxides.

In my future research, I plan to investigate processes to produce fuels and chemicals from biomass. The main challenge for the development of such processes is the discovery of a selective catalyst. This can be achieved by tuning the electronic properties of catalysts, via the use of active supports, bimetallic nanoparticles and the formation of well-controlled surface oxides. The efficiency of these approaches can be evaluated in reactions and the activity and selectivity of the catalysts correlated with a suitable descriptor. Such descriptors can be obtained from spectroscopic methods, including infrared and X-ray absorption and can be combined with input from computational chemistry to guide further catalyst design.

Teaching Interests:

During my PhD at UC Berkeley, I assisted in the teaching Thermodynamics and Process Design and I won the Dow Excellence in Teaching award for my work in teaching Thermodynamics. In my future teaching career, I plan on incorporating active learning in the undergraduate classroom, through group discussions and conversation, in order to engage more students and improve educational outcomes.

At the graduate level, I am planning to develop classes on Catalysis and Spectroscopy. In doing so, I envision discussing topics from my graduate and postdoctoral research in both classes. In my Catalysis class I will cover topics from the fundamental kinetics to the advanced methods of catalyst characterization, while in a Spectroscopy class I plan to further familiarize my students with the most advanced techniques in the field, emphasizing recent advances in X-ray and infrared spectroscopy.

505424 Computational Driven Strategies for the Rational Design of Novel Catalysts for Clean Energy Generation and Fuel Synthesis

Abstract:

Research Interests:

First principles density functional theory (DFT) calculations have been proven successful in predicting materials properties at the nanoscale for various applications, including those for energy related catalysis and electrocatalysis. Such calculations not only provide atomic details of the underlying reaction mechanisms that are not easily accessible in experiments, but also predict catalytic trends that can ultimately provide rational guidance toward the development of improved catalysts. The static DFT calculations performed at 0 K provide valuable insights to understand the fundamentals of a chemical reaction. However, they lack the accurate prediction of the kinetics of reactions occurring at experimental reaction conditions. This can be addressed by kinetic Monte Carlo (KMC) simulations performed at experimental reaction conditions using the DFT calculated energetics: activation energies, rate constants and so on. Therefore, KMC simulations provide a way to directly compare the theoretical results with experimental ones and validate the predictions made on the basis of DFT calculations.

In this work, combined DFT and KMC simulations are performed to gain atomic level understanding of the reaction mechanisms and guidance to tune the activity and selectivity of CO_2 hydrogenation to C1 (CO, CH_3OH and CH_4) compounds on metal/oxide catalysts. The DFT and KMC study show that the activation and conversion of CO_2 can be achieved on multifunctional catalytic sites available at the metal/oxide interface by taking advantage of the synergy between the metal nanoparticles and oxide support. The CO selectivity is found to be primarily controlled by the CO binding energy. In contrast, the CH_3OH selectivity can be tuned selectively by improving the stability of *CO, *HCO, and *H_2CO while destabilizing *HCOO and *H_3CO intermediates. Finally, to tune the selectivity of metal/oxide catalysts toward CH_4 the metal/oxide interfaces should strengthen the binding for H_xCO species to facilitate the C-O bond scission.

Teaching Interests:

Classroom teaching creates an environment where students learn, interact, challenge and respect others ideas. Most important to me, classroom teaching provides an opportunity to foster studentsâ€™ passion and curiosity, instill civic values and train the citizens of tomorrow which might not be possible in any other setting.

Mastering my students about science concepts and developing problem solving skills to the problems that are inherent to the nature are main goals of my teaching. Students tend to learn fundamental concepts better when they find themselves applying textbook concepts to some real world problems/examples. I found students more interested when the course materials and assignments to some extent are related to real life experiences. Furthermore, I experienced that students enjoy in a class even with little activities (in individual or group). Thus as an instructor, I try to integrate

scientific knowledge to real life experiences with examples and hands-on activities. My tools for successful teaching are student-centered teaching pedagogy, an interactive classroom environment, and hands-on demonstrations.

505469 Data Driven Catalyst Design and Optimization

Abstract:

Data Driven Catalyst Design and Optimization

*Yongchun Hong**

University of California at Berkeley, Berkeley, CA 94720

*yongchun.hong@berkeley.edu

Research Interests:

In some heavily documented areas in catalysis, such as CO oxidation, data (kinetic, theoretical, and spectroscopic) accumulated by our pioneers are beyond our capability to thoroughly summarize and survey, and new data are being reported by our peers at an unprecedented rate. It is even more challenging to consolidate those often contradictory observations. I am interested in developing a statistic based tools to analyze the data reported in the literature to allocate plausible mechanisms beyond those observations and to propose measurements or calculations to validate or invalidate those mechanisms with minimum efforts. Such efforts can be translated into standardized protocols that can be adapted by the catalysis community to facilitate the communication of observations. The statistical analysis of historical data and protocols that enable higher quality data acquisition will ultimately lead to a more comprehensive understanding of underlying mechanisms, that further enables a rigorous data-driven catalyst design and optimization that relies on the collective intelligence of the entire community instead of intuition and rigor of researchers in one of few groups.

Teaching Interests:

I have supervised 5 undergraduate students and mentored 1 graduate student during my academic training and I enjoy interacting with students with diverse backgrounds as well as sharing my knowledge and experience with them. I would like to develop and teach undergraduate courses on topics including catalysis, transport phenomena, and reaction engineering. I am especially interested in developing an undergraduate course that focuses on the basic concepts and applications of catalysis for students from Chemical Engineering and Chemistry backgrounds as well as other related disciplines because I found that such knowledge is essential for ChE and Chem students but also important for students from other related disciplines in the era of interdisciplinary cooperations. I would also like to develop and teach graduate courses on advanced topics including experimental and theoretical techniques in catalysis research because I think it is very important for graduate students to be aware of available tools to solve problems they encounter during their research, and more importantly, to explore new techniques by incorporating concepts and tools in other areas such as statistics and computer science.

Research Experience:

As a researcher in catalysis, I have 10 years' research experience in various areas of heterogeneous catalysis, including catalyst syntheses, characterization, kinetic assessment and modeling, and molecular simulation. My formal training in catalysis started in Tsinghua University, focusing on the syntheses of Au and Au-based bimetallic catalysts and the exploration of their applications in oxidative abatement of CO and formaldehyde and in the selective hydrogenation of a,ß-unsaturated carbonyl compounds. During my Ph.D. training in Washington State University and Pacific Northwest National Laboratory, I focused on the structure-performance relationship of Fe-based bimetallic catalyst in hydrodeoxygenation of bio-oil by correlating the structure information obtained via advanced *in-situ* characterization tools such as XAS, XRD, and XPS with the catalyst performance, and by accessing the mechanistic details based on kinetic inquiries and DFT simulation. I am currently working as a postdoctoral researcher at the University of California, Berkeley, and Lawerence Berkeley National Laboratory, pursuing a fundamental understanding of the catalytic consequence of oxygen binding properties of metal oxide on the abatement of N_2O in the engine exhaust via a combination of kinetic inquiries and DFT simulation.

Successful Proposals: XSEDE (PI), DOE (Participant)

Advisors and Collaborators: Prof. Enrique Iglesia (UCB, postdoctoral advisor); Prof. Yong Wang (WSU/PNNL, Ph.D. advisor); Prof. Bo-Qing Xu (THU, M.S. advisor); Dr. Andrew Getsoian (Ford); Prof. JeanSabin McEwen (WSU); Prof. Jun-Ming Sun (WSU); Prof. Franklin Tao (KU); Prof. Ayman Karim (VT); Mark Engelhard (PNNL); Dr. Janos Szanyi (PNNL).

Academic Services: Reviewer for *ACS Catalysis*, *Applied Catalysis B*, *Journal of Molecular Catalysis A*, *ChemSusChem*, *RSC Advances*, and etc.

Selected Publication:

1. **Hong, Y. C.**, Zhang, S. R., Tao, F., Wang, Y., *ACS Catal.* 7, 3639 (2017).
2. Wang, X., **Hong, Y. C.**, Shi, H., Szanyi, J., *J. Catal.*, 343, 185 (2016).
3. **Hong, Y. C.**, et al., *ACS Catal.*, 4, 3335 (2014).
4. Hensley, A. J. R., **Hong, Y.C.**, et al., *ACS Catal.*, 4, 3381 (2014).
5. **Hong, Y. C.**, Sun, K. Q., Han, K. H., Liu, G., Xu, B. Q., *Chem. Commun.*, 47, 1300 (2011).

505443 Designing Multicomponent Nanostructured Materials for Energy Storage and Conversion

Abstract:

Research Interests:

The development of energy storage and conversion systems to replace fossil fuels with renewable sources depends on the optimization of critical reactions. Though many reactions of interest are thermodynamically feasible, some of the best catalysts to date consist of materials that are simply too expensive to be practical when deployed on a wide scale. Much of the research to expand available catalysts for these reactions has focused on tuning composition. However, fully overcoming fundamental constraints to catalyst activity requires techniques beyond alloying alone, such as developing novel nanostructured catalysts. A prime example is in electrochemical H_2 generation, where the state-of-the-art catalyst is nanostructured Pt/C and much attention has been devoted to finding better materials. Since ideal catalysts are complex and must optimize both composition and structure, I propose to design novel multicomponent materials using a comprehensive approach which spans from theory and fundamental reactions in controlled conditions to synthesis and testing of catalysts under realistic conditions. As a faculty member, I am particularly interested in designing nanostructured and nanoporous multimetallic alloy and oxide catalysts to induce synergistic effects and tailored reactivity for a wide range of catalytic and electrocatalytic systems, including electrochemical CO_2 reduction to chemicals and fuels, electrocatalysts for air battery systems, and supported catalysts for biomass conversion.

In my research experiences, I have found opportunities to develop expertise in a wide range of experimental and computational techniques necessary to develop a complete understanding of structure-property relationships critical for design of novel materials. During my doctoral work, I focused on the design and characterization of nanostructured materials for energy conversion and storage, with a heavy emphasis on electrocatalysis. In a project on electrochemical H_2 evolution, I developed nanoporous Cu-M (M= first-row transition metal) bimetallic catalysts with activity approaching Pt/C and used density functional theory calculations to understand the experimental results.[1] To examine catalyst structure, I have used synchrotron X-ray absorption spectroscopy both *ex situ* and *in operando* to find critical properties of lithium-air battery electrocatalysts,[2] CO_2 reduction electrocatalysts,[3] and Co-based photocatalysts for water splitting.[4] Additionally, I have developed nanoporous ternary Cu-M-Al for vapor-phase hydrodeoxygenation of biomass derivatives.[5] In my post-doctoral work, I have expanded my technical knowledge to characterization with fundamental surface science techniques, where I have studied multimetallic thin films and the growth of two-dimensional aluminosilicate bilayers as zeolite analogs and atomically thin membrane materials.[6] In this session, I plan to provide an overview of my research in these areas and how this collected experience enables my future research and provides a complete basis for understanding catalyst properties and finding new materials to solve the most pressing issues in energy storage and conversion.

Graduate Advisor: Feng Jiao (University of Delaware, Department of Chemical and Biomolecular Engineering)

Post-Doctoral Advisor: Eric I. Altman (Yale University, Department of Chemical and Environmental Engineering)

Selected Publications (Total Papers: 21, Total Citations: 913, h-index: 11 [Google Scholar]):

(1) Lu, Q.; Hutchings, G. S.; Yu, W.; Zhou, Y.; Forest, R. V.; Tao, R.; Rosen, J.; Yonemoto, B. T.; Cao, Z.; Zheng, H.; Xiao, J. Q.; Jiao, F.; Chen, J. G. Nat. Commun. 2015, 6:6567.

(2) Hutchings, G. S.; Rosen, J.; Smiley, D.; Goward, G. R.; Bruce, P. G.; Jiao, F. J. Phys. Chem. C 2014, 118, 12617.

(3) Rosen, J.; Hutchings, G. S.; Lu, Q.; Forest, R. V.; Moore, A.; Jiao, F. ACS Catal. 2015, 5, 4586.

(4) Hutchings, G. S.; Zhang, Y.; Li, J.; Yonemoto, B. T.; Zhou, X.; Zhu, K.; Jiao, F. J. Am. Chem. Soc., 2015, 137, 4223.

(5) Hutchings, G. S.; Luc, W.; Lu, Q.; Zhou, Y.; Vlachos, D. G.; Jiao, F. Ind. Eng. Chem. Res. 2017, 56, 3866.

(6) Hutchings, G. S.; Jhang, J.-H.; Zhou, C.; Hynek, D.; Schwarz, U. D.; Altman, E. I. ACS Appl. Mater. Interfaces 2017, 9, 11266.

Teaching Interests:

I believe that the curriculum in chemical engineering should balance the need for fundamental knowledge with experience solving realistic problems using tools students will encounter in their careers. For example, when given the opportunity to give lectures during my teaching assistantships, I paired a walkthrough of fundamental derivations and problem solving with hands-on demonstrations of how to solve more complex problems with software used in practice. I found that students responded well to this approach and could apply the concepts effectively to their semester projects. Given my research background, I am most excited to apply this approach to courses in kinetics/reaction engineering as well as the introduction to chemical engineering. I would also like to develop courses in materials characterization and (electro)catalysis as graduate-level electives to provide a firm foundation for students interested in those areas. Additionally, I have experience mentoring summer undergraduates both during my graduate work and as part of a Research Experiences for Undergraduates (REU) program at Yale, where I have taught practical lab experience and provided guidance that has led to published results with the undergraduates as co-authors.

505369 Developing Fundamental Insights into Heterogeneous Catalytic Reactions for Selective Chemical Production and Sustainable Fuels

Abstract:

Research Interests:

The control of selectivity in complex chemical reaction networks has become the focus of heterogeneous catalysis in the 21st century where the goal is to decrease unwanted byproducts and to maximize cost effectiveness, energy efficiency and sustainability. My research interests focus on gaining fundamental insights into chemical reactions and selective chemical production by combining careful catalyst synthesis with state of the art *in-situ* spectroscopy in order to understand the dynamic nature of active sites under reaction conditions. The rational design of new catalysts through fundamental knowledge will be critical to meeting the demands of the changing landscape of the chemical and energy industries.

My Ph.D. dissertation work with Professor Phillip Christopher in the department of Chemical & Environmental Engineering at the University of California, Riverside focused on a multifaceted approach to investigate the impacts of local geometric and electronic structure on catalytic performance, in close collaboration with theoretical and *in-situ* microscopy techniques. Some examples included: exploiting *in-situ* infrared spectroscopy and rigorous reaction kinetic studies to reveal the active site in structurally dynamic reaction systems, utilizing photocatalysis to control selectivity and activate targeted strong metal-adsorbate bonds by matching photon wavelength with bond energies of hybridized metal-adsorbate bonds, and modifying metal-support interfaces to efficiently remove atmospheric contaminants for confined space pollution control applications.

My current postdoctoral work with Professor Mark E. Davis in the department of Chemical Engineering at the California Institute of Technology has involved the investigation of catalytic reaction mechanisms over microporous zeolite catalysts. For example, using controlled catalyst synthesis, reaction condition optimization and reaction kinetics to investigate the direct synthesis of important commodity chemicals from CO_2 and CH_4. I plan to continue working on the controlled design of zeolite based catalysts for investigating the fundamental mechanisms of complex reaction pathways.

In the future, I will unite the skills I have learned in my Ph.D. and postdoctoral work and focus on gaining fundamental insights into important chemical reactions, improving efficiency for the selective production of commodity chemicals, and sustainable fuel production. Specifically, I plan to use controlled catalyst synthesis, *in-situ* spectroscopy, detailed reaction kinetics, and photocatalysis as tools to promote the fundamental understanding of these reactions, and enhance catalytic selectivity in important reactions. It is my goal to move the field towards meeting the increasing demands of society for energy efficiency and sustainability through developing fundamental knowledge of reaction mechanisms and utilizing this knowledge to enhance efficiency and catalyst performance.

Teaching Interests:

During my time as an undergraduate and graduate student, I have served as a tutor for several chemistry and chemical engineering classes, been a TA for various chemical engineering laboratories, in addition to giving lectures in a Catalysis Reaction Engineering course. I have also served as a direct mentor for ten undergraduate students and four graduate students in the lab. In the future, I hope to help future chemical engineers develop a passion for catalysis, reaction engineering and energy research. I feel that I would be able to best accomplish this through teaching courses such as Kinetics and Catalysis & Reaction Engineering, but would be comfortable with core chemical engineering courses such as Thermodynamics and Transport Phenomena.

Selected Publications:

[1] Li, K.; Hogan, N. J.; **Kale, M. J.**; Halas, N. J.; Nordlander, P.; Christopher, P. "Balancing Near-Field Enhancement, Absorption, and Scattering for Effective Antenna-Reactor Plasmonic Photocatalysis", *Nano Letters* (2017) *17*, 3710-3717.

[2] Avanesian, T.*; Dai, S.*; **Kale, M. J.***; Graham, G. W.; Pan, X.; Christopher, P. "Quantitative and Atomic Scale View of CO-Induced Pt Nanoparticle Surface Reconstruction at Saturation Coverage via DFT Calculations Coupled with *in-situ* TEM and IR", *JACS* (2017), *139*, 4551-4558.

[3] **Kale, M. J.***; Gidcumb, D.*; Gulian, F. J.; Miller, S. P.; Clark, C. H.; Christopher, P. Evaluation of Precious Metal Catalyst Stability for Confined Space Pollution Control. *Applied Catalysis B* (2017), *203* 533-540.

[4] **Kale, M. J.**; Christopher, P. Utilizing quantitative *in-situ* FTIR spectroscopy to identify well-coordinated Pt atoms as the active site for CO oxidation on Al_2O_3 supported Pt catalysts. *ACS Catalysis* (2016), *6*, 5599-5609.

[5] **Kale, M. J.**; Christopher, P. Plasmons at the interface. *Science* (2015), *349*, 587-588.

[6] **Kale, M. J.**; Avanesian, T.; Xin, H.; Yan, J.; Christopher, P. Controlling Catalytic Selectivity on Metal Nanoparticles by Direct Photoexcitation of Adsorbate-Metal Bonds. *Nano Letters* (2014), *14*, 5405-5412.

[7] **Kale, M. J.**; Avanesian, T.; Christopher, P. Direct Photocatalysis by Plasmonic Nanostructures. *ACS Catalysis* (2014), *4*, 116–128.

**Co-first author*

505487 Efficient Catalytic Pathways for Carbon Utilization and Emission Control Technologies

Abstract:

Research Interests:

Development of emission control technologies for remediation of toxic gases from stationary and mobile combustion sources is of great interest due to the stricter regulations and changes in the emissions as a result of the improvements in burner and engine technologies. Within my previous research efforts, one of my main objectives was to understand pollutant formation mechanisms to develop better emission remediation technologies. In particular, I have worked on removal of trace metals from coal-fired power plants, and elucidated the heterogeneous reaction mechanism of mercury over activated carbon. Additionally, I have developed a low temperature hydrocarbon oxidation catalyst for diesel exhaust applications. Aside from emission control technologies, my research focused on spectroscopic investigation of CO adsorption and oxidation over single crystal surfaces at different pressures, developing a catalyst for cracking of hydrocarbons to valuable low hydrocarbons and designing analytical instruments for high-throughput experimentation. The wide range of research topics that I have studied has provided me with vast experience in catalysis, spectroscopy, and instrument design and development.

As a tenure-track faculty, I would like to continue my research on developing nano materials and novel processes in the area of low temperature exhaust remediation, trace metal capture, CO_2 utilization, methane reforming and alternative synthesis routes for the production of C_2-C_4 hydrocarbons. Currently, in most of the states, coal and natural gas power plants still remain the largest energy source for electricity production and play a crucial role to compensate the base loads in the electricity grid. Therefore, there is an urgent need for developing breakthrough technologies for CO_2 utilization and capture, and advanced combustion systems that can be implemented to the existing fossil fuel-fired power plants. I will initially develop my research program focusing on the aforementioned areas in carbon utilization and exhaust remediation and expand my research to create alternative energy pathways. My goal is to create a collaborative research environment interacting with national laboratories and industrial partners, and ultimately bridge the gap between fundamental science and industrial applications. In my poster presentation, I will discuss different projects to demonstrate my expertise in the area of catalysis and spectroscopy.

Teaching Interests:

Teaching has been my passion throughout my education and academic career. I gave lectures in various chemical engineering courses: Chemical Engineering Thermodynamics I, Chemical Kinetics and Modeling, Chemical Engineering Design, Chemical Plant Design Project, Mass Transfer, Kinetics and Reactor Design and Molecular Modeling. I am very interested in teaching core chemical engineering courses such as thermodynamics, heat transfer and chemical kinetics. I also plan to develop a special topics course on in-

situ spectroscopy for surface science applications. I will discuss my teaching vision during the poster presentation.

References

1. Sasmaz, E. (Corresponding Author); Wang, C.; Lauterbach, J. "Spectroscopic investigation of Pd local structure over Pd/CeO2 and Pd/MnOx-CeO2 during CO oxidation", Journal of Materials Chemistry A, 2017, 5, 12998.
2. Wang, C.; Wen, C.; Lauterbach, J.; Sasmaz, E. (Corresponding Author); "Superior oxygen transfer ability of Pd/MnOx-CeO2 for enhanced low temperature CO oxidation activity", Applied Catalysis B: Environmental, 2017, 5, 1.
3. Wang, C.; Binder, A. J.; Toops, T.J.; Lauterbach, J.; Sasmaz, E. (Corresponding Author) "Evaluation of Mn and Sn modified Pd-Ce based catalysts for low-temperature diesel exhaust oxidation" Emission Control Science and Technology, 2017, 3, 37.
4. Sasmaz, E.; Mingle, K.; Lauterbach, J. "High-throughput screening using Fourier transform infrared imaging" Engineering, 2015, 1, 234.
5. Wang, C.; Sasmaz, E.; Wen, C.; Lauterbach, J. "Pd supported on SnO2-MnOx-CeO2 catalysts for low temperature CO oxidation" Catalysis Today, doi:10.1016/j.cattod.2015.02.021.
6. Kim, S.; Mayeda, K.; Sasmaz, E.; Lauterbach, J. "One step process for the production of BTEX and LPG-like fuel from pentanediol" ACS Sustainable Chemical Engineering, 2015, 3, 381.
7. Kim, S.; Sasmaz, E.; Lauterbach, J. "Effect of Pt and Gd on coke formation and regeneration during JP-8 cracking over ZSM-5 catalyst" Applied Catalysis B: Environmental, 2015, 168, 212.
8. Galloway, B.D.; Sasmaz, E.; Padak, B. Binding of SO3 to Fly Ash Components: CaO, MgO, Na2O, and K2O" Fuel, 2015, 145, 179.
9. Saha, A.; Abram, D.N.; Kuhl, K.P.; Paradis, J.; Crawford, J.L.; Sasmaz, E.; Chang, R.; Jaramillo, T.F. Wilcox, J. "An XPS study of surface changes on brominated and sulfur-treated activated carbon sorbents during mercury capture: Performance of pellet versus fiber sorbents" Environ. Sci. & Tec., 2013, 47, 13695.
10. Bedenbaugh,J.; Kim, S.; Sasmaz, E.; Lauterbach, J. "High-throughput investigation of zeolite catalysts for military aviation fuel cracking to liquefied petroleum gas" ACS Combinatorial Science, 2013, 15, 491.
11. Sasmaz, E.; Kirchofer, A.; Saha, A.; Abram, D.; Jaramillo, T. F. Wilcox, J. "Mercury chemistry on brominated activated carbon" Fuel, 2012, 99,189.
12. Wilcox, J.; Sasmaz, E.; Kirchofer, A.; Lee, S-S. "Heterogeneous mercury reaction chemistry on activated carbon" J. Air & Waste Manage. Assoc. 2011, 61, 418.
13. Sasmaz, E.; Aboud, S.; Wilcox, J. "Mercury binding on Pd alloys and overlays" J. Phys. Chem. C, 2009, 113, 7813.
14. Aboud, S.; Sasmaz, E.; Wilcox, J. "Mercury adsorption on PdAu, PdAg and PdCu alloys" Main Group Chemistry, 2008, 7, 205.
15. Sasmaz, E.; Wilcox, J. "Mercury species and SO2 adsorption on CaO(100)" J. Phys. Chem. C, 2008, 112, 16484.

486649 Enabling New Reaction Pathways through Creation of Tailored Molecular Sieve Catalysts

Abstract:

Research Interests:

New catalysts enabled the creation of many important technologies in the 20th century. For example, processes such as ammonia synthesis for fertilizer, fluidized catalytic cracking for high-octane gasoline, olefin polymerization for plastics and elastomers, and three-way catalytic converters for automotive emissions control all have made very large, positive impacts on society. Effective use of heterogeneous catalysts in future applications for key areas of interest, such as energy, consumer goods, and infrastructure, requires both rational design of these nanomaterials at the molecular level and development of new, alternative routes to produce high-value chemicals from carbon-containing resources.

The increased accessibility of unconventional gas sources in shale formations throughout the world presents an unprecedented opportunity to devise new reaction pathways to transform lower alkanes, such as methane (CH_4), into a diverse range of suitable chemical precursors with enhanced selectivity and reduced environmental impact. One possible strategy for harnessing CH_4 as a chemical feedstock in the near-term is through formaldehyde (CH_2O): the simplest alkanal derived from CH_4 via methanol (CH_3OH) and synthesis gas ($CO/CO_2/H_2$) that can serve as a primary building block molecule (Scheme 1, blue path). Drawing from the concept of the formose reaction for sugar synthesis in prebiotic chemistry, I propose to combine CH_2O with biomass-derived glycolaldehyde (diose) through a series of intermolecular C-C coupling reactions over molecular sieve catalysts to produce a variety of useful oxygenated intermediates (e.g., *alpha*-hydroxy carboxylic acids, esters, ketones) and simple carbohydrates (e.g., trioses) in alcoholic and aqueous media. In addition to its viability for terrestrial chemicals production, CH_2O may also hold long-term promise for synthesis of edible, chiral sugars (e.g., D-glucose) from CO_2 and H_2 during extended missions in space and on Mars (Scheme 1, red path).

Scheme 1. Formaldehyde (CH_2O) as a chemical building block for CH_4 utilization (Earth) and carbohydrate synthesis (Space).

For this approach, I will use the catalytic functionalization of CH_2O, a model chemical building block, as a core research theme (Scheme 1, outlined region) to support the development of short- and long-term projects in areas such as:

1) *Design, synthesis, and application of tailored molecular sieves*

By combining *a priori* prediction of energetically favorable organic structure directing agents (OSDAs) and knowledge of organic-inorganic guest-host interactions, I will selectively produce targeted molecular sieves - the beginning elements of such a methodology have been successfully demonstrated in academia by the Davis research group at Caltech in combination with the Deem research group at Rice University. I plan to capitalize on their early successes and further develop this methodology using the reaction pathways outlined above as the initial system for implementation.

Additional research focuses will include:

- Experimentally deriving structure-property relationships to guide computationally-driven design and synthesis of new molecular sieve frameworks and compositions
- Examining effects of OSDAs and synthesis conditions on framework topology, intra-crystalline acid and metal site functionality, and enantioselectivity
- Investigating molecular sieve performance for adsorption and separation of oxygenates and carbohydrates
- Exploring catalytic functionality for chiral assembly of small molecules

2) *Assessment of catalytic active site function, tunability, and mechanistic implications*

Once the new structures are developed, I will utilize existing characterization methods to investigate the ability of acid and metal centers in these molecular sieves to manipulate CH_2O carbonyl functionality and catalyze reactions involving C-C and C-O bond formation as well as intramolecular carbon and hydrogen shift reactions.

Furthermore, I aim to:

- Understand the influences of molecular sieve pore structure, active site confinement, and local geometric and electronic structure on product selectivities and turnover frequencies for the reaction pathways outlined above
- Elucidate key reaction pathways and determine the relevant kinetic parameters for these systems by combining *in situ* and *operando* spectroscopic techniques with isotopic labeling experiments

Teaching Interests:

A student-focused teaching approach that values and facilitates individual learning is essential for more effective chemical engineering instruction. Students must not only become proficient at problem solving, but must learn to think for themselves, cultivate a sense of professionalism, and develop critical skills needed to succeed as practicing engineers. Engineering coursework is especially conducive to this type of multi-faceted learning approach because of the ability to integrate theory with practical applications. However, the critical link between principles and practice, which can be a powerful motivator for learning in many students, is often missing from the college classroom. Thus, one of my primary objectives as an instructor is to bridge this gap and help students organize theory and facts and apply these concepts to solve new problems.

Students will be introduced to theory through traditional, lecture-based instruction that will be supplemented with real-world examples to model scenarios that the students could encounter as practicing engineers. The appropriate mix of theory and application will be dependent upon the exact nature of the course as well as the audience. For instance, undergraduate students may be more interested and receptive to examples that are industrially-relevant, whereas graduate students may appreciate topics more closely tied to research and development in a laboratory environment. Potential teaching methods include:

- Conducting break-out sessions and solving short problems in small teams

- Demonstrating the use of software tools to solve more complex problems
- Performing small-scale experiments (if possible) to supplement course concepts
- Evaluating case studies on relevant incidents in industry and academia
- Discussions with invited guest speakers

Throughout my instruction, I will integrate themes of ethics, professionalism, and process safety into the curriculum that align with the course objectives. Having early exposure to these topics will help inform how students approach problem solving in their future work by enabling them to consider the potential impacts of decisions that engineers may face.

Core courses that I am most interested in teaching include kinetics and reaction engineering, thermodynamics, mass and energy balances, unit operations, and engineering materials. Additionally, I would also like to design new upper-level undergraduate and graduate courses in engineering ethics for safety and the environment, industrial catalysis and chemical technology, catalysis and spectroscopic methods, and engineering entrepreneurship.

503776 Enhanced Catalytic Capability through Controlled Reaction Environments: A Merger of Solvent Effects and Rational Catalyst Design

Abstract:

Research Interests:

Satisfying the demand for energy and chemicals via efficient and sustainable utilization of carbon resources is the defining challenge of the 21st century. Catalysis has the potential to satisfy these demands by enabling the use of non-traditional resources and enhancing the efficiency of those currently consumed. Essential to achieving this goal is the ability to predict and design new catalytic materials, which requires a fundamental understanding of catalytic surfaces and the associated chemical landscape.

My **PhD research** experience focused on understanding the impact of water on carbonyl reduction over supported Ru catalysts, which is one of the key chemistries in catalytic biomass upgrading and thought to be strongly modified by electronic effects. Through rigorous microkinetic modelling, a quantitative understanding was developed of the mechanism by which ketones are reduced to alcohols on a Ru surface. The microkinetic model then provided better understanding of the effects of water on the rate of hydrogenation; water increases the rate of hydrogenation through a preferential stabilization of the kinetically relevant transition state on the Ru surface via hydrogen bonding. This was confirmed by comparison with alkanes, which are incapable of participating in hydrogen bonding and showed no promotional effect on the rate.

My **postdoctoral research** at the University of Minnesota focused on the production of renewable chemicals (diolefins) from lignocellulosic biomass. Diolefins are key monomers in the manufacture of synthetic rubber for technologies such as renewable car tires. Saturated furans, derived from sugars, are converted to diolefins through an acid catalyzed mechanism called 'dehydra-decyclization' that was discovered as part of my research. This was particularly exemplified with tetrahydrofuran, where near quantitative yields to butadiene were achieved using an all-silica zeolite impregnated with phosphoric acid. The production of isoprene from 3-methyltetrahydrofuran was similarly demonstrated, which was optimized through experimental high throughput catalytic screening. The technology provides a renewable route to diolefins which serve as vital building blocks for most polymer products in the rubber industry.

My **future research** direction lies at the intersection of catalytic descriptors, hybrid reaction environments and data-driven discovery. Much of catalyst discovery has been guided through heuristics and inherited knowledge, where an ad-hoc approach has been adopted for selecting better catalysts. Computational efforts provide a guided approach to catalyst design, with volcano plots based on the Sabatier principle as a prominent example. A significant push is yet to be made with experiments, where large kinetic data sets of information will be required. Beyond simply understanding catalytic functionality in traditional reaction environments, a need exists to understand a catalyst's behavior at varying reaction conditions. The nature of the solvent environment in which a catalyst is placed is one such example, which

has been shown to drastically affect the rates of reaction. Leveraging solvent effects has found beneficial applications in acid catalyzed dehydrations, metal catalyzed reductions, selective oxidation and multiple other catalytic systems. A solvent based approach to tuning catalyst performance becomes especially attractive in areas such as catalytic biomass upgrading, where the inherently high water content of biomass can be utilized. Through a predictive understanding of how a solvent environment can affect a catalytic surface, reaction conditions can be better tailored to maximize desired catalytic transformations. **Teaching Interests:**

Research can inspire teaching. As part of my graduate degree program, I was involved in undergraduate course development, where I designed experiments, equipment and course material for undergraduate teaching labs. I designed and constructed an experiment where senior year students operated a vapor phase packed bed reactor, studying the kinetics of isopropanol dehydration over solid acid catalysts. I also designed an experiment to demonstrate the concept of residence time distribution (RTD) in reactors, where students measured and applied RTD fundamentals to a series of continuously stirred tank reactors using step tracers. I also served as a teaching assistant for graduate level mathematics for chemical, biomedical and mechanical engineering students. While I would especially enjoy teaching reaction engineering courses, both at the graduate and undergraduate level, I am comfortable teaching many of the core chemical engineering courses (Mass & Energy Balances, Heat & Mass transfer and thermodynamics) as well as upper-level courses including Process Design, Unit Operations and Product Design.

505454 Explaining Surface-Catalyzed Reactions in Electrochemistry

Abstract:

Research Interests:

Electrochemical surface-catalyzed reactions provide potent chemical tranformations which are promising to enable an alternative energy future. One quintessential electrochemical surface-catalyzed reaction is carbon dioxide reduction. Essentially, this reaction can close the carbon cycle because the carbon from carbon dioxide is passed to carbon-containing fuels. High-performance computing will continue to provide insight into the atomistic scale reaction mechanism, which is poorly understood. Reaction mechanisms studies have initiated, but many reaction pathways remain unknown.[1,2] The reaction occurs with a catalyst surface which experiences electrical charge. Recently[3], as many as sixteen chemical products of this reaction have been measured. The number of surface reactions is large and obtaining the most accurate transition states for each reaction can make a difference in which is the dominant reaction mechanism. Furthermore, a change in a transition state can influence the rate-controlling step.[4] To face this challenge, the growing string method is able to automatically search for reaction paths of elementary steps and their contained transition states in a rapid manner.[5] This automated reaction path discovery may be initiated from an initial geometry-optimized reactant and driving coordinates including bond breaks, bond additions and/or bond torsions.

Continuum-scale modeling, orders of magnitude longer length and time scales than atomic-scale density functional theory modeling, can offer the most realistic predictions of reaction performance. Continuum-scale modeling incorporates non-idealities such as concentration and pH gradients. For example, the pH of the layer of solvent near the surface of a copper catalyst changes in pH which affects the solubility of CO_2.[6] CO_2 must migrate to the catalyst surface for the reaction to activate. Also, the pH of the solvent region near the catalyst may be considered to drive reactions in the solvent, absent the catalyst.[7] Transferring uncertainty from ab initio calculations to the reactor-scale continuum offers a challenge to incorporate all uncertainties and non-idealities.[8] Figure 1 displays the length and time scales bridged by a multi-scale model. All together, this careful multiscale modeling of the electrochemical reduction of carbon dioxide will advance understanding of its mechanism and enable the improvement of the reaction performance.

Teaching Interests:

For teaching undergraduate courses, I am particularly interested in: Numerical Methods for Chemical Engineers, Mass and Heat Transfer, and Chemical Engineering Kinetics. For teaching graduate courses, I am interested in courses such as Chemical Reactor Design and Chemical Process Principles. In any course I teach, the incorporation of computer programming would ensure that students learn attention to detail and how to formulate engineering problems in an explicit manner. Furthermore, I believe visualizing data and technical writing are skills that chemical engineers should obtain in preparation for the labor force or academia.

Figure 1. Length and time scales of density functional theory to chemical reactor modeling.

[1] Goodpaster, J. D.; Bell, A. T.; Head-Gordon, M. *Phys. Chem. Lett.* **7** (2016) 1471-1477.

[2] Gattrell, M.; Gupta, N.; Co, A. *J. Electroanal. Chem.* ***594*** *(2006) 1-19.*

[3] Kuhl, K. P.; Cave, E. R.; Abram, D. N.; Jaramillo, T. F. *Energy Environ. Sci.* ***5*** *(2012) 7050-7059.*

[4] Jorgensen, M.; Gronbeck, H. *ACS Catal.* **6** (2016) 6730.

[5] Jafari, M.; Zimmerman, P. M. *J. Comput. Chem.* ***38*** *(2017) 645-658.*

[6] Singh, M. R.; Kwon, Y.; Lum, Y.; Ager, J. W.; Bell, A. T. *J. Am. Chem. Soc.* **138** (2016) 13006-13012.

[7] Birdja, Y. Y.; Koper, M. T. M. *J. Am. Chem. Soc.* **139** (2017) 2030-2040.

[8] Tinsley Oden, J.; Prudhomme, S.; Romkes, A.; Bauman. P. T. *SIAM J. Sci. Comput.* **28** (2006) 2359-2389.

495060 Insight and Applications of Pt-Bi Bimetallic Catalysts: A Combined Experimental and DFT Study

Abstract:

Research Interests: Catalytic Reaction Engineering

Teaching Interests: Chemical Thermodynamics and Chemical Reaction Engineering

Owing to environmental concerns and scarcity of known fossil reserves, biomass has been shown as an important renewable source, which can be converted into alternative fuels and value-added chemicals. Heterogeneous catalysis plays an important role in such processes, while the following two issues still remain challenging and limit their industrial applications: tuning selectivity of catalyst towards desired products and improving stability of catalyst against coking/carbon deposit. Both issues can be addressed by the use of bimetallic catalysts, containing a primary metal (e.g. Pt, Pd, Ru, Rh, etc.) and a promoter (e.g. Mo, Sb, Bi, etc.). In the present work, we report such an example with insight and applications of Pt-Bi bimetallic catalysts. It demonstrates that, over Pt-Bi catalysts, selectivity of 1,3-dihydroyacetone (DHA) from glycerol oxidation is tuned, and catalyst stability for reactions involving CH4 (guaiacol deoxygenation using CH4) is improved. Based on experimental observations and theoretical analysis, Bi locates in an adatom-like configuration Pt and a cooperative effect between Pt as the primary metal and Bi as a promoter is identified. Insight into Pt-Bi bimetallic catalysts also proposes additional catalytic applications, such as short-chain alcohol oxidation and hydrocarbon activation.

505609 Integrating Computational Chemistry Techniques to Understand Complex Chemical Reactions

Abstract:

Research Interests:

My research employs Density Functional Theory and Wave Function-based methods to understand catalytic reaction mechanisms, including heterogeneous, homogeneous, and/or photo-activated reactions. In addition, I have considerable knowledge regarding the modeling of hard matter-soft matter interactions. Building upon my wide range of research experience, I plan to develop a synergistic research program that encompasses the different subfields of computational catalysis, such as homogeneous, heterogeneous, and photocatalysis, by combining the valuable ideas, concepts, and techniques developed in each field and applying them to complex problems at the interfaces of these fields. I believe this approach is the key element for modeling and gaining valuable insight into important complex chemical reactions under realistic conditions. Examples of problems that can benefit from such an approach include the dynamical aspects of supported nanoparticle catalysis, modeling of photoelectrochemical cells, and interfacial phenomena at solid-liquid interfaces. Last but not least, I have a keen interest in developing computational tools and frameworks that can help to disseminate these new approaches to the research community.

Teaching Interests:

My envisioned contributions to teaching involves instilling in the younger generations the â€˜engineering approachâ €™ which allows one to break down larger problems into manageable smaller tasks, and eventually solving the questions that seemed unsolvable at first glance. It is my belief that the inclusion of state-of-the-art research findings in teaching is one of the best ways to foster an engineering mindset, as well as promote and maintain the curiosity and open-mindedness of students to novel ideas. Integrating research with teaching would allow students to develop their self-motivation and drive to search for new interesting problems and potential solutions on their own. For this reason, I would like to develop undergraduate and graduate Catalysis and Molecular Modelling electives that are related to my own research interests.

500142 Integration of Machine-Learning and Data Management Methods for Accelerated Catalyst Modeling and Exploration

Abstract:

Research Interests:

My primary research interest is the development of new machine-learning based techniques to help accelerate the discovery of increasingly effective catalysts. As the quality of life continues to improve on a global scale, the demand for energy resources continues to grow. However, use of current fossil fuels results in harmful CO_2 and NO_x emissions, contributing to dangerous global climate change. As a result, sustainable energy sources are becoming increasingly necessary and popular. However, these technologies are still in early stages of development, and there is still a need to improve their energy efficiency to support commercial viability. Thus, the rapid development of new catalysts which can help these technologies compete in the current energy economy is of critical importance.

Towards these ends, I have spent the last five years studying computational catalysis at Carnegie Mellon University. I worked with Professor John Kitchin to develop a new kind of atomic potential constructed from neural networks. These potentials have become increasingly popular in the last decade with the advent of fingerprinting schemes, proposed by Behler and Parrinello, which allow them to be trained to a wide variety of atomic input structures. This has unlocked their ability to reproduce potential energy surfaces (PES) of a multitude of chemical systems with levels of accuracy unrivaled by their physical potential counterparts[1]. They are also significantly faster to evaluate than a full quantum chemistry calculation resulting in the ability to study much larger unit cells and more complex chemistry. Utilizing these techniques, we have demonstrated that it is possible to model the high-dimension PES for the mobile surface of Pd with coverage dependence of oxygen[2]. It has also been used to make quantitative predictions of segregation for AuPd under vacuum conditions at all bulk compositions of the alloy[3]; an integral step for relating the active sites, which computationalists model, to the factors which we have control over when creating the catalyst.

Currently, I am a post-doc at Stanford working with Thomas Bligaard on developing grey-box models to accurately predict adsorbate coverage effects across transition metals. While here, I will gain additional insight into input parameter selection as well as inplementing unsupervised machine-learning for characterizing these parameters. Selecting the inputs for these problems is critical because it is only through these inputs that the algorithms are able to make any useful predictions. However, when there are so many possible input parameters available, it becomes challenging to decide which are most important for characterizing the system of interest. For example, many machine-learning potentials are based on descriptions of the unit cell geometry and compositions of the atoms. While this is quite tractable for modeling a PES, the inputs quickly become too numerous when considering all possible adsorbate and transition metal elements. Moreover, we would intuitively no longer expect the positions of the atoms to be an effective descriptor for the interaction energies of different species. Fortunately, based on known relationships of adsorption across metal surfaces, we know such effective input descriptors are likely to exist. It then follows that the accuracy can be improved using other inputs which are important to co-adsorption in a machine-learning framework.

By learning and testing these new aspects of machine-learning, I will acquire the tools I need for streamlining and automating the technical aspects of the search for new catalysts.

As available computational resources continue to increase, the rate of data generation in the field of computational chemistry is quickly beginning to outstrip the ability of those in the field to interpret it all. Furthermore, there is often also a great deal of usefulness in results beyond that intended by those who created it. To improve the ability of all those in the field of catalysis to continue making rapid scientific progress, and to achieve more with less, there is a powerful need to develop data management techniques which are capable of storing and searching this information; similar to the way the Materials Project and others have done for the field of material science. But for catalysts, this is significantly more challenging due to the large number of possible catalytic systems of interests thereby creating a greater need for machine-learning. Even with a robust database structure capable of storing a breadth of relevant information, there is a critical need for tools which are capable of intelligently querying that database. Machine-learning not only benefits from such large repositories of information, it can also assist in making such intelligent identifications which allow the user to find new and interesting results. All of these tools would provide an incredible resource to the field, and my work tailors these methods to best suit the field of catalysis.

Teaching Interests:

As an instructor, I teach: thermodynamics, kinetics, molecular simulation (and underlying quantum theory), reaction engineering, functional coding languages, and general chemical engineering. Chemical engineers fulfill a diverse range of demand in industry positions today. As modern tools become increasingly computationally intensive, there is an ever growing need to shift the emphasis of education to computational tools and techniques. As such, I strongly support the integration of coding into the classroom. This also ties into my interest in educational research which I have outlined below.

As an undergraduate, I spent a cumulative three years working in a general chemistry tutoring center, mentoring individuals and small groups. I often attribute my initial love of education to this time, and I look forward to a future as an educator. In my first year at CMU, I was a teaching assistant for the undergraduate introductory courses for general chemical engineering as well as thermodynamics. Since then I have also been the TA for the graduate level molecular simulation and reaction engineering courses. Most recently, I have been the math software TA, providing introductory level courses for MatLab to all students in the chemical engineering department. For all of these courses I have provided regular office hours and often additional recitations for the students. I have also given multiple guest lectures in graduate level courses in the mechanical and chemical engineering departments, for which I planned the subject material. For my educational contributions to the department, I received The Mark Dennis Karl Outstanding Teaching Assistant Award in 2015.

I have also pursued instruction on educational techniques for engineering by taking an evidence-based teaching course specific to STEM fields. I am a strong proponent of the flipped classroom as a pedagogical teaching model. This model requires students to study lecture material before the class and implement it on assignments given during class. Among other benefits, such a system allows the instructor to provide immediate feedback to students which is shown to improve student retention rates and longevity of student knowledge. Further details of my pedagogical teaching strategies and achievements can be found on my website.

I would like to study the potential benefits of teaching code-based chemical engineering tools in an immersive environment. The concept is similar to that of language immersion, in which various subject material is taught through the medium of a second language. Language immersion is known to dramatically improve student comprehension of the second language in technical communication with no discernible negative impact on primary language skills. Teaching students to solve chemical engineering problems can similarly be framed in the medium of a coding language, such as Python or MatLab.

Relevant Work

1. Jacob R. Boes, Mitchell C. Groenenboom, John A. Keith, and John R. Kitchin. Neural network and ReaxFF comparison for Au properties. International Journal of Quantum Chemistry, 116(13):979â€“987, 2016. DOI

2. Jacob R. Boes and John R. Kitchin. Neural network predictions of oxygen interactions on a dynamic Pd surface. Molecular Simulation, 43(5-6):346-354, 2017. DOI

3. Jacob R. Boes and John R. Kitchin. Modeling Segregation on AuPd(111) Surfaces with Density Functional Theory and Monte Carlo Simulations. The Journal of Physical Chemistry C, 121(6):3479-3487, 2017. DOI

Curriculum Vitae

ORCID

Website

Phone: (607) 342-1846

Email: jacobboes@gmail.com | jrboes@stanford

502910 Magnetic Polymer Nanocomposites for Giant Magnetoresistance and Electromagnetic Shielding

Abstract:

Due to the incomparable advantages such as light weight, optical, magnetic and enhanced mechanical properties, polymer nanocomposites have been widely studied by the researchers for different applications such as environmental remediation, energy (or information) storage and sensing.[1-4] Giant Magnetoresistance (GMR) phenomenon can be designed for information storage, our previous work shows that polymer nanocomposites show unique GMR behavior, indicating great potential applications for information storage.[5] Meanwhile, as the development of electrical devices, the electromagnetic wave radiation is a potential threat for human being's health.[1]Thus, the materials with electromagnetic wave absorption are urgently needed. In this project, we found the negative GMR in the magnetite polypyrrole nanocomposites synthesized by the surface initiated polymerization method. The temperature dependent resistance of the magnetite nanocomposites indicates a 3-D variable range hopping (VRH) electrical conduction mechanism. The positive GMR is observed in the synthesized pure polypyyole at room temperature and analyzed by the wave function shrinkage model. The negative MR is obtained in the synthesized magnetic PNCs at room temperature and analyzed by the orbital magnetoconductivity theory. Meanwhile, the enhanced electromagnetic wave shielding is observed in its epoxy nanocomposites, which is due to the increased interface area, eddy current loss and increased anisotropic energy. Moreover, the significantly reduced flammability is observed and analyzed as well.

Research Interests:

- Conductive Polymers
- Polymeric Giant Magnetoresistance Materials
- Environmental Remediation Polymers
- Electromagnetic Interference Shielding/Microwave Absorption Composites

Teaching Interests:

- Thermodynamics
- Fluid dynamics
- Advanced materials processing

Reference:

1. J. Guo, H. Song, H. Liu, X. Liu, C. Luo, X. Zhang, J. Kong, Z. Guo, Y. Ren, T. Ding, M. A. Khan and D. P. Young, *Journal of Materials Chemistry C*, 2017, DOI: 10.1039/C7TC01502J.

2. J. Guo, L. Guan, H. Wei, M. A. Khan, X. Zhang, B. Li, Q. Wang, B. L. Weeks, D. P. Young, T. Shen, S. Wei and Z. Guo, *Journal of The Electrochemical Society*, 2016, **163**, H664-H671.

3. H. Gu, J. Guo, H. Wei, S. Guo, J. Liu, Y. Huang, M. A. Khan, X. Wang, D. P. Young, S. Wei and Z. Guo, *Advanced Materials*, 2015, **27**, 6277-6282.

4. B. Qiu, J. Guo, X. Zhang, D. Sun, H. Gu, Q. Wang, H. Wang, X. Wang, X. Zhang and B. L. Weeks, *ACS applied materials & interfaces*, 2014, **6**, 19816-19824.

5. J. Guo, H. Gu, H. Wei, Q. Zhang, N. Haldolaarachchige, Y. Li, D. P. Young, S. Wei and Z. Guo, *The Journal of Physical Chemistry C*, 2013, **117**, 10191-10202.

505444 Making Renewables Chemicals and Biofuels Economical: Toward Complete Utilization of Lignocellulosic Biomass

Abstract:

Research Interests:

My research has focused on the production of biofuels and renewable chemicals. I believe that renewable energy sources are key for future development. My PhD research focused on the production of biodiesel. The goal was to replace homogenous catalysts (NaOH) by more sustainable heterogeneous catalysts. During this period, I was able to learn and become familiarized with the basic and fundamental tools utilized in the catalysis field (catalyst preparation and characterization). This work provided me with the background necessary for my next stage as a Postdoctoral researcher at UW-Madison under the supervision of Prof. Dumesic.

During my stay in Prof. Dumesic's group my research goal was to develop new processes to convert lignocellulosic biomass into biofuels and renewables chemicals. In my first year, I focused on the conversion of gamma-valerolactone (GVL), a biomass derived building block, into butene and high molecular weight olefins to be used as biofuel. We developed an integrated system to combine 3 reactions and two separation steps. Achieving that level of process integration was critical for the success of the project that was published in Science. This was the beginning of a journey working with GVL.

In the next years, I discovered that GVL can not only be used as platform molecule to produce high value chemicals and biofuels, but it also can be used as a solvent to process lignocellulosic biomass. In collaboration with other PhD students, within and outside the group, we demonstrated that GVL can be used as a solvent to produce key intermediates in biomass conversion, such as fermentable sugars, furfural, HMF and levulinic acid. We were able to show that dehydration and hydrolysis reactions are 30 and 100 times faster in GVL than in water leading to significant advantages, such as being able to carry out typical dehydration and hydrolysis reaction at much milder conditions than in water which significantly increases the yields obtained. These works were published in high impact journals (Science, Energy and Environmental Science, Angewandte Chemie International Edition and Green Chemistry). The success of the work led to the publication of 5 patents that were licensed by the start-up company, Glucan Biorenewables LLC, which hired me to commercialize the technology. In the past 4 years, I have divided my research efforts between my work at UW-Madison with Prof. Dumesic and my activities at Glucan Biorenewables LLC as Director of R&D. Working for a company gave me a new perspective for my research. Not only do we need to understand the fundamentals of the processes and provide innovative solutions, but these solutions need to make sense economically. When working with lignocellulosic biomass, it was obvious that better economics were necessary to move the technology from the lab to commercial scale. With that in mind, I started working towards the complete utilization of lignocellulosic biomass. The aim was to simultaneously obtain value from the three main fractions in biomass: cellulose, hemicellulose and lignin. By taking advantage of previous work performed using GVL as solvent, I designed a new process to fractionate lignocellulosic biomass. In the first step the hemicellulose and the lignin are dissolved in the GVL solvent at mild conditions. This step provides a high purity cellulose stream that can be used to produce high quality fibers and clothes. In a second step, the lignin is precipitated from the GVL solution. Because of the mild conditions used in the first step, the lignin retains its native β-O-4 linkages and it is more reactive than other lignins. A collaboration with the University of Tennessee demonstrated that because of its purity and reactivity this lignin could be used to produce battery anodes and carbon foams. After removal of the lignin, the hemicellulose solution can be concentrated and converted into furfural at excellent yields (95%) within the GVL solvent. Taking advantage of the furfural-water azeotrope (98 °C), furfural can be easily removed from the GVL solvent. The process converts more than 80% of the carbon present in the lignocellulosic biomass into high value products, obtaining revenues over $500 per metric ton of dry biomass and demonstrating that a biorefinery could be competitive with a petroleum refinery. The work was recently published in Science Advances.

Selected publications (total 46):

1. D. M. Alonso et al., "Increasing the revenue from lignocellulosic biomass: Maximizing feedstock utilization," Sci. Adv., vol. 3, no. 5, p. e1603301, May 2017.
2. J. Q. Bond, D. M. Alonso, D. Wang, R. M. West, J. A. Dumesic, "Integrated Catalytic Conversion of gamma-Valerolactone to Liquid Alkenes for Transportation Fuels", Science, 2010, 327, 1110-1114.
3. J.S Luterbacher, Jacqueline M Rand, D. M. Alonso, J. Han, J T.Youngquist, C.T Maravelias, B.F. Pfleger, J.A Dumesic, "Nonenzymatic Sugar Production from Biomass Using Biomass-Derived γ-Valerolactone," Science 2014, 343, 6168, 277-280
4. D. M. Alonso, J. Q. Bond, J. A. Dumesic, "Catalytic conversion of biomass to biofuels", Green Chemistry, 2010, 12, 1493-1513.
5. D. M. Alonso, S.G. Wettstein, J.A. Dumesic Bimetallic catalysts for upgrading od biomass to fuels and chemicals, Chem. Soc. Rev., 2012, 41, 8075-8098
6. D. M. Alonso, S. G. Wettstein, J.A. Dumesic "Gamma-Valerolactone, a sustainable platform molecule derived from lignocellulosic biomass." Green Chemistry. 2013, 15, 584-595
7. D. M. Alonso, S. G. Wettstein, M. A. Mellmer, E. I. Gurbuz and J. A. Dumesic, "Integrated conversion of hemicellulose and cellulose from lignocellulosic biomass" Energy & Environmental Science, 2013, 6 (1), 76-80
8. D. M. Alonso, S.G. Wettstein, J.Q. Bond, T.W. Root, J.A. Dumesic, "Production of biofuels from cellulose and corn stover using alkylphenol solvents," ChemSusChem, 2011, 8, 1078-1081
9. M.A Mellmer, C. Sener, J.M.R. Gallo, J. S. Luterbacher, D. M. Alonso, J.A. Dumesic" Solvent Effects in Acid-Catalyzed Biomass Conversion Reactions" Angewandte Chemie International Edition 53 (44), 11872-11875
10. M. L. Granados, M. D. Z. Poves, D. M Alonso., R. Mariscal, F. C. Galisteo, R. Moreno-Tost, J. Santamaria, J. L. G. Fierro, "Biodiesel from sunflower oil by using activated calcium oxide", Applied Catalysis B-Environmental 2007, 73, 317-326.

Funding: Obtaining funding from Government agencies has been part of my activities over the last few years. This responsibility will also be critical as a Faculty. NSF and DOE are the main funding agencies I will target to fund my research. By my previous interactions with these agencies I have realized that collaboration with other professors and other Universities will be critical to obtain funding for pioneering projects. I would also like to engage companies in research projects. Funding with companies will not only provide addition money for the research, but also a valuable interaction between the students and the companies so they are better prepared for their future work.

Successful proposals:

1. DE-FOA-0001619 DOE STTR: High Purity Cellulose for Low-Cost Nanocellulose and Biofuel Production (ref. 0000230723). Department of Energy. June 2017-January 2018. $150.000. Principal Investigator
2. NSF Phase II: SBIR Phase II: Catalytic Conversion of Lignocellulosic Biomass into Furfural and Dissolving Pulp using Green Solvents (ref. 1632394). National Science Fundation (USA). September 2016-August 2018. $750.000. Principal Investigator
3. NSF Phase IB: SBIR Phase I: Catalytic Conversion of Lignocellulosic Biomass into Glucose using Green

Solvents (ref. 1602713). National Science Fundation (USA). January 2016-July 2016. $30.000. Principal Investigator

4. NSF Phase I: SBIR Phase I: Catalytic Conversion of Lignocellulosic Biomass into Glucose using Green Solvents (ref. 1519869). National Science Fundation (USA). July 2015-December 2015. $150.000. Principal Investigator
5. NSF Phase IB: Green Solvent-Enabled Synthesis of Biobased Furans (ref. 1406606) National Science Fundation (USA). January 2014-June 2014. $29.999. Principal Investigator
6. NSF Phase I: Green Solvent-Enabled Synthesis of Biobased Furans (ref. 1315356). National Science Fundation (USA). July 2013-December 2013. $150.000. Principal Investigator

Teaching Interests:

My main goal as a professor will be to provide undergraduate and graduate students with the tools necessary to become successful engineers prepared for future challenges. To achieve that goal, I will provide them with a combination of fundamental and practical skills, which they will use to solve problems individually and in teams during their courses. Example of courses I would enjoy teaching are: mass and energy balances, reactor design and kinetics. I also consider it important to provide the students with knowledge about renewable energies and renewable products as it will be critical for a more sustainable future for the chemical engineering field. In this sense, my experience in biomass conversion can be critical to start new projects or courses.

505464 Mechanisms of Heterogeneous Catalysis for Clean Energy Conversion and Efficient Chemical Production

Abstract:

Luke M. Neal

Postdoctoral Research Scholar

Postdoctoral Research: Chemical Looping Redox Catalyst

Syngas Production: Chemical looping-reforming replaces the gaseous oxygen of oxidative reforming with the lattice oxygen of a redox-catalyst. This technology can improve the efficiency of Gas-to-liquid (GTL) processes. I formulated and characterized novel high-selectivity, high-oxygen capacity core-shell redox catalysts for this system. I reported that there were 4 distinct reaction regions for this type of catalyst, which have subsequently been observed in several related systems. My research also shows that the selectivity of a redox catalyst with high internal oxygen mobility is determined by the types of oxygen species available at the redox catalyst surface.

Chemical Looping-Oxidative Dehydrogenation of Ethane: I co-invented a licensed chemical-looping redox catalyst for oxidative dehydrogenation of ethane (ODH). My research demonstrates that this approach can give high yields and selectivity without the significant dilution typically needed in co-feed ODH. In follow-up work I developed kinetics for the design of a pilot unit currently under construction.

Thesis Research:

Ph.D. Dissertation: "Nano-Particle Supported Catalyst" Advised by Dr. Helena Hagelin-Weaver in the Department of Chemical Engineering at the University of Florida

The oxidative coupling of 4-methyl-pyridines to 4,4′-Dimethyl-2,2′-bipyridine is a green chemistry process and a useful probe reaction. Palladium on metal oxide supports had previously been reported as inactive for this reaction. By synthesizing and extensively characterized multiple palladium catalysts, I showed that "inert" nano-particle metal-oxide supports can induce similar behavior in platinum-group-metal catalysts typically associated with oxides exhibiting strong metal-support interactions resulting in high activity.

Research and Funding Successes: I have authored or co-authored 18 refereed articles on a variety of catalytic reactions. I have supervised multiple research projects, mentoring over 12 graduate and undergraduate researchers, and have participated in the writing of 6 funding proposals and multiple SOPOs, and reports leading to project successes. I co-invented a novel chemical looping ODH technology which secured a $3.8 million overall plus-up from DOE/ARPA-E. My work has led to 3 provisional and 2 nonprovisional patent applications 2 of which were licensed to a commercial partner.

Future Direction: Catalysis is the classic means of "process intensification" in reaction engineering, but new insights into catalyst performance by mechanistic studies of real, practical catalysts is needed to keep pace with growing demands for clean energy and chemical production worldwide. My research on both supported platinum-group metal- and mixed-metal-oxide redox- catalysts will contribute to improved mechanistic understanding of greener, low-emission catalytic processes. My future research will focus upon bridging the gap between real-world catalyst performance and fundamental studies on model systems. The goal of this emphasis is to build mechanistic tools to guide rational development of more efficient industrial catalysts. I will also actively pursue multi-dispensary collaboration. For example, system modeling is valuable for evaluating the interaction between process economics and catalyst performance parameters. Model catalyst development and characterization in collaboration with computer simulation for determining catalyst mechanisms is also important. I also welcome other opportunities to expand my knowledge by using my expertise in heterogeneous catalysis in support of my colleagues' research programs.

Research Interests: Building on my experience in practical reaction experimentation and mechanistic characterization of commercially relevant catalysts, I plan to conduct research in three areas: I) Oxygen storage materials for NO_x emissions reduction catalyst; II) Mechanisms of oxygenate formation in hydrocarbon systems; III) Unburned hydrocarbon oxidation over true earth-abundant materials.

Rational design of oxygen storage materials for enhanced NO_x reduction and advance combustion: Several emissions catalyst rely on "oxygen storage material" (OSM) supports. These materials allow the catalyst to oxidize hydrocarbons and CO in the presence of low pressures of oxygen and/or reducing agents needed for the suppression/reduction of NO_x. Rational design of OSMs has, so far, focused heavily on oxygen capacity at operating temperatures. Research on direct NO_x reduction catalysts has shown that other properties are impotent, but has tended to limit studies to unrealistically low N_2 and O_2 concentrations. My work will look at a full range of parameters, including oxygen-vacancy formation energies, internal oxygen transport rates and surface site pH. The principles developed will allow the rational design of improved emissions abatement catalysts.

Mechanistic study of oxygenate formation: The creation of alcohols and other oxygenates from alkanes is simultaneously a desired energy conversion process and a nuisance. Unwanted oxygenates can entrain water, poison catalysts, and add complexity to separations. However, oxygenate formation in processes such as gas-to-liquids and steam cracking is not well understood. I plan to use n-butane dehydrogenation as a model system. My research will probe a variety of catalysts with differing selectivities for oxygenate to determine underlying mechanisms. My studies will include in-situ/operando measurements, as well as surface characterizations. This project will yield mechanistic insights that guide suppression of oxygenates, while potentially developing improved catalysts for *on*-purpose oxygenate production and other processes.

True earth-abundant materials for deep oxidation of UHCs and CO: Novel, inexpensive mixed-metal-oxides will be used for deep oxidation of unburned hydrocarbons (UHCs) and carbon monoxide emitted in flue gases. Some redox catalysts, including Mg_6MnO_8 and $SrMnO_3$, have significant deep oxidation activity. These mixed oxides can exhibit better thermal stability and poisoning resistance than traditional platinum-group metal (PGM) catalysts, while potentially working well in very low partial pressures of oxygen. They can also be made more cheaply than typically proposed rare-earth oxides. Understanding the flexible properties of these true earth-abundant metal oxides can guide improvement of coal and biomass energy conversion efficiency, while lowering catalyst costs.

Teaching Interests: I welcome the opportunity to teach courses; I am capable of teaching any core undergraduate chemical-engineering course, and I have found that revisiting core chemical engineering concepts opens up deeper understanding of the subject. My background in applied, heterogeneous catalysis has given me particular skills and experiences well suited to teaching undergraduate courses relating to mass balances, thermodynamics, kinetics, and design. I am also interested in teaching graduate-level courses in kinetics and surface sciences, as well as special topic courses in catalysis and energy conversion. I am committed to studying and implementing teaching theory and techniques such as active learning. My approach to teaching has been molded by my experience as a TA and guest lecturer, as well as my supervision and mentoring of graduate and undergraduates in laboratory research. I have TA'd undergraduate students in thermodynamics and unit operations, and I have trained, mentored, and supervised over a dozen graduate and undergraduate students in their research, including setting research plans for three master students and a Ph.D. student. These mentoring interactions have included numerous international and four women and minority students. I am committed to pursuing further diversity through teaching and providing research opportunities.

Selected Publications:

S. Yusuf+, **Luke M. Neal+**, F. Li. "Effect of Promoters on Manganese Containing Mixed Metal Oxides for Oxidative Dehydrogenation of Ethane via a Cyclic Redox Scheme." ACS Catalysis, *Accepted* (2017); (+Co-first authors);

Luke M. Neal+, S. Yusuf+, F. Li. "Oxidative Dehydrogenation of Ethane: A Chemical Looping Approach." Energy Technology, **4**: 1200 (2016); (+Co-first authors);

Luke M. Neal, A. Shafiefarhood, F Li. "Effect of Core and Shell Compositions on $MeOx@La_ySr_{1-y}FeO_3$ Core–Shell Redox Catalysts for Chemical Looping Reforming of Methane." Applied Energy **157**, 391–398 (2015);

Luke M. Neal, A. Shafiefarhood, F. Li. "Dynamic Methane Partial Oxidation Using an $Fe_2O_3@La_{0.8}Sr_{0.2}FeO_{3-\delta}$ Core–Shell Redox Catalyst in the Absence of Gaseous Oxygen." ACS Catalysis, **4**: 3560–3569 (2014);

Luke M. Neal, M. Everett, G. Hoflund, H. Hagelin-Weaver. "Characterization of Palladium Oxide Catalysts Supported on Nanoparticle Metal Oxides for the Oxidative Coupling of 4-Methlypyridine." Journal of Molecular Catalysis A: Chemical, **335**: 210-22 (2011);

Luke M. Neal, S. D. Jones, M. L. Everett, G. B. Hoflund, H. E. Hagelin-Weaver. "Characterization of Alumina-Supported Palladium Oxide Catalysts in the Oxidative Coupling of 4-Methylpyridine." Journal of Molecular Catalysis A: Chemical, **325**: 25-35 (2010).

Nonprovisional Patent Applications:

F. Li, **Luke M. Neal**. "Ethylene Yield in Oxidative Dehydrogenation of Ethane and Ethane Containing Mixtures." US **15/422,743** (2017);

J.A. Sofranko, F. Li, F. Li, **Luke M. Neal**. "Oxygen Transfer Agents for the Oxidative Dehydrogenation of Hydrocarbons and Systems and Processes Using the Same." US **62/054,424** (2015).

505582 Molecular Modelling for Catalytic Reaction Engineering

Abstract:

Research Interests:

In course of my PhD and Post-Doctoral research I have maintained keen interest in fields like chemical vapour deposition synthesis of carbon nanostructures, photo-electro catalysis, biomass pyrolysis and so on and have been involved in research in molecular modelling for catalysis and reaction engineering in fields like methane activation and conversion and catalytic transformation of biomass derived feedstock.

Mechanistic insights on C-H bond activation and materials for methane chemistry

With recent discoveries of significant reserves of gas hydrates, my PhD work focussed on the chemistry of methane, as a step to contribute to realising a chemical supply chain around methane feedstock, replacing existing ones based on petroleum. In the thesis I showed that 1) high impact intermolecular collision of methane can result in their activation and direct bimolecular transformation to ethylene, unlike the unimolecular thermal dissociation and subsequent biradical coupling, 2) sub-nanometre clusters of copper and copper oxide are extremely active in dissociating methane unlike bulk copper which is inactive, with the former being effective in coupling reactions to form C_2 hydrocarbons, and 3) vacancy defects and dopants significantly alter reactivity of the surface of metal oxides, with Ni vacancies and Li as a low valent dopant significantly increasing reactivity of nickel oxide for methane dissociation. Insights on C-H bond activation of methane resulting from non-catalytic high impact collisions and the dissociation process on nano-clusters and oxide surfaces of abundant transition metals are crucial to develop new processes and to screen and design efficient catalysts for its conversion to value added chemicals.

Mechanistic insights on multifunctional catalysis for chemicals from biomass and solvent effects in such transformations

Biomass forms a carbon neutral feedstock for producing a variety of fuels and chemicals which are currently derived from petroleum feedstock. My research efforts also span the chemical transformation of biomass derived sugars and alcohols to value added chemicals with focus on unravelling mechanistic details of the transformations on selected multifunctional homogeneous and heterogeneous catalysts and explaining the contribution of different solvents in altering reactivity and product selectivity. The projects I worked on include 1) homogeneous Cr catalysis for glucose isomerization to fructose, 2) non-equilibrium solvation and solvent dynamics during hydride transfer in glucose catalysed by Sn-beta, 3) catalytic glycerol oxidation to dicarboxylic acids on copper (II) oxide and 4) glycerol hydrogenolysis to propane diol on ReOx modified Ir catalyst.

Expertise and skills

- Density Functional Theory (DFT), Møller–Plesset perturbation theory (MP), Coupled Cluster Theory, Classical Molecular Dynamics, *Ab Initio* Molecular dynamics (CPMD), Metadynamics

- Potential/Free energy surfaces, transition states, reaction paths, charge transfer, electronic structure, bonding, energy decomposition, vibrational analysis, band structure of solids, solvation, dynamics, first principles X ray spectroscopic characterization

Publications

- Jithin J. Varghese, Q. T. Trinh and Samir H. Mushrif, "Insights into the synergistic role of metal-lattice oxygen site pairs in four-centred C-H bond activation of methane: The case of CuO", *Catal. Sci. Tech.*, **6,** 3984 – 3996, 2016
- Jithin J. Varghese and Samir H. Mushrif, "First principles investigation of the dissociation and coupling of methane on small copper clusters: Interplay of collision dynamics and geometric and electronic effects", *J. Chem. Phys.*, 142, 184308, 2015
- Samir H. Mushrif, **Jithin J. Varghese**, Chethan B.K., "Solvation Dynamics and Energetics of Intramolecular Hydride Transfer Reactions in Biomass Conversion", *Phys. Chem. Chem. Phys.*, **17**, 4961-4969, 2015
- Samir H. Mushrif, **Jithin J. Varghese**, Dion G. Vlachos, "Insights into the Cr(III) catalyzed isomerization mechanism of glucose to fructose in the presence of water using ab initio molecular dynamics", *Phys. Chem. Chem. Phys.*, **16**, 19564-19572, 2014

Future direction

Computational catalysis over the years has become an integral part catalytic reaction engineering, where, the modelling work explains atomic and molecular phenomena and helps in computational screening of catalysts and solvents, together leading to design of more efficient catalytic reaction systems and processes. With thousands of new catalysts being discovered, their stability and recyclability are crucial for their commercial development and deployment. Computational effort in understanding and evaluating the stability and lifetime of catalysts in realistic operation conditions is lacking and this is a direction I would like to pursue. In this direction, I would like to work closely with experimental catalysis experts to develop multiscale models to understand catalyst deactivation and stability.

Teaching Interests:

Teaching Philosophy

'Teaching is a process of learning to inspire others'. This is my philosophy on teaching. A successful teacher shares existing knowledge, induces curiosity in the students, inculcates the spirit of questioning and inspires the quest for advancing the state-of-the-art in science and technology

University teaching experience

The 'University Teaching for Teaching Assistants' course at NTU helped me reflect on how we as students learnt best. At NTU, I had opportunities to test myself, where I engaged under-graduate students in laboratory sessions as well as tutorial sessions for the course modules as a graduate teaching assistant. I engaged students for two laboratory modules: Tensile testing of materials and Analysis of chemical reactors. Being a teaching assistant for tutorial sessions in the core Chemical Engineering modules of Unit Operations (Principles of separation), Chemical Reaction Engineering and Fluid systems gave me the real flavour of university teaching.

Teaching methods and practises and student feedback

Short questions, discussions, facts and scenarios to ponder about, helped in getting active class participation, and encouraged students to ask questions, making the sessions lively and useful to the students. My focus on presenting a methodical approach to solve the problems encouraged critical thinking by the students rather than rote learning. The use of technology like Youtube videos, power point presentations, Microsoft Excel spreadsheets with mathematical functions to solve problems in addition to the conventional speech and black/white board based approach to teaching greatly enhanced the classroom experience. Frequent citing of examples from the process industry helped students appreciate the relevance of the subject and get acquainted with situations they might encounter in their professional life ahead. A rating of 4.5 and higher (on 5) based on feedback from over hundred students who attended the tutorials sessions is testimony to my good performance as a teacher.

Courses

As a chemical engineering graduate, with professional experience in operations of a petrochemical unit, I am confident of teaching all the core chemical engineering topics among which preference would be to teach Unit Operations and Separations, Fluid Systems, Heat and Mass transfer, Chemical Reaction Engineering, Process Equipment and Plant Design. With research expertise in molecular modelling for a wide range of focus areas, I would like to develop undergraduate courses like Fundamentals of Molecular Modelling and Molecular Modelling for Catalysis.

505554 Novel Approaches for Carbon Neutral Energy Conversion

Abstract:

Postdoctoral Fellow

Research Interests:

Global energy needs are met by the combustion of fossil fuels such as oil, natural gas, and coal – finite resources while polluting the environment through carbon emissions. In order to achieve energy security in an environmentally friendly manner, the energy strategy must create alternative technologies based on renewable sources. A central challenge, however, is the development of new processes capable of tapping these sources and feeding the vast energy needs. One of the promising routes is to convert renewable energy into chemical energy, a process typically involves catalysis, speeding up and directing chemical reactions. My particular research interest roots in catalysis advance understanding of reactions essential to industrial production, health and the environment; harnessing catalysis to make chemical bonds in new ways; developing new materials that overcome the power density and stability limitations faced by existing catalysts in the conversion and storage of electrochemical energy.

Research Experience:

Postdoctoral research at UC Berkeley: My research focuses on the development of catalysts for CO_2 reduction as a route towards the storage of intermittent renewable energy sources. We are taking a unified approach to this small-molecule activation problem by developing molecular and materials catalysts for CO_2 reduction that can be used in parallel under environmentally green, aqueous conditions. Materials science provide concepts as well as components to develop new solar-to-chemical conversion processes. In particular, molecular materials derived from molecular-nanoparticle composites and molecular-bulk materials hybrids, complement heterogeneous catalyst designs. In particular, we are inspired by recent advances in organometallic chemistry, where careful design and modification of ligand structures has led to novel reactivity of small molecule organometallic complexes for homogeneous catalysis of organic reactions. As such, these molecular-material hybrids can be thought of as heterogeneous analogs to homogeneous organometallic complexes.

Ph.D research at Purdue University: we pursue several research fronts rooted in synthetic and physical inorganic chemistry, with emphasis on compounds of importance in electronic and opto-electronic materials, and energy science. We focus on improving the energetic efficiency of electronic devices through molecular design. We seek to functionalize state-of-the-art organic molecular wires through incorporation of redox-active transition metal building blocks. Our particular interest in the area of carbon-rich organometallic materials focuses on the synthesis and characterization of conjugated Ru2-alkynyl and 3d earth-abundant transition metal-alkynyl compounds. These wire-like single molecules are fabricated into electronic circuits to serve as electronic components, which exhibit undiscovered electron transport efficiency and reduced energy consumption.

Future Direction:

New catalysts for carbon-neutral energy conversion processes are essential to addressing climate change and rising global energy demands. The emphasis of my future research will be focused on the design and optimization of catalytic materials present new opportunities for overcoming current performance constraints, especially targeting energy inputs from renewable sources. Due to my multiple training background in chemistry, materials, and engineering, my primary research interest roots in the merging of different disciplines, discovering specific properties of known materials at new interfaces, whereas open up new routes for the design of novel materials with unprecedented interfaces. Particularly, creation of molecule/materials interface may further advance materials' robustness by integrating the strength from both molecular and materials side. These interfaces are capable of creating special synergistic effects between each component, leading to significantly improved efficiencies of the energy conversion/storage systems. The sparks from interfacing chemistry, materials science, and engineering are expected to enable new technological advancements in sustainable energy conversion and storage.

Grant Writing Experience:

I have received the Young Researcher Program of Synfuels China (YRPSC) Fellowship, which was awarded for my independent proposal of "Hybrid Molecular-Materials Catalysts for Carbon-Neutral Energy Conversion"

I have participated in drafting of six grant proposals and three beamline proposals for my graduate and postdoctoral supervisors. Successful proposals include:

1) The "Novel Carbon Rich Organometallics based on iso-Triacetylene: Effects of Cross Conjugation" funded by National Science Foundation (NSF).

2) "Cross-conjugated 3d and 4d Metal-alkynyl Frameworks: Synthesis and Properties" funded by National Science Foundation (NSF).

3) The "Molecular/Material hybrid catalysts for electrochemical N_2 reduction" funded by the Energy Biosciences Institute (EBI)

Teaching Interests:

I have broad research interests supported by my experience in various areas of chemistry. At Purdue, I was a teaching assistant for a sophomore inorganic chemistry course and a freshman general chemistry course. I worked closely with the professor to define the curriculum and develop problem sets and exams. This was a highly rewarding experience for me and helped me to formulate more clearly my teaching philosophy and to understand that I truly enjoy teaching and interacting with students both in the research laboratory and in the classroom. During my time at both Purdue and Berkeley, I have mentored the work of two graduate students and four undergraduates for their researches. Based on my experience, I hope to teach undergraduate General Chemistry, Inorganic Chemistry, Materials Chemistry, and Engineering Thermodynamics; a graduate course in Advanced Inorganic Chemistry, Materials Chemistry, Electrochemistry, Heterogeneous Catalysis, or Science and Engineering of Sustainable Energy.

Selected Publications: (40 Peer-Reviewed Papers Total, 2 Book Chapters)

Cao, Z.; Derrick, J. S.; Gao, R.; Gong, M.; Xu, J.; Nichols, E. M.; Smith, P. T.; Reimer, J. A.; Wen, X. and Chang, C. J. *Angewandte Chemie International Edition* 2017, submitted. "A Molecular Surface Functionalization Approach to Tuning Bulk Palladium Electrocatalysts for Carbon Dioxide Reduction"

Gong, M.†; **Cao, Z.**†; Liu, W.†; Liu, J.; Nichols, E. M.; Smith, P. T.; Derrick, J. S.; Xu, J.; Liu, Y. –S.; Guo,

J.; Wen, X. and Chang, C. J. *ACS Central Science* 2017, under revision. "Molecular Pocket Functionalized Metal Surfaces for Selective Electrochemical CO Reduction" (†Co-first author)

Gao, R.†; **Cao, Z.**†; Xie, L.†; Liu, X. –W.; Liu, X.; Ma, D.; Yang, Y.; Li, Y. –W. and Wen, X. *ACS Catalysis* 2017, accepted. "Theoretical Insights into Carbon Permeation of Iron-Catalyzed Fischer-Tropsch Synthesis" (†Co-first author)

Cao, Z.; Kim, D.; Hong, D.; Yu, Y.; Xu, J.; Lin, S.; Wen, X. D.; Nichols, E. M.; Jeong, K.; Reimer, J. A.; Yang, P. and Chang, C. J. *Journal of the American Chemical Society* 2016, 138, 8120-8125. "A Molecular Surface Functionalization Approach to Tuning Nano-particle Electrocatalysts for Carbon Dioxide Reduction"

Cao, Z.; Xi, B.; Jodoin, D. S.; Zhang, L.; Cummings, S. P.; Gao, Y.; Tyler, S. F.; Fanwick, P. E.; Crutchley, R. J. and Ren, T. *Journal of the American Chemical Society* 2014, 136, 12174-12183. "Diruthenium–Polyyn-diyl–Diruthenium Wires: Electronic Coupling in the Long Distance Regime"

505080 Rational Design of Material Interfaces for Electrochemical Energy Conversion and Storage

Abstract:

Ming Gong---Postdoctoral Fellow (University of California, Berkeley)

Research Interests:

Accelerating increase of energy demand and fossil fuel consumption has urged the need for sustainable energy solutions. Renewable energy has recently attracted significant attention as alternative energy sources, but the spatial and temporal intermittency of most renewable energy resources demands the conversion and storage of renewable energy into other energy forms when surplus. One of the most promising routes is to transform renewable energy into chemical energy by either upgrading abundant resources into high-value fuels or charging battery materials into high-energy states. Developing new materials is crucial to improve the efficiencies of these energy conversion systems, but is often limited by the lack of novel synthetic methodology. An alternative strategy is to interface different disciplines from chemistry, physics, biology and materials science, which may offer new opportunities of discovering unique properties of known materials at new interfaces as well as open up new routes of designing unprecedented materials with novel interfaces. During my academic career, I have been actively pursuing for bridging chemistry and materials science to create novel material/material or material/molecule interfaces for electrocatalysis or batteries. These interfaces are capable of creating special synergistic effects between each component, leading to significantly improved efficiencies of the energy conversion/storage systems. The sparks from interfacing chemistry and materials science have motivated me to further explore the interdisciplinary areas for diversifying functional materials using highly tunable chemical tools.

Research Experience:

Postdoctoral Project: ¡°*Supramolecular porphyrin cage assembly at molecular-materials interfaces for electrocatalysis*¡±

Under supervision of Prof. Christopher J. Chang, Department of Chemistry, University of California, Berkeley

PhD Dissertation: ¡°*Novel materials for high-performance water electrolyzer and ultrafast battery applications*.¡±

Under supervision of Prof. Hongjie Dai, Department of Chemistry, Stanford University

My academic career has span many fields related to energy conversion and storage. During the graduate studies, I gained expertise in nanomaterial synthesis, electrochemical/photo-electrochemical catalysis and battery assembly. During the postdoctoral research, I was further trained on molecule synthesis for chemically tuning the catalysts. The major achievements are listed as follow:

1) The discovery of a crystalline NiFe layered double hydroxide catalyst with high activity towards water oxidation and a NiO/Ni hetero-structure catalyst with high activity towards water reduction led to an alkaline water electrolyzer with <1.5 V operating voltage for the first time using non-precious metal-based catalysts. The durability could be further extended to 500 hours by introducing Cr_2O_3 overcoats to NiO/Ni catalysts.

2) The design of highly porous graphitic materials enabled an ultrafast Al ion battery operating in ionic liquids that was capable of delivering a voltage of ~ 2 V and lasting hundreds/thousands of cycles for the first time. Detailed mechanistic studies revealed that the battery operates through intercalation/de-intercalation of chloroaluminate anions into graphite on cathodes and deposition/dissolution of Al on anodes.

3) Supramolecular assembly of porphyrin cages on metallic substrates significantly increased the activity and selectivity of electrochemical CO reduction into C-C coupled products. The efficient catalysis is achieved by synergistic effect of stabilizing and transforming CO reduction intermediates inside the cage catalysts.

Future Direction:

Catalysis plays a central role in energy conversion and storage. Advancing towards more complicated catalytic reaction demands highly specific catalysts like the enzymes evolved by nature; however, it remains a grand challenge to either design enzyme-like catalysts with high specificity through bottom-up approaches or effectively screen a variety of catalysts for the optimal specificity of a target reaction through top-down approaches. As faculty I would like to take use of my expertise in molecular/materials catalysis and further diversify the search for novel catalysts. In particular, I would like to utilize molecule tools to atomically control the surface active states of materials catalysts as well as to facilitate synthesize novel materials with defined surface structures via bottom-up strategies. The systematically tuned surface structures by tunable molecules could enable high specificity and selectivity for catalysis. The catalytic reactions can involve gas-phase catalysis, liquid-phase small molecule activation, electrocatalysis and also photocatalysis/photoelectrochemical catalysis. In addition to the bottom-up approaches, I would also like to explore the dynamic combinatorial chemistry using molecular tools and nanomaterials catalysts to expedite the catalyst search. The reversible molecular binding/dissociation cycles together with in situ catalytic activity detection might provide feedback loops for high-throughput screen of catalysts, mimicking the directed evolution strategy in enzyme engineering. Therefore, in the future, I expect to combine different disciplines for advanced catalyst search and design.

Grant Writing Experience:

I have written a total of six grant proposals and three beamline proposals for my graduate and postdoctoral supervisors. Successful proposals include the zinc-air battery proposal funded by Precourt Institute of Energy (Stanford University) and the aluminum ion battery proposal funded by US Department of Energy.

Teaching Interests:

I have extensive teaching experience during my undergraduate, graduate and post-doctoral periods. I spent one month in summer teaching basic science to high school/middle school students in rural China during my undergraduate; I TAed three undergraduate courses and one graduate course (Chem 31X-General Chemistry, Chem 174/176-Physical Chemistry Lab A/B and Chem 277-Material Chemistry and Physics) during my graduate in Stanford University. Since topics of catalysis, electrochemical processes and sustainable energy are attracting more focus from Chemistry Engineering departments, my teaching interest lies in related courses (e.g. Science and Engineering of Sustainable Energy, Principles of Electrochemical Processes and Catalysis). Lastly, I have mentored four undergraduate students and five graduate students (including visiting graduate students) during graduate and post-doctoral periods.

Selected Publications: (#equal contribution)

Gong, M.#; Cao, Z.#; Liu, W.#; Nichols, E. M.; Smith, P. T.; Derrick, J. S.; Liu, Y. S.; Liu, J, J.; Wen, X. D.; Chang, C. J. Supramolecular porphyrin cage assembly at molecular-materials interfaces for electrocatalytic CO reduction, *submitted*

Wu, Y. P.#; **Gong, M.#**; Lin, M. C.#; Yuan, C. Z.; Angell, M.; Huang, L.; Wang, D. Y.; Zhang, X. D.; Yang, J.; Hwang, B. J.; Dai, H. J. 3D Graphitic Foams Derived from Chloroaluminate Anion Intercalation for Ultrafast Aluminum-Ion Battery, ***Adv. Mater.***, 2016, 9218-9222

Gong, M.#; Zhou, W.#; Kenney, M. J.#; Capusta, R.; Cowley, S.; Wu, Y. P.; Lu, B. A.; Lin, M. C.; Wang, D. Y.; Yang, J.; Hwang, B. J.; Dai, H. J. Blending Cr_2O_3 into NiO-Ni Electrocatalyst for Superior Water Splitting, ***Angew. Chem. Int. Ed.,*** 2015, 127, 12157-12161

Lin, M. C.#; **Gong, M.**#; Lu, B. A.#; Wu, Y. P.#; Wang, D. Y.; Guan, M. Y.; Angell, M.; Chen, C. X.; Yang, J.; Hwang, B. J.; Dai, H. J., An Ultrafast Rechargeable Aluminum Ion Battery, ***Nature***, 2015, 520, 324-328

Gong, M.#; Zhou, W.#; Tsai, M. C.; Zhou, J. G.; Guan, M. Y.; Lin, M. C.; Zhang B.; Hu, Y. F.; Wang, D. Y.; Yang, J.; Pennycook, S. J.; Hwang, B. J.; Dai, H. J. Nanoscale NiO-Ni Heterostructures for Ultra-Active Hydrogen Evolution Electrocatalysis, ***Nat. Comm.***, 2014, 5, 4695

Gong, M.; Li, Y. G.; Zhang, H. B.; Zhang, B.; Zhou, W.; Feng, J.; Wang, H. L.; Liang, Y. Y.; Fan, Z. J.; Liu, J.; Dai, H. J., Ultrafast High-Capacity NiZn Battery with NiAlCo Layered Double Hydroxide, ***Energy Environ. Sci.***, 2014, 7, 2025-2032

Kenney, M. J.#; **Gong, M.#**; Li, Y. G.#; Wu, J. Z.; Feng, J.; Lanza, M.; Dai, H. J. High-Performance Silicon Photoanodes Passivated with Ultrathin Nickel Films for Water Oxidation, ***Science***, 2013, 342, 836-840

Gong, M.#; Li, Y. G.#; Wang, H. L.; Liang, Y. Y.; Wu, J. Z.; Zhou, J. G.; Wang, J.; Regier, T.; Wei F.; Dai, H. J. An Advanced Ni-Fe Layered Double Hydroxide Electrocatalyst for Water Oxidation, ***J. Am. Chem. Soc.***, 2013, 135, 8452-8455

Reviews:

Gong, M.; Wang, D. Y.; Chen, C. C.; Hwang, B. J.; Dai, H. J. A mini review on nickel-based electrocatalysts for alkaline hydrogen evolution reaction ***Nano Res.***, 2016, 9, 28-46

Gong, M.; Dai, H.J. A mini review on NiFe-based materials as highly active oxygen evolution reaction electrocatalysts, ***Nano Res.***, 2015, 8,23-39

Selected Patents:

Ultra-fast rechargeable metal-ion battery, Hongjie Dai, Meng-Chang Lin, **Ming Gong**, Bingan Lu, Yingpeng Wu, US20150249261, Exclusively licensed

Strongly coupled inorganic-graphene hybrid materials, apparatuses, systems and methods, Hongjie Dai, Hailiang Wang, **Ming Gong**, US9237658

Plasmonic beads for multiplexed analysis by flow detection systems, Bo Zhang, Hongjie Dai, Yingping Zou, **Ming Gong**, Jiang Yang, US20160216252, Licensed

Heterostructures for ultra-active hydrogen evolution electrocatalysis, Hongjie Dai, **Ming Gong**, US20160017507, Exclusively licensed

505618 Renewable Bulk Chemicals Production Using Porous Catalytic Materials: A Mechanistic Perspective

Abstract:

Research Interests:

With the shale gas revolution, the availability of cheap and abundant natural gas may pose a threat to biofuels production, on the other hand, it opens exciting new opportunities for the renewable production of a selection of bulk chemicals. The shift to lighter shale gas feeds results in a tight global supply of C3/C4 olefins (propylene and butadiene) and aromatics benzene, toluene and xylene (BTX). Therefore, on-purpose renewable production routes for the aforementioned building blocks are desired to make up the shale-gas-related scarcity and shift to a more sustainable chemical industry in the future. In this regards, my postdoctoral research focused on the mechanistic understanding of renewable butadiene production from tetrahydrofuran in porous zeolite materials using a combined density functional theory (DFT) calculations, experiments and microkinetic modeling approach. On the basis of reaction pathway studies, important factors that affect reactivity and selectivity, specifically the zeolite pore size, acid type (Brønsted vs. Lewis) and strength were also investigated as part of an effort to develop structure-reactivity relationships that could lead to rational design of more efficient zeolite catalysts for the reaction. Future plans will utilize the knowledge obtained from the butadiene chemistry to find other applications in biobased olefins and aromatics production, to discover new biobased routes with higher transformation efficiency, and to design zeolite materials with superior performance in biobased bulk chemicals production.

Teaching Interests:

Thermodynamics, Chemical Kinetics, Chemical Reaction Engineering.

505423 Solar Energy Conversion Via Photovoltaics and Photocatalysis

Abstract:

Research Interests: Photocatalysis, Photovoltaics, Materials Science and Electrochemistry for Solar Energy Conversion and Storage

Teaching Interests: Catalysis, Photoelectrochemistry, Electrochemistry, Fundamentals of Photovoltaics, Chemical Reaction Engineering, and Teaching Method (I received "Kaufman Teaching Certificate from MIT Teaching & Learning Lab)

Due to the forthcoming shortage of natural resources, the demand for more efficient and eco-friendly chemical processes for the conversion of energy and substance, especially with respect to carbon management, is growing rapidly. Therefore, a search for high-performance solar energy conversion systems to end the current carbon economy era is of paramount importance in both academic and industrial sectors. In this regard, this talk aims to bring together experts from a variety of disciplines in order to discuss the latest fundamental progress in solar water splitting by doping-treated bismuth vanadate ($BiVO_4$) and printable and flexible photovoltaics by oCVD (Oxidative Chemical Vapor Deposition) polymers.

oCVD is a solvent-free conformal vacuum-based technique to enable thin-film fabrication of insoluble polymers at moderate vacuum (~ 0.1 Torr) and low temperature (25 â€“ 150 Â°C). Moreover, oCVD carries the well-cited vacuum processing benefits, such as easy patterning, well-defined thickness control, large-area uniformity, and inline integration with other standard vacuum processes (e.g., vacuum thermal evaporation). Based on these technical advantages from oCVD, polyselenophene, poly(3,4-dimethoxy-thiophene), and polythiophene have been successfully incorporated into organic photovoltaics as donor, neutral hole transporter, and photoactive neutral hole transporter, respectively.

Efficient, sustainable, and large-scale solar hydrogen harvest requires the development of suitable photocatalysts for photoelectrochemical (PEC) cells and artificial water photolysis systems, both of which are able to cost-effectively convert sunlight energy into gaseous hydrogen (H_2). To develop such appropriate photocatalysts, their atomic structure control is of primary importance since their functionalities (e.g., electronic band frame, electric properties, kinetics, etc.) are governed by their atomic structure. In this regard, $BiVO_4$â€™s atomic structure has been engineered via phosphorus, indium and molybdenum doping. The significantly enhanced photo-responsive characteristics of doping-treated $BiVO_4$ have been studied within experimental and theoretical domains.

505477 Multi-Scale Cellular and Protein Therapeutic Engineering for the Development of Novel Immunotherapies

Abstract:

Multi-Scale Cellular and Protein Therapeutic Engineering for the Development of Novel Immunotherapies

John James Blazeck

Postdoctoral Fellow, University of Texas at Austin

Successful Proposals:

American Cancer Society Postdoctoral Fellowship

NIH F32 Ruth L. Kirschstein Postdoctoral NSRA

Research Interests:

Novel immunotherapies have the capacity to cure cancers by activating anti-tumor T cell-driven immune responses and have shown remarkable efficacy in the clinic. However, clinical response is limited to a subset of patients because these therapies employ a single intervention strategy that can be rendered ineffective by redundancies within the immune system. To account for this issue, my research will aim to develop tools that precisely control cellular and immunological outputs through distinct, easily combined mechanisms for rapid therapeutic development. More specifically, I have accrued expertise in the fields of cellular, metabolic, and immunological engineering to implement large-scale, multi-level engineering programs to develop tumor-inhibition-resistant and tumor-activated T cell therapies while concurrently engineering immunological proteins with novel functionalities.

My graduate work provided me with an in-depth tool kit for engineering cellular response at the transcriptional, pathway, and whole-cell level that will be invaluable towards designing and engineering cellular therapies. In particular, I developed techniques to precisely control transcriptional regulation in response to multiple environmental inputs, and I utilized widespread metabolic pathway perturbations to drastically alter cellular phenotype. For instance, I developed an array of novel synthetic promoters in multiple eukaryotic organisms [1-3], and I engineered a nonconventional yeast strain to accumulate lipids at the highest reported values when published [4]. To complete the foundation for my own research program, I cultivated an immunological protein engineering skill set during my postdoctoral work that I will couple with my cellular engineering tool kit. Specifically, I developed an anti-cancer enzyme therapeutic that degrades kynurenine, an immune-inhibitory small molecule produced by cancers. There are currently >20 clinical trials evaluating small molecule inhibition of kynurenine production, but efficacy is limited because kynurenine is synthesized *in vivo* by three enzymes. Thus, we hypothesized that enzyme-mediated clearance of tumoral kynurenine would be an ideal route to relieve immune suppression. To this end, I undertook a large-scale protein engineering effort that improved the activity of the Human Kynureninase enzyme by 600 fold - to pre-determined levels suitable for kynurenine clearance. Administration of engineered kynureninase enzymes to mouse models of cancer depleted kynurenine levels, allowed for a robust T cell-driven anti-cancer immune response, and drastically slowed tumor growth [manuscript under review]. In the future, I plan to combine my expertise engineering immunological proteins and controlling cellular phenotypes to enable translational research to effectively target and kill cancer cells.

Future Directions:

A major challenge of developing therapies for immunological applications is that the complexity of the immune system can prevent desired outcomes. To address this challenge, my research program will merge my expertise in cellular manipulation and immunological engineering to develop novel anti-cancer cellular and protein therapies by engineering control of the immune response through multiple mechanisms. Specifically, by engineering immune cells at the transcriptional, enzymatic, and pathway levels, I will create robust autologous T cell therapies that are impervious to tumor-mediated immune suppression and are activated by the tumor microenvironment. To this end, I will design novel transcriptional control units activated by tumors (e.g. by lactic acid, kynurenine, hypoxia etc.), and I will rewire cellular pathways to resist the nutrient-limited tumor metabolic environment. Similarly, T cell inhibitory receptors can be co-opted to invert tumor-mediated signaling towards stimulatory outcomes, and engineered immune cells could even be designed to act as cellular sinks for tumor-produced suppressive small molecules or for metabolites required for tumor anabolic growth.

Additionally, I will engineer the basic architecture of immune proteins, such as T cell receptors and antibodies, to introduce novel functionalities that will enable new applications as single-entity anti-cancer therapeutics or for efficient combinatorial targeting of cancer cell mutations. For instance, by engineering physically separable antigen-specific and constant (cell-bound) regions of the T cell receptor (TCR), it will be possible to combine a clonal T cell population expressing the cell-bound TCR-region with multiple antigen specific TCR fragments to generate a well-defined, multi-target, 'drop-in' autologous anti-cancer T cell therapy. I will further engineer antibody backbones by introducing novel functional sites to create new classes of Bi or Tri-specific binding proteins designed to be ideal partners for cellular therapy treatments. While my initial work will focus on establishing independent control of distinct cellular levels within T cells, my long term vision is to synergize these control mechanisms within single cellular therapies. The capacity to rewrite the T cell regulatory and metabolic network will allow me to exploit the interconnected nature of the immune response and direct it towards beneficial outcomes against cancer, and engineered T cell therapies can be further combined with protein therapeutics to activate immune cells to fight cancer. The tools developed to achieve these goals will also be generalizable to many other immune-altered states (e.g. autoimmune disorders) and other biological networks with similar complexity. My foundation in synthetic biology and pathway engineering leaves me well-situated to undertake this research program in immunological engineering.

Teaching Interests:

I was fortunate to have an excellent experience teaching undergraduate course material before beginning my graduate research program. Specifically, as an MCAT organic chemistry and biology instructor for the Princeton Review, I designed and taught over forty hours of fully participatory lecture material. I kept my students interested and actively engaged in my lectures by consistently guiding them towards conclusions and relying on them to supply key pieces of information as I taught, i.e., by using the Socratic Method. I also had my students work together on short sample questions to let them apply their new knowledge set and gain confidence in their ability to solve MCAT-styled problems. In short, I learned how to effectively create lecture material that both led and collaborated with my students in their intellectual pursuits.

During graduate school, I happily served as a teaching assistant for two chemical engineering courses, Thermodynamics and Chemical Reactor Analysis and Design. I was able to see first-hand the benefit that a passionate instructor could have towards motivating and engaging young, bright minds. I learned techniques to improve course quality and provide additional support as students tackled difficult subject matter. For instance, I will begin my lectures with a brief overview of the previous and current classes' subject matter, employ peer-teaching opportunities and small-group efforts, and I will write my lecture notes by hand as class proceeds to pace delivery of material.

Core Courses: I am prepared to teach Kinetics/Reactor Design, Transport, and Mass/Energy Balances.

Advanced Courses: I am prepared to develop a Bioreactor and Cellular Engineering course that will be dedicated to understanding cellular conversion and metabolism at all scales, i.e. enzymatic, single-cell, and bioreactor levels.

Mentorship:

I have greatly enjoyed mentoring undergraduate and beginning graduate students as they start their research careers. To date, I have mentored more than ten undergraduate researchers, many of whom have gone onto graduate school, and three beginning graduate students. Six of my undergraduate mentees secured $1,000 Undergraduate Research Fellowships and three won 1st place in department or university poster competitions.

I look forward to mentoring graduate students, postdoctoral researchers, and undergraduate students in my own lab. As an advisor, I will strive to (1) encourage the development of my mentee's critical thinking and project development skills, (2) advance their understanding of their project and their field, and (3) cultivate their experimental expertise and communication skills. I will encourage my mentees' development by granting them greater domain over their projects as they progress, and I will host weekly rapid-fire round-robin journal clubs to establish the importance of keeping up with scientific literature. All incoming graduate students will 'intern' with a senior lab member for 3-6 months. In this manner, senior lab members will develop valuable teaching skills as beginning researchers learn experimental techniques. To aid in the development of communication skills, I will strongly encourage all students and postdocs to apply for fellowships, involve mentees in the grant writing process, and expect mentees to author scientific manuscripts and to present at group meetings and scientific conferences. Furthermore, all mentees will author a review article relevant to their project in their first 1-2 years to hasted development of their writing skills and understanding of their field. I will passionately urge all mentees to mentor undergraduate researchers, as it has been of such personal and scientific benefit to me. Finally, I will meet weekly with all graduate students and postdocs and help them achieve goals through bi-annual progress reviews that include individualized career and lab-specific development plans.

Referenced Publications:

[1] **Blazeck, J.** and Alper, H. (2013). Biotechnology Journal. 8(1): 46-58.

Times Cited = 103.

[2] **Blazeck, J.,** Garg, R., *et al.* (2012). Biotechnology and Bioengineering. 109(11) 2884-2995.

Times Cited = 122.

[3] **Blazeck, J.,** Liu, L. *et al.* (2011). Applied and Environmental Microbiology. 77(22): 7905-7914.

Times Cited = 105.

[4] **Blazeck, J.**, Hill, A., *et al.* (2014). Nature Communications. 5(3131) doi:10.1038/ncomms4131.

Times Cited = 138.

505055 Structure-Function Relations in Bifunctional Catalysis: Kinetic, Spectroscopic, and Theoretical Approaches

Abstract:

Research Interests:

Modern society relies on catalytic chemical transformations of simple molecules and varied feedstocks to agrochemicals, therapeutics, plastics, and fuels. Traditional heterogeneous catalysts are robust and easily recovered and regenerated but often suffer from poor selectivity that requires both energy-intensive separations for product streams and the disposal of the significant quantities of hazardous waste. Conversely, homogeneous catalysts often afford precise chemo- and regioselectivity, but they suffer from short lifetimes, difficult recovery, and poor regenerability. The theme central to my research interests is examining and developing materials interfacing homogeneous and heterogeneous catalysis that leverage the unique advantages of each (e.g., selectivity & specificity, stability & ease-of-processing, respectively) to investigate and develop green, atom-efficient chemical transformations relevant to the synthesis of bulk and fine chemicals.

My dissertation work with Professor Enrique Iglesia at the University of California, Berkeley focused on understanding mechanisms of alkane isomerization and Î^2-scission on bifunctional metal-acid catalysts. I used kinetic, spectroscopic, and computational techniques and developed reaction-transport formalisms, necessary to dissect the ubiquitous and unavoidable consequences of diffusion on reaction rates and selectivities, to establish structure-function relations for mesoporous and microporous solid acid catalysts. These solid acid catalysts had well-defined chemical connectivity and structure, which permitted rigorous kinetic analysis and comparison to theory (DFT). I will learn a suite of complementary skills in my postdoctoral research position, which will involve surface organometallic chemistry; I will rationally design, synthesize, and characterize hybrid organic-inorganic materials with well-defined surface species as catalytic sites. The skills and techniques I will learn here will allow me to synthesize catalysts

Teaching Interests:

In addition to my teaching experience as a graduate student instructor for three upper-division undergraduate core chemical engineering courses (Introduction to Chemical Process Analysis; Transport and Separations Processes; Transport Processes), I have been a research mentor for an undergraduate researcher at UC Berkeley. Teaching and mentorship, alongside scholarship in research, are of great importance to me. My teaching philosophy and methods combine insights from pedagogical research with a first-principles problem-solving approach to promote and assess student learning.

504815 Supported Molybdenum Dio-Oxo Catalysts for Acceptorless Aqueous Alcohol Dehydrogenation

Abstract:

H_2 is produced industrially via catalytic steam reforming of nonrenewable methane at temperatures of 400-1000 °C and at steam partial pressures near 30 bar. Hydrogen production from renewable biomass is challenging due to sluggish catalytic rates in water (the most common biomass contaminant), and catalyst deactivation. Furthermore, H_2 transportation and storage present safety issues that could be addressed using a carrier which releases H_2 on demand, such as biomass-derived methanol, ethanol, or formic acid, all of which are compatible with direct alcohol fuel cell technologies. The state-of-the-art for sustainable, environmentally benign methanol and ethanol production typically employs biomass (e.g., cellulose) fermentation to "bioalcohols" having ~10% alcohol in aqueous solution.

In this report we describe a heterogeneous catalytic system for base- and additive-free methanol and ethanol reforming that operates under mild conditions (40 - 90 °C and 1 atm) with an inexpensive supported Mo catalyst that is air- and moisture- stable. Additionally, this system is active for aqueous alcohols, exhibits no deactivation over days under these conditions, and is selective towards valuable aldehydes with negligible production of greenhouse gases or fuel cell poisons (CO_2 or CO), conforming to requirements for both commodity formaldehyde production and direct alcohol fuel cell applications.

Research Interests:

My primary research goals are to obtain multi-scale control of catalyst/support architectures for difficult reactions in energy research while providing a multidisciplinary approach for training students. My research aims include the development and integration of tandem/bifunctional catalyst systems for process intensification, the synthesis and deposition of organometallic precursors onto a range of surfaces to study fundamental support effects, and the controlled synthesis and preparation of metal phosphides, nitrides, and sulfides for catalysis. Concurrent work will also focus on the development of new designer supports and materials using renewables. The fundamental question my research plans to answer is: through molecular engineering of homo- and heterogeneous catalysts, can we obtain knowledge commensurate with that of homogeneous processes? Can we then use that knowledge to develop superior catalysts which promote historically challenging transformations? This interdisciplinary program will cover areas in chemistry, chemical engineering, and materials research, providing multiple avenues for cross department collaborations and teaching.

Teaching Interests:

My teaching interests include energy and catalysis, responsible research conduct, kinetics, and thermodynamics. I would also like to develop a course on the boundaries of both chemistry and engineering for freshman undergraduates. I am a graduate course guest lecturer (Responsible Conduct of Research) in the Department of Material Science and Engineering at Northwestern and have also taken 2 faculty teaching workshops on "Infusing Critical Thinking into Your Course Design" and "Developing Effective Learning Objectives" through Northwestern University's Searle Center for Advancing Learning and Teaching which has made me aware of these critical issues common in post-graduate teaching.

505462 Surface Interactions of High Performance Materials for Energy Efficient Technologies

Abstract:

Increasing energy demand is pervasive in our society. Expanding populations and economic growth rapidly escalate energy consumption. To keep abreast of demand, our society needs to intelligently and efficiently use the limited resources that remain, utilize renewable resources, and tap into previously untouched resources.

Research Interests: High Performance Materials, Catalysis, Energy, Surface Science, Interfacial Reactions

My research interests involve using high performance materials for use in energy efficient technologies. I intend to contribute to the development of these technologies by undertaking projects that are focused on the design, synthesis and characterization of advanced heterogeneous materials. My laboratory will investigate high performance materials to understand the fundamental relationships between atomic structure and characteristic properties/function for use in electrochemical, solar, and biomass derived syngas applications.

My graduate career was spent pursuing the advancement of clean fuels for energy utilization through an improved understanding of heterogeneous reactive systems for fuel reforming, with a particular focus on the molecular level surface characterization of sorbents/catalysts. Specifically, I have been involved in the systematic preparation and characterization of silver-titania adsorbents for the desulfurization of liquid phase logistic fuels. The ultimate goals of this project were to improve energy efficiency, to reduce exhaust pollutants thus improving the air quality and subsequently public health, and to promote environmental sustainability.

I have designed and modified surface analysis experiments in order to garner an improved fundamental understanding of ``real world" adsorbents/catalysts. Through this work, I have developed several experimental protocols using an atomic/molecular level approach in order to correlate structure with function in these real systems. My master's thesis presented a stepwise characterization of the Ag/TiO_2 adsorbent system using a multi-technique comparative approach. My Ph.D. thesis was focused on using oxygen chemisorption and XPS to find a robust means of particle size determination. These projects involved using fundamental studies from multiple techniques to unify molecular level descriptions at the surface of Ag/TiO_2 adsorbents.

Teaching Interests: Material/Energy Balances, Thermodynamics, Reaction Engineering.

My life has been tremendously impacted by dynamic teachers who possessed the ability to recognize and encourage talent development by motivating, instituting challenges, setting goals, and providing constructive feedback. I wish to create a similar impact through my own teaching. As an instructor, it is important to passionately impart knowledge to support reception and retention. While teaching methods

vary depending on the course, common teaching goals remain the same. These are---equip students with the necessary skills and inspire the confidence that through continued effort they will be able to understand even the most challenging topics. During my graduate career I have taught or tutored a number of different courses. I have become adept in small group instruction. In teaching larger groups, student engagement is key; incorporating response system technologies, such as Turning Point, facilitate immediate engagement. I was the recipient of the McLeod Outstanding Chemical Engineering Graduate Student Teaching Award for my efforts as teaching assistant in Unit Operations Lab. I have mentored undergraduate research assistants and helped international students to improve their technical writing skills. My students and peers describe me as hardworking, organized, approachable, resourceful and passionate to make a difference. My personal philosophy is to do everything ``with intent" of a deliberate and successful outcome. These are the traits with which I intend to pursue my teaching career. In addition to teaching the traditional undergraduate courses I would like to develop a special graduate course on 'Surface Analysis for Real World Systems'.

505587 Synthesis of Organometallic Single-Site Heterogeneous Catalysts for Sustainable Chemistry

Abstract:

Research Interests:

Amid environmental concerns associated with fossil fuels and the ever-increasing trend in worldwide energy consumption, interest in the industrial-scale production of fuels and products from biomass and carbon dioxide (CO_2) is rapidly growing. The primary challenges in current approaches include low reaction rates, low product yields and high costs. All of these challenges could be overcome through the development of active, selective, and robust catalysts. To this end, I am interested in studying the synergistic effect between active supports and homogeneous catalysts in single-site heterogeneous catalysts (SSHC) to achieve a more economically feasible and sustainable conversion of biomass. In contrast to traditional heterogeneous catalysts that are designed using rare transition metals, my research will focus, on: (1) developing SSHC from relatively abundant transition metals (Fe, Cu, Ni) and active supports such as zeolites, tunable acidic & basic clays (montmorillonite & hydrotalcite respectively), and sulfonated SiO_2 , (2) studying catalyst valorization of lignocellulosic sugars, lignin, chitin, and CO_2 to fuels and higher-valued products and (3) investigating novel base-catalyzed dehydration products of glucose. Computer simulations will complement experimental studies to probe mechanisms and rationally design catalysts with optimal activity, selectivity and lifetime.

Submitted Proposals:

National Research Council (NRC) Postdoctoral Research Fellowship, recommended for funding, unfunded

Arnold O. Beckman Postdoctoral Fellowship, unfunded

Awards and Fellowships:

Innovation & Entrepreneurship Prize, The George Washington University, Spring 2017

Eastman Kodak Scholars Program, SUNY University at Buffalo, Spring 2014

Honorable Mention, SUNY University at Buffalo School of Engineering and Applied Sciences,

Graduate Student Poster Competition, Buffalo, NY, April 23, 2014

Best professional poster, SUNY University at Buffalo Department of Chemical & Biological Engineering Symposium, Buffalo, NY, October 18, 2013

Mark Diamond Research Award: Non-Aqueous Production of Biofuel, Fall 2013

Postdoctoral Project: "*Glycerol Catalytic Transfer Hydrogenation of CO_2 using Ir and Ru Single-Site Heterogeneous Catalysts Immobilized on Hydrotalcites.*"

Under supervision of Adelina Voutchkova-Kostal, Chemistry, The George Washington University

PhD Dissertation: "*A Study of Selectivity During Acid-Catalyzed Dehydration of Cellulosic Sugars.*"

Under supervision of Carl R.F. Lund, Chemical and Biological Engineering,

Research Experience:

My graduate studies focused on fundamental reaction engineering and heterogeneous catalysis. More specifically, I pursued conversion of cellulosic sugars to value-added products (5-hydroxymethylfurfural, levulinic acid, and GVL) which could be used for fuels, polymers, and specialty chemicals. During this time I learned how to operate and fabricate high pressure batch and flow reactors, design and characterize heterogeneous catalysts, and perform ab-initio quantum chemistry calculations (Gaussian-3 (G3)). For my post-doctoral studies, I joined The George Washington University's chemistry department to expand my research breadth and knowledge in organic synthesis. I learned how to synthesize homogeneous catalysts that could be immobilized onto active supports to create SSHC with a synergistic effect between the support and immobilized organometallic catalyst.

Teaching Experience:

As a scholar and researcher, I believe it is extremely important to impart knowledge and enjoy mentoring, training, and working with researchers of all levels. Throughout my graduate studies, I served as a Teaching Assistant for an undergraduate/graduate course in Material Engineering and Science and for undergraduate courses in Organic Chemistry I and Organic Chemistry II in the Department of Chemical and Biological Engineering and in the Department of Chemistry, respectively. In addition, as a postdoc, I have given guest lectures for a graduate course in Green Chemistry and Engineering. Over the course of my Ph.D. studies I trained a postdoctoral fellow and mentored over 2 dozen undergraduates, 3 master students, and a high school senior. It was most valuable to see the progress students made towards becoming critical thinkers and independent researchers, integrating classroom knowledge with additional literature search and curiosity to form hypotheses, design experiments to test hypotheses, and interpret results.

Teaching Interests:

Given my research and educational background, I am most interested in teaching undergraduate and graduate level kinetics and reaction engineering, as well as undergraduate unit operations, product design, and introductory courses in chemical engineering that introduce students to fundamental concepts. I am interested in developing graduate and undergraduate elective courses in green chemistry and engineering, catalysis, and a course focusing on various commonly used instrumentations such as mass-spectroscopy. I am particularly drawn to the concept of flipped classrooms in which students study the course material outside the classroom while class time is devoted to more in-depth discussion of particular concepts, problem-solving, and engaging hands-on activities to further support course material. I was introduced to this concept during an undergraduate kinetics and reaction engineering course.

Selected Publications:

Karwa, S., Gajiwala, V., **Heltzel J.,** Patil S. K.R., and Lund C. R.F. *Reactivity of Levulinic Acid during Aqueous, Acid-Catalyzed HMF Hydration.* Catalysis Today., doi:10.1016/j.cattod.2015.06.020

Heltzel J, and Lund C. R.F. *Glucose Formate Conversion in gamma-Valerolactone. Catalysis Today.. 2016 26, 88-69*

Patil S. K.R., **Heltzel J.,** and Lund C. R.F. *Comparison of Structural Features of Humins Formed Catalytically from Glucose, Fructose and HMF.* Energy & Fuels., 2012. **26** (8), 5281-5293

Braden D. J., Henao C. A., **Heltzel J.**, Maravelias C. C., and Dumesic J. A. *Production of liquid hydrocarbon fuels by catalytic conversion of biomass-derived levulinic acid.* Green Chem., 2011.**13**:1755-1765.

Heltzel J, Patil S.K.R., and Lund C. R.F. *Temperature-Programmed Pyrolysis and Oxidation of Glucose-Derived Humins.* (manuscript under review at Industrial & Engineering Chemistry Research)

Book Chapters:

Heltzel, J., S. R. Patil and C. F. Lund (2016). Humin Formation Pathways. Reaction Pathways and Mechanisms in Thermocatalytic Biomass Conversion II. M. Schlaf and Z. C. Zhang, Springer Singapore: 105-118.

Patents:

Heltzel, J., Finn, M.T., Voutchkova-Kostal, A..Catalytic Transfer Hydrogenation of CO_2 to Potassium Formate using Glycerol as a Hydrogen source. (provisional application number 62/508,509)

505482 Understanding and Improving Heterogeneous Catalysis for Sustainable Production of Renewable Fuels and Chemicals

Abstract:

Research Interests:

The growing population has increased the global energy consumption and placed stress on the environment and society. Catalysis plays a critical role in the utilization of renewable energy resource, like wind, sun and biomass to provide energy and chemicals to meet the growing demand for a sustainable future. My research interests focus on the pursuit of understanding and development of heterogeneous catalysts towards production of fuels and chemicals using kinetics and insights from spectroscopy techniques.

In my Ph.D research with Prof. Johannes A. Lercher in the Chair of Technical Chemistry II at Technical University of Munich, I have developed effective catalytic strategies to convert biomass-derived molecules into aromatics, alkanes and alcohols. In particular, by exploiting the base metal catalysts in aqueous phase, we enabled the activation and cleavage of the C-O ether bond to upgrade biomass-derived oxygenates into renewable hydrocarbons and we also developed stable dual-function catalysts for the hydrodeoxygenation of biomass-derived oxygenates to bio-fuels by understanding the elementary steps through kinetic study and in-situ spectroscopy. For example, using Reactor IR, I was able to track the intermediates during the cascade reactions. In my current postdoctoral research with Prof. James A. Dumesic and Prof. George W. Huber, in the Department of Biological and Chemical Engineering at University of Wisconsin, Madison, my research interest is on the catalytic conversion of the lignocellulosic biomass into Î±,Ï‰-diols. Using the combination of kinetics and spectroscopic techniques I was able to draw correlations between the structure and catalytic activity of various transition metal oxide system and identify the active catalytic sites to gain a better understanding of chemistry and catalysis.

At the poster session I will present an overview of my past and present research, as well as my plan for future research activities which will continue to exploring fundamental insight towards catalytic conversion of renewable energy sources to fuels and chemicals. By understanding the chemistry behind the heterogeneous catalysis, more efficient and economic catalytic solution can be developed.

Teaching Interests:

My preference includes the entry-level courses in chemical engineering. With my background in chemistry, I am particularly well suited for teaching courses in chemical kinetics, materials balances and thermodynamics. I would also be interested in teaching more specialized courses in heterogeneous catalysis and/or microkinetics as elective classes at the graduate level.

505592 Structure-Function Correlations of Nanomaterials in Heterogeneous Catalysis

Abstract:

Research Interests:

"And yet it is only by studying function that we can understand function, so that the kinetic aspect must be allowed to retain its place and assigned its part in the unravelling of the great mystery"

– C. N. Hinshelwood, 1947

Catalysis is the science and technology of influencing the rates of chemical reactions, which are at the core of chemical industry. Practical success of catalysis needs the interdisciplinary intersection of chemistry, chemical engineering, and material science. My research interests lie on understanding the mechanism of chemical reactions and the role of catalysts under reaction conditions in order to transform qualitative observations and knowledge into predictive science.

In heterogeneous catalysis, under reaction conditions, active centers produce products. The presence of active centers depends on the local geometry of a catalyst particle, which can be varied by tuning the interactions between catalyst particle and the support.

As a first of its kind demonstration of this concept, in my Ph.D. work, the ruthenium-carbon interactions were varied by depositing Ru nanoparticles on carbons with different local geometry and functionalities. Using atomic level TEM analysis, we found that the local disorder of the support induces local disorder to the Ru particle[1] and influences catalysis as confirmed through the direct observation of the reaction intermediate (adsorbed dinitrogen) using *operando* XPS over Ru/C catalyst operating at 400 °C and 0.5 mbar ammonia.[2]

Identification of active centers requires interdisciplinary characterization techniques. For instance, the high H_2 production from ammonia over Mo_2C was due to the prevalence of highly energetic sites (twin-boundaries, stacking faults, steps and defects) which were confirmed with pre- and post TEM and EELS analysis.[3] Similarly, in my postdoctoral work on biomass conversion, by combining a series of characterizations of XPS, *ex/in situ* and *operando* EXAFS and TEM along with reactivity studies, we have discovered that the high yield of 2,5-dimethylfuran from 5-hydroxymethylfurfural benefits from dual functionality catalytic centers of strained mixed Ru_xO_y structure.[4,5],

The knowledge gained from realizing the active centers can lead to optimized catalysts. In a first instance, we have modified iron catalysts by alloying Co or Ni leading to enhanced activity and stability for ammonia decomposition.[6] Furthermore, we have developed a novel route to produce nanosized early transition metal carbide particles embedded in carbon, which show an improvement of ammonia cracking activity and

a promotion effect in PEM fuel cells.

Teaching Interests:

"Give a man a fish and you feed him for a day; teach a man to fish and you feed him for a lifetime"

- Lao Tzu

My teaching philosophy has been heavily influenced by this Chinese proverb. I strongly believe that teaching is more than delivering knowledge; it is the teacher's commitment to the society to raise a new generation able to surpass the "mountains" they may face in their lifetime. The general purpose of science and engineering education is to train independent and collaborative problem solvers. On one hand, it is critical for students to be able to gather, understand and utilize information independently because not only is it impractical to learn everything in class but more importantly, most innovative endeavors have no existing formula to follow. On the other hand, collaborative practice has increasingly been adopted as one of the fundamental mechanisms in the scientific community to tackle interdisciplinary challenges. I am also a candid believer of students' interest being a key prerequisite for any successful science and engineering education. I have adhered to this philosophy throughout my entire career, and would like to see it's fulfillment in making me a better researcher, teacher, and person. Therefore, inspiring interest, training independent learning ability and promoting collaborative practice are the three pillars of my teaching philosophy. I am confident that I can teach major course in an undergraduate chemical engineering curriculum. I very much look forward to the opportunity to teach and interact with students in Chemical Reaction Engineering, Kinetics, Material Sciences, Electron Microscopy and Interdisciplinary Methods in Heterogeneous Catalysis.

Reference

[1] W. Zheng, et al, *ChemSusChem*, **2010**, *3*, 226-230.

[2] W. Zheng, R. Schloegl, to be submitted.

[3] W. Zheng, R. Schloegl, et al, *J. Am. Chem. Soc*, **2013**, *135*, 3458-3464.

[4] J. Jae, W. Zheng, R. Lobo, D. Vlachos, *ChemSusChem*, **2013**, *6*, 1158-1162.

[5] J. Jae, W. Zheng, A. Karim, R. Lobo, D. Vlachos, *ChemCatChem*, **2014**, *6*, 848-856.

[6] J. Zhang, J. Mueller, W. Zheng, et al, *Nano Lett*, **2008**, *8*, 2738-2743.

505521 Advanced Functional Porous Materials As Heterogeneous Catalysts

Abstract:

Research Interests:

Advanced Functional Porous Materials as Heterogeneous Catalysts

Research Interest:

My research focuses on the interface of chemistry, chemical engineering and materials science and I am specifically interested in the areas of renewable energy, design of hetrogenous catalysts and advanced functional porous materials. In particular, my research focuses on the development of advanced functional porous materials and their use as supports for heterogeneous catalysts. Here, I presented a novel technology to convert biomass to high surface area carbon with tunable porosity, while in the past, I developed a method of fabricating zeolites and hydrotalcite with high selectivity to branched hydrocarbons with lower pour points and higher octane numbers. Moreover, I conducted research on complex fluid and flow behavior, the work which will have an impact on how to design the fluid for a better heat transfer performance. In this research, I used Computational Fluid Dynamics to have a better understanding of how the nanoparticles influence the thermal physical properties of nanofluids.

A) Carbon Based Catalyst from Biomass Conversion

During my postdoc, I presented a novel technology for conversion of biomass to functionalized carbon at ultra-fast time scales. The use of agricultural waste as carbon precursors is renewable and relatively less expensive. Highly porous carbon has many applications, including water treatment, separation, purification, catalysis, energy storage, batteries, fuel cells, electrodes for electric double layer capacitors, and others. Although there has been a large amount of research and interest in the conversion of biomass to carbons for various applications, the challenges remain, both in terms of processing time scale and also with the ability to tune the resulting carbon's properties. In addition, the functionalization schemes for using the resulting carbons toward catalyst schemes require additional development. I utilized a novel technology, plasma based technique to convert biomass to high surface area carbons with tunable porosity at ultrafast time scales of less than 1 minute. This high surface area carbons with tunable porosity, has a potential to be used as support for catalytic applications, which I demonstrated using them as supports for Pt catalysts. Specifically, I dispersed Pt nanoparticles on carbon supports using hydrogen plasma treatment and was able to obtain a uniform dispersion of Pt with narrow size distribution (less than 5 nm). The small size of Pt nanoparticles and tunable carbon supports were significant challenges in conventional methods. The final products obtained from plasma will be more economically competitive than those obtained via traditional techniques and have the ability to be produced in large industrial scales. I have filed a patent application.

B) Zeolites and Hydrotalcite Catalysts

During my PhD, I synthesized zeolites and hydrotalcite catalysts and I loaded Pt nanoparticles on these support. The fundamental idea of this catalyst program was to convert the waste oil to biofuels in a single catalytic process. This project represented significant advancement in the process of conversion of lipid biomass materials to 'drop-in' hydrocarbon transport fuels, such as motor and aviation gasoline. The deoxygenation of the fatty acid and subsequent isomerization/hydrocracking of the resulting C_{17}–C_{18} linear paraffins to branched, gasoline-range C_5–C_{10} hydrocarbons with lower pour points and higher octane numbers was accomplished over a bi-functional catalyst in a single step process. Such an investigation had not been reported before. This research led to two patent applications and several journal publications.

Specific Research Directions And Future Research Objectives:

In my current research, I use plasma technique to convert biomass such as hemp and soy-hulls to functionalized carbon, which are then used as catalyst support. For this, I disperse Pt nanoparticles on the fabricated high surface area carbon with tunable porosity and was able to demonstrate a uniform dispersion of Pt with narrow size distribution. Similar process can be applied to convert other types of biomass to functionalized carbons. In the future, I would like to apply plasma technique for dispersion of other noble metal on carbon supports and expand this approach to prepare bi-functional catalyst or bimetallic catalyst using plasma technique. These novel catalysts will be evaluate for their performance under hydrogenation reaction. Currently, I am using Pt/C in hydrogenation of xylose to xylitol.

My future research will also focus on converting the waste oil to biofuels in a single catalytic process. According to my results obtained for the zeolites, they are the nominate catalyst for this purpose because of the high percentage of decarboxylation and high selectivity they have for branched and aromatic compounds. I investigated the optimum reaction conditions in my previous research and I have high chance of collaborating with the industrial companies. For future research other non-noble metals, such as Cu, Al and Ni, which are much cheaper than Pt, could be used. Moreover, catalytic decarboxylation and conversion of oleic acid to parrafins and hydrocarbons over metal organic frameworks could be investigated. This research can be combined with activated carbon research, as well.

Resource Funding:

DOE(BES)/CSGB) Division- DOE(EERE)-NSF/ (CBET)/Particulate and Multiphase Processes/Computational and Data-Enabled Science and Engineering (CDS&E), Catalysis/Energy for Sustainability

Teaching Interests:

I have worked in a variety of academic institutions in a wide range of teaching areas. Teaching has always been a rewarding experience for me, mainly because of the great satisfaction that comes with sharing my knowledge with others. During my time at the University of Louisville, I received Teaching Assistant positions for 5 years. These included Introduction to Chemical Engineering, Material Science, Unit Operation Laboratory and Safety Health & Environment. At Kerman University, I taught Process Calculation, Fundamentals of Heat and Mass Transfer, Thermodynamics, Fluid Mechanics, Principles of Chemical Engineering and Chemical Reaction Engineering. Thus, I am confident that I could teach broad range of topics in Chemical Engineering. This is because of the fact that my research includes the core areas of Chemical Engineering. My teaching interest includes kinetics and reaction engineering, transport phenomena and fluid mechanics courses for undergraduate and graduate students, which naturally fit my research and teaching experience. Because of the fact that my research includes kinetics and reaction engineering, heat and mass transfer and fluid mechanics concepts, so I could give students team working projects and provide them the opportunity to improve their knowledge about the fundamental principles and technical details of basics of Chemical Engineering. Moreover, I always enjoy teaching hands-on laboratory classes such as Unit Operation laboratory or chemistry laboratory.

JOURNAL PUBLICATIONS, PATENTS AND CONFERENCE PROCEEDINGS

Patents:

M. Ahmadi, P. Ratnasamy, M.A. Carreon, "Single-step process for production of branched, cyclic, aromatic, and cracked hydrocarbons from fatty acids," United States Patent Application No. 14/489,852, filed in September 2014

M. Ahmadi, P. Ratnasamy, M.A. Carreon, "Single-step decarboxylation and further conversion of fatty acids over bifunctional catalysts," United States Patent Application No. 61/948,139, filed in March 2014

Journal Publications:

M. Ahmadi, J. Jasinski, J. Satyavolu, M. Sunkara, "A novel carbonization process to produce functionalized carbon from biomass," Under Review

M. Ahmadi, G. Willing, "Heat transfer enhancement in water-based CuO nanofluids," *International Journal of Heat and Mass Transfer*, Under Review

M. Ahmadi, A.Nambo, J.B. Jasinski, P. Ratnasamy, M.A. Carreon, "Decarboxylation of oleic acid over Pt catalysts supported on small-pore zeolites and hydrotalcite," *Journal of Catalysis Science & Technology*, 2015,5, 380-388

M. Ahmadi, E.E. Macias, J.B. Jasinski, P. Ratnasamy, M.A. Carreon "Decarboxylation and further transformation of oleic acid over functional Pt/SAPO-11 catalyst and Pt/chloride Al_2O_3catalysts," *Journal of Molecular Catalysis A*, 2014, 386,14–19

M. Ahmadi, M.R. Ghasemi, H. Hashemipour, "Study of different parameters in TiO_2 nanoparticles formation," *Journal of Material Science and Engineering*, 2011, 5, 87-93

Conference Publications:

M.Ahmadi, J. Jasinski, J. Satyavolu, M. Sunkara, "Biomass conversion to functionalized carbon," *2017 Renewable Energy & Energy Efficiency (RE3) Workshop,* May 2017, University of Louisville, Louisville, KY

M. Ahmadi, G. Willing, "Computational fluid dynamics modeling of momentum and heat transfer in nanofluids," *AIChE Annual Meeting*, Nov 2016, San Francisco, CA

M. Ahmadi, G. Willing, "Measurement of heat transfer mechanisms in water-based nanofluids," *AIChE Annual Meeting*, Nov 2015, Salt Lake City, UT

M. Ahmadi, E.E. Macias, J.B. Jasinski, P. Ratnasamy, M.A. Carreon, "Oleic acid decarboxylation and further transformation over pt/sapo-11 catalysts and pt/al_2o_3 catalysts," *AIChE Annual Meeting*, Nov 2014, Atlanta, GA

M. Ahmadi, A. Nambo, J.B. Jasinski, P. Ratnasamy, M.A. Carreon, "Decarboxylation and further isomerization of oleic acid over Pt-supported on small pore zeolites and hydrotalcite catalysts," *AIChE Annual Meeting*, Nov 2014, Atlanta, GA

M. Ahmadi, J. Jasinski, M. Carreon, "Decarboxylation of oleic acid over solid catalysts," *North American Catalysis Society Meeting*, June 2013, Louisville, KY

M. Ahmadi, M.R. Ghasemi, H. Hashemipour, "Investigating of operating conditions effects on TiO_2 nanoparticles properties," *International Conference on Nanoscience and Nanotechnology*, February 2010, India

505403 Designing Solid-Liquid Interphases and Polymer Composite Networks for Energy Storage and Carbon Capture

Abstract:

Research Experiences:

Advances in the basic science and engineering principles of electrochemical energy storage are imperative for significant progress in portable electronic devices. In this regard, metal based batteries comprising of a reactive metal (like Li, Na, Al) as anode have attracted remarkable attention due to their promise of improving the anode-specific capacity by as much as 10-fold, compared to the current state-of-art Li-ion battery that uses a graphitic anode. Perhaps their greatest advantage lies in the possibility of using of a Li-free high-capacity cathode like Oxygen that can improve the gravimetric energy density of batteries from ~0.3kWh/kg to ~12kWh/kg (i.e. comparable to the useful energy available from combustion of hydrocarbons). A persistent challenge with batteries based on metallic anodes, concerns their propensity to fail by short-circuits produced by dendrite growth during battery recharge, as well as by runaway of the cell resistance due to internal side reactions with liquid electrolytes. In my thesis research, I utilized multiscale transport modeling and experiments to fundamentally understand and to thereby develop rational designs for polymer electrolytes and electrode-electrolyte interphases that overcome these challenges (Tikekar, Choudhury et al. Nat. Energy 2016).

There have been several studies in literature dedicated to the prevention of dendrite growth by means of a high modulus physical barrier. However, electrolytes/separators with high mechanical strength tend to have low ionic conductivity, thus limiting their practical use. A recent theoretical study from our group reported a linear stability analysis of dendrite growth (Tikekar et al., Sci Adv. 2016) of metal electrodeposition and showed that the length-scale on which transport occurs near the electrodes could be as important as electrolyte modulus in stabilizing metals against dendrite formation. In a nutshell, this study concluded that dendrites can be prevented from crossing over to the counter electrode using battery separators with pore-diameter lower than critical (smallest) size of the dendritic nucleate. To evaluate this proposal, we designed cross-linked nanoparticle-polymer composite electrolytes with tunable pore size and quantified the stability of metal electrodeposition in these systems. In contrast to most previously reported polymer electrolytes, the crosslinked membrane simultaneously showed good mechanical strength (~1MPa) and high ionic conductivity at room temperature (~5mS/cm), which is a consequence of the high crosslinking node points in these membranes (Choudhury et al. Nat. Commn. 2015). Direct visualization experiments were performed to understand the effect of pore-size on dendrite growth, which showed remarkable agreement with the theoretical predictions. Furthermore, when operated in a battery, the crosslinked membrane showed the highest short circuit time compared to similar electrolytes reported in the literature.

Importantly, these studies showed that while the tendency for battery failure by dendrite-induced short-circuits can be reduced in polymer electrolytes, the issue of capacity-fading as a result of continuous reactions of the metal with liquid electrolyte persists. An additional striking fact in the electrodeposition literature not addressed by the linear stability analysis is that certain metals, including Magnesium, do not form dendrites. Further in my PhD thesis, I showed how multiscale analysis of transport at electrochemical interfaces enables design of stable solid-liquid interphases for reactive metal batteries. Recently, we used Density Functional Theory (DFT) calculations to quantify the diffusion energy barrier of ions on Mg, Li, Na surfaces and interestingly it seen that the diffusion barrier of Mg (0.02eV/atom) is several fold lower than Li (0.14eV/atom) or Na (0.16eV/atom) metals. In fact, the diffusion barrier of Li2CO3, Li2O (the commonly found compounds in lithium interface) is even higher, which is consistent with the dendritic electrodeposition in such batteries. However, in quest for finding stable interfaces, we observed that most metal halides (LiF, LiBr, NaF etc.) have much lower diffusion barrier. In other words, halide-rich interfaces on lithium or sodium can lead to stable electrodeposition similar to Mg deposition (Choudhury et al. Nat. Commn., in revision). We further utilized Classical Nucleation Theory (CNT) to understand implication of the ab-initio model in macroscopic metal deposition. The predictions from the coupled DFT-CNT model were validated using ex-situ scanning electron microscopy as well as in-situ optical microscopy. The nucleation pattern, indeed, showed a strike difference between usual (carbonate-rich) and halide-rich lithium interfaces. Based on these fundamental understanding, the solid electrolyte interphase in lithium metal batteries were artificially modified using organo-metallic reactions to enable enhanced reversibility in high energy density Lithium-Oxygen battery that demonstrated extended capacity retention and longer cycle life (Choudhury et al. Sci. Adv. 2017).

Research Interests:

The issues related to energy storage and climate change are often thought to be coupled. It is because, most renewable energy sources like solar or wind are intermittent, so they require efficient energy storage devices for full-time operation. On the other hand, although electric vehicles reduce vehicular emissions of carbon dioxide, the main sources of electricity generation are fossil fuel based. In this regard, a clever way to curb the CO2 emissions as well as store (and generate) electricity is designing a high energy density metal-CO2/O2 battery. In my future research career, I want to understand the fundamental barriers and find out clever solutions to successfully build high energy density batteries (or fuel cells) that provide multi-functional platform for energy harvesting and carbon capture, one example is aqueous Lithium-CO2 battery. There are at least three directions of research in the same broad area. (1) The first challenge is closely related to my PhD research, which is stabilizing the lithium metal under harsh conditions like aqueous media. This area of research will deal with strategies to reduce the chemical activity of the reactive fluids with lithium or sodium. It has been observed that the electrochemical stability window of water can be altered within a specific range by changing the salt concentration. It is hypothesized that excess salt in water forms a protective film on the electrode, thus changing the electric field at the interface. Another school of thought is the fact that excess salt binds the water molecule together, thus preventing their access to the electrode surface. Thus, a more detailed understanding is required to further push the limits of electrochemical activity in reactive conditions. I believe, my prior research work on understanding the surface diffusion of ions at the interface will provide me an important base to carry out my future work. Specifically, I will use techniques like in-situ AFM and Focussed Ion Beam milling, to understand the properties of the interface and based on that I will design the chemistry of electrolyte. Also, creating an artificial solid-electrolyte interphase using inert materials like Indium or Graphene can be a direction. (2) The second sub research area is on designing hybrid polymer membrane for separating the anode and cathode compartments in a secondary battery. Particularly, in a metal based battery, there are two major problems, one associated with the reactivity of the metal and dendrite formation, while the other is related to the cathode, involving high voltage stability or dissolution of reaction products by cathodic reactions. These two problems are decoupled and need to be solved using different electrolyte chemistries. I plan to synthesize polymeric composite membranes that can regulate the flow of ions, while restricting the solvent passage. Cylindrically aligned block copolymer or janus-particle films are some obvious approach, which can interact with solvent molecules differently from different sides. I would be interested in characterizing the structure of the membrane using cross-sectional scanning electron microscopy and in-situ optical microscopy. My previous experience in designing nanoporous crosslinked nanoparticle-based membranes for dendrite suppression can help me in this research area. (3) The third research area, I am going to focus on deals with the design of semi-solid cathode materials with high capacity loadings. Flow batteries are not suitable for consumer devices, owing to the requirement of a flow system and low volumetric density, while slurry-casted planar cathodes are limited in the active material loading due to the ion and electron transport-related issues in thick cathodes. I plan to synthesize gel-type building blocks of cathodes that can have good electron conductivity as well as high packing density. One pathway for this is synthesizing carbon based nanostructures, whose surface may be decorated with redox active species. The interconnected carbon structure can provide electron mobility, and a high grafting density of redox species on the carbon structure can improve the total loading capability. It will be important to understand the rheological characteristics as well as the self-assembly of the cathode building blocks using techniques like Rheo-SAXS, and Rheo-SANS. A major portion of my PhD research dealt with the understanding of structure and dynamics of polymer grafted hairy nanoparticles. I hope to bring in these experiences in my future research.

Experience in Grant writing:

Previously I have written at total of three grant proposals and six beamline proposals for doctoral advisor as well as for my graduate committee member. Successful proposal include Lithium metal battery interphases funded by ARPA-E of DOE and Polymer Battery Separators funded by NSF.

Teaching Interests:

I have prior experience of teaching two undergraduate level courses as a Teaching Assistant (TA) in Cornell University. The first course I served as a TA was for the junior year Chemical Engineering course – 'Heat and Mass Transfer' comprising of 96 students. My responsibilities in the course included a weekly recitation session, design of homework and exam questions as well as taking office hours and review sessions for students. My second TA appointment was in the Physics Department of Cornell University for the Freshman Level Course – 'Mechanics and Heat'. My task in this course was to instruct discussion sessions, conduct Physics Lab and take office hours as well as grade homework and exam papers. In addition, in my research group, I have guided several junior graduate (three M.S. and 1M.Eng.) students as well as three undergraduate students. I was also invited to give research talks in several research groups and R&D companies. Based on my PhD research, I am comfortable giving lectures in different areas including Polymer Physics, Electrochemistry, and Surface Science.

Publications:

1. Tuning surface diffusion barriers for high reversibility of lithium metal anodes S. Choudhury, Z. Tu, K. Fawole, S. Stalin, D. Gunceler, R. Sundararaman and L. A. Archer. Langmuir, in revision

2. Designing Solid-liquid Interphases for Sodium Batteries S. Choudhury, S. Wei, Y. Ozhabes, D. Gunceler, M. J. Zachman, Z. Tu, J. H. Shin, P. Nath, A. Agrawal, L. F. Kourkoutis, T. A. Arias and L. A. Archer. Nature Communications, in revision

3. Designing artificial solid-electrolyte interphases for single-ion, high-efficiency transport in batteries Z. Tu*, S. Choudhury*, M. J. Zachman, S. Wei, K. Zhang, L. F. Kourkoutis, L. A. Archer. Joule, in press [* First authorship shared]

4. Self-suspended Polymer Grafted Nanoparticles. S. Srivastava, S. Choudhury, A. Agrawal. Current Opinion in Chemical Engineering, 2017

5. Designer Interphases for the Lithium-Oxygen Electrochemical Cell. S. Choudhury, C. T. Wan, W. I. Al Sadat, Z. Tu, S. Lau, M. J. Zachman, L. F. Kourkoutis and L. A. Archer. Science Advances, E1602809 (2017) *†Highlighted in: Cornell Chronicle, Phys.org, Green Car Congress, Nanowerk Nanotechnology News*

6. Highly stable sodium batteries enabled by functional ionic polymer membranes. S. Wei*, S. Choudhury*, J. Xu, P. Nath, Z. Tu, and L. A. Archer. Advanced Materials, 29, 1605512 (2017) [* First authorship shared] *†Featured on the cover*

7. Nanoporous Hybrid Electrolytes for High Energy Batteries Based on Reactive Metal Anodes. Z. Tu, M. J. Zachman, S. Choudhury, S. Wei, L. Ma, Y. Yang, L. F. Kourkoutis, L. A. Archer. Advanced Energy Materials, 1602367 (2017) *†Featured on the cover*

8. Design Principles for Electrolytes and Interfaces for Stable Lithium-metal Batteries. M. D. Tikekar, S.

Choudhury, Z. Tu, and L. A. Archer. Nature Energy, 1:16114 (2016)

9. Lithium Fluoride Additives for Stable Cycling of Lithium Batteries at High Current Densities. S. Choudhury and L. A. Archer. Advanced Electronic Materials, 2, 1500246 (2016)

10. Hybrid Hairy Nanoparticles Stabilize Lithium Metal Batteries. S. Choudhury, A. Agrawal, S. Wei, E. Jeng and L. A. Archer. Chemistry of Materials, 28 (7), 2147-2157 (2016)

11. A Stable Room Temperature Sodium-sulfur Battery. S. Wei, S. Xu, A. Agrawral, S. Choudhury, Y. Lu, Z. Tu, L. Ma, L. A. Archer. Nature Communications, 7: 117222 (2016)

12. Molecular Origins of Temperature-Induced jamming in Self-Suspended Hairy Nanoparticles A. Agrawal, H.Y. Yu, A. Sagar, S. Choudhury, L. A. Archer. Macromolecules, 49 (22), 8738-8747 (2016)

13. Multifunctional Separator Coatings for High-Performance Lithium-Sulfur Batteries M. S. Kim, L. Ma, S. Choudhury, L. A. Archer. Advanced Materials Interfaces, 3 (22) (2016) *†Featured on the cover*

14. Fabricating Multifunctional Nanoparticle Membranes by a Fast Layer-by-Layer Langmuir-Blodgett Process: Application in Lithium-Sulfur Batteries M. S. Kim, L. Ma, S. Choudhury, S. S. Moganty, S. Wei, and L. A. Archer. Journal of Materials Chemistry A, 4, 14709-14719 (2016)

15. Interactions, Structure, and Dynamics of Polymer-Tethered Nanoparticle Blends A. Agrawal, B. M. Wenning, S. Choudhury. Langmuir 32 (34), 8698-8708 (2016)

16. Multiscale Dynamics of polymers in Particle-Rich Nanocomposites R. Mangal, Y. H. Wen, S. Choudhury, L. A. Archer. Macromolecules 49 (14), 5202-5212 (2016)

17. A Highly Reversible Room Temperature Lithium Metal Battery based on Cross-linked Hairy Nanoparticles. S. Choudhury, R. Mangal, A. Agrawal and L. A. Archer. Nature Communications, 6: 10101 (2015) *†Highlighted in: Cornell Chronicle, Phys.org, IEEE Spectrum, Engineering.com*

18. Self-suspended Suspensions of Covalently Grafted Hairy Nanoparticles. S. Choudhury, A. Agrawal, S A Kim, and L. A. Archer. Langmuir, 31 (10) 3222-3231 (2015)

19. A Highly Conductive, Non-flammable Polymer-nanoparticle Hybrid Electrolyte. A. Agrawal*, S. Choudhury*, and L. A. Archer. RSC Advances, 5, 20800-20809 (2015) [* First authorship shared]

20. Dynamics and yielding of binary self-suspended Nanoparticle Fluids. A. Agrawal, H. Y. Hsiu, S. Srivastava, S. Choudhury, S. Narayanan and L. A. Archer. Soft Matter, 11, 5224-5234 (2015)

21. Electronic and Chemical Properties of Germanene: The Crucial Role of Buckling. A. Nijamudheen, R. Bhattacharjee, S. Choudhury, and A. Datta. Journal Physical Chemistry C, 119 (7), pp 3802–3809 (2015)

Patents:

1. Dendrite Inhibiting Electrolytes for Metal-Based Batteries. L. A. Archer, T. A. Arias, Y. Lu, Z. Tu, D. Gunceler, S. Choudhury U.S. Patent Serial No. 15/128,635, Filed on September 23, 2016

2. Protective layers for Metal Electrode Batteries. Z. Tu, S. Choudhury, S. Wei, and L. A. Archer. U.S. Provisional Patent No. 62/436,248. Issued on December 19, 2016

3. Binder-Free and Multicomponent Layer-by-Layer Separator Coating for Lithium Batteries. M. S. Kim, S. Choudhury, L. Ma, and L. A. Archer U.S. Provisional Patent No. 62/265,539. Issued on December 10, 2015

504926 Electrodeposition-Based Additive Manufacturing: Combining Bipolar Electrochemistry with Scanning Probe Methodology for Freeform Fabrication

Abstract:

Recent advances in additive manufacturing technologies (such as fused deposition modeling or selective laser sintering) have been primarily driven by straightforward integration of computer software with hardware components, enabling simple 3D fabrication from computer aided design files. Despite these recent advances, a similar transition has yet to emerge for fully software-reconfigurable, electrodeposition-based prototyping. Typical industrial electrodeposition patterning techniques require a series of mask, deposition, etch, and planarization steps to develop fully formed 3D objects, limiting their customization. Localized direct-write electrodeposition techniques can bridge the advantages of electrodeposition-based micro/nanofabrication with user-friendly software intrinsic to present commercially successful additive manufacturing methods. Dimensions of the pipettes or electrodes used in localized electrodeposition dictate the spatial resolution of the material deposited (existing pipette/electrode fabrication techniques achieve sub 100nm resolution) while vertical resolution is controlled by the applied current and/or electrochemical waveform, providing methods for monolayer thickness control. Additionally, implementing bipolar electrochemistry in localized electrodeposition removes the need to electrically connect to the substrate during electrodeposition, providing improved automation and easier integration of software with hardware components.

Research Interests:

My research background is centered around electrochemical engineering; specifically in electrodeposition and electroanalytical characterization. My broader interests align in electrochemical design and integration of fundamental transport, kinetics, and thermodynamics for innovative solutions to industrially relevant problems. Research in the area of bipolar electrochemistry during my PhD provides me with a unique and comprehensive skillset, as bipolar electrochemistry involves intricate coupling of the following:

1. Thermodynamic principles similar to those seen in electroless deposition or nanoparticle syntheses.
2. Spatially segregated heterogeneous reduction and oxidation kinetics on a single substrate analogous to corrosion phenomenon.
3. Specifically designed ohmic distribution in an electrochemical cell to minimize shunt current losses akin to design of flow batteries.

During my postdoctoral studies at NIST, I have continued research in metal electrodeposition while adding expertise in various electroanalytical measurement techniques. The topics being probed include:

1. Effect of chemical adsorption on metal surfaces during electrodeposition for morphological control, directed growth, and negative differential resistance (S-NDR) generated spatial patterning.
2. Implementing localized electrodeposition with scanning probe technology for 3D freeform fabrication and combinatorial characterization of thin film materials.
3. Electrocatalysis of thin film metal alloys and metal oxides/hydroxides for energy applications.

A variety of electroanalytical measurement techniques are being utilized to characterize the above systems including: rotating disk electrochemistry, electrochemical quartz crystal microbalance, microelectrodes, scanning droplet cell electrochemistry, and scanning electrochemical microscopy.

Future Directions: I plan to combine aspects of my PhD and postdoctoral studies to create a research environment grounded in electrochemical fundamentals while engineering materials and processing challenges for electronics and energy applications. The rising need for more efficient energy technologies and the imminent transition to next generation computer processor manufacturing for quantum computing applications will require continued development, processing, and fabrication of thin film and 3D-structured materials. Integrating bipolar electrochemistry with scanning ion conductance microscopy and electrochemical impedance spectroscopy will create a new technique for in-situ measurement of film growth during 3D freeform fabrication, furthering customization capabilities for electrodeposition based additive manufacturing at the nanoscale. This technology can also be used for combinatorial characterization of materials for their electrocatalytic and corrosion resistant properties.

Teaching Interests:

My academic teaching interests align with my research background in studying the coupling between fundamental mass transport, thermodynamics, and kinetics for engineering electrochemical systems. Education in each of these core chemical engineering concepts individually is vital to the development of chemical engineering students with the requisite skill-set to succeed in industry. However, students commonly only put these fundamentals together during their capstone senior design project exploring design and optimization of a single chemical process at factory production scale. With the types of jobs that chemical engineering students are entering after graduation spanning a greater range of industries than traditionally, I plan to focus on incorporating core chemical engineering concepts across a broader range of commercial applications in class to better prepare students for their careers. Courses I have interest in teaching are a reflection of my past research and teaching experiences. These include core chemical engineering courses such as momentum/mass transport, reactor design, senior design, graduate level surface kinetics/catalysis, and an interdisciplinary elective course on electrochemical fundamentals. Within the classroom I aim to build an environment that fosters creativity, critical problem solving, and independent thinking while promoting group discussion and collaboration. This will be achieved with group exercises, example problems, and open-ended questions/discussions mixed into lecture periods and recitations. In my prior experiences as a teaching assistant, guest lecturer, and mentor I found that willingness and availability of the educator for individual or small group exchanges provides the students with a low-stress opportunity to ask questions, stimulate comprehensive discussion, and provide immediate feedback on lectures. Thus, office hours and group mentorship circles will be utilized by myself and teaching assistants to promote these activities.

505431 Engineering the Next-Generation of Electrochemical Energy Storage

Abstract:

Engineering the Next-Generation of Electrochemical Energy Storage

Kevin W. Knehr, Princeton University

2nd Year Postdoctoral Research Associate

Research Interests:

Batteries are complex, multidisciplinary, electrochemical energy storage systems that are crucial for powering our society. The increasing demand for better batteries across a diverse range of applications (*e.g.,* wearable tech, smartphones, electric vehicles, and grid-level storage) can only be met through the development of next-generation batteries employing new chemistries, new materials, and new designs. Identifying strategies for developing better batteries requires an understanding of the complex, coupled processes associated with mass transfer, reaction kinetics, and thermodynamics occurring within these systems. The combination of electroanalytical chemistry, materials characterization, continuum-level modeling, and engineering design offers a unique set of tools which can identify the performance limiting processes in batteries and provide solutions for enhancing the performance of next-generation systems.

Teaching Interests:

My objective as an educator is to have a heavy involvement with the undergraduate curriculum while maintaining a healthy research laboratory devoted to training the next generation of doctoral researchers. As an educator of undergraduates, I plan on leveraging my background as an electrochemical engineer to demonstrate how the fundamentals of chemical engineering (*e.g.,* thermodynamics, reaction kinetics, transport, and engineering design) intersect in real-world applications, like batteries. To accomplish this, I am prepared to teach any of the chemical engineering core courses. In addition, I am interested in establishing an elective course in electrochemical engineering based on my Ph. D. adviser's text book, which will contain a chapter on batteries I am helping to develop. As a mentor of graduate students, I anticipate a healthy laboratory of 4 to 6 Ph. D. students that I can personally train as electrochemical researchers. I believe direct involvement in all my students' projects is important to cultivate the best research mindsets and to develop sound research practices.

Postdoctoral Project: *"Development of a Minimal Architecture Zinc-Bromine Battery for Low Cost Electrochemical Energy Storage"*

Under supervision of Daniel A. Steingart, Department of Mechanical and Aerospace Engineering, Andlinger Center for Energy and the Environment, Princeton University

Ph.D. Thesis: *"Identification, Characterization, and Mitigation of the Performance Limiting Processes in Battery Electrodes"*

Under supervision of Alan C. West, Department of Chemical Engineering, Columbia University

Research Experience:

My academic training has been rooted in the use of electrochemical theory, engineering, and design to probe the mechanisms controlling battery performance. As a result of my investigations into several different battery technologies, I have developed skillsets in both experimental characterization and theoretical modeling of complex electrochemical systems. For example, in one project (which I conducted as an NSF-GRFP fellow), I coupled in situ, synchrotron X-ray techniques at Brookhaven National Laboratory with electroanalytical methods to understand and control the growth of lead-sulfate crystals in lead-acid batteries. In another project, I used continuum level modeling to probe the mass transport limitations across multiple length scales (nm to cm) in lithium-ion batteries containing nanosized materials. During my time as a postdoc, my main focus has been on the investigation and advancement of a novel, membrane-free, zinc-bromide battery. To accomplish this, I have employed diagnostic techniques using microcontrollers and electroanalytical methods to understand the sources of efficiency loss and to study approaches for improving the lifetime, efficiency, and energy density of the system.

In addition to many skillsets, the multidisciplinary nature of batteries has provided experience in many research areas, including: chemistry, materials science, and of course, chemical engineering. Furthermore, the majority of my work during by Ph.D. involved close collaboration with synthetic chemists and beamline scientists at Brookhaven National Laboratory and Stony Brook University as a result of a DOE-Energy Frontier Research Center involving eight partner institutions. These collaborations have yielded invaluable training on the management, organization, and execution of large, multidisciplinary research projects.

Teaching Experience:

Along with my academic training, I have had several opportunities to enrich my abilities as an educator. In a formal setting, I have served as a teaching assistant and a guest lecturer. Additionally, I am involved in writing a chapter on batteries for my Ph.D. adviser's text book: *Electrochemistry and Electrochemical Engineering: An Introduction*. In an informal setting, I have been a mentor to six undergraduate, one M. S. and five Ph. D. students throughout my career. I have also had the opportunity to work with middle school students through an outreach program I helped to develop as the President and Outreach Chair of the Chemical Engineering Graduate Organization at Columbia University. The program targets under-represented minorities in STEM, and consists of in-class activities and science fair projects at Frederick Douglass Academy II, a middle school in Harlem, NY.

Future Directions:

As a faculty member, I would like to continue on the path of characterizing and mitigating the performance limiting processes in current battery technologies and advancing novel battery systems and designs, which are tailored to modern applications. I am particularly interested in three main thrusts: *a)* development of electrochemical models for predicting and understanding battery degradation, *b)* advancement of batteries uniquely designed for the electric grid, where only cost and lifetime are the key design constraints, and *c)* investigation of the system-level performance of new, high-capacity materials that are being researched for portable applications. My current expertise in electrochemical theory, electroanalytical methods, and continuum-level modeling provides a useful framework for advancing new battery technologies because these tools make it possible *i)* to identify and mitigate the "rate-limiting" processes that are most detrimental to the performance of a specific battery chemistry and *ii)* to analyze the trade-offs associated with system-level design.

My overall approach is to leverage a deep understanding of electrochemical theory to develop experimental approaches and continuum-level simulations which probe and explain the underlying mechanisms controlling the performance of electrochemical systems. Therefore, my future lab will have both experimental and theoretical components. In addition, I anticipate close ties to collaborators which have access to advanced characterization techniques (*e.g.*, at national laboratories) that can provide supporting information, especially on the evolution of materials properties within these electrochemical system.

Selected Publications (21 in total, 2 in preparation):

Shaurjo Biswas, Aoi Senju, Robert Mohr, Thomas Hodson, Nivetha Karthikeyan, **Kevin W. Knehr**, Andrew Hsieh, Xiaofeng Yang, Bruce E. Koel, and Daniel A. Steingart, "Minimal Architecture Zinc-Bromine Battery for Low Cost Electrochemical Energy Storage," *Energy & Environmental Science*, 10, 114-120 (2017). *DOI: 10.1039/C6EE02782B*

K. W. Knehr, Christina A. Cama, Nicholas W. Brady, Amy C. Marschilok, Kenneth J. Takeuchi, Esther S. Takeuchi, and Alan C. West "Simulations of Lithium-Magnetite Electrodes Incorporating Phase Change," *Electrochimica Acta*, 238, 384-396 (2017). *DOI: 10.1016/j.electacta.2017.04.041*

K. W. Knehr and Alan C. West, "Theoretical Considerations for Improving the Pulse Power of a Battery through the Addition of a Second Electrochemically Active Material," *Journal of the Electrochemical Society,* 163, A1576-A1583 (2016). *DOI: 10.1149/2.0621608jes*

K. W. Knehr, Nicholas W. Brady, Christina A. Cama, David C. Bock, Zhou Lin, Christianna N. Lininger, Amy C. Marschilok, Kenneth J. Takeuchi, Esther S. Takeuchi, and Alan C. West, "Modeling mesoscale transport of lithium-magnetite electrodes using insight from discharge and voltage recovery experiments," *Journal of the Electrochemical Society*, 162, A2817-A2826 (2015). *DOI:10.1149/2.0961514jes*

K.W. Knehr, Christopher Eng, Jun Wang, and Alan C. West, "Transmission X-Ray Microscopy of the Galvanostatic Growth of Lead Sulfate on Lead: Impact of Lignosulfonate," *Electrochimica Acta*, 168, 346-355 (2015). *DOI: 10.1016/j.electacta.2015.04.022*

K. W. Knehr, Christopher Eng, Yu-chen Karen Chen-Wiegart, Jun Wang, and Alan C. West, "In situ transmission x-ray microscopy of the lead sulfate film formation on lead in sulfuric acid," *Journal of the Electrochemical Society,*162, A255-A261 (2015). *DOI: 10.1149/2.0141503jes*

K. W. Knehr and E. C. Kumbur, "Role of convection and related effects on species crossover and capacity loss in vanadium redox flow batteries," *Electrochemistry Communications*, 23, 76-79 (2012). *DOI: 10.1016/j.elecom.2012.07.008*

K.W. Knehr, Ertan Agar, C. R. Dennison, A. R. Kalidindi, and E. C. Kumbur, "A transient vanadium flow battery model incorporating vanadium crossover and water transport through the membrane," *Journal of the Electrochemical Society*, 59, A1446-A1459 (2012). *DOI: 10.1149/2.017209jes*

K. W. Knehr and E. C. Kumbur, "Open circuit voltage of vanadium redox flow batteries: discrepancy between models and experiments," *Electrochemistry Communications*, 13, 342-345 (2011). *DOI: 10.1016/j.elecom.2011.01.020*

Book Chapter:

"Batteries: Operation and Design," in *Electrochemistry and Electrochemical Engineering: An Introduction*, Alan C. West, *in progress*.

Successful Proposals: NSF-GRFP Fellowship (2012-2016), NDSEG Fellowship awardee (*Declined*).

502777 Stable Electrochemical Growth in Viscoelastic Electrolyte

Abstract:

Research Interests: Electrodeposition is used in various manufacturing processes for creating metal, colloid, and polymer coatings on conductive substrates. The process also plays an important role in electrochemical storage technologies based on rechargeable batteries, where it must be carefully managed to facilitate stable and safe operations at low operating temperatures, high rates, and over many cycles of charge and discharge. A successful electrodeposition processes requires fast transport of charged species (e.g. ions, particles, & polymers) and stable redox reactions at liquid-solid interfaces. In almost all used liquid electrolytes, deposition is subject to a variety of hydrodynamic and morphological instabilities that lead to complex transport phenomena in the electrolyte and unstable deposition, including formation of ramified structures known as dendrites on the substrate/electrode. This talk considers the stability of electrodeposition of reactive metals on planar electrodes with an emphasis on its role in enabling next-generation batteries in which metallic lithium or sodium serves as the electrode. Such batteries have been argued to offer step-change improvements in electrochemical storage technology over todayâ€™s state-of-the art lithium ion batteries and are under active investigation worldwide for high-energy, portable energy storage solutions in multiple fields. Beginning with analyses of ion transport processes and stability of electrodeposition in a structured electrolyte, the talk will discuss the fundamental fluid properties such as viscosity, elasticity, electro-convection etc. of the electrolytes that affect the stability of deposition and electrode-electrolyte interphases for stabilizing deposition of reactive metals. The talk will also discuss contemporary efforts to evaluate these approaches using electrochemical and visualization studies.

Teaching Interests: Mass Transport, Physics

505566 Designing Electrochemical Surfaces and Interfaces for Catalysis, Separation Membranes, and Sensors

Abstract:

Designing Electrochemical Surfaces and Interfaces for Catalysis, Separation Membranes, and Sensors

Jesse D. Benck

Postdoctoral Research Associate, Massachusetts Institute of Technology, benck@mit.edu

Research Interests:

My research interests focus on electrochemical technologies including electrocatalysts, separation membranes, and sensors. In particular, I am interested in studying the fundamental properties of surfaces and interfaces that control adsorption, reaction, and permeation processes relevant to these technologies. I aim to develop an experimental research program with the capabilities to fabricate materials with atomically-precise surfaces and interfaces, characterize these materials using *in situ* microscopy and spectroscopy, and test their electrochemical performance. Through these efforts, I hope to reveal physical insights that can be used to design new materials and devices that can address global challenges in energy, water, and health.

My research program will focus on the following areas:

1. Electrocatalysis for efficient energy conversion.

Understanding how electrocatalyst composition, chemical state, and surface structure determine catalytic activity will enable the design of improved catalysts for electrochemical energy conversion reactions such as water oxidation and carbon dioxide reduction.

2. Nanoporous membranes for selective and efficient electrochemical separations.

Studying the relationship between the size, shape, and functionalization of nanoscale pores in two-dimensional materials with the flux of ionic or molecular species through these pores will enable the design of high-performance membranes for water purification, desalination, and electrochemical energy conversion devices.

3. Selective adsorption at electrochemical interfaces.

Understanding how three-dimensionally structured inorganic materials and macromolecule-functionalized interfaces can form adsorption sites tailored to bind specific molecules will enable the design of sensors and electrocatalysts with precise molecular selectivity.

My previous research on electrochemical devices and gas separation membranes has prepared me to lead this proposed research program.

My PhD research with Professor Thomas F. Jaramillo in the Stanford University Department of Chemical Engineering focused on electrocatalysis and photoelectrochemistry for solar hydrogen production. I developed nonprecious metal sulfide and phosphide electrocatalysts for the electrochemical hydrogen evolution reaction. Using *in situ* microscopy and spectroscopy to study the catalyst structure and composition enabled me to identify the properties that controlled activity. I also designed and fabricated silicon and III-V semiconductor photoelectrodes with integrated catalyst and surface protecting layers for H_2 production via photoelectrochemical water splitting, focusing on improving device stability. I created a photovoltaic-electrolyzer system with world record solar-to-hydrogen efficiency. These experimental studies were all guided by technoeconomic analysis and numerical simulations of solar hydrogen production systems.

My postdoctoral research with Professor Michael S. Strano in the Massachusetts Institute of Technology Department of Chemical Engineering focused on nanoporous graphene membranes for gas mixture separations. Atomically thin graphene membranes have the potential to separate gas mixtures with extremely high flux and selectivity, leading to reduced energy consumption. I developed a procedure for fabricating suspended graphene membranes and measuring gas permeation and mixture separation selectivity across these membranes using on-line mass spectrometry. Through characterization and modeling of the nanopores, we took initial steps towards understanding the relationship between atomic-scale pore structure and gas permeance, which forms the foundation for the design of membranes with nanopores tuned for excellent separation performance.

At the poster session, I will provide a more complete account of my research experience and plans for my future research program.

Teaching Interests:

Throughout my education, I have been deeply devoted to developing my skills as a teacher and mentor. During graduate school, I was a teaching assistant for two years, and I received a Stanford Department of Chemical Engineering Outstanding Teaching Assistant award both years. I later served as the Stanford Chemical Engineering Department's head mentor teaching assistant to help younger graduate students develop their teaching skills. I founded and led an independent study course called Stanford ChemEng 432, "Introduction to Electrochemistry." In my final year at Stanford, I became a graduate teaching fellow and the primary instructor for an undergraduate elective course titled Stanford ChemEng 25E, "Energy: Chemical Transformations in Production, Storage and Use," and as a postdoctoral researcher, I gave a guest lecture in the MIT graduate-level course 10.585, "Engineering Nanotechnology." In addition, I have substantial experience mentoring many undergraduate and young graduate students in the lab.

Continuing to improve my skills as a teacher and mentor will be an important goal throughout my career as a faculty member. I am interested in teaching Chemical Engineering core courses including kinetics, thermodynamics, and separations, as well as developing elective courses related to my research interests, such as an "Electrochemical Energy Conversion" class. I also look forward to the many teaching opportunities that arise in a laboratory setting. As a research advisor I will focus on helping undergraduate and graduate students to develop their skills as independent scientists.

Selected Publications:

Full publication list on Google Scholar profile: http://scholar.google.com/citations?user=bM3D8a8AAAAJ&hl=en

1. **J.D. Benck**, Z. Chen, L.Y. Kuritzky, A.J. Forman, and T.F. Jaramillo. "Amorphous Molybdenum Sulfide Catalysts for Electrochemical Hydrogen Production: Insights into the Origin of their Catalytic Activity." *ACS Catalysis*, 2012. 2 (9): 1916-1923. http://dx.doi.org/10.1021/cs300451q

2. B.A. Pinaud, **J.D. Benck**, L.C. Seitz, A.J. Forman, Z.B. Chen, T.G. Deutsch, B.D. James, K.N. Baum, G.N. Baum, S. Ardo, H.L. Wang, E. Miller, and T.F. Jaramillo. "Technical and economic feasibility of centralized facilities for solar hydrogen production via photocatalysis and photoelectrochemistry." *Energy & Environmental Science*, 2013. 6 (7): 1983-2002. http://dx.doi.org/10.1039/C3EE40831K

3. **J.D. Benck**, S.C. Lee, K.D. Fong, J. Kibsgaard, R. Sinclair, and T.F. Jaramillo. "Designing Active and Stable Silicon Photocathodes for Solar Hydrogen Production Using Molybdenum Sulfide Nanomaterials." *Advanced Energy Materials*, 2014, 4: 1400739. http://dx.doi.org/10.1002/aenm.201400739

4. **J.D. Benck**,* T.R. Hellstern,* J. Kibsgaard, P. Chakthranont, and T.F. Jaramillo. "Catalyzing the Hydrogen Evolution Reaction (HER) with Molybdenum Sulfide Nanomaterials." *ACS Catalysis*, 2014, 4: 3957-3971. http://dx.doi.org/10.1021/cs500923c

5. T.R. Hellstern,* **J.D. Benck**,* J. Kibsgaard, C. Hahn, and T.F. Jaramillo. "Engineering Cobalt Phosphide (CoP) Thin Film Catalysts for Enhanced Hydrogen Evolution Activity on Silicon Photocathodes." *Advanced Energy Materials*, 2016. 6 (4). http://dx.doi.org/10.1002/aenm.201501758

6. S.C. Lee, **J.D. Benck**, C. Tsai, J. Park, A.L. Koh, F. Abild-Pedersen, T.F. Jaramillo, and R. Sinclair. "Chemical and Phase Evolution of Amorphous Molybdenum Sulfide Catalysts for Electrochemical Hydrogen Production." *ACS Nano*, 2016. 10 (1): 624-632. http://dx.doi.org/10.1021/acsnano.5b05652

7. J. Jia,* L.C. Seitz,* **J.D. Benck**,* Y. Huo, Y. Chen, J.W.D. Ng, T. Bilir, J.S. Harris, and T.F. Jaramillo. "Solar water splitting by PV-electrolysis with a solar-to-hydrogen efficiency over 30%." *Nature Communications*, 2016. 7: 13237. http://dx.doi.org/10.1038/ncomms13237

8. K.V. Agrawal,* **J.D. Benck**,* Z. Yuan, R.P. Misra, A. Govind Rajan, Y. Eatmon, S. Kale, X.S. Chu, D.O. Li, C. Gong, J. Warner, Q.H. Wang, D. Blankschtein, and M.S. Strano. "Fabrication, Pressure Testing and Nanopore Formation of Single Layer Graphene Membranes." *The Journal of Physical Chemistry C*, 2017. http://dx.doi.org/10.1021/acs.jpcc.7b01796

*Equally contributing authors

503337 Adsorption of Copper and NickelÂ from Wastewater in Fixed Bed Using Bentonite Clay

Abstract:

Research Interests: Water treatment

Teaching Interests: Chemical Engineering

Abstract:

Presence of heavy metals in wastewater and its land disposal are potential sources of environmental hazards. A bentonite clay from Saudi Arabia was used as an adsorbent for removing copper and nickel from wastewater in fixed bed. The effect of various experimental parameters such as bed depth (5-10cm) and initial concentration (20-100 mg L^{-1}) on the time of the bentonite clay bed and shape of the breakthrough curve was examined. The pore diffusion coefficient and the equilibrium parameters estimated from the batch adsorption data were used for modeling the adsorption of copper and nickel from wastewater by Saudi bentonite clay in an experimental fixed bed adsorber. The partial differential equations for pore diffusion and as well as copper and nickel mass balance in the packed bed were solved by regression lines to predict the breakthrough curves for fixed bed. The predictions from fixed bed pore diffusion models correlated well with the experimental data.

505066 Investigating Kinetics Under Extremely-Harsh Conditions for Energy and Food Processing

Abstract:

Wu, Xiao-Yu, PhD'17, Massachusetts Institute of Technology

Postdoctoral Associate, Reacting Gas Dynamics Laboratory/Center for Energy and Propulsion Research, MIT

Research Interests:

My research interests focus on the fundamental understanding of the kinetics in the thermo-electro-chemical processes under extremely-harsh operating conditions, i.e., high temperature/pressure, supercritical conditions and corrosive environment. Harsh operating conditions are inevitable in many energy and food processing applications such as fusion, combustion, supercritical fluid extraction and membrane filtration. The insights for both the chemical kinetics and process/reactor kinetics, i.e., the reaction kinetics, thermo-fluid transport, material stability and degradation under these conditions are essential for the optimal design of such systems in a safe and efficient manner. Both experimental and simulation tools are used to investigate these processes.

My PhD thesis with Professor Ahmed F. Ghoniem at MIT focused on development of oxygen permeable membrane reactor for enhanced H_2O/CO_2 splitting as an energy storage technology. Elevated temperatures are required to facilitate the kinetics of both surface reactions and bulk diffusions. Using both experiments and simulations, I was able to derive the chemical kinetic data for the heterogeneous surface reactions, and then model the oxygen transport process and design a membrane reactor. Additionally, the membrane stability under extremely reducing environment at elevated temperatures is one bottleneck for the implementation. Therefore, I characterized the membrane materials to understand the degradation mechanism. These insights from my PhD thesis can help engineers design better membrane reactor for this H_2O/CO_2 splitting process for energy storage.

My previous work in Zhejiang University was investigating heat and mass transfer under two extreme conditions: falling film evaporation in a vacuum and kerosene-nanofluids at supercritical conditions. Both projects were derived from industrial applications and the goal was to understand the transport phenomena and to design surface structures for enhanced heat/mass transfer.

Currently, I am in my first year of postdoctoral research with Professor Ahmed F. Ghoniem and Professor Bilge Yildiz at MIT. This project is sponsored by Exelon Corporation to investigate different hydrogen production technologies utilizing nuclear resources. The operating conditions for these technologies are different from the existing reforming technologies, and the understanding of the kinetics involved can help transform nuclear resources from solely an electricity producer into a versatile industry. Apart from studying the reaction and transport phenomena experimentally, I also plan to set up a techno-economic framework to compare the existing and developing technologies. This framework aims at identify the bottleneck for the development of certain technologies by comparing the kinetics in the processes. Following this, researches can investigate the improvements of these limiting steps to better develop the hydrogen production technologies.

In this poster session, I will give a more detailed presentation on my previous and current projects as well as the proposed research topics which will continue investigating the processes under extremely-harsh conditions for efficient energy and food processing. The goal of my research is to derive fundamental understanding of these processes, and guide the design and operation for a safe and clean future. These insights can also lead to new applications in related fields.

Teaching Interests:

Chemical kinetics, reactor design, transport phenomena, thermodynamics, or other related subjects

At MIT, I took the graduate course "Instruction in Teaching Engineering" and applied the knowledge on teaching when I TA'ed three courses at MIT: Thermal-Fluids Engineering II (undergrad core subject), Fundamental of Advanced Energy Conversion (both undergrad and grad) and Introduction to Heat Transfer (undergrad). Besides, I have mentored three undergraduates at MIT under my thesis work: one from MIT-Egypt Exchange Fellowship Program (3 months) and two from MIT Undergraduate Research Opportunities Program (UROP) (5 and 8 months). One of my supervisees received MIT UROP research funding for two terms under my thesis project.

Awards:

1. MIT-France Travel Fellowship, 2017

2. NSF Travel Award, 2017, 2016

3. Best paper initiative in AIChE Journal, 2016

4. MIT Golden Beaver Award – Institute award for excellence in student group leadership, 2016

5. Best Presentation in the 2015 AIChE Annual Meeting, 2015

6. Audience Award in MIT Mechanical Engineering Research Exhibition Poster Session, 2015

7. MISTI-MITEI fellow in Germany, MIT, 2015

Selected Publications:

1. X.Y. Wu, M. Uddi, A.F. Ghoniem, "Enhancing co-production of H_2 and syngas via water splitting and POM on surface-modified oxygen permeable membranes", *AIChE Journal*, **62 (12) 2016, 4427 - 4435, (invited contribution to AIChE Journal as "Best paper" initiative)**

2. X.Y. Wu, L. Chang, M. Uddi, P. Kirchen, A.F. Ghoniem, "**Toward enhanced hydrogen generation from water using oxygen permeating LCF membranes**", *Physical Chemistry Chemical Physics*, 17 2015 10093-10107

3. D. Huang, **X. Y. Wu**, Z. Wu, W. Li, H. T. Zhu, B. Sunden, "Experimental Study on Heat Transfer of Nanofluids in a Vertical tube at Supercritical Pressures", *International Communications in Heat and Mass Transfer*, 63 2015 54-61

4. D. Huang, B. Ruan, **X.Y. Wu,** W. Zhang, G.Q. Xu, Z. Tao, P.X. Jiang, L.X. Ma, W. Li, "Experimental Study on Heat Transfer of Aviation Kerosene in a Vertical Upward Tube at Supercritical Pressure", *Chinese Journal of Chemical Engineering*, 23 (2) 2015 425 - 434

Personal website:

http://xywu.scripts.mit.edu/xywu/

505561 Applying CVD Polymers in Membrane Separation, Biomedical Devices and Soft Electronics

Abstract:

Research Interests:

My current and future research interest lies in the chemical vapor deposition (CVD) of multifunctional polymer thin films, including microporous polymers, zwitterionic polymers, polymer based ion gels and conducting polymers, and their applications in gas separation, water desalination/purification, biomedical devices and soft electronics. Two CVD techniques, initiated CVD (iCVD) and oxidative CVD (oCVD) are chosen for their solvent-free nature. Due to this unique feature, iCVD and oCVD are distinct from solution-based or melt-processing methods for being able to directly form ultrathin (<10 nm), large area, defect-free and conformal polymer coatings on almost any substrates from vapor phase monomers in one step.

1. Previous and Current Research

1.1. Fabricating ultrathin polymer membranes for gas separations: Achieving both high flux and high gas selectivity is the ultimate goal in the membrane gas separation community. Reducing the thickness layer should increase the gas flux due to the inverse relationship between the membrane thickness and the gas flux. However, there is known challenge to achieve sub-100 nm pinhole-free gas-selective layer via conventional phase inversion technique. In this regard, three unique techniques have been employed to fabricate pinhole free sub-100 nm gas-selective layer. The use of these techniques also expands the library of gas-selective polymer membranes. These studies were carried out under the supervision of Prof. Karen Gleason at MIT and Prof. Steven Regen at Lehigh University.

1) Sub-100 nm gas-selective microporous polymer films synthesized via CVD.

Microporous polymers represent a class of most promising membrane materials for gas separation. However, two representative types of microporous polymers, polymers of intrinsic microporosity (PIMs) and hyper-crosslinked polymers (HCPs) both have their drawbacks. PIMs are subject to physical aging at low thickness due to their linear structure, while the highly crosslinked HCPs are difficult to form thin films. Therefore, an initiated plasma enhanced CVD technique was employed to solve this dilemma. The choice of monomer was a porphyrin derivative Zinc-tetraphenylporphyrin (ZnTPP). Porphyrin and its derivatives are promising monomers for synthesizing microporous polymer films due to their robust and rigid macrocycle structure. Nevertheless, there is no successful example of porphyrin derived polymeric membranes by conventional solution-based techniques. In contrast, ZnTPP derived ultrathin (<100 nm) and large area (175 cm^2) microporous polymer films were successfully synthesized via the iPECVD technique. These films displayed high selectivities for multiple gas pairs, including CO_2/N_2, CO_2/CH_4, O_2/N_2 etc. Gas-selective ultrathin films derived from CoTPP, MnTPP and TPP were also successfully synthesized by the same technique. Furthermore, such films can be crosslinked by adding co-monomer to the vacuum system to further enhance the film stability. This research represents the first effort on fabricating highly gas-selective

CVD polymer films.

2) Sub-10 nm Langmuir-Blodgett (LB) Bilayer as Highly H_2 Selective Membranes.

LB technique is known for its ability to deposit ultrathin films at nanometer scale; however, the defect (e.g., pinhole) control has been a persistent issue, which prevents the technique from applying in the (defect-sensitive) gas separation field. To overcome this difficulty, a strategy combining the organic synthesis and LB deposition was employed. In short, a series of charge bearing calix[n]arene based surfactants and polymeric surfactants have been synthesized and used as the building blocks of the LB bilayers. The defects were then removed by inserting polyelectrolyte in between the bilayer of surfactants to ionically crosslink the surfactant building blocks. Extremely high H_2/CO_2 selectivity (ca. 200) was achieved for a 7 nm LB bilayer by optimizing the surfactant/polyelectrolyte combination and the depositing condition. The intrinsic H_2/CO_2 separation performance of the 7 nm LB bilayer reached Robeson's Upper bound, which is the best performance reported for a 7 nm polymer membrane.

3) Sub-20 nm Polyelectrolyte Multilayers (PEMs) as Highly CO_2 Selective Membranes.

Polyelectrolyte Layer-by-Layer (LbL) deposition is known for depositing defect-free nanometer thick ultrathin films of different compositions. By combining the polyelectrolyte synthesis and the LbL technique, a series of sub-20 nm PEMs were fabricated to study their gas separation performances. Ultrahigh CO_2/N_2 selectivity (up to 120) was observed for a 14 nm PEMs membrane. Further study revealed that by optimizing the size of the polyelectrolyte side group, the permeance of CO_2 can be doubled while maintaining the CO_2/N_2 selectivity above 100. The research demonstrated the feasibility of applying polyelectrolyte LbL technique in the field of gas separation.

1.2. Fabricating Ultrathin Ion Gels via Initiated CVD (iCVD) and Their Application as Thin Film Transistor (TFT) Gate Insulator.

Ion gel is considered as a potential gate insulator replacement for oxides or polymer thin films in TFTs operated at low voltage, due to its large capacity (>1 μF/cm^2). However, due to the inherent slow polarization speed, current ion gel membranes suffer from significant capacity loss at higher operation frequency (e.g., > 20 KHz). Thinning down the membrane is considered to be the best strategy to overcome this weakness, but was not possible to obtain membranes at < 1 μm by solution-based techniques. In contrast, by employing an iCVD technique, a series of ultrathin crosslinked polymers were deposited as the base membranes, followed by injecting ionic liquid into the based membranes to form ion gel membranes with thicknesses from 20 nm to 2000nm. A clear thickness dependency of the ion gel capacity at high frequency is observed. By reducing the thickness of Ion gel membranes below 300 nm, the capacity of the ion gel can be maintained at 1 to 2 μF/cm^2 from 10 Hz to 1MHz. This capacity value is 1 order higher than the best polymer film insulator reported to date (Moon et al., Nat. Mater. 2015, 14, 628). At the same time, the capacity value at higher frequency (20 KHz to 1MHz) is far superior than the best performance reported to date for solution-processed ion gel membranes. (Cho et al., Nat. Mater. 2008, 7, 900). The study was carried out under the supervision of Prof. Karen Gleason at MIT.

1.3. Surface modification via Initiated CVD (iCVD) for applications in water desalination.

Biolfouling and scaling are two persistent issues in the field of water desalination. Biofouling affects membrane desalination, e.g., reverse osmosis (RO) desalination, while mineral scaling affects mostly the thermal based desalination processes, e.g., multi-stage flash distillation (MSF). To this regard, both ultrathin amphiphilic polymer and zwitterionic polymer films have been deposited onto RO membranes to render surfaces with desired fouling-resistant properties. As for scale control in the heat exchange tubes, the research has focused on the design of a new depositing technique to deposit on the internal wall of the heat exchange tube. UV lamp instead of conventional hot filament is chosen for better heat management capability. The use of UV lamp also simplified the deposition process by avoiding the use of initiator, because the UV can trigger the self-polymerization of the monomers. The studies were carried out under the supervision of Prof. Karen Gleason at MIT.

2. Future Research Plans:

2.1. Fabricating ultrathin microporous polymer membranes and their composite membranes with metal organic framework (MOF) via iCVD and oCVD, for gas separation, and potential applications in water purification, energy storage, catalysis and sensors.

2.2. Fabricating ultra-stable fouling-resistant zwitterionic polymers via iCVD, for their applications in water desalination/purification, oil/water separation and biomedical devices (e.g., electrode for bio-implants, endoscope etc.)

2.3. Fabricating ion gels via iCVD and conducting polymers via oCVD for their applications in soft electronics and in the field of renewable energy.

Teaching Interests:

I would like to teach polymer science and engineering core classes, including polymer chemistry, polymer physics, polymer science and engineering, polymer laboratory, plastics engineering etc. I can also teach some of the core courses in Chemical Engineering, Materials Science and Engineering and Chemistry. For instances, transport process, thermodynamics and kinetics, Laboratory Chemistry, Introduction to Chemical Engineering, Inorganic Chemistry, physical chemistry, organic chemistry and analytical chemistry etc.

I am also happy to design and teach new classes, such as CVD polymers, membrane separation processes, metallic/conducting polymers etc.

Selected Publications:

1. Andong Liu†, **Minghui Wang†**, Stefan Schroeder, Junjie Zhao & Karen K. Gleason, "Nanoscale Pinhole-Free Ion-Gel Film Enabled by Initiated Chemical Vapor Deposition: A Promising Material as Gate Insulator for Low-Voltage High-Speed Thin Film Transistors", (**†Equally contributed**, manuscript finished and to be submitted, patent application in process)
2. **Minghui Wang**, Peter Kovacik & Karen K. Gleason, "Conductive and Fouling-Resistant Zwitterionic Polymeric Films Synthesized by Chemical Vapor Deposition", Manuscript finished and to be submitted, MIT disclosure form filed, MIT Case 19732J)
3. Nicolas D. Boscher†, **Minghui Wang†**, Alberto Perrota, Katja Heinze, Mariadriana Creatore & Karen K. Gleason, "Metal Organic Covalent Network Chemical Vapor Deposition for Gas Separation", *Advanced Materials,* 2016, 28, 7479-7485. (**†Equally contributed. Related Patent Application, US20170158809A1**)
4. **Minghui Wang**, Nicolas D. Boscher, Katja Heinze & Karen K. Gleason, "Gas Selective Ultrathin Organic Covalent Networks Synthesized by iPECVD: Does the Central Metal Ion Matter?" *Advanced Functional Materials,* 2017, DOI:10.1002/adfm.201606652.
5. **Minghui Wang**, Junjie Zhao, Xiaoxue Wang, Andong Liu & Karen K. Gleason, "Recent Progress on Submicron Gas-selective Polymeric Membranes", *Journal of Materials Chemistry A*, 2017, DOI: 10.1039/C7TA01862B
6. **Minghui Wang†**, Xiaoxue Wang†, Priya Moni†, Andong Liu, Do Han Kim, Won Jun Jo, Hossein Sojoudi & Karen K. Gleason, "CVD Polymers for Devices and Device Fabrication", *Advanced Materials,* 2017, 29, 1604606. (**†Equally contributed**)
7. **Minghui Wang**, Vaclav Janout & Steven L. Regen, "Gas Transport Across Hyperthin Membranes", *Accounts of Chemical Research*, 2013, 46 (12), 2743-2754.
8. **Minghui Wang**, Song Yi, Vaclav Janout & Steven L. Regen, "A 7 nm Thick Polymeric Membrane With a H_2/CO_2 Selectivity of 200 That Reaches the Upper Bound", *Chemistry of Materials*, 2013, 25 (19), 3785-3787.
9. **Minghui Wang**, Vaclav Janout & Steven L. Regen, "Unexpected Barrier Properties of Structurally Matched and Unmatched Polyelectrolyte Multilayers", *Chemical Communications*, 2013, 49, 3576-3578.
10. **Minghui Wang**, Vaclav Janout & Steven L. Regen, "Hyper-thin Organic Membranes with Exceptional H_2/CO_2 Permeation Selectivity: Importance of Ionic Crosslinking and Self-healing", *Chemical Communications*, 2011, *47*, 2387-2389.

505550 Mechanistic, Spectroscopic and Theoretical Assessment of Porous Catalytic Materials

Abstract:

Research Interests:

Porous crystalline materials such as zeolites or metal organic frameworks provide a large and diverse pool of heterogeneous catalysts and catalyst supports. Thorough mechanistic investigations that employ characterization of material properties and surface structure, kinetic and isotopic studies, and computational modeling are needed in order to elucidate how these materials work as catalysts regardless of reaction or application. Further, this approach can be used in conjunction with specialized synthetic strategies in order to produce well-defined catalysts, imperative in understanding catalytic function.

Brønsted acidic zeolites, one class of these materials, catalyze a plethora of hydrocarbon and oxygenate reactions, all of which are mediated by ion-pair transition states. My graduate research with Prof. Enrique Iglesia investigated the factors that affect reactivity, specifically the size, shape and charge of the organic cation (derived from reactants) and inorganic anion (formed after deprotonation of the catalyst). These results were garnered from reactions that couple alkenes (oligomerization) as well as incorporate alkanes (alkylation via hydride transfer), but have been shown to extend to several other chemistries. We combined state-of-the-art theoretical methods (density functional theory) with rigorous kinetic measurements from a range of reactants and catalysts that led to a copious array of transition states and precursors. The synergy between theory and experiment exposed unprecedented mechanistic detail, such as the flexibility of zeolites to distort locally to increase van der Waals interactions with the cation that ultimately affect reactivity.

Part of my postdoctoral work with Prof. Christopher Jones has been with another class of porous materials, metal organic frameworks (MOFs). I have utilized MOFs, not as catalysts themselves, but as catalyst precursors to form iron-carbide materials for propane dehydrogenation reaction studies. My other postdoctoral project has involved design and synthesis of aminopolymers used in sorbent systems for CO_2 capture from dilute sources such as ambient air and flue gas. This project has broadened my skill set in materials and polymer synthesis and characterization, as well as separations and adsorption processes. Future plans for my own independent research group will incorporate kinetic investigations with theoretical methods and advanced synthetic techniques as a framework for studying porous catalytic materials from different catalyst classes and for other reaction pathways/applications. This framework can be extended to understand effects of important catalyst properties on reactivity and selectivity in order to advance rational catalyst design.

Postdoctoral Projects:

"Propane dehydrogenation on MOF-derived iron carbide catalysts" and *"Development of aminopolymer-based sorbents for CO_2 capture with improved capacity and oxidative stability."*

Under the supervision of Christopher W. Jones, Chemical and Biomolecular Engineering, Georgia Institute of Technology

PhD Dissertation: *"Alkene and Alkane Chain Growth on Solid Acid Catalysts."*

Under the supervision of Enrique Iglesia, Chemical and Biomolecular Engineering, University of California, Berkeley

Teaching Interests:

A key driving force for pursuing an academic career is my interest in teaching, both in the classroom and in the laboratory. One benefit of research in catalysis is that multiple fundamental pillars of chemical engineering (kinetics, transport and thermodynamics) must be mastered in order to understand and solve complex problems. Most of my experience comes from teaching chemical reaction engineering courses, and these would be the courses that I would prefer. As a graduate student instructor at UC Berkeley for the undergraduate and graduate level, I both prepared discussion sections used to advance the material as well as guest lectured in the main course. I have also guest lectured at Georgia Tech in the material and energy balances course. Further, I designed and taught an elective at UC Berkeley titled "Energy and Atom Efficient Catalytic Conversions," which covered both current and green(er), alternative industrial catalytic reactions for energy, fuel and chemical production.

505206 Membrane Separations for Clean Energy Conversions

Abstract:

Membrane Separations for Clean Energy Conversions

Simona Liguori – Department of Chemical and Biological Engineering, Colorado School of Mines, Golden, CO

Research Assistant Professor – *Colorado School of Mines* 2016 – present

Physical Science Research Associate – *Stanford University* 2014 – 2016

Post Doc – *Institute on Membrane Technology – ITM-CNR* 2012 – 2014

Motivation and **Research Interests:**

In the view of a significant reduction of greenhouse gases emissions amidst fossil fuel exploitation, the use of sustainable energy sources are a mandatory option as well as the use of alternative fuels for the transportation sector. In this context a strong interest has been addressed towards the possible development of a global Ôhydrogen economyÕ, by considering hydrogen as the energy carrier of the future and as reliable fuel source for transportation. Among all of the different technologies to produce green and sustainable hydrogen, inorganic membrane reactors are a promising alternative to enable distributed and centralized production and separation of hydrogen, respectively.

Hydrogen can be generated using feedstock from renewable sources, such as biogas, biomethanol, bioethanol, and bioglycerol or from fossil fuel such as natural gas. In both cases, two streams are produced by performing the reforming reaction in the membrane reactor. One stream contains high-purity hydrogen, which can be directly fed to a PEM fuel cell or internal combustion engine, while the other stream is mainly composed of CO_2, which can be easily separated and stored. An evident outcome of using a membrane reactor is emissions-free energy generation.

Inorganic membrane reactors are a clear example of process intensification and their utilization provides several other benefits when applied to chemical processes. Specifically, they are able to reduce the equipment footprint, energy consumption, and environmental impact of manufacturing processes. For these reasons, they may also be applied to NH_3 production. They allow for performing the process at milder operating conditions than the conventional Haber-Bosch process and to produce ammonia on-site eliminating the need for the U.S. import of foreign ammonia. Specifically, for the first time it has been demonstrated that N_2 permeates through metallic membranes by solution-diffusion and, by using the hydrogen as a sweep gas, it is feasible to produce NH_3 at a lower operating pressure.

This new method to produce NH_3 is highly significant as the current process is the second most energy-intensive chemical manufacturing process in the US and worldwide due to the harsh operating conditions.

In general, inorganic membranes have the potential to play a critical role in the area of clean energy conversion and the study of different materials through alloying is the key to reaching high membrane reactor performance.

In summary, my research is focused on:

á Hydrogen production by renewable sources or natural gas at lower operating conditions,

á NH_3 production via an alternate route from the conventional Haber-Bosch process,

á manufacturing of membranes and the study of different materials through alloying to produce high-performing membranes,

á the direct capture of CO_2 from a residue stream before it is released into the atmosphere.

Teaching Experience and **Teaching Interests:**

Sharing knowledge and communicating effectively is very important to me. I have served as a Teaching Assistant (TA) for Inorganic Chemistry courses for two years in Italy. Teaching is a very complex activity. It involves planning for learning, organizing materials, prioritizing ideas, interacting with students, learning to "monitor and adjust", "differentiate instruction" for students of various abilities, and learning how to accomplish goals that sometimes seem to be mutually exclusive, all while keeping "control" of a class and meeting the expectations of parents, administrators, and peers. Anyhow, I am always enthusiastic and passionate when I have to share my knowledge with student and when they are engaged during my lesson. I have had the opportunity to serve as mentor for two undergraduate students in Italy and one PhD student at Stanford and currently mentor a PhD student at the Colorado School of Mines.

Courses that I would be interested in teaching are fluid mechanics, transport phenomena, thermodynamics, energy conversion processes, and chemical kinetics.

Research Experience

I have been working on several fields on chemical engineering: reactor analysis and chemical kinetics, heat/mass transfer, and reaction/diffusion through catalysts and membranes.

During my PhD, my research focused on experimental and simulation analysis of reforming reactions of bio-fuels in both membrane and fixed-bed reactors for hydrogen production. Since completing my PhD, I have continued my research in membrane science and technology in my position as a Post-Doctoral Research Fellow at the Institute on Membrane Technology from 2012 to 2014 and as a Physical Science Research Associate at Stanford University from 2014 to 2016. I have been conducting research as a Research Assistant Professor with the Colorado School of Mines since June 2016 continuing my focus on membrane science and clean energy conversion processes.

Future Direction

As a faculty member I would like to continue applying membrane science and technology to different areas of engineering, as well as industrial applications. In particular, I would like to merge the knowledge that I obtained during my PhD and post-doctoral position: chemical engineering, and environmental and applied science. I believe that the synergy between the two fields can make feasible the reduction the greenhouse gases, above all CO_2, by producing H_2 and NH_3 by means of an alternative device including a membrane reactor.

Although I only started working with the Colorado School of Mines in 2016, I have hit the ground running in terms of improving my knowledge base associated membrane separations. I have had the opportunity while at Mines to continue my collaborations with Prof. Jennifer Wilcox in addition to beginning collaborations with membrane expert, Professor Doug Way. I am currently working on the evaluation of the performance of a membrane reactor to produce NH_3 at ambient pressure and by using a layer of catalyst deposited on the membrane surface (H_2 permeates through the membrane and N_2 is used a sweep-gas); NH_3 production in a membrane reactor at low operating conditions compared to the Haber-Bosch process (*opposite case*: N_2 permeates through the membrane and H_2 is used a sweep-gas); H_2 production and separation in Pd-alloys membrane reactor from reforming reactions; writing proposal for NSF, NineSigma, and RES Kaidi related to the H_2 production and separation in membrane reactor technology from renewable and natural gas reforming, respectively; writing scientific articles and chapters.

Selected Publications:

1. B. Anzelmo, J. Wilcox, **S. Liguori**, ÒNatural gas steam reforming reaction at low temperature and pressure conditions for hydrogen production via Pd/PSS membrane reactorÓ, Journal of Membrane Science, 522 (**2017**) 343-350. DOI: 10.1016/j.memsci.2016.09.029 (Corresponding Author)

2. J. Wilcox, P.C. Psarras, **S. Liguori**, ÒAssessment of reasonable opportunities for direct air captureÓ, Environmental Research Letters 12 (**2017**) 065001. DOI: 10.1088/1748-9326/aa6de5. (It was invited)

3. A. Iulianelli, **S. Liguori**, J. Wilcox, A. Basile, ÒAdvances on methane steam reforming to produce hydrogen through membrane reactors technology: a reviewÓ, *Catalysis Reviews Science & Engineering*, 58 (**2016**) 1-35, DOI:10.1080/01614940.2015.1099882.

4. A. Iulianelli, **S. Liguori**, Y. Huang, A Basile, ÒModel biogas steam reforming in a thin Pd-supported membrane reactor to generate clean hydrogen for fuel cellsÓ, *Journal of Power Sources*, 273 (**2015**) 25-32, DOI: 10.1016/j.jpowsour.2014.09.058.

5. **S. Liguori**, A. Iulianelli, F. Dalena, P. Pinacci, F. Drago , M. Broglia, Y. Huang, A Basile, ÒPerformance and long-term stability of Pd/PSS and Pd/Al_2O_3 membranes for hydrogen separationÓ, *Membranes*, 4 (**2014**) 143-162, DOI: 10.3390/membranes4010143.

6. **S. Liguori**, A. Iulianelli, F. Dalena, V. Piemonte, Y. Huang, A. Basile, ÒMethanol steam reforming in an Al_2O_3 supported thin Pd-layer membrane reactor over Cu/ZnO/Al_2O_3 catalystÓ, *International Journal of Hydrogen Energy*, 39 (**2013**) 18702-18710, DOI:10.1016/j.ijhydene.2013.11.113.

7. **S. Liguori**, M. Yuan, K. Lee, B. Anzelmo, N. Buggy, T. Fuerst, D. Way, S. Paglieri, J. Wilcox, ÒOpportunities and Challenges for Hydrogen Production/Separation via Metallic MembranesÓ, *Energy & Environmental Science*, in progress.

8. **S. Liguori**, K. Lee, J. Wilcox, ÒInnovative N_2-selective metallic membranes for improving CO_2 capture technologyÓ Ó, *Energy & Environmental Science*, in progress

505598 Membranes As Phase Contactors and Catalytic Interfaces

Abstract:

Research Interests:

Membrane science is a rapidly growing area in the field of separations and fundamentally is the study of mass transfer through nano- to micro-scale materials. The physical and chemical properties of permeating species and their interactions with the membrane material dictate the fluxes that are observed. Separation of a mixture of chemical species is possible when the fluxes of the various species are different. Applications for membranes range from water treatment and desalination to chemical separations to controlled drug delivery. Membrane reactors are a unique application where the membrane functions as a phase contactor and is utilized to either selectively remove a reaction product or to selectively supply a reactant into a reactive system. When acting as a reactant deliverer, parameters such as membrane composition, thickness, pore size, and catalyst location allow selective control of reactant flux and targeted delivery to catalytic sites if they are integrated with the membrane. Three-phase heterogeneous hydrogenation is an example of a system that benefits from such control. Hydrogen solubility in liquids, especially water, is inherently low, thus these systems often suffer mass transfer limitations or necessitate extremely high pressures to obtain appreciable reaction rates. Utilizing a membrane to deliver hydrogen directly from the gas phase to membrane-integrated catalytic sites avoids potentially rate limiting dissolution and diffusion in the liquid phase.

My doctoral research focused on the development of polymeric membranes with integrated noble metal catalysts for three-phase hydrogenation reactions. Integrally-skinned asymmetric polyimide membranes, porous polytetrafluorethylene (PTFE) membranes, and composite ceramic/polymer thin film membranes were explored. Porous PTFE membranes were proven highly effective in the aqueous phase hydrogenation of levulinic acid demonstrating reaction rates several factors higher than catalytically equivalent packed bed reactors under identical conditions. A notable challenge to effective utilization of the asymmetric polyimide membranes was solubility of the liquid phase solvent in the membrane material. Many organic solvents exhibited relatively high solubility in the polyimide thus significantly altering and decreasing hydrogen transport. This motivated an intensive investigation into the solubility and diffusivity of a variety of organic species in thin films of the polymer. Short chained alcohols (C1-C8) were identified as a class of solvents that produced only moderate swelling in the polymer and an additional study was conducted to further mitigate this effect by chemically cross-linking the polymer chains.

My postdoctoral research is exploring a variety of separation methods relevant to refining bio-oil from the fast pyrolysis of biomass. The unique fast pyrolysis reactor and fractional collection system developed at Iowa State University has created several novel bio-oil product streams and has thus required innovative separation approaches to separating and purifying the desired products from mixtures of hundreds of chemical species. One important separation is the recovery of sugars from a bio-oil "heavy-ends" organic phase. To accomplish this a combination of solvent extraction and liquid phase adsorption processes were investigated and pilot-scale studies are currently being developed.

Teaching Interests:

I have always thought of myself as an educator and thoroughly enjoy sharing my knowledge and experience with others. It is truly gratifying as an educator to help students overcome learning barriers and reach new heights in understanding. This may be as simple as stating ideas in different language or as complex as deconstructing complex or highly abstract ideas into fundamental concepts and helping the student to 'see' the pieces and understand how to rebuild them into complex scientific and engineering principles. I believe that making mistakes is integral to learning and developing oneself into an original and independent thinker. Failure is not something to fear, and the willingness to fail encourages the ability to take calculated risks and to self-correct when the results are not desirable.

My experience as an educator began during my undergraduate education where I tutored young students in a children's psychiatric unit and coached youth soccer. Concurrently I worked for several years at a company providing services for intellectually disabled individuals and worked closely with clients with a range of intellectual disabilities. In graduate school I was a teaching assistant for materials science and mechanical properties courses. I mentored three undergraduate research assistants in laboratory research over the course of three years and also mentored four REU students in the summers between academic terms. I encouraged the students to test their own ideas in the lab and to draw conclusions from their work. As a postdoctoral researcher I oversee several undergraduates in the lab and work closely with two associated with my research projects, while also mentoring graduate students in our research group. As a faculty I would be comfortable teaching undergraduate courses in any of the core chemical engineering curriculum and would be interested in developing specialized or graduate courses in membranes and advanced separations. I believe educators are students as well and that teaching is a two-way communication. To obtain the best outcome the teacher must be able and willing to understand concepts from the student's perspective and not just give the student the correct answer, but help him/her to develop the tools to find the answers on his/her own.

Future Directions:

As a faculty I plan to continue pursuing membrane based separations and more specialized applications of membranes integrated in reactive systems. There are many fundamental questions to investigate in this area, including catalyst integration, catalyst/membrane interactions, gas and liquid mass transfer through the membrane and at the membrane surface interface, and catalytic site location, functionality, and activity. Biomass based renewable chemicals is a burgeoning area where novel and innovative catalytic upgrading and separation strategies will help renewable chemicals become economically competitive with their fossil fuel counterparts. As my undergraduate degree is in biomedical engineering, I hope to expand membrane based investigations into biological systems. This may include enzyme immobilization on a membrane for a biologically reactive system, controlled release or selective delivery of drugs, removal of toxins or inhibitory species, or building cellular scaffolds that simultaneously offer controlled nutrient supply. Besides fundamental studies, I also greatly appreciate the need for practical solutions to current scientific and engineering challenges. As an adviser and mentor to undergraduate and graduate students, I will encourage my lab to conduct detailed and meticulous fundamental studies with an overarching constraint that the investigations support practical and rational solutions to real world problems regardless if the timeframe is the near or distant future.

Education

B.S. 2010 University of Iowa Biomedical Engineering
B.A. 2010 University of Iowa Music Performance
Ph.D. 2016 Kansas State University Chemical Engineering

Publications

Stanford, J. P., Soto, M. C., Pfromm, P. H., Rezac, M. E. "Aqueous phase hydrogenation of levulinic acid using a porous catalytic membrane reactor", Catalysis Today, (2016).

Stanford, J. P., Maier, A. L., McDonald, L. A., Pfromm, P. H., Rezac, M. E. “Kinetic and equilibrium sorption of organic liquids and vapors in Matrimid”, Journal of Membrane Science, (2016).

Stanford, J. P., Pfromm, P. H., Rezac, M. E. “Effect of vapor phase ethylenediamine cross-linking of Matrimid on alcohol vapor sorption and diffusion,” Journal of Applied Polymer Science, (2017).

Stanford, J. P., Hall, P. H., Rover, M. R., Smith, R. G., Brown, R. C. “Separation of fast pyrolysis derived sugars and phenolic species from an aqueous product stream using polymeric resin adsorbents”, (2017, in preparation).

Stanford, J. P., Rover, M. R., Smith, R. G., Brown, R. C. “Modeling and experimental validation of a continuous adsorbent based separation of sugars and phenolic species in an aqueous product stream in a pyrolysis biorefinery”, (2017, in preparation).

505373 Nanoporous Ultrathin Skinned Hollow Fiber Membranes

Abstract:

Research Interests:

Chemical separations consume roughly 50% of U.S. industrial energy use, and molecularly-selective membranes can significantly reduce this number. Numerous membrane materials have emerged to address unmet separation challenges in the past two decades; however, large-scale membrane processes are still limited to a handful of applications. The capability to economically form selective materials into thin separation (skin) layers on scalable membrane geometry is key to enable large-scale application of ANY membrane process. Many challenging separations in the chemical and energy industry require highly-rigid nanoporous molecular sieves to achieve precise differentiation (<0.02 nm) of closely-sized molecules. These materials often have low permeabilities, and ultrathin skin layers must be formed to provide practical productivity.

During my Ph.D. at Georgia Tech with Professor Bill Koros, I built a framework to develop zeolitic imidazolate framework (ZIF)/polymer hybrid membranes for challenging separations. I designed hybrid membrane materials with unprecedentedly-high olefin/paraffin selectivity enhancements, which were translated to high-loading nanocomposite hollow fiber membranes with excellent scalability. My research also advanced the fundamental understanding of molecular diffusion in flexible ZIFs, and developed new avenues to create nanoporous ZIFs with tunable transport properties. Currently, I am expanding my portfolio by working on carbon molecular sieve membranes in collaboration with industry to address the challenges and opportunities involved with processing aggressive natural gas. My most recent work introduced ultraselective carbon molecular sieve membranes showing excellent separation performance for purification of supercritical natural gas.

My future program aims to lead the next revolution in molecularly-selective membranes by uniquely integrating highly-selective nanoporous molecular sieves with ultrathin skinned hollow fibers. These â€œnanoporous ultrathin skinned hollow fibersâ€ represent a class of next-generation hollow fiber membranes comprising ultrathin skin layers relying upon rigid nanoporous molecular sieves. Design of rigid nanoporous materials with tunable transport properties will be guided by a molecular-level understanding of diffusion and material structure-property relationships. Economically-attractive module productivity is enabled by ultrathin skin layers and high packing density of the hollow fibers. With extensive experience in nanoporous materials and strong expertise in creating complex multi-layer composite hollow fibers, I am uniquely suited to undertake these challenging research efforts. The first-generation ultrathin skinned hollow fiber membranes will debottleneck separations of hydrocarbon vapor mixtures with significant energy savings, and enable membrane desalination of challenging feed water.

Teaching Interests:

Chemical Engineering Fundamentals, Mass Transfer, Separation Processes, Fundamentals of Membranes and Adsorbents

Selected Publications (18 total, 8 first author, 842 citations, H-index 11):

- **Zhang, C.**; Koros, W. J., Ultraselective carbon molecular sieve membranes with tailored synergistic sorption selective properties. *Advanced Materials* **2017**, DOI: 10.1002/adma.201701631.
- **Zhang, C.**; Koros, W. J., Tailoring the transport properties of zeolitic imidazolate frameworks by post-synthetic thermal modification. *ACS Applied Materials & Interfaces* **2015**, *7*, 23407-23411.
- **Zhang, C.**; Zhang, K.; Xu, L.; Labreche, Y.; Kraftschik, B.; Koros, W. J., Highly scalable ZIF-based mixed-matrix hollow fiber membranes for advanced hydrocarbon separations. *AIChE Journal* **2014**, *60*, 2625-2635.
- **Zhang, C.**; Lively, R. P.; Zhang, K.; Johnson, J. R.; Karvan, O.; Koros, W. J., Unexpected molecular sieving properties of zeolitic imidazolate framework-8. *The Journal of Physical Chemistry Letters* **2012**, *3*, 2130-2134.
- **Zhang, C.**; Dai, Y.; Johnson, J. R.; Karvan, O.; Koros, W. J., High performance ZIF-8/6FDA-DAM mixed matrix membrane for propylene/propane separations. *Journal of Membrane Science* **2012**, *389*, 34-42.

505623 Microporous Inorganic and Composite Membranes for Energy Efficient Separations

Abstract:

Research Interests:

Membranes made of microporous inorganic materials, e.g., carbon molecular sieve (CMS), zeolite, metal organic framework (MOF) are capable of surpassing the performance limit of polymer membranes for separations of many important gas or liquid mixtures. Moreover, their exceptional thermal, chemical and mechanical stability make inorganic membranes well suited for separations under industrially relevant harsh conditions that polymeric materials can not handle. However, high cost per unit area of membrane has limited their implementation to only medium-sized industrial plants for the dehydration of organic solvents. At current permeance levels, large surface areas and correspondingly high capital investment are required for many applications. To overcome this challenge, further improvement of membrane performance and developing cost-effective membrane synthesis strategies are needed.

One way forward is to increase the membrane permeance by reducing membrane thickness, without the formation of non-selective defects. One of my graduate research projects focused on the fabrication of ultra-thin CMS membranes. Through deliberate design of the polymer precursors and the porous structure of substrate, the thickness of CMS membranes was decreased down to 300 nm, and gas permeance was considerably enhanced without compromising selectivity. The major share of the cost of inorganic membranes is actually imposed by the underlying porous ceramic support. One approach to reducing membrane cost is to use more cost-effective and scalable supports, such as polymeric hollow fibers, which can be easily made into membrane modules of high packing density. Another approach is to suspend nano-sized microporous materials into polymer matrix to form defect-free mixed matrix membranes (MMMs). These two research areas are the focuses of my current postdoc research. An innovative chemical vapor growth approach is being developed to fabricate MOF membranes on polymeric supports. Without the use of any solvents, this vapor phase growth eliminates the homogeneous nucleation/growth of MOFs and the dissolution of membrane materials during membrane synthesis, enabling the formation of conformal MOF membranes on solvent-sensitive polymer supports. Another focus of my current research is to in situ form MOF nano crystals inside the polymer matrix to produce MMMs free of any MOFs/polymer interface.

My future research interests are in design and synthesis of novel inorganic/polymeric composite materials/membranes for separation applications and understanding their fundamental synthesis-structure-property relationships. The combination of microporous inorganic materials (zeolite, MOFs and carbon-based molecular sieves) and polymer materials is likely to result in a scalable, cost effective and high performance membranes for energy-efficient separations. Building on my interdisciplinary background and expertise, my research will involve: (1) synthesis of new microporous materials with strong molecular sieving effect and/or preferential adsorption towards the targeted molecules, as well as excellent stability. (2) Understanding and improving the interface interaction between inorganic and polymeric phases; (3) developing innovative processing strategies for better structure and morphology control of the composites systems; and (4)

Understanding the transport properties. MOFs will be the focus of the inorganic phase materials because of their high porosity, tunable pore structure and surface property. My goal is to develop novel hybrid materials as high-performing and cost-effective sorbents and membrane materials for the separations of gas or liquid mixtures, as well as water purification and desalination. Exploring these hybrid materials for other energy and environment related applications will also be pursued.

Research Experience:

My research experience has been a blend of chemistry, materials science and chemical engineering. Clean water technologies were the focuses of my undergraduate and master research. Novel functional polymers with powerful sorption capability to heavy metal ions were synthesized for water purification, and a new class of sensor based on an intrinsically conductive polymer was fabricated for efficient and reliable detection of leads ions in water. My PhD research was about membrane technology. Besides the aforementioned CMS membranes, I also worked on MOF-5, ZIF-8 and MFI zeolite membranes, with focuses on synthesis, structure characterization and studies on gas transport properties. I successfully developed positron annihilation spectroscopy (PAS) as an in-situ and non-destructive means to reveal the hierarchical microporous structure of zeolite membranes. This method is also applicable to other types of inorganic membranes materials, providing an opportunity to better understand the synthesis-structure-properties relationship of microporous inorganic membranes. My current research focuses on developing MOF/polymer composite membranes as the next-generation membranes for energy efficient separations.

Postdoctoral Project: "*MOF-based membranes for gas separation*"

Supervised by Dr. Michael Tsapatsis, Department of Chemical Engineering and Materials Science, University of Minnesota

PhD Dissertation: "*Synthesis and characterization of microporous inorganic membranes for propylene/propane separation*"

Supervised by Dr. Jerry Lin, School for Engineering of Matter, Transport and Energy, Arizona State University.

Teaching Interests:

I am interested in teaching heat and mass transfer, thermodynamics, transport processes, and materials related courses at both undergraduate and graduate levels. I am also interested in introducing and teaching a class at the interface between materials science and chemical engineering. This class will focus on the synthesis and characterization of materials that are particularly important to the field of chemical engineering, and the chemical engineering principles that can be used to develop and process novel materials.

Selected Publications:

X. L. Ma, X. T. Wei, J. Kniep, Y. S. Lin, Ultra-thin carbon molecular sieve membrane for propylene/propane separation, AIChE Journal, 2016, 62, 491–499.

X. L. Ma, H. Wang, H. B. Wang, J. O'Brien-Abrahama, Y. S. Lin, Pore structure characterization of supported polycrystalline zeolite membranes by positron annihilation spectroscopy, Journal of Membrane Science, 2015, 477, 41-48.

X. L. Ma, Jerry Y.S. Lin, Preparation Chemistry of Inorganic Membranes, Modern Inorganic Synthetic Chemistry, Elsevier, 2017.

D. F. Liu, **X. L. Ma**, H. X. Xi, Y. S. Lin, Gas transport properties and propylene/propane separation characteristics of ZIF-8 membranes, Journal of Membrane Science, 2014, 451, 85-93.

X. G. Li, **X. L. Ma**, J. Sun, M. R. Huang. Powerful reactive sorption of silver(I) and mercury(II) onto poly(o-phenylenediamine) microparticles. Langmuir, 2009, 25, 1675-1684.

505112 Molecular Design of Redox-Active Electrochemical Interfaces: Selective Separations and Beyond

Abstract:

Research Interests:

Molecular level design is crucial for the development of new and more efficient electrochemical processes for a range of industrial and environmental applications. In particular, selective separations remain among the most important steps in the chemical and biochemical industries, and are crucial for water purification and environmental remediation. Thermal and pressure-based separations often incur high energetic costs, while fixed-bed adsorption requires environmentally harmful solvents and regenerants. This is a particularly challenging task for the recovery of value-added products, and the mitigation of micropollutants of emerging concern. Although there has been great interest in electrochemical systems for energy storage and electrocatalysis, their use in separation science has been severely limited by the lack of molecular selectivity.

Redox-active species offer an attractive materials platform, especially organometallic compounds (e.g. metallopolymers and associated metal-organic complexes), due to the ability to control electronic and binding properties. My research interests focus on tuning the electrode interfaces at a molecular scale, to overcome the inherent limitations of current technologies. We engineer specific Faradaic-driven interactions to provide selectivity, and study electrochemical processes across several scales:

(i) We probe the intermolecular binding mechanisms through a combination of electronic structure (DFT) calculations and spectroscopy, and tune these interactions through synthetic functionalization of our electrodes. We enable the selective sorption and release of anions, cations, and even proteins, based solely on electrochemical control, in the presence of excess competing species.

(ii) At the nanoscale, we explore the physico-chemical behavior of our redox-active polymer films through in-situ measurements, especially during selective binding, and point to emerging directions beyond separations such as new materials development.

(iii) From an electrochemical cell perspective, we propose asymmetric Faradaic systems as a next generation configuration with higher charge efficiency and separation factors. We focus on counter-electrode design, in which the redox processes at the cathode and anode operate in tandem, lowering operating voltages, preventing parasitic side-reactions and ultimately enhancing ion-selective performance.

Fundamentally, the concepts explored shed insight on interfacial interactions of organic and inorganic ions with electrode surfaces; from a practical perspective, we propose electrochemical-swing processes as a key direction for sustainable separations. In the long-term, we expect integrated electrochemical systems to contribute greatly towards process intensification in the chemical industry.

Teaching Interests:

My teaching interests are in the areas of transport processes (in particular heat and mass transfer), separations science, thermodynamics/physical chemistry, and reaction engineering/kinetics. I am interested in focusing initially on the core classes in the classical chemical engineering curriculum, at both an undergraduate and graduate level. In parallel, I plan to develop an advanced class on electrochemical methods for chemical engineers. From a broader educational perspective, I would like to bring an aspect of scientific writing and presentation skills into more senior courses, to strengthen training on these aspects, and foster a degree of scientific inquiry for the students within a class environment.

505192 Bio-Mimetic Membranes for Energy Efficient Clean Water Processes

Abstract:

Research Interests: water-energy nexus, bio-mimetic materials, bio-inspired materials, membrane separations, water and wastewater treatment systems, surface functionalization

Over one billion people lack access to clean drinking water and membranes provide an energy efficient platform to purify impaired waters and fill that need. Unfortunately, treatment of impaired waters exposes membranes to biological and abiotic species, which leads to fouling and loss of membrane productivity over time. Since reduction in flux due to fouling is one of the largest costs associated with membrane processes in water treatment, new membrane modification methods that limit fouling would have significant economic and societal impacts.

In one project, a dual-mode membrane coating was developed that is capable of limiting and reversing membrane biofouling by switching reversibly between anti-fouling and anti-microbial states of biofouling control. This strategy differs fundamentally from most other surface modification strategies that rely solely on using chemical coatings designed only to weaken foulant adhesion and, thus, limit their accumulation or adding biocides. In the second project, a method to nano-pattern the surface of thin film composite membranes was developed to combine with chemical coatings to yield membrane surfaces that are more effective at fouling control than either method alone. Nature provides many examples where structured features can prevent accumulation of 'foulant' materials on the surface (i.e., shark skin, lotus leaves, etc.). We were the first to show that polyamide thin film composite membranes can be directly patterned.

Proposed future research will apply my expertise in membrane modification to mimicking bio-inspired materials to solve challenging separations problems and to improve the energy efficiency of membrane processes. One research project would be to mimic shark skin patterns by fabricating microfiltration membranes by phase inversion to reduce fouling in membrane distillation processes. Membrane distillation has a lot of promise in desalination, and water and wastewater treatment, however fouling on the membrane surface and in the membrane pores is still a major issue. A second project would be to modify membrane surfaces by mimicking a chemistry similar to what mussels use to exclude salt to stay attached to rocks to improve the salt rejection of nanofiltration membranes. Nanofiltration membranes offer the benefit of some salt rejection and higher fluxes than reverse osmosis membranes. However, if the salt rejection is increased while maintaining the flux, the energy efficiency would increase for desalination. A challenge of this research will be implementing nature's models into a membrane system. The results of this research will have implications in the field of membrane science, water treatment, and wastewater reuse.

Teaching Interests:

For 2 years, I have taught *Membrane Separations for Clean Water and Energy* through the Creative Inquiry program at Clemson University. I taught 12 students the basics of different types of membranes and guided them on designing and building two different membrane separation units to be used for unit operations lab and outreach events. The students presented two posters over the course of this class, including winning 2nd place in People's Choice at the Focus on Creative Inquiry event at Clemson in 2016. In conjunction with my NSF GRFP, I have participated in numerous outreach events through Clemson to engage K-12 students in science through filtration and separation activities. As a fundamentally trained chemical engineer with a focus on membrane separations, I am excited to teach courses related to reaction engineering, mass and energy balances, transport, polymers, and membrane separations.

505014 Renewable Transportation Biofuel and Value-Added Chemical Production from Wet Biowaste

Abstract:

Research Interests:

My research enthusiasm was influenced by my past and present research which encompasses biofuel conversion technologies, biowaste treatment, and systematical analysis of the nutrient and bioenergy recovery processes. Combining both laboratory and computational methods to link local processes to up-scaled system in our dynamic environment can help address the following questions:

(1) How to energetically harness available energy and nutrients from wet organic wastes?

Catalytic hydrothermal gasification (CHG) is an emerging technology for wet organic waste conversions with much faster reaction times (minutes), high conversion efficiencies (>90%), and a cost trajectory that is projected to be competitive with fossil fuels when the feedstock has a low or negative cost. However, there are no full-scale applications of this technology because commercial-scale equipment has not been developed, and the literature highlights several potential concerns (*e.g.*,

Anaerobic digestion (AD) is traditionally a dominant conversion technology used for wet organic wastes. In order to achieve a higher organic loading rate and energy recovery, a two-stage AD process is proposed in my research program and the results will be compared to those from a conventional AD. By optimizing the conditions of sludge loading rate, operating temperatures, and the types/dosage of adsorbents, the yields of hydrogen and methane as well as the energy recovery of the proposed two-stage AD system can be improved. Conventional AD process produces mainly methane, while two-stage AD can produce more hydrogen because it can better control the reaction environment and pathways by dividing the fermentation process into hydrolysis, acidogenesis, and methanogensis separately.

CHG and AD can both substantially recover the carbon nutrients in wet organic wastes, but not nitrogen and phosphate. On the other hand, aquatic biomass such as macrophytes and algae has been proven to have removal efficiencies of greater than 80% and 95% for nitrogen and phosphorous, respectively. In addition, crops that can grow in the hydroponic system also have the potential to further assimilate the residual nutrients in the CHG and AD effluents. Therefore, an algal treatment and/or a hydroponic farming system are proposed in my research program to simultaneously uptake the nutrients and remediate the wastewaters coming from AD and CHG. In this case, research efforts on algal biotechnology and hydroponic cropping development for concentrated wastewater treatment can be advanced to satisfy growing need in the FEW nexus, while the potential of these concentrated effluents as nutrient sources for aquatic biomass grown can be realized.

(2) How to selectively fine tune the biomass conversion processes to obtain valuable products?

No full-scale applications of the hydrothermal treatment technology have been realized so far, because the data available highlights several practical concerns that need to be resolved before large investments in the technology are justified.

(i) Catalysts development for effective CHG

Previous studies have extensively focused on using noble or transitional metal catalysts such as Pt and Ni for CHG. To make CHG a more affordable technology for the public, other economic catalysts such as activated carbons, zeolites, calcium carbonates and iron, will be explored and their catalysts performance will be evaluated. Further, synthesizing two catalysts with different functions would be explored for CHG reactions. Different synthesis methods (*e.g.*,

(ii) activity under CHG

Fouling of catalyst is another issue under hydrothermal processes and will be studied in my research program. One previous study using CHG to treat waste newspaper present that unless the catalysts can be reused and regenerated, CHG can produce more net-energy and show a more efficient energy consumption ratio than AD of waste newspaper. Thereby, a detailed physicochemical characterization of catalysts before/after CHG reactions would be valuable to provide information regarding catalyst deactivation mechanisms. In my research program, the catalysts will be characterized before and after the reactions to understand the deactivation mechanisms through SEM, TEM, TGA/DSC, XRD, and XRF analyses. Different regeneration methods for catalysts will be explored and the morphology of the regenerated catalysts will be investigated. Meanwhile, the lifespan of the catalysts will be understood. Ultimately, methodologies to minimize the catalysts deactivation will be established.

(iii) network for bioactive chemicals under CHG

Literature has demonstrated that CHG and HTL can effectively degrade bioactive chemicals, such as fatty acids, environment hormones and phenols, entrained in wet organic wastes. However, the reaction kinetics and network of these bioactive chemicals remained unknown. Understanding the reaction kinetics and network for bioactive chemicals would help design the up-scaled CHG reactor and make CHG a more versatile tool for treating different types of bioactive chemicals. In order to achieve this goal, I propose to conduct CHG with model bioactive compounds and isotopes. The products obtained at different reaction conditions will be analyzed by GC/MS, NMR, and FTIR assays to understand the reaction pathways of biowastes under CHG. Sensitivity analysis will also be conducted to explore significant factors governing CHG reactions.

(iv) Supercritical fluids (SCFs) applications to extract value-added products

The majority of recent research involving thermochemical conversion of biomass has been focused on fuels rather than on value-added products. However, analysis has shown that integrating co-products with biofuels offers a substantially higher return on investment while simultaneously meeting energy and economic goals. Supercritical fluids (SCFs) is a promising solvent for extracting value-added products from biomass, due to their relative lack of intrinsic hazard and robust tolerance of water content. For example, supercritical carbon dioxide ($scCO_2$) has been used to extract triacylglycerides (TAGs) from microalgae leading to a comparable yield as conventional extractions with organic solvents, but with the additional advantage of increased selectivity. Furthermore, solvation ability of SCFs can be tuned based on their density or by adding SCFs entrainers, such as ethanol, which have been shown to enhance solubility and increase yield of TAGs and value-added chemicals, such as astaxanthin (AX), from microalgae. As a consequence, I want to explore the possibility of using SCFs to develop viable and cost-effective process trains to selectively extract valuable products such as nutraceuticals from biomass, particularly from wet organic wastes.

(3) How do we combine new technologies with multiple scale of investigation to demonstrate an affordable system, minimize greenhouse gas emission, FEW nexus?

System analysis of the proposed biomass conversion systems can help understand and identify research gaps to establish a robust FEW nexus. For example, the necessity of reusing residual nutrients in the aqueous products (PHWW) obtained from hydrothermal treatment (HT) processes has been identified through a multi-component mass balance. Conducting the mass/energy balance and process design of the integrated system using the Aspen Plus®, STELLA®, SimaPro, and CHEMCAD software can elucidate and identify the techno-economic feasibility and life cycle loadings. A multi-objective optimization will also be carried out to create Pareto curves that aim to maximize the bioenergy production and minimize the environmental impact of the proposed FEW nexus.

In short, my research program aims to harness the chemical energy stored in wet organic biowastes. The ultimate goal of my research program is to demonstrate the feasibility of a novel concept for integrated bioenergy/value-added chemical production, biowaste treatment, as well as nutrient/water reuse via treatments of CHG, SCFs, and a two-stage AD of wet organic wastes.

Teaching Interests:

My teaching philosophy was molded by my education as an engineer. For me, education is significantly more than imparting knowledge found in textbooks. As an engineering student involved in a multidisciplinary research group, I had the opportunity to gain a great deal of know-hows by participating in different research activities. This experience also motivated and reinforced my teaching philosophy. I truly believe that by involving students with hands-on projects and real-world solid examples, learning new knowledge can be more effective while creative thinking and innovative solutions will be strongly encouraged. As Benjamin Franklin has said, "Tell me and I forget. Teach me and I remember. Involve me and I learn."

Teaching Philosophy

I am deeply committed to effective teaching and mentorship. I have been very fortunate to have had many positive and influential mentors during my education. This experience incentivizes and eagerly drives me to be a supportive and helpful mentor to my students. My objective when teaching and mentoring is to create a learning environment that simultaneously develops critical thinking skills and fosters an open mind to all possibilities by emphasizing the rich conceptual issues that defines sustainable energy science and engineering and how this discipline impacts our lives. By expanding not only the studentsâ€™ scientific understanding of engineering problems but most especially how they apply it to the real world, we can critically review more available sustainable engineering practices and/or stewardship. Advantages and limitations of the established methods would be identified during this critical thinking process, and valid alternative state-of-the-art methods can then be developed. Ultimately, I hope my students can think â€œoutside the boxâ€ and develop their confidence in proposing solution to existing and new problems.

Teaching experience

As I have served as a teaching assistant (TA) in the class of . As a mentor, I usually ask students some questions and give some solid examples to initiate their interest. Literature review and experimental design are generally required in class projects. Through lab projects, students can gain knowledge on bioenvironmental engineering design, particularly about biomass conversion and sustainable energy engineering, while I always benefit from the mentoring and discussion. Teaching and mentoring are not only a give-and-take process but also an interactive and mutual beneficial discussion with the mentored undergraduates at the same time.

Mentoring experience

By participating a multidisciplinary research programs in University of Illinois, joining the algae club in my home department, and involving with SWE (Society of Women Engineers) on campus, I was very fortunate to have ample opportunities to mentor or work with students from elementary school, high school, to different levels of undergraduate students. As providing technical training and helping them develop research topics, I always benefit from the mutual beneficial discussion. Students often inspire me from a different point of view and pay more attention to details than I do. This mutually beneficial discussion has been a cornerstone of my research career. For example, when I worked with my first undergraduate, sparked by her previous experience on centrifugation, we began to develop a more sustainable pretreatment technique, combining centrifugation and ultrasonic processes, for improved bioenergy conversion efficiency from algal biomass. Another instance is that I have further advanced my knowledge on gas-chromatography mass-spectrometry (GC-MS) by instructing several undergraduate and graduate students on using this equipment for analyzing biocrude oil and volatile aqueous productsâ€"mentoring itself is an intensive learning experience.

Teaching methods

When I served as a TA four years ago, I found it was drastically different between knowing the knowledge and transferring to others. For instance, demonstrating a practice example is likely more helpful to the novice than throwing a challenging problem at him/her, which may spoil a studentâ€™s interests and confidence. During my work with undergraduate students from different departments and culture background in University of Illinois, I realized that working on projects can significantly motivate students to learn a specific topic and inspire studentsâ€™ potential as well as establish their confidence. Therefore, I will combine lecture and lab projects in my classes. For fundamental classes (e.g. transport phenomenon and heat and mass transfer), I plan to apply in-class discussion to one of the weekly lectures with a worksheet. I will also separate students into different groups and assign different groups of students to design the worksheet every week. This type of student-centered learning method can further help students think about why the lecturer/textbook asks questions like that way and why those questions matter. For advanced classes, I will combine lectures, lab projects, field trips, term-paper writing, and practice proposal writing together. When I took classes and served as a TA in University of Illinois, I found out lab projects usually can make students effectively learn hands-on skills and apply what they learnt in classes to real-world problems while field trips can greatly initiate studentsâ€™ interest on a specific topic. In addition, term paper and practice proposal writing can help students synthesize their thoughts on specific topics and present their ideas in a professional way. Furthermore, I will conduct an early feedback, which will happen in the one third or half of the semester, in my classes so that I will have time to fine tune my teaching style or adjust course content during the semester. Overall, by using versatile teaching methods and techniques, I hope to give students a great incentive to learn different levels of courses and understand how they may change and impact the world.

505049 Organ-on-a-Chip and 3D-Printing Technologies: Applications in Nephro-Cardiovascular Diseases

Abstract:

Research Interests: organ-on-a-chip, 3D printing, nepho-cardiovascular diseases, personalized Medicine, stem cell engineering

Approximately 26 million Americans are diagnosed with kidney diseases, with an estimated collective cost to the healthcare system of over $40 billion. Chronic Kidney Disease (CKD) is considered the most fatal and expensive to treat disease since it leads to kidney failure and the only solution may be kidney transplantation. Although organ transplantation has saved millions of patient lives, significant challenges remain, such as the high cost of kidney donor transplant and the long waiting lists for donor organs. In addition, patients who receive a kidney transplant are subject to lifelong anti-rejection medications, which are accompanied with lifelong side-effects such as increased rates of infections, and diabetes. **Thus, it is necessary to engineer methods for optimizing organ transplantation and for kidney therapeutics, which can be applied to personalized medical treatment based on individual physiological conditions of each patient.**

My past and current research interests focus on the development of new tools to identify new biological targets involved in vascular diseases (1-7). Vascular involvement is a primary cause for acute renal failure (ARF) or acute kidney injury (AKI) which can result in prolonged loss of kidney function and vascular perfusion. Therefore, a deeper understanding of the complex bidirectional and interdependent interactions between the vascular and renal compartments in pathophysiological conditions is needed. My long term vision is the integration of biomedical and technological approaches within a single framework for potential translational and therapeutic purposes of nephro-vascular diseases. Such a vision requires a multidisciplinary research program which encompasses: (i) **Organ-on-a-chip technology** to build multicellular level functional units of kidney through micro-engineering approaches. This strategy will involve a functional biomimetic microfluidic unit that captures both the microvascular and renal tubular systems, where increasingly complex architectures with tunable biophysical and biochemical properties will be studied. This *in vitro* platform is ideal to mimic kidney filtration (GFR) and microvascular function (**kidney-on-a-chip**) and to identify novel drugs/targets involved in acute or chronic renal-vascular diseases. (ii) **Engineering stem cell therapy technologies** to model the appropriate cells involved in nephro-vascular diseases within the kidney-on-chip and establish such cells for regenerative applications. Engineering functional stem cells (iPS-driven cells/adult stem cells) in combination with molecular biology strategies including **CrispR technology** will be used to investigate the genetic basis for ARF. Additionally, the differentiated functional cells can be applied for cell therapy purposes needed for personalized medical treatment of CKDs. (iii) **3D printing technology** to build functional organ for transplantation. Integration of organ-on-a-chip and stem cells technologies with **3D printing technologies** will enable the 3D-printing autologous organ for transplantation in animal disease models, such as spontaneously hypertensive rats (SHR), which are developing abnormalities in vascular-renal functionality.

Aside from the exciting and challenging intellectual aspects of the technologies of organ-on-a-chip, stem cell therapy and 3D printing in treating nephron-vascular diseases, I view them as platforms to address hurdles in developing novel therapeutics for organ diseases and engineering whole organs.

References

1. **Alimperti**, S., Lei, P., Tian, J., and Andreadis, S. T., "A novel lentivirus for quantitative assessment of gene knockdown in stem cell differentiation", *Gene Therapy*, **19**, 1123-32 (2012).
2. **Alimperti**, S., Wen, Y., Lei, P., Tian, J., Campbell, A., and Andreadis, S. T., "Serum-Free Spheroid Culture maintains high proliferation and differentiation potential of MSCs", *Biotechnology progress*, **30**, 974-983 (2014).
3. **Alimperti**, S., You, H., George, T., Agarwal, S., Andreadis, S. T., "Cadherin-11 regulates both mesenchymal stem cell differentiation into smooth muscle cells and contractile function in vivo", *Journal Cell Science*, **127**, 2627-2638 (2014).
4. **Alimperti**, S., Andreadis, S. T., "CDH2 and CDH11 act as regulators of stem cell fate decisions", *Stem Cell Research*, **14**, 270-282 (2015).
5. Row, S., Liu, Y., **Alimperti**, S., Agarwal, S., Andreadis, S. T., "Cadherin-11 is a novel regulator of extracellular matrix and tissue mechanics", *Journal of Cell Science*, **129**, 2950-2961 (2016).
6. **Alimperti**, S., Mirabella, T., Bajaj, V., Polachek, W., Pirone, D., Duffield, J., Assoian, R., Chen, C.S., "A 3D biomimetic vascular model reveals a RhoA, Rac1, and N-cadherin balance in pericyte-regulated barrier function" (provisional accepted *in PNAS)*.
7. McCurley, A., **Alimperti**, S., et.al., "Inhibition of avb5 integrin leads to a pericyte-dependent attenuation of vascular permeability and protects against renal ischemia-reperfusion injury", *Journal of the American Society of Nephrology*, 2017

Teaching Interests:

The opportunity to educate future generations of chemical engineers is my principal motivation for pursuing an academic career in Chemical and Biological Engineering departments. During my Ph.D. in chemical and biological engineering department at State University of New York at Buffalo and as well, during my postdoctoral training at Wyss Institute at Harvard University, I established expertise in engineering fundamentals, which I applied them to develop new biomedical engineered tools and ultimately, to investigate new biological questions. My philosophy on teaching in a chemical engineering department is built from this experience, and i**n the laboratory**, I will emphasize in integrating engineering design principles with biology. **In the classroom,** my major focus will be to establish primary principles that build intuition to enable rational problem solving in biological sciences independently of a student's chosen career path. Establishing a solid background in chemical engineering courses including mass and heat transport (Transport phenomena), unit operations, tissue engineering, material engineering, and chemistry, I would be comfortable teaching similar or related courses at the undergraduate or graduate level. Finally, one of my key interests is in participating in senior design projects designed to introduce undergraduates to department faculty and their research.

500267 Metal Oxide Redox Materials for Energy Applications

Abstract:

My research interests lie at the interface of materials science and chemically reacting systems for applications in energy and sustainable engineering. Most industrial processes involve gas-solid heterogenous reactions and the interplay between the solid material and the process provides a platform for interesting alternative processing pathways and enhanced performance. My future research vision involves engineering novel production processes for fuels and commodities that are of significant positive impact on society and the environment.

Research Interests:

- High-temperature thermochemical processes for fuels and energy storage
- C1 activation chemistry
- Materials for high-temperature chemistry, photocatalysis, advanced semiconductors, and sensors

Teaching Interests:

- Material/energy balances
- Process design
- Applied data analysis
- Energy and sustainability

505220 Atomistic Modeling of Energy Storage Materials

Abstract:

The lithium-ion (Li-ion) battery has revolutionized the consumer electronics industry in the past two decades, achieving moderate battery lifetimes and higher energy storage densities. However, we have begun to reach the limits to improvements in Li-ion batteries. Further, in order to reach the energy demands posed by the transportation industry, batteries that possess energy densities much larger than current state-of-the-art Li-ion technologies are highly desirable. This research investigates two â€˜beyond Li-ionâ€™ battery technologies that boast theoretical energy densities more than five times larger than Li-ion batteries: batteries employing (i) metallic Mg anodes and (ii) metallic Li anodes. We used density functional theory to probe atomistic interactions on each of these metallic surfaces. For the metallic Mg anode, we identified likely reactions between the anode surface and the battery solvent. The composition of the model Mg anode surface was modified to mimic realistic electrode surfaces. We found that solvent decomposition products and reaction energies were highly dependent on the model anode surface composition, with the pristine Mg anode displaying a greater tendency for solvent decomposition. For the metallic Li anode, I will show our recent efforts to model the interface between Li and its native oxide. These studies provide a more complete picture of the atomic-scale factors governing the successful implementation of metallic anodes for beyond Li-ion batteries.

Research Interests:

I am broadly interested in modeling processes/phenomena related to energy storage materials. Some specific examples from my research background and interests include:

- Analyzing interfacial effects on electrochemical phenomena
- Multi-scale modeling of electrochemical processes
- Predicting energy storage capacities of current and new materials
- Applying machine learning to the discovery of new materials
- Employing sampling algorithms to reveal complex reaction pathways

Teaching Interests:

I am interested in teaching all types of courses including introductory, core, and elective courses. However, based on my interests and expertise I would be best-suited to teach Calculus I, II, and III, General Physics I and II, Engineering Thermodynamics, Chemical Reaction Engineering, Scientific Computing, and Battery Fundamentals.

501751 Convergence As a Chemical Engineering Career

Abstract:

My research experiences include deep dives in the areas of applied environmental study, energy in highly political settings, and sustainability education. My future interests invovle nanotechnology that can be applied to energy, environmental applications, and many aspects of life science. The core intersection of these topics embodies convergance. For example, making use of citizens and affordable analytical technology require skills beyond preparing engineers for chemical manufacturing plant settings. Further, SMART technologies to improve energy use and at human technology interfaces are in nascent stages of development and deployment. Both of these examples are areas that I am intersested in exploring and working with creative teams to tackle.

Research Interests: Data - materials convergence, sensors & networks, energy, nanomaterials, next generation catalysis, water, biomedical devices

Teaching Interests: Sustainability education, student centered education, experiential learning, action research, policy

505234 Developing Energy Materials through New Material Synthesis and Advanced Optoelectronic Characterization

Abstract:

Research Interests: and **Teaching Interests:**

The design of functional materials is a rapidly growing research area, particularly with concern to identifying new materials for energy applications. Research in materials-by-design is led by computational modeling, where thousands of new materials have been identified and screened for potential applications in areas such as photovoltaics, thermoelectrics, and energy storage. However, high-throughput synthesis and experimental characterization of theoretically identified materials is needed to realize the development of these materials and accurately verify their optoelectronic properties. Here, research is proposed for the development of unique chalcogenide and perovskite semiconductors â€“ with particular focus on photovoltaic materials â€“ comprised of facile solution-based processing and advanced optoelectronic characterization techniques.

My doctoral and postdoctoral research at Purdue University and Helmholtz-Zentrum Berlin has focused on the development of chalcogenide and perovskite semiconductors through nanoparticle-, solution-, and vacuum-based processing techniques. Control of material defects and transport properties has been demonstrated through unique thermal processing techniques, nano-structured device architectures, and the development of new photovoltaic materials such as Ge- and Ag-based kesterite absorbers. My current research also involves the development of solution-based ABX_3-type chalcogenide perovskites, recently identified as a promising photovoltaic absorber through computational design of materials.

In addition to material synthesis, the successful development of materials-by-design requires characterization of their intended optoelectronic properties. Such characterization is often non-trivial for early-stage/non-ideal materials, in particular when bare materials are probed in lieu of completed electronic devices. My doctoral and postdoctoral research has also focused on the development of optoelectronic characterization techniques and methodologies for non-ideal semiconductors. These techniques include electronic device measurement and simulation, as well as non-destructive probing of absorber layers for defect properties, homogeneity, and the spatial mapping of optoelectronic properties. Past research has focused on advanced quantum-efficiency, current-voltage, and impedance spectroscopy techniques for materials with non-ideal defect properties; current research is focused on developing unique photoluminescence and time-resolved photoluminescence techniques utilizing supercontinuum and multiphoton excitation.

My experiences in semiconductor material synthesis and characterization offer a unique advantage in the proposed research area of developing unique chalcogenide and perovskite semiconductors, where feedback from material characterization will be used to direct material synthesis. This research area is supported by international research initiatives, including the Materials Genome Initiative (US) and European Materials Modeling Council. Additionally, this work benefits student learning and collaboration as a multi-disciplinary approach bridging chemical engineering, materials science, chemistry, physics, and electrical engineering is required. As a faculty member I plan to continue my research to establish a world-class semiconductor research program developing energy materials, while inspiring and educating students in this area.

505597 From Fundamental Understanding Towards Materials Design of High Energy Battery Materials

Abstract:

Research Interests:

Electrochemical energy conversion and storage is critical for vehicle electrification and storing clean energy from intermittent sources (i.e. wind, solar) on the grid. Towards these goals, my work as a doctoral student has encompassed the broad spectrum of stages necessary for technological innovation: (1) developing new tools to discover fundamental understandings, (2) designing and synthesizing unique materials to address intrinsic failure modes, and (3) patenting the technology for real world applications.

As a doctoral student, I have developed [1] and patented [2] a novel materials design approach that enables silicon to cycle as a lithium-ion battery anode with over four times the theoretical capacity of commercial graphite (372 mAh g^{-1}). Using *in situ* transmission electron microscopy (TEM), I showed that a conformal graphene cage surrounding the silicon is electrically conductive and mechanically robust enough to overcome the dramatic failure modes (i.e. particle rupture, moving interface) of silicon during battery operation. This work is important because it demonstrated for the first time that fractured particles could still be recharged in a battery, something that the battery community believed to be previously impossible. Furthermore, the concept of a conformal graphene coating shows that simple physical mixing of two materials may be insufficient to achieve the desired properties of a composite materials system.

Looking beyond silicon, I established strategies to investigate fundamental aspects of pure lithium metal, which can potentially enable batteries with energy density approaching that of gasoline. Previous techniques were unable to study lithium metal at the nanoscale due to its extreme reactivity to the ambient environment and sensitivity to the electron beam. Using environmental [3] and cryogenic [4] TEM techniques that I developed, my research revealed the nanoscale corrosion behavior and detailed crystallography of lithium metal dendrites. The new insights guided the design of a protective nitride layer that enabled improved cycling stability of lithium metal. My results represent a simple but telling demonstration of the practical applications that can be initiated by fundamental studies on reactive battery materials. These newly developed techniques open up many excellent opportunities for future scientific discovery and invention.

My future research seeks to develop fundamental understandings of the working mechanisms and failure modes in electrochemical devices at both the macroscopic and atomic scale. Guided by these insights, I will design, synthesize, and characterize materials systems that can overcome intrinsic challenges and enable devices for practical application. Using the advanced characterization tools and chemistry expertise developed in my doctoral studies, I intend to further study the atomic structure of electrocatalysts at reaction conditions to understand their behavior during operation [5].

Selected Publications:

[1]: **Y. Li**, K. Yan, H.W. Lee, Z. Lu, N. Liu, Y. Cui. ***Nature Energy*** 1 (2016): 15029.

[2]: **Y. Li**, K. Yan, Z. Lu, Y. Cui. ***US Patent Application No***. 15/178,366. (2016).

[3]: **Yuzhang Li**, Yanbin Li, Y. Sun, B. Butz, K. Yan, A.L. Koh, J. Zhao, A. Pei, Y. Cui. ***Nano Letters*** (in review)

[4]: **Yuzhang Li**, Yanbin Li, A. Pei, K. Yan, Y. Sun, C.L. Wu, L.M. Joubert, R. Chin, A.L. Koh, Y.Yu, J. Perrino, S. Chu, Y. Cui. ***Science*** (in review)

[5]: H. Wang, S. Xu, C. Tsai, **Y. Li**, C. Liu, J. Zhao, Y. Liu, H. Yuan, F. Abild-Pedersen, F.B. Prinz, J.K. Norskov, Y. Cui. ***Science*** 354.6315 (2016):1031-1036.

Teaching Interests:

My teaching audiences have spanned middle school students to post-doctoral scholars. These experiences have trained my ability to clearly explain difficult concepts in a simple or detailed manner. Coming from a chemical engineering undergraduate background, I am excited to teach a variety of courses at the undergraduate and graduate levels. In particular, I am interested in teaching materials chemistry, transport phenomena (heat, mass, momentum), and atomic structure and arrangements. Creating a great educational experience for young scientists and researchers is critical for the continued advancement of science and it is with this view that I will approach teaching as a future faculty member.

Proposal Experience:

- Office of Energy Efficiency and Renewable Energy Materials Research Program (2016)
- Department of Energy Batteries for Advanced Transportation Technologies (2016)

499131 Investigation and Implementation of Adsorption Models in Nuclear Energy

Abstract:

Adsorption is a complex physical-chemical process by which molecules are attached to surfaces of solid particles. The type of adsorption that occurs may often depend on the media the phenomenon is occurring in, making the design of models for various adsorption systems an arduous task. Regardless of the media, however, the basic mechanisms of the adsorption process are the same. Therefore, a plausible approach to the development of adsorption models in different systems would be to design a generalized mathematical framework with all the necessary methods built in that will be used as a platform to develop system-specific adsorption models. In this work, the investigation and development of such a structure will be discussed and a host of system-specific adsorption models that have been developed on top of that framework will be detailed. The problems of interest are all related to nuclear energy and specifically the availability of uranium in the Nuclear Fuel Cycle via recycling spent uranium fuel rods and capturing uranium fuel from seawater. In recycling spent uranium, the reprocessing procedure produces numerous gas pollutants that must be removed from the off-gases before emission to the atmosphere. To facilitate the design of that capture system, adsorption models have been developed to predict isothermal equilibria of complex gas mixtures and to quantify the rates of adsorption for various adsorbent materials. For recovering uranium from seawater, two different models were produced: (i) a predictive, multi-ligand adsorption model to incorporate effects of pH, ionic strength, and competing metals and (ii) an analytical model for quantifying the impact of current velocity on the mass transfer limitations of braided fiber adsorbents. The culmination of these adsorption models will provide tools for scientists and engineers to better understand adsorption phenomena in the applications of interest and subsequently design and/or control the necessary capture systems at both the front and back ends of the Nuclear Fuel Cycle.

Research Interests:

Throughout my academic career, my research focus has been on the development and application of adsorption models for engineered and natural systems. In environmental sciences and engineering, it is not very common to have researchers who have focused primarily on modeling and mathematics as opposed to experimentation. However, I felt drawn to modeling as I have a natural affinity towards mathematics, especially the applications of mathematics to solving complex problems. In addition, having expertise in computer modeling and simulation is becoming increasingly important today as computers have become cheaper and faster than ever before.

Part of my future plan includes exploring new and innovative numerical modeling techniques that I have learned over these last several years in graduate school. The current modeling framework I have developed is excellent for large, sparse systems of equations, but could be improved further by incorporating Message Passing Interface (MPI) aware/safe code. This would then allow the framework to be scaled onto massively parallel computing architectures. In addition, I would also like to expand some preliminary work I have done in developing my own types of numerical solvers, such as a recursive Krylov method for sparse linear systems and a variational polynomial technique for partial differential equations.

My work in the design and implementation of adsorption process models has given me a breadth of knowledge in both air and water quality engineering, as well as numerical methods and computer simulation. Beyond the nuclear fuel cycle application, the modeling framework for adsorption has broad functionality in numerous other areas of environmental science and engineering including ground water remediation, carbon capture, and activated carbon filtration. It is my hope to continue to add to this modeling framework while pursuing new research opportunities.

Teaching Interests:

Teaching environmental science and engineering has been a passion of mine ever since I started college. I have always enjoyed working with other students and have often taken the lead on many group activities, as I was extremely motivated to learn and to help others around me learn. As a teacher, I would work to communicate my passion of environmental science and engineering to my students and strive to instill in them the same motivation for learning as I had when I was a student.

A healthy balance between research and teaching is necessary to foster a successful career in academia. However, these two areas need not be treated entirely as separate entities. It has been my experience that the best college professors are the ones who can successfully connect classroom concepts to real world challenges. By bringing research concepts and problems into the course material, one can help foster interest and engagement in the subject by demonstrating its applicability outside of the classroom. I plan to regularly apply my own research in lectures whenever possible and would even want to design special homeworks or assignments based on problem solving exercises from research.

One course that I would be especially motivatd to teach is a numerical methods course with an emphasis on applying/designing algorithms to solving complex problems in environmental science. In todayâ€™s modern computing world, there is increasing demand for the use of computer simulations to predict how different environmental and engineered systems will behave under different stimuli or scenarios. For this course, I want to introduce students to the various numerical techniques developed for solving these very complex, coupled systems, then have them apply that knowledge through coding exercises in a computer language like C++ or Matlab.

504806 Mechanical Principles of Biofilm FormationÂ

Abstract:

Research Interests: The overarching goal of my future research is to understand how bacteria, once thought to be solitary and asocial, cooperate in space and time to behave like multicellular organisms. By combining genetic tools, single-cell resolution live imaging, and *in situ* measurements of mechanical properties, I aim to connect the genetic and biochemical programs underlying the construction of the biofilm architecture to its mechanical properties. This multidisciplinary approach will provide a comprehensive understanding of how bacteria build multicellular communities cell by cell, what unique materials they use to do so, and what mechanical features emerge at the level of the collective. With this understanding, I will discover engineering principles to promote or discourage bacterial community formation in relevant medical and industrial settings.

Teaching Interests: Biomaterials, Soft Matter Physics, Polymer Chemistry and Physics, Rheology, Microbiology.

505162 Multi-Level Systems Modeling

Abstract:

In this poster session, I will discuss my Ph.D. research in the area of Process Systems Engineering (PSE) at Purdue University, my Postdoctoral research at MIT, the research topics and projects I am proposing to pursue, and my teaching interests and experience. For more information, please visit my website: http://www.emregencer.com/.

Research Interests:

Current energy and chemical sectors predominantly depend on fossil resources. Limited fossil resources and significant increase in atmospheric greenhouse gases urge to develop and implement low-carbon energy conversion processes for a sustainable economy.

The overarching goal of my research is to meet essential human needs including transportation fuel, chemicals, electricity and water while cost effectively reducing carbon foot print and improving resource utilization. My research focuses on multi-level systems analysis of energy conversion technologies for synthesis of energy efficient, synergistic processes incorporating process intensification for optimal utilization of resources.

During my PhD, I have focused on determining the optimal synergistic process design to produce liquid fuel from biomass and natural gas, developing a methodology to synthesize efficient solar thermal hydrogen production processes, identifying and optimizing efficient solar thermal hydrogen production process designs based on the developed methodology, synthesis of integrated thermal desalination processes, creating efficient and intensified processes to convert solar thermal energy into electricity and to continuously supply to the grid by integrated energy storage solutions. Moreover, I have worked on policy analysis of large-scale energy storage options for uninterrupted renewable power supply, particularly on closed loop CO_2-driven energy storage solutions.

For my postdoctoral research, I am working on developing a process superstructure based modeling framework to systematically analyze techno-economic indicators of integration of carbon capture, utilization, and storage technologies with various CO_2 sources such as fossil fuel-fired power plants and industrial processes. I am also involved with performing life cycle analysis of major energy conversion technologies.

Teaching Interests:

I had extensive teaching experience as a teaching assistant at Purdue University. I gave several lectures and prepared several teaching materials including a new curriculum and a complete set of lab instructions of the newest version of Aspen Plus for the *Design and Analysis of Processing Systems* course. I am passionate about teaching and I am particularly interested in enhancing students' understanding on energy conversion technologies, process design and advanced mathematical modeling. I am also interested in teaching thermodynamics, separations, heat and mass transfer.

I received several teaching awards including 2016 Graduate School's Excellence in Teaching Award, Purdue University's highest award in recognition of graduate student teachers, Excellence for Undergraduate Teaching Assistant Award as voted by the graduating seniors for Teaching Excellence in Chemical Engineering and 2013 CETA (Committee Education of Teaching Assistants) Award.

References

1. Gençer E, Miskin C, Sun X, Khan R, Bermel P, Alam A, Agrawal R (2017). Directing solar photons to sustainably meet food, energy and water needs, *Scientific Reports*, 7, 3133, DOI: 10.1038/s41598-017-03437-x.
2. Gençer E, Agrawal R (2017). Strategy to synthesize integrated solar energy coproduction processes with optimal process intensification. Case study: Efficient solar thermal hydrogen production, *Computers and Chemical Engineering,* DOI: 10.1016/j.compchemeng.2017.01.038.
3. Gençer E, Agrawal R (2017). Synthesis of Efficient Solar Thermal Power Cycles for Baseload Power Supply, *Energy Conversion and Management*, 133, 486-497.
4. Gençer E, Agrawal R (2016). A commentary on the US policies for efficient large scale renewable energy storage systems: Focus on carbon storage cycles. *Energy Policy,* 88, 477-484.
5. Gençer E, Mallapragada DS, Maréchal F, Tawarmalani M, Agrawal R (2015). Round-the-clock power supply and a sustainable economy via synergistic integration of solar thermal power and hydrogen processes. *Proceedings of the National Academy of Sciences (PNAS),* 112(52), 15821-15826.
6. Parsell T, Yohe S, Degenstein J, Jarrell T, Klein I, Gençer E, Hewetson B, Hurt M, Kim J I, Choudhari H, Saha B, Meilan R, Mosier N, Ribeiro F, Delgass W N, Chapple C, Kenttamaa H I, Agrawal R, Abu-Omar M M, (2015). A synergistic biorefinery based on catalytic conversion of lignin prior to cellulose starting from lignocellulosic biomass. *Green Chemistry*, 17(3), 1492-1499.
7. Gençer E, Tawarmalani M, Agrawal R (2015). Integrated Solar Thermal Power and Chemical Coproduction Process for Continuous Power Supply and Production of Chemicals. *Computer Aided Chemical Engineering, 12th International Symposium on Process Systems Engineering and 25th European Symposium on Computer Aided Process Engineering,* 37:2291-2296.
8. Gençer E, Mallapragada DS, Tawarmalani M, Agrawal R (2014). Synergistic biomass and natural gas conversion to liquid fuel with reduced CO_2 emissions. *Computer Aided Chemical Engineering, Proceedings of the 8th International Conference on Foundations of Computer-Aided Process Design*, 34:525-530.
9. Gençer E, Al-musleh E, Mallapragada DS, Agrawal R (2014). Uninterrupted renewable power through chemical storage cycles. *Current Opinion in Chemical Engineering*, 5, 29-36.

Reports

1. Gençer E, O'Sullivan F (2017). CCUS Technologies Baselines Report. *MIT Energy Initiative Center for Carbon Capture, Utilization, and Storage.*

505138 Ion Transport in Charged Porous Media: From Porous Electrodes to Geological Flows

Abstract:

Research Interests: Electrokinetics, Transport Phenomena, Computational Science and Engineering, High Performance Computing

Teaching Interests: Transport Phenomena, Thermodynamics, Numerical Analysis, Mathematical Physics

Understanding ion transport phenomena and electrochemical processes in porous media is relevant to many applications. Some examples include charge transport and storage in batteries and supercapacitors, water desalination through capacitive deionization, electrokinetic remediation of contaminated soil, and monitoring of fluid flow in underground reservoirs. Although these problems span many orders of magnitude in both length and time scales, the fundamental picture of ion transport and electrokinetic phenomena is remarkably similar. Here, I will discuss some of the underlying physical mechanism and their implications for optimizing existing technologies and possible novel applications.

For my doctoral thesis at UCSB (under the supervision of Prof. Frederic Gibou and Prof. Todd Squires), I studied the effects of pore microstructure on ion transport in porous electrodes. In most existing models, pore morphology is neglected and transport is described using homogenized equations. Despite their popularity and success, these models cannot correctly describe heterogeneous transport which occurs in ordered porous materials, e.g. in mesoporous carbon or CNT-based electrodes. Therefore, pore microstructure can significantly impact the charging kinetics. To test this hypothesis, I developed multi-resolution numerical algorithms for solving the Poisson-Nernst-Planck equations at the pore-scale. Specifically, I showed that surface conduction, which was ignored in previous models, can significantly enhance the charging kinetics by providing additional transport pathways. This discovery may be used for designing better performing electrodes for high power density applications.

During my postdoctoral research at MIT with Prof. Martin Bazant, I have been investigating novel applications of electrokinetic phenomena to underground fluid flow problems. For instance, our recent theory shows that electro-osmotic flows can control, and under certain conditions suppress, viscous fingering. This interfacial instability can lead to residual trapping of oil during secondary recovery process and is hard to control. However, we show that by adjusting the injection ratio of electric current to fluid flow, viscous fingering can be manipulated such that stable displacement is possible. Electro-osmotic flows are also more efficient in submicron pores where hydraulic resistance is very large. Therefore, they can lead to a more uniform displacement even in reservoirs with very large permeability variations. Finally, it might be possible to leverage the large electro-osmotic pressures, generated in tight and dead-end pores, as a new and efficient fracturing technique.

505493 Modeling of Light-Driven Heterogeneous Catalysis and Other Excited-State Processes at the Nanoscale

Abstract:

Research Interests:

The ever-increasing drive to use solar power to generate renewable electricity and fuel has resulted in a boom in studies on photo-driven processes in light-harvesting materials. These include photo-induced chemical reactions and charge carrier generation and transport in semiconductors, metals, and their interfaces. The need for better models and therefore understanding of reactions and processes involving electronically-excited states is essential to further drive this research forward. The goal of my future research is therefore to use state-of-the-art computational tools in fields where understanding of light-matter interactions and excited-state physics at the atomic-scale delivers unparalleled design principles to accelerate fabrication of useful solar-driven technologies and applications. My research teamâ€™s areas of expertise will include (*i*) visible light-driven photocatalysis with emphasis on free-radical polymerization, (*ii*) nanoscale photovoltaics, and (*iii*) charge carrier transport across novel nanoscale junctions.

Teaching Interests:

I believe I am most qualified to teach quantum mechanics, solid-state physics and chemistry, chemical thermodynamics and kinetics, and theoretical and computational materials science and chemistry (with emphasis on first-principles atomistic modeling).

505083 Transitional Solutions Towards Decarbonized Economy

Abstract:

Research Interests:

Record-breaking rise in atmospheric greenhouse gas (GHG) emissions and consequent climate change in recent decades have induced public and private sectors to implement targets for reducing GHG emissions by moving towards sustainable technologies. Conventional energy industries (e.g. oil & gas and coal) are well established, characteristically conservative, and have inertia against intense changes mandated by environmental regulations. Alternative energies on the other hand are gradually rolling but have shown their technological and economic feasibility limitations. Wise integration of renewables with fossil fuel energy systems not only can play a systematic transformative role in transition towards low-carbon economy, but also would provide process-level synergistic enhancements. I like to utilize my research background - both in renewable process engineering and system-level life cycle analysis – and participate in this social and technological transition.

Successful Research Proposals:

- NSERC PDF, the *Natural Sciences and Engineering Research Council* of Canada, 90,000$, 2016-2018.
- Vanier Canada, NSERC, 150,000$, 2010-2013.
- 4YF, University of British Columbia (UBC), 100,000$, 2009-2013.

Postdoctoral Projects:

- "*Life cycle environmental and energetic productivity assessment of fossil fuel energy systems for quantitative policy & decision making*"
- "*Can CO_2 enhanced oil recovery catalyze gigatonne carbon capture and storage deployment?*"

Advisor: Prof. Adam R. Brandt, Energy Resources Engineering Department, Stanford University, CA.

- "*Biorefinery in western North America: Technology maturation platform to enable commercial scale production of biojet and renewable diesel fuels*"

Advisor: Prof. Tony X. Bi, Boeing-UBC joint project.

PhD Dissertation:

- "*Biomass/Fossil fuel co-thermochemical conversion with & without integrated CO_2 capture*"

Advisor: Profs. John R. Grace, Chemical & Biological Engineering Department, University of British Columbia, Vancouver, Canada.

Research Experience:

My research has been a composite of many fields in *Energy* and *Environment* for the last ten years, and this has granted me potent interdisciplinary insight. I always believed that experimentation and theoretical models are two necessary wings for a mature research and I consistently followed this path. I have worked and published in different experimental scales:

"fluidization and multi-phase gas-solid systems, solid-state reaction kinetics, thermogravimetric analysis, carbon (in biomass/biosolids & wastes/coal) thermochemical conversion (pyrolysis and gasification) in lab and pilot scales, catalytic renewable processes, and carbon capture via solid sorbents"

by using various techniques:

"micro/meso/macro surface area measurement via N_2 and CO_2 adsorption, SEM/EDX, XRD, TPR, TPO, TPD, GC, Fuel characterization (ultimate, proximate, and ash analysis), European standard tar sampling, laser particle size analysis, and hydrodynamics cold model study."

On the modeling side, I have experience in: "system-level environmental-economic policy assessment, life cycle assessment, energetic productivity analysis, quantitative policy & decision making, instability analysis and perturbation methods, and multiphase flows computational fluid mechanics."

Throughout my past experience in industrial sponsored and collaborative projects, I have demonstrated my capability and effectiveness in tactfully reaching the objectives of the projects within the given timeline. I have also attempted to improve my leadership skills by entering to UBC management sub-specialization program and taking some business courses like "business decision for technology ventures" and "management fundamentals for engineering enterprises".

Teaching Interests:

1. **Philosophy:** Education transforms one's viewpoints, horizons, and habits of mind. I have a strong desire to be contributory in that transformation. My principle teaching philosophy is to provide an atmosphere of caring and respect in order to encourage self-governing learning, to be honest about potential confusions, and unambiguous in my explanations. I de-emphasize memory work, and stress the importance of integration of knowledge by turning the classes into model of opportunity for *inquiry-based learning*. My goal is to present topics by drawing on both work and research experiences to make material relevant. Once students realize the value of their marketable skills and knowledge, it rises their desire to master those skills and concepts.
2. **Experience:** I recently participated in an advanced training and received a certificate of Stanford postdoctoral scholars teaching workshop. I acted as the teaching assistant of the following courses at UBC: Unit operation I (undergrad), Chemical engineering laboratory (undergrad), Elements of physics (undergrad), Environmental engineering laboratory (undergrad), Kinetics and reactor design (undergrad), Engineering management (grad), Business decision for technology venture (grad). I also served as guest lecturer at UBC: Kinetics and reactor design (undergrad), Advanced reactor design (grad), Business decision for technology venture (grad), Green engineering principles and industrial applications (grad). Since summer 2016, I have been mentoring two PhD and Masters students from Imperial College and Stanford on their thesis research. I also performed research mentorship of six visiting graduate/undergraduate students from Stanford, UCSB, Morehouse College, UBC, University of Lyon, Technical University of Munich and University of Nottingham.
3. **Interests:** The world is not on course for a sustainable future and we need global citizens with the creativity to solve complex multidisciplinary challenges. Thus, regardless of the course I am assigned to teach, I try to embed "environmental" and "sustainability" orientations in my lectures. I am in particular interested to offer the following courses:

- *Optimization of Energy Systems*: an introductory course in mathematical programming and optimization, with examples and problems drawn from the energy industries.
- an interdisciplinary graduate course in *Energy, Environment and Sustainability* by including global exergy resources, thermodynamics for renewables, solar thermal, biomass energy systems, and

geothermal systems.
- *Engineering Life Cycle Assessment of Energy Systems*: covers methods of full-fuel-cycle LCA, including mass and energy balances in complex systems analysis, systems boundary considerations, co-product allocation methods, and embodied energy calculations for capital investment and infrastructure. It will also cover secondarily the complexities associated with broad-scale engineering analysis, including economic factors (so called "consequential" LCA).

Future Direction:

My research record shows my broad range of experience and interest in sustainable energy, from analytical and numerical multi-phase studies, to fundamental and large scale experimental approaches, and ultimately towards big-picture system-level environmental assessment analyses. This could not have been grown without collaboration, knowledge transfer and team working via effective communications and partnership with others.

My main research objective as faculty is to work on transitional solutions towards low-fossil-fuel economy by founding a Laboratory of Energy and Environment. Research in this lab is aimed to apply a fundamental understanding gained from lab scale experiments, together with appropriate mathematical modelling, to the industrial challenges of having more sustainable energy systems. The system-level life cycle studies would be our compass to strategies the lab's future research directions. The main themes of the lab though not limited to are:

- **Conversion:** to enhance multi-phase conversion processes, e.g. biomass/waste pyrolysis and gasification to produce bio oil and syngas;
- **Upgrading:** to investigate different pathways towards liquid fuel (e.g. jetfuel) production from bio-syngas and bio oil upgrading;
- **Synthesis:** to develop cost-effective materials and techniques for CO_2 capture and storage;
- **Design:** to introduce novel reactor designs according to new process specifications and criteria, e.g. auto-thermal multiple bed reactor systems for simultaneous multiple conversion processes (e.g. combustion and thermochemical conversion);
- **Macro-Analysis:** to embrace system-level sustainability studies for quantitative decision and policy making by incorporating environmental assessment and optimization methodologies to evaluate our understanding of the environmental impacts of fossil-fuel-based energy systems and to determine how these impacts can be reduced.

Selected Publications:

1. **Masnadi MS**, Brandt AR, Climate impacts of oil extraction increase significantly with oilfield age, *Nature Climate Change,* 2017 (in press).
2. **Masnadi MS,** Brandt AR, Energetic productivity dynamics of global super-giant oilfields, *Energy & Environmental Science*, 10, 1493-1504, 2017.
3. Yu MM, **Masnadi MS**, et al., Co-gasification of biosolids with biomass: thermogravimetric analysis and a pilot scale study in a bubbling fluidized bed reactor, *Bioresource Technology*, 175, 51-58, 2015.
4. **Masnadi MS**, Grace JR, et al., Single-Fuel steam gasification of switchgrass and coal in a pilot-scale bubbling fluidized bed: A comprehensive parametric reference for co-gasification study, *Energy,* 80, 133-147, 2015.
5. **Masnadi MS**, Grace JR, et al., From fossil fuels towards renewable: inhibitory and catalytic effects of carbon thermochemical conversion during co-gasification of biomass and fossil fuels, *Applied Energy,* 140, 196-209, 2015.
6. **Masnadi MS**, Grace JR, et al., From coal towards renewables: Catalytic/Synergistic effects during steam co-gasification of switchgrass and coal in a pilot-scale bubbling fluidized bed reactor, *Renewable Energy,* 83, 918-930, 2015.
7. **Masnadi MS**, Grace JR, et al., Biomass/coal steam co-gasification integrated with in-situ CO_2 capture, *Energy,* 83, 326-336, 2015.
8. Ellis N, **Masnadi MS**, et al., Mineral matter interactions during co-pyrolysis of coal and biomass and their impact on intrinsic char co-gasification reactivity, *Chemical Engineering Journal,* 279, 402-408,

2015.
9. **Masnadi MS**, Habibi R, et al., Fuel characterization and co-pyrolysis kinetics of biomass and fossil fuels, *Fuel*, iSGA-3 Special Issue, 117, 1204-1214, 2014.
10. James AK, Helle SS, Thring RW, Rutherford MP, and **Masnadi MS**, Investigation of air and air-steam gasification of high carbon wood ash in a fluidized bed reactor, *Energy and Environment Research*, 4(1), 15-24, 2014.
11. Habibi R, Kopyscinski J, **Masnadi MS**, et al., Co-gasification of biomass and non-biomass feedstocks: synergistic and inhibition effects of switchgrass mixed with sub-bituminous coal and fluid coke during CO_2 gasification, *Energy & Fuels*, 27(1), 494–500, 2013.
12. **Masnadi MS**, Elyasi S, et al., Gas-Solid flow distribution through identical vertical passages: modeling and stability analysis, *AIChE*, 56(8), 2039-2051, 2010.
13. **Masnadi MS**, Grace JR, et al., Distribution of multi-phase gas-solid flow across identical parallel cyclones, *Separation & Purification Technology*, 72(1), 48-55, 2010.
14. Bi XT, **Masnadi MS**, Chapter 11: Multiphase reactors for biomass processing and thermochemical conversions, In: Cheng Y, Wei F, Jin Y, editors, Multiphase reactor engineering for clean and low-carbon energy applications, *John Wiley and Sons*, 331-376, 2017.

505330 Pore-Level Multiscale Simulation of SAGD

Abstract:

A numerical framework is presented to simulate SAGD in multi-scale porous media images. To leverage pore-scale data sets and evaluate residual oil saturation and performance of the SAGD as a complex EOR method, it is essential to provide comprehensive computationally cheap algorithms for multi-phase and multi-physics micro-scale studies. Here, a hybrid pore-scale simulation approach called DyMAS is utilized to simulate SAGD in a reliable and time efficient manner applying two quasi-static and CFD-based techniques to micro- and macro-pore domains. Quasi-static simulations are conducted in micropores, where the capillary force is dominant, and a dynamic CFD solver is applied to macropores taking into the account the gravity and viscous forces, simultaneously. The workflow is applied to three-dimensional digital rock samples representing the McMurray formation of the Athabasca Oil Sands Deposit, and a detailed sensitivity analysis is performed on operational conditions leading to minimum residual oil. The post-processing results, including capillary pressure curves, residuals, and effective permeability/thermal conductivity curves, are then predicted and discussed in terms of pore-scale phenomena. According to the results the condensed water imbibition plays a crucial role in decreasing the residual oil within micropores, and buoyancy-driven flows are determining in intergranular spaces demonstrating the scale-dependent nature of gravity, viscose and capillary forces.

Research Interests:

- Multiphase flow and heat transfer in porous media.
- Pore-level simulations and characterizations.
- Dynamic and static reservoir simulation.
- Coding and algorithm design.
- Geostatistics.
- Well testing and stimulation.
- Production surveillance and optimization in gas condensate reservoirs.
- Characterization of Shale and tight reservoirs.
- Hydrate-bearing sediments.

Teaching Interests:

- Reservoir Simulation.
- Geostatistics.
- Well testing.
- Reservoir Rock and Fluid Properties.

References:

- Mohammadmoradi, P., Kantzas, A. (2017). DyMAS: A Direct Multi-Scale Pore-Level Simulation Approach.

  SPE Western Regional Meeting, https://doi.org/10.2118/185720-MS
- Mohammadmoradi, P., Pore Morphological Multi-Phase Digital Rock Physics Models, PhD Thesis, University of Calgary, http://hdl.handle.net/11023/3485.

SCREENING IMPROVED RECOVERY METHODS IN TIGHT-OIL FORMATIONS BY INJECTING AND PRODUCING THROUGH FRACTURES

Harpreet Singh[1]
[1]National Energy Technology Laboratory, Morgantown, WV, USA

Abstract

Tight-oil reservoirs exhibit two characteristic behaviors that limit potential for improved recovery: i) limited fluid movement from the unfractured matrix limits the potential to sweep additional oil towards the production wells with injection of a displacing fluid, ii) the wettability of these reservoirs tend to be oil-wet which holds oil in relatively smaller pores, as compared to gas or water, making it difficult to mobilize that trapped oil. Primary oil production from these formations can decline to half of initial rates in the first year due to low permeability in the unfractured reservoir matrix, and a large amount of unrecovered oil remains in smaller pores that is not able to move out. Declining oil rates and low ultimate recoveries from oil-rich shale formations could benefit from application of secondary fluid injection schemes.

To address these challenges, this study investigates the potential to improve oil recovery from tight-oil reservoirs by a method that includes injecting a fluid into the fractured reservoir to produce hydrocarbons from adjacent fractures intersecting the same wellbore. The injection and production through fractures has a potential economic advantage over huff-n-puff scheme such that there is no lag-period between injection and production. We compare the estimates of incremental oil recovery obtained by injecting water, immiscible CO_2, and surfactant over the oil recovered by primary depletion (with no injection). The results suggest that injection of fluids may not always improve recovery from tight-oil reservoirs, especially in a heterogeneous reservoir (having different rock types) whose average matrix permeability is lower than 0.01 mD. In cases where recovery is improved over primary depletion, water flood and surfactant flood perform almost equally well with no noticeable difference between the two. Although, surfactant results in favorable fluid properties to mobilize oil, the reason surfactant did not perform well in tight-oil formation studied here is because the unswept oil is inaccessible due to small permeability, and not because of high capillary pressure.

Keywords: tight-oil; shale; EOR; water; surfactant; immiscible CO_2;

1. Introduction

Oil and gas production from unconventional tight-oil resources has become an important source of energy security in the United States (US) and it is emerging to be as important in other countries, especially Russia, Argentina, and Canada will see significant tight-oil production by 2040 (Aloulou, 2016). In the U.S., the contribution of oil produced from unconventional shale resources has jumped from 0.5 million barrels per day (BPD) in 2005 to ~5 million BPD in 2015, i.e. ~52% of the total US crude oil production (US EIA, 2016). This increase in production from unconventional shale resources has been mainly due to advancements in drilling efficiency and well configuration (Rystad Energy, 2014). However, similar improvements have not been achieved for well performance; it has been observed (Rystad Energy, 2014) that year after year wells in mature plays (e.g. Eagle Ford and Bakken) have shown similar initial rates and decline trends. Production from tight-oil resources has two drawbacks with a potential for improvement: i) low ultimate recovery factor (Gaswirth and Marra, 2015; Gaswirth et al., 2013), and ii) decline in daily production rate by more than 50% within the first year of production from an average well in the Bakken (Lund, 2014; Wachtmeister et al., 2017) and Eagle Ford (Delaihdem, 2013; Gong, 2013) plays. For instance, estimated oil resources in the Bakken formation is estimated to be 300 billion barrels (Flannery, 2006), out of which 7.38 billion barrels is estimated to be technically recoverable using current best-available technology (Gaswirth and Marra, 2015; Gaswirth et al., 2013), i.e. just ~2% of the original oil in place (OOIP). In addition to low recovery factor, the production rate of a new well from tight-oil reservoirs have been reported to decline by more than half within a year (Delaihdem, 2013; Gong, 2013; Lund, 2014; Wachtmeister et al., 2017), primarily because of fast depletion from fractures and slow fluid movement from low matrix permeability.

To arrest the declining production rates and recover additional oil from tight-oil resources, traditional methods of improved oil recovery (IOR) can be considered (e.g. water/gas/surfactant flooding etc.). However, unlike IOR in conventional reservoirs with relatively high permeability, unconventional tight-oil resources present two major challenges to IOR: i) very limited fluid movement outside the fractures limits the possibility of an efficient sweep between the wells, ii) the wettability of these reservoirs tend to be oil-wet that keeps the oil in the smallest pores compared to gas or water, making it difficult to drive oil out.

We investigate the potential of IOR from tight-oil reservoirs by a method that includes numerical reservoir simulation-based evaluation of the performance of fluid injection through a set of fractures to stimulate additional hydrocarbon recovery from a different set of fractures intersecting the same wellbore (Dombrowski et al., 2015) as shown in Figure 1. This injection-production scheme is less expensive compared to IOR using several injection and production wells in tight-oil reservoir (e.g. Ghaderi et al., 2017; Hustak et al., 2017), and unlike huff-n-puff, does not require shutting-in the well.

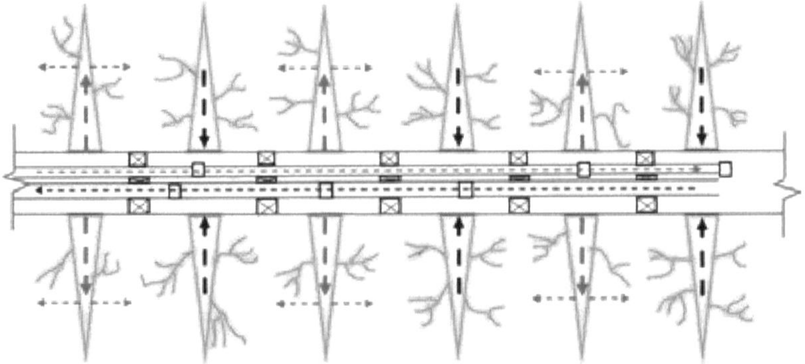

Figure 1: Schematic for fluid injection and production through hydraulic fractures in tight-oil reservoir. Dashed red and black colored arrows represent the direction of the injected and produced fluids, respectively.

To investigate the potential of improving recovery in tight-oil reservoirs, we study the injection of three different fluids (water, immiscible CO_2, and surfactant) in a single wellbore that is intersected by three injection-fractures and three recovery-fractures. This representation used for simulation of what is very likely a network of engineered fractures and pre-existing natural fractures. We compare the estimates of oil recovery obtained by injecting water, immiscible CO_2, and surfactant with the oil recovered by primary depletion (from all six fractures) with no injection. The impact of fluid injection on oil recovery is assessed for three different average permeabilities (very low, low, and relatively high), three different injection rates (low, high, and very high), and when injection is constrained by constant bottom-hole pressure in a homogeneous and heterogeneous medium with rock characteristics representative of a typical tight-oil reservoir.

2. Method

Below, we discuss the development of the reservoir model and the fluid model used in simulations. The reservoir model is composed of discrete hydraulic fractures that are connected with the matrix grid by a nearest neighborhood connection approach (Lie, 2016). The flow simulations are performed using Matlab Reservoir Simulation Toolbox (MRST), a full-physics reservoir simulator (Lie, 2016), with fully implicit solutions of blackoil model.

2.1. Reservoir Model

2.1.1. Injection and Production in Tight-Oil Reservoirs

A schematic of a single horizontal well (henceforth referred to as *lateral*) intersecting six hydraulic fractures in a tight-oil reservoir is depicted in Figure 2. The conventional approach of recovering hydrocarbons from tight reservoirs would lead to low injection rates and very limited fluid movement due to extremely low permeability, therefore this drawback can be addressed by either having extremely long fractures between injection and production wells (which is not technically possible at present) or drilling additional infill wells that are closely spaced. The former option is difficult to achieve with current state of fracturing technology where fracture half-length is on the order of a few hundred feet (Zuloaga-Molero et al., 2016), while the option of drilling closely spaced additional wells may prove uneconomical for reservoirs with low rates of production. Dombrowski et al. (2015) proposed an economical approach that addresses these concerns by a method that includes injecting a fluid through a set of fractures and producing hydrocarbons from a different set of fractures within the same wellbore. As the gravity forces have relatively small impact on light oil in tight reservoirs compared to viscous forces, the hydrocarbons can be recovered from both above and below the fractures.

In this study, numerical simulation of a single horizontal wellbore intersecting six hydraulic fractures is used to investigate the injection of fluids as a measure to improve recovery after production rate drops significantly. Initially, all six fractures are used for production by primary depletion (without any injection) for ~1800 days when the production rate is significantly lower (< 50 %) than its initial value.

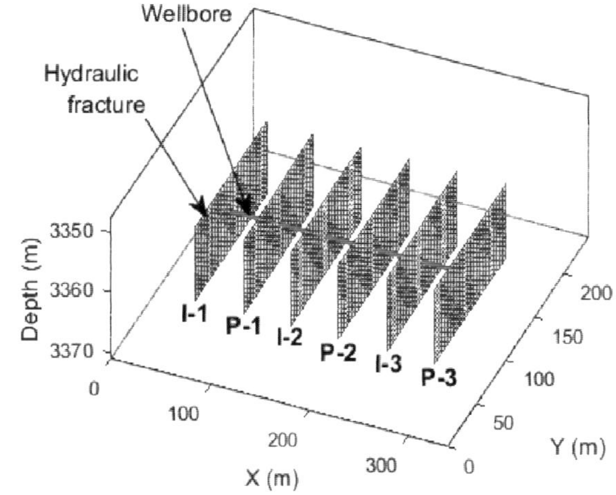

Figure 2: Computational domain for fractured tight-oil reservoir with six hydraulic fractures intersecting the lateral. After 1800 days (~ 5 years) of production by primary depletion, three existing hydraulic fractures are used to inject fluid (I-1, I-2, I-3) and the remaining three fractures are used for production (P-1, P-2, P-3).

Although complex fracture geometries (including interaction of hydraulic and natural fractures) are common in tight-oil reservoirs, such geometries are not considered here in order to control the number of variables and to avoid the complexity beyond necessary.

2.1.2. Reservoir Properties

The reservoir and fracture properties are taken from other studies (Sanchez-Rivera et al., 2015; Zhu et al., 2017; Zuloaga-Molero et al., 2016) on Bakken tight-oil. The parameters for reservoir including well, and hydraulic fractures (shown in Figure 2) are given in Table I. The hydraulic fractures in Figure 2 are modeled using embedded discrete fracture and are linked to the matrix grid through nearest neighborhood connection approach (Lie, 2016).

Table I: Parameters for (a) reservoir including well, and (b) hydraulic fractures.

(a)

Reservoir parameters	Value	Unit
Initial oil saturation (S_{oi})	80	%
Initial reservoir pressure	8000	psi
Production pressure (BHP)	1800	psi
Reservoir temperature	240	F
Reservoir depth	11000	ft
Formation compressibility	5e-06	psi^{-1}
Time when injection begins	1800	$days$
Time when injection ends	20	$years$
Mean permeability (matrix)	0.1, 0.01, 0.005	mD
Mean porosity (matrix)	8	%

(b)

Hydraulic Fracture Parameters	Value	Unit
Aperture	5e-03	ft
Permeability	10	D
Porosity	50	%
Spacing	160	ft
Half-length	200	ft
Height	40	ft

2.2. Fluid Model

2.2.1. Fluid Properties

In-situ fluid properties

The PVT analysis conducted on in-situ fluids from different tight-oil reservoirs (Ghaderi et al., 2017; Hawthorne et al., 2017a; Li and Luo, 2017; Mavor, 2013) reveal the crude oil from these reservoirs to be light, with low density and viscosity. The density, viscosity, compressibility, and formation volume factor of light oil and brine in a typical tight-oil formation like Bakken at initial reservoir conditions (Zhu et al., 2017; Zuloaga-Molero et al., 2016) are shown in Table II. Only brine and oil are considered as the in-situ fluid present at initial conditions in the reservoir. The bubble point pressure of Bakken oil is ~1480 psi (Ghaderi et al., 2017), therefore, it is safe to assume that no free gas exists at initial reservoir condition (8000 psi, 240 ^oF), with all the gas dissolved in oil. Another reason to assume no initial free gas in the reservoir is due to limited capability of surfactant flooding module in MRST that does not provide an option for the gas phase; therefore, having no initial gas in the reservoir would allow for consistent comparison from all four types of recovery processes, i.e. primary depletion, water flooding, gas flooding, and surfactant flooding.

Table II: Properties of formation (a) light oil, and (b) brine, with fluid properties reported at in situ conditions.

(a)

Oil Properties	Value	Unit
Density ($\rho_{o,i}$)	40.47	lb/ft^3
Viscosity ($\mu_{o,i}$)	0.142	cp
Compressibility (c_o)	6.65e-05	bar^{-1}
Formation volume factor	$0.94e^{c_o(p-p_{ini})}$	–

(b)

Brine Properties	Value	Unit
Density ($\rho_{w,i}$)	1000	kg/m^3
Viscosity ($\mu_{w,i}$)	0.61	cp
Compressibility (c_w)	4.28e-05	bar^{-1}
Formation volume factor	$0.99e^{c_w(p-p_{ini})}$	–

Injection fluid properties

Three different fluids are injected to study their impact on improving recovery from tight-oil formations, which are water, immiscible CO_2, and surfactant. The properties of CO_2 (Andersen, 2017) at initial reservoir conditions (8000 psi, 240 F) are given in Table III. We assume CO_2 is immiscible with reservoir oil, i.e. both CO_2 and oil exist as separate phases.

Table III: Properties of CO_2.

Gas Properties	Value	Unit
Density ($\rho_{g,i}$)	803.5	kg/m^3
Viscosity ($\mu_{g,i}$)	0.075	cp
Compressibility (c_g)	6.25e-04	bar^{-1}
Formation volume factor	$e^{c_g(p-p_{ini})}$	–

Although surfactant flooding has been studied extensively in conventional reservoir rocks (Sheng, 2015) to improve recovery, only few studies (He et al., 2014; Mirchi et al., 2015; Shuler et al., 2016; Wang et al., 2011, 2012, 2016; Xu et al., 2011) exist about their application in tight reservoir rocks like shale, especially on real reservoir rock. Surfactant's ability to reduce the interfacial tension between oil and water (σ_{ow}) results in changing the relative permeability and decreasing the residual oil saturation. The decrease in interfacial tension (IFT) by adding surfactant results in increasing the capillary number (N_c), which is a measure of viscous force relative to the capillary force, i.e.:

$$N_c = \frac{|k\nabla p|}{\sigma_{ow}} \quad (1)$$

Therefore, the consequential effect of an increased capillary number is the reduction in residual saturation of oil. The reduction in residual saturation is a function of miscibility condition, which is defined in terms of miscibility factor ($m \in [0, 1]$) that is calculated by interpolating the logarithmic value of the capillary number (Jørgensen, 2013).

$$m(N_c) = \frac{\log_{10}(N_c) - \log_{10}(N_c^{misc})}{\log_{10}(N_c^{immisc}) - \log_{10}(N_c^{misc})} \quad (2)$$

The surfactant adsorption is assumed to be negligible such that it does not result in any change in porosity or permeability of the rock. Relevant details on the surfactant model implementation can be found in the Appendix or in Jørgensen (2013) for specific details. The parameters used for surfactant flooding are taken from Xu et al. (2011) who conducted surfactant flooding experiments for a tight-oil formation (Yanchang) in China with various concentrations of surfactant. The IFT and viscosity as a function of surfactant concentration, and the magnitude of miscibility for different capillary numbers (function of σ_{ow}) as reported in Xu et al. (2011) are given in Table IV:

Table IV: (a) interfacial tension and viscosity as functions of surfactant concentration, and (b) miscibility factor as a function of capillary number.

(a)

$C_s\ (kg/m^3)$	$\sigma_{ow}\ (N/m)$	$\mu_w\ (cp)$
0	0.05	0.61
0.1	5e-04	-
0.5	1e-05	-
1	1e-06	-
30	1e-06	0.8
100	1e-06	1

(b)

$log_{10}(N_c)$	$m(N_c)$
-10	0
-5.5	0
-4	0.5
-3	1
2	1

2.2.2. Relative Permeabilities

Relative permeability data govern the flow movement and relative ease of flow for multi-fluids in tight-oil reservoirs. Unfortunately, there are not many experimental data for relative permeability in tight-oil reservoirs, however, there is one notable study by Honarpour et al. (2012) that published relative permeability data for tight-oil Eagle Ford samples categorized as calcite-rich, organic-rich, and fractured. Besides this study, Liu et al. (2010) published the relative permeability data for a tight-oil reservoir, Yanchang oilfield, in China. The data from these two studies suggest that end-point relative permeability values in tight-oil reservoirs are lower than generally higher values in conventional reservoir rocks (Moghadasi et al., 2015) and the range of saturation within which fluids are mobile is tighter than conventional rocks. To the authors' knowledge there are no experimental data available tight-oil reservoir in Bakken, but many simulation studies for Bakken (Sanchez-Rivera et al., 2015; Zhu et al., 2017; Zuloaga-Molero et al., 2016) have used data from Yu et al. (2014) who assumed Brooks-Corey model for relative permeability and estimated its parameters by history-matching pressure and production data from the Middle Bakken. However, the relative permeability parameters estimated by Yu et al. (2014) for the Middle Bakken show two contradictions:

i) their estimated end-point relative permeability values are highly optimistic ($k_{rw}^{o} = k_{row}^{o} = k_{rog}^{o} = 1; k_{rg}^{o} = 0.85$) when compared to experimental observations (Honarpour et al., 2012; Liu et al., 2010) from other tight-oil formations, e.g. Eagle Ford in Texas (Honarpour et al., 2012) and Yanchang oilfield in China (Liu et al., 2010; Xu et al., 2011).

ii) their estimated parameters (end-point and residual saturations) suggest that Middle Bakken is water-wet ($S_{wr} > S_{or};\ k_{rw}^{o} = k_{row}^{o}$), however, experimental survey for Bakken (Wang et al., 2012) reveals that it is slightly oil-wet. A review on wettability of various shale formations around the world by Singh (2016) revealed that most shales tend to be oil-wet.

Therefore, we use the relative permeability data that are representative of a slightly oil-wet medium ($k_{row}^{o} \le k_{rw}^{o};\ S_{wr} = S_{or}$) and a tight-oil formation ($k_{row}^{o}, k_{rw}^{o} < 1$). The relative permeability data for oil-water flow are taken from an experimental surfactant flooding study (Xu et al., 2011) on tight-oil cores from Yanchang oilfield in China, while for oil-gas flow the relative permeability parameters from Yu et al. (2014) are modified slightly to account for oil-wet and tight-oil medium properties as discussed in Zhang et al. (2016). The Brooks-Corey relative permeability model for oil-water and gas-liquid flow can be defined as:

$$k_{rw}(S_w) = k_{rw}^{0}.(S_{wn})^{n_w} \quad (3)$$

$$k_{row}(S_w) = k_{row}^{0}.(1 - S_{wn})^{n_{ow}} \quad (4)$$

$$k_{rg}(S_g) = k_{rw}^{0}.(S_{gn})^{n_g} \quad (5)$$

$$k_{rog}(S_g) = k_{rog}^{0}.(1 - S_{gn})^{n_{og}} \quad (6)$$

Where, k_{ri} is the relative permeability of phase $i(= o, w, g)$, k_{ri}^{o} is the end-point relative permeability of phase i, n_i is the Corey curve exponent for k_{ri}, and S_{in} is the normalized saturation of phase i. The parameters in above relative permeability model for different fluids are defined in Table V. The

Table V: Relative permeability model parameters.

Water-Oil					
S_{wr}	S_{or}	k_{rw}^{0}	k_{row}^{0}	n_w	n_{ow}
0.2	0.2	0.6	0.5	3	3
Gas-Oil					
S_{gr}	S_{or}	k_{rg}^{0}	k_{rog}^{0}	n_g	n_{og}
0.0	0.2	0.85	0.7	3.15	5.75
Water-Surfactant-Oil					
S_{wr}	S_{or}	k_{rw}^{0}	k_{row}^{0}	n_w	n_{ow}
0.05	0.05	1	1	1.5	1.5

2.3. HETEROGENEITY

Based on laboratory (Dong et al., 2017; Hawthorne et al., 2017b; Tuero et al., 2017) and field (Kenomore et al., 2017; Steiner et al., 2016; Witkowsky et al., 2012) observations of shale rocks, it can be broadly divided into three rock types, i.e. inorganic (poor quality), organic (good quality), and intermediate (fair quality) that lies between the inorganic and organic. A spatially heterogeneous reservoir model based on this concept is developed using sequential indicator simulation algorithm (Deutsch and Journel, 1997) by considering marginal probabilities of 0.6, 0.2, and 0.2 for the inorganic, intermediate, and organic rock types, respectively. These marginal probabilities are chosen to reflect the volumetric percentages of organic pores found in many shales, which is about 20% by volume (Witkowsky et al., 2012) based on 1.1 g/cc kerogen grain density. The spatial correlation of each rock type is defined using three nested spherical variograms along 60 degrees to 120 degrees.

The rock properties of inorganic, organic, and intermediate type are considered based on laboratory-scale and field-scale observations of shale rock (Hawthorne et al., 2017b; Kenomore et al., 2017; Tuero et al., 2017; Witkowsky et al., 2012) that show organic rock type, which are also associated with micro-fractures (Hawthorne et al., 2017b), have larger pore sizes (Dong et al., 2017) in comparison to intermediate and inorganic rock types that occupy larger pore volume of the rock (Tuero et al., 2017) with significant saturation of hydrocarbons (Tuero et al., 2017). Therefore, these qualitative observations are used to assign rock properties (porosity, permeability, and initial oil saturation) to each rock type as shown in Figure 3 .

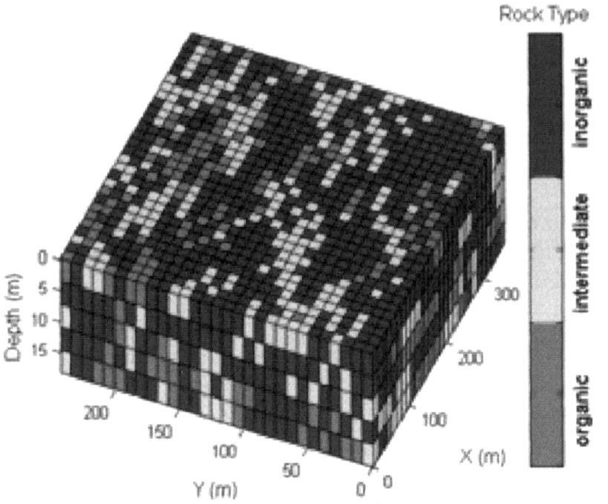

	Geology	Φ (%)	k (mD): Scenario I	II	III	S_{oi} (%)
Heterogeneous	organic	3	0.4	0.04	0.02	60
	intermediate	6	0.1	0.01	0.005	70
	inorganic	10	0.01	0.001	0.0005	90
	Mean	8	0.1	0.01	0.005	80
Homogeneous		8	0.1	0.01	0.001	80

Figure 3: Reservoir model is composed of three different rock types (shown by colors) for a spatially heterogeneous model, where each rock type has a distinct porosity, permeability, and initial oil saturation value. The average values of porosity, permeability, and initial oil saturation for the heterogeneous model are equal to their corresponding values in a homogeneous model. Three matrix permeability scenarios considered are very low (mean 0.005 mD), low (0.01 mD), and relatively high (0.1 mD).

2.4 Injection Rates

Three injection rates are chosen and denoted as low (~7 rb/d), high (32 rb/d), and very high (48 rb/d). The magnitude of each injection rate is denoted as low, high, or very high based on range of injection rates adopted in other studies (Sanchez-Rivera et al., 2015; Zhu et al., 2017) for tight-oil reservoirs with different permeability and fracture spacing.

3. Results

The reservoir is depleted naturally without any fluid injection for ~1800 days when a fluid (water, surfactant, or immiscible CO_2) is injected for the remaining life of the reservoir (~20 years) to arrest the declining production rates and recover additional oil.

The impact of injected fluid on oil recovery is assessed for three different average permeabilities (very low: 0.005 md, low: 0.01 mD, and relatively high: 0.1 mD) and different injection rates (~ low: 7 rb/d, high: 32 rb/d, and very high: 48 rb/d) in reservoir barrels (rb/d) for both homogeneous and heterogeneous scenarios. A similar set of results is obtained to see the impact of injecting fluids constrained by a constant bottom-hole pressure (BHP ~ 8100 psi) instead of a constant injection rate.

3.1. Oil Rates

Figure 4 shows oil rates for four types of recovery processes, i.e. by primary depletion (no injection), by water injection, by surfactant injection, and by immiscible CO_2 injection. Each plot on left column of Figure 4 is for different reservoir matrix permeability in a homogeneous medium, whereas a corresponding plot on the right side of this figure is for heterogeneous medium with an average permeability equal to its corresponding homogeneous value. The results presented in Figure 4 are for an injection rate of ~7 rb/d for all three injection fluids. Figure 5 and Figure 6 show similar results as presented in Figure 4, but for higher injection rates of ~32 rb/d and ~48 rb/d, respectively. Figure 7 shows similar set of results as presented in Figure 4, but these results are obtained when the fluid injection is constrained by a constant BHP (~8100 psi) instead of a constant injection rate.

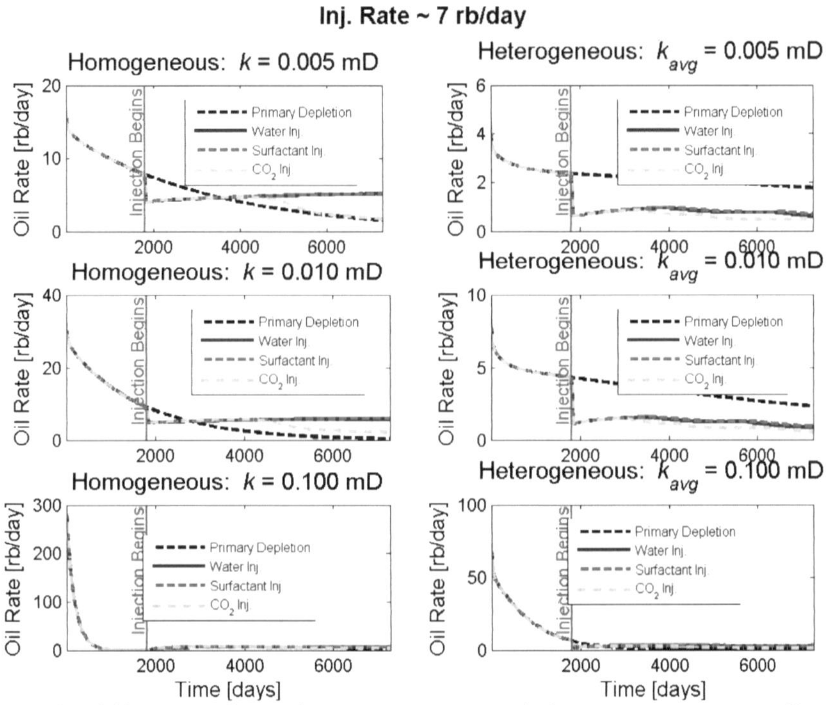

Figure 4: Oil rates from homogeneous and heterogeneous media with three improved recovery methods (injection rate ~ 7 rb/day). Primary depletion represents the base case where oil is produced without any fluid injection.

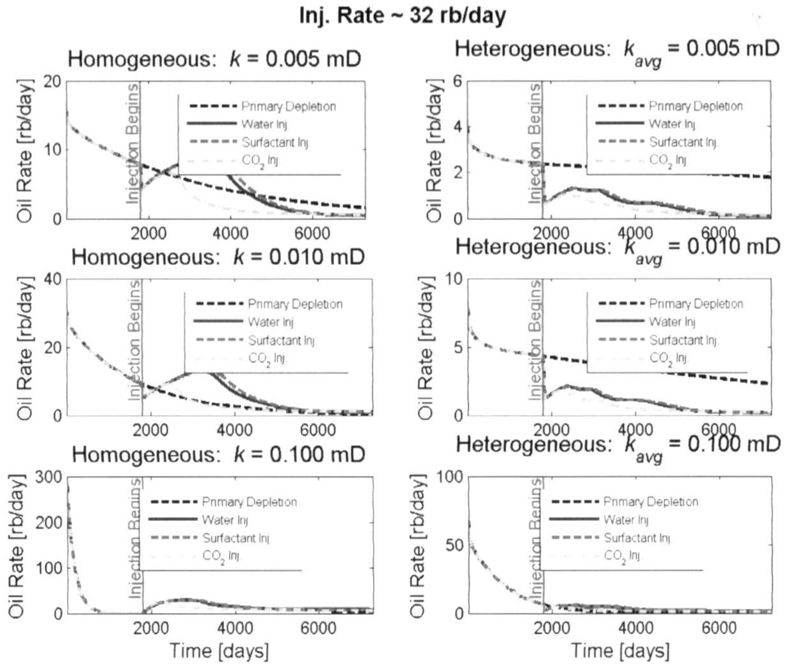

Figure 5: Oil rates from homogeneous and heterogeneous media with three improved recovery methods (injection rate ~ 32 rb/day). Primary depletion represents the base case where oil is produced without any fluid injection.

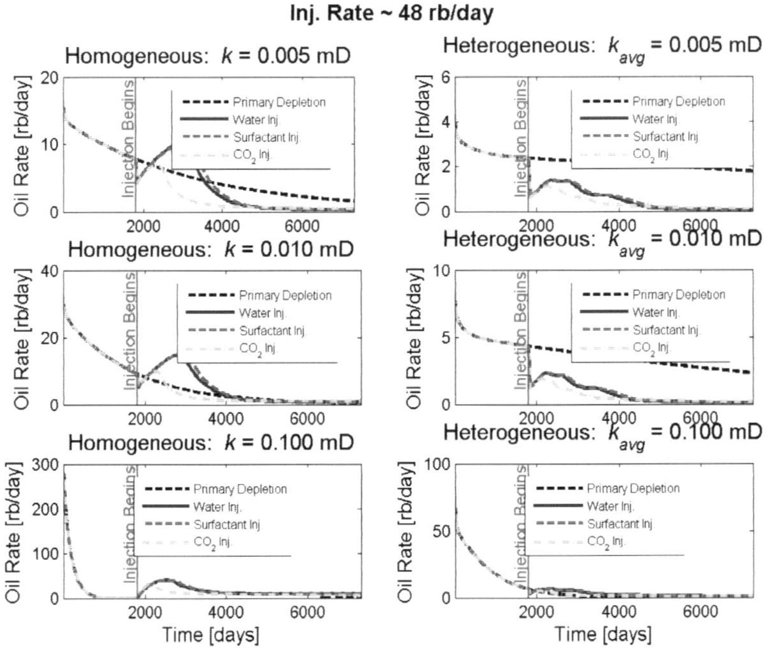

Figure 6: Oil rates from homogeneous and heterogeneous media with three improved recovery methods (injection rate ~ 48 rb/day). Primary depletion represents the base case where oil is produced without any fluid injection.

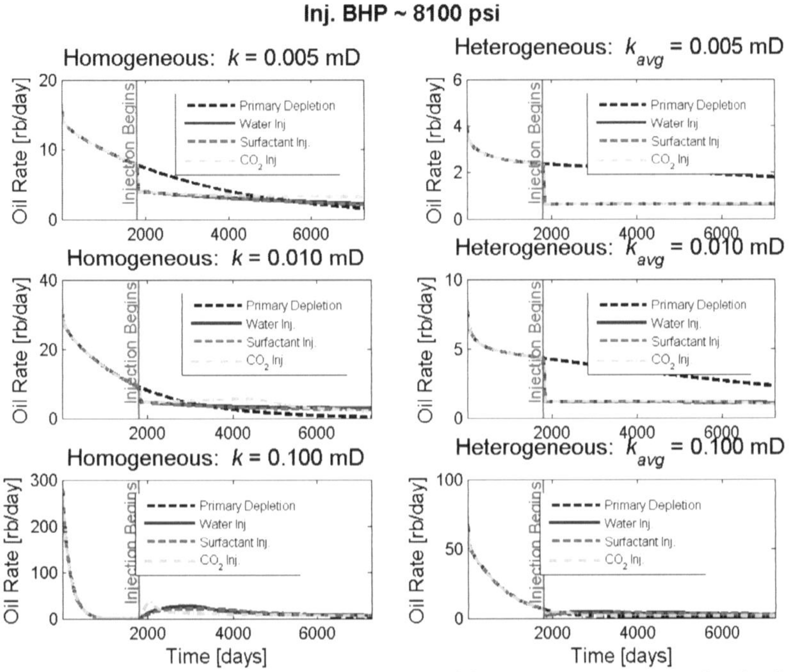

Figure 7: Oil rates from homogeneous and heterogeneous geological models with three improved recovery methods (injection BHP ~ 8100 psi). Primary depletion represents the base case where oil is produced without any fluid injection.

3.2. Oil Recovery

The oil rates plotted in Figure 4 through Figure 7 are used to find oil recovered as percent of the original oil in place (OOIP).

Figure 8, Figure 9, Figure 10, and Figure 11 show oil recovered as percent of the original oil in place (OOIP) for the corresponding oil rates plotted in Figure 4 through Figure 7, respectively. Among the recovery values from all scenarios, maximum oil is recovered (~1.2 % of OOIP; bottom left plot in Figure 10) by injecting surfactant (or water) at the largest rate of ~48 rb/d in a homogeneous medium with 0.1 mD permeability. The maximum oil recovered is ~65% more than the primary depletion (bottom left plot in Figure 14) by injecting surfactant (or water, with no noticeable difference compared to surfactant), whereas the oil recovered by immiscible CO_2 is about 25% less than that by water or surfactant, but it is 40% more than the primary depletion. This is because, as shown in Figure 6, the oil rates by water (or surfactant) get a larger bump in rates than by immiscible CO_2 initially, and the rate bump provided by water (or surfactant) is able to be sustained for a longer period than by immiscible CO_2. The reason for the higher bump provided by water (or surfactant) as opposed to immiscible CO_2 is that earlier CO_2 breakthrough (~2,200 days) is predicted in comparison to water (or surfactant) breakthrough (~2800 days). Minimum oil is recovered (~0.04 % of OOIP; top right plot in Figure 10) with the largest injection rate of ~48 rb/d in a heterogeneous medium with 0.005 mD average permeability. In this case, there is no substantial difference in oil recovered between the three injected fluids (top right plot in Figure 14), whereas the oil recovered by primary depletion (no injection) is almost 50% greater than the oil recovered by injection. The corresponding oil rates for the minimum oil recovered are shown by Figure 6, where it can be observed that the oil rates by water (or surfactant) get a larger bump than by immiscible CO_2 initially, however, the magnitude of oil rates ($< 2\ rb/d$) with all three fluids is not substantial enough to make any difference in the percent of oil recovered.

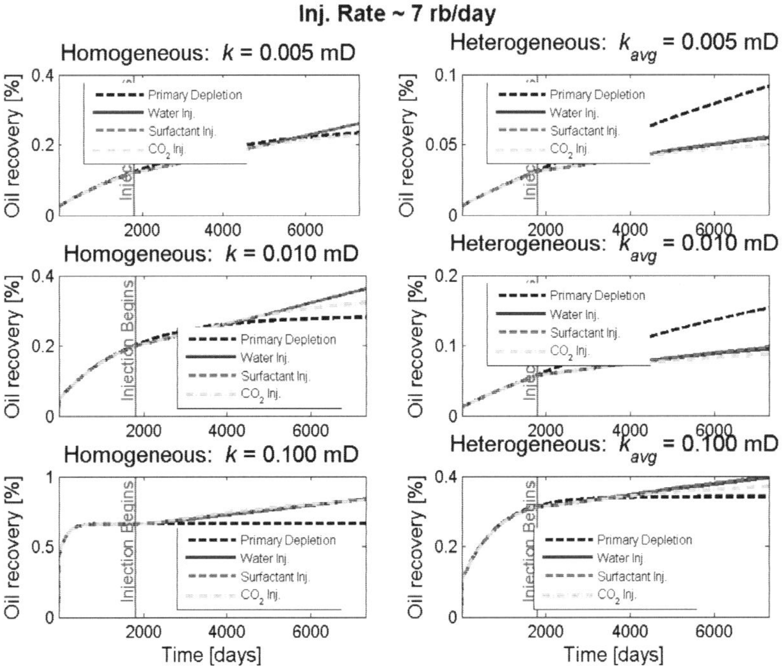

Figure 8: Percent cumulative recovery from homogeneous and heterogeneous geological models with three improved recovery methods (injection rate ~ $7\ rb/day$). Primary depletion represents the base case where oil is produced without any fluid injection.

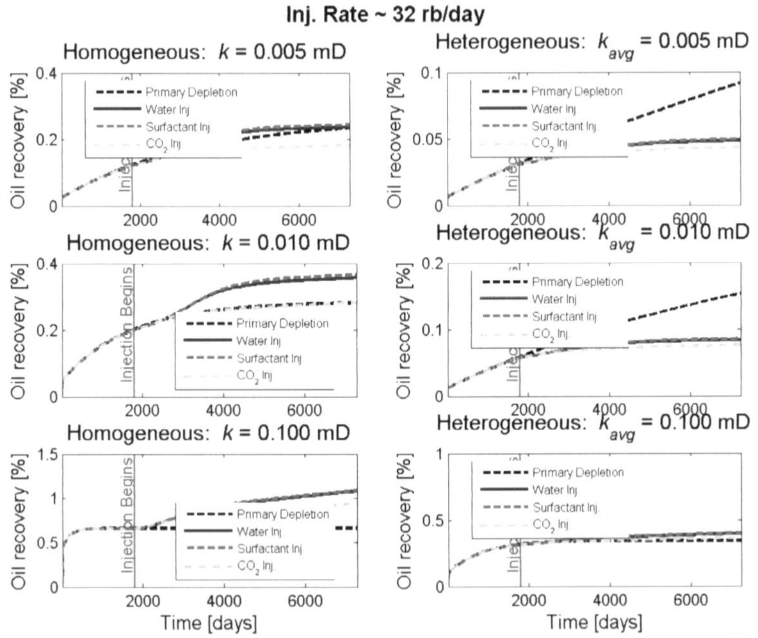

Figure 9: Percent cumulative recovery from homogeneous and heterogeneous geological models with three improved recovery methods (injection rate ~ 32 rb/day). Primary depletion represents the base case where oil is produced without any fluid injection.

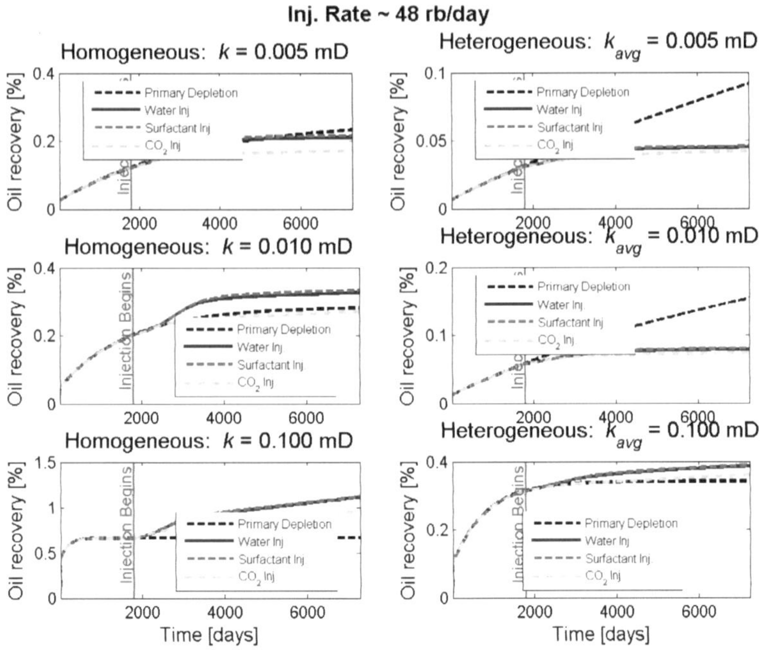

Figure 10: Percent cumulative recovery from homogeneous and heterogeneous geological models with three improved recovery methods (injection rate ~ 48 rb/day).

Primary depletion represents the base case where oil is produced without any fluid injection.

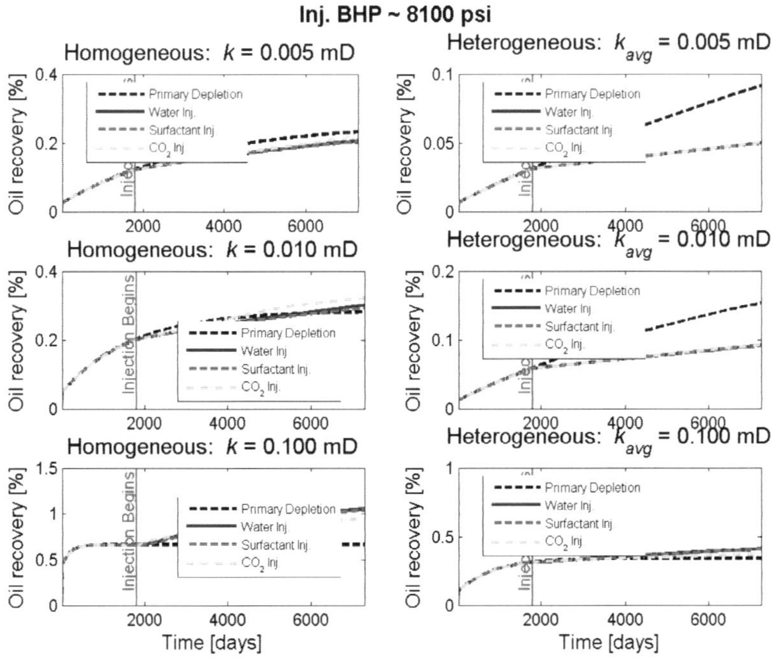

Figure 11: Percent cumulative recovery from homogeneous and heterogeneous geological models with three improved recovery methods (injection BHP ~ 8100 *psi*). Primary depletion represents the base case where oil is produced without any fluid injection.

3.3. Incremental Oil Recovery

To show the impact of injecting different fluids on recovery, incremental recovery (IR) with respect to primary depletion is calculated as shown:

$$Incremental\ recovery\ [\%] = \left(\frac{recovery_{inj} - recovery_{PD}}{recovery_{PD}}\right) \times 100 \qquad (7)$$

Where subscript inj represents the method of recovery by fluid injection (water, surfactant, CO_2), and PD is the method of recovery by primary depletion with no injection (base case). Figure 12, Figure 13, Figure 14, and Figure 15 show relative change in oil recovered as percent with respect to primary depletion (PD) for the corresponding oil recoveries plotted in Figure 8, Figure 9, Figure 10, and Figure 11, respectively.

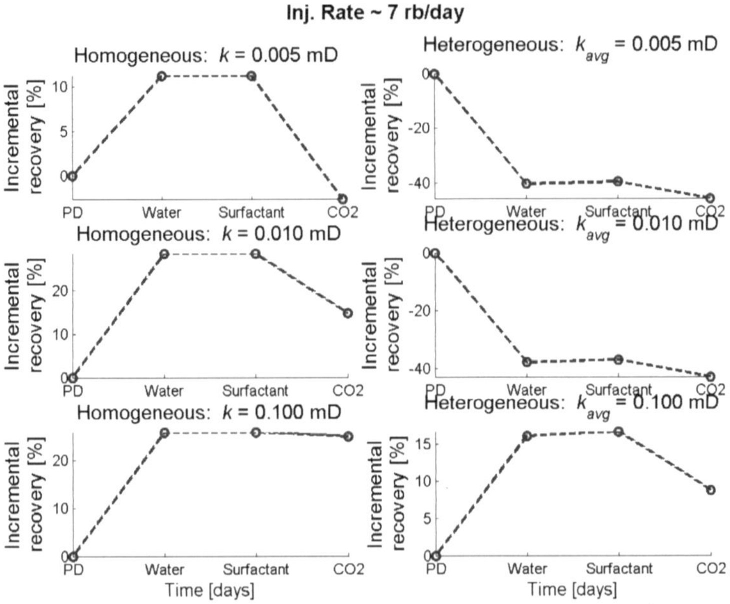

Figure 12: Incremental recovery (IR %) with respect to primary depletion (base case) for the recovery values plotted in Figure 8.

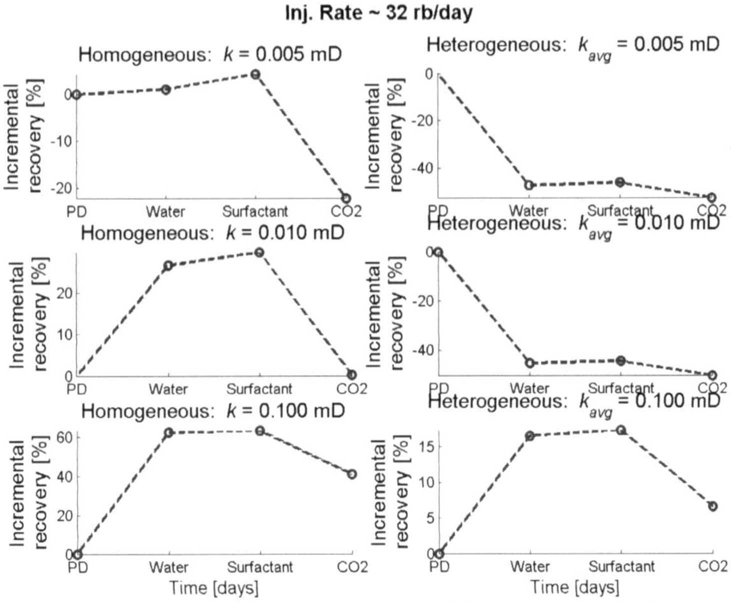

Figure 13: Incremental recovery (IR %) with respect to primary depletion (base case) for the recovery values plotted in Figure 9.

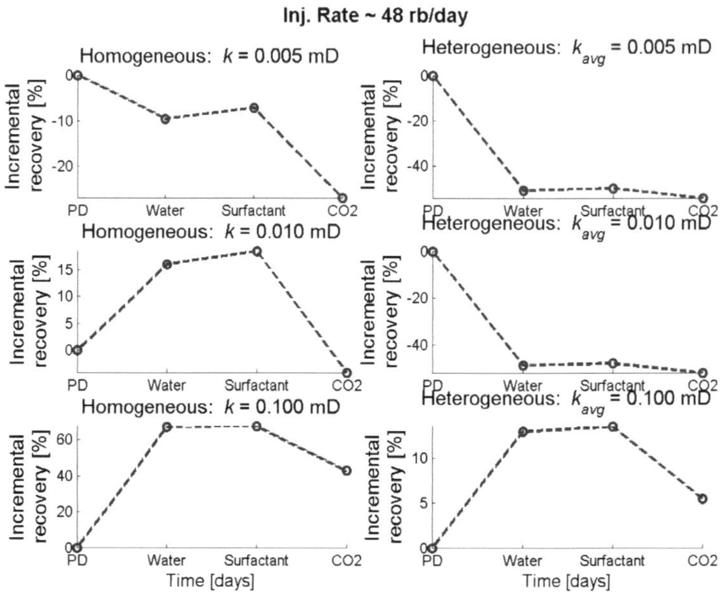

Figure 14: Incremental recovery (IR %) with respect to primary depletion (base case) for the recovery values plotted in Figure 10.

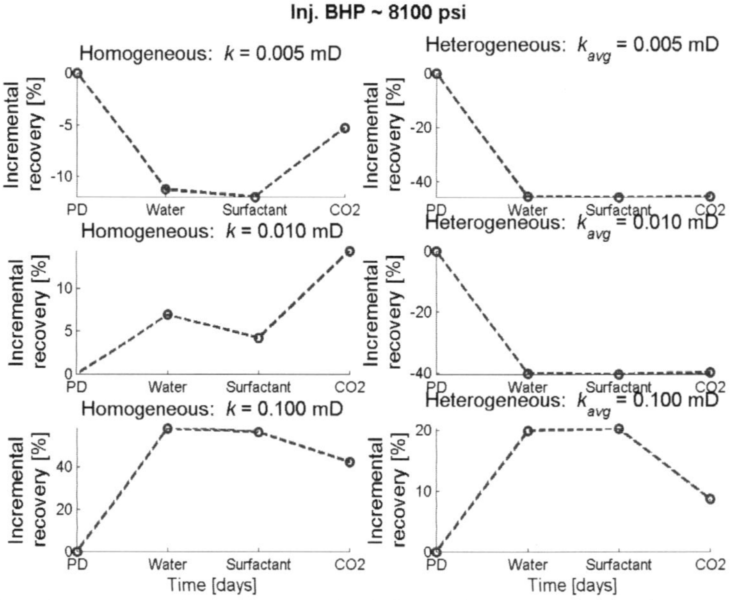

Figure 15: Incremental recovery (IR %) with respect to primary depletion (base case) for the recovery values plotted in Figure 11.

Impact of an Additional Lateral

To study the impact of number of laterals and hydraulic fractures on the incremental recovery, we replace a single lateral in the middle with two similar laterals (each intersecting six hydraulic fractures) at equal distance from the boundaries. For detailed results with two laterals, readers may refer Singh and Cai (2018). The results presented in Singh and Cai (2018) show that the percent of incremental recovery due to fluid injection is not affected by having an additional lateral in a homogeneous medium. Although the percent of IR is affected in heterogeneous medium due to relative change in spatial heterogeneity with two laterals as compared to a single lateral in the middle (Figure 2), the behavior of incremental recovery due to injection of each fluid (water, surfactant, or immiscible CO_2) is still the same; that is, the relative performance exhibited by each fluid in increasing fluid recovery with respect to the base case still shows the same behavior despite the relative change in heterogeneity.

3.4. Effect of Surfactant Concentration

The surfactant concentration used for this study is 50 kg/m^3 (or 5% by mass of water), which provides minimum surface tension and maximum miscibility factor of 1 as shown by Table IV. The minimum surface tension and maximum miscibility is obtained with surfactant concentration of 30 kg/m^3 (or 3% by mass of water). Any surfactant concentration above this value did not change the recovery values.

3.5. Effect of Capillary Pressure

Capillary pressure curve for water-oil and gas-liquid was obtained by using the laboratory-scale capillary pressure data from middle Bakken cores (Karimi and Kazemi, 2015) and extrapolating them by fitting a curve over the data. The expression obtained after extrapolating the data beyond the water saturation range of ~62-81% is shown as following:

$$p_c(psi) = \begin{cases} (-9.34 \times 10^{-42}).\, e^{143.3.S_w}, & water\ displacing\ oil \\ (7.78 \times 10^{3}).\, e^{-4.91.S_o}, & gas\ displacing\ oil \end{cases} \tag{8}$$

The capillary pressure at in-situ oil saturation of 80% as given by this curve is negligible in comparison to the initial reservoir pressure of ~8000 psi. Even after producing for 20 years, the pressure did not drop enough to be comparable to the highest possible value of capillary pressure by the above expression. Therefore, using the above curve as capillary pressure for the reservoir fluids in a model (mimicking properties of Bakken) did not affect the recovery estimates. The reason why capillary pressure did not play a role in affecting recovery is because the pore space that is inaccessible is due to its small permeability, and not because of low pressure support. The average pressure support in the reservoir even by the end of 20 years of production is much higher that can overcome capillary forces.

It should be noted that, although, there are some studies (e.g. Zhang et al., 2017) that show numerically that capillary pressure can have significant impact on oil recovery (upto 20 % higher), their methodology uses Young-Laplace equation for the estimates of capillary pressure, which usually gives impractical results for shales as discussed in a review by Singh (2016).

4. Discussions

Results suggest that fluid injection can improve recovery over the base case (with no injection), but, the recovery is not improved universally across all the studied scenarios. Fluid injection did not improve recovery (negative IR %) in 10 out of 24 cases studied, where 8 out of those 10 unsuccessful cases were from heterogeneous media with very low (0.005 mD) to low (0.01 mD) average matrix permeability. The remaining 2 unsuccessful cases are for very low (0.005 mD) matrix permeability in homogeneous media with very high injection rate (~48 rb/d), or when the production is constrained by a constant BHP, respectively. In the 14 successful or improved recovery (positive IR %) cases where injecting fluids improved the recovery, surfactant injection performed the best in 3 cases while immiscible CO_2 injection performed the best in 1 case. In the remaining 10 out of the 14 improved recovery cases, there was no clear winner among the three fluids; in most cases, surfactant and water performed almost equally, but better than immiscible CO_2.

4.1. Effect of Matrix Permeability

An increase in matrix permeability results in increasing oil recovery for all the recovery methods, including primary depletion, and for the three methods of recovery by fluid injection. This impact is consistent irrespective of injection rate or well constraint, and the relative change in oil recovery between different recovery methods increases with the increase in matrix permeability as shown in Figure 12, Figure 13, Figure 14, and Figure 15.

4.2. Effect of Injection Rate

An increase in injection rate leads to decrease in oil recovery matrix with very low (0.005 mD) to low (0.01 mD) matrix permeability, but it leads to increase in oil recovery for matrix with relatively higher permeability (0.1 mD). This can be seen in in Figure 12, Figure 13, Figure 14, and Figure 15, which show that at small injection rate (~ 7 rb/d) the recovery by injection for reservoir with very low (0.005 mD) to low (0.01 mD) matrix permeability is greater than recovery by primary depletion; but, with increase in injection rates (~ 32 rb/d and 48 rb/d) there is a relative drop in recovery for reservoir with very low (0.005 mD) to low (0.01 mD) matrix permeability. However, for reservoir with relatively high (0.1 mD) matrix permeability, the recovery increases relatively with increase in injection rate from 7 rb/d to 48 rb/d. This is possibly due to the fact that high injection rate (equivalent to relatively high convective flow and low diffusion) is useful only when the rock is sufficiently permeable that it allows the fluid flow; whereas low injection rate (equivalent to relatively low convective flow and high diffusion) is useful when the rock is extremely tight that the primary mode of flow is through diffusion.

4.3. Effect of Injection Constrained by Constant BHP

An injector constrained by a constant BHP is relatively more effective in improving recovery in a reservoir that has a relatively high matrix permeability (0.1 mD) than in a reservoir having a very low (0.005 mD) to low (0.01 mD) matrix permeability. Injector constrained by a constant BHP is most effective for immiscible CO_2 injection, which tends to perform better with constant BHP than with constant injection rate. An injector constrained by a constant BHP performs better with the injection of immiscible CO_2 than with water or surfactant when the reservoir matrix permeability is very low (0.005 mD) to low (0.01 mD); whereas, effectiveness of immiscible CO_2 in case of reservoir with relatively high (0.1 mD) matrix permeability is not as good as water or surfactant, but it's still relatively better than its performance in the cases where injection is constrained by a constant rate.

4.4. Effect of Heterogeneity

Injecting fluids in reservoirs with very low (0.005 mD) to low (0.01 mD) average matrix permeability does not lead to a better recovery. In fact, switching three fractures from producers to injectors for the purpose of injecting fluids in these very low (0.005 mD) to low (0.01 mD) permeability reservoirs lead to a net decrease in recovery relative to primary depletion; that is, producing through all six fractures with no injection leads to a higher recovery than by injecting fluids through half of the fractures. However, injecting fluids in reservoirs with relatively larger average matrix permeability of 0.1 mD does lead to a better recovery compared to primary depletion; lower injection rates are relatively more effective than a higher injection rate in increasing the recovery, whereas production constrained by a constant BHP is even better than low injection rates to increase recovery rate.

4.5. Effect of Type of Fluid

Results suggest that fluid injection improved recovery (positive IR %) in 14 out of 24 cases studied, where surfactant injection performed the best in 3 cases while immiscible CO_2 injection performed the best in 1 case. In the remaining 10 IR cases, there was no clear winner among the three fluids; in 9 cases out of 10, surfactant and water performed almost equally, but better than immiscible CO_2, while only in one case all the three injection fluids (water, surfactant, and immiscible CO_2) performed equally.

The 10 cases where injecting fluids did not improve recovery relative to primary depletion are:

1) Heterogeneous reservoirs with very low (0.005 mD) to low (0.01 mD) average matrix permeability irrespective of well constraint or injection rate [8 cases]
2) Very high injection rates (~48 rb/d) in reservoir with very low (0.005 mD) matrix permeability [1 cases], and
3) Production constrained by a constant BHP in reservoir with very low (0.005 mD) matrix permeability [1 case].

The 14 cases where injecting fluids improved the recovery relative to primary depletion are:

1) Surfactant injection performed the best in reservoir with very low (0.005 mD) matrix permeability with high injection rate (~32 rb/d) [1 case], and in low (0.01 mD) matrix permeability with high to very high injection rates (~32 rb/d and 48 rb/d) [2 cases]. The improvement in recovery by surfactant ranged between 1-3% higher than water, relative to primary depletion.
2) Immiscible CO_2 injection performed the best in reservoir with low (0.01 mD) matrix permeability when the production is constrained by a constant BHP [1 case].
3) In the remaining 10 IR cases, there was no clear winner among the three fluids; in 9 cases out of 10, surfactant and water performed almost equally, but better than immiscible CO_2, while only in one case all the three injection fluids (water, surfactant, and immiscible CO_2) performed equally.

In general, the oil rates by water (or surfactant) get a larger bump in rates initially than by immiscible CO_2 (as shown by Figure 4 through Figure 7) because of earlier breakthrough of immiscible CO_2 in comparison to water (or surfactant) by few years in most cases. This is because gas being a lower viscosity fluid than the water (or surfactant) will have a higher velocity (by Darcy's law) and that results in its earlier breakthrough.

A brief summary of above results is presented in form of matrix (Table VI and Table VII) to compare the performance of methods used for oil recovery in a homogeneous and heterogeneous medium, respectively. Scenarios marked as X is shown in red colored box to indicate that injection of fluid did not help improve recovery for that particular scenario. Also, scenarios marked with 'Water or Surf' (or 'Water or Surf or Immiscible CO_2') indicate that there was no single injection fluid that performed the best in that particular scenario, and that any one of the injection fluids mentioned in that scenario will result in a similar recovery as obtained by the rest of the fluids mentioned for that scenario.

Table VI: Recovery technique for each scenario that recovered the most oil in a homogeneous medium. Scenarios marked as X is shown in red colored box to highlight that injection of fluid did not improve recovery.

HOMOGENEOUS	*Low Injection Rate*	*High Injection Rate*	*Very High Injection Rate*	*Constrained by BHP*
Very Low Permeability	Water or Surf	Surfactant	X	X
Low Permeability	Water or Surf	Surfactant	Surfactant	Immiscible CO_2
High Permeability	Water or Surf or Immiscible CO_2	Water or Surf	Water or Surf	Water or Surf

Table VII: Recovery technique for each scenario that recovered the most oil in a heterogeneous medium. Scenarios marked as X is shown in red colored box to highlight that injection of fluid did not improve recovery.

HETEROGENEOUS	*Low Injection Rate*	*High Injection Rate*	*Very High Injection Rate*	*Constrained by BHP*
Very Low Permeability	X	X	X	X
Low Permeability	X	X	X	X
High Permeability	Water or Surf	Water or Surf	Water or Surf	Water or Surf

4.6. Relating Recovery with Mechanism of Injected Fluid

Recovery with Water and Immiscible CO_2

In general, the oil rates by water (or surfactant) get a larger bump in rates initially than by immiscible CO_2 (as shown by Figure 4 through Figure 7) because of earlier breakthrough of immiscible CO_2 in comparison to water (or surfactant) by few years in most cases. This is because gas being a lower viscosity fluid than the water (or surfactant) will have a higher velocity (by Darcy's law) and that results in its earlier breakthrough. Both water flooding and immiscible CO_2 flooding work by increasing the pressure that drives the oil out by immiscible displacement. Although, the relative permeability of gas with oil is different than the relative permeability of water with oil, it does not unilaterally dictate the oil recovery; this is because, relative permeability of oil with gas favors higher mobility of both the gas and the oil, which is good for driving the oil out faster, however, it also simultaneously results in faster breakthrough of gas due to its higher mobility. This is especially important in a tight-oil reservoir where the tight pore space dominates the reservoir, and even a slightly higher permeable pathways can lead to preferential pathway for the gas that may result in its breakthrough. In the current injection-production scheme, the gas is injected through the fractures, and it tends to breakthrough from the other fractures that are working as production zones. However, water being a heavier fluid with about ten times higher viscosity than the CO_2, is not likely to breakthrough as fast as gas because of its much lower velocity compared to CO_2 ($\propto \mu_{gas}/\mu_{water}$~0.1). Therefore, without early breakthrough, water is able to sustain higher pressure than CO_2 for a longer period of time.

Recovery with Surfactant

Unlike water and immiscible CO_2 flooding, surfactant flooding works by mixing with the oil to form a miscible solution that reduces the surface tension of the oil and helps bring down the capillary forces; therefore, reduced capillary forces push the oil out from the small pores that hold the oil due to high capillary pressure. Therefore, the success of a surfactant flooding in extracting oil depends not just on its chemical properties that can reduce the surface tension of oil (also referred as wettability alteration), but also on the amount of oil trapped by capillary forces. If the average reservoir pressure is sufficiently high, such as in the studied scenario (and also the scenario with most tight-oil formations, which are over pressured) the amount of oil that can be displaced by surfactant depends on the capability of the surfactant to first access the tight rock to mobilize additional oil for recovery. In conventional reservoirs, surfactant flooding generally performs better over water flooding because the surfactant is able to access the pores that are otherwise inaccessible due to high capillary pressure with water. Although, surfactant results in favorable fluid properties to mobilize oil, the reason surfactant did not perform well in tight-oil formation studied here is because the unswept oil is inaccessible due to small permeability, and not because of high capillary pressure. The average pressure support in the reservoir even by the end of 20 years of production is much higher that can overcome capillary forces. Even though the surfactant performs better in 2 cases out of 24 cases, the increase in recovery (1-3% higher than water) is not sufficiently higher to warrant its supremacy over water in tight-oil formations, primarily because it cannot access the tight pore space.

5. Summary and Conclusions

Production rates in tight-oil reservoirs decline rapidly within a year after the well is put to production. To arrest the declining production rates and recover additional oil from tight-oil resources, traditional methods of IOR (water flooding, immiscible CO_2 flooding, surfactant flooding) were investigated. To overcome the challenge of negligible fluid movement and sweep outside the fractures, the fluid was injected through a set of fractures and produced through another set of fractures intersecting the same lateral. We compare the estimates of oil recovery obtained by injecting water, surfactant, and immiscible CO_2, with the oil recovered by primary depletion (with no injection).

Results suggest that fluid injection did not improve recovery in 10 out of 24 cases studied, where 8 out of those 10 unsuccessful cases were from heterogeneous medium with very low (0.005 mD) to low (0.01 mD) average matrix permeability. The remaining 2 unsuccessful cases are for very low (0.005 mD) matrix permeability in homogeneous media with very high injection rate (~48 rb/d) and when the injection is constrained by a constant BHP, respectively. In the 14 successful or improved recovery cases, surfactant injection performed the best in 3 cases while immiscible CO_2 injection performed the best in 1 case. In the remaining 10 improved recovery cases, there was no clear winner among the three fluids; in 9 cases out of 10, surfactant and water performed almost equally, but better than immiscible CO_2, while only in one case all the three injection fluids (water, surfactant, and immiscible CO_2) performed similarly.

In conclusion, the results summarized in Table VI and Table VII suggest that an enhanced recovery method (e.g. water, immiscible CO_2, or surfactant) may not always improve recovery from tight-oil reservoirs, especially in a heterogeneous reservoir whose average matrix permeability is less than a threshold of 0.01 mD. Although, surfactant results in favorable fluid properties to mobilize oil, the reason surfactant did not perform well in tight-oil formation studied here is because the unswept oil is inaccessible due to small permeability, and not because of high capillary pressure. Therefore, it can be said that reservoir heterogeneity coupled with the average matrix permeability of a tight-oil reservoir seems to play the most significant role in exploring the potential of enhanced recovery.

There have been some lab-scale EOR studies (Li and Sheng, 2016, 2017; Li et al., 2017) on tight cores, but we should be cautious in directly adapting the results from lab-scale studies; that is because, unlike EOR from conventional reservoirs where reservoir heterogeneity is the primary variable that decides the performance of the recovery, there are multiple factors that may decide the performance of an EOR in tight-oil formations. One of those factors, besides other factors (e.g. fractures), is reservoir heterogeneity coupled with the average matrix permeability, which seems to play the most important role in the feasibility of an EOR method as discussed above. Most of the lab-scale studies look at performance of a certain EOR method on a core that is fairly homogeneous (with no permeability variation) and do not account for other important parameters (e.g. fractures) that may impact the performance of some EOR methods, specifically the EOR by gas injection. Therefore, we must be careful in adapting the results of those studies to support field scale response.

6. Acknowledgment

This research was supported in part by an appointment to the National Energy Technology Laboratory Research Participation Program, sponsored by the U.S. Department of Energy and administered by the Oak Ridge Institute for Science and Education.

7. References

Aloulou, F. (2016). World tight oil production to more than double from 2015 to 2040.

Andersen, O. (2017). Simplified models for numerical simulation of geological CO2 storage. The University of Bergen.

Delaihdem, D.K. (2013). Decline curve analysis and enhanced shale oil recovery based on Eagle Ford Shale data. Thesis. University of Alaska Fairbanks.

Deutsch, C.V., and Journel, A.G. (1997). GSLIB: Geostatistical Software Library and User's Guide (Oxford University Press, USA).

Dombrowski, Fonseca, Karanikas, Moya, and Lamantia (2015). Fluid injection in light tight oil reservoirs. US Patent US20130228337.

Dong, T., Harris, N.B., Ayranci, K., Twemlow, C.E., and Nassichuk, B.R. (2017). The impact of composition on pore throat size and permeability in high maturity shales: Middle and Upper Devonian Horn River Group, northeastern British Columbia, Canada. Marine and Petroleum Geology *81*, 220–236.

Flannery, J. (2006). Integrated Analysis of the Bakken Petroleum System, U.S. Williston Basin. pp. 138–145.

Gaswirth, S.B., and Marra, K.R. (2015). U.S. Geological Survey 2013 assessment of undiscovered resources in the Bakken and Three Forks Formations of the U.S. Williston Basin Province. AAPG Bulletin *99*, 639–660.

Gaswirth, S.B., Marra, K.R., Cook, T.A., Charpentier, R.R., Gautier, D.L., Higley, D.K., Klett, T.R., Lewan, M.D., Lillis, P.G., Schenk, C.J., et al. (2013). Assessment of undiscovered oil resources in the Bakken and Three Forks Formations, Williston Basin Province, Montana, North Dakota, and South Dakota, 2013 (Reston, VA: U.S. Geological Survey).

Ghaderi, S.M., Clarkson, C.R., Ghanizadeh, A., Barry, K., and Fiorentino, R. (2017). Improved Oil Recovery in Tight Oil Formations: Results of Water Injection Operations and Gas Injection Sensitivities in the Bakken Formation of Southeast Saskatchewan. (Society of Petroleum Engineers), p. SPE-185030-MS.

Gong, X. (2013). Assessment of Eagle Ford Shale Oil and Gas Resources. Thesis. Texas A&M University.

Hawthorne, S.B., Miller, D.J., Grabanski, C.B., Sorensen, J.A., Pekot, L.J., Kurz, B.A., Gorecki, C.D., Steadman, E.N., Harju, J.A., and Melzer, S. (2017a). Measured Crude Oil MMPs with Pure and Mixed CO2, Methane, and Ethane, and Their Relevance to Enhanced Oil Recovery from Middle Bakken and Bakken Shales. (Society of Petroleum Engineers), p.

Hawthorne, S.B., Jin, L., Kurz, B.A., Miller, D.J., Grabanski, C.B., Sorensen, J.A., Pekot, L.J., Bosshart, N.W., Smith, S.A., Burton-Kelly, M.E., et al. (2017b). Integrating Petrographic and Petrophysical Analyses with CO2 Permeation and Oil Extraction and Recovery in the Bakken Tight Oil Formation. (Society of Petroleum Engineers), p.

He, K., Xu, L., Gao, Y., Neeves, K.B., Yin, X., Bai, B., Ma, Y., and Smith, J. (2014). Validating Surfactant Performance in the Eagle Ford Shale: A Correlation between the Reservoir-on-a-Chip Approach and Enhanced Well Productivity. (Society of Petroleum Engineers), p.

Honarpour, M.M., Nagarajan, N.R., Orangi, A., Arasteh, F., and Yao, Z. (2012). Characterization of Critical Fluid PVT, Rock, and Rock-Fluid Properties - Impact on Reservoir Performance of Liquid Rich Shales. (Society of Petroleum Engineers), p.

Hustak, C., Dieva, R., Baker, R., MacIsaac, B., Frankiw, K., and Clark, B. (2017). Waterflooding a Multi-layered Tight Oil Reservoir Developed with Hydraulically Fractured Horizontal Wells. (Society of Petroleum Engineers), p. SPE-185021-MS.

Jørgensen, K. (2013). Implementation of a Surfactant Model in MRST With Basis in Schlumberger's Eclipse. Norwegian University of Science and Technology.

Karimi, S., and Kazemi, H. (2015). Capillary Pressure Measurement using Reservoir Fluids in a Middle Bakken Core. (Society of Petroleum Engineers), p.

Kenomore, M., Hassan, M., Dhakal, H., and Shah, A. (2017). Total organic carbon evaluation of the Bowland Shale Formation in the Upper Bowland of the Widmerpool Gulf. Journal of Petroleum Science and Engineering *150*, 137–145.

Li, L., and Sheng, J.J. (2016). Experimental study of core size effect on CH4 huff-n-puff enhanced oil recovery in liquid-rich shale reservoirs. Journal of Natural Gas Science and Engineering *34*, 1392–1402.

Li, L., and Sheng, J.J. (2017). Numerical analysis of cyclic CH4 injection in liquid-rich shale reservoirs based on the experiments using different-diameter shale cores and crude oil. Journal of Natural Gas Science and Engineering *39*, 1–14.

Li, S., and Luo, P. (2017). Experimental and simulation determination of minimum miscibility pressure for a Bakken tight oil and different injection gases. Petroleum *3*, 79–86.

Li, L., Sheng, J.J., Xu, J., and others (2017). Gas Selection for Huff-n-Puff EOR in Shale Oil Reservoirs Based upon Experimental and Numerical Study. In SPE Unconventional Resources Conference, (Society of Petroleum Engineers), p. SPE-185066-MS.

Lie, K.-A. (2016). An introduction to reservoir simulation using MATLAB: User guide for the Matlab Reservoir Simulation Toolbox (MRST). SINTEF ICT. http://www.sintef.no/Projectweb/MRST/Publications.

Liu, R., Liu, H., Li, X., Wang, J., and Pang, C. (2010). Calculation of Oil and Water Relative Permeability for Extra Low Permeability Reservoir. (Society of Petroleum Engineers), p.

Lund, L. (2014). Decline curve analysis of shale oil production: The case of Eagle Ford. Uppsala University.

Mavor, M. (2013). Reservoir Fluid Properties Required for Low Permeability Oil Reservoir Analysis, by Matt Mavor; #90186 (2013). In Hydrocarbon Charge Considerations in Liquid-Rich Unconventional Petroleum Systems, (Canada: AAPG), p.

Mirchi, V., Saraji, S., Goual, L., and Piri, M. (2015). Dynamic interfacial tension and wettability of shale in the presence of surfactants at reservoir conditions. Fuel *148*, 127–138.

Moghadasi, L., Guadagnini, A., Inzoli, F., and Bartosek, M. (2015). Interpretation of two-phase relative permeability curves through multiple formulations and Model Quality criteria. Journal of Petroleum Science and Engineering *135*, 738–749.

Rystad Energy (2014). SHALE COMPANIES HAVE PROVEN MORE EFFICIENT BUT NOT MORE PRODUCTIVE (Rystad Energy).

Sanchez-Rivera, D., Mohanty, K., and Balhoff, M. (2015). Reservoir simulation and optimization of Huff-and-Puff operations in the Bakken Shale. Fuel *147*, 82–94.

Sheng, J.J. (2015). Status of surfactant EOR technology. Petroleum *1*, 97–105.

Shuler, P.J., Lu, Z., Ma, Q., and Tang, Y. (2016). Surfactant Huff-n-Puff Application Potentials for Unconventional Reservoirs. (Society of Petroleum Engineers), p.

Singh, H. (2016). A Critical Review of Water Uptake by Shales. Journal of Natural Gas Science and Engineering *34*, 751–766.

Singh, H., and Cai, J. (2018). Screening improved recovery methods in tight-oil formations by injecting and producing through fractures. International Journal of Heat and Mass Transfer *116*, 977–993.

Steiner, S., Ahsan, S.A., Raina, I., Dasgupta, S., and Lis, G.P. (2016). Interpreting Total Organic Carbon TOC in Source Rock Oil Plays. (Society of Petroleum Engineers), p.

Tuero, F., Crotti M, M., and Labayen, I. (2017). Water Imbibition EOR Proposal for Shale Oil Scenarios. (Society of Petroleum Engineers), p.

US EIA (2016). Annual Energy Outlook 2016 (Government Printing Office).

Wachtmeister, H., Lund, L., Aleklett, K., and Höök, M. (2017). Production Decline Curves of Tight Oil Wells in Eagle Ford Shale. Nat Resour Res *26*, 365–377.

Wang, D., Butler, R., Liu, H., and Ahmed, S. (2011). Surfactant Formulation Study For Bakken Shale Imbibition. (Society of Petroleum Engineers), p.

Wang, D., Butler, R., Zhang, J., and Seright, R. (2012). Wettability Survey in Bakken Shale With Surfactant-Formulation Imbibition. SPE Reservoir Evaluation & Engineering *15*, 695–705.

Wang, D., Zhang, J., Butler, R., and Olatunji, K. (2016). Scaling Laboratory-Data Surfactant-Imbibition Rates to the Field in Fractured-Shale Formations. SPE Reservoir Evaluation & Engineering *19*, 440–449.

Witkowsky, J.M., Galford, J.E., Quirein, J.A., and Truax, J.A. (2012). Predicting Pyrite and Total Organic Carbon from Well Logs for Enhancing Shale Reservoir Interpretation. (Society of Petroleum Engineers), p.

Xu, F., Guo, X., Wang, W., Zhang, N., Jia, S., and Wang, X. (2011). Case Study: Numerical Simulation Of Surfactant Flooding In Low Permeability Oil Filed. (Society of Petroleum Engineers), p.

Yu, W., Lashgari, H., and Sepehrnoori, K. (2014). Simulation Study of CO2 Huff-n-Puff Process in Bakken Tight Oil Reservoirs. (Society of Petroleum Engineers), p.

Zhang, K., Seetahal, S., Alexander, D., Lv, J., Hu, Y., Lu, X., Zhang, D., Wu, K., and Chen, Z. (2016). Effect of Confinement on Gas and Oil Relative Permeability During CO2 Flooding in Tight Oil Reservoirs. (Society of Petroleum Engineers), p.

Zhang, Y., Yu, W., Sepehrnoori, K., and Di, Y. (2017). A Comprehensive Numerical Model for Simulating Fluid Transport in Nanopores. Sci Rep *7*.

Zhu, P., Balhoff, M.T., and Mohanty, K.K. (2017). Compositional modeling of fracture-to-fracture miscible gas injection in an oil-rich shale. Journal of Petroleum Science and Engineering *152*, 628–638.

Zuloaga-Molero, P., Yu, W., Xu, Y., Sepehrnoori, K., and Li, B. (2016). Simulation Study of CO2-EOR in Tight Oil Reservoirs with Complex Fracture Geometries. Sci Rep *6*.

Appendix

SURFACTANT MODEL (FROM JØRGENSEN, 2013)

$$f(N_c) = \begin{cases} N_c^{misc};\ k_\alpha^{misc}(s_\alpha);\ s_{\alpha,res}^{misc}, & max(N_c) \\ N_c^{immisc};\ k_\alpha^{immisc}(s_\alpha);\ s_{\alpha,res}^{immisc}, & min(N_c) \end{cases} \quad \text{(A-1)}$$

Where, $\alpha \in \{w, o\}$

The residual saturation values for the relative permeability are obtained by weighted average of residual saturation values at immiscible (without surfactant) and fully miscible conditions.

$$s_{\alpha,res}^{N_c} = m.\, s_{\alpha,res}^{misc} + (1-m).\, s_{\alpha,res}^{immisc} \qquad \text{(A-2)}$$

In addition to reducing the IFT and residual oil saturation, surfactant also changes the relative permeability due to miscible flow such that the effective relative permeability after adding surfactant is the weighted average of the relative permeabilities with immiscible (without surfactant) and with fully miscible conditions.

$$k_{r\alpha}(s_\alpha) = m.\, k_{r\alpha}^{misc}\left(\bar{s}_\alpha^{misc}, N_c\right) + (1-m).\, k_{r\alpha}^{immisc}\left(\bar{s}_\alpha^{immisc}, N_c\right) \qquad \text{(A-3)}$$

The discrete relative permeability values cannot be directly scaled (a continuous function) to match the new end-points, therefore, it require modified saturation values that are calculated as following.

$$\bar{s}_\alpha^{misc}(N_c) = \left(\frac{s_\alpha - s_{\alpha,res}^{N_c}}{1 - s_{\alpha,res}^{N_c} - s_{\alpha',res}^{N_c}}\right).\left(1 - s_{\alpha,res}^{misc} - s_{\alpha',res}^{misc}\right) + s_{\alpha,res}^{misc} \qquad \text{(A-4)}$$

$$\bar{s}_\alpha^{immisc}(N_c) = \left(\frac{s_\alpha - s_{\alpha,res}^{N_c}}{1 - s_{\alpha,res}^{N_c} - s_{\alpha',res}^{N_c}}\right).\left(1 - s_{\alpha,res}^{immisc} - s_{\alpha',res}^{immisc}\right) + s_{\alpha,res}^{immisc} \qquad \text{(A-5)}$$

Also, addition of surfactant to the water increases its viscosity in a way that the effective viscosity of water after adding surfactant is given as:

$$\mu_w^{surf}(c_s, p) = \frac{\mu_w(p)}{\mu_w(p_{ref})} \cdot \mu_s(c_s) \qquad \text{(A-6)}$$

505232 Aerosol Synthesis of Materials for Sunlight Harvesting Applications

Abstract:

Research Interests:

Research abstract:

With the increase in the greenhouse gas emission from conventional energy sources and concerns about global warming, the demand for renewable energy is growing rapidly. Sunlight is one of the cheapest and abundant sources of energy available in nature. However, the high cost of the most efficient and commercialized silicon solar cells makes it challenging to take full advantage of the freely available sunlight. Therefore, it is important to develop cheap, efficient, and stable devices to harvest solar energy. My research is focused on the synthesis of materials via novel aerosol generation techniques (at Aerosol and Air Quality Research Laboratory), advanced material characterization (at Argonne National Laboratory) and their application in sunlight harvesting.

Aerosol-based techniques are simple, scalable, and operate at atmospheric pressure. Moreover, they can be operated continuously and allow precise control of material properties (size, morphology, crystal structure, composition), which are the challenges for earlier solution-based techniques. I will discuss one such aerosol-based technique, electrohydrodynamic atomization (i.e., electrospray), for fabricating materials with desired properties for efficient sunlight harvesting. Specific examples in brief are:

1. Directed-assembly of Thylakoid membrane: Thylakoid membrane contains light harvesting complexes, Photosystem (PS) I and II. This project focuses on re-engineering the naturally occurring photosynthetic thylakoid membrane using electrospray to fabricate an artificial photo-electrochemical cell for solar energy conversion.

- **Kavadiya, S.**, Chadha, T. S., Liu, H., Shah, V. B., Blankenship, R. E., and Biswas, P., *"Directed assembly of the thylakoid membrane on nanostructured TiO_2 for a photo-electrochemical cell"*, Nanoscale, 8, 1868, 2016

2. Improving stability of perovskite solar cells under humid environment: Recently developed highly efficient and low-cost perovskite ($CH_3NH_3PbI_3$) solar cells (PSCs) are vulnerable in ambient humid environment. Our electrospray technique makes stable PSCs by controlling the rate of perovskite formation precisely and under ambient conditions. The perovskite film fabricated using electrospray is highly smooth and moisture-resistant.

- **Kavadiya, S.**, Niedzwiedzki, D. M., Huang, S., and Biswas, P., *"Electrospray-assisted Fabrication of Highly Stable and Efficient Perovskite Solar Cells at Ambient Conditions"*, Advanced Energy Materials, doi: 10.1002/aenm.201700210, 2017

- Carvalho, B. A. de*, **Kavadiya, S.***, Huang, S., Niedzwiedzki, D. M., and Biswas, P., *"Highly stable perovskite solar cells fabricated under humid ambient conditions"*, IEEE Journal of Photovoltaics, 7, 532, 2017 (*equal contribution)

3. Crumpling of 2D graphene oxide (GO) sheet: Graphene oxide has numerous applications in almost every field, because of its remarkable properties. However, the 2D sheets tend to restack and lose the surface area, hence hinder their efficient utilization. Using electrospray technique, we crumple the 2D sheets into stable nanoscale 3D structure at room temperature, preserving the surface area to a significant extent. We also synthesize composite materials with CGO and utilize these nanomaterials for efficient conversion of carbon dioxide to useful fuels (CO, CH_4) and sensing application.

- **Kavadiya, S.**, Raliya, R., Schrock, M., and Biswas, P., *"Crumpling of graphene oxide through evaporative confinement in nano-droplets produced by electro-hydrodynamic aerosolization"*, Journal of Nanoparticle Research, 19: 43, 2017
- Lin, L-Y., Nie, Y., **Kavadiya, S.**, and Biswas, P., *"N-doped reduced graphene oxide promoted nano TiO_2 as a bifunctional adsorbent/photocatalyst for CO_2 photoreduction: Effect of N species"*, Chemical Engineering Journal, 316, 449, 2017

Overall, based on PhD research work, I gained experience in material synthesis, advance level characterization, and device fabrication. For the above-mentioned work, I received a doctoral student research award from the Department of Energy, Environmental and Chemical Engineering at Washington University in St. Louis.

Research interest:

In future, as a faculty, I would like to continue applying aerosol-based methods to synthesize materials for various applications. I believe that aerosol-based techniques have large potential to prepare materials with desired properties (self-assembled structure in the first project, moisture-resistance smooth film in the second project and crumpled GO structure in the third project). Furthermore, aerosol-based techniques are scalable for large-scale production and efficiently use the precursor material, reducing material loss significantly. I will also try to advance the technique I develop from lab-scale to larger-scale for commercial use. I will describe my detail research plan during the poster session. Below are few proposed future research projects in brief:

i. Hydrophobic crumpled graphene-perovskite composite: I have experience with synthesis and application of both graphene-based composite materials and perovskite. Furthermore, aerosol based techniques are exceptionally good in synthesizing crumpled graphene-based composite materials. As mentioned earlier, perovskite are prone to moisture. The aim is to cover the perovskite with waterproof graphene sheet and make the perovskite hydrophobic. My expertise on aerosol-based synthesis of crumple graphene composites will be unique to execute this project.

ii. Large area perovskite solar cells: I have successfully fabricated stable PSCs using electrospray technique at ambient condition. It is very important to make efficient PSCs on larger scale. Other than the excellence of aerosol-based techniques in precisely controlling the material properties, these techniques are scalable for commercial application. My aim for this project is to scale up the electrospray deposition system to make large area stable perovskite solar cells under ambient condition.

Finally, I will combine the synthesis with the advanced characterization techniques through my collaboration at Argonne National Lab, to understand their formation mechanism and specific properties. Other than experimental characterization of the materials, I believe that the use of quantum mechanical modeling provides a fair idea about the material properties at atomic scale and optimizes the experimental efforts. Therefore, I foresee myself collaborating closely with computational groups and/or having a computational component in my own laboratory.

Teaching Interests:

Besides doing research work, I have also been interested in teaching. Through teaching, I want to ignite curiosity in students, teach them to think critically and develop essential skills. I started gaining teaching experience by doing teaching assistantship (TA) since my undergraduate education. During the past years in my PhD, I have TAed courses such as Engineering Analysis of Chemical Systems, Advanced thermodynamics and Material science, in which the first two included classroom teaching, organizing homework and exam material. A challenge to me in teaching is to teach undergraduate class from the base level while maintaining the flow. Therefore, I always be careful in choosing examples and try to connect the scientific principles to the daily life examples. I am enrolled in teaching citation program at the Washington University teaching center, which is focused on improving teaching skills through workshops, developing new teaching methods, applying them in class and coming up with the teaching philosophy. In future, as a faculty, I would like to teach fundamental courses on Chemical Processes (consisting mass, energy and momentum balance), Aerosol Science, and Material Science and Engineering.

Lastly, I also mentored three undergraduate students with very diverse background in their summer intern research projects in our lab and have journal publications with two of them. I am always enthusiastic to teach them the importance, capabilities and useful outcome from doing novel research work. I see myself getting better in mentoring students each year. I will continue to develop my teaching skills in both classroom teaching and in-person student mentorship.

505556 Harvesting, Coversion, and Direct Utilization of Solar Energy

Abstract:

Research Interests:

The emergence and success of photovoltaics for the conversion of solar to electrical energy has inspired increasing interest in the harvesting, conversion, and direct utilization of solar energy. Research in these areas typically requires using/developing materials that can effectively collect the energy of sunlight, studying and understanding the various pathways through which the collected energy can be dissipated, and exploiting or engineering a highly selective energy dissipation pathway through which some useful function can be performed. Tackling such projects requires a multi-faceted approach combining materials science, solid state physics, chemistry, and engineering. I am primarily interested in the conversion of solar/light energy into chemical energy (i.e. solar fuels) and the direct use of solar/light energy to drive chemical transformations (photocatalysis).

My doctoral research is focused on a specific class of light-harvesting materials known as plasmonic metal nanoparticles (nanoparticles of mainly Cu, Ag, and Au). I am studying the ability of these materials to collect the energy of visible light and channel that energy into the vibrational modes of molecules or into other metals for photocatalysis. Using surface-enhanced Raman spectroscopy, we showed that positioning strongly-interacting molecules near the surface of plasmonic metals unlocks a highly selective channel through which light energy collected by the plasmonic metal can be funneled into the vibrations of the molecule [1]. The new physical insights into energy harvesting and dissipation gained from this work inspired us to conceptualize the design of multimetallic plasmonic nanostructures in which light energy could be collected by the plasmonic metal and selectively transferred to catalytic sites on another metal to perform light-driven chemical reactions on a non-plasmonic metal surface [2]. Through working on these projects and others, I have become very well equipped with materials development and characterization skills, chemical analysis and characterization skills, and have developed a solid foundation of knowledge in catalysis and light-matter interactions.

[1] C. Boerigter, **U. Aslam**, S. Linic, *ACS Nano*, 10, 2016.

[2] **U. Aslam**, S. Chavez, S. Linic, *Nature Nanotechnology*, Accepted June 2017.

Teaching Interests:

My educational background is in chemical engineering and I am most-suited for and interested in teaching classes in reactions engineering, heterogeneous catalysis, separations, unit operations lab, thermodynamics, and mass and heat transfer. My teaching experience includes acting as a teaching assistant for a Calculus II class with recitation lectures taught by me twice a week, as a teaching assistant for a unit operations course in chemical engineering, as a teaching assistant for a fuel cells and fuel processes course, and as an informal teaching assistant for a fundamentals in heterogenous catalysis class. In addition, I have mentored 2 graduate students and 8 undergraduate students during my

time as a doctoral student.

I personally find teaching to be interesting in it of itself. It is challenging to construct easy-to-follow lectures that seamlessly flow from class-to-class. I believe students learn best when they are truly engaged during class (paying attention, thinking as they go, asking questions) and the best teachers make it as easy as possible for the students to do these things. The teacher must be energetic, reading the faces of the students, pausing to allow students to think and ask questions, openly asking questions to get the students to think, and finding ways to make the material as tangible and relatable as possible.

504408 Solution Processed Optoelectronics. Materials to Devices

Abstract:

Research Interests:

Research Experience. Solution processing has the potential to become a transformative technique for producing photovoltaics at both the low cost and large scale needed to meet energy demand, yet many challenges remain for both materials and devices. During my Ph.D. research, under the direction of Prof. Prashant Kamat and the University of Notre Dame, I applied ultrafast transient spectroscopy to probe charge recombination and charge transfer across semiconductor heterojunctions with specific relevance to solution-processed photovoltaics. In addition to these spectroscopic studies, I led our group's efforts investigating lead halide perovskite photovoltaics, developing novel device architectures and elucidating the mechanism of moisture-induced degradation. This work led to the publication of my thesis entitled, "Mesostructured Thin Film Solar Cells: Examining Hole Transfer Mechanisms and Device Stability" which earned me the Eli J. and Helen Shaheen Graduate School Award in Engineering.

During my postdoctoral work as an EERE Postdoctoral Fellow at the National Renewable Energy Laboratory (NREL), I have worked to address the challenges associated with solution processed halide perovskite solar cells. While these materials have already reached 22% efficiency, their stability remains a hurdle on the path toward commercialization. At NREL, I have utilized a suite of characterization techniques and formed numerous fruitful collaborations to investigate degradation processes in halide perovskite materials while simultaneously working to fundamentally improve their stability and performance. This has led to a greater understanding of degradation process and material properties, and led to the development of perovskite solar cells which remain largely unchanged despite 1000 hours of continual, unencapsulated operation.

Future Direction. I am excited by the potential of solution-processed optoelectronic devices, including photovoltaics, light emitting diodes, and dynamic windows, and the potential these devices have for transforming the energy sector. However, bringing solution-processed optoelectronic devices from the lab to the market requires a better understanding of the materials themselves, including the development of new materials, and solution-based formation/deposition processes that enable material formation and scalable device fabrication. As a faculty member I intend to develop a research program that addresses both of these materials challenges, materials and their formation, and applies this understanding to the fabrication of devices. Specifically, I intend to work with materials, particularly halide perovskites because of their GaAs-like optoelectronic properties, across length scales, from nanoparticles to large-area thin films and single crystals, to better understand the effects of surface chemistry, grain boundaries, and defect chemistry on material properties. This will enable the rational design of materials for optoelectronic devices. In tandem with this work, I aim to enable the scalable manufacturing of these materials and devices my moving beyond traditional material synthesis and deposition techniques. I will work to develop inherently scalable material formation techniques, including blade coating and spray coating, by understanding and controlling the kinetics and mechanisms of film formation. These two thrusts are highly complementary and will allow for iterative feedback within the research group as well as exposing new areas for investigation.

Teaching Interests:

During the 2014-2015 academic year, I served as the instructor for Introduction to Engineering Systems I & II at the University of Notre Dame, an introductory level engineering course of 40+ students. As one of seven instructors across all of the sections of this course, I collaborated with senior Notre Dame faculty members to develop lecture and exam materials, gaining valuable insight into both teaching and course design. In addition to this teaching experience, during my doctoral education I received guidance through teaching workshops hosted by the University of Notre Dame Kaneb Center for Teaching and Learning and was awarded the Advanced Teaching Scholar Certificate.

I enjoy devising unique examples and problems to help students learn difficult concepts. I frequently use in-class examples and activities to help students learn from their peers and help me better gauge student learning, allowing me to better address specific weaknesses in future lectures and avoid wasting time on material where they already show proficiency. In addition, I find this keeps lectures interactive, engaging, and unique from the textbook.

Beyond the classroom, I have participated in mentoring five undergraduate students and two high school teachers in their research. This has helped me develop my skills as a research mentor and has led to two second author research papers for undergraduates I mentored.

Selected Publications:

Christians, J. A.; Schulz, P.; Tinkham, J. S.; Schloemer, T. H.; Harvey, S. P.; Tremolet de Villers, B. J.; Sellinger, A.; Berry, J. J.; Luther, J. M. Tailored Interfaces of Unencapsulated Perovskite Solar Cells for >1000 Hours of Ambient Operational Stability. **2017**, *under review*.

Swarnkar, A.; Marshall, A. R.; Sanehira, E. M.; Chernomordik, B. D.; Moore, D. T.; **Christians, J. A.**; Chakrabarti, T.; Luther, J. M. Quantum Dot-Induced Phase Stabilization of α-$CsPbI_3$ Perovskite for High-Efficiency Photovoltaics. *Science*, **2016**, 354, 6308, 92-95.

Manser, J. S.; **Christians, J. A.**; Kamat, P. V. Intriguing Optoelectronic Properties of Metal Halide Perovskites. *Chem. Rev.* **2016**, 116, 12956-13008.

Nenon, D. P.†; **Christians, J. A.**†; Wheeler, L. M.; Blackburn, J. L.; Sanehira, E. M.; Dou, B.; Olsen, M. L.; Zhu, K.; Berry, J. J.; Luther, J. M. Structural and Chemical Evolution of Methylammonium Lead Halide Perovskites During Thermal Processing from Solution. *Energy & Environmental Science*, **2016**, *9*, 2072-2082.

†These authors contributed equally

Christians, J. A.; Miranda Herrera, P. A.; Kamat, P. V. Transformation of the Excited State and Photovoltaic Efficiency of $CH_3NH_3PbI_3$ Perovskite upon Controlled Exposure to Humidified Air. *J. Am. Chem. Soc.* **2015**, *137*, 1530-1538.

Christians, J. A.; Leighton Jr., D. T.; Kamat, P. V. Rate Limiting Interfacial Hole Transfer in Solid-State Solar Cells. *Energy Environ. Sci.* **2014**, *7*, 1148-1158.

Christians, J. A.; Fung, R. C. M.; Kamat, P. V. An Inorganic Hole Conductor for Organo-Lead Halide Perovskite Solar Cells. Improved Hole Conductivity with Copper Iodide. *J. Am. Chem. Soc.* **2014**, *136*, 758–764.

505385 Integrated Modeling for Solutions in Carbon Management

Abstract:

Research Interests:

Carbon management encompasses those technologies whereby carbon dioxide is prevented from release (carbon capture) removed from the atmosphere (carbon dioxide removal) and coupled to utilization or reliable storage. The advantage of an integrated approach to complex system modeling lies in the ability to simultaneously consider components from separate disciplines, such as material design and techno-economics. Expansion of model boundaries to include interdisciplinary constraints facilitates the engineering of solutions with greater precision than those produced in single-discipline or limited-discipline approaches. For example, advances in separation technologies typically focus on the lowering of energy and water intensity. However, considerations to regional air quality (humidity, contaminants), CO_2 source (purity, feasibility of waste-heat integration), scale, proximal reliable storage and utilization opportunities may produce a more accurate merit-order ranking of compatible technologies.

Research Experience:

My introduction to modeling came by way of a NASA fellowship involving *in-situ* resource utilization and the assessment of promoted-Ni and Fe surface analogs for the conversion of Martian CO_2. This project blossomed into a terrestrial examination of CH_4 selectivity in Fischer-Tropsch synthesis via density functional theory, both in cluster (Gaussian-type) and later periodic (plane-wave) calculations. This led to my involvement with the Clean Energy Conversions Laboratory (then at Stanford University, now with the Colorado School of Mines) which had also led some DFT investigations on Fe/Co Fischer-Tropsch processing. Though I began my career as a chemist, my experience has been largely rooted in engineering principles and approaches. I find it beneficial to have experience on "both sides", as it allows me to approach problems from different vantage points. At Stanford, I worked with the Graduate School of Business to increase my understanding of the techno-economic factors associated with small and large-scale CCS deployment. Further, I have worked alongside experimentalists to bolster the functional understanding of materials at the molecular level. These collective projects have brought me experience in surface chemistry and catalysis, simulations, molecular modeling, geographic systems modeling, techno-economic modeling, high-performance computing and coding. More importantly, they have helped shape my desire to use these approaches symbiotically and have solidified the importance of collaborations within and outside of my field.

Teaching Interests:

In addition to my research, I have significant experience in teaching. I enjoyed early success as a graduate student, earning an annual TA teaching award (2011) and career TA teaching award (2014). As a result of this success, I was offered an opportunity to teach two undergraduate lectures as a graduate student: Physical Chemistry I (16 students) and Survey of Physical Chemistry (7 students). Since, I have been invited back as a part-time lecturer, and have been responsible for up to 300 students in a semester. Those duties included General Chemistry II (160 students), Preparatory Chemistry (200 students) and Physical Chemistry I and II (15 and 80 students, respectively). I have also mentored undergraduate and graduate students in research, which has led to publication in the *Journal of Physical Chemistry C.*

Future Direction:

As faculty, my vision is to continue to examine the more relevant and pressing barriers associated with carbon management. It is my opinion that in the absence of drastic changes in carbon policy, we must make detailed, localized, compelling cases that are not vulnerable to the terminal uniqueness which often befalls broader, generalized assessments. Specifically, I would like to define industry-specific solutions, which would consider the appropriate separation technology given exhaust conditions, process design, and regional considerations such as local CO_2 demand (current and feasible future), reliable storage opportunities, and land management (renewable energy potential, algae cultivation potential, etc.). The goal is to distill our available knowledge into a specific pathway for deployment, available to any region wishing to advance carbon management yet unsure of how to proceed.

I also recognize the need to remain flexible. In my early career, crude-oil prices drove interest in FT. Since, the price has dropped significantly, yet I have found new life in FT by way of carbon utilization and C1 chemical processing. I believe the diversity of my skillset makes me especially capable to accommodate such changes, though I firmly believe that research in carbon management is healthy and burgeoning, even in the face of current skepticism. I am excited to bring my approach to these problems, to collaborate with departmental and interdepartmental partners to strengthen these solutions, and – most of all – to make a contribution.

Selected Publications:

Psarras, P., Wilcox, J. Molecular Simulations of Nitrogen-Doped Hierarchical Carbon Adsorbents for Post-Combustion CO_2 Capture. *Physical Chemistry Chemical Physics,* 2016, DOI: 10.1039/C6CP05865E

Psarras, P. et al. Effect of water on the CO_2 adsorption capacity of amine-functionalized carbon sorbents. (invited, accepted, issue cover, *Industrial Engineering and Chemistry Research,* May, 2017 **DOI:** 10.1021/acs.iecr.6b05064) – This paper was identified by Session Chair William J. Koros (Georgia Institute of Technology, United States) as the Best Presentation from the session "Novel Material for Gas Separation, Storage & Utilization" from the 2016 ACS Fall Meeting in Philadelphia, PA.

He, J., To, J., **Psarras, P.** et al. Natural Gas Sweetening Using a Tunable Polyaniline-Based Porous Carbon with Ultrahigh Surface Area. *Advanced Energy Materials*, 1502491 (2016)

Psarras, P., Ball, D., Wilcox, J. Effect of Ag and Pd Promotion on CH_4 Selectivity in Fe(100) Fischer-Tropsch Catalysis *Physical Chemistry Chemical Physics*, 2017, DOI: 10.1039/C6CP07116C

Wilcox, J., **Psarras, P.**, Liguori, Simona. Assessment of Reasonable Opportunities for Direct Air Capture. (invited, accepted, *Environmental Research Letters*, 12(6) 2017)

Psarras, P. et al. Slicing the Pie: How Big Could Carbon Dioxide Removal Be? *WIREs Energy and Environment,* (2017) (invited by associate editor, Mark Barteau, accepted) – this article has been selected for promotion on the Wiley/Wires science news site, http://www.advancedsciencenews.com

495241 Advanced Control for Next-Generation Materials Synthesis and Smart Manufacturing

Abstract:

Research Interests:

Strong trends in chemical and biomolecular engineering have led to a rapid increase in complexity of the design and manufacturing of high-tech products. Systems engineering (including computer modeling, simulation, optimization, and control) encompasses a powerful set of tools that can be used to help overcome the many challenges of these complex systems that include: (1) nonlinear (multi-scale) physical phenomena occurring on a large range of length and time scales, (2) mixed continuous and discrete dynamics due to the presence of time delays, interconnected units, and component faults/failures, (3) process uncertainties, disturbances, and noise, and (4) tight product specifications and safety-critical constraints.

During my Ph.D. research at the Massachusetts Institute of Technology (MIT) (with advisors: Prof. Richard Braatz and Prof. Michael Strano), I developed practical optimization-based methods that are able to address many of these challenges for a wide-range of complex systems using first principles models in combination with advanced statistical techniques. I have applied these methods to a number of emerging applications including continuous pharmaceutical manufacturing, advanced manufacturing of biological drugs, carbon nanotube-based solar cells, and the templated assembly of nanocrystals. Specifically, I have shown the impact of uncertainty on standard model-based control methods and established ways to systematically account for this uncertainty in the design of system-wide control structures.

Throughout my postdoctoral work at the University of California, Berkeley (with advisor: Prof. Ali Mesbah), I have been expanding the scope of my research toward autonomous â€œfault-tolerantâ€ model predictive control schemes. The main application focus has been on plasma processes for both medical therapy and semiconductor fabrication. The large number of reactions taking place in the plasma and their non-equilibrium nature make the models far too complicated to be directly used for optimization. In this work, more tractable models can be derived by applying identification and machine learning methods to data generated from the complex first principles calculations.

In my future research program, I am interested in using my systems engineering expertise to bridge the gap between the synthesis/design of next-generation materials (e.g., pharmaceuticals, or catalysts) and the development of processes for their smart manufacturing. This involves cooperative research toward (1) materials design by explicit modeling of molecular- and atomic-scale interactions, (2) efficient material screening through machine learning and reduced order modeling, and (3) the development of tractable methods for simultaneous process optimization and advanced control (across multiple spatial and temporal scales) in the presence of uncertainty. These methods can be applied to a wide-class of complex chemical and biomolecular systems including battery systems, (bio)pharmaceuticals, power/smart-grid networks, advanced biofuel production, and catalytic chemical processes. Based on my background in applied mathematics, I also intend to develop application-oriented theory and numerical methods.

Teaching Interests:

I have extensive experience as a teaching assistant (TA) both as an undergraduate at UT Austin and as a graduate student at MIT. At MIT, I TA'ed the Numerical Methods in Chemical Engineering graduate course. For my efforts as a TA, I received the Outstanding Graduate Teaching Assistant Award from the Chemical Engineering Department and the campus-wide 2015 School of Engineering Award for Extraordinary Teaching and Mentoring from the Dean of Engineering. I am particularly interested in advancing the computing, numerical methods, and control skills of chemical engineering students. In addition, I am very interested in teaching transport phenomena, thermodynamics, unit operations, and applied mathematics.

505524 Data Driven Modeling and Control for Engineering Next-Generation Processes

Abstract:

Research Interests:

Keywords: Dynamic systems and control, multiscale modeling, machine learning for system identification, nonlinear systems analysis

Application areas: bioengineering and biomanufacturing, photovoltaics and semiconductor manufacturing

Process systems engineering has enabled routine high performance and high reliability in modern chemical manufacturing. As the interests of chemical engineers have diversified to new domains including biology, materials science, and alternative energy, new tools are needed to bring fundamental advances to bear. The systems that govern these new domains are *complex* and *nonlinear*, which makes conventional systems analysis and process engineering challenging or impossible. In my work, I use theory and computation for *understanding* complex systems, *developing* effective dynamic models, and *designing* new processes and robust control systems so that these new technologies can be deployed. Here, I present two two example of my research:

1. Postdoctoral Project: Data-driven analysis and control of novel photobioreactors

Advised by Yannis G. Kevrekidis at Johns Hopkins University and Princeton University

Recent advances in machine learning, data science, and synthetic biology enable a new biomanufacturing platform where high-resolution process inputs and outputs can be used for online optimization and control. In particular, biosensors and optogenetic "switches" allow for measuring and manipulating the state of a bioreactor with more precision than ever before. In this "photobioreaction" process, the reactor state is observed using biosensors measured with an IR detector and actuated with LEDs that turn metabolic pathways "on" and "off" using optogenetics. Abundant state information from biosensors can be used for data-assisted modeling and data-driven control system design. We are developing new machine learning tools for dynamical system identification, and we will use tools from control systems theory (optimal control, model predictive control) to optimize photobioreactor performance. The effectiveness of these methods will be demonstrated through optimization of a process to produce the advanced biofuel isobutanol. We envision that this platform will also be used for myriad processes that produce pharmaceuticals, chemicals, and other products, all from renewable feedstocks.

2. PhD Dissertation: Rapid thermal processing for production of chalcopyrite thin films for solar cells: Design, analysis, and experimental implementation

Advised by Babatunde A. Ogunnaike and Robert W. Birkmire at the University of Delaware

Due to their potential for low cost and high efficiency, thin film solar cells are a strong candidate for utility-scale electricity production. One attractive manufacturing route for thin film solar cells is rapid thermal processing. For my PhD research, I designed a rapid thermal processing (RTP) reactor for producing thin film solar cells and deployed a novel control system using an advanced model-based controller and a nonlinear observer. With the reactor in place, I developed a novel stochastic model to describe thin film deposition in the reacting system. I derived expressions to relate simulation parameters to physical properties, showed that the model could predict film composition—including composition gradients that affect solar cell performance—as observed in the lab, and used the stochastic nature of the model to characterize lateral heterogeneity in the film. I also showed that the model predictions correspond with experimental data. Finally, I applied statistical design of experiments to probe the effects of process variables on the material properties of the films and optimize the efficiency of solar cells.

Teaching Interests:

Experience:

2016, University of Delaware, Assistant Instructor
Courses: Engineering Statistics (graduate), Chemical Engineering Laboratory (undergraduate)

2012, University of Delaware, Graduate Teaching Assistant
Course: Process Dynamics and Control (undergraduate)

I plan to treat teaching as a vital component of my position. I have experience as the instructor of computer labs (as a TA) and experimental labs (as an instructor), as well as lecturing experience at the graduate level. Beyond classroom experience, I helped develop an undergraduate laboratory project where students produced biodiesel from soybean oil with bench-scale and pilot-scale reactors.

My teaching philosophy is to to lead the class through experiential learning. After core concepts have been introduced, I believe that most effective way to cement student understanding is through a series of projects, hands-on demonstrations, and "virtual" demonstrations. I will use the programming skills I have developed through my research to build educational applications and demonstrations that students can use to reinforce the material.

I am capable of teaching any core chemical engineering course at the undergraduate or graduate level. Given the option, I would choose to teach courses with a substantial mathematical modeling component, such as Process Dynamics and Control, Computational Methods, or Differential Equations for Engineers. I am also interested in developing/teaching elective courses, possibly in fields such as Advanced Process Control and Machine Learning for Chemical Engineers.

Publications:

2017, AIChE Journal, "Hierarchical monitoring of industrial processes for fault detection, fault grade evaluation and fault diagnosis," L. Luo, **R. J. Lovelett**, B. A. Ogunnaike.
www.dx.doi.org/10.1002/aic.15662

2016, Journal of Process Control, "Design and experimental implementation of an effective temperature control system for thin film Cu(InGa)Se_2 production via rapid thermal processing," **R. J. Lovelett**, G. M. Hanket, W. N. Shafarman, R. W. Birkmire, B. A. Ogunnaike.
www.dx.doi.org/10.1016/j.jprocont.2016.07.005

2016, AIP Advances, "A stochastic model of solid state thin film deposition: Application to chalcopyrite growth," **R. J. Lovelett**, X. Pang, T. M. Roberts, W. N. Shafarman, R. W. Birkmire, B. A. Ogunnaike.
www.dx.doi.org/10.1063/1.4948404

2016, University of Delaware Ph.D. Dissertation "Rapid thermal processing for production of chalcopyrite thin films for solar cells: Design, analysis, and experimental implementation," **R. J. Lovelett**,
http://udspace.edu/handle/19716/21450

2016, IEEE Photovoltaic Specialists Conference, "Growth of Cu(In,Ga)(S,Se)$_2$ films: Unravelling the mysteries by in-situ x-ray imaging," B. West, M. Stuckelberger, L. Chen, **R. J. Lovelett**, B. Lai, J. Maser, W. Shafarman, M. Bertoni. www.dx.doi.org/10.1109/PVSC.2016.7749650

2015, IEEE Photovoltaic Specialists Conference, "A stochastic model for Cu(InGa)(SeS)$_2$ absorber growth during selenization/sulfization," **R. J. Lovelett**, W. N. Shafarman, R. W. Birkmire, B. A. Ogunnaike. www.dx.doi.org/10.1109/PVSC.2015.7356226

505449 Discrete and Hybrid Dynamics, Cyber-Physical Systems, and Formal Methods in Chemical Engineering

Abstract:

Research Interests:

My research falls into the broad categories of systems, control, cyber-physical systems, and process systems engineering. Specifically, I work on problems related to analyzing and designing the automation logic in a modern chemical plant. This logic is the part of the control system that's responsible for making decisions on the order of every second in response to measurements from the plant and input from operators to ensure that the plant operates safely and efficiently. Some of the key challenges that arise in this field are modeling the behavior of a chemical plant's existing automation logic, specifying the desired behavior of the control and automation system, formally determining whether the existing logic meets the specification, and designing logic that's guaranteed to meet a specification. Each of these challenges is magnified by the large size and high complexity of a typical chemical processing plant.

A cyber-physical system (CPS) is a system that involves interaction between a "cyber" component (such as the control system in a chemical plant) and a "physical" component (such as the chemical process itself). This type of system is not new, but, over roughly the past 10 years, interest in dealing directly with CPSs (as opposed to decomposing the "cyber" and "physical" parts of the systems) has increased significantly in diverse disciplines such as aerospace, computer, electrical, and mechanical engineering. Applications include air-traffic control, software and hardware verification, energy grid management, and self-driving cars. The growth of CPS-related research has not been as strong in chemical and biomedical engineering, despite the fact that examples of CPSs abound on scales from entire chemical plants down to medical devices. The main objective of my research to date, and goal of my future research, is to fill this gap by identifying the tractable problems in the field of chemical engineering that existing CPS theory can be applied to, and by extending the existing theory to address problems that are outside its reach. Such problems that I have worked on (or continue to work on) in my Ph.D. and Postdoctoral research include verification and falsification ("bug finding") of chemical plant automation systems and addressing the interaction between low-level automation logic and high-level scheduling decisions that occur in batch processing plants.

CPS research in chemical engineering will serve to accelerate the application of smart manufacturing and automation (which require a tight integration between physical processes and digital/cyber control and scheduling) in the chemical-processing industry. Smart manufacturing is one of the most promising frontiers in the chemical-processing industry because it seeks to address all or most of the primary manufacturing objectives, including safety, reliability, efficiency, flexibility, and profitability. The research problems that need to be addressed to enable its application include:

- managing discrete/continuous interactions in process control and automation (analysis and design)
- bridging the gap between high-level scheduling decisions (such as how much of a particular product to produce, made by human operators, often with the aid of optimization tools, at time scales on the order of hours and days) and low-level discrete automation decisions (such as whether or not to open a safety valve, made by an embedded control system, at time scales less than a second)
- synthesizing discrete automation logic that is provably correct with respect to a given specification (for example, a safety or operability requirement) when applied in a given plant
- detecting errors in existing automation logic that cause it to violate a given specification

The fact that each of these problems involves CPSs demonstrates the potential impact of pushing forward CPS-related research in chemical engineering.

Teaching Interests:

I believe that the main reason why I am in the position to make impactful contributions to systems research in chemical engineering is that, as an undergraduate student at the University of Texas and a Ph.D. student at Carnegie Mellon, I was fortunate enough to learn from professors in chemical engineering who are leading experts in the areas of systems, control, modeling, and optimization. My main objective in teaching is to pass along the same core skills in mathematics, science, and also in computer programming and software development that I rely on every day in my own work. I feel that these skills are critically important for a chemical engineer to be successful today, as engineering in general increasingly involves writing software to apply sophisticated algorithms to complex problems; there can be no smart manufacturing in the chemical-processing industry without engineers who have at least a basic understanding of the underlying theory and the ability to apply that theory, and the same is true of chemical engineers working in other industries.

Given my background, I believe that I'm qualified to teach graduate courses in control, optimization, and mathematics, and all core undergraduate courses in chemical engineering (thermodynamics, transport phenomena, chemical kinetics and reaction engineering, mathematics, introduction to chemical engineering, computer programming, control, separations, etc.) with a preference for control, mathematics, and computer programming. In addition, I'm interested in developing and teaching a graduate-level course in cyber-physical systems and their application to chemical engineering, and extending how control is taught in undergraduate lab courses to include automation, the associated low-level discrete logic that needs to be developed, and how to implement it.

505416 Novel Strategies for Quantification of Model Uncertainty and Real-Time Optimization of Batch Operations

Abstract:

Research Interests:

In the last few decades, there has been a steady growth in application of optimization techniques to the solution of problems of industrial importance, e.g. supply-chain management, operational planning/scheduling and optimal process control/dynamic optimization. This trend has been motivated by both increasing global competition and tighter environmental regulations, which are driving industry towards reducing costs while also minimizing environmental impacts.

Although extensive process systems engineering research has been reported so far, most of it has addressed continuous processes and only limited attention has been devoted to batch operations, which are widely used to produce high value-added products (specialty chemicals, pharmaceuticals, cosmetics, etc.). Given the opportunities that this application domain provides, I have focused my research on methodologies/applications belonging to four main areas:

- Operational planning and scheduling of batch processes;
- Deterministic and stochastic model predictive control and/or dynamic optimization of batch processes;
- Static and dynamic estimation of model uncertainty from experimental measurements (model uncertainty is represented as the probability distribution of an appropriate set of uncertain parameters);
- Dynamic identification of the optimal uncertainty set in stochastic dynamic optimization and model-based control problems (the optimal uncertainty set is simply intended as the ensemble of the most relevant uncertain parameters of the process model).

In the past, I have also addressed supply-chain problems[1], and continue to have interests in that enterprise-level research area as well.

More recently, I have focused my research efforts on developing methods for the optimal, real-time management of both single batch operations and multi-unit/multi-step batch processes. In particular, I have formulated and implemented two dynamic optimization and/or model-based control algorithms applicable to single batch operations (BSMBO&C [2,3] and RBSMBO&C [4,5]) as well as a novel strategy aiming at the integration of scheduling, dynamic optimization and model-based control of multi-unit batch processes (MUBSMBO&C [6]). The BSMBO&C and MUBSMBO&C algorithms assume that model mismatch is negligible while RBSMBO&C explicitly accounts for model uncertainty, through a novel dynamic scenario selection procedure. These approaches are very flexible, and will accommodate virtually any batch system and/or performance metrics.

My latest research work involves the estimation of the probability distributions (PDFs) of the uncertain parameters of

ODE/DAE models as well as the real-time selection of the optimal uncertainty set in stochastic online optimization and control problems. I have already developed a methodology for estimating the PDF of the uncertain parameters of an ODE/DAE model, which relies on a novel back-projection concept (PDFE&U [7]), and provides satisfactory accuracy at a low computational cost. In fact, PDFE&U appears to be more efficient that many existing state-of-the-art random sampling-based approaches, such as MCMC. I am currently working on a strategy for dynamic estimation of the optimal set of uncertain parameters to be used in a stochastic real-time optimization and/or model-based control problem, which exploits and combines PDF estimation techniques, e.g. PDFE&U, multi-point sensitivity analysis, and novel raking indices. Both of these frameworks show interesting synergies with RBSMBO&C.

In the near future, I am planning to apply these algorithms both to solve drug delivery problems (individualized drug dosing) and to improve the efficiency and reliability of food processing technologies, active pharmaceutical ingredients (APIs) production and specialty chemicals synthesis. Moreover, I am also interested in augmenting MUBSMBO&C to make it capable of explicitly handling model uncertainty, as well as in further improving the computational efficiency of all of the aforementioned strategies. Further algorithmic improvements may also involve incorporation of goal programming techniques into some of the aforementioned strategies in order to solve multi-objective scheduling, dynamic optimization and model-based problems under uncertainty.

Teaching Interests:

My research work and graduate studies have provided me with deep knowledge of control theory, numerical methods, dynamic and steady-state modelling, coding/parallel computing (especially in C++), and statistics. Therefore, I would be interested in teaching any course broadly related to these subjects, e.g. applied statistics, process control, numerical methods, dynamic and steady-state modelling and process optimization. I would also like to introduce and teach two new classes, whose topics lie at the boundaries of Chemical Engineering, Statistics and Computer Science:

- Applied statistics for process optimization/control under uncertainty;
- Parallel programming techniques applied to process simulation and optimization.

Since optimization strategies have become an essential tool for the solution of many problems of industrial relevance in the last few years, I strongly believe these interdisciplinary courses would provide new chemical engineers with useful and valuable knowledge.

References:

1. Rossi F, Manenti F, Reklaitis G. A general modular framework for the integrated optimal management of an industrial gases supply-chain and its production systems. *Computers and Chemical Engineering*. 2015;82:84-104.
2. Rossi F, Manenti F, Buzzi-Ferraris G. A novel all-in-one real-time optimization and optimal control method for batch systems: Algorithm description, implementation issues, and comparison with the existing methodologies. *Industrial and Engineering Chemistry Research*. 2014;53:15639-15655.
3. Rossi F, Copelli S, Colombo A, Pirola C, Manenti F. Online model-based optimization and control for the combined optimal operation and runaway prediction and prevention in (fed-) batch systems. *Chemical Engineering Science*. 2015;138:760-771.
4. Rossi F, Reklaitis G, Manenti F, Buzzi-Ferraris G. Multi-scenario robust online optimization and control of fed-batch systems via dynamic model-based scenario selection. *AIChE Journal*. 2016, DOI: 10.1002/aic.15346.
5. Rossi F, Manenti F, Pirola C, Mujtaba I. A robust sustainable optimization & control strategy (RSOCS) for (fed-)batch processes towards the low-cost reduction of utilities consumption. *Journal of Cleaner Production*. 2016;111:181-192.
6. Rossi F, Casas-Orozco D, Reklaitis G, Manenti F, Buzzi-Ferraris G. A computational framework for integrating campaign scheduling, dynamic optimization and optimal control in multi-unit batch processes. *Computers and Chemical Engineering*. 2017;DOI: http://dx.doi.org/10.1016/ j.compchemeng.2017.05.024.
7. Rossi F, Manenti F, Buzzi-Ferraris G, Reklaitis G. A novel back-projection approach to the estimation or update of the probability distribution of the uncertain parameters of an ODE/DAE model. *In preparation. Estimated submission at the end of 2017.*

505396 Development and Assessment of New Processes for the Production of Bio-Products

Abstract:

Development and assessment of new processes for the production of bio-products

Sampath Gunukula

University of Maine Chemical and Biological Engineering and Forest Bioproducts Research Institute, Orono, ME, 04469-5737, USA

Research Experience:

My academic research is a blend of the fields of process engineering, economics, and sustainability. My academic training is in process simulation, process research, techno-economic analysis, economic risk assessment, and life-cycle analysis. In my research I have worked on quantifying the financial risk of investments for the development of platform technologies to produce biobased products, introducing a new methodology for the economic and environmental assessment of early stage renewable chemical technologies, and developing general design rules that guide the design of biocatalytic technologies and chemical catalytic pathways for the transformation of platform biorenewable chemicals. As a result of these projects, I have expertise in the areas of chemical process development, cost benefit analysis, biofuel and biorenewable chemical production, and sustainability assessment.

Postdoctoral Projects: Economic and process analysis of biorefineries, Assessment of the viability of the platform chemical concept, the effect of biomass densification technologies on the economics of biofuel production processes, and finding new applications of bio-char.

Supervisors: Dr. M. Clayton Wheeler, Dr. William DeSisto, and Dr. Hemant Pendse, Chemical and Biological Engineering, Forest Bioproducts Research Institute, University of Maine at Orono.

Ph.D. Dissertation: â€œ*Techniques for Evaluating and Guiding Development of Renewable Chemical Technologies for Sustainable Products*â€

Under the supervision of Dr. Robert Anex, Biological Systems Engineering, University of Wisconsin at Madison

Research Interests:

1. **a. Technology assessment**

In a faculty position I would continue my work using process modeling, economic analysis, and sustainability assessment in the development of new renewable technologies. Additionally, I would like to focus on incorporating social costs, policy interventions, and policy uncertainty into the techno-economic modeling for the evaluation of renewable chemical technologies. Moreover, I would like to explore the utility of life cycle assessment for assessing food, energy, and water security.

1. **b. Development of separation and purification processes**

The lack of processes to purify low-value added biorenewable products from clarified fermentation broths that contain a large number of low concentration impurities such as amino acids have limited the commercialization of biorenewable technologies. I plan to develop less energy intensive, sustainable processes to overcome this limitation. I plan to work on developing new sustainable techniques for the separation and purification of biorenewable products. In the near term, my focus will be on modifying bio-char for use as an adsorbent for removing minor impurities such as amino acids from a clarified fermentation broth. Additionally, I will work on finding inducers to lower pH of the clarified fermentation broths for the easy separation of organic acids.

1. **c. Process research**

The use of sulfuric acid as an acid catalyst in the production of biorenewable products from lignocellulosic feedstock is problematic because it is difficult to recover for reuse. Neutralizing sulfuric acid with calcium salts instead of recovering it may not be a viable option for the large-scale production of biorenewable products, as the storage, handling, and transportation of tonnes of sulfuric acid can be hazardous. Thus, future research is necessary to provide an acid environment for the production biorenewable products without the addition of an acid catalyst. In collaboration with Dr. Wheeler, I will work on replacing an acid catalyst with organic acids that can be recovered easily.

1. **d. Valorizing biorefinery waste**

The high capital and operating costs of new biorefinery technologies are creating barriers for attaining commercial relevance. Such barriers can be overcome by generating additional revenue from the biorefinery waste streams by converting them to value added products. Many potential research opportunities exist in the area of valorizing biorefinery waste. One such opportunity is the production of in situ hydrogen from the aqueous waste of fast pyrolysis process without using a catalyst to improve the overall economics of renewable gasoline and diesel production via fast pyrolysis pathway. I will collaborate with Dr. Wheeler and Dr. Desisto to work on concentrating organics from the aqueous waste and use these organics as a feedstock for the production of hydrogen via fermentation.

Teaching Interests:

I have observed while I was mentoring undergraduate students as well as co-supervising graduate students that students try to memorize scientific concepts. I also noticed among these students that there is a fear of learning new knowledge as well as difficulty with communicating their work properly. Having desire to teach how to learn science has motivated me to purse career as a faculty member. As faculty member, I will help students to develop their creative abilities and critical analysis skills through class assignments. I will put effort to improve studentâ€™s communication skills through assigning team projects as well as asking them to give presentations very often. In my lectures, I will include application of chemical engineering concepts in studying other fields of science to encourage students to learn new fields and to promote interdisciplinary understanding. As a faculty member, I plan to teach chemical process design, biological unit processes, engineering economics, and separation and purification techniques in food and pharmaceutical industry.

Publications:

1. S. Gunukula, P. Keeling, R. Anex, Risk advantages of platform technologies for biorenewable chemical production, Chem. *Eng. Research Design*. 2016.
2. S. Gunukula, R. Anex, Evaluating and guiding the development of sustainable biorenewable chemical technologies, *Biochem. Eng. J.* 2017
3. J.L. Carrasco, S.Gunukula, A.A. Boateng, C.A. Mullen, W.J. DeSisto, M.C. Wheeler, Pyrolysis of forest residues: An approach to techno-economics for bio-fuel production, *Fuel*.2017

4. S. Gunukula, R. Anex, Comparative economic analysis of multiple bio-based routes to adipic acid. *Biofuels Bioprod Bioref*, 2017

Under review

1. S. Gunukula, T. Runge, R. Anex, Assessment of Biocatalytic Production Parameters to Determine Economic and Environmental Viability, 2017. (*ACS Sustainable Chem. Eng.*)
2. H. AlMohamadi, S. Gunukula, W.J. DeSisto, M.C. Wheeler, Formate Assisted Pyrolysis: An economic analysis, 2017 (*Biofuels Bioprod Bioref*)

Awards:

1. Student-led Research Competitive Grant: S. Gunukula, J.A., and Lopez-Ruiz, Determining the different reaction engineering factors that affect the economic feasibility of the production of alpha olefins from carboxylic acids through techno-economic analysis. CBiRC Student-led Research Grant, Iowa State University, 2012.

505177 Single Cell Analysis Using Droplet Microfluidics

Abstract:

Research Interests:

The complexity of biological systems suggests the existence of a whole new universe of novel catalysts, functions and mechanisms. Dissection and analysis of this complexity at single cell level will lead to the study of these novel catalysts, functions and mechanisms, hence, largely enhance our understanding of biological systems and drastically expand the capacity of bioengineering. Droplet microfluidics emerged as a versatile tool for single cell screening and analysis in recent years. In this poster, I will highlight several of recent efforts to demonstrate the promise of droplet based single cell analysis for bioengineering and biomedical research.

My graduate research focused on applying synthetic biology and metabolic engineering strategies to engineer nonconventional yeast *Yarrowia lipolytica* for high lipid production. *Y. lipolytica*, as an oleaginous yeast, possesses intrinsic capacity of accumulating large amount of lipid, which can be converted into numerous value-added chemicals including biofuels and nutrients. To expand its genetic toolbox, I first developed a set of hybrid promoters and centromere regulated plasmids to allow efficient manipulation of gene expression in this yeast. I then employed combinatorial, evolutionary and inverse metabolic engineering strategies to drastically increase the lipid production titer. Finally, to gain insights over the lipogenic phenotype, I also performed comparative genomic analysis to identify beneficial genomic and transcriptomic changes in high lipogenic strains. With these works, I applied synthetic biology concepts and genome scale analysis to achieve the success of engineering for value-added chemical production and to gain insights of mechanism of lipogenesis.

While I was engineering *Y. lipolytica* for high lipid production, I started to appreciate the existence of cellular heterogeneity. Characterization at population level, rather than single cell level, greatly limited our ability to understand and engineer biology. During my postdoc, I applied droplet microfluidics to develop novel bioengineering methods and to gain better understanding of biological systems. First, I applied droplet microfluidic workflows for effective screening of secretory phenotype. I applied flow cytometric sorting of double emulsion for improving riboflavin production in *Y. lipolytica*. I also worked on developing RAPID (RNA aptamers in droplets) by taking advantage of the quantification function of RNA aptamers and droplet sorting ability to improve tyrosine and streptavidin production in *S. cerevisiae*. Second, I developed a novel workflow for high throughput yeast colony RNA-seq by combining yeast colony formation in hydrogel microdroplets and Drop-seq (barcoded single cell RNA-seq). I further applied this method to study the white-opaque transition in *C. albicans*. Third, I employed Drop-seq to collect and analyze single cell genome wide expression data from dissociated mouse lung tissues and revealed an important subgroup of macrophage with mouse model of pulmonary fibrosis. We also started to look into the relation between gut microbiome to pulmonary fibrosis in mouse using single cell RNA-seq. In addition, I have also gained experience with droplet microfluidic techniques such as PACS (PCR-activated cell sorting), In-Drop single cell RNA-seq and droplet MDA for single cell genomic sequencing.

Looking ahead, the expertise and experience I gained through my training will enable my future lab to fully utilize the power of droplet microfluidic based single cell analysis to accelerate the engineering and discovery of biology. I will start by using droplet microfluidic platform to screen for novel industrial relevant enzymes and strains from environmental samples through *in situ* cultivation method and cultivation free method. I will then apply my expertise in synthetic biology to conduct protein engineering and strain development with promising candidates. In the meantime, I am also looking for establishing fruitful collaborations in the field of single cell sequencing, such as characterization of gut microbiome (single cell genome sequencing) and host response (single cell transcriptome sequencing).

Teaching Interests:

I have served as teaching assistant for biochemistry laboratory course and general microbiology course. Through the process, I gained valuable experiences such as interacting with students from different backgrounds and leading discussion and review section. I have also mentored several undergraduate students teaching them basic lab techniques, experimental design and critical thinking. Based on my academic background, I am willing to teach undergraduate lecture courses such as Introduction of Chemical and Biomolecular engineering and Biochemical Engineering. I also plan to develop course focusing on introducing core concepts and methods in the fields of metabolic engineering, synthetic biology and microfluidics.

505600 Investigating Continuous Biochemical Processing Strategies Utilizing Process Systems Engineering Fundamentals

Abstract:

Research Interests:

Much of current research is driven toward developing a platform for a sustainable future by determining cost-effective alternatives to manufacturing various products traditionally made from fossil fuels. While many studies have shown the promise of biochemical platforms to yield a variety of products, including alternative fuels, biopolymers, and a variety of specialty chemicals, multiple challenges exist in their process development. Many of these technologies are currently in their infancy and lack the productivity and economic viability to compete with traditional petroleum-based processes. In addition, the use of complex, often growth-associated biochemical reactions result in unit operations that must be monitored and controlled to counteract batch-to-batch variability and maximize productivity. These challenges can be overcome by catalyzing the transition of commercial-scale bioprocesses from traditional batch methods to continuous processing strategies that offer the potential to exploit the cost reduction effects of economies-of-scale. To this point, much of the focus in advancing bioprocessing applicability has been on improvement through biochemical means, resulting in novel recombinant strains with higher conversion and yield, rather than the development of novel processing strategies that could provide further answers to these challenges. However, the fundamentals of process systems engineering, typically employed for the modeling, synthesis and control of traditional chemical processes, can be utilized to develop feasible, cost-effective solutions for the continuous, commercial implementation of bioprocessing methods.

To this end, my research plan is to focus on the implementation of a process systems approach to develop novel processes in key areas of biochemical manufacturing, with applications in a variety of areas including sustainable energy, bio-plastics, targeted cell therapies, and pharmaceuticals. My doctoral work at Texas A&M University, under the guidance of Dr. M. Nazmul Karim, has provided me with extensive experience developing methods for the modeling, process synthesis, and optimal control of novel, continuous processes for the biochemical production of fuels and chemicals. I will look to extend this knowledge to my group to afford students a multi-disciplinary environment capable of providing opportunities to develop both experimental and computational skills in the areas chemical and biochemical engineering. My research will facilitate the bench-to-commercial scale development of new continuous bioprocesses to increase their reliability and economic viability. The application of process synthesis methods to the knowledge of biochemical reaction pathways in a variety of organisms will direct process development to novel cocultures capable of achieving the desired conversion of a selected substrate-product combination. Work in developing integrated systems to simultaneously produce pharmaceuticals and liquid transportation fuels will providing a sustainable platform to reduce dependence on non-renewable liquid fuel sources and increase availability of necessary products in the health care industry. The investigation of intensified production schemes for biological production of pharmaceuticals will utilize the cost-reducing practices of continuous processing and process intensification as well as guard against risks such as contamination and product variability. The interdisciplinary nature of the group will also necessitate strong collaborations that will further enrich this training environment.

Teaching Interests:

I am comfortable teaching all general subjects within the chemical engineering curriculum for both undergraduate and graduate students. However, I would prefer to teach process economics, process design, process control, or numerical analysis courses based on my experience teaching these subjects as a co-instructor under the Graduate Teaching Fellow program at Texas A&M University.

Based on my own academic background and my experience in the Texas A&M chemical engineering department, I would like to design two new courses that can be offered to either the graduate students or advanced undergraduate students: a process synthesis and optimization course and a biochemical processing course.

The first course would cover the various aspects of process synthesis, such as the fundamentals of nonlinear and mixed-integer optimization, the modeling paradigm consistent with the construction of process superstructures, and the use of process integration and intensification strategies to improve upon superstructures. In recent years, process systems engineering has become a special focus in many universities, therefore various topics in this field will be a focus of this course.

The second course would cover the basics of biochemical processing, both upstream and downstream, and serves to teach essential topics to developing alternative sustainable processes. This course would cover the fundamentals of biochemical reactions (kinetic modeling and bioreactor design), bioseparations (chromatography, membrane extraction, liquid-liquid extraction, etc.), and the applications and challenges of designing large scale bio-systems. Specific applications to the areas of alternative fuel, pharmaceuticals and biologics, and cellular therapies would be explored.

Select Publications:

1. **Raftery, J.P.** and Karim M.N. (2017) Economic viability of consolidated bioprocessing utilizing multiple biomass substrates for commercial-scale cellulosic bioethanol production, *Biomass and Bioenergy* 103, 35-46.
2. **Raftery, J.P.** and Karim M.N. (2017) Economic Improvement of Continuous Pharmaceutical Production via the Optimal Control of a Multi-Feed Bioreactor, *Biotechnology Progress,* doi: 10.1002/btpr.2433
3. Ordoñez, M.C., **Raftery, J.P.**, Jaladi, T., Chen, X., Kao, K., and Karim, M.N. (2016). Modelling of batch kinetics of aerobic carotenoid production using *Saccharomyces cerevisiae*, *Biochemical Engineering Journal* 114, 226â€"236.

505517 Process Systems Engineering in Pharmaceutical Process Development

Abstract:

Research Interests:

Firstly, I firmly believe that mathematical tools are the best scientific languages to understand the complex and dynamic chemical engineering processes. My background in Process Systems Engineering (PSE) has helped me to gain knowledge of a broad category of engineering tools for pharmaceutical process modelling, design, optimisation, and control. Hence, I am very interested in developing a systems-based pharmaceutics development methodologies, which involves a holistic approach to drug formulation and the design of the drug manufacturing process that makes use of a mechanistic model-based framework and targeted experiments. We have conducted our research framework based on PSE tools and contributed to the development of new crystallization and manufacturing process, e.g. the plug-flow crystalliser, continuous oscillatory baffled crystalliser, continuous mixing blender, dry impact milling. My current research project in Purdue University is funded by and working with the FDA. Our main goal is to advance regulatory science to support the implementation of continuous solid dose manufacturing systems, equipped with control systems that are capable of handling raw material variability and assuring product quality in real time. My specific work is to determine the risk levels from the different sensing configuration and evaluated them and use them to enable control systems capable of assuring product quality, which is of great concern to the real-time release strategy of the continuous manufacturing.

My future research interests would continue exploring the implementation of these process systems engineering tools in pharmaceutical continuous manufacturing and crystallization, and seeking collaborating opportunities with other faculty members in this area.

Teaching Interests:

In terms of teaching, which requires the experience and capability to interpret the rigid scientific terminologies in understandable language, is usually a challenge for new faculty members. During the course modules for the Doctoral Training Centre (DTC) in CMAC Loughborough University, I have had the chances to experience for myself this hurdle. And I find I could overcome this and enjoyed the teaching, especially when I was preparing the course structure and assignment. In addition, having laid a solid foundation in Chemical Engineering, especially the advanced course modules in NUS (Ranked 1st in Asia and 5th in the world universities, QS World Univerisites), I would feel comfortable teaching across a broad range of subjects at undergraduate and postgraduate levels.

505129 Scientific Computing and Mathematical Modelling for Multiscale Nonlinear Systems

Abstract:

Research Interests: Optimization, Nonlinear dynamics, Stability, Biological/chemical processes.

Process synthesis, optimization, and design of energy-production systems are becoming increasingly challenging as they rapidly grow in size and complexity. Successful and sustainable development of energy infrastructures hinges on integrated energy and water systems, in which all safety, environmental, and economical considerations are accounted for. Polygeneration of energy can significantly improve efficiency and reduce capital investment cost by combining processes that involve similar intermediate products and operation units. Strategic planning for projects of such magnitude requires accurate and predictive mathematical models and a systematic multi-scale decision-making framework to provide the best plan among numerous alternatives (e.g. feedstocks, products, and technologies) that satisfies a prescribed set of objectives. Formulation of such problems often leads to large-scale nonlinear dynamic problems that are computationally challenging to solve for systems of practical importance. My overall approach is to tackle these problems from two fronts: (i) Developing robust and fast computational algorithms for the optimization of process-level models and (ii) leverage advanced mathematical techniques to derive macro/microscopic unit-process models, which are simple enough to be computationally tractable and realistic enough to capture the essential underlying physics.

Research Experiences: I am currently a postdoctoral fellow in the Process Systems Engineering Laboratory at MIT under the supervision of Prof. Paul Barton (**successful proposal: NSERC Postdoctoral Fellowship**). I am currently working on the multi-scale design and analysis of carbon capture, utilization, and storage networks, focusing on algae cultivation for fuel production and carbon sequestration. Since I started, I have worked on (i) fast and robust multi-parametric programming algorithms for the flux balance analysis of genome-scale metabolic networks and (ii) superstructure optimization for techno-economic analysis of carbon sequestration networks. In the first, I have achieved more than six-fold improvement in computational speed of large-scale parametric linear and quadratic programs over current parametric programming tools. In the second, using explicit solutions from the foregoing parametric study, I incorporated detailed metabolic models into superstructure models, demonstrating the computational viability of integrated metabolic-network and process-level optimization models.

I obtained my PhD from the Chemical Engineering Department at McGill under the supervision of Prof. Reghan Hill and Theo ven de Ven (**successful proposal: NSERC Strategic Green Fibre Network**), where I studied the stability and dynamics of wood-fibre collapse upon drying. I developed an idealized model that captures the underlying physics of drying-induced collapse, demonstrating the connection between elastocapillary stability and dry-state conformations. I rigorously established a connection between stability loss and the shape of bifurcation diagrams, providing an alternative means to identify the stability of elastocapillary systems, which is much less costly than the eigenvalue computations in conventional linear stability analyses.

Future Directions: As faculty, I would like to continue working on the development of mathematical-models and computational tools for the simulation and optimization of multi-scale biological and chemical processes with emphasis on polygeneration systems. I am particularly interested in mass production of renewable fuels, where algal bio-refineries are integrated with power and steam generation, water treatment, carbon capture, and enhanced oil recovery (EOR) plants within a large-scale carbon-sequestration networks. My future efforts are classified into three categories: (i) Improving mathematical models for existing technologies (e.g. EOR), (ii) developing new theories, models, and computational tools for studying biological systems to uncover their potential application in renewable-fuel production, and (iii) developing robust global optimization tools for superstructure modelling and techno-economic analysis of polygeneration networks. Among these, the first and last are short-term projects, where fast progress can be made by extending the research efforts from my postdoctoral training. The second is a long-term and challenging project with potentially high impact on energy systems. The potential of microbial communities in the production of clean fuels and other value-added products is not fully understood. This is due the lack of mathematical tools to characterize their complex metabolism that can quantify the behavior of micro-organisms in response to environmental stimuli. A comprehensive mathematical framework must be developed to coherently describe steady-states and transients of metabolic networks and the dynamics of the transition between these two modes.

Teaching Interests: Optimization, nonlinear dynamics, fluid mechanics, transport phenomena, computational fluid dynamics.

One of my motivations for pursuing an academic career besides research is teaching. A successful research career can be complemented by satisfying teaching responsibilities. My teaching philosophy is based on three principles: (i) maintaining studentsâ€™ interest, (ii) being accessible and accountable, and (iii) valuing concepts and fundamentals over case studies and memorization. During my PhD, I was a teaching assistant and lab instructor for several undergraduate- and graduate level courses on most chemical-engineering core subjects. My experiences include preparing course material, handling lab sessions, holding tutorials and group discussions, and lecturing small (fewer than 20 students) and medium-sized classes (around 80 students).

Selected Publications:

A. Akbari, R.J. Hill, Liquid-bridge stability and breakup on surfaces with contact-angle hysteresis, *Soft matter*, 12, 6868-6882, 2016 (preprint arXiv:1507.06549).

A. Akbari, R.J. Hill, T.G.M. van de Ven, Stability and folds in an elastocapillary system, *SIAM J. Appl. Math.*, 76(1):87-109, 2016 (preprint arXiv:1505.07315).

A. Akbari, R.J. Hill, T.G.M. van de Ven, Catenoid stability with a free contact line, *SIAM J. Appl. Math.*, 75(5):2110-2127, 2015 (preprint arXiv:1505.07159).

A. Akbari, R.J. Hill, T.G.M. van de Ven, Liquid bridge breakup in contact-drop dispensing: Liquid bridge stability with a free contact line, *Phys. Rev. E*, 92:022404, 2015**.**

A. Akbari, R.J. Hill, T.G.M. van de Ven, An elastocapillary model of wood-fibre collapse upon drying, *Proc. Roy. Soc. A*, 471:20150184, 2015.

A. Akbari, M. Akbari, R.J. Hill, Effective thermal conductivity of two-dimensional anisotropic two-phase media, *Int. J. Heat Mass Transfer*, 63:41-50, 2013.

505440 Chemical Thermodynamics of Aqueous Atmospheric Aerosols: Modeling and Microfluidic Measurements

Abstract:

Lucy Nandy, Advisor: Cari Dutcher, *University of Minnesota-Twin Cities*

4th Year PhD Candidate

Research Interests:

Accurate predictions of gas-liquid-solid equilibrium phase partitioning of atmospheric aerosols by thermodynamic modeling and measurements is critical for determining particle composition and internal structure at conditions relevant to the atmosphere. Organic acids that originate from biomass burning, and direct biogenic emission make up a significant fraction of the organic mass in atmospheric aerosol particles. In addition, inorganic compounds like ammonium sulfate and sea salt also exist in atmospheric aerosols, that results in a mixture of single, double or triple charged ions, and non-dissociated and partially dissociated organic acids. Statistical mechanics based on a multilayer adsorption isotherm model can be applied to these complex aqueous environments for predictions of thermodynamic properties.

In this work, thermodynamic analytic predictive models are developed for multicomponent aqueous solutions (consisting of partially dissociating organic and inorganic acids, fully dissociating symmetric and asymmetric electrolytes, and neutral organic compounds) over the entire relative humidity range, that represent a significant advancement towards a fully predictive model. Moreover, the model has been applied to determine the activity of ions to predict pH of acidic aerosol particles. The model is also successful in predicting hygroscopicity of amino acids and other organic compounds in the atmosphere. In addition to the modeling approach, water loss of multicomponent aerosol particles is measured by microfluidic experiments to parameterize and validate the model. In the experimental microfluidic measurements, atmospheric aerosol droplet chemical mimics are generated in microfluidic channels and stored and imaged in passive traps until dehydration to study the influence of relative humidity and water loss on phase behavior.

These efforts together will enhance understanding of atmospheric aerosol phase, solid/liquid/gas partitioning, and liquid-liquid morphologies found in the troposphere. Additionally, although this research emphasizes atmospherically relevant systems, the measurements and modeling are useful to any application that requires thermodynamic predictions of water content, like in soil permeability, drying processes in food science, waste water treatment and industrial discharge.

Teaching Interests:

My teaching activities are focused on fluid mechanics and thermodynamics. I am currently designing a course on aerosol chemistry.

Course interests - Fluid Mechanics, Multiphase Flows (Macro- and Micro- scale), Thermodynamics, Heat Transfer

PhD Dissertation: "*Thermodynamic Properties of Water Soluble Organics in Atmospheric Aerosols*"

Under supervision of Prof. Cari S. Dutcher, Mechanical Engineering, University of Minnesota

Research Experience:

My research projects have been a combination of fluid mechanics, thermodynamics, aerosol chemistry and heat & mass transfer. My current project is thermodynamic modeling and measurements of atmospheric aerosol properties and phase behavior. The model is based on adsorption isotherm by statistical mechanics, and the measurements are done by biphasic microfluidic experiments. Collaboration with other research groups has aided me to find various applications of the model, like finding pH and hygroscopicity of aerosol particles. Prior to my doctoral studies, I have worked on projects in a steel industry that required a heat transfer model for dynamic secondary cooling control for uniform cooling of steel slabs, and a mixing model for alloy dissolution in gas-stirred steel ladle for complete homogenization. I have also worked on a small project during my masters' on development of microchannel assisted cooling devices that involved image-analyzing interferometry. My masters' dissertation project was numerical simulation of evaporative processes using ANSYS (FLUENT) that had applications in design process of two-phase cooling systems.

Teaching Experience:

In addition to my research career, my teaching role as a teaching assistant has helped me to be more analytical in my approach to learning and teaching. I have led discussion sessions for both fluid mechanics and thermodynamics for undergraduate students, and have also TAed applications of thermodynamics in fluid flows for undergraduate and graduate students. For all the TA positions, my duties also included holding office hours, grading assignments and exams, and very few guest lectures. Aside from that, I have mentored an undergraduate student in a mentor program in the college of science and engineering, and have mentored another undergraduate student in our laboratory who works on microfluidic droplet experiments. Additionally, I have taken a course over a semester in Preparing Future Faculty for teaching in higher education.

Future Direction:

Since I want to pursue my career in academia after completing the PhD program, I plan to continue working in the fields of aerosol science, chemical thermodynamics and multiphase microfluidics to gain new insights and future research responsibilities. I would also like to merge the knowledge with the topics that I worked on during my Masters that involved microscale heat transfer and computational fluid dynamics (CFD). I believe CFD will play an important role in multiphase microfluidics that uses water-in-oil droplets of volumes in the range picolitres to nanoliters.

I would like to take the current research further by developing thermodynamic models for complex liquid-liquid phase separated atmospheric aerosols to study the amount and composition of each phase and the degree of wetting. This will help predict properties of the liquid phases, and what factors are important to determine the equilibrium morphology of aerosols. Furthermore, liquid-liquid phase separation has applications in maintaining the physical stability of protein formulations. I would also like to explore this field that will enhance maintain stability across the protein therapeutics shelf life for desirable products.

Since microfluidic multiphase flows have a variety of applications like emulsions in food industry, mixing by mass transfer in a microreactor, micro separation, and particle production and synthesis, I would like to combine the heat transfer studies to understand microfluidic emulsifications for emulsion stability that can influence processes. This will address long-term storage stability of food-relevant compounds. A computational fluid dynamics (CFD) model can also be developed to study the flow field, heat and mass transfer to study internal flow patterns and dissolved chemical transfer across fluid interfaces. This can be extended to determine chemical reactions like nitration of toluene in the industries.

As a faculty, I will ensure that I collaborate with modelers and experimentalists for future research possibilities, and get realistic findings that can have various applications in aerosol and water studies. I will also impart my knowledge to students and help them see themselves having ability to understand science and engineering problems and how these problems exist in real concrete ways.

Publications:

Nandy, L. and Dutcher, C. S., Isotherm-Based Thermodynamic Model for Solute Activities of Asymmetric Electrolyte Aqueous Solutions; *J. Phys. Chem. A* (under review)

Craig, R. L., **Nandy, L.**, Axson, J. L., Dutcher, C. S., Ault, A. P.; Spectroscopic Determination of Aerosol pH from Acid-Base Equilibria in Inorganic, Organic, and Mixed Systems; *J. Phys. Chem. A* (under review)

Marsh, A., Miles, R. E. H., Rovelli, G., Cowling, A. G., **Nandy, L.**, Dutcher, C. S., and Reid, J. P.; Influence of Organic Compound Functionality on Aerosol Hygroscopicity: Dicarboxylic Acids, Alkyl-Substituents, Sugars and Amino Acids; *Atmos. Chem. Phys.*, 17, 5583–5599, **2017**. https://doi.org/10.5194/acp-2016-1051

Nandy, L., Ohm, P. B., and Dutcher, C. S.; Isotherm-Based Thermodynamic Models for Solute Activities of Organic Acids with Consideration of Partial Dissociation; *J. Phys. Chem. A*, 120 (24), 4147–4154, **2016**. http://doi.org/10.1021/acs.jpca.6b01904

Rindelaub, J. D., Craig, R. L., **Nandy, L.**, Bondy, A. L., Dutcher, C. S., Shepson, P. B., and Ault, A. P.; Direct Measurement of pH in Individual Particles via Raman Microspectroscopy and Variation in Acidity with Relative Humidity; *J. Phys. Chem. A*, 120 (6), 911-917, **2016**. http://doi.org/10.1021/acs.jpca.5b12699

Nandy, L. and Dutcher, C. S., Phase behavior of ternary ammonium sulfate and dicarboxylic acids mixture solutions in aqueous aerosol mimics (in preparation)

505211 Molecular Modeling and Simulation for Energy, Environment and Life Science

Abstract:

Research Interests:

My postdoctoral research in the group of Prof. A. Z. Panagiotopoulos at Princeton University has been focused on the design of molecular force fields for water, CO2 and electrolytes, as well as the development of algorithms for molecular dynamics simulations of interfacial wettability and nucleation phenomena. In particular, I have designed new polarizable force fields that substantially improve the accuracy of simulations relative to prior fixed-point charge force fields for thermodynamic (e.g. phase equilibrium) and transport properties (e.g. transport coefficient, viscosity) of water/CO2/electrolytes mixtures, which are crucial for CO2 geological sequestration process.[1,2] I also developed a novel free energy based algorithm in molecular dynamics simulations to study the wettability, i.e. contact angle, of heterogeneous fluid/solid systems. The algorithm can be used to obtain the contact angles of fluids (e.g. water, hydrocarbon) on both non-polar (e.g. Lennard-Jones) and polar (e.g. clay minerals) surfaces without using microscopic fluid droplet in simulations. In addition, in collaboration with Prof. Pablo Debenedetti, I have studied the homogeneous nucleation of NaCl in its supersaturated aqueous solutions, and I was able to obtain the nucleation rates directly from simulations without resorting to classical nucleation theory. During my PhD studies at University of Wyoming, I improved the statistical associating fluid theory using Monte Carlo simulations to predict the phase equilibria of natural gas hydrates under the inhibition of aqueous ionic liquid solutions.[3]

In the future, I plan to apply my expertise in molecular modeling and simulation to problems relevant for energy, environment and biophysics. First, I plan to develop advanced force fields in order to predict accurately both the thermodynamic (phase equilibrium) and kinetic (crystallization) properties of natural gas hydrate, which is a flow assurance challenge for oil & gas production, with the aim to help design novel low dosage kinetic gas hydrate inhibitors. Second, I plan to apply molecular dynamics simulations to study the heterogeneous nucleation and growth of nanoparticles (e.g. calcium carbonate, iron oxide) at mineral interfaces (e.g. quartz, clay, etc.), which is essential to understand the transport of aqueous species in underground aquifers. Third, I plan to develop coarse-grained molecular models for protein aggregation and assembly and apply these models with advanced free energy methods (e.g. umbrella sampling, metadynamics) to study the kinetics and pathways of protein self-assembly.

Teaching Interests:

My teaching interests include undergraduate and graduate level chemical thermodynamics, statistical mechanics, transport phenomena, molecular modeling and simulation methods, and scientific computing.

[1] Jiang, H.; Economou, I. G.; Panagiotopoulos, A. Z. Molecular Modeling of Thermodynamic and Transport

Properties for CO2 and Aqueous Brines. Acc. Chem. Res., 2017, 50, 751.

[2] Jiang, H.; Economou, I. G.; Panagiotopoulos, A. Z. Phase Equilibria of Water/CO2 and Water/n-Alkane Mixtures from Polarizable Models. J. Phys. Chem. B 2017, 121, 1386.

[3] Jiang, H.; Adidharma, H. Thermodynamic Modeling of Aqueous Ionic Liquid Solutions and Prediction of Methane Hydrate Dissociation Conditions in the Presence of Ionic Liquid. Chem. Eng. Sci. 2013, 102, 24.

492934 Solvation Behavior of Self-Assembled Systems: Investigating the Colloidal Interface Via Molecular Simulations

Abstract:

Single walled-carbon nanotubes (SWCNTs) possess unique properties that make them attractive for a number of applications. However, the difficulty in obtaining uniform samples in terms of size, chirality, and handedness inhibits their widespread use. Various separation techniques address this problem through use of surfactants in order to suspend the tubes in aqueous media and then apply other approaches such as aqueous two phase extraction (ATPE) or ion-exchange chromatography (IEX) to sort the dispersed SWCNTs by their physiochemical properties. Single-stranded DNA (ssDNA) is a very effective dispersant that shows sequence-specific behavior during separation. Previous experimental work has shown that this specificity can be harnessed to tune the separation in favor of particular nanotube chiralities. Currently, the nature of this specificity is not yet well understood and optimal ssDNA/SWCNT pairs must be searched by trial and error. In this study, we use replica exchange molecular dynamics (REMD) simulations to assemble the ssDNA-SWCNT complexes and then apply free energy perturbation methods to obtain an estimate for the solvation energy of the complex. Such systems require the construction of a novel free energy pathway to access the quantity of interest. Additionally, the structure and dynamics of the hydrating water has been analyzed to better understand the solvation behavior on a molecular level and to provide insight into the effects of ssDNA sequence. These investigations work towards our goal to provide better insight into the sequence/chirality specific separation mechanism, and to eventually develop a model that allows for the prediction of other pairs leading to the efficient sorting of a desired nanotube chirality.

Research Interests:

My experience with the tool of molecular simulations has allowed me to explore a number of problems and to increase the breadth of my research interests. Molecular self-assembly, solvation behavior, membrane transport, and non-equilibrium thermodynamics all provide interesting topics which are accessible via various MD schemes. I would particularly be interested in continuing my current work dealing with the solvation of self-assembled systems and to extend this to other assembled species and molecular environments. This would help to focus on the application of such self-assembles complexes as they are often constructed to behave as sensors or delivery mechanisms and the local environments are crucial to their behavior and function.

Teaching Interests:

A strong background in engineering and a wide range of research experiences would allow me to be comfortable teaching many fundamental courses in chemical engineering. Of particular interest are undergraduate courses such as thermodynamics, numerical/computational methods, and transport phenomena; as well as graduate courses: thermodynamics, advanced heat and mass transfer, engineering mathematics. Additionally, I would be interested in developing a course more closely related to my research interests that would serve as an introduction to molecular simulations. This course would not only teach the basic concepts behind this technique, but also focus on application via projects and provide a survey of new directions within the field. Such a course was not available during my time in the classroom and would have been highly beneficial as I began my research career.

505715 Chemistry and Physics of Biological Fluids on the Mesoscopic Scale

Abstract:

Research Interests: Chemistry and Physics of Biological Fluids on the Mesoscopic Scale. The complex behavior of fluids such as the cytoplasm of the cell and the blood in veins encompasses rich viscoelastic phenomena, interesting rheology and dynamics. We use state-of-the-art computer modeling and simulations to bridge the atomistic, mesoscopic and continuum length scales to study these fluids in health and disease.

Teaching Interests: Thermodynamics, computational chemistry and physics, continuum physics, statistical mechanics, numerical methods, transport

Interfacial Transport Phenomena with Applications to Environment Remediation and Biomedical Engineering

Interfaces between two distinct phases are ubiquitous in nature and many engineering processes. From the fundamental studies in physics, material science and biology, to applications in various fields, the "interface" between two fluid phases includes complex molecular and particulate structures. The mechanical, chemical, thermal and transport properties of such complex interfaces are crucial to the responses of many natural and engineering systems. **My research interests lie in understanding the fundamental interfacial transport phenomena with complex fluids and structurally compound interfaces at microscopic scales, with an emphasis on establishing multiple platforms that shed light into biomedical engineering and environmental science.** I utilize a combination of experimental, analytical and numerical tools to investigate a wide range of systems inspired by nature and industrial observations, and I am also dedicated to applying what we learn from laboratory discoveries to tackling practical problems in different fields. Currently, my research includes the following thrusts: interfacial mixing for multi-functional and bioavailable nanocluster assembly, complex structured microparticle generation and sorting using microfluidics, thin-film dynamics and instability and materials characterization using laser spectroscopy (Fig. 1, upper panels).

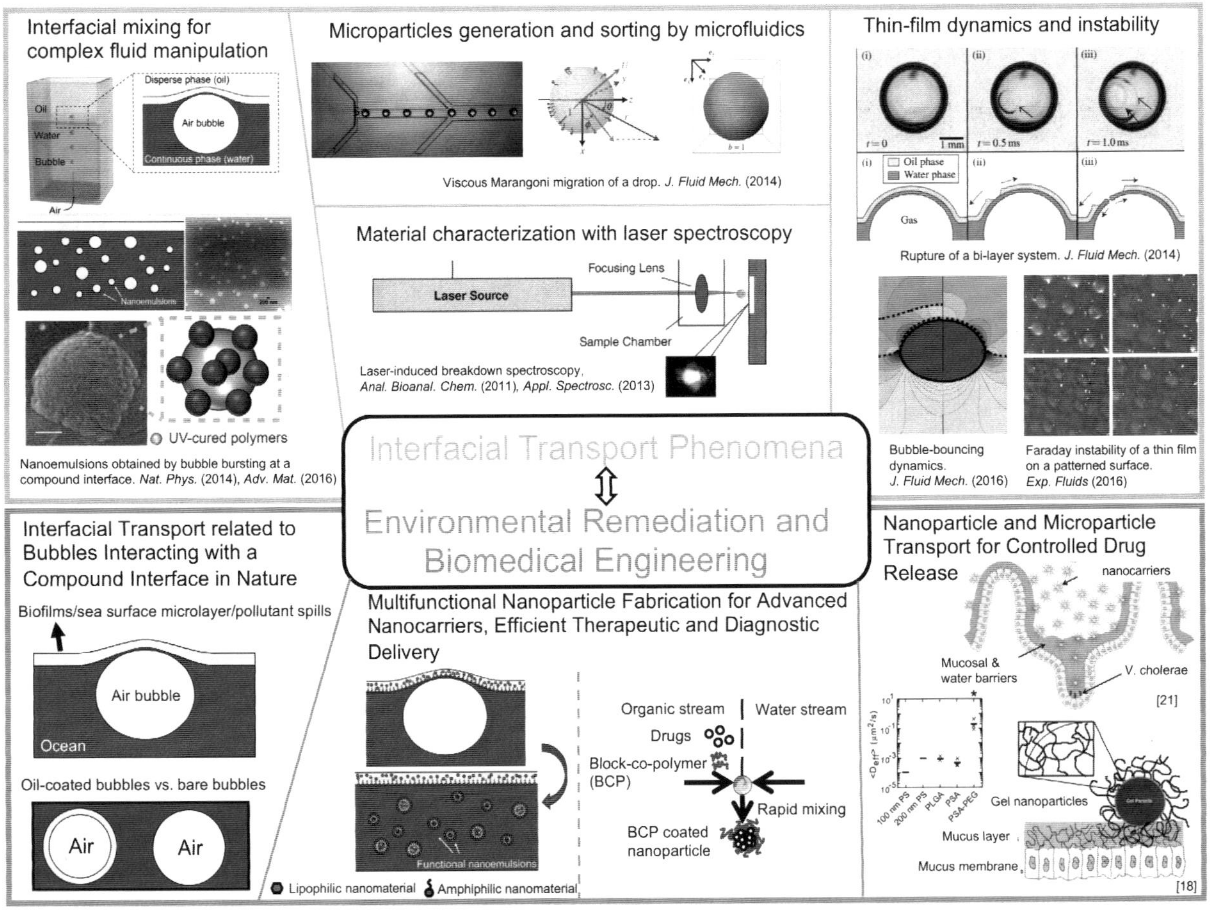

Figure 1: Overview of the proposed research plan.

Recent Research Accomplishments and Progress

In recent years, I have been focusing on understanding multiple interfacial transport phenomena and working to develop novel experimental and analytical techniques for environmental science and materials engineering. One such platform for functional nanomaterials that I recently discovered involves the production of nanoemulsions by bubbles bursting. This technique was developed based on my study of the dynamics and colloid science of bubbles bursting at an aqueous surface coated by a layer of oil. Through systematic experiments, I showed how the bursting of air bubbles in this complex system can produce nanometer-sized droplets in the water phase (published in **Nat. Phys.** [1] and featured in "News and Views" [2]). This result is surprising because earlier work of bubbles bursting at an air-water interface produced aerosol drops in the air, while our discovery reveals a new mechanism of mass transport across such a compound interface.

These results inform the design of new nanomaterial fabrication using bubbles. This dispersal system has various advantages for nanoemulsion fabrication compared with conventional emulsification methods, including enhanced energy-efficiency, as well as providing a continuous and scalable emulsification approach. Furthermore, this technique is compatible with a variety of nanomaterials (published in **Adv. Mat.** [3]). Considering the diversity of materials that can be encapsulated, along with the potential for mass production, this one-step encapsulation system can offer a new toolbox for generating (multi-)functional nanoemulsions and nanoparticles.

Another challenge in the development of polymeric nanoparticles is the inability to control the mixing processes required for stable nanoparticle physicochemical properties. Currently, I am working on developing Flash Nanoprecipiation, which utilizes interfacial mixing in a confined volume of two opposing streams to kinetically control the production of nanoparticles with low polydispersity and uniform properties stabilized by amphiphilic block copolymer. This technique provides a versatile platform for structured nanomaterial synthesis by applying interfacial transport phenomena.

Additionally, in many cases functional nanomaterials require further encapsulation into micron carriers for stability and controlled release. One important toolbox that allows complex fluid manipulations on the micron scale is microfluidics. I have been intensively involved in the research of particle movements and migration mechanism in microfluidics [4], which enables me to further fabricate multi-functional and structurally complex polymeric particles on a micron scale. Currently, I am also working on generating polymeric microparticles to encapsulate nanoparticles for the delivery of peptides, biologics and soluble therapeutics. Furthermore, many nanofabrication processes involve thin-film dynamics, and such dynamics are also relevant to the investigation of fundamental problems of polymer diffusion and adsorption, which contribute to a better control for nanofabrication related to polymer dynamics. I have also developed my analytical skills through the modeling of the thin-film instability and dynamics in bi-layer systems [5, 6] or at the presence of a designed surface topography [7]. Finally, laser spectroscopy is widely applied as a non-contact technique of characterizing various nanomaterials, and I have been recently involved in the application of laser spectroscopy to analyzing the elemental information of materials [8, 9, 10].

Throughout these experiences, I have acquired relevant experimental and analytical techniques useful for investigating interfacial transport phenomena and explore the potential for environmental science and materials engineering in multi-scale systems. With such capabilities, my group will use various ideas from investigating interfacial transport phenomena and bridge these understandings to address important questions in environmental science and materials engineering. In the following sections, I will outline the proposed research agenda that my group will pursue (Fig. 1, lower panels).

Proposed Research

1. Multi-functional nanoparticle fabrication for advanced nanocarriers, efficient therapeutics and controlled delivery

Based on my experience with both Bubble-Bursting Nanofabrication and Flash Nanoprecipitation, my group will focus on engineering nanoparticles not only to develop advanced nanocarriers for applications such as bioimaging, but also for the development of efficient therapeutics and controlled drug delivery.

The advantages of Bubble-Bursting Nanofabrication lie in its high energy efficiency, low surfactant requirements, wide compatibility with various nanomaterials and the ease of scaling-up. This interface transport phenomena can be utilized to produce a variety of composite particles. My group will explore the potential of such a bubble-bursting platform for advanced nanocarriers, such as for multimodal imaging (Fig. 2). Bioimaging is important since it can provide necessary information about the transport of biomaterials. Multimodal imaging is particularly useful in the case when the signal of one imaging agent may be interfered by other noisy emissions. A significant obstable for a wide application of multimodal and theranostic nanoparticles has been the complexity of reported synthesis approaches, and the multistep and complex nanoparticles synthesis also presents a major problem in being scaled-up [14]. However, the bubble-bursting platform provides a one-step, cost-effective and continuous approach. We have demonstrated that this technique can produce nanoemulsions encapsulating two species of quantum dots as a promising preliminary study [3] (Fig. 2).

More specifically, we will try to incorporate magnetic-resonance-imaging and near-infrared-active agents into nanoparticles by our single-step platform. Such advanced nanocarriers enables a multimodal imaging, which provides a robust imaging probe in various environments. Besides bioimaging, we will also investigate the possibility of producing structural nanomaterials by this platform, such as Pickering nanoemulsion with a good stability or liposome-stabilized nanomeulsions for vaccine delivery [15].

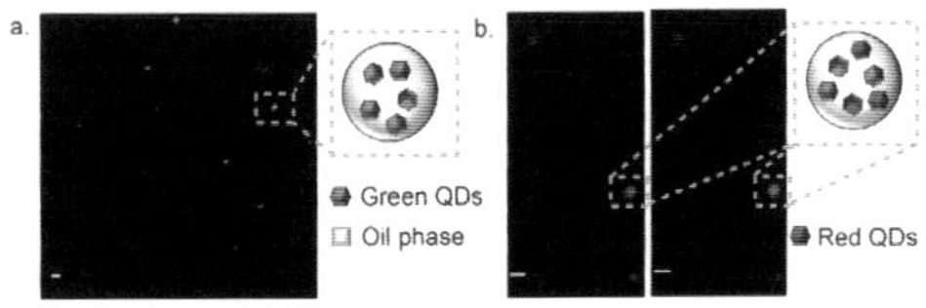

Figure 2: Simultaneous sample images of nanoemulsions encapsulating two different quantum dots, with detectors limited in different wavelength ranges [3].

On the other hand, with significant attention focused on nanoscience in recent years, nanomaterial-based drug delivery has been propelled to the forefront by researchers from both academia and industry. A lot of new developed therapeutics for diseases cannot be moved into clinical trials since the new compounds are highly hydrophobic and difficult to dose. Using Flash Nanoprecipitation, my group will develop a low-cost and scalable protocol to prepare difficult-to-deliver therapeutics into nanoparticles for drug delivery applications. My group will work closely with people from pharmaceutical chemistry and focus on the hydrophobic therapies, which are difficult to delivery in tranditional conditions. The rules for relating encapsulation efficiencies, yields, dissolution kinetics, and *in-vivo* pharmacokinetics to drug molecular properties will be formulated, which will inform the selection of excipients and processes for improving the bioavailability of the drug compounds.

Furthermore, with these "nanoparticles", the ultimate goal is to prepare dry powder compounds with high bioavailability. In particular, high bioavailability is enabled by two characteristics: high surface-area-to-mass ratio and amorphous solid structure. Flash Nanoprecipitation achieves high supersaturations in time scales of $O(1$ ms), which drives high nucleation rates that ensure nanoparticle sizes. The rapid assembly of $O(10$ ms) traps compounds in amorphouse states, which increases their bioavailability. In addition, after the particle formation process, my group will further demonstrate how the materials should be collected in dry powder form for oral drug delivery and test the storage stability of the powders. For instance, direct precipitation and filtration, spray drying and lyophilization will be tested based on various formulations to evaluate suitable and optimized conditions. In addition, with microfluidics, we are capable of generating composite microparticles encapsulating nanoparticles with hydrophobic or hydrophilic polymer scaffolds to provide an additional barrier to prolong the release of the therapeutic and decrease the burst release. Such a procedure is favored for injectable polymeric depots with peptide and protein [16].

2. Nanoparticle and microparticle transport for controlled drug release

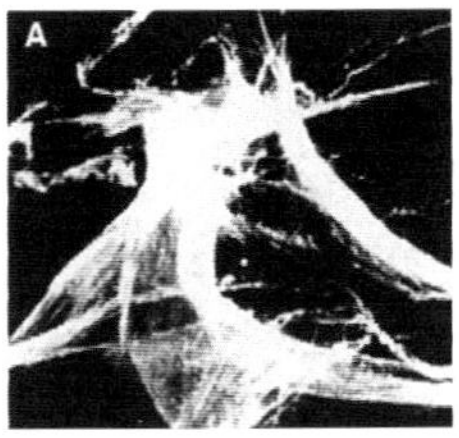

Figure 3: Fluoresecently labeled polystyrene particles formed thick cables with mucin fibers in human cervicovaginal mucus [17].

After obtaining multi-function nanoparticle and microparticles, my group will further study how these nanoparticles or microparticles transport through different biosystems. In particular, recently the development of controlled release systems for drug and gene delivery to the mucosal surfaces, such as those of the lung airways, gastrointestinal tract, female reproductive tract, nose and eye, is of widespread interest [17]. However, the viscous, elastic and sticky mucus layer that lines all mucosal tissues has evolved to protect the body by rapidly trapping and removing foreign particles and hydrophobic molecules [18]. The limited permeability of drug delivery particles and many hydrophobic drugs through the mucus barrier leads to their rapid clearance from the delivery site, often precluding effective drug therapies at non-toxic dosage.

Nanoparticles have been proposed as an efficient drug delivery system, but in order to avoid rapid mucus clearance and/or reach the underlying epithelia, nanoparticles must quickly traverse at least the outmost layers of the mucus barriers. Therefore, a key question is to understand and control how polymeric nanoparticles transport through viscoelastic mucus structures as a drug cargo.

In order to penetrate the mucus layer, synthetic nanoparticles must avoid adhesion to mucin fibers and be small enough to avoid. Recent researchers have tried to modify the surface chemistry of the nanoparticles and study the interface transport phenomena related to the parameters of the particle size and the mucus properties [19] (Fig. 3). However, there is a vast array of *in vitro* systems and animal models that have been used, which has produced discordant results regarding the optimum characteristics for efficient nanoparticle delivery. Also, how the unique rheological and adhesive properties of the mucus respond to the foreign particles needs to be further addressed [20]. Since we are capable of engineering composite nanoparticle or microparticle with various surface properties and imaging agents, we would like to study how the drug-loaded particles transport through the mucus layer. The wide compatibility for various materials and the multimodal imaging properties of the nanoparticles by our platforms enable us to observe the dynamics. In particular, we will search for a cost-effective way to increase delivery efficiency.

In order to provide insights into the issues discussed above, my group will 1) create a model system in the laboratory to isolate the physical properties of the mucus layer, through which we will identify the important controlled parameters during particle transport and build up a quantitative physical model. The coupling of the surface chemistry of the particles and the mucus properties will be identified; 2) fabricate mucus-penetrating particles based on the quantitative model. We will learn from nature itself, such as how the virus is able to penetrate the musus layer [18], and we will also try to create "active" particles for new delivery strategy. For instance, we can apply a potential field, such a magnetic force, to manipulate the nanoparticle penetration if the particles are engineered appropriately. We can also design nanoparticles with releasing dynamics, such that the released agents can adjust the properties of the mucus towards an efficient drug delivery; 3) perform *in vitro* studies to test the capability of the proposed composite nanoparticles and the related drug release dynamics. If possible, my group will work closely with researchers from biology to carry out *in vivo* experiments and understand how our nanoparticles behave in a real biological environment. Furthermore, this study can be extended to the nanoparticle transport through different biofilm systems, especially with applications to a targeted drug delivery [21].

3. Interfacial transport related to bubbles interacting with a compound interface in nature

In recent years, we have learned that there are important fluid dynamics problems involving bubbles interacting with complex interfaces. Bubbles are fundamentally important in many natural physical processes and in a host of industrial and man-related activities. In particular, they play a key role in many cross-interface approaches. For instance, sea spray, which transports dissolved gases, salts, surfactants, and biological materials to the atmosphere, is largely attributed to aerosols produced by the estimated 10^{18} to 10^{20} bubbles that burst every second across the oceans (Fig. 4).

In many complex fluid systems, the interface is not "clean". It may be complicated by a variety of particulate and bulk structures. For example, most bubbles formed in nature (e.g. in oceans, lakes, etc.) are coated by an oil layer with a thickness of 1-100 μm because the ecosystem is filled with organic materials [12]. When two such coated bubbles collide with each other, a compound liquid/liquid/gas interface consisting of air, oil and water is present at the interaction, leading to more complex dynamics than the collision of bare bubbles. In addition, most of the atmosphere/ocean interface is covered by the sea surface microlayer, an organic film containing lipids, proteins and hydrocarbons from the ecosystem [13]. This indicates that such compound interfaces exist widely in nature, but the physics of bubbles interacting with a compound interface are not fully understood.

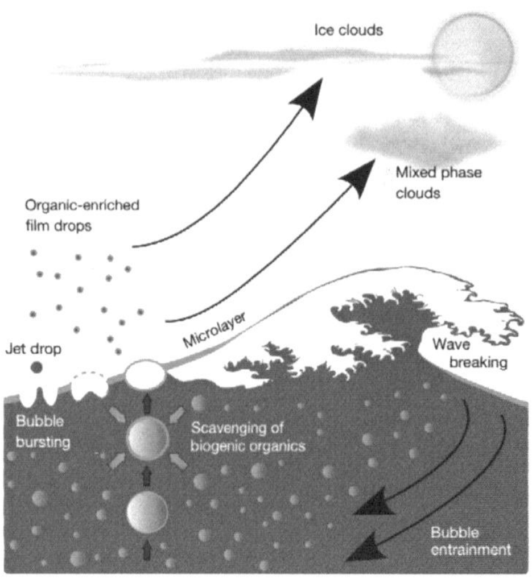

Figure 4: Sea-spray aerosol particles enriched in organic material are generated when bubbles burst at the air/sea interface [11].

To address the remaining challenges concerning bubble and particulate dynamics at compound interfaces, my group will 1) explore experimentally and theoretically the unrecognized role that the dispersal mechanism of bubbles bursting (from upper phase to the lower phase) may play in the fate of the sea surface micro-layer and of pollutant spills. In particular, we will apply biofilms and petroleum layers for the respective scenarios and observe how these materials are redistributed in the water column by bubble-bursting. The effects of various surfactants on the fluid motion and the dispersal properties will be studied in detail. We will pay close attention to the wetting state between the dispersed phase and the continuous phase since the wide presence of the surfactant-like biomaterials can significantly change the interfacial properties; 2) extend the understandings of the mass transport to environmental remediation, since this mechanism not only has implications for mass transfer processes in the oceans, but also suggests possibilities regarding the clean-up and degradation of oil spills. Small oil droplets may help bacteria or algae to degrade pollutants faster because of the high surface-to-volume ratio of the droplets, and thus, directly generating oil-in-water nanoemulsions using bubble bursting at the locations of oil spills may enhance the rate of environmental remediation, especially when other well-identified dispersal mechanisms, such as wave breaking, are limited; 3) investigate how the bubble dynamics is modified by the compound interface. For instance, we will study how both oil-coated and bare bubbles interact and coalesce with compound interfaces including organic material layers. These results will provide insights into the role of compound interfaces on particle dynamics and multiphase physics important to natural systems.

In summary, my group will combine laboratory experiments and theoretical modeling to study various interfacial transport phenomena while pay attention to insights and applications in environmental science and materials engineering. Through my research, I hope to contribute to not only the development of fundamental science but also solving practical problems for human and the society.

References

[1] J. Feng, M. Roché, D. Vigolo, L. N. Arnaudov, S. D. Stoyanov, T. D. Gurkov, G. G. Tsutsumanova, H. A. Stone. Nanoemulsions obtained via bubble-bursting at a compound interface, *Nat. Phys.* 2014, *10*, 606-609.

[2] J. Eggers. Fluid dynamics: Getting the drops in, *Nat. Phys.* 2014, 10, 548-549.

[3] J. Feng, J. K. Nunes, S. Shin, J. Yan, Y. L. Kong, R. K. Prud'homme, L. N. Arnaudov, S. D. Stoyanov, H. A. Stone. A scalable platform for functional nanomaterials via bubble-bursting, *Adv. Mat.* 2016, *28*, 4047-4052.

[4] O. S. Park*, J. Feng*, H. A. Stone. Viscous Marangoni migration of a drop in a Poiseuille flow at low surface Peclet numbers, *J. Fluid Mech.* 2014, *753*, 535-552. (*equal contribution)

[5] J. Feng, M. Muradoglu, H. Kim, J. T. Ault, H. A. Stone. Dynamics of a bubble bouncing at a liquid/liquid/gas interface, *J. Fluid Mech.* 2016, *807*, 324-352.

[6] P. S. Stewart, J. Feng, L. S. Kimpton, I. M. Griffiths, H. A. Stone. Stability of a bi-layer free film: simultaneous or individual rupture events? *J. Fluid Mech.* 2015, *777*, 27-49.

[7] J. Feng, I. Jacobi, H. A. Stone. Experimental investigation of the Faraday instability on a patterned surface, *Exp. Fluids* 2016, *57*, 1-7.

[8] J. Feng, Z. Wang, L. West, L. Li, Z. Li, W. Ni. A nonlinearized multivariate dominant factorbased partial least squares (PLS) model for coal analysis by using laser-induced breakdown spectroscopy, *Appl. Spectrosc.* 2013, *67*, 291-327.

[9] J. Feng, Z. Wang, L. West, Z. Li, W. Ni. A PLS model based on dominant factor for coal analysis using laser-induced breakdown spectroscopy, *Anal. Bioanal. Chem.* 2011, *400*, 3261-327.

[10] J. Feng, Z. Wang, Z. Li, W. Ni. Study to reduce laser-induced breakdown spectroscopy measurement uncertainty using plasma characteristic parameters, *Spectrochim. Acta B* 2010, *65*, 549-556.

[11] T. W. Wilson. A marine biogenic source of atmospheric ice-nucleating particles, *Nature* 2015, *525*, 234-238.

[12] B. D. Johnson, R. C. Cooke. Bubble populations and spectra in coastal waters: A photographic approach, *J. Geophys. Res.* 1979, *84*, 3761-3766.

[13] D. J. Carlson. Surface microlayer phenolic enrichments indicate sea surface slicks, *Nature* 1982, *296*, 426-429.

[14] M. Akbulut, P. Ginart, M. E. Gindy, C. Theriault, K. H. Chin, W. Soboyejo, R. K. Prud'homme. Generic method of preparing multifunctional fluorescent nanoparticles using flash nanoprecipitation, *Adv. Funct. Mater.* 2009, *19*, 718-725.

[15] R. L. Richards, M. Rao, T. C. Vancott, G. R. Matyas, D. L. Birx, C. R. Alving. Liposome-stabilized oil-in-water emulsions as adjuvants: increased emulsion stability promotes induction of cytotoxic T lymphocytes against an HIV envelope antigen, *Immunol. Cell Biol.* 2004, *19*, 531-538.

[16] R. F. Pagels, R. K. Prud'homme. Polymeric nanoparticles and microparticles for the delivery of peptides, biologics, and soluble therapeutics, *J. Control. Release* 2015, *219*, 519-535.

[17] S. K. Lai, Y. Y. Wang, J. Hanes. Mucus-penetrating nanoparticles for drug and gene delivery to mucosal tissues, *Adv. Drug Deliv. Rev.* 2009, *61*, 158-171.

[18] R. A. Cone. Barrier properties of mucus, *Adv. Drug Deliv. Rev.* 2009, *61*, 75-85.

[19] S. K. Lai, D. S. Harrold, S. T. Man, Y. Y. Wang, R. Cone, J. Hanes. Rapid transport of large polymeric nanoparticles in fresh undiluted human mucus, *Proc. Natl. Acad. Sci. U.S.A.* 2007, *104*, 1482-1487.

[20] L. M. Ensign, R. A. Cones, J. Hanes. Oral drug delivery with polymeric nanoparticles: The gastrointestinal mucus barriers, *Adv. Drug Deliv. Rev.* 2007, *64*, 557-570.

[21] H. D. Lu, A. C. Spiegel, A. Hurley, L. J. Perez, K. Maisel, L. M. Ensign, J. Hanes, B. L. Bassler, M. F. Semmelhack, R. K. Prud'homme. Modulating Vibrio cholerae quorum-sensing-controlled communication using autoinducer-loaded nanoparticles, *Nano Lett.* 2015, *15*, 2235-2241.

505507 Modeling Liquid Crystals, Active Matter and Other Non-Equilibrium and Nonlinear Soft Materials

Abstract:

Research Interests:

Soft Matter comprises of a wide range of materials, which are deformable or structurally switchable by thermal or mechanical stress of the magnitude of thermal fluctuations. Because of the richness and complexity in its structures and properties, soft matter is promising for design of materials with novel features. Examples are liquid crystal-based biosensors, actuators, soft robotics, and kirigami-based mechanic metamaterials. Theory and modeling of soft matter are not only important for discovering functional material by design, but also helpful for understanding the biophysics of life, for instance, protein folding, cell morphogenesis, and animal flocking; This knowledge can be in return used to help design biomimetic, autonomous materials. The equilibrium structures of many soft matter can be predicted by free energy considerations. Their non-equilibrium behaviors and responses to driven forces, are however nonlinear and therefore difficult to model due to their intrinsic microstructures. I am interested in pursuing a career on modeling non-equilibrium and nonlinear soft matter, including but not limited to liquid crystal flows, active matter, and mechanic metamaterials.

Specifically, I propose to work in the following research directions as a faculty:

1) Liquid crystal flows. Liquid Crystals exhibit versatile mesophases, including nematic, smectic A, smectic C, columnar phases and their chiral counterparts. Nematic phase refers to a system when rod-like particles exhibit long-range orientational ordering but remain spatially disordered. The other mesophases encompass partial spatial ordering. These mesophases can engender distinct, periodic defect structures, for example blue phases and focal conic domains, which are useful for novel photonic devices and organizing and sorting microparticles and even surfactants. The above liquid crystal phases can be found in many dense suspensions of aspherical particles, such as carbon nanotubes, DNAs, virus, and certain drugs. Modeling their hydrodynamic behaviors could help us better understand the rheology of dense suspensions with emerging ordering. One could also use liquid crystals to transport cargos and separate particles. Therefore, a robust simulation technique on liquid crystal flows is beneficial. During my postdoc, I have developed a simulation tool for flowing nematics. I envision my future research direction to be modeling the flow of chiral nematics and other higher ordered liquid crystal mesophases.

2) Active Matter. Active matter converts other forms of energy into local mechanic work and drive the system out of equilibrium. Consequently, a spontaneous flow and structural ordering may emerge. Active matter can be biopolymer networks, bacteria suspensions, self-propelling colloids, and even vibrating rods. Active matter could be the candidate for biomimetic, autonomous materials. The major reason that hinders its wider and further applications is that it is very difficult to control and tailor the spontaneous flows. A special type of active matter is namely active nematics, which comprises of elongated particles and exhibit nematic phases. In such systems, topological defects may persistently move. If one is able to manipulate the motion of defects, one could transport defect-carried particles. As my second research direction, I plan to explore different active matters and find a uniform model to better elucidate their dynamics.

3) Mechanic Metamaterials. Elastic materials show linear response to small stresses. At higher stress, they may undergo buckling instabilities. This nonlinear behavior has been exploited to engineer mechanic metamaterials, which exhibit for example negative Poisson ratio. Another interesting direction for elastic materials is to design foldable three-dimensional structures out of two-dimensional materials. This simple, hierarchical material design philosophy could fundamentally advance our current manufacturing technique. However, the possible structures that lead to useful functions are not fully exploited. I plan to develop a robust simulation tool to handle buckling instabilities and able to predict the mechanic response of the structure under certain mechanic stresses.

During my Ph. D, I worked with Prof. Joel Koplik at Levich Institute in CCNY, on multi scale simulations of microfluidic phenomena. During my postdoc experience with Prof. Juan de Pablo at The University of Chicago, I developed a parallelized hybrid lattice Boltzmann code to simulate flowing nematic LC system, and applied it to the biopolymer systems to understand active nematics phenomena.

Teaching Interests:

(Computational) Fluid Mechanics; Thermodynamics; Statistical Physics; Transport phenomena; Liquid Crystals; Advanced Simulation Method for Soft Matter

487435 Multiphase Interactions to Create Designer Material

Abstract:

Multiphase Interactions to Create Designer Material

3rd Year Postdoctoral Fellow

PhD Dissertation: Dynamics of droplet shedding and coalescence under the effect of shear flow.

Under supervision of Prof. Nabil Esmail and Prof. Ali Dolatabadi, Department of Mechanical Engineering, Concordia University, Montreal, Canada.

1st Postdoctoral Project: Study of droplet impact and shedding on laser treated substrates.

Under supervision of Prof. Anne Kietzig, Department of Chemical Engineering, McGill University, Montreal, Canada.

2nd Postdoctoral Project: Study of unit operations and material characterization involved in continuous manufacturing of solid dose pharmaceutical products.

Under supervisor of Prof. Fernando Muzzio. Department of Chemical and Biochemical Engineering, Rutgers University, NJ, USA.

Research Experiences:

During my academic career, I was involved in various engineering fields from fluid dynamics to pharmaceutical unit operations. My former training in PhD and first postdoctoral fellowship was in experimental and computational fluid dynamics and surface engineering. Where I learned various simulation techniques such as Volume of Fluid and Smoothed Particle Hydrodynamics in addition to experimental techniques in shadowgraphy and laser micromachining. During my second postdoctoral fellowship, I was involved in characterizing pharmaceutical unit operations where I acquired invaluable experiences in pharmaceutical manufacturing, unit operations, and material characterization.

Research Interests:

The general theme running through my research activities is my great interest in multiphase interactions and surface engineering. My goal is to establish a topnotch multiphase flow and surface engineering research program emphasizing their practical applications in various engineering fields. I have found that I greatly benefit from working on multidisciplinary projects. I enjoy bringing together individuals with different backgrounds and approaches, to allow them to spark off novel perspectives. In this respect, I have collaborated with researchers from many different fields. My research objectives are to take advantage of theoretical, computational, and experimental tools for better understanding of the multiphase interactions. My emphasis is always to convert the knowledge obtained into successful designs with a wide range of engineering applications.

My immediate research plans that follow from my Ph.D. and postdoctoral research work can be categorized as follows.

Droplet dynamics:

Droplets interactions with substrates involve in variety of industrial applications from ink-jet printing to rain impingement (and consequential icing phenomenon) on structural systems like power-lines, wind turbines and airfoils. During my PhD and first postdoctoral fellowship, I performed a fundamental experimental and numerical study on droplet dynamics. I investigated the effect of surface wettability and air flow speed in predicting the drop behavior in terms of shedding and coalescence. I showed that on a hydrophilic substrate (providing high enough air speed), narrow streams of liquid were formed from merging droplets. In contrast, on a superhydrophobic substrate coalesced droplets rolled on the surface and detached from it when the air speed was sufficiently high. In addition, the results indicated a contrast in the mechanism of the coalescence and subsequent detachment between a single and two droplets on a superhydrophobic surface. At low air speeds, the two droplets coalesced by attracting each other before detaching with successive rebounds on the substrate, while at higher speeds the detachment occurred almost instantly after coalescence, with a detachment time decreasing exponentially with the air speed. The experimental results were also compared with the results of numerical simulation based on Volume of Fluid method to give a better understanding of the phenomena by analyzing the aerodynamic forces, velocity vectors, and stream lines on the droplets.

In continuation of this work, I would like to investigate the effect of liquid viscosity and surface tension in droplet behavior in terms of shedding, impact and coalescence. I am also interested in studying the effect of surface and ambient temperature on droplets dynamics, specifically, in freezing conditions. There, the heat transfer equations will play an important role in modelling the system's dynamic. Solidification has various potential applications from icing to surface coating with thermal spray techniques where molten particles with high velocity hit the target surface and solidified on it. Investigating the effect of droplet size (sub-micro droplets) can be another interesting topic for research. There, the pillar size (micro and nano structures) existed on substrates becomes comparable with the size of the droplet. Therefore, the effect of surface roughness on the droplet dynamics should also be included in the experimental and numerical analyses.

Thermal spray process to create durable superhydrophobic surfaces:

Thermal spray techniques are used extensively in aerospace, automotive, and power generation industries to provide protective coatings on engine, landing gear, and turbine components that are exposed to heat, corrosion, and wear. It is also found that the suspension plasma spray technique has the potential to be used as a suitable method for creating durable superhydrophobic coatings. This approach is much more affordable compared to other techniques such as laser etching. Superhydrophobic coatings can be used to generate waterproof, anti-corrosive, and anti-icing surfaces. These surfaces can also be used as protection in circuits and grids due to their self-cleaning properties. In aerospace industries, these surfaces are used to reduce the drag force and increasing the speed which lead to higher efficiency. However, the durability of the superhydrophobic coatings is generally weak today due to their delicate micro and nano structures. This is the main reason that limit their applications. During my collaborative work in "Thermal Spray Lab" at Concordia University, I grasped knowledge in thermal spray processes and their applications. Therefore, due to my interest in wetting science, I am interested in a detailed study on applying thermal spray techniques to generate durable superhydrophobic coatings.

Wettability of granular materials:

Granular materials are widely used in various industries such as pharmaceuticals, food, agricultural chemicals, detergents, and catalysts. Wetting properties of both raw and processed material is one of the significant characteristics involved in various process steps from formulation development to the design of manufacturing pathway. It is known that granules' wetting properties will change significantly while undergoing different manufacturing processes. It is also known that granules' wettability affects their dissolution performance which can be a significant aspect for industries such as pharmaceuticals. Therefore, knowing the wetting properties of granular material is inevitable for quality control in advanced pharmaceutical manufacturing. Having closely worked with pharmaceutical industries during my postdoctoral fellowship at Rutgers, and considering my background in wetting science, I am passionate to carry on research on developing experimental and numerical methods to characterize granular material wettability. Together with the wettability analysis, I am also interested in doing research on droplet impact dynamics on porous material.

Teaching Experiences:

In the middle of my third year of undergraduate studies I became an instructor for teaching engineering software. It was initially a huge challenge for me to prepare the lecture, and develop exam questions. However, I truly enjoyed teaching and seeing how my students learned from me. I received good feedback from the attendees. This first experience encouraged me to continue working as a teaching assistant and guest lecturer for six years through my master as well as during my PhD. Again, I received good feedback not only from the main instructors of those courses, but also from the students. Currently, as an adjunct lecturer and postdoctoral fellow at Rutgers University, I am supervising various students, who are working towards their thesis projects. Due to my passion for teaching, alongside my research, I teach one course each semester.

Teaching Interests:

Having a strong background in engineering, I have the experience and knowledge to teach basically all fundamental chemical and mechanical engineering courses. Particularly, if assigned to me, I would be excited to teach: at the undergraduate level: unit operations, reaction engineering, fluid mechanics, heat and mass transfer, thermodynamics, statics, dynamics, computer fundamentals, numerical methods, and engineering mathematics; at the graduate level: advanced fluid mechanics, computational fluid dynamics, advanced heat and mass transfer.

Furthermore, I would be thrilled to develop a course on the graduate level that is more closely related to my field of research including but not limited to the theory and application of multiphase flows which introduces students into the various techniques and processes that emerged recently at the forefront of research. Apart from familiarizing students with different techniques, they would be required to come up with their own application ideas as part of a course project. This course would not only encourage interdisciplinary thinking and team work, but also will attempt to trigger new research and applications in one of the core engineering research fields.

505445 Spherically Confined Colloidal Suspensions of Hydrodynamically Interacting Particles: A Model for Intracellular Transport

Abstract:

Research Interests:

We study the diffusion in and rheology of hydrodynamically interacting colloids confined by a spherical cavity via dynamic simulation, as a model of intracellular and other confined biophysical transport. The modeling of transport and rheology in such confined inhomogeneous soft materials requires an accurate description of the microscopic forces driving particle motion, such as entropic and hydrodynamic forces, and of particle interactions with nearby boundaries. Previous models of such micro-confined transport behavior had been limited primarily to a single particle inside a spherical cavity. Although attempts had been made to extend such models to more than one confined particle, none had yet successfully accounted for the effects of hydrodynamics, owing to the difficulties of modeling many-body long-ranged interactions. To accurately model spherically confined suspensions, we derived new far-field mobility functions and, together with the appropriate near-field resistance functions, implemented them in a Stokesian-dynamics like approach. The method fully accounts for all many-body far-field interactions and near-field interactions both between the particles themselves and between particles and the enclosing cavity. Utilizing our newly developed method, we study short- and long- time self-diffusion at equilibrium, with a focus on the dependence of the former on particle positions relative to the cavity, and of both on volume fraction and size ratio. We find the cavity exerts qualitative changes in transport behavior, such as a position dependent and anisotropic short-time self-diffusivity and anisotropic long-time transport behavior. Such qualitative changes suggest that careful interpretation of experimental measurements in 3D confined suspensions requires accounting for such confinement induced behaviors. To elucidate the effects of confinement on inter-particle hydrodynamic interactions, we utilize our method to determine the concentrated mobility of particles in the spherically confined domain. We find that confinement induces qualitative changes in the functional dependence of particle entrainment with inter-particle separation. For widely separated particles, the functional dependence on inter-particle separation can be predicted via a Greenâ€™s function, we discuss how this behavior can be utilized to develop a more accurate framework for two-point microrheology measurements near confining boundaries.

Teaching Interests:

My primary teaching interests are in topics related to fluid mechanics, transport phenomena, soft matter physics and mathematical methods at both the undergraduate and graduate level. In teaching, I am interested in instilling in students the need to understand these subjects from first principles, and in clearly communicating both the subject matter and its importance in practical applications.

505519 Computational and Experimental Investigation of Membrane Biomechanics

Abstract:

Research Interests:

Potentially toxic drugs and compounds from endogenous or exogenous sources constantly interact with the human body. Many of these compounds are small molecules that alter the physicochemical properties of physiological membranes, including changing permeability, fluidity, packing, electrostatic potential, rheology, and barrier function. In many instances, this may lead to the development of various pathophysiological states. The molecular mechanisms underlying these changes and their effects on proteins and other membrane structures are not entirely clear and depend strongly on the chemical identities of the offending compounds. My research interests focus on elucidating these molecular mechanisms through a combination of computational and experimental techniques and revealing their role in pathogenesis.

As a faculty member, I will build upon the experience and skills I have gained during my postdoctoral and graduate school training. In my lab, I look forward to applying a combination of: (i) coarse-grained molecular dynamics simulations, (ii) atomistic simulations, (iii) spectroscopic methods, and (iv) cell biology techniques. By utilizing these methods I aim to elucidate the mechanisms underlying the biophysical and biomechanical consequences of molecules involved in various pathophysiological states, leading to cardiovascular, metabolic, and immunological diseases. My research expertise, experience authoring scientific publications and writing grant proposals, as well as my experience mentoring both graduate and undergraduate students will enable me to establish a competitive research program that focuses on topics at the interface of Chemical Engineering and Biomedicine.

In my current position as a National Institutes of Health (NIH) Pulmonary T32 Postdoctoral Research Fellow at the University of Illinois College of Medicine, under the supervision of Dr. Irena Levitan, I investigate the influence of bioactive oxidized species on model bilayer and cell membrane biomechanics, as well as the relationship between of cell volume changes and membrane tension. Oxidized species have been implicated in the development of atherosclerosis. In this role, by utilizing a combination of experimental and computational methods, I aim to uncover the molecular mechanisms underlying the paradoxical impact of pro-atherogenic compounds on endothelial cell stiffening and membrane biomechanics.

As a PhD student in the Chemical Engineering Department at the University of Illinois, Chicago, I undertook coarse-grained molecular dynamics simulations to probe the effects of small molecules on the structural and dynamic properties of phospholipid layers. Specifically, I investigated the membrane altering capability of toxins as applied to: (i) the differential cytotoxicity of anticoagulant species, (ii) the role played by endogenous metabolites in causing consciousness impairment, and (iii) the association of various drugs with lipid emulsions acting as detoxification agents. Using my simulation approach, I was able to obtain detailed molecular-level information on the absorption, transport, and partitioning of toxin molecules into model membranes, which provided useful insight and predictions.

Teaching Interests:

In addition to my research experience, I have extensive teaching experience, from elementary through college level. My teaching interests are Numerical and Computational Methods in Chemical Engineering, Biochemical Engineering, Material & Energy Balances, Transport Phenomena, Thermodynamics, and other special topics related to my research background. Overall, my expertise in interdisciplinary science will serve as a framework for innovative study using both computational and experimental techniques.

505502 Engineering Metal Surfaces Via Electrochemical Reactions for Advanced Functionalities

Abstract:

Research Interests:

My research interests are surface modifications of metals by electrochemical reactions to attain advanced functionalities. Due to beneficial combination of mechanical strength and ductility, metals and metal alloys are widely used in many industries ranging from home appliances and infra structures to marine, petrochemical and biomedical. Therefore, the extensive and varied applicability of metals often require exposure to various environment including liquids, ice, seawater, hydrocarbons and physiological solutions. My research focuses on understanding phenomena at metal-environment interfaces and design of the metal surfaces to advance their functionalities.

To establish rational strategy to control the interfacial behavior and improve their functionalities, controls over surface structure and surface chemistry are critical. Electrochemical methods enable tuning surface structures from micrometer to nanometer scale as well as chemical composition. My first research goal is to attain relationship between electrochemical variables (i.e. potential, current density, and electrolyte) and metal surface characteristics. Second step is to investigate interactions between modified metal surfaces and exposed environment such as wettability, phase change (i.e. condensation and icing), fouling, and corrosion. Based upon the knowledge I would acquire from my first and second goals, design strategy to improve functionalities can ultimately be suggested. I believe that fundamental studies on metal surface engineering can also be applied to energy related fields such as electrode designs in battery and fuel cells.

PhD Dissertation: Modification of Metal Surfaces for Advanced Functionalities

Research Experiences:

My academic trainings have been focused on surface/interface engineering of metals and polymers to improve performance of applications. During my Ph.D. study in Georgia Tech, I have mainly focused on surface modification of metals by electrochemical reactions. I have developed new strategies to control topography of metals and studied effect of structures in multi-length scale on wettability, corrosion resistance, and interaction with bio-organisms such as protein, bacteria and cells. My M.S. study in Seoul National University in South Korea focused on engineering nano-morphology of polymer blend thin films in organic solar cells. For this study, I utilized small angle neutron scattering to scrutinize the phase separated morphology and correlated the morphological information to performance of organic solar cells. In addition, I have research experience in developing remote control fuel cell airplane during my internship in Korea Institute of Science and Technology. Bevy with diverse research experience ranging from surface/interface engineering to device/system development, I can be a mentor to students performing diverse research projects.

Teaching Interests:

During my teaching carrier development, I enjoyed serving as a teaching assistant in "Thermodynamics" and "Senior Design" at Georgia Tech. Thermodynamics is one of the core course in chemical engineering and I was fortunate to teach recitation class every week. During recitation sessions, I learned how to clarify concepts step by step to help students' understanding. Senior design requires students to tackle real industrial problems with their chemical engineering knowledge. As a TA, I needed to be prepared to guide their problems with multiple aspects such as thermodynamic feasibility, kinetical concerns, and process designs and economics. These experiences made me confident in teaching chemical engineering stem courses. Also, from my research experience, I am confident in teaching specific courses "Electrochemistry", "Surface and Interface Engineering", and "Scattering Theory and Applications". Also, I have experience mentoring undergraduate students, and helped them to develop and trim their creative thought processes. Based on these teaching and mentoring experiences, I learned PI's roles how to help students to overcome obstacles and encourage them to proceed.

Future Direction:

I would like to continue my academic career on modification of metal surfaces and understanding interfacial behaviors. Particularly, I am interested in surface modifications of metals and metal alloys used in chemical and petrochemical industry such as heat exchangers and hydrocarbon cracking reactors. Surface characteristics play a pivotal role in interfacial behavior and/or reactions at metal surfaces. Therefore, proper control of metal surfaces with mechanical stability can improve performance of their applications. For example, surface wettability of heat exchanger has much to do with condensation rate and mobility of condensate at the metal surface, which can largely influence on heat transfer efficiency. Like this, I would like to focus on understanding role of surface characteristics on interfacial behavior and developing functional metal surfaces that have practical relevance. Besides, I have interests in predicting interfacial behavior depending on surface characteristics. In order to achieve this goal, I would like to collaborate with researchers who have expertise in simulations and modeling, which I think will shed light on in-depth understanding on interfacial phenomena and future pathways on my research.

Selected Publications:

(1) **W. T. Choi**, K. Oh, P. M. Singh, V. Breedveld, and D. W. Hess, Hydrophobicity and improved localized corrosion resistance of grain boundary etched stainless steel in chloride-containing environment, *Journal of The Electrochemical Society*, 164, C61-C65 (2017)

(2) **W. T. Choi**, K. Oh, P. M. Singh, V. Breedveld, and D. W. Hess, Wettability control of stainless steel via evolution of intrinsic grain structures, *Journal of Materials Science,* 51, 5196-5206 (2016)

(3) **W. T. Choi**, J. Song, J. K. Ko, Y. Jang, T. Kim, Y. Han, J. Lim, C. Lee, and K. Char, Effect of solvent additives on bulk heterojunction morphology of organic photovoltaics and their impact on device performance, *Journal of Polymer Science Part B: Polymer Physics,* 54, 128-134 (2016)

(4) **W. T. Choi**, K. Oh, P. M. Singh, V. Breedveld, and D. W. Hess, Systems and methods for producing anti-wetting structures on metallic surfaces, *US Patent* No. 2017/0114472 A1, Assignee: Georgia Tech Research Corporation (2017)

505628 Explore Colloidal and Interfacial Phenomena in Complex Fluids: From Isolated Fluid Particles to Their Close Packing Structures

Abstract:

Research Interests:

Complex fluids are aqueous solutions of dispersed colloids (solids, drops, and bubbles) and solute species such as polymers and surfactants. The complexity is often characterized by diverse interactions among components in the solution and manifested in a wide range of colloidal and interfacial phenomena. While typically occurring at nano or micro scales, many of these phenomena can further impact macroscopic fluid properties, presenting topics that are of interest to both scientific studies and practical applications. The goal of my research is to obtain fundamental understanding of colloidal and interfacial phenomena in complex fluids and formulate rationales for engineering bulk material properties.

My research training involves utilizing microfluidic systems to study electro-kinetics and fluid dynamics of biopolymers and particle dispersions. As a graduate student advised by Prof. Victor Ugaz, I performed DNA gel electrophoresis in microchip, seeking enhanced performance of DNA separation by studying interactions between DNA molecules and hydrogel pore morphologies. My postdoctoral studies with Prof. Todd Squires examined particle migration in chemical gradients. I have learned and designed microfluidic devices, which enable versatile chemical gradients to be imposed and colloidal migration to be observed and measured. In particular, I am now studying, both experimentally and theoretically, convective fluid flows induced by surfactant gradients.

My future research will focus on behaviors of fluid particles (drops and bubbles) in dilute and concentrated emulsions. First of all, for individual particles in diluted emulsions, I will explore how interfacial mass transport and fluid flows are coupled at particle surfaces, particularly with the presence of multiple surfactants and anisotropic surface properties. Second, concentrated fluid particles in confined channels start to pack into denser configurations, forming unique structures involving islands of carrier fluids connected by thin liquid bridges. My particular research thrust is 1) to develop the capability to produce such structurers with robust properties by exploring the parameter space dictating the fluid particle packing process and 2) to exploit their applications. For example, thin bridges in the structure can be used to probe colloid dynamics (in confined spaces) and to enforce diffusion-dominated solute transport.

Teaching Interests:

As a future instructor, I see my job not only in imparting materials from textbooks, but also in helping students to develop the capability of acquiring knowledge independently and continuously. This requires a considerable shift from the conventional teacher-dominated concept to a more student-centered methodology. My teaching philosophy is therefore formulated to reflect my efforts to achieve this by focusing on three essential elements in teaching activities: 1) recognizing students as individual learners, 2) encouraging self-evaluations, and 3) promoting active learning.

My mentoring/teaching experiences include advising undergraduate in summer research program and serving as teaching assistant (numerical methods and heat transfer). In particular, I have developed considerable interests to integrate programming with teaching and research activities, thus addressing broader audiences. As an example, I programed an iPad app (â€œBrownian Dynamics,â€ available on Apple App Store) that allows people to interact with and visualize polymer structures in various environments.

505634 Tailoring Functionality from Disorder : Complex Nonequilibrium Phenomena at Biological and Nanomaterial Interfaces

Abstract:

Research Interests:

Materials design is at the forefront of pioneering technology to address some of our most challenging global issues. One notable direction has been to leverage unique behavior that arises at the interface of dissimilar material constituents and phases, e.g., nanomaterials, complex fluids, and soft condensed materials. The additional interactions at these interfaces typically induce molecular structuring that deviates from the characteristic disorder found in bulk counterparts. Intriguingly, such subtle microscopic changes tend to propagate to macroscopic changes in material properties. Furthermore, the use of external forces, such as mechanical strain or electrical bias, introduce a dynamic means of tuning interfacial phenomena. Hence, identifying the key molecular mechanisms at play, which can span multiple length and time scales, will facilitate the design of hybrid materials with tailored functionality. To pursue this goal, I am particularly interested in (1) understanding the principles used by biological building blocks (i.e., proteins, lipids, and nucleic acids) to regulate cellular processes and (2) adapting these principles to create functionally diverse nanomaterials. Given the crowded environment within cells, it is fascinating that many concurrent yet distinct biological activities proceed with high specificity. I anticipate that the fundamental structure-function relationships and mechanisms that regulate these inherently dynamic processes can be leveraged toward materials design in applications including energy storage and conversion, drug targeting and delivery, and water purification. To this end, I envision a research program that utilizes my extensive background in multiscale computational simulations and unites my interests in nanotechnology, biophysics, and sustainability.

Successful Proposals: Ruth L. Kirschstein National Research Service Award (NIH F32)

Postdoctoral Project: "Multiscale Simulations of HIV-1 Assembly, Budding, and Maturation" under the supervision of Gregory A. Voth, Department of Chemistry, Institute for Biophysical Dynamics, James Franck Institute, The University of Chicago

PhD Dissertation: "First-Principles Investigation of Carbon-Based Nanomaterials for Supercapacitors" under the supervision of Gyeong S. Hwang, McKetta Department of Chemical Engineering, The University of Texas at Austin

Research Experiences:

My training, which has been highly interdisciplinary, has culminated in a hierarchical computational skillset that ranges from nano- to micro-scales. As a graduate student, I developed a combined quantum mechanical and classical molecular dynamics framework to study low-dimensional material interfaces. I primarily focused on understanding charge storage mechanisms at the interface of carbon nanomaterials and ionic liquid electrolytes for supercapacitor applications. As a postdoctoral scholar, I am now extending my expertise in atomistic simulations with coarse-graining (CG) theory and enhanced sampling techniques. In my present work, I am utilizing CG approaches to investigate viral assembly and budding at cell membrane interfaces to discover new avenues for anti-viral treatment. I have also had the benefit of many fruitful collaborations with my experimental colleagues, especially through my participation in ARPA-E, NSF and NIH-funded interdisciplinary centers. Through these connections, I have been privileged to engage with experts that have broadened by scientific perspectives and have allowed me to explore new directions for my research. I am confident that these comprehensive experiences will enable me to address unique scientific questions at the interface of synthetic materials and biology.

Teaching Interests:

Throughout my education and early career, my mentors have been invaluable for my growth as an independent scientist and engineer. As a faculty member, I eagerly look forward to my responsibility as a mentor and teacher for the next generation of young scientists. My core teaching philosophy can be described by three tenets. My foremost goal is to instill a passion for learning by fostering the curiosity and confidence to be inquisitive without fearing failure. One of my most rewarding experiences has been to design interactive activities that highlighted interesting concepts in nanotechnology and computational research, which I further showcased to local students ranging from elementary to high school levels. My second goal is to create tangible connections between what students learn in academic settings and real-world problems. For this reason, I firmly believe in the value of immersive, project-based learning as a supplement to lectures. Not only do these projects help translate engineering skills toward problem solving, but also help establish communication and collaborative skills. During my undergraduate studies, I appreciated the value of such courses, one of which eventually inspired me to pursue computational research in graduate school. Finally, I hope to create an environment for my students to practice a critical-thinking mindset, e.g., to question underlying assumptions and to develop physical intuitions. In my experience as a teaching assistant (for undergraduate courses on fluid mechanics and computational methods), I regularly held recitation sessions with time allocated to stimulate such discussions; many of my students have told me that they found this approach refreshing and helpful for understanding the material. I plan to continue to evolve my teaching philosophy during my academic career. While I am prepared to teach any course, I am most interested in fluid mechanics, thermodynamics, or numerical methods. I am also looking forward to developing new courses based on my core competencies, such as a broad introduction to the growing field of computational sciences.

Selected Publications:

1. **A. J. Pak**, J. M. A. Grime, P. Sengupta, A. K. Chen, A. E. P. Durumeric, A. Srivastava, M. Yeager, J. A. G. Briggs, J. Lippincott-Schwartz, G. A. Voth, "Immature HIV-1 lattice assembly dynamics are regulated by scaffolding from nucleic acid and the plasma membrane," under review (2017).
2. **A. J. Pak** and G.S. Hwang, "Molecular insights into the complex relationship between capacitance and pore morphology in nanoporous carbon-based supercapacitors," *ACS Appl. Mater. Interfaces*, 8 (2016), 34659-34667.
3. W. Wei, L. Chang, K. Sun, **A. J. Pak**, E. Paek, G. S. Hwang, Y. H. Hu, "The bright future for electrode materials of energy devices: Highly conductive porous Na-embedded carbon," *Nano Lett.*, 16 (2016), 8029-8033.
4. **A. J. Pak** and G. S. Hwang, "Charging rate dependence on ion migration and stagnation in ionic liquid-filled carbon nanopores," *J. Phys. Chem. C*, 120 (2016), 24560-24567.
5. **A. J. Pak** and G. S. Hwang, "Theoretical analysis of thermal transport in graphene supported on hexagonal boron nitride: The importance of strong adhesion due to p-bond polarization," *Phys. Rev. Appl.*, 6 (2016), 034015.
6. D. Wu*, **A. J. Pak***, Y. Liu, X. Wu, Y. Ren, Y. Zhou, Y. Tsai, M. Lin, H. Peng, G. S. Hwang, and K. Lai, "Thickness-dependent dielectric constant of few-layer In2Se3 nano-flakes," *Nano Lett.*, 15 (2015), 8136-8140.
7. **A. J. Pak**, E. Paek, and G. S. Hwang, "Impact of graphene edges on enhancing the performance of electrochemical double layer capacitors," *J. Phys. Chem. C*, 118 (2014), 21770-21777.
8. **A. J. Pak**, E. Paek, and G. S. Hwang, "Tailoring the performance of graphene-based supercapacitors

using topological defects: a theoretical assessment," *Carbon*, 68 (2014), 734-741.
9. E. Paek, **A. J. Pak**, and G. S. Hwang, "A computational study of the interfacial structure and capacitance of graphene in [BMIM][PF6] ionic liquid," *J. Electrochem. Soc.*, 160 (2013), A1-A10.

* indicates co-first authorship

Full list can be found in my Google Scholar profile (https://scholar.google.com/citations?user=J19dfCQAAAAJ&hl=en)

505077 Computational Micro/Nanofluidics

Abstract:

Research Interests:

I am interested in understanding transport phenomena in fluid medium at micro/nanoscales using high-performance computing. These transport phenomena have broad applications in energy storage, water/air purification, and particle/cell focusing and separation, just to name a few. The success of micro/nanofluidic devices hinges on the precise control of the fluid, ion, thermal, and particulate transport in these devices. The powerful codes I developed (http://miccomcodes.org) for simulations of particulates suspended in micro/nanosystems governed by electrostatic, magnetostatic, and hydrodynamic interactions open up numerous opportunities to explore and understand the fundamental physics of transport phenomena in micro/nanosystems for diverse applications.

Teaching Interests:

At both undergraduate and graduate level, I can teach any course in the transport phenomena and thermodynamics area. I can also develop graduate courses in extreme-scale scientific computing by leveraging my expertise in high-performance computing and scalable scientific software.

Successful Proposals:

Length-scale bridging computational scheme for structure and transport, assisted writing annual renewal proposal for postdoctoral supervisor Dr. Olle G. Heinonen

Postdoctoral Projects:

Length-scale bridging computational scheme for structure and transport, under the supervision of Dr. Olle G. Heinonen at the Materials Science Division, Argonne National Laboratory

Continuum-Particle Simulation Software in Midwest Integrated Center for Computational Materials, under the supervision of Prof. Juan J. de Pablo at the Institute for Molecular Engineering, University of Chicago

PhD Dissertation:

Nonequilibrium Transport of Ionic Liquids in Electrified Nanosystems, under the supervision of Prof. Rui

Qiao at the Department of Mechanical Engineering, Virginia Tech

Research Experience:

My academic career has been focused on developing and utilizing high-performance computer simulations, that span multiple length scales (nanometer to micrometer) and time scales (picosecond to second), for scientific research in the forefront of multiple physical domains (electrostatics, magnetostatics, hydrodynamics, ion transport, to name a few). I have developed powerful mesoscale computational tools for computing electrostatic/magnetostatic interactions between arbitrary-shaped particles, and for modeling Brownian dynamics of hydrodynamically interacting objects in general geometries. I have made the code publicly available through the http://miccomcodes.org site. I have also discovered novel transport of ionic liquids in electrified nanosystems using molecular and continuum simulations, which laid important theoretical foundations for improving the performance of supercapacitors.

Teaching Experience:

Aside from my academic research, I have spent two semesters teaching lectures and experiments for an undergraduate core course "Mechanical Engineering Laboratory" in the Department of Mechanical Engineering at Clemson University. I have also mentored graduate students and am the "Go-to Expert" on using scalable libraries and high-performance computational tools in research groups at Argonne National Laboratory and The University of Chicago.

Future Direction:

The short-term research projects that I will pursue focus on exploring and understanding task-oriented manipulation of suspended objects, such as macromolecules and colloidal particles, in microfluidic systems and ion transport of novel electrolytes, such as novel polyelectrolytes and ionic liquids, in nanofluidic systems. The long-term research that I will pursue aims to explore and obtain an integrated understanding, in close collaborations with experimentalists, of thermal, ion, and mass transport in biological systems (e.g., blood vessels), energy systems (e.g., supercapacitors) and mechanical systems (e.g., magnetorheological/electrorheological devices). The outcomes of my research are the better understanding and new insights for designing advanced technologies for water/air purification, macromolecule manipulation and sequencing, and energy storage/transport. In the ultimate level, my research could help address critical demands for clean environment and energy, and benefit human health.

Selected Publications:

Xujun Zhao[†], Jiyuan Li[†], **Xikai Jiang**[†], Dmitry Karpeev, Olle Heinonen, Barry Smith, Juan P. Hernandez-Ortiz, Juan J. de Pablo, *Parallel O(N) Stokes' solver towards scalable Brownian dynamics of hydrodynamically interacting objects in general geometries*, J. Chem. Phys., 146, 244114, **2017** (†: the authors contributed equally)

Xikai Jiang, Jiyuan Li, Xujun Zhao, Jian Qin, Dmitry Karpeev, Juan Hernandez-Ortiz, Juan J. de Pablo, and Olle Heinonen, *An O(N) and parallel approach to integral problems by a kernel-independent fast multipole method: Application to polarization and magnetization of interacting particles*, J. Chem. Phys., 145, 064307, **2016**

Xikai Jiang, Ying Liu, and Rui Qiao, *Current rectification for transport of room-temperature ionic liquids through conical nanopores*, J. Phys. Chem. C, 120 (8), 4629-4637, **2016**

Guang Feng, **Xikai Jiang**, Rui Qiao, and Alexei Kornyshev, *Water in ionic liquids at electrified interfaces: the anatomy of electrosorption*, ACS Nano, 8 (11) 11685-11694, **2014**

Xikai Jiang, Jingsong Huang, Hui Zhao, Bobby Sumpter, and Rui Qiao, *Dynamics of electrical double layer formation in room-temperature ionic liquids under constant-current charging conditions*, J. Phys. Condens. Matter, 26 (28), 284109, **2014**

Xikai Jiang, Jingsong Huang, Bobby Sumpter, and Rui Qiao, *Electro-induced dewetting and concomitant ionic current avalanche in nanopores*, J. Phys. Chem. Lett., 4 (18), 3120-3126, **2013**

Xikai Jiang, and Rui Qiao, *Electrokinetic transport in room-temperature ionic liquids: amplification by short-wavelength hydrodynamics*, J. Phys. Chem. C, 116 (1), 1133-1138, **2012**

505293 Imaging the Structure and Dynamics of Soft Materials

Abstract:

Research Interests:

Soft materials are ubiquitous in our daily life with examples ranging from plastics in the manufacturing sector to particle suspensions in cosmetic, food and medical products. Although soft materials exhibit a wide range of fascinating characteristics, they share a few common organizing principles that arise from their unique mesoscopic structures. Thus, the key to advance soft materials research is to understand the interplay between mesoscopic structures and macroscopic properties of soft materials and, furthermore, to realize controls over the assembly of these mesoscopic structures. Current state-of-the-art microscopy techniques allow us to image the mesoscopic structures of various soft materials under different external stimuli with unprecedented spatiotemporal resolutions, providing the direct link between structures and material properties. Within this exciting context, I would like to establish an experimental program, where I will employ various cutting-edge imaging tools to investigate the structural origin of different soft-material properties and explore the structural transformation of soft materials under external stimuli.

Specifically, my research focuses on phase transitions in colloids and the transport dynamics of active fluids. In my PhD work, I combined video microscopy with a new local heating technique, which allows me to explore the kinetics of phase transitions in colloidal crystals and study the influence of confinement and external flows on the nucleation. The results from my research reveal the nucleation mechanisms at single particle scales and are insightful for materials engineering. In my postdoctoral work with Prof. Xiang Cheng, I studied the transport and rheology of active fluids and explore the onset kinetics of the swarming motion of active fluids. These studies deepen our understanding on transport processes in microbiological systems and provide new insights on the emergent collective behaviors of active fluids.

In my future research plan, I will focus on soft materials out of equilibrium. These include the self-assembly and phase transitions of colloidal suspensions under external fields, glassy dynamics and glass transition of colloidal suspensions, and transport and rheology of active matter. I will combine the video microscopy and rheology/microrheology techniques to understand the detailed dynamics of these nonequilibrium processes. Furthermore, I aim to apply the understanding in practical applications such as coating process, bacterial infection, biomedicine and design of smart materials.

Postdoctoral project: Collective dynamics of bacterial suspensions. Department of Chemical Engineering and Materials Science, University of Minnesota, under supervision of Xiang Cheng.

PhD Dissertation: Phase transitions in thin-film colloidal crystals. Department of Physics, Hong Kong University of Science and Technology, under supervision of Yilong Han.

Selected Publication:

1. Yi Peng, Wei Li, Feng Wang, Tim Still, Arjun G. Yodh, and Yilong Han, Diffusive and martensitic nucleation kinetics in solid-solid transitions of colloidal crystals, ***Nature Communications***, 8, 14978 (**2017**).
2. Yi Peng, Lipeng Lai, Yishu Tai, Kechun Zhang, Xinliang Xu and Xiang Cheng, Diffusion of ellipsoids in bacterial suspensions, ***Physical Review Letters*** 116, 068303 (**2016**) (Highlighted by Editorsâ€™ Suggestion).
3. Yi Peng, Feng Wang, Ziren Wang, Ahemd Alsayed, Zexin Zhang, Arjun Yodh and Yilong Han, Two-step nucleation mechanisms in solid-solid phase transitions, ***Nature Materials***, 14, 101â€"108 (**2015**). (Focus Article with Cover Image)
4. Yi Peng, Ziren Wang, Ahemed Alsayed, Arjun Yodh and Yilong Han, Melting of colloidal crystal films, ***Physical Review Letters*** 104, 205703 (**2010**) (featured by Physical Review Focus)

Teaching Interests:

My teaching philosophy rests on the belief that education is not the filling of a pail, but the lighting of a fire. Teaching is not only about conveying facts or techniques written in textbooks, it is also about showing students how we can tackle new problems using the knowledge and skills we already have. I was a teaching assistant for Fundamentals of Physics, Electrical and Magnetism, Energy and Environment and the related laboratory courses in my graduate study. I have also mentored several graduate and undergraduate students. In the future, I would like to teach Thermodynamics and Statistical mechanics, Fluid Mechanics and Rheology classes, as well as to develop new courses on Soft Materials

505434 *In silico* Design of Ionic Liquid Adducts for Biomedical and Electrochemical Applications

Abstract:

Research Interests:

Ionic liquids (ILs) are materials with a low degree of toxicity that can be incorporated in a variety of drug formulations. Due to their non-flammability and high ionic conductivity, they can also be used in energy storage and electrochemical applications. My research will focus on the computational design of ILs for these applications including the creation of efficient numerical methods for computing the strong electrostatics interactions inherent to these systems.

Design ILs that increase the bioavailability of drugs: The biggest obstacles facing drug development today are limited bioavailability and water solubility, low dissolution rate, and polymorphic crystalline structure of a drug. Polymorphism of the drug leads to different physical properties, such as water solubility. The goal of this research is to increase the water solubility and stability of drugs, such as Brivudine and trifluridine, which are anti-HIV drugs with low bioavailability by means of ILs. A drug or active pharmaceutical ingredient (API) can be combined physically or chemically with an IL to produce a modified drug (prodrug) in a form of an active liquid salt that increases the solubility and limits undesirable polymorphic crystalline structures. Simulation tools, such as molecular dynamics (MD) and density functional theory (DFT) will be employed to characterize the interaction of different anions and cations with water molecules and to elucidate the microstructure and spatial characteristic of pure ILs and IL modified drugs (API-IL). An atomistic force field will be developed as a benchmark to determine the physicochemical properties of API-ILs systems. Further, a coarse-grained model of API-ILs will be proposed in order to remove the limitations of short time scales in atomistic simulations.

Design ILs that build a uniform solid-electrolyte interface (SEI) in batteries: Batteries operating at ambient operating temperatures have low ionic conductivity ($10^{-3}\ S\ cm^{-1}$) and metal cycling efficiency, and have raised safety concerns due to the incompatibility of the metal with the electrolyte. ILs are of a great interest as electrolytes because of their high ionic conductivity, excellent thermal stability, and non-volatility. For example, ILs can increase the conductivity of poly (ethylene oxide) (PEO) electrolytes by a factor of 10 in batteries. Thus, the conductivity of the electrolyte can be tuned by changing the ratio of the PEO to IL. The SEI, which is the result of the reduction of the electrolyte solvent and the lithium salt at the anode surface, controls the lifetime of a battery. To optimize battery performance a clear understanding of the composition and structure of the SEI is needed for alkali metal batteries. Simulation techniques, such as quantum chemistry and atomistic MD simulations using a reactive force field (ReaxFF) will be incorporated to characterize the reaction pathways that take place in various SEIs. These tools will also be used to find ILs that suppress the growth of dendrites at the electrodes because they lead to formation of a more uniform SEI, which improves the cycling behavior. Furthermore, the effect of the aqueous electrolyte, which has been used in experiments, on the structure of the electrolyte at IL-electrode interface and its impact on the charge storage ability are open questions that will be addressed in my future research.

Devise polymerised-ILs that improve ion transport in the fuel cells: In proton exchange membrane fuel cells (PEMFCs), the platinum catalyst is poisoned by carbon monoxide in the hydrogen feed and the addition of the extra pure hydrogen significantly increases the cost of the process. Poisoning of the platinum catalyst can be reduced by elevating the temperature above 140ËšC. However, this reduces the efficiency of the PEMFCs due to the reduction in proton conductivity. Protic ILs can be combined with different polymers to enhance the ionic strength of the electrolyte, interfacial stability, and inflammability. Furthermore, various polymerised-IL electrolytes can be designed to meet specific requirements of the fuel cells electrolytes. In order to design and optimize a new generation of electrolytes, a detailed understanding of the nano- and microscale properties of the hybrid membranes is required. With this goal in mind, the volumetric, structural and dynamic properties of the membranes composed of different types of IL, IL/polymer mixture, and polymerised-IL electrolytes will be characterized.

Devise polymerised-ILs that improve ion transport in the fuel cells: In proton exchange membrane fuel cells (PEMFCs), the platinum catalyst is poisoned by carbon monoxide in the hydrogen feed and the addition of the extra pure hydrogen significantly increases the cost of the process. Poisoning of the platinum catalyst can be reduced by elevating the temperature above 140ËšC. However, this reduces the efficiency of the PEMFCs due to the reduction in proton conductivity. Protic ILs can be combined with different polymers to enhance the ionic strength of the electrolyte, interfacial stability, and inflammability. Furthermore, various polymerised-IL electrolytes can be designed to meet specific requirements of the fuel cells electrolytes. In order to design and optimize a new generation of electrolytes, a detailed understanding of the nano- and microscale properties of the hybrid membranes is required. With this goal in mind, the volumetric, structural and dynamic properties of the membranes composed of different types of IL, IL/polymer mixture, and polymerised-IL electrolytes will be characterized.

New computational methods for ionic liquids: Studying these problems using simulations requires significant computational time due to the atomistic and charged nature of the systems. To tackle this challenge I will develop a new simulation method to speed up the electrostatic computations. Coarse-grained models of ILs in a single phase that reduces the number of charge interaction points will increase the efficacy of the simulations. For simulations of electrodes at a constant electrical potential in presence of ILs, the electrodes will be treated with a fixed uniform surface charge density with no explicit atomic sites, which will decrease the computation time.

Teaching Interests:

I am excited to teach any core course in the undergraduate curriculum of chemical engineering, and I want to develop graduate elective courses on simulation. I will develop an interactive teaching method for students through using a computer-based teaching method. Using programming languages such as Julia and C++, students will learn how to apply different computational tools to solve not only problems that they face in the class, but also various intricate modeling problems that emerge in chemical engineering.

In particular, I look forward to teaching numerical analysis, process control and design courses that involve mathematical modeling and simulation. An advanced project-based course that covers molecular and Monte-Carlo simulations with a focus on complex fluids, such as ILs and biomolecules will be created as an elective course.

Research Background:

I have been fortunate to work on various research topics during my graduate studies in the Khare research lab at Texas Tech University. I started my graduate research by investigating the structural and rheological properties of dilute polymer solutions of different chain architectures. Later on, I created rigorous atomistically-detailed models for determining the volumetric, structural, and rheological properties of neat and polymer modified asphalts. Specifically, I showed that master curve can be constructed from simulation values of viscoelastic properties, thus this increases the range of time and frequency scales accessible to simulations by a few decades and facilitates a comparison with experiments. In addition, the simulations were able to capture the sub-glass transition temperature relaxation of the model structures that has been reported in experimental studies of glass former liquids. I have also closely collaborated with an experimental group in order to develop novel polymeric materials that can be utilized as the membrane in pervaporation applications. During the latter project, I demonstrated that the dynamics of solvent molecules (water and ethanol) are closely correlated with dynamics of cross-linked polymer regardless of the solvent concentration and this dynamic coupling could be used as an additional factor in designing polymeric materials for pervaporation based separations. In addition, in a collaborative study (Zhang *et al.*, J. Phys. Chem. B, **2015**, 119 (47)), we showed that MD simulations can capture the experimental trend of the viscosity of imidazolium-based ILs and is able to connect the macroscopic properties to the microstructure of ILs.

During my post-doctoral appointment in the Bonnecaze research group at the University of Texas at Austin, I am developing models and simulations of magnetically-driven flows of dilute suspensions of magnetite particles. In this study, we aim to model and simulate the drug delivery to a fully occluded blood vessel using a magneto-rheological fluid that is caused by a rotary magnetic field with a gradient. I also study the rheological properties of soft particle glasses. In particular, I found that the polydispersed soft particles glasses, which are jammed beyond the random close packing volume fraction, can form crystalline structures in the presence of shear flow. I am also creating atomistic models to study the interface of calcite and water in the presence of oil and surfactant molecules. This project is of interest to oil recovery applications in which the interfacial properties of a reservoir play a crucial role in determining the efficiency of the process.

In summary, I will apply these skills (atomistic and coarse-grained MD simulations of polymers/ionic liquids, microscale simulations of suspensions, and continuum mechanics) to other problems in the field of engineering to understand the mechanisms underlying different phenomena that occur at nano- and micro-length scales.

505065 Modeling across Disparate Spatiotemporal Scales – Enabling Answers to Grand Engineering Challenges

Abstract:

Research Interests:

My broader research interest is to conduct theoretical/simulation studies to develop a fundamental understanding about transport phenomena at the solid/metal surface or interface relevant to energy and electronics applications. My training in graduate and postdoctoral research is in developing surface mass transport models across disparate spatiotemporal scales and will serve as the foundation for pursuing my future research interests. Surface transport has ubiquitous applications in Chemical Engineering, from thin film growth to heterogeneous catalysis. Alongside traditional fields there are emerging areas of interdisciplinary nature, like the plasma-material interface in a fusion reactor where surface transport plays an important role and the underlying physics is mostly unexplored. The structural and thermomechanical response of the plasma-facing materials that are exposed to the harsh plasma environment inside the nuclear reactor and damaged by the high fluxes of helium atoms produced by nuclear fusion – is one of the major limitations toward realizing nuclear fusion. Considering its future potential and current impediments in realizing that potential, nuclear fusion is widely recognized as one of the grand challenges for engineering and applied science in the 21st century. Another grand challenge in the field of energy application is to improve the solar cell efficiency for making solar energy economically viable; an idea for enhancing efficiency, which is worth exploring, involves developments of nanocrystalline active photovoltaic materials and integrating these nanostructures into a system that can transmit the energy into a circuit. Transferring core Chemical Engineering knowledge to another field to address such grand challenges is an exciting research proposition for me.

Theory, computation, and experiment taken together to probe the structure, chemistry, and function of new and complex materials is a comprehensive approach to address such a grand problem. However, connecting an atomic-scale (fusion reaction) or quantum-scale phenomenon (photon absorption and emission of charge carriers in photovoltaics) to experimentally/operationally relevant spatial and temporal scales will require O(300 Million years) wallclock time of atomic-scale simulations, using current state-of-the-art supercomputing facilities. And modeling/simulation challenges will increase by several folds if we make the models realistically accurate by taking materials heterogeneity and imperfection into account. Systematic development of atomistically-informed multiscale mass and energy transport model(s) can bridge the gap between atomistic modeling and experimental data, and my research will focus precisely on the development of such an approach.

Teaching Interests:

I am comfortable with teaching any of the core Chemical Engineering undergraduate and graduate courses, including courses of interdisciplinary nature. The only exception is courses which require substantial background in bioengineering or the life sciences. I will particularly be interested in teaching transport phenomena and fluid mechanics to graduate students. In addition to core courses, based on my academic and research experience, I would like to design two specially structured courses for first-year graduate students, one on *Mathematical and Statistical Methods for Chemical Engineering (including design of experiments)*: an introductory graduate-level course to support the needs of students interested in experimental research, and another one on *Computational Methods in Materials Science*: an introductory graduate-level course to prepare engineering students for computational research in the interdisciplinary arena of materials science and engineering.

Postdoctoral Project: *"Modeling of Plasma-Surface Interactions and Radiation Damage Dynamics;"* Advisor: Prof. Brian D. Wirth, University of Tennessee Knoxville and Oak Ridge National Laboratory.

PhD Dissertation: *"Analysis of External-Field-Driven Surface Stabilization and Patterning;"* Advisor: Prof. Dimitrios Maroudas, University of Massachusetts, Amherst.

Research Experience:

My academic and research career path has traversed through many fields of Chemical Engineering and covered the entire spectrum of simulation scales, from the atomic level to the chemical plant scale. Nevertheless, my graduate and postdoctoral research have been focused on development of multiscale models, hierarchically parametrized coarse-grained models, using atomistic information through molecular-dynamics (MD) simulations and/or first-principles density functional theory (DFT) calculations, which are experimentally validated, for surface mass transport in diverse systems of interest and applications. For example, a part of my Ph.D. thesis research was focused on developing a model to facilitate external-field-driven complex nanopattern formation strategies starting from single-layer islands of a conducting material on crystalline conducting substrates, including current-driven formation of nanowires. In my postdoctoral research, I focus on developing an atomistically informed model to predict the structural and morphological evolution of plasma-facing material components in the fusion reactor. My current projects involve extensive collaborations with research institutes and national laboratories including close collaborations with experimentalists from Oak Ridge and Sandia National Laboratories. Also, as a part of a DOE-funded Scientific Discovery through Advance Computing (SciDAC) project, I have acquired proficiency in high-performance computer simulations in leadership computing facilities. Before joining the UMass Amherst Chemical Engineering Ph.D. Program, I have worked on a collaborative research project with Honeywell UOP to develop system identification and model predictive control strategies for an industrial fluidized bed polymerization reactor.

Teaching Experience:

I have extensive teaching experience, mostly in the capacity of teaching assistant (TA) for undergraduate and graduate level courses, including some laboratory courses. I have TAed Chemical Engineering Analysis (graduate level), Chemical Reactor Engineering (graduate level), Introduction to Statistics and Design of Experiments (undergraduate level) with regular responsibilities of holding office hours, grading homework assigfnments, and preparing solution sets for assignments and exams. In addition to these, I have guest-lectured a graduate-level Chemical Reactor Engineering course at UMass Amherst (I was nominated for the departmental TA award by the students). Moreover, for Introduction to Statistics and Design of Experiments offered to ~650 freshman students, I had special responsibility as the head TA to host regular training sessions for 32 other TAs for the course. Finally, I have mentored graduate students in the research groups where I conducted my Ph.D. research and currently conducting my postdoctoral research.

Publications:

1. **D. Dasgupta**, R. Kolasinski, L. Du, D. Maroudas, and B. D. Wirth, "Modeling of nano-fuzz formation in helium-ion-irradiated tungsten," (Manuscript in preparation).

2. **D. Dasgupta**, A. Kumar, and D. Maroudas, "Oscillatory dynamics of current-driven single-layer epitaxial islands," (Manuscript in preparation).

3. A. Kumar, **D. Dasgupta**, and D. Maroudas, "Complex pattern formation from current-driven dynamics of

single-layer homoepitaxial islands on crystalline conducting substrates," *Physical Review Applied* (Accepted; will be published in July 2017).

4. A. Kumar, **D. Dasgupta**, and D. Maroudas, "Surface nanopattern formation due to current-induced homoepitaxial nanowire edge instability," *Applied Physics Letters* **109**, Article No. 113106, 4 pages (2016).

5. A. Kumar, **D. Dasgupta**, C. Dimitrakopoulos, and D. Maroudas, "Current-driven nanowire formation on surfaces of crystalline conducting substrates," *Applied Physics Letters* **108**, Article No. 193109, 5 pages (2016).

6. L. Du, **D. Dasgupta**, and D. Maroudas, "Weakly nonlinear theory of secondary rippling instability in surfaces of stressed solids," *Journal of Applied Physics* **118**, Article No. 035303, 14 pages (2015).

7. L. Du, **D. Dasgupta**, and D. Maroudas, "Stabilization of the surface morphology of stressed solids using simultaneously applied electric fields and thermal gradients," *Journal of Applied Physics* **116**, Article No. 173501, 12 pages (2014).

8. L. Du, **D. Dasgupta**, and D. Maroudas, "Stabilization of the surface morphology of stressed solids using thermal gradients," *Applied Physics Letters* **104**, Article No. 181901, 4 pages (2014).

9. **D. Dasgupta** and D. Maroudas, "Surface nanopatterning from current-driven assembly of single-layer epitaxial islands," *Applied Physics Letters* **103**, Article No. 181602, 4 pages (2013).

10. **D. Dasgupta**, G. I. Sfyris, and D. Maroudas, "Current-driven morphological evolution of single-layer epitaxial islands on elastic substrates," *Surface Science* **618**, L1-L5 (2013).

- This work was highlighted by the editor and a perspective article on this work was written by Prof. Yannis G. Kevrekidis in the same issue of the journal.

11. G. I. Sfyris, **D. Dasgupta**, and D. Maroudas, "The effect of a thermal gradient on the electromigration–driven surface morphological stabilization of an epitaxial thin film on a compliant substrate," *Journal of Applied Physics* **114**, Article No. 023503, 13 pages (2013).

12. **D. Dasgupta**, G. I. Sfyris, M. R. Gungor, and D. Maroudas, "Electromigration- driven complex dynamics of void surfaces in stressed metallic thin films under a general biaxial mechanical loading," *Journal of Applied Physics* **112**, Article No. 083523, 10 pages (2012).

13. **D. Dasgupta**, G. I. Sfyris, M. R. Gungor, and D. Maroudas, "Surface morphological stabilization of stressed crystalline solids by simultaneous action of applied electric and thermal fields," *Applied Physics Letters* **100**, Article No. 141902, 4 pages (2012).

14. **D. Dasgupta** and S. C. Patwardhan, "NMPC of a continuous fermenter using Wiener-Hammerstein model developed from irregularly sampled multi-rate data," *IFAC Proceedings Volumes* **43**(5), 637-642 (2010).

15. U. Sarkar, **D. Dasgupta**, T. Bhattacharya, S. Pal, and T. Chakroborty, "Dynamic simulation of activated sludge based wastewater treatment processes: Case studies with Titagarh Sewage Treatment Plant, India," *Desalination* **252**(1-3), 120-126 (2010).

16. S. Pal, U. Sarkar, and **D. Dasgupta**, "Dynamic simulation of secondary treatment processes using trickling filters in a sewage treatment works in Howrah, West Bengal, India," *Desalination* **253**(1-3), 135-140 (2010).

505544 Spin-Segregation of Active Spinners

Abstract:

Research Interests:

In the following, I briefly explain the projects I am currently involved as postdoctoral researcher, and my past studies as a graduate student:

Postdoctoral Research:

(i) Reversibility of a Colloidal System (Prof. Arratia's lab): I am currently studying a colloidal system adsorbed at an oil-water interface, which is subject to cyclic shear. The colloidal particles are sub-micron size, such that Brownian motion has significant contribution to the single particle dynamics. Through this study, we found that thermal noise significantly affects the reversibility of material in a nontrivial manner.

(ii) Predicting Reversible Rearrangements by Machine Learning Techniques (Collaboration with Prof. Liu and Prof. Yodh): Through this study, I am interested in finding the structural signatures which determine the dynamics (here, reversibility of cycles). This turns out to be a very challenging problem for disordered materials, and remains as an active research area. We have recently developed techniques based on machine learning to predict the reversibility dynamics based on the structure of a disordered matter.

(iii) Dynamics of Active Systems (Prof. Durian's lab): In active particulate systems, in contrast to passive systems, single particle dynamics is independently present even without external driving. We have designed two species of active granular systems: spinners, and translators. The intra-structure of particles (which are 3D printed) is designed such that they either spin or translate in the presence of an upward air flow. In this ongoing study, we aim at describing the collective dynamics, and in particular, we search for an effective temperature which governs the dynamics. Subsequently, we will be able to extract an equation of state for these systems.

PhD Research:

My PhD research was an experimental study on the jamming and rheology of granular materials (Prof. Behringer's lab), with a specific focus on elliptical-shaped particles. I performed continuous Couette shear (rheology) and cyclic compression (jamming-unjamming) on both systems of circular and elliptical shaped particles. The identical experiment, on both circles and ellipses, enabled us to understand the role of particle anisotropy on jamming and the bulk rheology.

Teaching Interests:

My teaching experience during graduate studies involved teaching lab sessions, and helping students with their homeworks, as well as the opportunity of teaching a full course (Introductory Electricity, Magnetism, and Optics) as

an instructor. Even though my graduate studies were in the field of Physics, however, I have a BSc. degree in Chemical Engineering, which provides me sufficient tools and familiarity with subject matter, to teach undergrad engineering courses.

Emulsions and Microcapsules for Food, Carbon Capture and Beyond

Dr. Srinivas Mettu (smettu@unimelb.edu.au), The University of Melbourne, Melbourne, Australia

ABOUT ME

- Education

  PhD (Lehigh University, USA) with Prof. Manoj Chaudhury

  27 Scholarly works (17 publications, 8 patents, 2 theses)

- Research Experience

1. Wetting, Colloids and Surface Science
2. Printing, Microfabrication
3. Surface Force Measurements, Emulsion formulations and Ultrasound Encapsulation
4. Biofuels from Microalgae, Interfacial Rheology and Nanofabrication

- Teaching Experience

  Physical Chemistry

  Fluid Mechanics

Referees:

1. Prof. Manoj K Chaudhury
2. Prof. Raymond Dagastine
3. Prof. Muthupandian Ashokkumar

WETTING AND SURFACE SCIENCE RESEARCH AT LEHIGH (PHD) AND XEROX RESEARCH

Drop Motion with Vibration

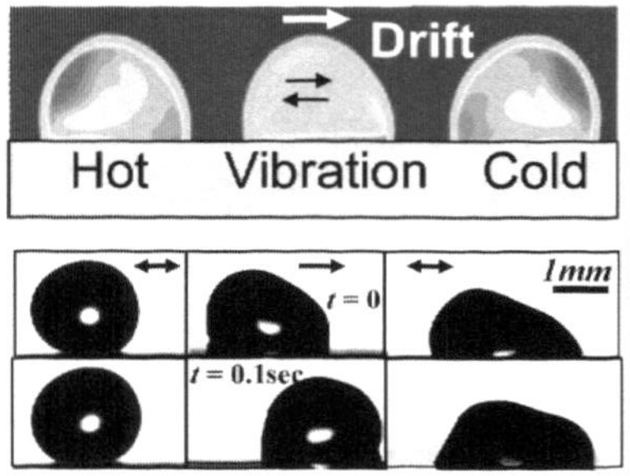

Wetting on Print Surfaces Xerox Research

SURFACE FORCE MEASUREMENTS

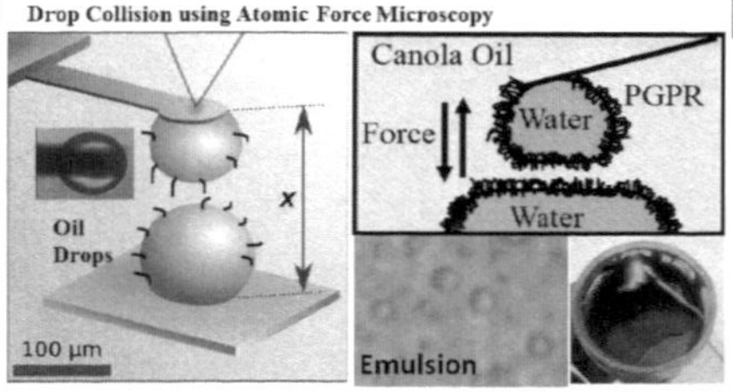

20% Water Chocolate

Measured surfaces forces and formulated emulsions for chocolates with 20% water

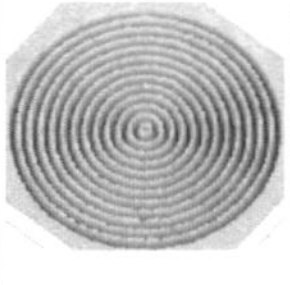

Nanofabrication (using NanoFrazor®)

Fabricated Nanochannels and measured surfaces forces on them

ULTRASOUND ENCAPSULATION

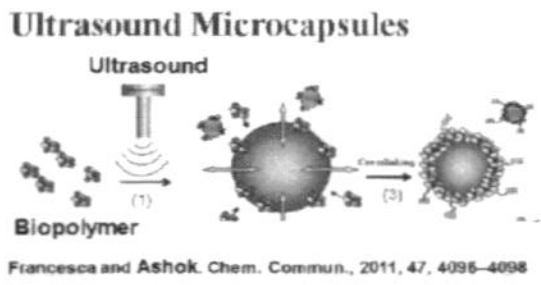

Fabricated microcapsules for food applications, tested their strength through AFM Indentation experiments

BIOFUELS EXTRACTION FROM MICRO ALGAE

Current Work

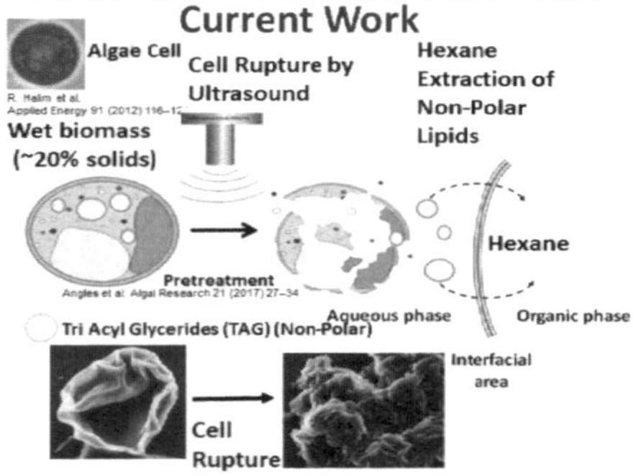

Improving the Extraction Efficiency of Lipids using Ultrasound

Probing interfacial properties of microbial emulsions using AFM, Interfacial Rheology and Tensiometry

FUTURE (AT YOUR GREAT UNIVERSITY!!)

- **Surface Forces for Emulsions**

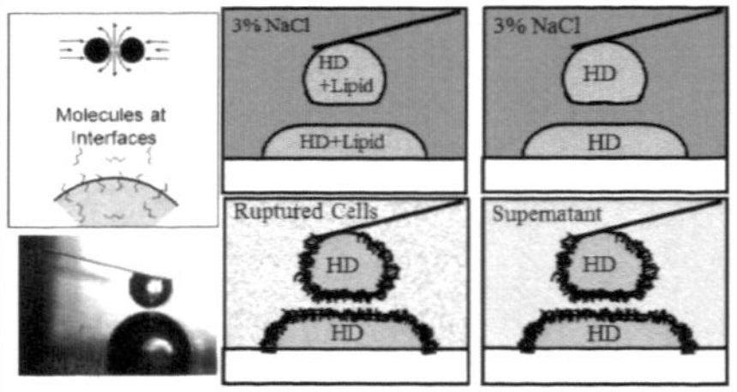

- **Solvent Free Extraction of Lipids (Using Single Cell Experiments)**

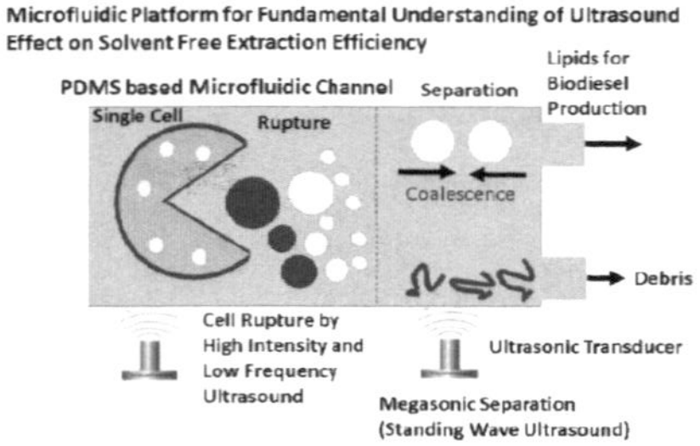

- **Microcapsules for Carbon Capture and Beyond**

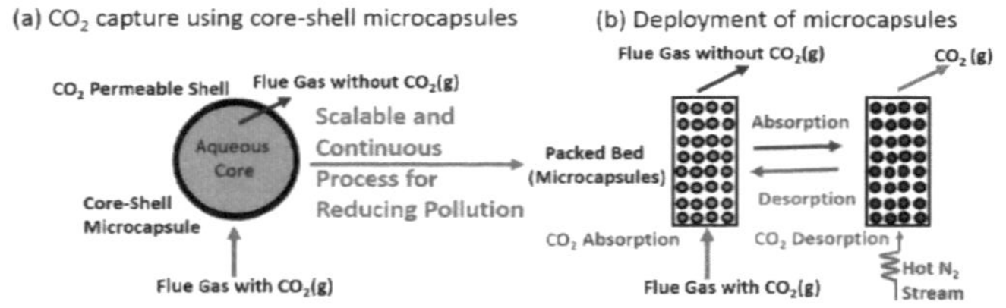

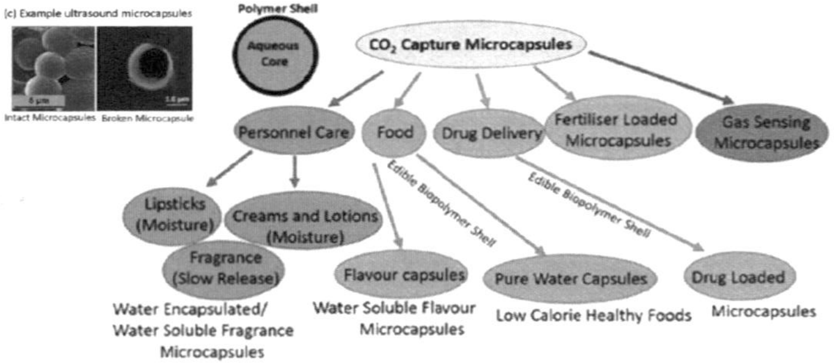

505366 Curvature Matters. Reconfigurable Materials from Anisotropic Colloid Interactions

Abstract:

Research Interests:

My research interests are centered on the fundamental phenomenological behavior and applications of colloidal suspensions. In particular, I am interested in suspensions composed of anisotropic particles, and suspensions with a polarized behavior generated by the influence of an external field. These kinds of colloids have directional interactions, which depend on the relative position and orientation while undergoing a translational and rotational Brownian motion. Different applications of colloidal suspensions take advantage of the precise control over the individual and the collection of particles using different fields, such as gravitational, electric, magnetic, acoustic, etc. The directional behavior generated by anisotropic particles will impact, lead advances in the colloid science, and improve the current applications, such as controlling the directed assembly, tunable/responsive (soft) materials, active suspensions, colloidal delivery systems, heat transfer fluids, photonic materials, separation processes, etc.

My research combines my expertise in synthesis, characterization, real-time imaging/measurements, and modeling at different scales of approximation, to fundamentally understand the interactions and collective response of suspensions composed of colloidal particles. The future research plan combines the fundamental understanding and quantification of anisotropic particle interactions, and its applications which include, but not limited to, the next topics:

- **Polarized soft matter** (*electromagnetic fields, active suspensions, ferrofluids and tunable suspensions*): Improve the actuation of colloidal suspension by the effect of the orientational dependence of anisotropic particles and highly concentrated suspensions.

- **Asymmetry for hierarchical assembly** (*colloidal crystals, hierarchical structures, photonic/phononic materials*): Production of colloidal tunable/multifunctional crystals accomplished by the multipolar interactions of anisotropic particles under the influence of electric/magnetic/acoustic fields.

- **Shape-enable delivery systems** (*encapsulation, fragrance delivery, drug delivery and theranostics*): Use the directional and orientational surface and field interaction of anisotropic particles to improve the delivery/transport of encapsulated fluids.

Teaching Interests:

I have had the experience to teach "Fluid Mechanics" as Assistant Professor. Additionally, I have previous experience as an instructor of "Chemical Engineering Laboratory I", and as a Course Coordinator of "Thermal Sciences Laboratory", in Chemical and Mechanical Engineering. Furthermore, I have had several experiences as Teaching Assistant in the last ten years. I have taken this role in courses such as "Fluid Mechanics", "Thermodynamics", and "Transport Phenomena" at the undergrad level, and "Transport Phenomena/Continuum Basis" at the graduate level.

I would be interested in teaching "Fluid Mechanics", and the introductory course of transport phenomena commonly offered to undergraduates. Likewise, due to the nature of my work and academic interests, I would be interested in teaching "Transport phenomena/Continuum Basis" at the graduate level. From my research experience, I would be interested in teaching courses related to modeling of anisotropic particles, as well as on particle-particle interactions.

504596 Computational Design and Discovery of Materials

Abstract:

Research Interests:

The technological advances that will address the grand challenges of our generation share the same physical basis: materials. Advanced materials enable water purification, drugs to be delivered with precision, industrial catalytic and separation processes to take place efficiently, the capture of toxic compounds; advanced materials have and will continue to impact our standard of living. Computer simulations enable faster discovery and informed design of materials, guiding experimental efforts towards promising structure candidates and therefore saving resources and efforts. For example, large scale high-throughput screening of materials allows for identification of promising candidates for targeted applications and for a birdâ€™s-eye view of the material landscape, revealing structure-property relationships unavailable to small scale studies. On the other hand, using enhanced sampling techniques to model self-assembly provides fundamental understanding that can address the synthetic viability of a material in particular conditions. Further, it can give insights into conditions necessary to exploit or avoid particular physical properties. When used in concert, these computational techniques allow for the *in silico* discovery and design of materials.

Teaching Interests:

During the 5th year of my PhD I took part in the Teaching Apprenticeship Program at Northwestern University which allowed me to teach half of the separations course under the supervision of Professor Linda Broadbelt. I was also a teaching assistant for various courses: mass transport, molecular simulations, and separations. Additionally, I taught a summer algebra class to incoming high school students who speak English as a second language. I also tutored high school students from Evanston, IL in science and math. Lastly, I have mentored graduate and undergraduate students during my PhD and postdoc.

Research Experience:

During my PhD, I developed algorithms to quickly and accurately assess the performance of porous materials in energy-related applications: hydrogen storage, methane storage, and xenon-krypton separations. I produced a code which generates crystalline structures for a topologically-diverse database. Working with experimental collaborators, we were able to synthesize promising candidates of a rare topology. The studies led directly to the discovery of new materials and structure-property relations that will guide future synthetic efforts. Additionally, I used *ab initio* methods to calculated binding energies of hydrogen and solvent molecules onto open metal sites. As a postdoc, I have led efforts to produce a new software package, Software Suite for Advanced General Ensemble Simulations (SSAGES), that implements enhanced sampling techniques for molecular dynamics simulations. I used these enhanced sampling techniques to elucidate the self-assembly process of a metal-organic framework, MOF-5. I have also performed atomistic molecular dynamics simulations to investigate ion transport in diblock copolymer systems.

Future Direction:

As faculty, my research efforts will focus on targeted design of materials for engineering applications. Specifically, my first efforts will focus on the design of porous materials (metal-organic frameworks, porous cages, porous liquids, etc.) for applications dealing with water (desalination, water purification). This will require molecular simulations (grand canonical Monte Carlo and enhanced sampling techniques) to understand water uptake in these materials and to assess selectivities for separation processes. Then, enhanced sampling techniques will be used to evaluate the mechanical stability of the materials for the operating conditions required, a crucial metric for the viability of these materials. Further, the use of hybrid materials (porous materials-polymers, porous materials-nanoparticles) in water applications will be assessed using a combination of atomistic molecular dynamics and enhanced sampling techniques. As promising materials are identified, experimental collaborations will be needed to synthesize and characterize candidate structures. Lastly, I will seek to understand the molecular self-assembly pathways of porous materials in views to manipulate them and avoid or exploit physical properties like catenation and polymorphism.

505053 Computational Modeling of Catalytic Reactions and Nanomaterials: Mechanisms and Structure-Function Relationships

Abstract:

Research Interests:

My experiences of computationally-guided chemical discovery fall into two major directions: (1) understanding new energy-relevant catalytic materials, underlying structure-function relationships that would guide further catalyst discovery, and the mechanisms of reactions on the catalytic materials that are relevant to the energy and environments; and (2) understanding the fundamental properties of atmospheric aerosols, providing new molecular-level insight into the composition and physicochemical properties of aerosol surfaces and their interactions with water, and elucidating of how the arrangement of molecules at the surface of sea spray aerosols influences the chemical reactivity of these particles in the lower atmosphere. Theoretical models and computational techniques were developed to facilitate the discoveries. In this poster, I will first present several related projects that have been published and then illuminate in detail of several proposing research projects in the field of catalysis and materials.

Teaching Interests:

I deeply value the vocation of teaching and am excited about the prospect of contributing to the undergraduate and graduate programs. Although I academic training background is mainly in chemistry, but I am willing to teach at all levels of courses including thermodynamics and courses advanced electives aligned with my own research and the needs of the department. I would develop a new *Computational Catalysis and Computational Materials* course for graduate students and in addition to contributing to the computational curricula for undergraduates. So that students can gain the first-hand experience of computational chemistry/chemical engineering.

505594 Correlating Structure and Performance of Heterostrcuted Materials for Energy Generation and Storage

Abstract:

Research Interests:

Given the crucial role of catalysis science to our society, understanding of how to design catalyst structure for desired catalytic activity and selectivity becomes a grand challenge. Hetero-structured materials provide extra degrees of freedom for tuning. My PhD research was dedicated to develop a fundamental and detailed understanding of the relationship between the structure of nanoscopic metallic catalysts and their function using first participial simulation. Instead of quantifying absolute catalytic activity; rather, I am able to calculate how reactivity descriptors (which are known to correlate with the activity) vary with the particle structure and composition. In my post-doc at Stanford and U Penn, I then extended my research to a boarder pool of materials with higher complexity, such as metal oxide, metal carbides, 2D materials.

My future research direction will be focused, but not limited to the followings

1) Development of a general picture for structure-reactivity correlation of different classes of hetero-structures.

2) Development of dynamic understanding of catalytic processes on nanoparticles. The descriptor-based screening only provides a static and mean-field description of catalysis on nanoparticle. In reality, the morphology of the nanoparticle is varying during reaction, where a dynamic understanding will be crucial to have an accurate access of the catalytic function of a nanoparticle.

3) Development of long-time scale simulation method suitable for distributed computing environment and its application on catalytic processes.

Teaching Interests:

I have two years' teaching experiences as a research educator of Freshman Research Initiative(FRI) program at UT-Austin, where undergraduate students are exposed to authentic theoretical chemistry research. In addition to lecture classes, I have TAed "Physical chemistry lab" classes at UTAustin. Spring 2017, I have TAed and guest lectured "Computational Science of Energy and Chemical Transformations", where I helped to setup and develop the course.

I like to take the advantage of exposing students to the frontier research during the lecture, which not only largely attracted students' attention on the course but also received some interesting feedback on your own research.

505591 Materials and Methods for Sustainable CO_2 Conversion Towards Hydrocarbon Generation

Abstract:

Research Interests:

The current energy and environmental scenario demands immediate attention towards mitigating atmospheric CO_2 and repurposing it for hydrocarbon generation. Carbon capture and sequestration (CCS) routes for CO_2 reduction attracted sufficient investment over the last few years. However, its scale of operation (about 35 Mtpa, in early 2017) is far below the CO_2 emissions (about 35 Gtpa, in 2015). Probing other methods for sustainable conversion of CO_2 is thus a prime necessity. Amongst the different routes of CO_2 conversion, solar energy based options like solar photocatalytic or solar thermochemical cycles (STC) garnered a lot of interest. Photocatalysis can be a low temperature process, however the current materials are mostly limited to using UV light for the purpose and the rates of CO_2 conversion are still very low. STC can achieve superior rates of CO_2 conversion, but at the cost of high temperatures of operation (more than 1000 °C). Reverse water gas shift – chemical looping (RWGS-CL) presents the benefits of these processes without facing their limitations. It can operate at low temperatures of about 500 °C with even better CO_2 conversion rates.

Mixed metal oxides are key elements in RWGS-CL process. Perovskite oxides (ABO_3) in particular have shown promising results for this purpose. RWGS-CL operates in two-steps: the first one being generation of oxygen vacant materials ($ABO_{3-\delta}$) by heating in presence of hydrogen. In the second step, these oxygen deficient materials regain their stoichiometric forms (ABO_3) while CO_2 is converted to CO. Oxygen vacancy formation is the key step that governs RWGS-CL process. Hence, DFT-calculated oxygen vacancy formation energy (E_{vac}) serves as an appropriate descriptor for this process. Since perovskite oxides are perfect platforms for material composition tuning, we screened several perovskite oxides based on our descriptor (E_{vac}). The predicted materials were subsequently synthesized via Pechini method and tested for CO_2 conversion performance. We could demonstrate highest CO_2 conversion at lowest temperatures (~450-500 °C). The role of intrinsic material parameters towards CO_2 conversion has been studied as well.

On the photocatalytic frontiers, the use of materials with capability of harvesting solar light is encouraging. Oxy-nitrides, transition metal dichalcogenides and perovskite oxides are some potential candidates for this purpose. We hereby computationally investigate the role of heterojunctions, strain and vacancy defects for tuning the band gap and band edges of these materials. Investigating all these material properties provides the backbone for material prediction and unravelling the catalysis mechanism for a vast range of chemical reactions

Teaching Interests:

Basic chemical engineering principles are not restricted to any particular discipline. They can be applied to a lot of common natural phenomena happening around us. These basic principles are key to any materials and methods development and discovery. Over the years, I had the experience of being Teaching Assistant in several courses (Chemical Engineering Thermodynamics, Transport Phenomena, Numerical Methods, and Process Dynamics and Control). Along with these, two other courses (Material and Energy Balance and Chemical Reaction Engineering) form the core of chemical engineering. I am interested in teaching any of these courses. During my graduate studies I also enjoyed the role of mentoring undergraduate students. In future, I would like to advise and mentor students in their various academic endeavors.

505596 Molecular Modeling and Machine Learning for Catalysis and Separations

Abstract:

Research Interests:

As earth's population grows and the standard of living in developing nations improves, demand for energy, water, and agricultural products will increase. These will interact in complex ways as production increases to meet demand; the NSF has even prioritized these challenges in their Innovation at the Nexus of Food, Energy, and Water Systems (INFEWS) initiatives. Human and animal wastes have been viewed as environmental and sanitary liabilities, but they could be viable renewable resources instead. Technologies that economically convert biomass in wastewater to useful fuels and chemicals could conserve energy and clean water, making fresh water available for human or animal consumption or irrigation. In addition, as new biomass pyrolysis, dehydration, and hydrodeoxygenation processes approach commercialization, new water treatment processes will need to be developed to clean up process water, which often contain compounds toxic to microbes and resistant to traditional wastewater treatment.

Chemical transformation and solute-substrate interactions are at the heart of any such future technologies, creating a significant opportunity for applying catalytic science to water treatment. Moreover, in this age of computer-aided catalyst design, progress toward optimal catalyst and reactor designs can be accelerated by utilizing modeling tools to identify and alleviate bottlenecks in reaction networks and in processes, as well as to aid in the search for high-performance materials. As vast libraries of materials are created in lab and *in silico*, extracting actionable information from these big data can be difficult. Techniques from machine learning and data science can be harnessed to identify critical trends and descriptors, as well as to build predictive models for discovery.

Developing these technologies will require an interdisciplinary and collaborative approach, bringing together the insights and methodologies of chemical, environmental, and civil engineers. During my PhD, I developed expertise in reaction engineering, catalysis, and computational chemistry for processing biomass into fuels and chemicals, and in my current post doc, I am acquiring new skills for modeling separation processes and designing new materials *in silico*. Leveraging these techniques in collaboration with experimentalists will accelerate progress toward a more sustainable future.

Research Experience

During my Ph.D. with Prof. Dion Vlachos, I applied molecular dynamics and electronic structure calculations to investigate fundamental interactions between solvents, solutes, and catalysts in biomass catalysis. In my first project, I discovered how co-solvent interactions protect biomass-derived furans from side reactions in fructose dehydration [1] [2]. In my second project, I developed structure-activity relationships for homogeneous Lewis acids to probe mechanistic details in glucose isomerization [3] [4]. In my third project, I exhaustively screened site geometries for Sn-Beta, identifying the most stable active geometry and proposing a source for Brønsted acidity [5]. In my final project, I developed the reaction mechanism for fructose etherification on the hierarchical Sn-SPP zeolite [6].

In my current post doc with Prof. Ilja Siepmann, I am developing and applying Monte Carlo simulation techniques in three projects: 1) predicting thermophysical properties of hydrocarbons, 2) exploring aggregation mechanisms of biorenewable surfactants and developing *in silico* design procedures to predict novel surfactant properties, and 3) developing molecular modeling and machine learning techniques to interpret zeolite characterization experiments.

Teaching Interests:

Excellence in teaching has the potential to multiply progress made toward a more sustainable future through the engineers who will come after us.

I have served as a teaching assistant in both undergraduate and graduate thermodynamics courses, and I have been a guest lecturer in kinetics and reaction engineering at 3 institutions. In addition, I organized a computational chemistry workshop for the graduate reaction kinetics course, to guide students through exercises using the same software I use in my research.

I am particularly excited about incorporating machine learning and data science into undergraduate and graduate education. By broaching these topics in core courses and going deeper in elective courses, students will become acquainted with cutting-edge data analysis tools and equipped for careers in data science and its intersection with traditional chemical engineering careers.

Courses of expertise: kinetics and reaction engineering, thermodynamics and statistical mechanics, numerical methods and linear algebra, computational chemistry, and engineering statistics.

First-Year Postdoctoral Fellow

Postdoctoral Project: *Monte Carlo Simulations for Novel Materials and Process Design*

Advisor: Prof. J. Ilja Siepmann, Department of Chemical Engineering and Materials Science, University of Minnesota

PhD Dissertation: *Leveraging First-Principles Simulations for Deeper Insights in Biomass Processing*

Advisors: Prof. Dionisios Vlachos and Dr. Stavros Caratzoulas, Department of Chemical and Biomolecular Engineering, University of Delaware

AIChE Conference Talk:

[6] "Mechanism for Selective Fructose Etherification on Hierarchical Sn-SPP Zeolite" to be presented at 2017 AIChE Annual Meeting, Catalysis and Reaction Engineering division, Minneapolis, MN

Selected Publications

[5] **Josephson, T. R.**, Jenness, G. R., Caratzoulas, S., Vlachos, D. G., **2017**. "Distribution of open sites in the Sn-Beta zeolite." Microporous and Mesoporous Materials. doi: 10.1016/j.micromeso.2017.02.065

[4] **Josephson, T. R.**, Brand, S. K., Caratzoulas, S., Vlachos, D. G., **2016.** "1,2-H versus 1,2-C-shift on Sn-silsesquioxane." ACS Catalysis. doi: 10.1021/acscatal.6b03128

[3] Brand, S. K., **Josephson, T. R.**, Labinger, J. A., Caratzoulas, S., Vlachos, D. G., Davis, M. E., **2016**. "Methyl-ligated tin silsesquioxane catalyzed reactions of glucose." Journal of Catalysis. doi: 10.1016/j.jcat.2016.06.013

[2] **Josephson, T. R.**, Tsilomelekis, G., Bagia, C., Nikolakis, V., Vlachos, D. G., and Caratzoulas, S., **2014**. "Solvent-induced frequency shifts of 5-hydroxymethylfurfural deduced via infrared spectroscopy and *ab initio* calculations." Journal of Physical Chemistry A. doi: 10.1021/jp508340p

[1] Tsilomelekis, G., **Josephson, T. R.**, Nikolakis, V., and Caratzoulas, S., **2013**. Origin of 5-hydroxymethyl furfural stability in water/dimethyl sulfoxide mixtures." ChemSusChem. **Cover article**. doi: 10.1002/cssc.201300786

505508 Molecular Modeling of Anti-Microbial Peptides at Water-Membrane Interface

Abstract:

Research Interests:

In the last 20 years, interest in peptide folding mechanisms at water-membrane interfaces and peptide-membrane interactions has increased in diverse biological fields. An example is the interaction of anti-microbial peptides (AMP) against the bacterial membranes. Since the resistance of bacteria against conventional antibiotics is increasing, the interest in developing of new antibiotics with high level of performance has arisen. Research studies show future potential of AMPs as an applicable antibiotic drug based on two main reasons: The unique mechanism of actions and rapid killing of bacteria. AMPs mechanism of action involves its interaction with the lipid membrane of bacteria. The experimental studies suggested that the structure of peptide, the kind of amino acid sequences and the structure of membrane play the most significant roles in this mechanism. Hence, the study of 3D structure of peptide and membrane is the key to understand this mechanism. In spite of experiments that explored the details of peptide-membrane interactions, there is still much uncertainty in the exact mechanism of cell death by AMPs and the relationship between the structure of AMPs and its antibacterial functionality. Spectroscopic techniques especially NMR help toward a better understanding of the structure of peptide and membrane; however, these methods just provide a different points of view of peptide activities and cannot completely determine the AMPs-membrane interaction. Molecular modeling provides atomistic details of peptide-membrane interaction which contributes to investigation of this mechanism and also can help to find which biophysical parameters and AMPs characteristics can affect the performance of AMPs. Prior computational studies demonstrate the crucial effect of hydrophobicity, charge and the helical structure of peptides during interactions with membranes. These studies also point out the important role of aromatic amino acids such as Tryptophan (TRP) and Tyrosine (TYR) in peptide-membrane interactions.

As a general aspect of my research, within the group of Professor Michael Greenfield, I attempt to understand how changes in amino acid residues have an effect on peptide folding and peptide-membrane interactions. To that end, I am focusing on the structures of novel hybrid AMPs LM7-1 and LM7-2 at water-membrane interface. These peptides were designed previously using combinations of naturally occurring AMPs, and they differ in sequence only at the 15th residue. Prior experimental results indicate different performance of these two AMPs against diverse bacteria. One of the potential explanation for this issue can be the different folded structure of these AMPs at water-membrane interface. Molecular dynamics (MD) simulation is used as an appropriate tool to investigate the details of AMP and membrane structure exactly in the moment of interaction. The hydrophobicity distribution on the folded structure of peptide, the deviation of peptide from helical structure at the moment of binding with membrane, and the pore formation structure on the membrane are all key points that can be analyzed using the results of MD simulation. However, all of these kinds of studies depend on the accuracy of the simulation.

The accuracy of biomolecular modeling results is based in part on force field (FF) parameters. Several simulation studies have proved that applying different FFs leads to various peptide folded structures and consequently results in different functionalities of peptide. The results from the first steps of these simulations show that the structures of aromatic rings (TRP and TYR) have deviations from planarity. These aromatic amino acids have an effective role in the peptide-membrane interaction. Hence, I am investigating the geometry and inherent nonplanar vibrations of these aromatic rings in more detail. I applied all-atom Normal Mode Analysis (NMA) on the results of the MD simulation to calculate the inherent vibrations of atoms. On the basis of this analysis, some deficiencies in calculated nonplanar mode frequencies from reality have been observed. Geometry analysis and NMA have been applied to develop more accurate FFs parameters.

Investigating the journey of peptide from coil structure in solution to folded structure at a water-membrane interface can help us to find how the one different residue in the sequence of LM7-1 and LM7-2 leads the peptide to a different path of the folding process. MD simulation can provide the conformational space for a peptide folding process. Since large-scale motion in peptide are a consequence of small-amplitude fluctuations, it is also essential to investigate the individual atomic fluctuations over MD simulation. NMA is applied on the configuration results of MD simulations of peptide in solution or at water-membrane interface. This calculation provides the vibrations in a peptide and also provides the surface energy map that shows which specific vibrational modes lead a peptide to be folded or unfolded. Comparing analysis for both LM7-1 and LM7-2 can help us understand how one difference in the peptide sequence affects their performance against a bacterial membrane. The outcome of the computational studies will contribute to find atomistic detail in the process of peptide folding at water-membrane interfaces, and also provides a proper perspective in understanding of biophysical characteristics in the peptide-membrane interactions.

Teaching Interests:

Beside my research career, I have extensive teaching experience of about 12 years that shows my intensive interest in teaching. I believe teaching helps me to achieve the deeper level of understanding. I started teaching at high school and institutes when I was an undergraduate student. As a PhD student, I continued teaching as a teaching assistant (TA) at University of Rhode Island (URI) for 4 years. During these four years, I have been chosen to be a TA for diverse courses. Even as a Ph.D student, I had the opportunity to be the instructor of "Chemical Engineering Thermodynamics 1" course (CHE313-spring 2017) at URI.

My previous experiences in teaching:

1- Instructor of " Chemical Engineering Thermodynamics 1" at URI – Spring 2017

2- TA for Senior Chemical Engineering Lab – 7 semesters at URI

3- TA for courses: Chemical Kinetics and Reactor Design, Thermodynamic 2, Engineering Material, Chemical Process Calculations.

4- Teacher in Iran of Chemistry and Geometry, Analytical Geometry, Linear algebra for 8 years at high school and university entrance exam preparation institutes.

In the future, I am interested in teaching:

1- Applied mathematics in chemical engineering

2- Thermodynamics

3- Mass transfer

4- Heat transfer

5- Molecular modeling

504946 Multi-Scale Modeling of Liquid Solutions and Solid/Liquid Interfaces

Abstract:

Research Interests: One of the grand challenges faced by modern society is the unquestionable quest for clean, renewable and sustainable sources of energy. Innovations in new materials have historically paved the way for significant breakthroughs tackling the looming energy crisis. However, design and optimization of materials for applications in energy storage devices (such as supercapacitors and batteries) and energy conversion devices (such as solar cells) is confronted by multitude of materials related limitations. The proliferation of computer speed and accuracy of algorithms are enabling the rational design of novel materials for diverse technological applications with unprecedented fidelity. However, even advanced computational methods tailored for tackling problems at a specific length and time scales, are incapable of studying the properties and functions of materials and processes which span over wide spatial (angstroms to micron) and temporal scales (femtosecond to second). To bridge the gaps between these extremely wide length and time scales, I propose a high-throughput multi-scale simulation approach to provide the missing link between the processes occurring on atomistic length and time scales that affect properties on mesoscopic scales exceeding far beyond atomistic ones by predicting and understanding the physicochemical behavior of liquid solutions and solid/liquid interfaces. The development of such a predictive understanding of solutions and interfacial properties is fraught with complications, but crucial to the design and optimization of energy storage devices and material syntheses methods for which experiments alone are insufficient due to difficulties associated with distinguishing between different phases, limitations for experimental resolution and the high cost associated with experimental work. Thus, the synergetic fusion of multi-scale simulation and experimental input will be a key factor in achieving a fundamental, comprehensive understanding of the delicate interplay between the macroscopic properties and events occurring at much smaller length scales.

Teaching Interests: My goal as an educator and mentor is to facilitate advancement of my students in their career, skills and personality and not mere transfer of equations and facts. I believe that my students should be critical and independent thinkers with skills to work in teams. Teaching and mentoring has been an integral part of my training both as a teaching assistant at the Louisiana State University and a graduate student mentor at the Lawrence Berkeley National Laboratory. I have gradually built my pedagogical skills, confidence and gained experience to be a thoughtful educator and mentor for students at both undergraduate and graduate levels.

505509 Multiscale Simulations of Nonequilibrium Mechanisms in Aqueous Solutions

Abstract:

Personal Information:
I am a recent Alexander von Humboldt Research Fellow of the Max Planck Institute for Polymer Research. I am a Chemical Engineer by education, having received my Ph.D. at the University of California in Santa Barbara and my B.Eng. at McGill University in Montreal. I currently work as a data scientist at a private firm.

Research Interests:

My scientific research revolves around aqueous mechanisms on the molecular level (changes in polymer conformations, colloid aggregations, etc. in water). Importantly, such systems exhibit multiscale behavior: For example, a minor mutation in a protein sequence can cause a major error in a cell network. Aiming at clarifying the nuances of such issues, I employ my computational skills in multiscale simulations, which are algorithms for correctly and efficiently connecting microscopic and macroscopic models in soft matter. My most distinguished scientific achievements are the following:

- While working with Prof. M. Scott Shell, I resolved a paradox between numerical predictions and empirical observations that has existed for decades, convincingly showing that the force between small hydrophobes is weak oscillatory and between large hydrophobes is strong monotonic.
- While employed by Prof. Kurt Kremer, I formulated a multiscale strategy that gains an efficiency of as much as an order of magnitude, while describing the correct static and dynamic behavior of molecular liquids across their phase diagram.

I am currently moving in the direction of nonequilibrium systems relevant for environmental and pharmaceutical nanotechnologies, particularly focusing on adjusting their microscopic features for optimizing their macroscopic functions. At the onset of my career, I will pursue two specific routes in my research. The first one deals with impurity filtration via graphene nanochannels: My main goal here will be in optimizing a nanochannel in being permeable for water but selective against contaminants. The second one deals with drug delivery via liposome nanocarriers: My main goal here will be in optimizing a nanocarrier for releasing the bulk of a certain medicine in nominal time. Over time, I will incorporate these two projects in a research program that also explores multiscale phenomena in biology: In particular, I am interested in how microscopic mutations in residue sequences influence macroscopic errors in signal networks. Finally, I also have plans of collaborating across the globe (i.e., Prof. Jing Yu and Prof. John M. Frostad), as well as on the campus.

Teaching Interests:

I have a unique background in pedagogy. At university, I volunteered earlier on as a Teaching Aide, and I worked later on as a Teaching Assistant and as a Teaching Associate. While the former position involved the regular duty of a student, the latter position comprised the entire role of an instructor. In fact, it was because of my distinguished performance as a Teaching Assistant that I received a prestigious award that provided me with the opportunity of being a Teaching Associate. Consequently, I have correspondingly completed an official certificate in instruction. In my spare time, I am also employed as a mathematics instructor at a local academy.
I can lead many courses in Chemical Engineering. Considering my mathematical expertise, I am especially suitable for the fundamental courses (e.g., Thermal Physics, Transport Phenomena, Reaction Kinetics, Quantum Physics, etc.). Via my scientific experience, I also intend to develop two elective courses (i.e., one about molecular simulations and one about aqueous systems). Importantly, the cognitive sciences find that there is a mismatch between the teaching styles of most instructors and the learning styles of most students, and in all of my classes, I correspondingly intend to combine various pedagogical approaches.

Annual Meeting:

I would be happy of meeting you in person at my poster on Oct. 29 after noon. Besides, you may be interested in the two lectures which I am giving at the annual meeting:

- My work on "Employing a Multipole Approximation in a Hybrid Fluid via Relative Resolution" can be found at the "Faculty Candidates in CoMSEF I: Biomolecules, Soft Materials, and Algorithms" session, which is on Oct. 30 before noon.
- My work on "Molecular Simulations of the Force Spectroscopy of a Biomimetic Polymer" can be found at the "Thermodynamics of Biomolecular Folding and Assembly" session, which is on Nov. 1 before noon.

497355 Predictive Bottom-up Design of Nanomaterials for Biomimicking Applications

Abstract:

Research Interests: The goals of my research program are to design of adaptive, programmable nanomaterials that are of potential use in a host of nanotechnology applications including, but not limited to, optical devices, drug delivery, biosensing and energy storage and conversion. For these goals, I am interested in exploring design rules for a wide range of molecular and colloidal systems via bottom-up approaches such as self- and directed-assembly techniques. The design rules cover not only the assembling building block geometry and interactions, but also assembly pathways, at equilbrium and far-from-equilbrium. The target nanostructures are judiciously chosen for specific applications in optical devices, drug delivery, biosensing and energy storage and conversion. As such, the assembly processes may occur either in bulk or at the interfaces and the interactions between the assembling building blocks require mutiscale modeling. To efficiently address the fundamental challenges with bottom-up approaches, I am particularly intersted in employing and developing tools and methods that have been actively used across multidisciplines, ranging from large-scale GPU-accelerated simulation codes to enhanced sampling methods to data mining techniques.

Teaching Interests: I am interested in teaching Chemical Engineering core courses including Thermodynamics and Transport, and developing my own courses on Polymer Physics and Computational Nanoscience at graduate level. I would like to facilitate cooperative learning in classroom and incorporate practical aspects with fundamental problems into lecture materials. My objectives are to explain to the students why the coursework benefit them with their career and how they can succeed with the courses.

505530 Wave Function-Based Framework for Computational Catalyst Discovery

Abstract:

Research Interests:

In heterogeneous catalysis, the primary unsolved problem is determination of the most optimal catalyst type, structure, and composition for an arbitrary chemical reaction. Breakthroughs in the critical areas of precious metal-free catalysis and carbon-neutral production of chemicals depend on the knowledge of "ideal" catalyst site properties. For my faculty research, I propose a methodological advancement in the area of computational chemistry, aimed at tackling the catalyst discovery problem. Previously, I have demonstrated as a proof of concept that systematic reduction of Kohn-Sham equations of the density functional theory (DFT) that employ the notions of Harris energy functional and chemical pseudopotential theory, lead to Huckel theory-type equations governing energetics of surface processes. Reformulation of the method in terms of local molecular orbitals indicate that a limited set of atom-specific parameters (3 per each interacting atom pair) plus the interatomic connectivity uniquely defines the energy of a "catalyst + reaction intermediate" system. Consequently, systematic variations of parameters describing the catalyst for a given known reaction mechanism allow determination of the local catalyst site structure and composition that should yield the highest reaction rates and/or product selectivities. The method provides a chemistry-inspired alternative to widely used linear scaling relationships in catalyst design and holds promise for simultaneous optimization of reaction pathways and heterogeneous catalysts.

Teaching Interests:

I have always been enthusiastic about teaching science in the classroom setting, be it an explanation of difficult concepts to students, demonstration of a "big picture" relevant to material being studied, or going through homework solutions. I am a strong believer that problem solving is the crucial skill that students should acquire in the class, which not only help them advance the research, but also facilitate understanding of seemingly abstract ideas, such as entropy. In addition, due to the fact that students internalize just 10% of all information presented in the traditional lecture format, I am a proponent of the interactive lecture format, in which it is critical for a professor to stimulate class discussions and ask questions. My ultimate goal is to educate the future generation of chemical engineers and equip them with the knowledge for tackling 21st century problems our society faces. I am very motivated to develop Quantum Theory for Engineers elective at the graduate level.

505126 Data Analytics for Complex Systems

Abstract:

Research Interests:

In my research, I create, modify, and adapt machine learning algorithms for applications in chemical and biological engineering. Machine learning is an exciting field due to improved sensor systems yielding new types of datasets, decreasing data storage costs, and increasing computational power. This combination of resources gives engineers unprecedented access to data to answer questions about complex systems.

During my PhD work, I have participated in several collaborations. A key aspect of my work is the translation of modeling goals to algorithms. For instance, in a collaboration with a biomanufacturing company, I designed a methodology to learn a sparse model such that the most important control variables could be identified [1]. The resulting approach was able to achieve 95% prediction accuracy for the desired tolerance on an unseen test dataset. In an academic collaboration, I developed an algorithm that could use high-dimensional bio-assay data with missing entries by simultaneously performing imputation and model building [2]. The resulting models were at least as accurate and sparser than competing methods. More recently, I have collaborated with industrial and academic partners on energy applications. I collaborated with an oil and gas company to predict anomalous well conditions using a semi-supervised likelihood ratio test [3] and have started work on data-driven capacity monitoring for lithium ion batteries. These projects use a combination of statistics, optimization, numerical methods, and chemical engineering to improve decision making and increase system understanding.

Data analytics can help solve many open problems and chemical and biological engineers have an important role to play in designing techniques that will work for these applications. My future research will use data-driven techniques to improve process monitoring [4], better characterize parametric uncertainty, and build predictive models. I will focus on solving chemical and biological engineering problems by developing and employing techniques that handle issues specific to chemical processes such as missing data [2,5], unstructured data, and heterogeneous measurements as well as biological data issues such as high dimensionality and small sample size [6]. I will establish collaborations with manufacturing companies in the areas of time series analysis, unsupervised datasets, feature engineering, and manifold learning. I will also work with other faculty members to help develop methods to use data from new sensor systems and high throughput analysis.

Teaching Interests:

I am particularly interested in the important role software and computational tools play in engineering and would hope to integrate coding into any class I teach. I would prefer teaching applied mathematics, process design, unit operations, control, and transport phenomenon.

References:

[1] K. Severson, J.G. Van Antwerp, V. Natarajan, C. Antoniou, J. Thömmes, and R.D. Braatz. Elastic net with Monte Carlo sampling for data-based modeling in biopharmaceutical manufacturing facilities. *Computers & Chemical Engineering*, 80:30-36, 2015.

[2] K.A. Severson, B. Monian, J.C. Love, and R.D. Braatz. A method for learning a sparse classifier in the presence of missing data for high-dimensional biological datasets. *Bioinformatics*, in press. doi: 10.1093/bioinformatics/btx224

[3] K.A. Severson, P. Chaitwatanodom, M.C. Molaro, R.S. Bailey, and R.D. Braatz. Anomaly detection and diagnosis using semi-supervised models for industrial time-series data, in preparation.

[4] K. Severson, P. Chaiwatanodom, and R. D. Braatz. Perspectives on process monitoring of industrial systems. *Annual Reviews in Control*, 42:190-200, 2016.

[5] K.A. Severson, M.C. Molaro, and R.D. Braatz. Methods for applying principal component analysis to process datasets with missing values. Special Issue on Process Data Analytics, *Processes*, accepted.

[6] K.A. Severson, J.G. VanAntwerp, V. Natarajan, C. Antoniou, J. Thömmes, and R.D. Braatz. A systematic approach to process data analytics in pharmaceutical manufacturing: The data analytics triangle and its application to the manufacturing of a monoclonal antibody. In *Multivariate Analysis in the Pharmaceutical Industry*, edited by A. P. Ferreira, J. C. Menezes, and M. Tobyn, Elsevier, in press.

505004 Dynamic Systems Spanning Engineering to Medicine

Abstract:

Dynamic Systems Spanning Engineering to Medicine

Anwesha Chaudhury, PhD

John Higgins' group (February 2015-present)

Center for Systems Biology, Massachusetts General Hospital; Pathology, Massachusetts General Hospital; Systems Biology, Harvard Medical School

Email: achaudhury@mgh.harvard.edu, anwesha_chaudhury@hms.harvard.edu

Education

Doctor of Philosophy

December 2014
Department of Chemical and Biochemical Engineering

Rutgers-The State University of New Jersey, NJ, USA

Masters of Technology

August 2010

Department of Chemical Engineering

Indian Institute of Technology, Kharagpur, India

Bachelors of Technology

June 2008

Department of Chemical Engineering

West Bengal University of Technology, India

Research Interests:

The overarching theme of my research is to define and delineate complex dynamic systems -- both biological and synthetic -- in terms of tractable yet accurate mathematical models. My diverse training in engineering and medicine puts me in a strong position to make significant progress toward these goals. I firmly believe that we are at an inflection point in our ability to minutely quantify enough aspects of complex systems and to computationally use vast troves of such data to make increasingly accurate predictions about even systems that are not completely understood. The ongoing advancements in clinical measurements and rapid growth in the availability of computational resources presents a unique opportunity to develop inexpensive tools that circumvent inefficient manufacturing process operations and streamline the diagnosis and treatment of diseases in patients. My current research is working toward such a clinical tool that takes advantage of fine-grained but readily available blood cell measurements, and predicts risk for a set of specific diseases for individual patients. In the short-term, I aim to continue developing data-driven population dynamics models for blood cells to predict onset and risk of several important diseases. Using this approach, I also intend to develop hypotheses for the underlying molecular and cellular mechanisms of disease conditions, which can thereafter be investigated using experiments. I plan to integrate these models for the various lineages of white blood cells in order to investigate the crosstalk in the immune system - a subject with vast potential applications.

From a process systems perspective, I intend to continue developing mathematical models for particulate systems, which can then be utilized for process control and optimization. I have significant experience with developing numerical techniques to solve complex partial differential equations, which I would like to continue incorporating into my research. One of my primary goals is to develop realistic models, which can actually be used for practical applications. With this intent in mind, I plan to continue collaborating closely with experimentalists in order to use systems engineering with a practical application.

Research experience:

I am a Chemical Engineer by training, and have worked on a wide variety of research projects within the field, and in several other realms of basic science and engineering. The common denominator across all my research experiences is mathematical modeling and systems engineering. From working on non-linear system dynamics pertaining to reaction engineering, to developing mechanistic mathematical models for improving processes operation in pharmaceutical drug manufacturing, and now working with clinical data to understand human pathophysiology, I have been interested in different fields where I feel I may be able to make a significant contribution using the core skill sets of systems modeling. While working as a theoretician during my PhD tenure, I collaborated closely with my experimental colleagues in order to develop realistic mathematical models and better understand the underlying assumptions required for the models. Towards the end of my PhD, I was drawn into the enigma of systems biology, which led me to switch fields and broaden the applications of my mathematical modeling skills. In my current position, I am working in an extremely multi-disciplinary environment where I am leveraging my engineering training and expertise to develop models that can improve disease prognosis and alleviate the shortcomings of a delayed therapeutic intervention while also testing hypotheses and revealing details about molecular and cellular disease mechanisms. A specific disease condition -- sickle-cell disease -- has physical and chemical mechanisms with deep relevance to my chemical engineering training, and I am studying the biorheological properties of blood precipitating a clinical sickle cell vaso-occlusive crisis. Our work motivates a collaboration with single-cell assay experts to develop a novel method to identify biomarkers for sickle cell disease. During my PhD and postdoctoral career, I have received several awards and grants, demonstrating an ability to secure external funding and interest in my research.

Teaching Interests:

In addition to research, I have a keen interest in mentoring students and teaching core as well as advanced engineering topics. During the tenure of my PhD I was a teaching assistant (TA) for several courses including: Transport phenomenon-I, Junior design and economics, and Process control. As a postdoctoral research fellow, I have co-instructed a Health Science and Technology course (Matlab for Medicine) for medical students, which is a required core course for the Harvard-MIT MD curriculum. I have derived immense pleasure from teaching and helping undergraduate students learn the chemical engineering curriculum. During my PhD tenure, I have mentored more than five undergraduate and masters students, who have successfully performed research and have co-authored publications in leading peer-reviewed journals. During my postdoctoral training, I have helped supervise two graduate rotation students and an undergraduate summer intern, as well. I have always encouraged undergraduate students to pursue research in addition to their coursework. I am very comfortable with teaching the core chemical engineering subjects, and I would be very excited to design additional courses that combine the knowledge of chemical engineering, mathematics and statistics.

Awards and Honors:

- October 2016, Life Science Research Foundation (LSRF) postdoctoral fellowship, Good Ventures fellow.
- May 2015, Awarded the prestigious Outstanding Graduate Student Award by the School of Engineering, Rutgers University in recognition of the achievements as a graduate student.
- June 2014, Awarded the Baden-Wuerttemberg Scholarship to pursue research at
- University of Konstanz, Germany.
- July 2012, Awarded the Marshall Plan Fellowship to pursue research at TU Graz, Austria.
- May 2012, Awarded the best poster prize at the New Jersey Pharmaceutical Association for Science and Technology graduate student poster competition.
- September 2010, School of Engineering Fellowship, Rutgers University, USA.

Selected publications: (total 22 published, 9 first author publications)

- **A. Chaudhury,** L. Noiret, J. M. Higgins, White Blood Cell Population Dynamics for Risk Stratification Of Acute Coronary Syndrome, Under Review.
- **A. Chaudhury,** A. Tamrakar, M. Schöngut, D. Smrčka, F. Štěpánek, R. Ramachandran, Multidimensional population balance model development and validation of a reactive granulation process, Industrial Engineering & Chemistry Research, *2015*, 54(3).
- **A. Chaudhury**, A. Armenante, R. Ramachandran, Compartment based population balance modeling of a high shear wet granulation process using data analytics, Chemical Engineering Research and Design, *2015*, 95.
- **A. Chaudhury**, H. Wu, M. Khan, R. Ramachandran. A mechanistic population balance model for granulation processes: Effect of process and formulation parameters. Chemical Engineering Science, *2014,* 107.
- **A. Chaudhury**, I. Oseledets, R. Ramachandran. A computationally efficient technique for the solution of multi-dimensional PBMs of granulation via tensor decomposition. Computers and Chemical Engineering *2014*, 61.
- M. Sen, **A. Chaudhury**, R. Singh, J. John, R. Ramachandran, Multi-scale flowsheet simulation of an integrated continuous purification-downstream pharmaceutical manufacturing process, International Journal of Pharmaceutics, *2013*, 445(1-2).
- **A. Chaudhury**, A. Niziolek, R. Ramachandran. Multidimensional Mechanistic Modeling of Fluid Bed Granulation Process: An Integrated Approach. Advanced Powder Technology, *2012*, 24.
- **A. Chaudhury**, S. Chakraborty, Dynamics of Mixing-Limited Pattern Formation in Non-Isothermal Homogeneous Autocatalytic Reactors: a Low-Dimensional Computational Analysis, Industrial Engineering and Chemistry Research, *2011*, 50 (8).

Book chapters:

- **A. Chaudhury**, D. Barrasso, Daniel A. Pohlman, James D. Litster, R. Ramachandran, Mechanistic modeling of high shear and twin screw mixer granulation processes, Modeling of pharmaceutical unit operations in solid dosage forms, Elsevier, *2016*, In Press.
- **A. Chaudhury,** M. Sen, D. Barrasso, R. Ramachandran, Population balance models for pharmaceutical processes, Handbook in Process Simulation and Data Modeling in Solid Oral Drug Development and Manufacture, Humana Press, *2016*, 43-83.

Teaching and research mentoring experience:

September 2011-June 2012; September 2014-December 2014, Teaching assistant, Department of Chemical and Biochemical engineering, Rutgers University

(Courses taught: Transport phenomenon I, Junior design, Process control).

August 2016; August 2017, Teaching assistant, Health Science and Technology, Harvard Medical School (Course: Matlab for Medicine).

505483 Global Optimization Techniques for System Identification and Green Engineering Applications

Abstract:

Research Interests:

My research focuses on linear programming (LP) and non-linear programming (NLP) techniques for process intensification applications. To this end, I have explored interval analysis-based branch-and-bound techniques for parameter identification, assessment/synthesis of sustainable systems, graphical/analytical techniques for the global solution of the total annualized cost problem for a series of compressors, and the introduction of the novel concept of the attainable region for process networks. The latter concept has been applied to identifcation of the attainable region for VLE separator networks in which the process network's outlet flowrates are allowed to vary. Throughout the course of my work, I have experienced first-hand the â€œcurse of dimensionalityâ€ prevalent in non-linear optimization algorithms. Thus, I have applied analytical techniques, dimensionality reduction, convexification strategies, and functional analysis to identify global optima and to enhance the efficiency of their computation.
My global optimization expertise has prepared me well to pursue new research avenues in several areas, particularly optimal system identification, initial project feasibility assessment via attainable region concepts, and sustainability assessment and decision making. My strong mathematical and computational background will also allow me to contribute to collaborations that can benefit from this kind of expertise.

Teaching Interests:

Once I become an assistant professor, I will disseminate all developed concepts and lessons learned throughout my career into course-friendly materials to provide students opportunities to apply basic chemical engineering and optimization concepts to real-world green engineering problems. The opportunity to observe how thermodynamic concepts, process control, and optimization techniques combine to foster decision-making in process systems design will enrich studentsâ€™ education in chemical engineering, and further prepare them in industry and academia.

505335 Multi-Scale Optimization in Process Systems Engineering

Abstract:

Research Interests:

Energy Systems, Mathematical Modeling(Optimization), Multi-scale Systems, Process Design and Control

Multi-scale optimization:
As multi-scale modeling and advanced simulation technologies mature, it is desirable to utilize their predictive power in process design and optimization. However, these models present unique challenges for traditional optimization frameworks. These models, such as those for complex fluid flow and molecular dynamics, span a very wide range of length and timescales. The highly specialized methods required to simulate these systems are incompatible with popular process optimization methods.
In this project, we are developing a framework for embedding complex simulations in optimization problems. The framework contains two methodologies that build on each other: Îµ-exact models can be used to get close to an optimum and a trust region algorithm that sequentially updates surrogate models. The trust region algorithm is supported by convergence theory to guarantee finding an optimal solution of the truth model. Although our case studies involve the integration of process models and CFD-like continuum models, the use of surrogate models can bridge the gap between other length and timescales. With this optimization framework, insights from the molecular scale can help the decision process at the device and process scale, in turn impacting the decisions at the enterprise scale. We have applied this method to solve flowsheet optimization problems in carbon capture systems. However, this framework could be applied anywhere surrogate models are used, from detailed property calculations to enterprise planning models.
Energy/Process Integration:
Heat integration is a classic and ever-challenging problem in process systems engineering. Pinch methods remain highly popular and practical for typical heat integration problems, as well as mass exchange networks, work exchange networks, and more. Mathematical programming has also become and indispensable tool to handle the many tradeoffs in these systems. However, the highly combinatorial nature of these problems usually requires models that sacrifice a certain degree of accuracy or computational tractability. In this work, we strive to combine these two approaches by using thermodynamic insight from pinch-based methods to build tractable optimization models. This strategy has been effectively applied to organic Rankine cycle design in refineries and to sub-ambient system integration in the oxycombustion process. The optimization results offered new insights into important trade-offs such as cycle efficiency and heat recovery, while greatly improving efficiency compared to previous studies. As energy costs become more important for competitiveness, these integration methods will become more important throughout the chemical industries.

Teaching Interests:

I am able to teach all core chemical engineering courses, with particular interest in teaching control, design, thermodynamics, and engineering math/programming. I also look forward to the opportunity to develop graduate courses in optimization and mathematical modeling of both dynamic and steady state processes.

Please visit my personal page: www.johneason.me

505105 Multiscale Processes Intensification and Optimization of Process Systems

Abstract:

Research Interests:

For years, industrial and manufacturing processes have relied on scaling-up to reduce production costs and increase competitiveness. Advances in Process Intensification, which aims to substantially reduce capital cost, energy consumption, environmental footprint, while increasing safety and production output, have the potential to change the industrial and manufacturing landscape in the upcoming years.

My research aims to systematically search for intensification opportunities – enhancement of intrinsic effects, combination of synergetic processes, novel equipment designs, among others – by using rigorous mathematical approaches, fundamental laws, modeling, and computational simulation to explore a wide range of scales (multiscale-integrated) in which process intensification opportunities can be sought, from molecular, to equipment, to the systems level.

During my Ph.D. I have developed research projects that involved rigorous mathematical modeling and simulation of reaction-diffusion processes in porous media in a multiscale fashion. In addition, I have developed models to systematically intensify reactive separation processes that guarantee global optimality. The set of skills I have learned includes the development of mathematical models from fundamental principles – mass, momentum, and energy balances; reaction kinetics, thermodynamic principles, and advanced transport models such as Stefan-Maxwell and Dusty-Gas Model – for multiscale, multiphase physical systems. In addition, I had formal training in optimization of large-scale process networks and process synthesis, which gives me the ability to find innovative conceptual designs. This combined with my previous industry experience as a senior mechanical engineer in designing and implementing large-scale industrial projects, in which I performed several activities from proposal writing to conceptual design. I am sure that my background and experience qualifies me to develop a strong research program in intensified chemical/manufacturing processes and novel equipment designs.

Proposal experience: NSF-EAGER, EERE-Industrial Assessment Centers, CEC L.A. Regional Energy Innovation Cluster.

Awards: UCLA Graduate Division Doctoral Dissertation Year Fellowship; UCLA JP Lemann Fellowship.

Ph.D. Dissertation: *"Multiscale performance limits quantification of chemical processes networks"*

Supervised by Prof. Vasilios I. Manousiouthakis, University of California, Los Angeles.

Teaching Interests:

I have over three years of experience as a teaching fellow for chemical engineering courses in both classroom and laboratory settings. I am prepared to teach general undergraduate and advanced graduate level courses in Chemical Engineering and Mechanical Engineering, such as transport phenomena, thermodynamics, heat transfer, fluid mechanics, controls, separations, numerical and mathematical methods, and process design. My previous experience as a senior engineer at a project design company can be beneficial to students of all levels, enriching the learning processes of cross-disciplinary classes such as process economics, process design, and entrepreneurship. I am also excited to teach graduate courses in optimization and process design, and look forward to developing new courses related to energy systems and process intensification.

Future direction:

As a faculty, I would like to apply my research efforts on the development of novel processes, equipment design, and production schemes that will help to shape an intensified, safer, greener, data-driven process industry. Through a combination of rigorous modeling, multiscale computational simulation, and optimization/control techniques, a diverse range of both traditional and new processes/manufacturing systems (reactive distillation, membrane reactors, additive manufacturing, among others) can be analyzed, modified, and/or synthesized in order to maximize performance. I envision myself collaborating closely with smart manufacturing/process intensification research centers, as well as companies in the manufacturing and processing industries.

Additional details are available at www.flaviodacruz.com

Publications:

da Cruz, F. E., and Manousiouthakis, V.I. "Process intensification of reactive separator networks through the IDEAS conceptual framework." *Computers & Chemical Engineering* (2016).

da Cruz, F. E., and Manousiouthakis V.I. "Parametric studies of steam methane reforming using a multiscale reactor model." *Ind. and Eng. Chemistry Research* (Submitted).

Karagöz, S., **da Cruz, F. E.**, Tsotsis, T. T., and Manousiouthakis, V.I. " The multiscale (pellet-reactor scale) membrane reactor (MR) modeling and simulation: Molecular sieve MR for hydrogen production by low-temperature water gas shift reaction." *Ind. and Eng. Chemistry Research* (Submitted).

da Cruz, F. E., and Silvio de Oliveira Junior. "Petroleum refinery hydrogen production unit: exergy and production cost evaluation." *International Journal of Thermodynamics* 11, no. 4 (2008): 187-193.

505272 Optimization-Based Control of Complex Process Networks: Application to Medicine and Energy Systems

Abstract:

Research Interests:

Regarding my future research efforts, I intend to extend my computational work to modeling, estimation, control, and optimization of advanced chemical and biological processes with application to energy systems and medicine. Specifically, the intellectual objective of my future research is to develop a systematic model predictive control and optimization framework tailored for large-scale complex process systems with unknown transport-reaction mechanisms when the system's uncertainty plays a vital role. For example, the source of parametric uncertainties and unmeasurable unknown parameters in chemical and living systems, which necessitates applying such novel control and optimization approaches, is reaction rates, activation energies, fouling factors, and microbial and cell growth rates which are only approximately known via complex experiments and/or high-fidelity multiscale simulations. The proposed research in this category will cover the controller parameterization, regulation of the estimation error, adaptive law synthesis, closed-loop system stability and robustness for different types of desired system performances. I plan to investigate the nonlinear adaptive model predictive control problem by introducing extra degrees of conservativeness considering robust performance and closed-loop stability. The proposed research will introduce systematic computationally efficient process identification methods to estimate system parameters required by the model predictive control to compute the optimal control action subject to a continuous decrease in the identification error.

Teaching Interests:

I view teaching as an integral and very important part of my overall academic activity. I plan to actively participate in the enhancement of the entire undergraduate and graduate curriculum. Given my Chemical Engineering background and my experience as a teaching assistant, lecturer and lab supervisor in several undergraduate courses, labs and workshops at Sharif University of Technology, the Pennsylvania State University, University of Delaware, and University of Minnesota, I feel confident that I can effectively teach every course in the undergraduate Chemical Engineering curriculum, and I will be happy to do so. Given my expertise in the areas of process dynamics, control, design and applied mathematics, my immediate teaching plans concern the improvement and development of courses pertaining to process dynamics and control, process design, and analytical/numerical methods for solving chemical engineering problems.

505413 Process Systems Engineering for Transforming Industrial Flares into a Source of Energy By Managing Uncertain Abnormal Situation

Abstract:

Meet the Faculty Candidate Poster Session

Computing & Systems Technology

Process Systems Engineering for Transforming Industrial Flares into a Source of Energy by Managing Uncertain Abnormal Situations

Monzure-Khoda Kazi

Department of Chemical Engineering, College of Engineering, Qatar University, P.O. Box-2713, Qatar

Email: kazi0001@qu.edu.qa

Advancement in technology always influences the world energy market by making excess to the valuable untapped resources. Profound research on potential unexploited/waste energy can lead the world to a new direction of energy sources; shale gas production is a very recent example of modern technological advancement and cutting-edge research efforts. Flaring is a very wasteful practice and a significant source of GHG emissions from oil, gas and petrochemical plants. The World Bank recently has identified flare gas as a key untapped resource to meet the world's increasing demand of fossil fuel energy while reducing harmful emissions and has announced an ambitious goal of 'Zero routine gas flaring by 2030'[1-3]. Therefore, further investigation on flare recovery and mitigation techniques/tools has become eminent and can open the doors of huge opportunity to turn the unproductive non-renewable flare streams into valuable commodity [4].

This recovered flare gas can be an excellent source of heat or power generation using energy alternative tools. This approaches will be viable economically if an eco-industrial park is developed to exchange by-products and wastes. The resources and wastes of multiple plants may be integrated in different ways. It is also possible to use common utility and treatment systems. This park can provide clear economic and other advantages over the current stand-alone processing model [5, 6]. However, the application of industrial ecology at factory level is absent from the literature.

Companies are now motivated to invest highly on reducing gas flaring and industrial complexes to optimize the utilization of this valuable resource and reduce its environmental impact. It is now clear that flare management has good financial and environmental benefits. But, it is still quite challenging to fit the flare recovery technologies with the proven process design of the plants. The main challenge is the lack of coordination among the industrial producers, the commercial vendors and the process designers. There are hardly any guidelines, framework or software which can handle the variable nature of industrial problems and specifications. It is understandable that no two industries are alike; so flare gas recovery systems should be tailored for individual applications both for large and small scale. Predefined size of the alternatives cannot assure optimal performance of the integrated process, which does not reflect the proper economic/environmental benefit or trade-offs. Moreover, in most cases the uncertain nature of the flaring events is compensated by overdesign of the equipment or by inefficient utilization of the energy stream.

Preliminary findings in flare recovering and utilization during abnormal situations from my previous works/publications have been proven promising in terms of important economic, environmental and energy conservation benefits obtained [7, 8]. It indicates that substantial fractions of the available flare streams generated during abnormal industrial situations can be used for heat and power generation through flare management tools.

Furthermore, industrial ecology is a young but growing multidisciplinary field of research which combines aspects of engineering, economics, sociology, toxicology and the natural sciences. Industrial ecology is concerned with the shifting of industrial process from linear (open loop) systems, in which resource and capital investments move through the system to become waste, to a closed loop system where wastes can become inputs for new processes. Existing industrial complexes can also evolve towards symbioses when corporations are motivated by waste reduction and environmental policy. The creation of an EIP entails technical and economic challenges that are best addressed through a process integration framework. Ultimately, it will help the decision makers to compare different available commercial and evolving products from different companies and choose their best fitted solution depending on which the corresponding users can make their critical decisions.

Figure 1. Concept of Agro-Industrial Eco Park combining with industrial flare utilization technologies (i.e., COGEN, TMD, water & heat recovery from flue gas and seawater greenhouse)

Research Goal

To develop application-oriented theory, computational tools, algorithms, and optimization methods for complex and multi-scale systems

Research Interests:

Modeling and optimization of complex systems

- Multi-scale systems engineering for energy and environment
- Nonlinear, nonconvex and stochastic optimization
- Sustainable process design through integration and intensification
- Grey-box and black-box systems

I am interested in developing application-oriented theory, computational tools, algorithms, and optimization methods for complex and multi-scale systems. My research background of process integration, process control and specially process optimization assist me for developing new concepts and tools to systematically enhance and reconcile various process objectives, such as cost effectiveness, yield enhancement, energy efficiency, and pollution prevention. I am working last few years to develop such methodical sustainable approaches for multi-scale systems engineering starting from small scale production (producing valuable pharmaceutical drags using liquid chromatography) to large scale facility (managing industrial flares during abnormal situations). I am interested to solve complex systems which are difficult to optimize due to their complexity, uncertainty, and the inability to generate accurate data using physical and/or computational experiments in reasonable time. I would also like to face the challenges from process uncertainty and from the black box systems which need to be resolved for accurate model prediction, control and optimization.

COMPLETED PROJECTS

- Control and Optimization of Continuous Chromatographic Separation Process (NTU, Singapore)
- Large-Scale Chromatography with green Solvents: Fundamentals and Novel-processes (GSK-Singapore Partnership project with NTU by Economic Development Board (EDB))
- An Integrated Approach to the Simultaneous Design and Operation of Industrial Facilities for Abnormal Situation Management via industrial flare management (National Priority Research Project in Qatar | Advisory board: Qatar Chemical Company Ltd (Q-Chem))
- A Life-Cycle Integrated Approach to the Incorporation of Safety in the Design, Operation, and Optimization of Industrial Supply Chains in Qatar (Preparing final report)

CONTRIBUTION AREAS

Process Control | Process Optimization | Process Integration | Process Design | Process Intensification | Process Modeling & Simulation | Process Safety | Ad hoc Reporting

QUALIFICATION SUMMARY

1. PhD degree in chemical engineering
2. 9 years of R&D experience focusing on: green technologies for pharmaceutical separation, GHG reduction strategies/tools, wastewater treatment facility, water networking, combined heat-power technologies, thermal membrane distillation
3. Possesses experience of developing modeling and simulation using deterministic and stochastic models; the developed models were based on the combination of finite volume and genetic algorithm
4. Expert in different optimization techniques, algorithm and software
5. Expert on different numerical methods and control theory
6. Developed several codes for optimizing control and for the solving of real life optimization problem based on MATLAB, C and LINGO. I am also expert in writing codes on FORTRAN and different problem solving tools/software.
7. 'Hands-on' experience with different chromatography technologies: LC-UV/Vis spectroscopy, GC-MS, SFC
8. Experience of conducting experimental study for chiral drug separation using in-house developed novel separation technique
9. Experience of characterization using advanced analytical techniques
10. Developed several codes using C and MQL programming combining with Monte Carlo simulation results

Teaching Interests:

I am comfortable teaching the *Chemical Engineering core courses* offered within the department at both the *undergraduate and graduate* level. Based on my academic, PhD work and research background, I would be well suited to teach fundamental course such as:

- **Process Control**
- **Process Design**
- **Process Modeling and Simulation**
- **Process Integration**
- **Engineering Economics**
- **Chemical Thermodynamics**
- **Numerical Methods**
- **Separation Technology**
- **Process Intensification**

I appreciate the need for faculty flexibility and embrace the opportunity to cover a broad spectrum of courses. I look forward to incorporating fundamental engineering concepts with cutting-edge technologies. I can also teach different programing language and software such as: MATLAB, FORTRAN, C, LINGO, GAMS, LabVIEW, Aspen HYSYS, ASPEN Plus Dynamics, ANSYS, Thermolib, Simulink, ProMAX, Auto CAD, Latex, Origin, EndNote, XLSim – Excel base risk solver etc.; which will provide the students with working experience and applications on computer-aided modelling, simulation and optimization of chemical engineering systems.

SUMMARY OF SCHOLARLY ACHIEVEMENTS

Total Peer Reviewed Publication: 10; Peer Reviewed Journal articles: 10 published (4 first- and 6 co-authored) and 2 (first-authored) in review; Conference Presentations: 14; Poster Presentations: 8

GRANT WRITING EXPERIENCE

1. "Addressing Design Challenges for Effective Implementation of Flare Recovery Systems", Source: QNRF – National Priority Research Program, Budget: US$ 599,422, PI: Assoc. Prof. Mohammad Aman Ullah, 2017-2020 *(*I assisted to write this proposal as this was the continuation of my current postdoctoral works)*
2. "Optimization of a single column continuous chromatographic separation process for the separation of optical isomers", Source: 13th cycle of the QNRF's undergraduate research experience program (UREP), Budget: US$ 58,850, Lead PI: Assoc. Prof. Mohammad Aman Ullah, 2013 *(*I helped to write this proposal as this was the continuation of my PhD thesis works)*

PROFESSIONAL AFFILIATION

Reviewer of Journals: Current Opinion in Chemical Engineering

Senior Member: American Institute of Chemical Engineers (AIChE)

Member: The Institution of Engineers, Bangladesh (Qatar Chapter), Chemical Engineering Alumni Association in BUET, NTU Alumni Club

Volunteer: Summer Youth Olympics, Singapore (2010)

Organizing Committee member: 5th Pacific Basin conference on Adsorption Science and Technology, Singapore (2009)

References:

1. The World Bank Feature Story. *Gas Utilization in Kuwait Reaps Economic and Environmental Benefits*. 2014; Available from: http://www.worldbank.org/en/news/feature/2014/08/25/gas-utilization-in-kuwait-reaps-economic-and-environmental-benefits.

2. The World Bank Press Release. *Initiative to Reduce Global Gas Flaring*. 2014; Available from: http://www.worldbank.org/en/news/feature/2014/09/22/initiative-to-reduce-global-gas-flaring.

3. The World Bank Press Release. *Bid to End Routine Gas Flaring Gets Huge Boost at COP21*. 2015; Available from: http://www.worldbank.org/en/news/press-release/2015/12/07/bid-to-end-routine-gas-flaring-gets-huge-boost-at-cop21.

4. Fleisch, T.H. *Associated Gas Monetization via miniGTL : Conversion of Flared Gas into Liquid Fuels and Chemicals, Report III. World Bank, Washington, DC. © World Bank*. Available from: https://openknowledge.worldbank.org/handle/10986/23609 License: CC BY 3.0 IGO.

5. El-Halwagi, M.M., *Process integration*. Vol. 7. 2006: Academic Press.

6. El-Halwagi, M.M., *Sustainable design through process integration: fundamentals and applications to industrial pollution prevention, resource conservation, and profitability enhancement*. 2012: Elsevier.

7. Kazi, M.-K., et al., *Integration of Energy and Wastewater Treatment Alternatives with Process Facilities To Manage Industrial Flares during Normal and Abnormal Operations: Multiobjective Extendible Optimization Framework*. Industrial & Engineering Chemistry Research, 2016.

8. Kazi, M.-K., et al., *Multi-objective optimization methodology to size cogeneration systems for managing flares from uncertain sources during abnormal process operations.* Computers & Chemical Engineering, 2015. **76**(0): p. 76-86.

504262 Computational Design and Characterization of Nanoscale Materials for Energy Applications

Abstract:

Research Interests:

Transitioning the global infrastructure from reliance on fossil fuels to clean and sustainable sources of energy is the defining technological challenge of our age. Solving this problem will require the development of more efficient novel materials for energy production, transport, and storage. My research interests involve using a combination of first principles and statistical methods to predict mechanical, thermodynamic, and electronic properties of materials to accelerate the discovery of new materials with broad applications related to energy. I am also interested in methodological developments to increase the speed and efficiency of these calculations.

High-performance computing coupled with a variety of theoretical frameworks, including density functional theory and Monte Carlo methods, allows us to design and characterize new materials much faster than could be achieved using only experiments. Theoretical calculations can screen tens of thousands of materials to identify top candidates for further examination. This type of screening can also be used to investigate trends between structure and function, which can be used to facilitate the design of new materials for specific applications.

My graduate research focused on using density functional theory to predict vibrational spectra for Si and Ge nanoparticles with impurities and heterojunctions. By understanding how changes to the geometry and electronic structure affect the Raman response, we can interpret experimental spectroscopic data to infer the size, dopant concentration, and interfacial strain in these nanoparticles. The calculation of vibrational spectra from first principles for a system containing hundreds of atoms is computationally expensive; therefore, we also introduced an efficient high-order integration scheme for calculating interatomic forces. This method provides good accuracy at a substantially lower computational cost and reduces memory requirements.

My postdoctoral work involves using high-throughput screening methods to discover new nanoporous materials for hydrogen storage, concentrating specifically on metal-organic frameworks (MOFs). MOFs are porous materials consisting of metal nodes connected by organic linkers. They are appealing candidates for gas storage and separations due to their porosity and high surface area. Due to the diversity of available nodes, linkers, and functional groups that can be combined to create a MOF, there are practically an infinite number of hypothetical MOFS. Using high-throughput computational methods to generate new structures and predict their textural properties and capacity for hydrogen storage, we have sifted through huge numbers of MOFs to identify those which are most promising and should be studied in more detail. We also identified structure-property trends and design principles for what an "ideal" MOF for hydrogen storage should look like, which can inform the creation of new MOFs tailored to a specific function.

In the future, I plan to build on my background in molecular modeling, materials science, and high-performance computing to discover new materials with applications in efficient chemical separations and energy production and storage. I am also interested in studying the behavior of MOFs at interfaces with other materials, such as metal nanoparticles or graphite oxide.

Teaching Interests:

My goal as an instructor in chemical engineering is to give students necessary foundational knowledge, while also imparting a sense of the broader connections between various courses they will take throughout their education. Therefore, I am interested in teaching any of the fundamental courses such as thermodynamics, kinetics, transport, and mass and energy balances. Since my background is primarily in molecular modeling and computation, I would also be interested in teaching courses such as numerical methods or MATLAB programming. I am also intrigued by the opportunity to teach more applied courses such as chemical engineering design, which draw from—and synthesize—all elements of chemical engineering.

In addition to these courses, I would be interested in designing and teaching an elective course for graduate students taken from my own research, which focuses on computational materials science and molecular modeling. The course would offer an introduction to density functional theory, molecular dynamics, Monte Carlo methods, and high performance computing. This would be a valuable primer for graduate students who plan to pursue computational research and also provide an accessible overview for students in experimental research who want to gain a sense of how modeling is done and what advantages and limitations various techniques offer.

505547 High-Performance ComputingÂ Approaches to Large-Scale Stochastic Programming andÂ Data Analysis

Abstract:

High-Performance Computing Approaches to Large-Scale Stochastic Programming and Data Analysis

Research Interests: High-performance computing, stochastic programming, machine learning, energy systems, robust/stochastic MPC

This poster highlights my recent work on algorithms and software implementations to solve nonlinear stochastic programming problems of unprecedented complexity to local and global optimality. SchurIpopt and PIPS-NLP are powerful local solvers for continuous problems, which exploit the arrowhead block representation of the linear algebra system using a distributed-memory Schur decomposition strategy. However, the scalability of the Schur decomposition approach is hindered by the number of first-stage variables. IPCluster is a solver that overcomes this fundamental bottleneck by performing adaptive clustering of scenarios to create effective preconditioners for the linear algebra system. SNGO is a global optimization solver that uses a specialized branch and bound strategy to enable parallelism and convergence for both continuous and mixed-integer problems. We also present PLASMO, a Julia-based open-source modeling framework that uses a graph-based approach to facilitate the expression of complex stochastic programming problems and scenario trees. We have applied these tools in the three applications.

Robust/Stochastic Control

Robust/Stochastic Control is an approach to design controllers that explicitly addresses uncertainty either from uncertain parameter or disturbance. A robust/stochastic nonlinear model predictive control (NMPC) strategy optimizes the worse-case/expected-value performance at each sampling instance. The size of problem is often too large to be implemented in real time. In a collaboration with the experimental research group of Prof. Zoltan Nagy at Purdue, we have found that unseeded batch crystallization processes reach optimization formulations of over half a million variables and constraints. The serial solver Ipopt takes 7 min to solve the problem and by using parallel implementations in SchurIpopt we are able to reduce the time to 30 seconds, allowing for real-time applications. We have also found that a robust NMPC strategy guarantees satisfaction of all state and input constraints for a set of uncertainty realizations, and also provides better robust performance. In particular, the standard deviation of crystal size and worst case mismatch from desired crystal size using robust NMPC is 30% smaller than that using nominal NMPC and is 50% better than that using open-loop optimal control.

It is also possible to use robust/stochastic NMPC formulations to optimize the parameters of controller laws. In a collaboration with General Electric, we have recently proposed a stochastic optimization formulation to identify optimal parameters for pitch and torque controllers in wind turbines considering historical data of wind speed. This approach seeks to extract maximum power and to satisfy extreme load requirements by design. Compared with nominal controller parameters, the optimal settings increase the revenue of a wind turbine by 18%, reaching revenue improvements of hundreds of thousands of USD per year. The optimization problem involves up to 7.5 million variables and constraints and can be solved in less than 1.3 hours using PIPS-NLP. The serial solver Ipopt takes over one day to solve the problem.

Optimization of Energy Systems

Many energy system suffers from significant uncertainty arising from climate conditions and electricity demands. Moreover, conflicting objectives must often be taken into account. We have recently developed a multi-objective stochastic optimization framework to solve these types of problems. We have applied this framework to the design of a combined heat and power plant that must take cost, emissions, and water usage under consideration. The design problem has more than 1 million variables and can be solved with PIPS-NLP in less than 6 minutes.

We have also applied our techniques to stochastic market clearing models in large power grid. A clearing problem for the Illinois power grid system has more than 64,000 first stage variables and more than 1.2 million total variables. The challenge in this problem is the large dimensionality of the first-stage. While Ipopt and PIPS-NLP both take more than 8 hours to solve the problem, a preconditioned strategy implemented in IPCluster takes only 11 minutes to solve the problem.

Experimental Data Analysis

High-performance computing and optimization algorithms can also be used to break existing limitations in estimation and learning problems. In a collaboration with the experimental group of Prof. Ophellia Venturelli at UW-Madison, we have demonstrated that we can estimate parameters in highly nonlinear microbial growth models to global optimality. Our solver SNGO solved this problem in less than 9 minutes to an optimality gap of 1%, while the state-of-the-art global solver SCIP cannot solve the problem (it takes one day to close the optimality gap to 60%). The advantage of using a global solver is that we can provide a certificate to experimental researchers that no better set of parameters can be found.

Machine learning problems from either support vector machines (SVM) or neural network can also be solved using the proposed algorithms. In a collaboration with Prof. Nicholas Abbott group at UW-Madison, we have applied SVM to classify the response of liquid crystals (LCs) under analytes such as DMMP (similar structure to sarin nerve gas) and water. We have used our algorithms to classify more than 70,000 images from experiments of LCs and we have achieved accuracies of 99%.

Teaching Interests:

I had gain extensive teaching experience during my PhD as a teaching assistant in modeling, high-performance computing, optimization, and control. I am qualified to teach undergraduate chemical engineering core courses such as design, control, and modeling. I am also interested in developing new courses related to applications of high performance computing, optimization, and machine learning in chemical engineering.

505229 Conducting Flow-Induced Crystallization Studies on Flexible and Semi-Rigid Polymers: A Facilitator of Education in Polymer Physics

Abstract:

Research Interests:

Training students and researchers on classic topics of conformation and molecular dimensions of polymer chains remains challenging. As a postdoc at Penn State with my supervisor, Prof. Ralph Colby, I was able to assemble a 17-student team of superb undergraduate researchers spanning across different aspects of polymer science. These undergraduate students worked on the following projects:

1. Flow-induced crystallization of PEEK, nylon 66, iPP, and PET.

2. Rheology of ionic liquids/cellulose/chitosan solutions.

3. Investigation of polyolefin-based interpenetrating polymer networks for collecting oil spills.

4. Renewable lignocellulose-based foams via Pickering emulsion templates.

To provide these undergraduates with opportunities for continuing research experience, I submitted two separate proposals to the Fall 2017 University Park Research Experience for Undergraduates Funding Program through the College of Engineering at Penn State. These grants would enable two of our students to work on a project that aims to develop biomass-based films.

Through my Penn State experience, I came to realize that newcomers to the polymer field are easily attracted to studying relationships between structure and function in polymeric systems. In particular, our newly purchased rheometers made it easier for us to have our students experience and understand how conformational ordering under shear in a molten polymer can significantly change its crystallization rate and resulting crystal morphology.

My interest in "green materials" builds upon my Ph.D. work, undertaken at UMaine under the guidance of Prof. Doug Bousfield, which elucidated the interactions between rheology of cellulose nanofibers (CNFs) suspensions and their processability. I focused on the development of new applications for CNFs as an alternative platform for petroleum-based products. That work contributed to the feasibility of utilizing waste biomass, and therefore has huge environmental benefits. As the PI, I wrote a proposal entitled "Renewable lignocellulose-based foams via Pickering emulsion templates" to USDA's Fiscal Year 2017 Biomass Research and Development. If successful, this grant will help me with funding my own research hopefully as a faculty member during 2018 and beyond.

I have taken my interest in controlling polymer physics and its effect on processing and final properties of the product to my current postdoctoral position at MIT with Prof. Greg Rutledge, where we seek to utilize structure-property relationship of polymer configuration and its molecular orientation to significantly alter the mechanical properties of polymeric fibers. For instance, by improving the molecular orientation and degree of crystallinity from gel-electrospinning ultra high molecular weight polyethylene, our group has successfully fabricated ultrathin fibers with high stiffness that rivals the commercial protective vest such as Spectra. By defining a fundamental connection between the predicted nanostructure and its functionality, we can expand much further on the capabilities of electrospun nanofibers to significantly impact the polymer processing and its application as a whole.

Teaching Interests:

During 2011, I was a fulltime teacher of Chemical Reaction Engineering, Numerical Methods, Unit Operations (+Lab), Aspen Plus Workshop, Heat Transfer (+Lab), and Fluid Mechanics (+Lab). At UMaine, I TAed in Biological Transport Phenomena. More recently at Penn State, I taught Polymer Rheology and Processing. In the future, I would welcome teaching these courses, but I would also like to teach Polymer Physics, Chemical Engineering Thermodynamics and Transport Phenomena. In 2015, I published a textbook entitled "Numerical Methods for Process Engineers" with Shiraz University Press. I believe effective teaching empowers the students' ability to think in a critical way and create their own knowledge, and this ultimately benefits the research community. I also think other legitimate purposes of teachers include the need to make a useful contribution to society by helping induct younger human beings into the society's thrive process.

506370 Utilization of Lignocellulosic Biomass to Value-added Bio-products

Abstract:

Research Interests:

Lignocellulosic biomass, the most abundant renewable resource produced by photosynthesis on Earth, has been considered a promising alternative fuel and chemical source. It is widely accepted that biomass-derived products can reduce the current dependence on fossil-based products; however, there are still many technical and economic challenges facing the utilization of this material. My research interest in lignocellulosic biomass stems from such challenges. The long-term goal of my work is to improve the utilization of lignocellulosic biomass by understanding the physicochemical properties of this material and developing effective conversion technologies and value-added biomass-derived products.

1. Understanding the physicochemical characteristics of biomass and bio-products

Significance: Lignocellulosic biomass is a complex biopolymer mainly composed of cellulose, hemicellulose, and lignin, with other minor components. The characteristics of biomass vary depending on its species, growing environments, harvesting periods, and region. In addition, biomass conversion methods, reaction conditions, and other factors can change characteristics of the products and residues produced from biomass. Therefore, characterizing biomass and its products is essential for understanding both biomass feedstock and the effects of different conversion methods.

Current Research and Accomplishments: I have conducted a range of biomass characterization research to study the effects of different pretreatments on biomass, understand biomass recalcitrance factors in natural variants, observe the characteristic changes in genetically modified biomass, and interpret the inhibitory effects of pretreated lignins on cellulose hydrolysis. My contribution to biomass characterization has not only been limited to the collection of data or interpretation of characteristic results. I have designed a bi-solvent system to enhance biomass handling and the resolution of NMR spectra.

Future Research Plan

Based on my research experience, I believe we can find many answers for the current challenges in biomass utilization by elucidating the properties of biomass and its products. In particular, I will focus on lignin characterization because it is the key factor of biomass recalcitrance and is not a well-explored component compared to the carbohydrates fraction in lignocellulosic biomass. The following plans detail my immediate research aims for characterizing lignin, including (a) to study the relationship between the physicochemical and thermochemical properties of lignin and (b) to develop new ways of rapidly characterizing valuable technical lignins.

2. Development of effective biomass conversion processes and value-added bio-products

Significance: The conversion of lignocellulosic biomass is challenging due to its rigid and compact structure and its recalcitrant nature due to various chemical and physical features. Therefore, effective and innovative technologies are needed to convert biomass into fuels and industrial chemicals. Also, due to the technical and economic barriers of current biofuels, such as cellulosic ethanol and biodiesel, researchers have turned their attention to developing new types of biofuels, chemicals, and biomaterials from biomass.

Current Research and Accomplishments: Various pretreatment and fractionation methods have been studied for biomass utilization, but the aforementioned barriers in biomass conversion have not been fully addressed yet. For these reasons, I have tried to consider the challenges for biomass utilization by lignin valorization to mono-aromatic compounds and dicarboxylic acids. Another approach I have developed is an efficient biomass conversion process for high-value chemicals and precursors using a biphasic system that consists of an aqueous phase of molten salt hydrate and an organic phase for extracting furan-based chemicals.

Future Research Plan

My future research will aim to design practical biomass utilization strategies for industrial applications. I will approach this goal by improving biomass conversion efficiency and/or by producing high-value lignin-based products/co-products. In my previous work, I learned about the technical limitations and cost-barriers in current biomass conversion processes, thus I will focus my research group's efforts on designing efficient biomass conversion strategies to overcome these challenges. In addition, I will develop an economic biorefinery process by valorizing co-products like lignin from current biomass conversion processes. To achieve these goals, the aims of my future research group will include: (a) to synthesize new graphene oxide-based acid catalysts for the efficient and cost-effective conversion of biomass; and (b) to produce green and high-value lignin-based biomaterials with direct commercial applications.

Research accomplishments

- Published 32 peer-review journal articles (16 articles as a first author and 16 articles as a co-author) in *Green Chemistry, ChemSusChem, ACS Sustainable Chemistry & Engineering, Bioresource Technology, Biotechnology for Biofuels*, and other journals.
- Published 3 book chapters about biomass utilization and characterization.
- Made 20 oral and 31 poster presentations at the professional conferences including *AIChE annual meeting, ACS annual meeting, Symposium on Biotechnology for Fuels and Chemicals*, and others.
- Made two invited presentations at *Kongju National University* and *University of Tennessee-Knoxville*.
- Contributing as a Topic Editor in the *Frontiers in Energy Research* journal.

Teaching Interests:

My experience as a student and an instructor in academia has led me to believe that a well-prepared teacher not only delivers knowledge to students, but also utilizes a diverse number of approaches to help *all* students comprehend the material. In particular, higher education aims to foster professionals who have both fundamental knowledge and problem-solving skills. Therefore, effective teaching should go beyond just the delivery of knowledge. My primary goal in teaching is to provide a confident environment for students to learn fundamental knowledge and to cultivate their abilities to apply their learning and problem-solving skills at future positions. To provide this type of quality instruction, I have three key approaches: 1) motivate students; 2) interact and communicate with them in diverse ways; and 3) build their logical reasoning strategies from learning by application.

My field of research, the characterization and utilization of biomass, has attracted considerable attention due to recent global energy and environmental issues. Compared to the amount of research activity on biorenewable materials, there are generally few related courses on the subject available to students. I believe more courses are necessary to introduce biorefinery concepts and their applications, and with this in mind, I was able to contribute to the development of a "Bioprocessing and Bioproducts" course, preparing an array of course materials that spanned from the syllabus to the content for weekly lectures. It was a great experience that allowed me to think about the connection between my research and teaching. I have also taught in several lab classes designed for undergraduate students. With these experiences, I would like to develop a new course introducing and explaining biorefinery. Having written two book chapters about lignocellulosic biomass pretreatment and cellulosic ethanol production and one book chapter on biomass analysis technologies, I feel well prepared to undertake this endeavor. In particular, one of these book chapters that targeted junior and senior college students to provide fundamental knowledge about the utilization of lignocellulosic biomass is now being used in "Biorefining" course at University of Wisconsin-Madison.

I believe a well-prepared course is an excellent medium through which an instructor can provide guidance to students who can in return share their own new and creative ideas. My educational experiences have provided a valuable stepping-stone and foundation for my career to serve as a future faculty member.

Education accomplishments

- Mentored two graduate, six undergraduate, and one high school students.
- Gave lectures as a TA in three courses.
- Gave lectures as a guest lecturer in two courses.
- Organized Summer Biofuel Workshops (supported by NSF) for local high school students in two years.

Other accomplishments

- Contributed two grant proposals (SIRE-REU; $4,865 per each project).
- Recommended *USDA 2015 AFRI NIFA Fellowship* as PI ($148,515; The fellowship was not awarded because of relocation to ORNL).
- Reviewed 3 grant proposals for USDA Small Business Innovation (SBIR) Phase I from 2015 to 2017.
- Awarded *Supplementary Performance Award ORNL* (2016), *Research Excellence Award* (2012), *Techno Dormitory Scholarship* (2006), *Science and Technology Scholarship* (2006-2007), *Hanyang Fusion Materials Scholarship* (2006-2007).
- Member of the honor societies, *Gamma Sigma Delta* and *Alpha Epsilon*.

506504 Leveraging Physiological Microenvironment to Transport across Biological Barriers

Abstract:

Research Interests: Each tissue in our body is composed of specific biological components that constitute a defined niche, and these characteristic compositions can be exploited to rationally design drug delivery systems to improve drug efficacy and reduce systemic toxicity. Tissue compositions alter significantly under pathological conditions compared with normal tissue at the genetic, molecular and cellular levels. Approaches that employ key biochemical or biophysical cues at the site of disease may effectively respond to endogenous stimuli in the biological milieu to cross biological barriers for targeted drug delivery and on-demand release. I am highly interested in understanding the biological changes associated with disease types and states, and how to apply these pathophysiological features as cues in systems for drug delivery and tissue regeneration.

I work at the interface of chemical engineering, biomedical engineering, material science, and immunological research, seeking innovative solutions for safer and more effective treatment of skeletal and gastrointestinal disorders. My research experience covers drug delivery systems for treatment of inflammation-associated diseases, including inflammation-targeting hydrogel microfibers for local drug delivery in inflammatory bowel disease (1), and bone-seeking bisphosphonate-modified nanoparticles targeting the bone microenvironment for treatment of multiple myeloma (2). I have also studied disulfide-linked bisphosphonate-conjugated proteins for triggered release (3), and protein-encapsulated polymeric nanoparticles for controlled release at the implantation site for de novo bone formation (4, 5).

Future Research: As a faculty member, I will focus on employing engineering strategies to tackle biological and immunological problems for medical innovations and applications. The gastrointestinal tract contains the most complex human immune system, which closely interacts with the gut microbiota. The healthy or diseased state of the gut is also connected with skeletal fitness. I will direct my research in the following areas to translate biological findings into therapies, and to provide clinically relevant solutions for medical applications:

(1) Targeting inflammation for autoimmune gastrointestinal disorders
(2) Promoting mucosal healing for intestinal homeostasis
(3) Modulating microbiome for integrative skeletal regeneration

Teaching Interests: My teaching philosophy focuses on inspiring students to understand fundamental concepts and apply knowledge to solve practical problems. I promote interactive teaching and learning activities, and appreciate the diversity of students' background and academic goals. I focus on teaching and evaluating the class in a manner that centers on effective communication, conceptual understanding, and creativity. At different stages of my academic training, I sharpened my pedagogical skills and cultivated my passion for teaching and mentoring through being Assignment Grader, Teaching Assistant, and Lecturer. As a graduate student, I was a teaching assistant for Mass Transfer and Heat Transfer. I was a lecturer in the Department of Chemical and Materials Engineering at the University of Alberta after I completed my Ph.D. studies. I developed and delivered lectures, designed midterms and finals, and held office hours for students. I am well prepared to teach core chemical engineering courses and special topic courses that I plan to develop, including Transport Phenomena in Living Systems, Biomaterial Science and Nanomedicine, and Drug Delivery Fundamentals and Design Principles. My mentoring experience also included working with seven undergraduates, five visiting students, and three technicians in a laboratory environment. I am open to fresh and effective teaching ideas that facilitate the teaching and learning processes, and dedicated to the training and development of young students.

Grant Writing:

1. Crohn's and Colitis Foundation – I was awarded a Research Fellowship Award from the Crohn's and Colitis Foundation (2015 – 2017).

2. NIH K99/R00 – I composed a K99/R00 to the National Institute of Diabetes and Digestive and Kidney Diseases – an impact score of 27 in 2013.

3. NIH R21 – I assisted in writing an R21 to the National Institute of Dental and Craniofacial Research – funded in 2013 (2013 – 2015).

4. NSERC – I was awarded the Natural Sciences and Engineering Research Council of Canada (NSERC) Postdoctoral Fellowship (2010 – 2012).

References (Selected Publications):

1. S. Zhang*, J. Ermann*, M.D. Succi, A. Zhou, M.J. Hamilton, B. Cao, J.R. Korzenik, J.N. Glickman, P.K. Vemula, L.H. Glimcher, G. Traverso, R. Langer, J.M. Karp. An Inflammationtargeting Hydrogel for Local Drug Delivery in Inflammatory Bowel Disease. Science Translational Medicine. 2015, 7, 300ra128. (* denotes equal contribution) Highlighted in: Science, 2015, 349, 702; Nature Reviews Drug Discovery, 2015, 14, 678; Nature Reviews Gastroenterology & Hepatology, 2015, 12, 551.

2. A.Swami, M.R. Reagan, P. Basto, Y. Mishima, N. Kamaly, S. Glavey, S. Zhang, M. Moschetta, D. Seevaratnam, Y. Zhang, J. Liu, T. Memarzadeh, J. Wu, S, Manier, J, Shi, N. Bertrand, Z. N. Lu, K. Nagano, R. Baron, A. Sacco, A. M. Roccaro, O. C. Farokhzad, I. M. Ghobrial. Engineered Nanomedicine for Myeloma and Bone Microenvironment Targeting. Proceedings of the National Academy of Sciences of the United States of America. 2014, 111(28): 10287-10292.

3. S. Zhang, J. E.I. Wright, G. Bansal, P. Cho, H. Uludağ. Cleavage of disulfide-linked fetuinbisphosphonate conjugates with Three Physiological Thiols. Biomacromolecules, 2005, 6: 2800-2808.

4. S. Zhang, C. Kucharski, M. R. Doschak, W. Sebald, H. Uludağ. Polyethylenimine-PEG Coated Albumin Nanoparticles for BMP-2 Delivery. Biomaterials. 2010, 31(5): 952-963.

5. S. Zhang, M. R. Doschak, H. Uludağ. Pharmacokinetics and Bone Formation by BMP-2 Entrapped in Polyethylenimine-Coated Albumin Nanoparticles. Biomaterials. 2009, 30(28): 5143-5155.

Hydrogeoxygenation of Long-Carbon Oxygenates to Jet and Diesel Fuels: Probing the Reaction Network

Saikat Dutta, Postdoctoral Researcher (Adviser: Dion G. Vlachos, Basu Saha)
University of Delaware, Newark, Delaware

About Me

Project-Proposal Experience

- Hydrodeoxygenation of high-carbon oxygenates using noble metal catalysts (US DOE)
- 2D-graphenic materials for condensation of biomass-derived oxygenates (US DOE)
- Metal-organic frameworks (MOFs) derived porous functional materials for charge-storage applications.
- Biolubricants from biomass and furnas (NSF-SBIR 2017) (National Science Foundation (Directorate for Engineering Industrial Innovation and Partnerships) (Phase-I $225,000) (*submitted*)
- Sustainable sugar and biofuel production from craft lignin and cellulose (Innovation Corps-National Innovation Network Program) (NSF-17-559) (Phase-I $75,000) (*submitted*)
- "2-Dimensional Functional Materials for Biofuel Cells and Biofuel Production" Ramanujan Fellowship, DST, New Delhi ($57,000 phase-I) (*submitted*)

Research Experience

- Graphenic materials catalysis
- Biomass processing and chemical conversions
- Enzymatic catalysis and enzyme-functionalized porous materials
- Porous hierarchical heteroatom doped carbon materials
- Electrochemical charge storage
- Algal biomass processing and enzymatic conversions
- Enzymatic biomass transformations and enzyme functionalized porous solids

Current HDO Strategies

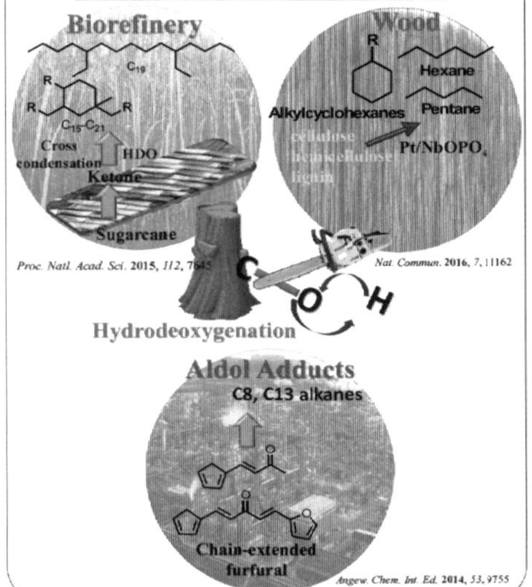

Proc. Natl. Acad. Sci. **2015**, *112*, 7645

Nat. Commun. **2016**, *7*, 11162

Angew. Chem. Int. Ed. **2014**, *53*, 9755

Our Methods: HDO of Furylmethane

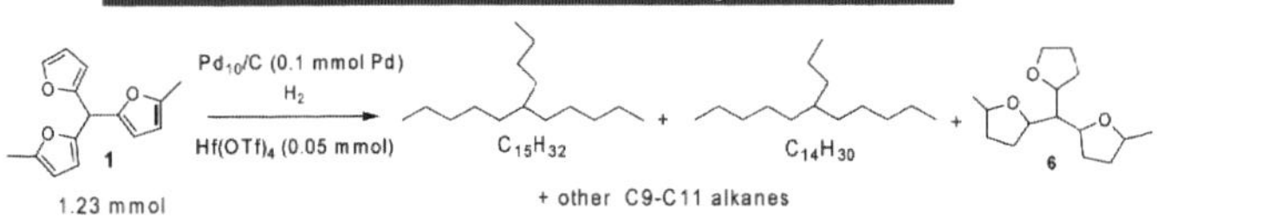

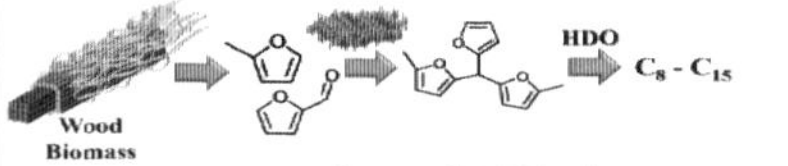

Dutta et al. *ACS Catal.* **2017**, *7*, 3905-3915

- Rapid hydrogenation of **1** at room temperature to form **6**
- **6** undergoes ring-opening by C–O hydrogenolysis
- C–O hydrogenolysis is promoted by $Hf(OTf)_4$

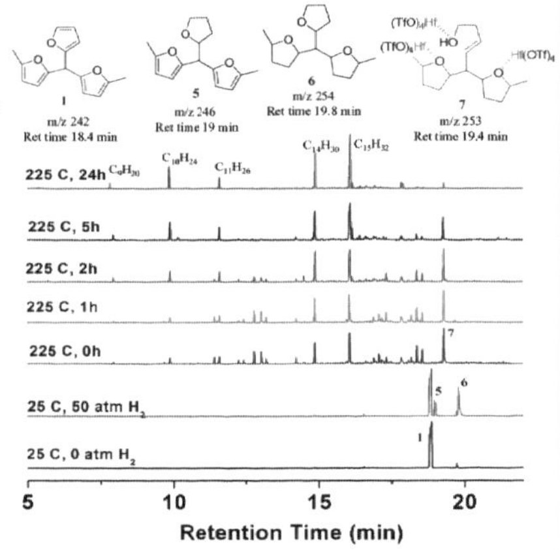

Figure 1. GC chromatograms of HDO of 1 with $Pd_{10}/C\text{-}HF(OTf)_4$; major compounds and retention times are shown over the plot. Controlled experiments:1 (0.300 g, 1.23 mmol), $Hf(OTf)_4$ (0.05 mmol, 0.040 g, H_2 (50 atm), cyclohexane (25 mL), 25-225 °C.

Reaction Network

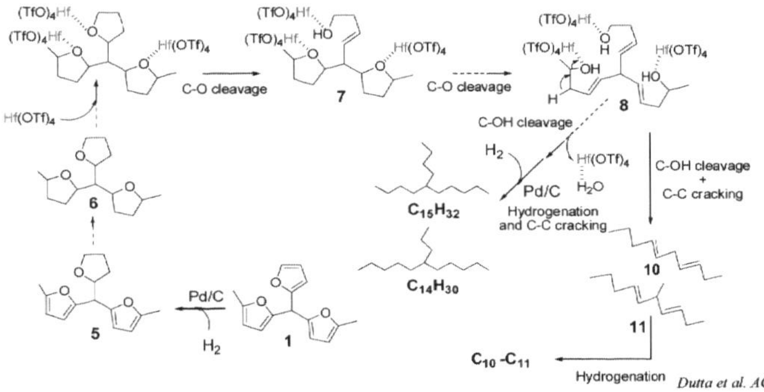

Dutta et al. ACS Catal. **2017**, *7*, 5491-5499

Scheme 1. Proposed reaction network for the HDO of **1** to alkanes using $Pd_{10}/C\text{-}Hf(OTf)_4$.

Alkane Selectivity

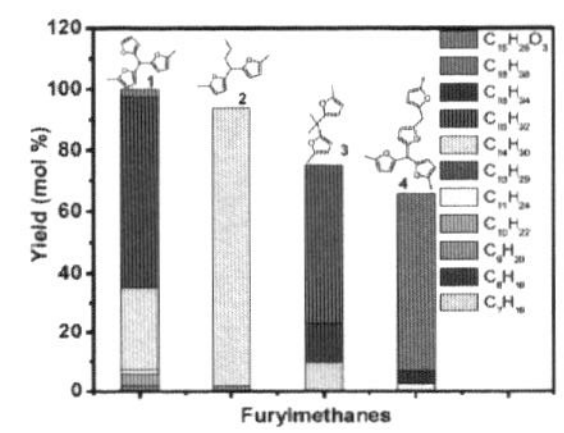

Figure 2. Yields of alkanes from HDO of furylmethanes using $Pd_{10}/C\text{-}Hf(OTf)_4$ under identical conditions in Table 1, entry 7.

Summary

- Deoxygenation of ring-opened oxygenates is faster for single ring-furan surrogates compared to trifurylmethane.
- Maximum 97% alkane products with 93% selectivity in C15 and C14 alkanes are achieved at optimal conditions
- $Hf(OTf)_4$ facilitates ring-opening of furan via formation of an adduct between Hf^{4+} and nucelophilic 'O' of the ring by activating C–O bonds.
- Reaction progresses successively through furan hydrogenation, ring-opening, deoxygenation and hydrogenation steps.

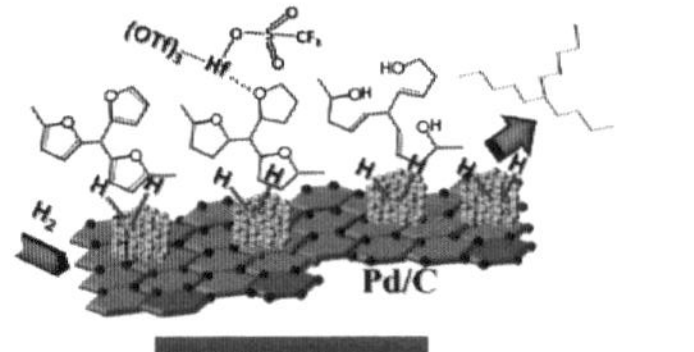

Future

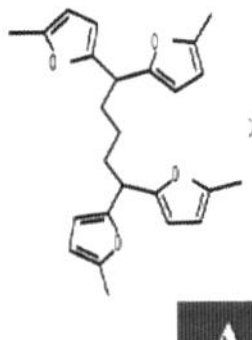

- Kinetics of hydrogenation and ring-opening

Acknowledgement

U.S. Department of Energy, Office of Science

U.S. Department of Energy, Office of Science, Office of Basic Energy Sciences, under Award No. DE-SC0001004

505421 Tissue-Engineered Models for Lymphatic and Blood Vascular Biology

Abstract:

Research Interests:

My research interest is to understand lymphatic and blood vessel morphogenesis, homeostasis and disease pathogenesis. Toward this end, I am working to develop novel experimental systems and cellular and molecular tools to better understand the mechanisms by which cells regulate and respond to biological and mechanical cues. During my doctoral research in Aleksander S. Popel's research group at Johns Hopkins University, I explored tumor-lymphatic vessel crosstalk in cancer metastasis using a novel mouse model for rapid onset of metastasis and biochemistry and cell biology methods. In addition, I generated new small peptides to inhibit lymphangiogenesis and angiogenesis using bioinformatics-guided methods, which was confirmed in diverse in vitro and in vivo cancer models. As a postdoctoral fellow in Christopher S. Chen's research group at Boston University and Harvard Wyss Institute, I am working to complement my expertise in lymphatic and cancer biology with novel organs-on-chip platforms to further extend the focus of my research to biomimetic 3D models of lymphatic and vascular biology. For my postdoctoral research, I am integrating novel 3D microfluidic platforms and cutting-edge molecular and cellular techniques, including cloning, CRISPR-mediated gene editing with in vivo mouse models to develop a novel and unique approach to investigating the mechanisms that contribute to lymphatic and vascular diseases and cancer. Specifically, we have discovered a previously undescribed mechanism by which inflammation impairs lymphatic drainage of fluid and dendritic cells by abnormally tightening the lymphatic junction. The molecular signaling pathway involved in this mechanism is known to be a master regulator of cell and matrix interaction, and I am currently working to investigate this pathway as a mechanism by which inflammation contributes to lymphatic tissue morphogenesis and pathogenesis. In addition, I focus on pancreatic cancer studying how pancreatic cancer cells invade and replace blood endothelium in our 3D tumor-on-chip and in mouse tumor models. This training and my scientific contributions have built a foundation for my long-term research goal of building new tools for investigating the effects of biochemical and physical cues on cell and tissue biology with the eventual goal of informing new and better treatments for lymphatic and vascular diseases and cancer.

SUCCESSFUL PROPOSALS

[1] LE&RN Postdoctoral Fellowship ($94,536.00), Lymphatic Education & Research Network; 2016-2018

[2] TL1 Postdoctoral Fellowship ($47,844.00), NIH/NHLBI; 2015-2016

[3] MOGAM Science Scholarship ($10,000.00), Green Cross (a pharmaceutical company); 2013

PENDING PROPOSAL

[1] K99/R00 NIH Pathway to Independence Award ($1,004,634.00), NIH/NHLBI; (Pending)

POSTDOCTORAL PROJECT

"Lymphatic and blood vessel-on-chip to study lymphatic and blood vessel related diseases such as lymphedema, inflammation, and cancer"

Under the supervision of Christopher S. Chen, Biomedical Engineering, Boston University; The Wyss Institute for Biologically Inspired Engineering, Harvard University

PH.D. DISSERTATION

"Novel role of lymphatic and blood vasculatures in breast cancer growth and metastasis and peptide agents with anti-lymphangiogenic and anti-angiogenic activity"

Under the supervision of Aleksander S. Popel, Department of Biomedical Engineering, Johns Hopkins University School of Medicine.

EDUCATION

2014 Ph.D. Chemical & Biomolecular Engineering, Johns Hopkins University (Mentor: Aleksander Popel)

2008 M.S. Pharmacy, Seoul National University, Korea (Mentor: Youngro Byun)

2006 B.S. (summa cum laude) in Chemical Engineering, Seoul National University, Korea

SELECTED PEER-REVIEWED PUBLICATIONS

(Among 12 first or co-first author publications, 18 total, h-index=11)

[1] Nguyen DHT*, **Lee E***, Alimperti SA, Wong A, Eyckmans J, Stanger BZ, Chen CS, "Pancreatic ductal adenocarcinoma replaces endothelium during tissue invasion", *Nat Biomed Eng*, 2017, Accepted. ***Equal contribution**

[2] **Lee, E.**, Song, H.H.G., Chen, C.S., "Biomimetic on-a-chip platforms for studying cancer metastasis", *Curr Opin Chem Eng*, 11:20-7, 2016

[3] **Lee, E.**, Fertig, E.J., Jin, K., Sukumar, S., Pandey, N.B., Popel, A.S., "Breast cancer cells condition lymphatic endothelial cells within pre-metastatic niches to promote metastasis", *Nat Commun*, 5:4715, 2014

[4] **Lee, E.**, Lee, S.J., Koskimaki, J.E., Han, Z., Pandey, N.B., Popel, A.S., "Inhibition of breast cancer growth and metastasis by a biomimetic peptide", *Sci Rep*, 4:7139, 2014

[5] **Lee, E.**, Pandey, N.B., Popel, A.S., "Lymphatic endothelial cells support tumor growth in breast cancer", *Sci Rep*, 4:5853, 2014

PATENTS

[1] Popel, A.S., Pandey, N.P., **Lee, E.**, Green, J.J., Shmueli, R.B., "A biomimetic peptide and biodegradable delivery platform for the treatment of angiogenesis- and lymphangiogenesis-dependent diseases", *US Patent 0122390*, 2016

[2] Byun, Y., **Lee, E.**, Jeon, O.C., Kim, S.Y., Park, R.W., "Heparin conjugates and methods", *US Patent 8088753*, 2012

Teaching Interests:

My previous teaching experiences have given me a sense of how to teach classes and mentor students in the future. As a professor, I will give students necessary tools for developing analytical thinking and to solve problems beyond idiomatic solutions. I will structure classes using a combination of hands-on learning through lab projects as well as reading the current literatures for students to gain an understanding of cutting-edge research. I will also expose students to problems with unknown solutions to challenge them to think critically. I can teach undergraduate students with Chemical Reaction Engineering and Fluid Dynamics or Transport Phenomena in Chemical Engineering. These courses would foster critical thinking, and help students learn the key engineering concepts and their applications in their research. I would be happy to teach topic-oriented courses in Cell and Tissue Engineering, Regenerative Medicine, Cancer, Nanomedicine, and Vascular and Lymphatic Biology for graduate level students.

TEACHING EXPERIENCE

2011-2012 Chemical Kinetic Processes, Johns Hopkins University, MD

2007 Advanced Physical Pharmacy, Seoul National University, Korea

2006 Physical Pharmacy Laboratory, Seoul National University, Korea

MENTORING EXPERIENCE

2017 – Keith Gagnon (M.D., Ph.D. student), Wyss Institute at Harvard University, MA

2015 – 2017 Bryan Hassell (Ph.D. student), Wyss Institute at Harvard University, MA

2013 – 2014 Ryan Walter (undergrad), Biomedical Engineering, Johns Hopkins University, MD

2012 – 2014 Seung-Jae Lee (undergrad), Biomedical Engineering, Johns Hopkins University, MD

Expanded Research Interests - Gur Pines

Summary:

My primary research interests are in sequence-to-activity mapping and the core principles for the ability to evolve and gain new functionalities. My experience spans from studying mutations that promote cancer to isolating mutations that confer new properties such as resistance to various antibiotics in microbes. My short-term goals include the investigation of two extremes of the mutational spectrum, namely a single mutation with a large phenotypic effect, and multiple mutations resulting with no impact at all. My longer term plans include a deeper study of mutational tolerance and to design overlapping, functional genes. I also plan to combine my two fields of expertise to implant mammalian signaling pathways into bacteria with the purpose of increasing the pace of drug discovery and expand the understanding of those pathways in an orthogonal environment.

Past research:

My graduate studies at the Weizmann Institute of Science (Israel) focused on a set of mutations found in the epidermal growth factor receptor (EGFR) in human glioblastoma tumors. A thorough investigation led me to uncover the mechanism by which these mutations bypass the tight regulation that typically characterizes this receptor[1]. Using mass spectrometry, I documented the oncogenic signaling pathways that are being activated by these mutants. Finally, I found that the mutations are sustained by the HSP-90 chaperone and that upon its inhibition, the levels of these oncogenic proteins are significantly reduced. The understanding of the autoinhibitory mechanism that is defective in these mutants led to another study in which we further elucidated the dimerization and autoregulation processes within the EGFR family[2].

Current research:

My postdoc research at the University of Colorado, Boulder, introduced me to the other side of mutations, namely harnessing the power of evolution rather than fighting it (as in cancer treatments), and doing so by using synthetic biology and bioengineering approaches. I developed a codon compression method for saturation mutagenesis that reduces library size without affecting its diversity. This method is superior to other methods since it is dynamic, taking into account codon usage and allows to determine the level of stringency it should follow. I also demonstrated the method's efficiency using a novel CRISPR-CAS9 method for bacterial genome mutagenesis[3]. Soon after, I led the development of a dedicated website for the codon compression tool, making it accessible to a wider audience[4] (http://www.dynamcc.com). In line with my support of open source coding, all codes are freely available for use, modification, and implementation in other tools.

I am also part of a team developing methods for generating large genomic mutational diversity, rationally and systematically. Applying our CREATE approach (CRISPR-Enabled Trackable

Genome Engineering) resulted (among other things) in isolating novel mutations conferring resistance to both trimethoprim[5] (a DHFR inhibitor) and fosmidomycin (a DXR inhibitor, also serves as an antimalarial drug, Pines et. al., *manuscript in preparation*). Mutants that confer resistance to fosmidomycin are of importance for the development of future DXR inhibitors. Together with my host Prof. Ryan Gill, I wrote and received an NIH R21 grant with the aim of expanding the search for fosmidomycin (and other drugs) resistant mutants targeting the complete DXP metabolic pathway (NIH grant #1R21AI128296-01A1).

A closer look at the resistance-conferring mutations revealed that a significant proportion of those cannot be obtained by popular methods such as error-prone PCR and directed evolution since they require more than a single nucleotide replacement within the codon, an extremely rare event. Together with James Winkler from our group, I designed theoretical alternative genetic codes that will alleviate this limitation by increasing the amino acid accessibility with a single point mutation (Pines and Winkler et al., *in revision*[6]). An organism employing such an altered genetic code will show increased evolvability and might prove an ideal platform for protein and genome engineering.

I believe that exciting scientific discoveries are laid in the borders of different disciplines and perspectives. This is the reason I chose to have my postdoc training in a different discipline than my Ph.D., and this is also the reason I try to collaborate as much as possible. A good example for this is the method for tracking genomic mutations developed in our lab. While it was designed for tracking bacterial genomes, I applied it to identify mutations in cancer cells, significantly increasing the impact of this method[7].

As an independent researcher, I plan to continue my studies of sequence to activity relationships on different levels. I aim at using synthetic biology approaches for achieving basic science and bioengineering-oriented discoveries. I am an experimentalist by training, but I do appreciate the power of computing, as shown by my current research. Hence my lab, while mainly focused on experimental work, will always have computational elements associated with it.

For more detailed information on my research interests, please contact me at: gur.pines@colorao.edu, or come visit me at the "meet the faculty candidate" poster session poster # **7ja** Methods for Efficient Sequence to Activity Mapping.

References

1. Pines, G., Huang, P. H., Zwang, Y., White, F. M. & Yarden, Y. EGFRvIV: a previously uncharacterized oncogenic mutant reveals a kinase autoinhibitory mechanism. *Oncogene* **29,** 5850–5860 (2010).
2. Bublil, E. M. *et al.* Kinase-mediated quasi-dimers of EGFR. *FASEB J.* **24,** 4744–4755 (2010).
3. Pines, G. *et al.* Codon compression algorithms for saturation mutagenesis. *ACS Synth. Biol.* **4,** 604–614 (2015).
4. Halweg-Edwards, A. L., Pines, G., Winkler, J. D., Pines, A. & Gill, R. T. A Web Interface for Codon Compression. *ACS Synth. Biol.* (2016). doi:10.1021/acssynbio.6b00026
5. Garst, A. D. *et al.* Genome-wide mapping of mutations at single-nucleotide resolution for protein, metabolic and genome engineering. *Nat. Biotechnol.* (2016). doi:10.1038/nbt.3718
6. Pines, G., Winkler, J. D., Pines, A. & Gill, R. T. Refactoring the Genetic Code for Increased Evolvability. *bioRxiv* 128058 (2017). doi:10.1101/128058
7. Zeitoun, R. I. *et al.* Multiplexed tracking of combinatorial genomic mutations in engineered cell populations. *Nat. Biotechnol.* **33,** 631–637 (2015).

507198 Colloidal and Interfacial Phenomena Involving Anisotropic Fluid

Abstract:

Research Interests:

Either bulk or interfaces of isotropic solvents have been widely used to direct assembly of molecules and colloidal particles, leading to a range of assemblies, elastomers and gels with complex mechanical, electronic and optical properties. Anisotropic fluid, such as liquid crystal (LC), possesses orientational ordering of constituent molecules which results in orientation-dependent properties (optical, mechanical and transport) that differentiate from isotropic liquids. My research aims to resolve the question how the orientational order in LC affects assembly of molecules and colloids. Here I present works on (i) LC topological defect-mediated molecular self-assembly, (ii) LC emulsion-templated colloidal assembly and particle synthesis, and (iii) non-monotonic responsive polymeric materials based on liquid crystal elastomer (LCE), which together provide the platform for realizing my goal to produce a remarkably diversity of colloidal and interfacial phenomena that are distinct from isotropic liquid.

Research Experience:

My first research topic is molecular self-assembly directed by topological defects in LCs. We found that nanoscopic environments defined by LC topological defects can selectively trigger processes of molecular self-assembly. By using fluorescence microscopy, cryogenic transmission electron microscopy and super-resolution optical microscopy, key signatures of self-assembly of amphiphilic molecules in topological defects are observed - including cooperativity, reversibility, and controlled growth of the molecular assemblies. By using amphiphiles that can be photocrosslinked, we also demonstrate preservation of molecular assemblies templated by defects, including nanoscopic “o-rings” synthesized from “Saturn-ring” disclinations. Our results reveal that topological defects in LCs are a versatile class of three-dimensional, dynamic and reconfigurable templates that can direct processes of molecular self-assembly in a manner that is strongly analogous to other classes of macromolecular templates (e.g., polymer—surfactant complexes).

The second part of my research discovers that LC droplets in LC-in-water emulsions can be utilized as templates for synthesis of anisometric microparticles with chemical surface patches. Remarkable control over the shape and surface morphology of polymeric particles templated from LC microdroplets can be achieved through chemical manipulation of configuration of the droplets. In addition, reversible adsorbate-driven ordering transitions in LC droplets are used to localize colloids at surfaces of water-dispersed LC droplets in a programmable manner. This approach can be used to synthesize either “Janus-like” microparticles or magnetically-responsive patchy microdroplets of LC with either dipolar or quadrupolar symmetry that exhibit distinct optical responses upon application of an external magnetic field.

My third research topic is development of a chemical platform for synthesis of a liquid crystalline material assuming a non-monotonic response with respect to the monotonic dynamics of a single stimulus. In this study, we synthesized a LCE polymer exhibiting two phase transitions – LC—isotropic and crystal—LC. Surprisingly, two deformations executed by such a LCE microplate proceeded in the opposite direction! By selecting the temperature range for thermal cycles, we programmed the LCE system to realize a rich palette of stimulus-dependent modes. Overall, these results demonstrate that LCEs consisting of multiple phase transitions can serve as a versatile class of adaptive materials for design of dynamic microstructures adopting non-monotonic stimulus-dependent responsiveness, which are unprecedented in traditional LCEs reported in previous studies.

Teaching Interests:

Chemical engineers applies fundamental physical sciences to design and produce chemicals and materials. This is exemplified in my PhD and postdoctoral research about colloidal and surface phenomena and polymer science involving LCs. My research experience has prepared me to teach a wide range of curriculum in Chemical Engineering (such as Thermodynamics, Polymer Science, Colloidal and Surface Science, and Transport Phenomena). In addition, I would like to work with undergraduate and graduate students in short-term projects in my course to deepen their understanding learned in the classroom, and promote student interest in scientific research and further graduate education.

Aside from my research, I also have been involved in a wide variety of educational and outreach activities that provided me with valuable experience to start an academic career. As a teaching assistant in my PhD study, I taught two intermediate level courses of Chemical Engineering at University of Wisconsin-Madison (UW). In my postdoctoral study at Harvard University, I was invited to give a lecture about "Intermolecular Interactions and Liquid Crystals" in a senior applied physics course of Chemistry in Materials Science and Engineering. Besides teaching, as a member of UW Materials Research Science and Education Center (MRSEC), I took roles and responsibilities for designing simple methods to gather public interest in science and engineering. As a consequence, I believe I have had valuable experience and gained confidence to advice students, and to influence public audience to gather interest to Chemical Engineering.

Selected Publications (23 total, 10 first author):

1. Wang, X.;* Zhou, Y.;* Kim, Y.-K.; Miller, D. S.; Zhang, R.; Martinez-Gonzalez, J. A.; Bukusoglu, E.; Zhang, B.; Brown, T. M.; de Pablo, J. J.; Abbott, N. L. "Patterned Surface Anchoring of Nematic Droplets at Miscible Liquid—Liquid Interfaces". Soft Matter DOI: 10.1039/C7SM00975E 2017. (*contributed equally)

2. Wang, X.; Bukusoglu, E.; Abbott, N. L. "A Practical Guide to the Preparation of Liquid Crystal- Templated Microparticles". Chemistry of Materials 29, 53-61, 2017.

3. Bukusoglu, E.;* Wang, X.;* Zhou, Y.; Martinez-Gonzalez, J. A.; Rahimi, M.; Wang, Q.; de Pablo, J.J.; Abbott, N. L. "Positioning Colloids at the Surfaces of Cholesteric Liquid Crystal Droplets". Soft Matter 12, 8781-8789, 2016. (*contributed equally)

4. Wang, X.; Bukusoglu, E.; Miller, D. S.; Pantoja, M. A. B.; Xiang, J.; Lavrentovich, O. D.; Abbott, N.L. "Synthesis of Optically Complex, Porous and Anisometric Polymeric Microparticles by Templating from Liquid Crystalline Droplets". Advanced Functional Materials 26, 7343-7351, 2016.

5. Wang, X.; Kim, Y.-K.; Bukusoglu, E.; Zhang, B.; Miller, D. S.; Abbott, N. L. "Experimental Insights into the Nanostructure of the Cores of Topological Defects in Liquid Crystals". Physical Review Letters 116, 147801, 2016.

6. Wang, X.; Miller, D. S.; Bukusoglu, E.; de Pablo, J. J.; Abbott, N. L. "Topological Defects in Liquid Crystals as Templates for Molecular Self-Assembly". Nature Materials 15, 106-112, 2016.

7. Wang, X.;* Yang, P.;* Mondiot, F.; Li, Y.; Miller, D. S.; Chen, Z.; Abbott, N. L. "Interfacial Ordering of Thermotropic Liquid Crystals Triggered by the Secondary Structures of Oligopeptides". Chemical Communications 51, 16844-16847, 2015. (*contributed equally)

8. Wang, X.; Miller, D. S.; de Pablo, J. J.; Abbott, N. L. "Organized Assemblies of Colloids Formed at the

Poles of Micrometer-Sized Droplets of Liquid Crystal". Soft Matter 10, 8821-8828, 2014.

9. Wang, X.; Miller, D. S.; de Pablo, J. J.; Abbott, N. L. "Reversible Switching of Liquid Crystalline Order Permits Synthesis of Homogeneous Populations of Dipolar Patchy Microparticles". Advanced Functional Materials 24, 6219-6226, 2014.

10. Wang, X.; Luo, Y.; Li, B.; Zhu S. "Ab Initio Batch Emulsion RAFT Polymerization of Styrene Mediated by Poly(Acrylic Acid-b-Styrene) Trithiocarbonate". Macromolecules 42, 6414-6421, 2009.

507088 Experimental Interrogation of Polymer Material Structure-Property Relationships

Abstract:

Research Interests:

My core areas of interest include **polymer science, surface science and metrology, hard-soft material interfaces, and functional composites**. Recently, it is important for a material to bring more to an application than its basic properties, like modulus or fracture toughness. Materials can be functional, whether the function is environmental sensing, self-healing, self-assembly, or stimulus response. I aim to create new polymeric and composite materials that take advantage of multiple phases and mechanically activated chemical ("mechanophore") dopants to create designed functional material components that are resistant to damage and fatigue. My interests align with this aim to make it a promising central focus for me in my future investigations.

One unifying theme of my research has been using model polymer materials to test the bounds of applicability for published structure-property relationships and thermodynamic models. I feel this is important for the health of the polymer science community, as there are strong temptations to claim agreement between data and currently accepted theories. Aside from continuing to perform these deliberate tests of hypotheses, I aim to develop new, improved, and more efficient methods for executing or replicating tests where critical gaps exist between current theories and practice for materials with designed properties.

For example, I have been attracted to the metrology of surfaces due to the wealth of information that can be extracted with high precision and spatial resolution. Although nanometer length scales of colloidal systems can be probed with scattering techniques, I aim to create two-dimensional mimics of interesting colloids and other heterogeneous systems to deeply understand the behavior of the simpler system. The variety of physical and chemical data that can be gathered in that geometry can then be fed back into understanding of the more complex system, helping for example to overcome questions of inhomogeneity in the scattering ensemble.

Also, there is inevitably an extreme gradient in the chemical environment at the interface between hard and soft materials, leading to important macroscopic effects like adhesion, adsorption, and interphase formation. By understanding this region, it is possible to better predict the behavior of composite materials and heterogeneous systems. I aim to explore this combinatorial space to identify novel interfaces or coatings that outperform existing approaches or even have unique new capabilities.

Research Experience:

Currently, I am working in a collaboration between Northwestern University, Owens Corning, and NIST to improve **measurement science of the micro- and nanomechanical properties of glass fiber**

coatings for fiber reinforced polymer composites. We are combining interphase modulus measurements with complementary interfacial shear strength measurements with complimentary fluorescent probe and interphase modulus measurements for the purpose of mimicking larger-scale material tests as well as to better predict macroscopic composite coupon strength and fatigue with multi-scale material models.

Previously, during my NRC postdoc at NIST. I was studying the **physics of end-tethered polymer films,** at all levels from synthesis to metrology with a variety of techniques available at NIST, from microscopy to neutron scattering, to develop new metrologies for these films. I measured (4) the swelling of dense polymer brushes in response to solvent vapor with X-ray reflectivity, as well as the surface interaction parameters (1) of sub-brush end-tethered chains swollen in solvent with liquid solvent. This project required rigorous thin film synthesis protocols (3) and comparison of optical, X-ray, and neutron scattering effects within the thin film while dry and swollen in solvent.

The focus of my graduate work at CU was on the **synthesis, rheology, and applications thermally reversible resins**, specifically Diels-Alder networks, as a mendable and recyclable thermoset material. I studied their chemical, mechanical (7), and viscoelastic properties (8), and demonstrated (5) that my model networks required a more advanced theoretical structure-property relationship than is typically applied to polymers containing reversible bonds.

Successful Proposals:

GAANN Fellowship

NRC Postdoctoral Research Associateship at NIST

NCNR neutron beam time requests (4)

Postdoctoral Projects:

Novel Metrology Approaches Utilizing the Swelling Physics of End-Tethered Polymers, under **Kate Beers**, Materials Science and Engineering Department, National Institute of Standards and Technology.

Multi-Scale Metrology for Visualization and Characterization of Polymer Composite Interphase, under **Jeff Gilman**, Materials Science and Engineering Department, National Institute of Standards and Technology, and **Cate Brinson**, Department of Mechanical Engineering, Northwestern University.

PhD Dissertation:

Rheology and Application of Thermoreversible Polymer Networks Based on the Diels-Alder Cycloaddition, under **Chris Bowman**, Department of Chemical and Biological Engineering, University of Colorado at Boulder

Teaching Interests:

The courses I am best prepared to teach are of course the ones most directly related to my research: **General chemistry, reaction kinetics, thermodynamics, numerical methods, statistics, and polymer science**. Additionally, I look forward to the opportunity to teach some courses that I enjoyed but have not had a reason to employ directly in my post-graduate work, like **process controls, unit operations, and optimal design of experiments**. I believe that one does not truly understand a topic until having taught it successfully to a student. Teaching serves as a valuable check of my own assumptions and conclusions, because I have to take the nebulous feelings and intuitions in my mind and render them into a concrete form that is coherent and logical to at least one other human being. As a chemical engineer turned polymer scientist, my teaching is influenced by the need to balance tactile and practical types of understanding with quantitative modeling and data analysis. Personally, mathematical expressions make far more sense to me once I can connect them to a physical process, or at least some kind of visualization. Different students will have different responses to these directions, but I believe in all cases that both parts, practical and quantitative, are necessary for successful learning. The outcomes I expect from teaching the course go beyond fulfillment of the stated syllabus – although that is certainly important! I'll know I have prepared the students successfully when the students themselves report that they have a mastery of the course material, they feel like the course has been worthwhile, and they are engaged to learn more.

Publications:

(1) **Sheridan, R. J.**; Orski, S. V.; Jones, R. L.; Satija, S. K.; Beers, K. L. Surface Interaction Parameter Measurement of Solvated Polymers via Model End-Tethered Chains. *Macromolecules* **2017** (accepted).

(2) Yang, Y.; Urbas, A.; Gonzalez-Bonet, A.; **Sheridan, R. J.**; Seppala, J. E.; Beers, K. L.; Sun, J. A Composition-Controlled Cross-Linking Resin Network through Rapid Visible-Light Photo-Copolymerization. *Polym. Chem.* **2016**, *7* (31).

(3) **Sheridan, R. J.**; Orski, S. V.; Muramoto, S.; Stafford, C. M.; Beers, K. L. Ultraviolet/Ozone as a Tool to Control Grafting Density in Surface-Initiated Controlled-Radical Polymerizations via Ablation of Bromine. *Langmuir* **2016**, *32* (32).

(4) Orski, S. V.*; **Sheridan, R. J.***; Chan, E. P.; Beers, K. L. Utilizing Vapor Swelling of Surface-Initiated Polymer Brushes to Develop Quantitative Measurements of Brush Thermodynamics and Grafting Density. *Polymer* **2015**, *72*. (*equal credit first authors)

(5) **Sheridan, R. J.**; Bowman, C. N. Understanding the Process of Healing of Thermoreversible Covalent Adaptable Networks. *Polym. Chem.* **2013**, *4* (18).

(6) Chatani, S.; **Sheridan, R. J.**; Podgórski, M.; Nair, D. P.; Bowman, C. N. Temporal Control of Thiol-Click Chemistry. *Chem. Mater.* **2013**, *25* (19).

(7) **Sheridan, R. J.**; Bowman, C. N. A Simple Relationship Relating Linear Viscoelastic Properties and Chemical Structure in a Model Diels-Alder Polymer Network. *Macromolecules* **2012**, *45* (18).

(8) **Sheridan, R. J.**; Adzima, B. J.; Bowman, C. N. Temperature Dependent Stress Relaxation in a Model Diels-Alder Network. *Aust. J. Chem.* **2011**, *64* (8).

508864 Level Set Algorithms for Polymer Field Theory

Abstract:

Abstract:

I will present how level set methods can be used to design and develop new algorithms for self-consistent field theory (SCFT) in polymer physics. I first present a computational framework, encoded on a forest of quad/oct trees in a parallel environment. I then introduce the concept of functional level-set derivative into SCFT which is then used to embed SCFT into a variable shape simulation. Finally an algorithm for the inverse geometric problem is presented: it finds a shape in order to obtain a desired structure.

Research Interests:

My research interests are but not limited to: computational and applied mathematics, soft matter, polymer field theory, fluid mechanics, shape optimization and high performance and parallel computing. During my master I developed a computational model to study the collective stochastic motion of biological cells during wound healing. In my PhD I developed algorithms for variable shape polymer field theory simulations using level set methods. Polymer field theory allow to find polymeric material properties and structure at thermodynamic equilibrium. Level set methods allow to vary the domain simulation in a simple way. When the geometry of the domain is a free variable of the polymeric material design, efficient algorithms are required to find the optimal shape. Currently during my postdoc I develop algorithms for massive parallelization of Stokesian dynamics to simulate large scale complex fluids where particles interact in a viscous fluid through hydrodynamic interactions.

In the future I intend to work on polymer field theory, fluid mechanics and numerical shape optimization methods. I am interested in electro active polymers which exhibit a change of shape and size when stimulated by an electric field. These are called artificial muscles and are used in soft robotics. I will develop a polymer theory which takes into account the electrical field, the mechanical stress and the change of shape. As well I will develop numerical methods and algorithms to solve the resulting equations. In fluid mechanics I am interested in soft robots with a continuous and transient shape immersed in a fluid. I am as well interested to develop numerical methods toward efficient shape optimization algorithms and applications. Indeed shape optimization methods have to rely on numerical shape derivatives which are highly complex objects and require special care either in finite difference or in finite element.

Teaching Interests:

My teaching interests are numerical methods, programming, finite element methods, mathematics, physics, solid mechanics, fluid mechanics, statistical mechanics and heat transfer. I enjoy teaching on the board, interact with students and answer questions.

509820 Fundamental Studies and Engineering Modeling of Industrially Relevant Systems

Abstract:

PHD DISSERTATION

"Fundamental Studies and Engineering Modeling of Hydrogen Bonding"
Under the supervision of Carl T. Lira, Department of Chemical Engineering and Materials Science, Michigan State University

SUCCESSFUL PROPOSALS

National Science Foundation (NSF) GOALI: Improved Association-based Models for Separations in the Bioeconomy

RESEARCH EXPERIENCE

I have worked on a few different areas in my time as a researcher. My PhD project is inspired in part by collaborations with chemical engineers in industry who expressed a need for a flexible, robust thermodynamic model that can accurately represent associating systems they frequently struggle to model. To do this, I developed a general expression for the Wertheim association contribution to the activity coefficient () which can be attached to any physical and combinatorial -model. Currently, I am parameterizing the model by leveraging spectroscopic tools (including IR and NMR spectroscopy) informed by quantum and molecular mechanics (QM/MM) simulations to capture the effects of hydrogen bonding at a microscopic level. Using a novel quantum calculation technique for hydrogen bonded clusters, I was able to decipher where different types of hydrogen bonds appear in IR spectra using QM simulations and prove that monomers and end groups with free hydrogen atoms ('beta' bonds) overlap significantly. This work has allowed me to develop my skills in experimental design, MATLAB coding, simulation (using several software packages) and Linux scripting. Upon completion of my PhD degree, I aim to completely execute and combine findings of the three legs of the project and deliver an AspenPlus user model to be utilized by our industrial colleagues and the broader engineering community.

As an undergraduate, I studied the effects of microwave-enhanced degradation of phenol (a common byproduct of industrial processes and component of their resultant waste streams) in the presence of nickel loaded zinc oxide nanocatalysts. Through this project, I learned how to design reaction experiments, uncovered the dominant reaction pathways under several conditions and authored the first of my research publications.

In addition to disciplinary research, I have also designed and implemented a research project in engineering education. I developed and implemented a collaborative in-class module on energy changes during phase transitions in a junior-level chemical engineering thermodynamics class to determine whether active learning can improve students' grades on an assessment of the topic compared to traditional lecturing. The intervention lead to over 30% more correct responses in the assessment.

TEACHING EXPERIENCE

I am passionate about understanding how people learn. For this reason, I applied for and was accepted into the Future Academic Scholars in Teaching (FAST) fellowship at MSU where I conducted a teaching-as-research project (see Research Experience for more information) and interacted frequently with a diverse learning community. My rewarding experience as a fellow lead me to accept the position of FAST committee graduate leader for the 2017-2018 academic year. In this role, I am leading biweekly discussion sessions on teaching, learning and assessment and supporting fellows as they define and conduct their own education projects in a range of STEM fields. Additionally, I am currently in the process of obtaining certification of college teaching in engineering from Michigan State University.

My teaching experiences also include serving as the teaching assistant (TA) for two courses (CHE311-Fluid Flow and Heat Transfer and CHE321-Chemical Engineering Thermodynamics) where my responsibilities included leading recitations, holding office hours and grading assessments. To better understand the theories that underlie effective teaching and learning, I took a course on the 'Fundamentals of Engineering Education'. This was instrumental in familiarizing me with pedagogical vocabulary and concepts.

I have also had several opportunities to serve as a mentor to chemical engineering undergraduate students outside the classroom. A particularly enlightening experience for me has been mentoring an undergraduate student for the last two years as he navigates various aspects of my research project. His work has ranged from experimental data collection and analysis to writing codes to process large amounts of data from molecular simulations.

Research Interests:

The design and optimization of separation processes in the chemical industry requires the utilization of accurate thermodynamic models. These allow engineers to predict solution properties under a variety of conditions. However, when models are deficient, more time and capital must be invested in pilot plant experiments, increasing the costs associated with separation processes. This issue is encountered frequently in many chemical industries. As a result, the development of robust models has therefore become an active area of research, generating a significant number of equations of state and activity coefficient models of varying accuracy and applicability. In my doctoral research, I approached this area by improving the representation of hydrogen bonding between molecules in liquid mixtures (see Research Experience). In the future, I would like to continue focusing on applied thermodynamics, specifically aimed at solving problem encountered in industry by deriving molecular insight from ab initio calculations. I will begin by extending the ideas and methods developed during my PhD to solid-liquid systems to better represent interactions occurring on catalyst and membrane surfaces.

Additionally, I aim to continue doing pedagogical research in chemical engineering. I am interested in developing a database of modules designed according to established principles of backward design and collaborative learning on important concepts within the discipline, including the Navier-Stokes equation, material balances and process control fundamentals. These units will be designed to be relatively short (1-3 class periods) to encourage faculty participation and require students to undergo authentic tasks representative of those they would carry out in industry. Educational research has indicated that these types of sessions boost student performance on assessments and help instructors identify common misconceptions.

Teaching Interests:

I have experience teaching and assessing courses on thermodynamics, fluid flow and heat transfer. While I am comfortable teaching these courses at both the undergraduate and graduate level, I am also interested in teaching several other areas such as reactions and separations. Through my various teaching experiences, I have recognized the importance of creating a learning environment based on active learning principles and therefore aim to do this in my own classroom.

PUBLICATIONS

- Bala, Aseel M., Liu, R., Neuroth, G., Mathias, P. M., Patel, N. C., Frank, T. C., Vu, D. T., Cheluget, E.L. and Lira, C. T., "Applications of a Wertheim association activity coefficient model", In Progress
- Bala, Aseel M., Killian, W. G., Storer, J. A., Killian W., Jackson, J. E. and Lira, C. T., "Integration of quantum calculations and spectroscopy for Wertheim alcohol association", In Progress
- Bala, Aseel M. and Lira, C. T., "Relation of Wertheim association constants to concentration-based equilibrium constants for mixtures with chain-forming components", Fluid Phase Equilibria, 430, pg. 47-56 (2016)
- Bala Ahmed, A., Jibril, B., Danwittayakul and S., Dutta, J., "Microwave-enhanced degradation of phenol over Ni-loaded ZnO nanorods catalyst", Applied Catalysis B: Environmental, 156-157, pgs. 456-465 (2014)

Research Program: Dr. Jai A. Pathak.

Theme: "Fundamental Molecular Engineering with Biophysics, Rheology and Thermodynamics: Elucidating Protein Stability in Flow Fields, Protein-Protein Interactions in Concentrated Solutions and Glassy Dynamics/Bulk Processing Flow Instabilities"

Peer Recognition, Key Strengths, Originality and Leadership: The PI has (co)-authored 32 peer-reviewed publications in respected journals. The impact of the PI's recent research, published with his industry post-doctoral fellow, Dr. Prasad Sarangapani, [1] at MedImmune/AstraZeneca, was recognized by Prof. John Prausnitz (Dept. of Chemical Engineering, UC Berkeley), who reviewed the paper and was invited by the editors of *Biophysical Journal* to write a "New & Notable" Comment. Prof Prausnitz wrote about [2] the significance and impact of the work of Sarangapani *et al.*:

"Many like me will be secretly unhappy about the demise of the colloidal theory of globular protein solutions. Although we knew that this theory was "sick," we hoped that it might "recover." But now after the report of Sarangapani *et al.*, the colloid-like theory is dead. We can take comfort in the remark of Sarangapani *et al.*, that while scientifically erroneous, the colloid-like theory may nevertheless be useful for some purposes in biotechnology. While we mourn with sadness, we also owe much thanks to Sarangapani *et al.*, that when describing nature, yes, by all means seek simplicity, but, with respect for complexity, don't forget to mistrust it."

The PI is a recognized expert in the solution rheology and bio-physics of proteins: he has delivered an invited plenary lecture at the XVII International Conference of Rheology (Kyoto, Japan, Aug. 2016). He has identified these research problems during his career, and has trained a post-doctoral fellow and a few graduate students for former academic collaborators. He has 11 publications in biophysics, and leverages his knowledge of biophysics, fluid mechanics and rheology extensively. He has an extensive network of domestic and international collaborators in academia, industry and in NIST and NIH, which he will rely upon to get federal funding and to involve appropriate subject matter experts in the inter-disciplinary research program proposed here. He is the lead author on a proposal with Prof. Christopher Roberts (Univ. of Delaware), Dr. Michael Tarlov (NIST) and Dr. Linda Narhi (Amgen) to initiate a new Gordon Research Conference (GRC) in Development of Bio-therapeutics and Vaccines; the revised proposal is currently under final review with the GRC board.

Research Program

The 21st century poses global challenges to mankind, especially in harnessing alternate sources of clean and renewable energy, responding to climate change, ensuring abundant food and clean water supply and revolutionizing healthcare/medicine. In healthcare/medicine, bio-technology has unleashed a revolution since the discovery of recombinant DNA technology, which has unleashed the discovery, development and commercialization/regulatory approval of numerous biologics (bio-pharmaceuticals): large molecule drugs expressed from cells, microbes etc., for use as medicines (therapeutics) and vaccines (prophylactics). These proteins are formulated at high concentrations ($\geq$100 mg mL^{-1}), due to low in-vivo potency (target binding). Protein aggregation, elevated viscosity and opalescence are common observed phenomena in crowded conditions, which are both fundamentally elusive and have technological for the bio-pharma industry

While Chemical engineers have made prolific contributions in bio-technology to the science of upstream technology (protein expression) and downstream technology (protein purification), numerous scientific/technical challenges abound. Protein stability towards pathways of degradation, e.g., fragmentation and aggregation, are encountered ubiquitously in bio-manufacturing unit operations: flow during cell culture

[1] "Critical Examination of the Colloidal Particle Model of Globular Models" Prasad S. Sarangapani, Steven D. Hudson, Ronald L. Jones, Jack F. Douglas and Jai A. Pathak. *Biophysical Journal*, 108, 724 – 737 (**2015**).
[2] "The Fallacy of Misplaced Concreteness" John Prausnitz, *Biophysical Journal*, 108, 453-454 (**2015**).

in bio-reactors, flow in chromatography columns for purification, flow in formulation vessels and filling piston/peristaltic/time-pressure pumps and fill nozzles during aseptic (sterile) fill/finish of drug products, flow during shipment of finished liquid drug product in vials/syringes and flow in syringes and other novel devices for self-administration or clinical administration. It is impossible to expresses, purify, process, ship and deliver protein drug products to patients without subjecting them to various flow fields: shear flow, extensional flow etc., wherein protein molecules in concentrated solution encounter a wide range of shear, dilatational and extensional rates. **Protein aggregation during these steps poses serious challenges to biologics development and bio-pharmaceutical manufacturing, slowing the rate at which medicines and vaccines are developed and brought to market and increasing development cost and cycle time.**

1) Protein Stability in Flow Fields and at Interfaces: The proposed research theme is distinct from quiescent protein aggregation (e.g. during storage), which has been studied extensively by Prof. Christopher Roberts (Univ. of Delaware) and his research group; **the work proposed aims to study the interplay of thermodynamics, flow fields, protein solution rheology and surfaces/interfaces on protein stability during flow.** The highly non-Newtonian rheology of concentrated protein solutions stems from hydrodynamic and thermodynamic interactions, and is also determined by the presence of solid/solution and air/solution interfaces, where electrostatic and hydrophobic interactions become important, and can lead to partial unfolding of proteins. The partially unfolded state is the precursor of aggregation, as protein molecules organize their exposed hydrophobic patches to shield them from unfavorable hydrophobic patch-water interactions. **This research will bridge the molecular, mesoscopic and *macroscopic/continuum domains.*** Bulk rheology (macro and micro) and interfacial rheology (interfacial shear and dilation) will be applied in conjunction with scattering (light and neutrons), *in-situ* circular dichroism, intrinsic/extrinsic fluorescence, optical microscopy (flow visualization) etc. to pinpoint **how the unfolding of protein molecules air/solution and solid/solution interfaces during flow leads to formation of microscopic aggregates (~ 10 nm – ~ 1 μm), followed by larger microscopic "sub-visible particles", as the FDA refers to them, (~1 – ~ 100 μm) and finally visible particles**. The effects of flow fields on thermodynamic phase behavior, which are well documented in synthetic polymer solutions, have not been scrutinized for protein solutions, and will also be investigated in the context of opalescence.

Impacts, Stakeholders and Funding Agencies: This fundamental phenomenon has practical consequences for bio-manufacturing. The formation of aggregates and particles poses a roadblock to therapeutic/vaccine development by increasing development cycle time and added costs. For ensuring patient safety against adverse immunogenic reactions, regulatory agencies such as the FDA, EMEA etc. have strict regulations for the maximum aggregate content and proteinaceous particle size/count that can be injected into patients. Statutory compliance with these regulatory requirements ensures that the bio-pharmaceutical industry will be a key stakeholder in this basic research, as development scientists do not have the time, resources and expertise to tackle these challenges at a fundamental scientific level; strict development timeline requirements only allow for engineering quick-fix solutions.

This work will proceed on two parallel tracks: with biophysically well-characterized model proteins for proof of concept development and rapid publication, and with proprietary proteins in collaboration with bio-pharma companies. A collaboration with protein expression and purification experts in academia will be initiated to leverage the upstream/downstream expertise required for supply of proteins, antibodies etc. The National Institute for Innovation in Biopharmaceuticals (NIIMBL), a public-private consortium dedicated to advancing Biopharmaceutical Manufacturing Innovation (NIST/US Dept. of Commerce), the Biomolecular Interaction Technology Center (Univ. of Delaware) and major bio-pharmaceutical companies will be approached for funding. This research will be disseminated journals like *Journal of Pharmaceutical Sciences*, *Biotechnology Bioengineering*, *Journal of Rheology*, *Soft Matter* etc.

2) The Physics of Protein-Protein Interactions in Concentrated Solutions: Research dedicated exclusively to the *molecular/microscopic domain* will probe the delicate interplay of hydrodynamics and thermodynamics in determining the stability of protein molecules, protein-protein interactions and viscosity/mutual diffusion/sedimentation behavior of protein solutions. While theoretical predictions from colloidal models, generalized Stokes-Einstein models and mode coupling theory are available for the inter-relationship amongst hydrodynamic and thermodynamic properties, they have not been interrogated in sufficient depth to elucidate the correct physics in the context of protein solutions. Solution physical hydrodynamic and thermodynamic properties will be measured and interrogated against available literature theoretical models. Static light scattering (SLS) will be used to measure the osmotic second virial coefficient B_{22}, dynamic light scattering (DLS) for quantifying the diffusion interaction parameter k_D, sedimentation coefficient k_s measurements using Analytical Ultracentrifugation (AUC) and viscometry to determine protein intrinsic viscosity $[\eta]$ and the Huggins coefficient k_H. The work on Huggins constant aims to revive the use of the Huggins equation to understand the solution thermodynamics and hydrodynamics of crowded protein solutions, and to use literature theoretical models such as the adhesive hard sphere model of Russel to interpret the observed behavior. The Huggins constant is mostly ignored or discarded for proteins solutions, although it is a The Huggins model is a powerful and accurate descriptor of the volume fraction dependence of relative viscosity: k_H contains hydrodynamic and thermodynamic information, and is influenced my macromolecular shape. Careful measurements and robust interpretation of the Huggins coefficient for diverse molecular constructs like antibodies, fusion proteins, antibody-drug conjugates etc. will enable use of crowded solution viscosity models for understanding crowded behavior, and will transform the current literature paradigm of (incorrectly) extrapolating dilute solution k_D to predict crowded solution interactions and viscosity. The Huggins constant will be related to solution stability with AUC, which is the gold standard for detecting reversible self-association (RSA) in protein solutions. RSA is detrimental to bio-therapeutic development owing to patient safety concerns, and quantitatively relating RSA (size of the equilibrium self-assembled species) to k_H will be a transformative contribution to protein solution biophysics

Impacts, Stakeholders and Funding Agencies: This work will be done strictly in the hydrodynamics/rheology and thermodynamics realm, with diverse protein constructs of clinical importance and with globular proteins in the public domain. The funding agencies for this work will be NIH (primary) and NSF - CBET (secondary). The PI will collaborate with a leading protein thermodynamics expert, Prof. Robin Curtis (Univ. of Manchester, UK), who will develop theoretical models to interpret k_H and B_{22}, in low ionic strength solutions, where charge-charge interactions are significant and for which no models are currently available. While this investigation will be at a fundamental level, its significance and impacts will have useful consequences for industry formulations scientists who perform these measurements routinely to determine protein-protein interactions, stability and high solution viscosity during the development of high concentration liquid protein formulations. Biophysical and thermodynamic measurements are done in dilute solution conditions, while HCLFs are far from dilute: inter-molecular hydrodynamic and thermodynamic interactions must be accounted for, and this work will bridge the gap from dilute to concentrated solutions, so no extrapolations are required in understanding the thermodynamic and hydrodynamic contributions to the viscosity of concentrated protein solutions. A wealth of knowledge will be extracted by performing these measurements on purified well-characterized proteins, and will be organized using rigorous theoretical principles. The learning from this research will be disseminated in journals like *Biophysical Journal*, *Journal of Chemical Physics*, *Journal of Physical Chemistry B*, *Physical Review Letters,* etc.

3) **Molecular Dynamics & Rheology of Protein Solutions: Protein Stability in the Glassy State (at and near the glass transition temperature T_g') and with Flow Instabilities above T_g':** Recall that protein molecules are often prone to aggregation and are hence unstable in concentrated solutions. Thus, a typical short-term fix for Phase I clinical trials, where it is urgent to get the molecule into the clinic for First Time in Human (FTIH to generate human tolerability and safety profiles, is to lyophilize (freeze dry) the protein

formulation. Lyophilization requires the development of a formulation with appropriate bulking agents and cryo-stabilizing preservatives, and development of a freeze-drying cycle. After freezing, nucleation and primary/secondary drying. the protein and other excipients are left in the form of an elegant cake inside the glass vial. The cake is reconstituted by addition of specified quantity of sterile Water for Injection in the clinic. While the molecule is in Phase I trials in the freeze-dried solid dosage form, a stable liquid dosage form formulation is developed in the interim for Phase II clinical trials.

While lyophilization removes water and stabilizes proteins, thus enabling rapid translation of the molecule into the clinic for FTIH milestone, trace quantities of moisture act as a plasticizer that reduces the T_g' of the lyophilized formulation. These trace quantities of water fasten the β-relaxations (secondary relaxations associated with side chains of protein molecules), which are related to the shelf-life of the lyophilized formulation, thanks to the pioneering work of Marcus Cicerone *et al.* at NIST, who used elegant inelastic neutron scattering techniques to measure the temperature dependence of β-relaxation timescales. Most academic and industrial researchers cannot easily access beam lines that are available at cold neutron reactor sources. The PI therefore proposes to use high frequency shear rheology and dielectric/fluorescence correlation spectroscopy to quantify the temperature and frequency dependence of these β-relaxations, as is done for polymeric glasses. High frequencies can be easily accessed with oscillatory shear rheology and fluorescence correlation, thus making these measurements of the β-relaxations and α-relaxation (glass transition) timescales over a broad temperature range. The temperature dependence of this β-relaxation timescale will help elucidate the storage shelf life of protein therapeutics. Typically, these liquid formulation therapeutics are stored in cold chain at 5 ^{o}C, while accelerated stability tests are typically performed at 40 ^{o}C, to kinetically accelerate the instability pathway. **The PI and his team will probe universality in the temperature dependence of the α-relaxation, which is a reasonable expectation from protein glasses, since they form fragile glasses. Such universality will also help guide the rational design of protein solutions for optimal stability.**

Finally, the behavior of therapeutic protein solutions as they flow out of filling nozzles will be visualized using high speed microscopy to observe the different regimes that are associated with non-Newtonian fluids: dripping, jetting, etc. The delicate balance between a plethora of physical forces, such as surface tension, gravity, inertia, elasticity etc. demarcates these various regimes observed in high speed video microscopy. This investigation has not been done in the context of concentrated protein solutions, while anecdotal evidence collected by the PI during supervision of filling of therapeutic formulations in Contract Manufacturing Organizations (CMOs) has revealed fascinating flow instability regimes. This research will be conducted on a strictly fundamental basis, and explained in terms of the flow field and protein solution rheology and physical characteristics. This work has tremendous technological importance too, as improper filling before lyophilization can lead to splashing etc., which leads to rejection of costly drug product during mandatory regulatory visual inspection. Consequently, bio-pharma companies suffer yield hits, and sometimes entire batches are rejected, leading to increased costs.

Impacts, Stakeholders and Funding Agencies: The PI will try to fundamentally study the physics of protein glasses, and translate the learning to NIH and to interested bio-pharma companies that are prospective collaborators. The learning from this research will be disseminated in journals such as *Molecular Pharmaceutics, Pharmaceutical Research etc.* The agencies that will be approached for funding the glass transition work include the NIH, the Burroughs-Wellcome Fund and the Bill and Melinda Gates Foundation: this work will be translated to improve the stability of vaccines in developing countries and extended to the development of room temperature stable vaccines by addition of approved excipients that raise the T_g' above room temperature, thus kinetically retarding the protein degradation pathway. The funding agencies for flow visualization of protein solutions include the NSF-CBET and bio-pharmaceutical companies. That research will be published in *Journal of Non-Newtonian Fluid Mechanics*, *Physical Review Fluids* and *Physics of Fluids*.

509945 Transport Properties of Polymers and Nanoparticles having Complex Mor-phologies: A Computational Modeling Study

Abstract:

Research Interests:

My scientic research focuses on the use of computational tools to study the structural, transport, electromagnetic, hydrodynamic, and thermodynamic properties of soft materials at different time, length, and energy scales. In particular, I use molecular dynamics simulations, Monte Carlo simulations, path-integration techniques, and nite element calculations to study the self-assembly of DNA-based nanostructures, the property-structure relationships of complex macromolecules and polymers in solution, and the mechanic and electromagnetic properties polymer-based systems. My goal is to expand my collaboration network among the scientic community and foster the formation of new scientists in order to form an internationally recognized scientic group centered on the design, characterization, and property optimization of advanced soft materials.

Research Experience:

During my B.Sc. studies, I worked under the supervision of Prof. Arvids Stashans at the National Polytechnic School, Quito, Ecuador, where we performed quantum mechanic calculations to determine the structural, electronic, and optical properties of F{centers (two electrons in one oxygen vacancy) in cubic and tetrahedral lattices of SrTiO3, a material of interest due to its superconductivity properties. Additionally, I worked in Prof. Alberto Celi's lab developing software to automate the data extraction and analysis from an X{ray diractometer. We applied this computational tool to study the relationship between the structural and dielectric properties of TiZrSrO3 with dierent dopants.

My Ph.D. studies in Prof. Francis W. Starr's group, Wesleyan University, Middletown, CT, USA, centered on the use of Monte Carlo, molecular dynamics, and free{energy calculations to explore the self{assembly of nanoparticles decorated with DNA strands. In particular, we studied how the structural parameters of the DNA{coated nanoparticles affect the topology of the resulting self{assembled network, and the assemblies thermodynamic properties. These networks are currently being applying for sensing, imaging, and therapeutics. Our calculations and theoretical results allowed us to establish collaborations with researchers at Brookhaven National Laboratory.

After graduating, I continued expanding my computational skills and theoretical background during my two postdoctoral positions at the the National Institute of Standards and Technology (NIST), both working under the supervision of Dr. Jack F. Douglas. At NIST, we combined novel multi{scale computer modeling calculations to study transport properties of macromolecules, electromagnetic properties of carbon{based composites, and the hierarchical self{assembly of DNA{based nanostructures. All these projects were performed in collaboration with experimental scientists at NIST, which familiarized me with the state{of{the{art experimental nanoparticle characterization techniques.

Teaching Interests:

I have had the opportunity of teaching mathematics and physics to students ranging from middle school up to college. Additionally, during my Ph.D. program, I worked as teaching assistant for a variety of undergraduate and graduate physics and computational courses. I also mentored undergraduate students in performing research.

I am qualied to teach introductory physics and mathematics courses, classical mechanics, quantum mechanics, statistical mechanics, thermodynamics, and also more ap-
plied and interdisciplinary courses related to my computational and soft matter background, such as polymer physics, multiscale modeling, materials science related courses.

Selected Publications:

9. F. Vargas-Lara, A. Hassan, M. L. Manseld, and J. F. Douglas, \ Knot Energy, Complexity, and Mobility of Knotted Polymers", Accepted, Scientic Reports, (2017)

8. F. Vargas-Lara, M. L. Manseld, and J. F. Douglas, \Universal Interrelation Between Measures of Particle and Polymer Size ", The Journal of Chemical Physics, 147, 014903 (2017)

7. C. Long, N. Orlo, K. Twedt, T. Lam, F. Vargas-Lara , M. Zhao, B. Natarajan,K. Scott, E. Marksz, T. Nguyen, J. F. Douglas, J. McClelland, E. Garboczi, J. Obrzut, A. Liddle, \Giant Surface Conductivity Enhancement in a Carbon Nanotube Composite by Ultraviolet Light Exposure", ACS Appl. Mater. Interfaces, 8, 35, 23230-23235, (2016)

6. F. Vargas-Lara, A. Hassan, E. Garboczi, J. F. Douglas, \Intrinsic Conductivity of Carbon Nanotubes and Graphene Sheets Having a Realistic Geometry", The Journal of Chemical Physics, 143, 204902, (2015)

5. F. Vargas-Lara, S. M. Stavis, E. A. Strychalski, B. J. Nablo, J. Geist, F. W. Starr and J. F. Douglas, \Dimensional reduction of duplex DNA under connement to nano uidic slits", Soft Matter, 14, 8273-84, (2015)

4. F. Vargas-Lara, J. F. Douglas, \Confronting the Complexity of Commercial Carbon Nanotube Materials", Soft Matter, 11, 4888-4898, (2015)

3. S H. Ko, F. Vargas-Lara, P N. Patrone, S. M. Stavis, F. W. Starr, J. F. Douglas, and J. A. Liddle, \High{Speed, High{Purity Separation of Gold Nanoparticle{DNA Origami Complexes using Centrifugation: Theory and Experiment", Soft Matter, 10, 7370{7378, (2014)

2. C. Cheng, F. Vargas-Lara, A. Tchanvchenko, F. Starr, O. Gang \The Internal Structure of Nanoparticle Dimers Linked by DNA", ACS Nano, 6 (8), 6793-6802, (2012)

1. F. Vargas-Lara and F. W. Starr, \Stability of DNA{linked Nanoparticle Crystals I: Effect of Linker Sequence and Length", Soft Matter 7, 2085{2093, (2011)

Energy Management and Sustainability in Chemical Engineering and Beyond

Research Interests:

- Energy generation, utilization, conversion, storage, transmission, conservation, management and sustainability
- Mitigation of environmental pollutants, and sustainable energy systems, CO2 capture & renewable energy
- Catalysis for sustainable energy and Fischer- Tropsch Synthesis (FTS)

Successful Proposals:

Advanced Research Projects Agency Energy (ARPA-E), U.S. Department of Energy (DOE), the Climate Change and Emissions Management Corporation (CCEMC) of Alberta, Canada, and the Advanced Conversion Technologies Task Force in Laramie, Wyoming.

Postdoctoral Project: *"Novel Solvent Exchange Distillation System- a Comparison Study with Batch & Flash Separations"*
Under supervision of Professor Phil C. Wankat, Davidson School of Chemical Engineering, Purdue University

PhD Dissertation: *"Dynamic Liquefied Natural Gas (LNG) Processing with Energy Storage Applications"*
Under supervision of Professor Larry L. Baxter, Chemical Engineering Department, - Brigham Young University

Research Experience:

My academic career path has been a blend of many fields of science and engineering. During Bachelor, I was working on simulation/process design and optimization of different industrial and petrochemical unites which I have published my results in four peer-reviewed journals. As a master student, I was working on catalysis, kinetic modeling and experimental reactor design. I am greatly interested in catalysis for sustainable energy and Fischer-Tropsch synthesis. As a result of my research, I have published and presented more than thirty technical articles in the fields of catalysis for renewable energy, kinetic modeling, chemical reaction engineering and Fischer-Tropsch synthesis. As a PhD candidate, I developed new technology, Cryogenic Carbon Capture (CCC) that aims to decrease energy consumption and the costs of separating CO_2 from other gases in flows including power plant exhausts, natural gas flows and most industrial streams and optimum heat exchanger design of natural gas liquefaction model. The fruit of my research projects have been published in famous journals and presented in prestigious conferences. Last but not the least, as a Postdoc; I have been doing detailed simulations of new configurations for batch/flash and dynamic distillation and developing low energy methods in pharmaceutical/food or chemical manufacturing environment. I have more than six papers ready to submit on my postdoc studies.

Teaching Experience:

I am interested in teaching all the courses I have served as a teaching assistant and lecturer. Specifically, I am qualified to teach undergraduate-level courses such as Heat Transfer, Chemical Reaction Engineering, Chemical Engineering Principles, Engineering Mathematics, Fluid Mechanics, Physical Chemistry, Industrial Equipment and Principles of Gas Purification and graduate-level courses such as Energy Engineering, Statistical Analysis, Heterogeneous Catalysts and Transport Phenomena. In addition, I teach CHE 20500: Chemical Engineering Calculations (to B.S. students), CHE 37700: Momentum Transfer, CHE 54000: Transport Phenomena and CHE 597: Capstone Design (to undergraduate and graduate students) as an Assistant Professor and serve as a director for Professional Graduate Student Organization (PGSO) in Purdue University. In general, I am very versatile with respect to other courses I would be interested in teaching at your respected university.

Future Direction:

As faculty, I plan to collaborate with researchers in both academia and industry to improve and develop new processes and mechanisms in FTS, Water Gas Shift (WGS) and Reverse Water Gas Shift (RWGS). Reaction mechanisms and catalytic effects on the enhancement of product efficiency for each process and hydrocarbon synthesis reactions will be presented. By considering the existing data, the best reactor configuration, which is fit with the reaction mechanism, will be proposed. I would like to combine WGS and WGS-FTS at low and high temperatures along with different catalysts and operating conditions. This will include mathematical reactor modeling and experimental test. I plan to expand my current CCC work on solar fuels to also address broader solar energy concerns and the impacts on environment and markets and utilize and develop my dynamic heat exchanger technology –patent pending- in different industries including oil, gas, pharmaceutical, food, beverage and so on to maximize energy savings and efficiency.

I have developed methods to greatly increase the efficiency of solvent exchange by distillation in industry. Many energy manufacturers have entered a new era where the efficiency of production has become important. I plan to improve this novel configuration and collaborate with not only oil and gas industries but also pharmaceutical, food and beverage industries.

Based on the nature of the research projects I have carried out and propose to pursue, I anticipate funding from agencies, such as the NSF, DOE, and Advanced Research Projects Agency-Energy (ARPAE) in the divisions and programs listed below. In addition, there is a great potential for collaboration with National Laboratories (e.g., NREL, NETL, Sandia). I also expect establishing collaborations and funding from industrial companies, such as Dow Chemical, ExxonMobil, Chevron, Valero Energy, Shell Corporation, BP, Phillips 66, Eli Lily, Archer Daniels Midland (ADM), PepsiCo, the Kraft Heinz Company and E. & J.

Gallo Winery, and others. Having an industrial collaborator was a very enriching experience, both research-wise and personally, during my Master, Ph.D. and Postdoc.

Selected Publications:

- Farhad Fazlollahi, Phil C. Wankat "Novel solvent exchange distillation system- a comparison study with batch separations" Chemical Engineering Science (In press).
- Farhad Fazlollahi, Samrand Saeidi, Mohammad Saeed Safdari, Majid Sarkari, Jiri Klemes and Larry L Baxter "Effect of operating conditions on cryogenic CO2 removal" Energy Technology 2017; 5: 1588-1598.
- Farhad Fazlollahi, Alex Bown, Edris Ebrahimzadeh and Larry L Baxter "Transient natural gas liquefaction and its application to energy storage with cryogenic carbon capture (CCC-ES)" Energy 2016; 103: 369–384.
- Farhad Fazlollahi, Alex Bown, Samrand Saeidi, Edris Ebrahimzadeh and Larry L Baxter "Transient natural gas liquefaction process comparison- dynamic heat exchanger under changes in flow" Applied Thermal Engineering 2016; 109:775-788.
- Farhad Fazlollahi, Alex Bown, Edris Ebrahimzadeh and Larry L Baxter "Design and analysis of the natural gas liquefaction optimization process- Energy Storage of Cryogenic Carbon Capture (CCC-ES)" Energy 2015;90: 244–257.
- Farhad Fazlollahi, Majid Sarkari; Hamed Gharebaghi; Hossein Atashi; Mohammad. M Zarei; Ali Akbar Mirzaei and W.C Hecker "Development of kinetic expression for a sol-gel Fe-Mn/Al2O3 Fischer-Tropsch Synthesis catalyst" Chinese Journal of Chemical Engineering 2013; 21 : 507-519.
- Farhad Fazlollahi, Majid Sarkari; Akbar Zare; Ali. A. Mirzaei and Hossein Atashi "Development of a kinetic model for Fischer-Tropsch Synthesis over Co/Ni/Al2O3 catalyst" Journal of Industrial and Engineering Chemistry 2012; 18: 1223–1232.

Peer- Reviewed Magazine:

- Farhad Fazlollahi and Larry L Baxter "Modeling and analysis of natural gas liquefaction process- Energy Storage of Cryogenic Carbon Capture (CCC-ES)" EM magazine, Journal of the Air & Waste Management Association, Aug 2015.

Peer- Reviewed Chapter in Book:

- Farhad Fazlollahi, Mohammad Mehdi Zarei, Maryam Seied Habashi, and Larry L Baxter "Method for removal of mercury from oil field brine with calcium carbonate co-precipitation" TMS Annual Meeting & Exhibition, Characterization of Minerals, Metals, and Materials, San Diego, Californian, USA, 2014 (http://onlinelibrary.wiley.com/doi/10.1002/9781118888056.ch16/summary).

- Farhad Fazlollahi , Majid Sarkari and Hossein Atashi "LaCoO3: The efficient catalyst to purify pollutant gases" TMS Annual Meeting & Exhibition, Characterization of Minerals, Metals and Materials, Characterization for Environmental Applications, San Antonio, TX, USA, 2013 (http://onlinelibrary.wiley.com/doi/10.1002/9781118659045.ch21/summary).

Reviews:

- Samrand Saeidi, Sara Najari, Farhad Fazollahi, Maryam Khoshtinat Nikoo, Fatemeh Sefidkon, , Jiří Jaromír Klemeš, Larry L Baxter "Mechanisms and kinetics of CO2 hydrogenation to value-added products: a detailed review on current status and future trends" Renewable and Sustainable Energy Reviews 2017; 80: 1292-1311.
- Samrand Saeidi, Farhad Fazollahi, Sara Najari, Davood Iranshahi, Jiří Jaromír Klemeš, Larry L Baxter "Hydrogen production: perspectives, separation with special emphasis on kinetics of WGS reaction: A state-of-the-art review" Journal of Industrial and Engineering Chemistry 2017; 49: 1-25.

Plasma Biomedicine and Plasma-Fabricated Nanomaterials for Energy, Health, and Electronics

Daniel T. Elg, 2nd-Year Postdoc, Department of Chemical & Biomolecular Engineering, University of California, Berkeley

Research Interests:

Two of the most promising areas of plasma-material interaction research are plasma biomedicine and nanomaterial fabrication/modification, both of which are very dependent on gas-phase plasma chemistry. Plasma biomedicine can enable processes such as sterilization, wound healing, and cancer treatment; nanomaterials have a myriad of uses, such as enabling catalysis, the continuation of Moore's Law, and the production of low-cost, efficient photovoltaics. Both of these areas have only scratched the tip of the iceberg in terms of their potential. Despite their seeming differences, both are plasma-material interaction techniques that are dependent on the same principles of plasma physics, nonthermal charged gas-surface interactions, and plasma chemistry. My previous research experience has been focused on plasma-material interactions, utilizing both the low-pressure and high-pressure cold, non-thermal plasmas seen in these fields. This background has prepared me for a career in advancing the science of plasma-material interactions for applications that will attract funding from both government and industry and are relevant to materials, energy, and health. **My goal as a professor is to build a group that contributes fundamental knowledge and new, innovative diagnostic and synthesis techniques to plasma biomedicine and nanomaterials.**

I intend to build upon my postdoctoral work and use my expertise in isolating O atom trends to identify reaction pathways for the production of biologically-active OCl^- in saline and calculate the rate constant for the dominant pathway. In the area of nanomaterials synthesis, I intend to use pulsed biasing and *in-situ* diagnostics to control and measure the energy per atom in dusty SiH_4 plasmas, allowing for tailored deposition of nanocrystalline and even epitaxial Si thin films for solar energy. An innovative design built upon my Sn etching experience will also be used to produce synthesize SiSn thin films from dusty plasmas. SiSn films have tunable direct bandgaps for photovoltaics and can also be used to fabricate transistors with low leakage currents. Additionally, deposition of nanomaterial catalysts at atmospheric pressure can be accomplished at a high rate by combining nanopulses with a continuous RF plasma. This has possible ramifications for sustainable energy by means of catalyst deposition for CO_2 reforming. Finally, I will maintain my reputation in the EUV lithography industry (and procure industrial funding) through projects involving the fabrication and hydrogen blistering of EUV mirrors. A thermal method will be used to measure sputtered neutral atom energies in order to tailor deposition conditions for smooth interfaces in alternating nanolayers of Mo and Si, and an ion gun and radical source will be used to investigate hydrogen blistering of these nanolayered structures.

Teaching Interests:

My pursuit of an academic career stems from a twin desire to pursue excellent research and to teach science in a way that maximizes its accessibility and prepares the next generation for success. Over the course of my time as both a student and a teacher, I came to the conclusion that **the unifying link between effective teaching in all styles is: structure**. The professor must understand the mindset of somebody without prior knowledge (and often without great interest in) the material, predict which directions such a mind will take, and structure the progression of the course to present the material in a manner that is logical to that mind. This is the approach I take in the classroom. As a graduate student at Illinois, I actively sought out teaching opportunities even though I was fully funded through my research. I taught gratis every year and was eventually listed as Co-Instructor for the course NPRE 101, "Introduction to Energy Systems". Even as an undergraduate, I was deeply interested in teaching and pedagogy, choosing to attend Olin College because of its commitment to reinvent engineering education and emphasize hands-on project-based learning.

My interdisciplinary background has given me the experience necessary to teach a wide variety of subjects. Additionally, I would be very interested in creating plasma courses, such as a course on plasma processing and plasma chemistry. Topics covered could include ambipolar diffusion, conservation equations, collisional processes, reaction rates, plasma chemistry modeling, plasma-surface interactions, sputtering, etching, and various types of plasma sources and their applications. Because of my commitment to preparing students for real-world applications of science, I especially appreciate the kind of hands-on experience that a lab class can give, and I would like to create a plasma lab course that would serve as an introduction for anybody looking to perform experimental plasma work.

Postdoctoral Research: *Selective Diagnostics for Plasma-Liquid Interactions in Plasma Biomedicine*

Supervisor: Prof. David B. Graves; Chemical & Biomolecular Engineering; University of California, Berkeley

At Berkeley, I have focused mainly on the area of plasma-liquid interactions, which are crucial for plasma biomedicine, since medical effects of plasmas are caused largely by plasma-produced reactive oxygen and nitrogen species dissolved in biological liquids. During my postdoctoral work, I used a spin trap to selectively isolate trends due to plasma-produced O atoms dissolved in water. O atoms are very important for plasma-induced reactions in biological material, and this was the first time their effect had been separated out from those of other plasma-produced species in the aqueous phase. Current research involves the use of gas and liquid-phase diagnostics to quantify the improvement of plasma-liquid interactions in a perfluorocarbon solvent. Additionally, I have also worked on both hardware and software for a project on advanced control of atmospheric pressure plasmas. Advanced control is necessary in order to provide reliable plasma treatments on living subjects.

PhD Research: *Removal of Tin from Extreme Ultraviolet Collector Optics by an In-Situ Hydrogen Plasma*

Advisor: Prof. David N. Ruzic; Nuclear, Plasma, and Radiological Engineering; University of Illinois

At Illinois, I developed, measured, and simulated plasma techniques for the extreme ultraviolet (EUV) lithography industry. EUV lithography sources enable enhanced resolution and the continuation of Moore's law by creating 13.5nm photons (an order of magnitude lower wavelength than currently-used 193nm photons) with high-energy Sn plasmas. The main factor now preventing high-volume commercial implementation is the fact that the Sn plasma debris can sputter, implant in, or deposit on the collector optic that focuses the EUV light. High-energy debris must be slowed down or deflected, and deposited Sn must be removed without turning the source off. My research provided multiple advances towards mitigating collector contamination. Important contributions include: (1) Demonstration of EUV reflectivity restoration by an *in-situ* technique to remove Sn from EUV collector optics with an H_2 plasma, (2) Isolation of ion energy flux as the main factor limiting the etch rate, (3) Exploitation of ion bombardment to achieve per-radical etching efficiencies orders of magnitude higher than those reported with traditional atomic H sources, (4) Demonstration of the fact that, contrary to previous thought, the unstable etch product SnH_4 does not significantly reduce etch rates by decomposing, and (5) Successful deflection of keV-range ions from a z-pinch EUV source with magnetic fields. This research was partially funded by an NSF GOALI grant which I co-authored.

Awards & Honors:

Mavis Future Faculty Fellowship

Co-Author of Successful NSF GOALI Proposal #14-36081: "In-Situ Plasma Cleaning of Optics: Building a Fundamental Understanding of the Etch Process in a Complex Plasma Environment"

International EUVL Symposium Student Scholarship (2014, 2015)

Selected Publications:

(Submitted) **D.T. Elg**, I. Yang, D.B. Graves, "Production of TEMPO by O Atoms in Atmospheric Pressure Non-Thermal Plasma-Liquid Interactions", *Journal of Physics D: Applied Physics* (2017).

D.T. Elg, G.A. Panici, S. Liu, G. Girolami, S.N. Srivastava, D.N. Ruzic, "Removal of Tin from Extreme Ultraviolet Collector Optics by Etching with an *In-Situ* Hydrogen Plasma", *Plasma Chemistry & Plasma Processing* (2017).

Q. Yang, K. Kalathiparambil, **D.T. Elg**, D.N. Ruzic, W.M. Kriven, "Microstructural Damage of α-Al_2O_3 by High Energy Density Plasma", *Acta Materialia* **132**, 479-490 (2016).

D.T. Elg, G.A. Panici, J.S. Peck, S.N. Srivastava, D.N. Ruzic, "Modeling and Measurement of Hydrogen Radical Densities of *In-Situ* Plasma-Based Sn Cleaning Source", *Journal of Micro/Nanolithography, MEMS, & MOEMS* **16**(2) (2016).

D.T. Elg, J.R. Sporre, G.A. Panici, S.N. Srivastava, D.N. Ruzic, "In-Situ Collector Cleaning and EUV Reflectivity Restoration by Hydrogen Plasma for EUV Sources", *Journal of Vacuum Science & Technology A* **34**(2) (2016).

J.R. Sporre, **D.T. Elg**, Kishor K. Kalathiparambil, D.N. Ruzic, "Modeling and Measuring the Transport and Scattering of Energetic Debris in an EUV Plasma Source", *Journal of Micro/Nanolithography, MEMS, & MOEMS* **15**(1) (2016).

D.T. Elg, J.R. Sporre, D. Curreli, I.A. Shchelkanov, D.N. Ruzic, K.R. Umstadter., "Magnetic Debris Mitigation System for EUV Sources", *Journal of Micro/Nanolithography, MEMS, & MOEMS* **14**(1) (2015).

Contact:

Daniel Elg
D75 Tan Hall
University of California
Berkeley, CA 94720
(630) 272-8503
daniel.elg@berkeley.edu

509894 Reinforced anion exchange membrane (AEM) Separators Based on Triblock Copolymers for Electrode-decoupled redox flow batteries (RFBs)

Abstract:

Research Motivation:

Energy consumption is one of the barometers of economic and human well-being and the equitable development of humankind requires sustainable energy equity. Achieving this overarching goal requires us to address issues such as the intermittency of ecologically non-disruptive renewable energy technologies (solar and wind), decarbonize the transportation sector, and find alternative chemical pathways for large scale, energy intensive industrial processes. *I aim to help address these key issues by leveraging my training and experience in multi-scale theoretical and experimental investigations of electrochemical systems.*

Postdoctoral Project – *"Reinforced anion exchange membrane (AEM) Separators Based on Triblock Copolymers for Electrode-decoupled redox flow batteries (RFBs)"*

Supervisor: Prof. Vijay K. Ramani, Department of Energy, Environmental and Chemical Engineering, Washington University in St. Louis

PhD Dissertation – *"Investigation of the oxygen reduction reaction at the lithium-oxygen cell cathode"*

Supervisor: Prof. Jai Prakash, Department of Chemical and Biological Engineering, Illinois Institute of Technology

Research Interests:

My research journey began with my undergraduate work on continuum scale modeling of the capacity fade mechanisms in lithium-ion batteries with Prof. Balaji Krishnamurthy. An internship at Tata Power Solar, where I assisted in formulating a new silicon texturing bath composition (resulting in a ~1% improvement in final cell efficiency), whetted my appetite for translational research.

During my PhD research with Prof. Prakash at Illinois Institute of Technology, I: 1) evaluated novel lithium-ion cell chemistries for use in sub-zero temperatures (in conjunction with Hydro-Quebec), 2) performed the first ever accelerating rate calorimetry study of the thermal safety of lithium-sulfur cells (again with Hydro-Quebec), and 3) performed a multi-year theoretical and experimental investigation of catalysts, solvents and kinetics of the oxygen reduction reaction (ORR) in lithium-oxygen cells in collaboration with Toyota Motors. Over the course of these projects, I extensively employed density functional theory (DFT) simulations, continuum scale kinetic modeling, and classical electrochemical and electroanalytical experimental techniques.

As an integral part of my postdoctoral research project, I am investigating anion exchange membranes for use in electrode-decoupled redox flow batteries (RFBs). Herein, I am applying my knowledge of electrochemistry and transport phenomena to understand the transport of ions across a functionalized membrane. The final goal of this project, funded by ARPA-E, is to develop selective membranes that allow for the deployment of low-cost, electrode-decoupled RFBs such as the Fe-Cr system.

Future Research Directions:

I plan to leverage my training and experience in multi-scale theoretical and experimental investigations of electrochemical systems to carry out transformative, translational research into next generation, beyond lithium-ion batteries, grid scale energy storage using redox flow batteries, and the electro-synthesis of bulk and fine chemicals.

Project 1: *Metal- (metal superoxide) batteries* – Lithium-oxygen batteries have been the holy grail of next generation battery systems due to their unmatched theoretical specific energy (3458 Wh kg^{-1}). Despite years of extensive work on catalysts, redox mediators and ever-more-stable electrolytes, a practical lithium-oxygen battery has not been realized. As part of this project I will examine sodium- and potassium- oxygen batteries whose reduction products are superoxides instead of the peroxides, allowing me to circumvent the columbic efficiency and degradation problems of lithium-oxygen batteries. I will combine insights from electrochemical kinetics studies with oxygen impermeable membranes and water tolerant ionic liquids to demonstrate long life alkali metal- (metal superoxide) batteries. *The end goal will be the development of a true metal-ambient air battery.*

Project 2: *Low-cost, grid scale energy storage* – The adoption of intermittent renewable energy sources calls for the large-scale deployment of energy storage solutions to maintain grid reliability. The DOE estimates that $100/kWh is the cost at which storage solutions become cost competitive. This cost target precludes both the Li-ion battery and the all-vanadium redox-flow battery. I will investigate alternative, low cost, earth abundant chemistries for redox-flow batteries. The use of catalytic inhibitors to suppress the hydrogen evolution and oxygen evolution reactions, in conjunction with electrolyte systems that increase the nominal redox potential of the redox couple of choice, will pave the way for >2V RFBs utilizing couples such as Cr-Ce and provide an unparalleled cost advantage compared to existing technologies. Further, I will develop highly perm-selective, high anion transference number anion-exchange membranes for this system both in-house and in collaboration with polymer scientists and material chemists.

Project 3: *Electro-synthetic alternatives to energy intensive chemical production* –

Ammonia production – Ammonia is produced to the tune of 140 million tons a year via the energy intensive Haber-Bosch process with a massive carbon footprint. Biomimetic or electro-catalytic ammonia production is a low CO_2 alternative to this process that will allow for low capital cost and distributed production of ammonia at the demand site. Further, the elimination of steam reforming to produce hydrogen and the avoidance of transportation costs and associated emissions makes this an extremely tantalizing proposition. I will examine metal, metal oxide, carbide, nitride and alloy catalysts for electrochemical ammonia synthesis from water and air. Combining DFT screening of catalyst candidates (nitrides, carbides, Pt and Pd alloys) with robust half-cell and device-level testing, I will identify candidate catalysts at the top of the activity pyramid, evaluate their nitrogen reduction kinetics and design, test and optimize larger-scale cells to produce ammonia.

CO_2 capture, concentration and conversion (C3)- Any realistic timeline of a transition to renewables envisions significant dependence on fossil fuels in the near-term. Thus, it is imperative that efficient CO_2 capture technologies be developed. In collaboration with other groups working on CO_2 capture, I will develop alkali metal-O_2/CO_2 batteries for direct conversion of CO_2 to electricity. An intriguing idea that will be investigated is a combined N_2 electro-reduction/ alkali metal-O_2/CO_2 system wherein the ammonia produced is used to capture and concentrate the CO_2 before use in the battery.

Teaching Interests:

The experience I have gained assisting in the teaching of 9 courses across the spectrum of ChE, from the sophomore level to the advanced PhD level, has prepared me well to serve as an instructor teaching the traditional chemical engineering core. I have served as Assistant to instructor and Support to instructor respectively for courses on Electrochemical Engineering and Energy Conversion and Storage at Washington University, where I have been actively involved in class planning and laboratory sessions in addition to traditional teaching assistant roles. During my PhD, I served as a Teaching assistant for 7 courses including thermodynamics, transport phenomena, undergraduate laboratory and electives in energy technology. The experience of assisting in the teaching of thermodynamics as an introductory first course, an intermediate course for 1st year master's students and as an advanced course for PhD students provided invaluable insights into the nuances of balancing depth and clarity without sacrificing technical accuracy. The mentor-mentee relationships I formed with high school, undergraduate and graduate students in the lab redoubled and renewed my commitment to excellence in teaching and training. I look forward to bringing this passion to my future role as a ChE faculty member.

Invention and Technology Translation:

I have always wanted to see my work changing society in a reasonable timeframe. My PhD work on Pd and Pt alloy catalysts for lithium-oxygen batteries with Toyota resulted in a patent and is the basis of their continued research in this field. My on-going postdoctoral work on reinforced membranes for flow batteries has already resulted in a $50,000 Leadership in Entrepreneurial Acceleration Program (LEAP) grant from Washington University to serve as a pre-seed fund for a start-up. My future independent research career will maintain that same strong practical, inventive and translational emphasis.

Publications (9 published, 1 in review and 4 under preparation)

1. **Sankarasubramanian,** and B. Krishnamurthy, A Capacity Fade Model for Lithium ion batteries including kinetics and diffusion, *Electrochim. Acta* **70**, 248-254 (2012).
2. Seo, **S. Sankarasubramanian,** C-S Kim, P. Hovington, J. Prakash, and K. Zaghib, Thermal characterization of Li/sulfur, Li/ S-$LiFePO_4$ and Li/S-LiV_3O_8 cells using Isothermal Micro-Calorimetry and Accelerating Rate Calorimetry, *J. Power Sources* **289**, 1-7 (2015).
3. **Sankarasubramanian*,** N. Singh, F. Mizuno, J. Prakash, *Ab initio* investigation of the Oxygen Reduction Reaction activity on noble metal (Pt, Au, Pd), Pt_3M (M=Fe, Co, Ni, Cu) and Pd_3M (M=Fe, Co, Ni, Cu) alloy surfaces, for Li-O_2 cells, *J. Power Sources,* **319**, 202-209 (2016).
4. **Sankarasubramanian*,** J. Seo, F. Mizuno, N. Singh, J. Prakash, Rotating ring-disc electrode investigation of the aprotic superoxide radical electrochemistry on multi-crystalline surfaces and correlation with Density functional theory modeling – implications for Lithium-air cells, *J. Electrochem. Soc.*, **163**(10), A2377-A2384 (2016).
5. **Sankarasubramanian*,** J. Seo, F. Mizuno, N. Singh, K. Takechi, J. Prakash, Enhancement of oxygen reduction reaction rate by addition of water to an oxidatively stable ionic liquid electrolyte for lithium-air cells, *Electrochem. Commun.*, **73,** 55-57 (2016).
6. **Sankarasubramanian*,** J. Seo, F. Mizuno, N. Singh, J. Prakash, Elucidating the Oxygen reduction reaction kinetics and the origins of the anomalous Tafel behavior at the Li-O_2 cell cathode, *J. Phys. Chem. C*, **121,** 4789-4798 (2017).
7. Zhang, J. Parrondo, **S. Sankarasubramanian,** V. Ramani, Detection of Reactive Oxygen Species in AEM Fuel Cells using *in situ* Fluorescence Spectroscopy, *ChemSusChem*, **10**, 3056-3062 (2017).
8. Seo[#], **S. Sankarasubramanian*, #,** F. Mizuno, N. Singh, K. Takechi, J. Prakash, Effect of cathode porosity on the Lithium air cell oxygen reduction reaction – a rotating ring-disk electrode investigation, *Electrochim. Acta*, **248**, 570-577 (2017).
9. He, G. Wang, J. Parrondo, **S. Sankarasubramanian**, V. Ramani, Pt/RuO_2-TiO_2 electrocatalysts exhibit excellent hydrogen evolution activity in alkaline media, *J. Electrochem. Soc.*, **164**(12), F1234-F1240 (2017).

510365 Understanding and Controlling Electro-Chemo-Mechanical Phenomena in Advanced Materials for Energy Storage & Harvesting

Abstract:

Research Interests
Advanced materials in energy-related application areas, such as batteries, fuel cells, electrolysis, and catalysis, are exposed to high temperatures, and/or harsh chemical and electrochemical environments during routine operation conditions. Performance and lifetime of these materials are directly associated with the interplay between the electrical, chemical and mechanical properties. Also, many functional materials in energy-related applications take large quantities of ions in and out, such as Li+ in batteries, H+ in hydrogen storage, and O2â€€€ in solid-state fuel cells during routine operation conditions. Repeated ion insertion and removal from the materials causes considerable volumetric changes and associated signicant stress development, eventually leads to severe performance loss and shorten the lifetime of the devices. Understanding and optimizing electro-chemo-mechanical responses of advanced materials is required in order to design new materials with synergistic properties, which oer the possibility of fabricating high performance energy conversion and storage devices at low cost.

Research Experience
In my graduate studies with Kurt Hebert and Pranav Shrotriya at Iowa State University, I investigated the electrochemical driving force for vacancy diusion-induced tensile stress and plasticity during stress corrosion cracking in aluminum. I developed substantial expertise in electrochemistry and mechanics of materials with an emphasis on defect chemistry and novel characterization methods. The most impactful accomplishment was the development of a rational basis for the formation of porous anodic oxide, which will help overcome barriers to commercialization of energy harvesting and energy storage devices made from these materials [1-5, 9, 10]. I also demonstrated that corrosion processes produce defective and presumably weakened surface layers that may be precursors for crack initiation [6].

My postdoctoral work at the University of Illinois at Urbana-Champaign under the mentorship of Nancy Sottos and Scott White has focused on the understanding and controlling of chemomechanical deformations in Li-ion battery electrodes. A key nding was that electrochemical stiness provides new insights into the origin of degradation mechanisms and desired properties of advanced battery electrodes for high-rate and -power application [8]. Furthermore, I have demonstrated that dynamic changes in the electrode surface due to interaction between electrode and electrolyte results into irreversible electrode deformations and capacity fade [7,11]. Finally, I have also shown that Au surface coating can improve mechanical stability and lifetime of the electrode materials [13].

Proposal Contributions
Interfacial Control for Self-Stabilizing Solid State Batteries, Shell Energy

Future Research Directions

My proposed research program will apply my expertise to gain a fundamental understanding and controlling of electro-chemo-mechanical responses of advanced materials for engineering the next-generation energy harvesting and storage devices. I will also develop novel characterization methods for probing in situ dynamic mechanical response of materials and simultaneous changes on the surface of materials. This research merges the fields of electrochemistry, defect chemistry, and solid mechanics. In the short term, I will work on solid-oxide fuel cells, sodium-ion batteries and lithium-air batteries. These studies will complement each other with respect to the understanding and controlling electro-chemo-mechanical deformations in materials for energy applications, therefore making a big step towards engineering any advanced and robust materials. In the long run, I
would like to extent my knowledge and expertise to other energy harvesting and storage systems such as hydrogen fuel cells and Li-sulfur batteries.

Teaching Interests
I strongly believe that teaching is an essential and enjoyable part of being a faculty member. My approach has been developed through the serving as teaching assistant, the mentoring of several graduate students as well as through the Preparing Future Faculty Program. I would like to teach core chemical engineering courses such as uid dynamics, heat & mass transfer, thermodynamics, reaction engineering, separation process, numerical methods and advanced mathematics for chemical engineers at undergraduate and graduate level. I have also had several enriching and rewarding advising experiences during my Ph.D. training and post-doctoral years.

Publications

[13] O. O. C apraz, S. Rajput, K.L. Bassett, A. Gewirth, S. White, N. Sottos, Controlling the Volumetric Expansions in the Lithium Manganese Electrode via Surface Modication,
in preparation, 2017, (Manuscript is available upon request)
[12] L. Zhao, E. Chenard, O. O. C apraz, N. Sottos, S. White, Direct Detection of Manganese Ions in Organic Electrolyte by UV-vis Spectroscopy, under review, 2017
(Manuscript is available upon request)
[11] O. O. C apraz, S. Rajput, S. White, N. Sottos, Strain Generation Mechanisms in Lithium Manganese Oxide Electrode, under review, 2017 (Manuscript is available upon request)
[10] O. O. C apraz, Q. van Overmeere, P. Shrotriya and K. R. Hebert, Stress Induced by Electrolyte Anion Incorporation in Porous Anodic Aluminum Oxide, Electrochim. Acta,
238, 368, 2017
[9] Shinsuke Ide, O. O. C apraz, P. Shrotriya and K. R. Hebert, Oxide Microstructural Changes Accompanying Pore Formation During Anodic Oxidation of Aluminum,
Electrochim. Acta, 232, 303, 2017
[8] O. O. C apraz, K. L. Bassett, N.R. Sottos, and A.A. Gewirth, Electrochemical Stiffness of Lithium Manganese Oxide, Adv. Energy Mater.,1601778, 2016
[7] E.M.C. Jones, O. O. C apraz, S.R. White and N.R. Sottos, Reversible and irreversible deformation mechanisms of composite graphite electrodes in lithium-ion batteries,
J.Electrochem. Soc., 163, 9, 2016
[6] O. O. C apraz, P. Shrotriya and K. R. Hebert, Tensile stress and plastic deformation in aluminum induced by vacancy diffusion during corrosion, Acta Materialia, 115, 434, 2016
[5] O. O. C apraz, P. Skeldon, G. Thompson, P. Shrotriya and K. R. Hebert, Role of Oxide Stress in the Initial Growth of Self-Organized Porous Aluminum Oxide, J. Electrochim. Soc., 167, 404, 2015, Editor Choice
[4] O. O. C apraz, P. Skeldon, G. Thompson, P. Shrotriya and K.R. Hebert, Factors Controlling Stress Generation during the Initial Growth of Porous Anodic Aluminum
Oxide, J. Electrochim. Soc., 159, 16, 2015
[3] O. O. C apraz, The Role of Stress in Self-Ordered Porous Anodic Oxide Formation and Corrosion of Aluminum, Dissertation (Ph.D.) 2014, Iowa State University
[2] O. O. C apraz, P. Shrotriya and K. R. Hebert, Measurements of Stress Changes During Growth and Dissolution of Anodic Oxide Films on Aluminum, J. Electrochem. Soc.,
161, D256, 2014
[1] O. O. C apraz, K. R. Hebert and P. Shrotriya, In situ Stress Measurement During Aluminium anodizing using, Phase-Shifting Curvature Interferometry, J. Electrochem. Soc.,
160, D501, 2013

510376 Microfabricated Devices for Drug Delivery and Tissue Engineering Applications

Abstract:

Research Interests

My main research interest is in the use of biomaterials for tissue engineering and controlled drug delivery with a focus on translational and platform technologies. Specifically, I plan to leverage the exceptional resolution of microfabrication tools originally developed for the integrated circuit industry to create microscalescale devices with the potential to improve patient health. Drug delivery devices have the potential to improve therapeutic outcomes and patient compliance by providing prolonged and/or targeted release that maintains drug concentrations within the therapeutic window. My research will focus on creating material- and geometry-dependent release devices that can be tuned to achieve different release kinetics for applications ranging from cancer to infectious disease. In addition, I also plan on using a similar set of techniques to create tissue engineering constructs such as a microfluidic retina-on-a-chip. This device will serve as a valuable drug screening tool, platform for the controlled study of genetic disease, and precursor to implantable tissue engineering scaffolds.

Postdoctoral Research

My postdoctoral research under the supervision of Robert Langer at the Massachusetts Institute of Technology has largely focused on three main projects. My first project aims to develop microfabricated particles with pulsatile release kinetics for single-injection vaccination, the second focuses on the use of microneedle-delivered ultrastable NIR quantum dots for on-patient medical records, and the third aims to improve the evaluation of in vivo release kinetics using fluorescence imaging and positron emission tomography.

PhD Research

As a graduate research assistant under the supervision of Magali Saint-Geniez at Harvard Medical School I worked on two projects. My first project involved the development of nanoporous polymeric scaffolds for retinal tissue engineering and the second used finite element modeling of oxygen distribution in the retina as a predictive tool for the progression of age-related macular degeneration.

Selected Publications

McHugh KJ*, Nguyen TD*, Linehan AR, Yang D, Behrens AM, Rose S, Tochka ZL, Tzeng SY, Norman J, Anselmo AC, Xu X, Tomasic S, Taylor MA, Lu J, Guarecuco R, Langer R, Jaklenec A. Fabrication of fillable microparticles and other complex 3D microstructures. *Science* 2017;357:1138-42.

Anselmo AC, McHugh KJ, Webster J, Langer R, Jaklenec A. Layer-by-layer encapsulation of probiotics for delivery to the microbiome. *Adv Mater* 2016;28:9486-90.

McHugh KJ, Tao SL, Saint-Geniez M. Porous poly(Îμ-caprolactone) scaffolds for retinal pigment epithelium transplantation. Invest Ophthalmol Vis Sci 2014;55:1754-62.

Cao H, McHugh K, Chew SY, Anderson JM. The topographical effect of electrospun nanofibrous scaffolds on the in vivo and in vitro foreign body reaction. J Biomed Mater Res A 2009;93:1151-9.

*Indicates equal contributions

Patents

McHugh K, Jaklenec A, Langer RS. Microdevices with complex geometries. Filed September 13, 2017. Application Number: US 62/558,172.

Jaklenec A, McHugh KJ, Langer RS. Microneedle tattoo patches and use thereof. Filed July 16, 2017. Application Number: US 62/533,081.

Jaklenec A, Gates W, Eckhoff PA, Nikolic B, Wood LL Jr., Langer RS, Nguyen TD, Tzeng SY, Norman JJ, McHugh K. Micromolded or 3-D printed pulsatile release vaccine formulations. Filed December 16, 2014, Application Number: US 14/572,631.

McHugh KJ, Saint-Geniez M, Kim L. Predicting retinal degeneration based on three-dimensional modeling of oxygen concentration. Filed April 24, 2013. Application Number: US 14/786,952.

Successful Proposals:

NIH (NIBIB) Ruth L. Kirschstein National Research Service Award (F32) awarded for "fluorescence-based molecular imaging of in vivo release kinetics," The Grimshaw Foundation grant received for "patient-specific modeling of oxygen distribution for predicting retinal degeneration," and NIH LRP awarded for "micromolded particles for single-injection pediatric vaccines."

Teaching Interests

As a PhD student I spent two semesters as a Graduate Teaching Fellow at Boston University teaching Biomedical Instrumentation II (ENG BE 492), a core laboratory class primarily for juniors majoring in biomedical engineering. Between this experience and my research expertise, I am particularly well suited for teaching courses that focus on micro- and nano-fabrication processes, drug delivery, polymeric biomaterials, bioreactor design, biomanufacturing, microfluidics, and thermodynamics, but I am capable of teaching other graduate and undergraduate core courses as well. Further, in addition to my teaching duties, I also look forward to the opportunity to mentor students. I have been fortunate to train under some of the leading researchers in biomaterials, biocompatibility, drug delivery, and tissue engineering. Their mentorship has allowed me to not only gain new technical skills, but also appreciate the positive effect that mentorship can have on a student's professional development. Since graduate school I have mentored seven undergraduates, eight technicians, and several graduate students. My interactions with these junior level have been overwhelmingly positive, so I look forward to continuing to grow my mentee network as an independent investigator.

510516 Towards Stronger and Smarter Materials via the Hybridization and Engineering of Dimensionality and Topology

Abstract:

Research Interests:

Macroscopic preparation of materials with precise structure control at the atomic and molecular level is an ultimate goal of chemical engineering, and this could lead to the production of materials with versatile properties and tangible technological applications. Despite our ability to synthesize molecules and nanoparticles, atomic control by introducing or compositing single atom defects or discontinuities in materials at a large scale remains a dream. However, the emergence of large-area, two-dimensional (2D) materials which have the thickness of a single atom or unit cell, such as graphene, molybdenum disulfide, or hexagonal boron nitride (hBN), may help to realize this dream. My research interests lie in the fundamental task of synthesizing 2D materials, polymers, nanoparticles, and their molecular composites and nanocomposites with precise structural control and the process of scaling up. The basic idea is to use material hybridization, dimension transformation, and topology optimization to achieve stronger and smarter materials. Hybridizing layers of 2D materials or intercalating/interfacing 2D layers with composite matrix materials such as polymers, is a simple, yet efficient, way to achieve atomic control, and can scale up to the macroscopic level. Simultaneously, these 2D materials are intriguing nanoinclusions that can be engineered to impart exotic mechanical reinforcements and various other functions as matrix composites. A further dimension transformation from 2D to 3D (i.e., by creating a 3D composite body comprised exclusively of 2D layers) can induce new and exotic functions that depend on the topology of the 3D object.

Postdoctoral Projects:

- Under supervision of *Michael S. Strano*, Chemical Engineering Department of MIT:

"Design And Synthesis Of Carbon-based Chainmaille Structures for Flexible, Ultra-lightweight Protection", funded by U.S. Army Research Office (ARO);

A Conceptual Model of Ambient Armor and Issues Relating to Experimental Realization, funded by ARO;

"FATE: Foldable and adaptive Two-Dimensional Electronics", Thrust II – Folding, MURI project, funded by Air Force Office of Scientific Research;

Thermal Management Technologies for Low Temperature Undersea Dive Persistence: a Novel Arctic Diving Suit, funded by U.S Navy.

Ph.D. Dissertation:

"Study on Tailoring Chain Topologies of Ethylene Copolymers and Their Functionalization"

- Under supervision of *Wen-Jun Wang* (College of Chemical and Biological Engineering, Zhejiang University, China), *Shiping Zhu* (Department of Chemical Engineering, McMaster University, Canada), and *Zhibin Ye* (Bharti School of Engineering of Laurentian University, Canada)

Fellowships:

2013 Fellowship for Outstanding Doctoral Dissertation of Zhejiang University, China;
2011 Fellowship for Excellent Doctoral Student granted by Ministry of Education, China;

Research Experience:

As a scientist trained with chemical engineering fundamentals, my research interfaces polymer science, nanotechnology, material mechanics, catalysis, and sensors. Specifically, I have focused on the precise synthesis and characterization of polymers, 2D materials, nanoparticles and their nanocomposites, and the experimental and theoretical study of their mechanical, electrical, optical, catalytic properties and sensing behaviors:

As a postdoc, I have worked on: *(1) Layered and scrolled polymer nanocomposites with aligned semi-infinite 2D inclusions,* which show substantial mechanical reinforcement and interesting electrical and optical properties at highly reduced volume fraction of 2D inclusions such as graphene. *(2) Autoperforation of 2D materials for generating two terminal memresistive Janus particles.* This *autoperforation* technique can guide Griffith facture of 2D materials, which provides a means of spontaneous assembly for surfaces comprised of 2D molecular surfaces. With this technology, I have created colloidal microparticles that can function as particulate electronic devices capable of collecting and storing digital information in their environment. *(3) Ambient armor (AA) concept for stand-off projectile defense and its mathematical modeling.* I have explored the theoretical possibility of new materials, particularly lightweight 2D nanocomposites with high toughness and elongation, for projectile protection in a stand-off way. (4) *Collaborative work with other lab members.* I have also involved in a diverse of projects including polymer-wrapped single-walled carbon nanotube sensors for carbohydrates detection, nitroaromatic detection with plant nanobionics, chemical doping of carbon nanotube fibers for electrical energy generation, 2D heterostructures and their photoresponse, strain engineering of 2D materials, thermal diodes, self-healing and growing hydrogels, synthetic state machine particles, and others.

As a PhD student, I mastered the synthesis of late transition metal catalysts, the catalytic polymerization of olefins, and "living"/controlled radical polymerization of functional monomers. Combining different polymerization techniques, I achieved a synergistic control of the chain sequences, topologies, and functionalities of ethylene polymers. I synthesized and characterized a library of functional polyethylenes (PEs) hybrids with various topologies including tree-like, hyperbranched star, miktoarm star and cross-linked mesh, and well-designed functionalities like tertiary amines, ethylene glycols, and amino acid groups. The characteristic chain architectures and functionalities endow the polymers with fascinating solution and melt properties and gas-sensitive behavior in aqueous solution, and I intensively characterized these properties with triple-detection gel permeation chromatography, rheometer, thermal analysis, and other techniques. I further demonstrated the new applications of these functional PEs as efficient supports for metal nanoparticles and organocatalysts to achieve a highly active and selective catalysis with good recyclability.

Teaching Interests:

Personally, I believe that one of the greatest pleasures of being a professor is teaching and mentoring students. During my Ph.D. and postdoc, I have mentored more than 17 students from different nations and cultural backgrounds with fruitful achievements, including 6 undergraduates, which have received awards and honors such as the "Excellent Graduation Thesis of Zhejiang University", and the "National University Student Innovation Program" of China. In addition, one student's work was highlighted by the newsletter of Material Processing Center of MIT. I have had valuable experience in teaching various courses in the fields of polymer science and reaction engineering, and I have been invited as a lecture in the Nano-Engineered Composite Aerospace Structures Consortium of MIT. With an interdisciplinary background, I can teach different curriculums in the field of polymer science, nanotechnology, nanocomposites, and catalysis and would like to develop new interdisciplinary courses and international teaching programs in the future.

Future Directions:

In future, I want to build an interdisciplinary research group in the fields of (I) *lightweight and high-strength materials for protective and space applications and* (II) *active and intelligent materials for catalytic and sensing applications.* In particular, I plan to work on: (1) The precise synthesis of large-area, high quality 2D materials and their molecular composites and nanocomposites via wet-chemistry methods. (2) The fundamental study of the reinforcing mechanism of these composites in aspects of mechanical, optical, electrical, and catalytic properties. (3) The topology design and optimization of the 3D composite body and the development of new methods and processes to realize the dimension transformation from 2D to 3D. An ultimate goal is to achieve a scalable production of 3D composite bodies with outstanding mechanical performance, chemical stability, and an integration of multiple functions such as energy generation and storage, sensing, memorizing, communicating, actuating, catalysis, and others. This composite can be a microparticle that can perform as a smart and autonomous device for biomedical and Internet of Things applications.

Selected Publications:

- 19 peer-reviewed journal articles (11 first-authored papers), 3 patents, 1 book chapter.

(1) **Liu, P.;** Jin, Z.; Katsukis, G.; Drahushuk, L. W.; Shimizu, S.; Shih, C.-J.; Wetzel, E. D.; Taggart-Scarff, J. K.; Qing, B.; Van Vliet, K. J., Layered and scrolled nanocomposites with aligned semi-infinite graphene inclusions at the platelet limit. *Science* **2016**, 353, 364.

(2) **Liu, P.**; Liu, A. T.; Kozawa, D.; Dong, J.; Saccone, M.; Koman, V. B.; Wang, S.; Son, Y.; Wong, M. H.; Strano, M. S. Autoperforation of 2D Materials for Generating Two Terminal Memresistive Janus Particles. Submitted to *Nat. Mater.*, out for review.

(3) **Liu, P.**; Cottrill, A. L.; Kozawa, D.; Koman, V. B.; Kaplan, Liu, A. T.; Tran, T.; Wong, M. H.; Strano, M. S. Emerging trends in 2D nanotechnology that are redefining our understanding of Nanocomposites. *Nano Today, invited review, prepared.*

(4) **Liu, P.;** Strano, M. S., Toward Ambient Armor: Can New Materials Change Longstanding Concepts of Projectile Protection? *Adv. Funct. Mater.* **2016**, 26, 943.

(5) Wong, M. H.; Giraldo, J. P.; Kwak, S.-Y.; Koman, V. B.; Sinclair, R.; Lew, T. T. S.; Bisker, G.; **Liu, P.**; Strano, M. S., Nitroaromatic detection and infrared communication from wild-type plants using plant nanobionics. *Nat. Mater.* **2016**, 16, 264.

(6) Liu, A. T.; Kunai, Y.; **Liu, P.**; Kaplan, A.; Cottrill, A. L.; Smith-Dell, J. S.; Strano, M. S. Electrical Energy Gene-ration via Reversible Chemical Doping on Carbon Nanotube Fibers. *Adv. Mater.* **2016**, *28*, 9752.

(7) Son, Y.; Li, M.-Y.; Cheng, C.-C.; Wei, K.-H.; **Liu, P.**; Wang, Q. H.; Li, L.-J.; Strano, M. S., Observation of switchable photoresponse of a monolayer WSe2–MoS2 lateral heterostructure via photocurrent spectral atomic force microscopic imaging. *Nano Lett.* **2016**, 16, *3571.*

(8) Wang, S.; **Liu, P.**; Wang, W.-J.; Zhang, Z.; Li, B.-G., Hyperbranched polyethylene-supported L-proline: a highly selective and recyclable organocatalyst for asymmetric aldol reactions. *Catal. Sci. Technol.* **2015**, 5, 3798.

(9) **Liu, P.**; Lu, W.; Wang, W.-J.; Li, B.-G.; Zhu, S., Highly CO_2/N_2-switchable zwitterionic surfactant for Pickering emulsions at ambient temperature. *Langmuir* **2014**, 30, 10248.

(10) **Liu, P.**; Dong, Z.; Ye, Z.; Wang, W.-J.; Li, B.-G., A conveniently synthesized polyethylene gel encapsulating palladium nanoparticles as a reusable high-performance catalyst for Heck and Suzuki coupling reactions. *J. Mater. Chem. A* **2013**, 1, 15469.

(11) **Liu, P.**; Ye, Z.; Wang, W.-J.; Li, B.-G., Hyperbranched Polyethylenes Encapsulating Self-Supported

Palladium (II) Species as Efficient and Recyclable Catalysts for Heck Reaction. *Macromolecules* **2012**, 46, 72.

(12) **Liu, P.**; Lu, W.; Wang, W.-J.; Li, B.-G.; Ye, Z.; Zhu, S., Synthesis and Characterization of PE-b-POEGMA Copolymers Prepared by Linear/Hyperbranched Telechelic Polyethylene-Initiated ATRP of Oligo(ethylene glycol) Methacrylates. In *Progress in controlled radical polymerization: mechanisms and techniques, American Chemical Society:* **2012**; Vol. 4, 39.

(13) **Liu, P.**; Landry, E.; Ye, Z.; Joly, H.; Wang, W.-J.; Li, B.-G., "Arm-first" synthesis of core-cross-linked multiarm star polyethylenes by coupling palladium-catalyzed ethylene "living" polymerization with atom-transfer radical polymerization. *Macromolecules* **2011**, 44, 4125.

510702 Functional 2D Material Heterostructures and Bio-Interfacing for Sustainable Energy Generation

Abstract:

Research Background:

The quantum confinement of charge carriers in two-dimensional nanomaterials (2DNs) evolve new and in several cases superior electronic and optical properties with applicability in solar energy conversion, quantum transport-based nanoelectronics, and bio-nano interfacing. The challenge is how to control the nucleation of these 2DNs and their quantum heterostructures for engineered properties. Owing to the Van Hove singularity in the electronic density of states, the 2D semiconductors exhibit strong light-matter interactions and can be coupled with bulk semiconductors or biological entities to create sustainable energy generations. For the past 8 years of my doctoral and post-doctoral training and research assistant professorship positions, I have built a solid foundation by addressing these challenging questions.

My graduate research was a result of research exposure from three countries with two scholarships including Commonwealth scholarship for doctoral exchange research at The University of Saskatchewan, Canada and second one is a Department of Science and Technology sponsored travel grant for visiting Technical University of Denmark for Ph.D. summer school on challenges in 2D materials. In the group of Prof. Qiaoqin Yang at The University of Saskatchewan, Canada, I primarily focused on developing strategies for synthesis of vertical graphene films with enhanced electron field emission characteristics. I have developed a novel strategy to produce high aspect ratio silicon nanowires directly on SiO_2/Si surfaces via oxidation and reduction in hydrogen environment. These nanowires exhibited quantum confinement effect as confirmed by Raman spectroscopic analysis.

My postdoctoral research with Prof. Vikas Berry at The University of Illinois at Chicago (UIC) involved nanoscale designing and bottom-up synthesis of 2D material heterostructures for low-temperature quantum transport and optoelectronics. Currently the quantum heterostructures are assembled by transferring the 2D materials produced on transition metal catalytic surfaces. These polymer-assisted transfer approaches introduce scattering centers for charge carrier transport and recombination centers for photovoltaic solar cells. To address this challenge, I have developed novel surface-chemical-interaction chemistry to nucleate graphene, h-BN and their heterostructures directly on oxide and nitride-based gate-dielectric substrates and investigated low-temperature quantum transport phenomena. I have developed mechanistic understanding of epitaxial graphene formation with ultra-high Raman I_{2D}/I_G ratio on Co (1 1 1) surfaces. I have also designed experiments to comprehend the photovoltaic and quantum efficiency characteristics in advanced 2D layer-interfaced-3D bulk semiconductor photojunctions for efficient photovoltaic sensors.

In my current role as Research Assistant Professor at UIC, I have focused on understanding the structure-property correlations of 2-atom, 3-atom, 4-atom thick 2D materials and interfacial photophysics of 2D-on-3D material junctions. With collaboration with Dr. Anirudha Sumant (Argonne National Laboratory) and Prof. Gajendra Shekhawat (Northwestern University), I am currently investigating two important experimental phenomena on piezoelectricity in 2D semiconductors and role of nanocrystalline diamond as an electron blocking inter-layer in graphene photovoltaic sensors. In the last 11-month period, I as a Co-PI received a grant of US$ 1,500 for a project on 'printing graphene-based solar cells' under UIC-Women in Engineering Summer Program 2017 and as a PI submitted several proposals including: (i) Samsung Global Research Outreach Program (US$ 100,062) and (ii) Centre for Nanoscale Materials, Argonne National Laboratory (User Research Proposal). I have also organized two graduate research and career symposiums in UIC Department of Chemical Engineering during fall of 2016 and 2017 and reviewed about 40 papers.

Research Interests:

With a solid foundation in the physics of two-dimensional (2D) materials for opto-plasmonic-based heterojunction-photovoltaics, low-temperature quantum transport, molecular-surface-functionalization, and bio-nano interfacing, research projects with high possible throughput are being targeted. My goal is to establish an independent research program that aims in developing four major projects: (A) New 2D Materials and Quantum Heterostructures Development: mechanistic understanding of growth chemistry for 1-atom, 2-atom, 3-atom, 4-atom, 5-atom thick inorganic 2D materials and complex heterostructure circuits and their fundamental structure-property correlations, (B) Energy Harvesting and Quantum Electronics: 2D materials-based emerging plasmonic-heterojunction-photovoltaics and low-temperature quantum transport in nano(opto)electronic circuits, (C) Materials Chemistry: non-covalent and vapor-phase surface functionalization and doping/bandgap engineering of the 2D materials to control the electronic structure towards efficient device functionality; and (D) Bio-Nanotechnology: quantum mechanical interfacing of 2D materials with biological entities (such as: bacteria, cells etc.) for functional bio-interfaces and new phenomena.

Teaching Interests:

The broad prospective of my research interests are further shaped by my recent teaching experiences. At UIC, I have developed an advanced nanoscience course 'Graphene and Two-Dimensional (2D) Materials (ChE 494)' from Spring 2017 and also teaching in Fall 2017. In this course, I have employed both knowledge-centered (lectures about 2D concepts) and learning-centered (2D laboratory, interactions or collaborations in a mini-2D-symposium) techniques to ensure long-term retention of fundamental concepts. In both Spring and Fall 2017, total of 40 graduate and undergraduate students are trained on emerging 2D materials theory, processing and applications. The learning goals are: 2D materials science, synthesis chemistry, nanoscale characterizations, and advanced device functionality. This course also includes laboratory demonstrations: bottom-up synthesis of graphene, its chemical transfer, and Raman spectroscopic characterizations. The student evaluation is based on conceptual quizzes, homework problems, writing a review article, and a research/project presentations. To inculcate idea of collaborations among students, I have started organizing a mini-symposium, where students have to present and interact with others fellow students. For Spring 2017, the overall teaching rating for this course is 5.00/5.00, which motivated me to further enrich the course program. For Spring 2018, I am selected to teach 'Science and Technology of Graphene and 2D Materials (Hon 201)' for UIC Honors College and also developing a new course 'Photovoltaic Science and Engineering (ChE 494)' for Spring 2018 in UIC Department of Chemical Engineering. Further as part of my outreach activity, I have delivered 8 invited lectures and 15 contributed talks at various institutions/conferences in USA, Denmark, Canada, and India. I recently received an invitation to present our '2D heterostructure nano-optoelectronics' work at 2017 MRS Fall Meeting, Boston in the 'Symposium NM02-Anisotropic Carbon Nanomaterials-Frontiers in Basic and Applied Research'.

Future Direction:

As an independent investigator, I would like to continue the idea of using experimental techniques and theoretical analysis to understand the structure-property correlations and diverse functionalities of 2D materials and their nanoscale attributes. In this regard, I believe the functional 2D material system I am currently focusing provides a rich experimental toolbox, with a number of fascinating and currently unexplored phenomena to reconnoiter. Potential and immediate future plan includes: (i) mechanistic understanding of charge transport at the interface of biological entities and 2D materials for bio-photovoltaics; (ii) Interfacing of 2D layers with perovskites for novel photojunctions; (iii) origami and kirigami of 2D nanocomposites and device functionality.

The distinctiveness of my methodology is the ability to develop complex 2D heterostructure circuits, characterize their structure-property correlations, and design novel energy junctions via nano-bio interfacing. Due to the highly interdisciplinary nature of my work, I envision a strong and active collaboration with fellow faculty in the university to setup specific techniques in my lab.

Selected Publications (32 total, 20 first/corresponding author, 220 citations, and 9 H-index):

S. K. Behura, P. Nguyen, R. Debbarma, S. Che, M. R. Seacrist, and V. Berry, "Chemical interaction-guided, metal-free growth of large-area hexagonal boron nitride on silicon-based substrates," ACS Nano, Vol. 11, p. 4985-4994 (2017). Highlighted in Nature India.

S. K. Behura, P. Nguyen, S. Che, R. Debbarma, and V. Berry, "Large-area, transfer-free oxide-assisted synthesis of hexagonal boron nitride films and their heterostructures with MoS2 and WS2," Journal of the American Chemical Society, Vol. 137, p. 13060-13065 (2015).

S. K. Behura and V. Berry, "Interfacial nondegenerate doping of MoS2 and other two-dimensional semiconductors," ACS Nano, Vol. 9, p. 2227-2230 (2015).

S. K. Behura*, S. Nayak, I. Mukhopadhyay, and O. Jani, "Junction characteristics of chemically-derived graphene/p-Si heterojunction solar cell," Carbon, Vol. 67, p. 766-774 (2014).
S. Che, K. Jasuja, S. K. Behura, P. Nguyen, T. S. Sreeprasad, and V. Berry, "Retained carrier-mobility and enhanced plasmonic-photovoltaics of graphene via ring-centered η6 functionalization and nanointerfacing," Nano Letters, Vol. 17, p. 4381-4389 (2017). Highlighted in Science News.

S. Deng, E. Gao, Y. Wang, S. Sen, T. S. Sreeprasad, S. K. Behura, P. Král, Z. Xu, and V. Berry, "Confined, Oriented and Electrically Anisotropic Graphene Wrinkles on Bacteria," ACS Nano, Vol. 10, p. 8403-8412 (2016). Highlighted in The Economist.

Selected Patents (2 filed, 1 under processing of filing, 6 invention disclosures):

V. Berry, S. K. Behura, P. Nguyen, and M. R. Seacrist, "Epitaxial growth of defect-free, wafer-scale single-layer graphene on thin films of cobalt," US Patent WO2017058928 A1, Priority Date: Oct. 1, 2015, Publication Date: Apr 6, 2017.

V. Berry, S. K. Behura, P. Nguyen, and M. R. Seacrist, "Direct formation of hexagonal boron nitride on Si-based dielectrics," US Patent Serial Number: 62/335, 149, Filing Date: May 2016.

510707 Techno-Economic and Life Cycle Analysis of the Renewable Energy Conversion Pathways

Abstract:

Research Interests:

Sustainable energy has a promising future to replace petroleum to power the world as well as providing a cleaner environment. Transdisciplinary knowledges and perspectives combining engineering, statistics, economics and environments are crucial to push the new renewable energy technologies to the real world. My research experiences focused on evaluating new energy production technologies using a combination method of process modeling simulation, statistics regressions, techno-economic analysis and life cycle analysis. I designed and modeled several novel chemical processes to produce carbon-negative energy and chemicals, involving both biological and thermochemical conversion processes, such as fast pyrolysis & hydroprocessing, solvent liquefaction, fermentation, anaerobic digestion etc. Based on the chemical process models built in Aspen Plus, I further conducted economic models to evaluate the estimated total investments, payback period, break-even cost etc., to comprehensively evaluate the economic feasibility of the new energy production process. Besides the economic analysis, I also assessed how the technology might interact with air, water, and human health, quantifying the environmental benefits of different sustainable energy production technologies through life cycle analysis. I also applied statistics methodology to building regression models, further investigating correlations among process related inputs and economic/environmental related outputs, as well, evaluating the sensitivity and uncertainty of various parameters.

Modeling the world to better understand and improve the real world is the fundamental concept of my research. My future interests will be combining the global economic markets with the potential sustainable energy process, to conduct deeper analysis of the mutual correlation between the real economic markets and the engineering technology. As well, I would like to build models to evaluate the global GHG emissions reduction potentials, helping accelerate the adoption of different renewable energy technologies.

Publications

- **Wenqin Li**, Dang, Q., Smith, R., Brown, R. C. & Wright, M. M. Techno-economic analysis

of the stabilization of bio-oil fractions for insertion into petroleum refineries. ACS Sustain.

Chem. Eng. (2017). (Special Issue Paper)

- **Wenqin Li**, Qi Dang, Robert C. Brown, David Laird, M. M. W. The impacts of biomass

properties on the pyrolysis yields, economic and environmental performance of the pyrolysis-

bioenergy-biochar platform to carbon negative energy. Bioresource Technology. (2017).

(Front Cover Paper)

- **Wenqin Li**, Denis Bbosa, Arpa Ghosh, Robert C. Brown, M. M. W. Techno-economic analysis of

ethanol production from solvent liquefaction of lignocellulosic biomass. (Submitted)

Teaching Interests:

I have co-instructed a mechanical engineering design class (Introduction to Mechanical Engineering Design) in fall 2013. I got a valuable chance to closely work with engineering undergraduates, helping them developing their brilliant ideas into real mechanical products, with a goal of improving peopleâ€™s lives in developing countries of Africa. I also participated in organizing an Expo for all 150 students at the end, connecting the students with industrial leaders, professors, graduate students and their undergraduate peers. I was very proud that one team from my class got the first prize in the Expo competition. My philosophy of teaching is using different projects and activities to help students better understand basic theoretical concepts/principles. As well, connecting the theoretical world with practical applications or experiences will motive studentsâ€™ interests as well as helping them accumulate real problem-solving skills.

511297 Synthesis and Characterization of Novel Hierarchical Porous Materials with Functional Properties

Abstract:

Research Interests:

Porous materials are essential components for many applications because they must facilitate mass transfer, provide active surfaces for chemical reactions, and conduct electrons and heat. Additionally, they must maintain long-term stability under operating conditions, resistance under mechanical stresses, and low-weight. Novel energy storage and conversion devices use porous materials, which drastically affect the performance. In fuel cells, porous carbon substrates are responsible for accurately controlling the complex two-phase mass transfer phenomenon through diffusion layers and facilitating triple-phase-boundaries in the catalyst layers. In redox flow batteries, porous electrodes are responsible for liquid electrolyte distribution and provide active surfaces for electrochemical reactions. Solid-state diffusion of lithium is a major performance limitation in lithium-ion batteries. The lack of sophisticated control over the microstructure and surface properties limits the performance of state-of-art materials. Additionally, the interfacial processes, such as wetting, have been largely overlooked.

The rational design of porous materials requires an understanding of structure-function relationships at multiple scales. During my future faculty position, I intend to fabricate advanced porous materials for a variety of applications relevant to chemical engineering and electrochemistry. I anticipate using simulations, such as Lattice Boltzmann and pore network modeling, to inform the experimental design of these novel materials. In this presentation, I will first briefly overview my past and current research. Then, I will discuss the future research directions that I plan to pursue, mainly: (1) synthesis of engineered porous materials with different degrees of hierarchization; (2) modification of surfaces using thin films to impart functional properties, such as wetting, electrochemical activity, or stability; (3) development of advanced characterization platforms and techniques to visualize and investigate porous materials and devices.

Research experience

PhD Thesis Project: Novel gas diffusion layers with patterned wettability for advanced water management strategies in polymer electrolyte fuel cells. *Advisor:* Prof. Thomas J. Schmidt (Department of Chemistry and Applied Biosciences, ETH Zurich)

(e-collection.library.ethz.ch/eserv/eth:50310/eth-50310-02.pdf).

My PhD research at ETH Zurich/Paul Scherrer Institute focused on the development of novel porous materials for fuel cell applications. During this time, I developed substantial expertise in materials science of porous media, microfluidics, imaging methods and electrochemical engineering. I invented a method, based on radiation grafting, to synthesize gas diffusion layers with patterned wettability. Monte Carlo simulations were used to theoretically predict the electron dose distribution throughout the porous materials, leading to notable improvements in resolution. The choice of monomer and reaction conditions allowed for the precise regulation of the hydrophilicity. Correlations between material microstructure, surface chemical composition, and wetting properties were elucidated, which provided a fundamental understanding about wetting in complex systems and capillary pressure transport. Additionally, we used a combination of advanced imaging techniques (e.g. neutron radiography and x-ray tomographic microscopy) and electrochemical diagnostics to understand the limiting performance factors and to inform the design of optimized materials. The modified diffusion layers with patterned wettability have improved the water management inside operando cells, thereby significantly increasing the fuel cell power density. The technology was patented and has attracted significant attention from industry. Presently, this technology is heading towards commercialization.

Postdoc Project: Development of advanced porous electrodes for non-aqueous redox flow batteries. *Advisor:* Prof. Fikile Brushett (Department of Chemical Engineering, MIT)

My ongoing postdoctoral work at the Massachusetts Institute of Technology focuses on the development of advanced porous electrodes for redox flow batteries. During this time, I investigated the effect of the electrode microstructure on flow battery performance. To deconvolute the role of electrode properties performance, specifically its impact on kinetics, ohmics and mass transport, we used platform redox couples, with well-defined electrochemical properties, in tandem with diagnostic flow cell configurations which enable individual electrode characterization in isolation, but at near, practical operating conditions. A key finding was that hierarchically organized microstructures, featuring a bimodal pore size distribution, greatly improve flow battery performance, informing the design of novel microstructures. To accurately understand transport within these highly anisotropic carbon-fiber electrodes, we are using X-ray tomographic microscopy to obtain detailed microstructural information and running computational fluid dynamics simulations. Furthermore, I am synthesizing metallic porous electrodes with a well-defined microstructure and composition using the dealloying method. Additionally, a variety of polymer thin films have been electrografted onto the surfaces of porous electrodes to impart desired properties, such as specific wettability, increased surface area or electrochemical activity. During this period, I am developing substantial expertise in electrode fabrication, energy storage, surface modifications through electrochemical methods, organic synthesis and numerical simulations.

Awards and Honors

- ETH Medal for Outstanding PhD Thesis, ETH Zurich, 2017
- Energy Technology Graduate Student Award, The Electrochemical Society, 2017
- Early Postdoc Mobility, Swiss National Science Foundation, 2016
- Chemical Engineering Valedictorian, University of Alicante, 2013

Selected Publications

1. **Forner-Cuenca** & F. Brushett, "Carbonaceous porous electrodes for non-aqueous redox flow batteries: understanding the role of microstructure", *in preparation*.
2. **Forner-Cuenca**, et al. (2017), "Mask-assisted electron radiation grafting for localized through-volume modification of porous substrates", ***Radiation Physics and Chemistry***, 135, 133-141.
3. **Forner-Cuenca**, et al. (2016). "Advanced water management in PEFCs: diffusion layers with patterned wettability. Part III: In-situ PEFC characterization with neutron imaging", ***Journal of The Electrochemical Society***, 163(13), F1389-F1398.
4. **Forner-Cuenca**, et al. (2016), "Advanced water management in PEFCs: diffusion layers with patterned wettability. Part II: Measurement of capillary pressure characteristic with neutron and synchrotron imaging", ***Journal of The Electrochemical Society***, 163(9), F1038-F1048.
5. Conder, **A. Forner-Cuenca**, et al. (2016), "Performance-enhancing asymmetric separator for lithium-sulfur batteries", ***ACS Applied Materials & Interfaces***, 8(29), 18822-18831.
6. **Forner-Cuenca**, et al (2015), "Engineered water highways in fuel cells: radiation grafting of gas diffusion layers", ***Advanced Materials***, 27, 6317-6322.

Teaching Interests:

Transport phenomena, Surface Science, Fluid Mechanics, Energy Conversion and Storage, Porous Media, Electrochemistry

511434 Chemically-Modified Biomolecules & Nanosystems to Sense & Modulate Biology

Abstract:

Research Interests:

My research focuses on engineering new light-responsive tools to control biological processes and on developing nanotechnology- based platform to deliver large biomolecules for diagnostics and therapeutics. During my graduate research in Prof. Simon H. Friedman's lab at UMKC, I developed a set of photochemical tools to achieve light activated RNA interference by modifying the ends of siRNA with photocleavable groups and applied the technology to pattern gene expression in cells using light. Later, I created a light activated insulin depot that can release insulin on demand for type I diabetics and reduce painful injections by developing a photocleavable crosslinker containing clickable functionality to link insulin and resin via photocleavable groups. As a part of my postdoctoral research in Prof. Sangeeta Bhatia's lab at MIT, I have been interested in advancing the light-controllable systems for in vivo applications and in developing nanoparticle systems for the delivery of peptides, proteins, and nucleic acids. I developed a light-activatable CRISPR/Cas9 system using photocleavable oligonucleotide complements of sgRNAs and applied them to control genome editing of multiple genes concurrently using light with improved specificity. In another project, we developed a photoresponsive protease nanosensors carrying synthetic peptide substrates for more reliable detection of secreted protease activity in the tumor microenvironment in a mouse model of colorectal cancer in the urine. More recently, we have successfully developed an efficient targeted nanoparticle approach to deliver CRISPR/Cas9 in vivo. Given my experience with the delivery and modification of nucleic acids and proteins, including CRISPR/Cas9, I am interested in engineering new genome editing tools with higher specificity and efficiency as well as in developing delivery technologies using biochemical approaches. I would like to then apply these tools for a range of applications including drug delivery, diagnostics, and treatment of genetic diseases including cancer. My long-term goal is to translate these systems into patients for better disease diagnosis and treatment.

Teaching Interests:

With my training and experience in chemistry and biomedical engineering, I will be enthusiastic to teach following core courses:

Transport phenomena
Chemical reaction engineering
Polymers materials
Materials chemistry

In addition, I will be passionate to teach as well as develop the following electives and graduate courses:

Biomaterials
Biochemical, molecular, and cellular engineering
Nanotechnology
Genetic engineering

Courses taught during my career as trainee:
Medicinal chemistry I (as a guest instructor for Pharm D. and undergrad students as a part of preparing future faculty fellowship)
Medicinal chemistry I & II (as a teaching assistant for Pharm D. and undergrad students)
Organic chemistry II (as an official peer tutor for undergraduates appointed by Prof. Kathleen Kilway, UMKC)
Chemistry, Physics, & Mathematics (as a professional tutor for high school students for 6 years)
Mentoring:
Hired, supervised, and mentored 3 undergraduates and 1 summer student at MIT and all of them are pursuing higher degrees in the STEM.
Hired, trained, and mentored 2 graduate students at UMKC on various techniques.
Teaching certificates received:
Kaufman Teaching Certificate Program, MIT (June 2016)
Graduate Certificate College Teaching & Career Preparation, UMKC (May 2013)
Teaching Fellowship
Preparing Future Faculty Fellowship, UMKC (2008-2011)
o Awarded to only 5 PhD students in the University for three consecutive years. Learned and implemented innovative ways to assess students' success by applying Bloom's Taxonomy. The fellowship provided me with a unique opportunity to teach a course on medicinal chemistry to a class of over 100 students spread over two campuses as a guest instructor teaching four review classes. Received excellent feedbacks from students and scored 95% positive comments as a teacher.
Grantsmanship
Awarded competitive personal fellowships
Dissertation Research Fellowship, UMKC (2011-2012)
Preparing Future Faculty Fellowship (PFF), UMKC (2008-2011)
Contribution to Funded Awards & Grants Reporting
NSF award 1052871, (2011-2015)
DARPA Cooperative Agreement HR0011-13-2-0017 (2013-2016)
NIH DP3 NIDDK award 1DP3DK106921-01 (2015-2019)
Education:
Interdisciplinary PhD in Pharmaceutical Sciences and Chemistry, 2013, University of Missouri-Kansas City, USA
Bachelor of Pharmacy, 2006, Dr. H.S. Gour University, India
Research/Work Experience:
• Massachusetts Institute of Technology (MIT), Institute for Medical Engineering and Science, 2013-Present
o Postdoctoral Associate, 2013-2017, Advisor: Prof. Sangeeta N. Bhatia, MD, PhD
• University of Missouri-Kansas City (UMKC), 2006-2013
o Graduate Research Assistant, 2006-2013, Advisor: Prof. Simon H. Friedman,PhD
o Graduate Teaching Assistant, 2007-2008, Medicinal Chemistry I & II
o Guest Instructor, Fall 2009, Medicinal Chemistry I (as a part of the Preparing Future Faculty Fellowship)
Google Scholar link
https://scholar.google.com/citations?user=qtpo58sAAAAJ&hl=en&oi=ao
Publications (8 published, 1 manuscript under revision, 3 manuscripts under preparation):
1. Jain PK, Ramanan V, Schepers AG, Dalvie N, Panda A, Fleming HE and Bhatia SN. Development of light-activated CRISPR using photocleavable protectors of guide RNAs. Angewandte Chemie International Edition, 55(40), 1521-3773 (2016).
• Highlighted on the 100+ scientific media and Altmetrics ranked it #32 over 20,000+ articles ever published by the journal.
2. BR Sarode, PK Jain, and SH Friedman. Polymerizing Insulin with Photocleavable Linkers to Make Light-Sensitive Macropolymer Depot Materials, Macromolecular Bioscience, 16 (8), 1138-1146 (2016).
• Highlighted on the back cover of the journal.
3. Dudani JS, Jain PK, Kwong GA, Stevens KR and Bhatia SN. Photoactivated Spatiotemporally-Responsive Nanosensors of in Vivo Protease Activity. ACS Nano, 9(12), 11708-11717 (2015).
• Highlighted as an important approach for Moonshot Initiative by Prof. Paula Hammond at MIT in an editorial.
4. Kala A, Jain PK, Karunakaran D, Shah S and Friedman SH. The synthesis of tetra-modified RNA for the

multidimensional control of gene expression via light-activated RNA interference. Nature protocols, 9(1), 11-20 (2014).
5. Kala A, Jain PK and Friedman SH. Patterning of cells through patterning of biology. Molecular BioSystems, 10, 1689-1692 (2014).
6. Jain PK, Karunakaran D and Friedman SH. Construction of a Photoactivated Insulin Depot. Angewandte Chemie International Edition, 52, 1404-1409 (2013).
• Highlighted on the frontispiece and covered by numerous scientific media and local news.
7. Jain PK, Shah S and Friedman SH. Patterning of gene expression using new photolabile groups applied to light activated RNAi.
Journal of the American Chemical Society, 133(3), 440-446 (2011).
8. Shah S, Jain PK, Kala A, Karunakaran D and Friedman SH. Light-activated RNA interference using double-stranded siRNA precursors modified using a remarkable regiospecificity of diazo-based photolabile groups. Nucleic Acids Research, 37(13), 4508-4517 (2009).
• Highlighted on the front cover of the journal.

Patents (2 filed patents and multiple disclosures)

US Patent App. 14/410,606, 2013, "Photocleavable drug conjugates." Friedman SH, Jain PK, Karunakaran D and Sarode B (2013).

U.S. Provisional Application No.: 62/332,096, "Methods and uses for remotely triggered protease activity measurements." Bhatia SN, Kwong GA, Jain PK, Dudani JS and Schurle S (2016).

511494 Modeling of Polymer Material Processing from Molecular Basis

Abstract:

Research Interests:

My research group will concentrate on linking different modeling techniques and tools for hierarchical polymer material design. Polymeric materials have been empowering industrial growth for decades. Polymers are present in everyday consumer products, and are rapidly seizing new roles in bio-medicine and hi-tech applications, like drug delivery and computer chip fabrication. However, our understanding of the fundamental relations between polymer chemical composition and the properties of the final product is rudimentary.
The emphasis of my research plan will be on the development of comprehensive connections between different levels of description for dynamics predictions. My group will employ common tools for molecular modeling, such as atomistic and molecular dynamics simulations, mesoscopic/coarse-grained models, dynamic/kinetic Monte-Carlo, analytical theory, and computational fluid dynamics. However, transparent connections will allow to easily switch to a coarse-grained representation of the system, thus facilitating accessible prediction built upon detailed chemical structure. Building upon my experience with rigorous coarse-graining and polymer modeling, with focus on dynamic properties, the group will create a hierarchical framework, where the most detailed members explicitly contain full information about chemistry, while the least detailed members are suitable for macroscopic simulations of final products for direct manufacturing optimization. Currently I propose to address challenges in the following three research areas:
â€¢ Multi-level polyelectrolyte complexation modeling
â€¢ Coarse-grained model for dynamics of topological constraints
â€¢ Non-equilibrium processes in polymer manufacturing with molecular models and computational fluid dynamics

Teaching Interests:

Rapidly growth computational research increases role of programming for engineering, however there are no classes in a chemical engineering curriculum that cover it thoughtfully. The absence of such a class is a frequent complaint among PhD students. To address this problem, I would like to apply extensive experience with high-performance computations to develop High-Performance Computing in a Molecular Modeling class. In this class, I will cover common computational methods for molecular modeling, like coarse-grained molecular dynamics, atomistic simulations and self-consistent filed theory. Specific algorithms and their implementations on GPU and other parallel environments will be discussed. The course will also provide general skills for a large-scale software development and cover modern software optimization techniques. I was involved with chemical engineering departments throughout my career and I am prepared to teach common core classes, like Transport Phenomena, Thermodynamics, Statistical Mechanics, both at graduate and undergraduate levels. In addition, my research was in the field of polymer physics, and can teach a Rheology or Polymer Science class.

Selected Publications:

1. M Andreev, A Chremos, JJ de Pablo and JF Douglas, Coarse-Grained Model of the Dynamics of Electrolyte Solutions, J. Phys. Chem. B, 2017
2. S Srivastava, M Andreev, AE Levi, DJ Goldfeld, J Mao, WT Heller, V Prabhu, JJ de Pablo and MV Tirrell, Gel Phase Formation in Dilute Triblock Copolyelectrolyte Complexes, Nature Communications, 8, 14131, 2017
3. M Andreev, JD Schieber, Accessible and Quantitative Entangled Polymer Rheology Predictions, Suitable for Complex Flow Calculations, Macromolecules, 48 (5), 1606-1613, 2015
4. JD Schieber, M Andreev, Entangled polymer dynamics in equilibrium and flow modeled through slip links, Annual Review of Chemical and Biomolecular Engineering (5), 367-381, 2014
5. M Andreev, RN Khaliullin, RJA Steenbakkers, JD Schieber, Approximations of the discrete slip-link model and their effect on nonlinear rheology predictions, Journal of Rheology, (57), 535-557, 2013

Research statement

Having caught the soft-matter bug as an undergraduate student at the University of Pennsylvania, I developed a particular interest in the ordering and assembly of colloids, and in the elasticity and rheology of amorphous solids — including glasses, emulsions, foams, and colloidal pastes and gels. These two broad topics are incredibly rich and diverse, and lend themselves both to very practical and applied research, as well as to more fundamental work. This balance between applied and fundamental aspects is what I find most appealing about these topics, as whatever work we do in these areas will be guaranteed to be both interesting and relevant: which, in addition to being personally fulfilling, also seems to help attract industrial collaborators and funding. Here I summarize some of my background in these two fields, and describe two ready-to-go projects that students and post-docs in my lab would be able to start working on immediately.

During my Ph.D. I focused on the elasticity and rheology of emulsions and foams[1], and studied the emergence of elasticity and yielding of particle- and surfactant-stabilized emulsions. My proudest scientific accomplishment as a graduate student was the discovery that the liquid-to-solid transition in barely-compressed and quiescent emulsions is controlled by the balance between the elasticity of the compressed solid and the thermal fluctuations of its constituent droplets[2]. The conclusion of this experimental *tour de force* — which combined diffuse-light-scattering microrheology, high-resolution MRI, mechanical rheology, monodisperse particle and emulsion synthesis, and high-performance GPU computing — was that when the osmotic pressure of the emulsion is less than a critical value, $\Pi^* \approx \epsilon \bar{T}^{1/3}$, where $\epsilon = \frac{\sigma}{R}$ is the ratio of interfacial tension to droplet radius and $\bar{T} = \frac{k_B T}{4\pi\sigma R^2}$ is a reduced temperature equal to the ratio thermal to interfacial energies, the thermal agitation of individual droplets is strong enough to yield the solid without any externally applied stress. This work also confirmed that the shear modulus of the emulsion is proportional to its osmotic pressure, which contradicts $\approx$20 years of results from numerical simulation, but is consistent with well over 30 years of experimental work.

Most interestingly, however, this critical confining pressure is extraordinarily large: for droplets with a ~10 μm diameter Π^* ~ $10^5 \cdot \frac{3k_BT}{4\pi R^3}$, which is several orders of magnitude larger than the osmotic pressure required to turn a dispersion of similarly-sized hard particles into a colloidal glass. Moreover, the essential features of emulsion rheology used to derive the expression for Π^* — the fact that shear moduli and yield strains appear to vanish as the osmotic pressure is lifted — are shared by many kinds of pastes and other soft-solids.

[1] Copy of thesis available at https://www.dropbox.com/s/tgywlp7vb0tgzsj/Guerra_thesis.pdf?dl=0 .

[2] This research is covered in chapters 2–4 of thesis, and copies of an in-preparation manuscript and SI are available here https://www.dropbox.com/s/vu7oomgp2plfor7/dws_nmr_g_093017.pdf?dl=0 , and here https://www.dropbox.com/s/sii90hsy5t1po2t/dws_nmr_SI_093017.pdf?dl=0 .

We should thus revisit the critical elastic properties of these ubiquitous materials as well, and I would like to make this one of my primary research directions as PI. During the last year of my Ph.D. I began to develop techniques to study this liquid-to-solid transition and the nature of the strongly-compressed-yet-fluid in emulsions, microgel suspensions, and colloidal glasses using confocal microscopy and particle-displacement correlation measurements: which have yielded encouraging results, and are primed for students and post-docs to adopt and build upon.

The most important practical realization from this part of my Ph.D. work was that it is much easier to modulate the behavior of an amorphous solid by controlling its osmotic pressure instead of its concentration. Filling vertical tubes with colloidal particles or droplets that are slightly denser than the surrounding fluid results in a sedimented pile with an osmotic pressure that starts at zero at the very top of the sediment, and increases nearly linearly with vertical depth. Because the osmotic pressure of any thin slice of material is set by the total weight of the material above it, the aging of any volume element will occur under practically isobaric conditions. Moreover, because the density difference between particle and liquid is intentionally small, small changes in temperature significantly change the particle's buoyant mass, and thus the applied pressure gradient: so we can implement Struik's aging protocol by heating and cooling the system.

Together with Prof. Thomas Kodger, a fellow graduate student at the time, we extended this approach to sediments composed of fluorescent emulsion droplets, microgels, or hard-sphere particles, which we loaded into hermetically sealed glass chambers and imaged using a home-built horizontal spinning-disk confocal microscope. We analyzed time-lapse movies of the 3-D structure of each of these samples taken over days or weeks to detect particle positions and displacements using high-throughput, GPU-accelerated image processing software that I wrote during that time. Using a generalization of Prof. John Crocker's two-point displacement correlation function that I too developed — that projects all pairs of particle displacement onto all admissible spherical-vector harmonics — we converted these microscopic data into a reconstruction of the average displacement field of particles that results from the infinitesimal displacement of a reference particle.

This preliminary work shows that, unlike the smooth, dipole-like displacement field expected for a linear and isotropic continuum, the displacement field of a dense, supercooled liquid displays strong oscillations that correlate with oscillations in the pair-distribution function, $g(r)$. The equivalent measurement for a glassy configuration — measured deeper into the sediment of the same tube — shows a similar structure, but displacements are effectively zero in a ring that roughly corresponds to the first minimum of $g(r)$:

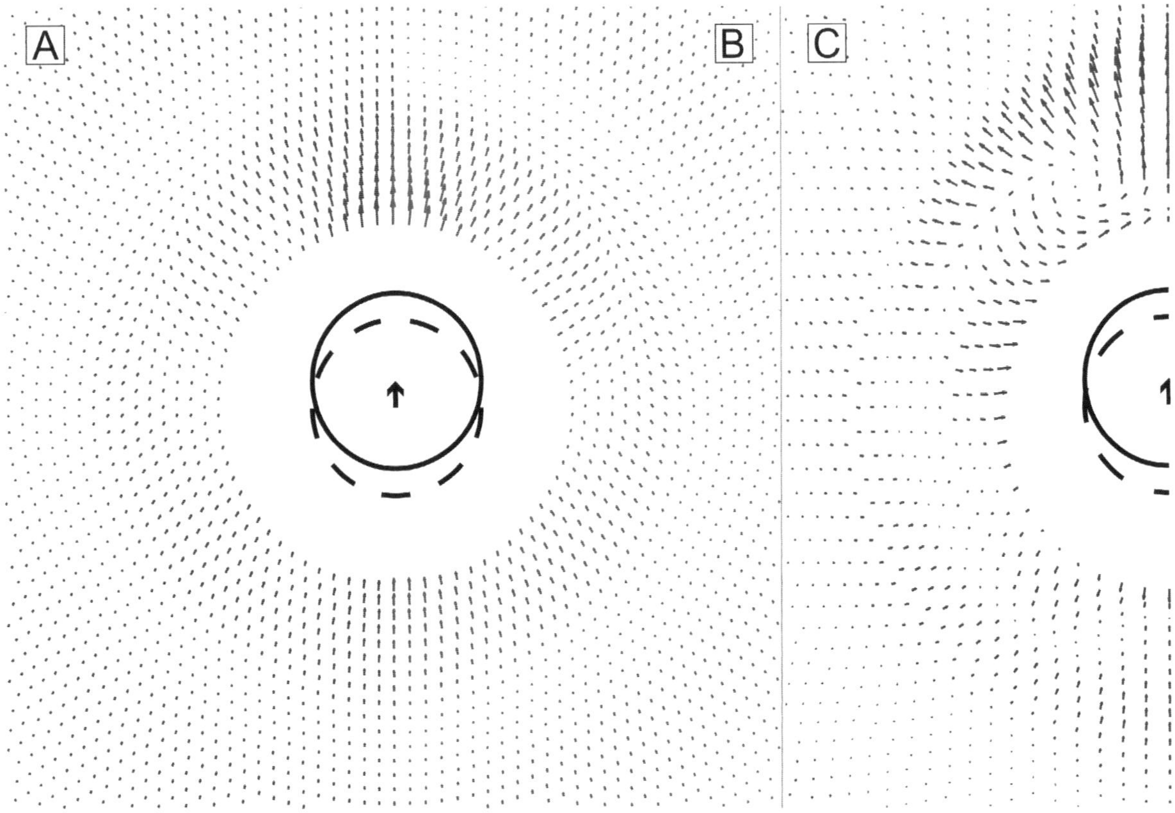

Figure 1. Two-point displacement fields for supercooled liquid and glass Infinitesimal displacement fields computed for a dense binary fluid (A, red) and colloidal glass (B, blue) using my generalize two-point displacement projection method. (C) Rescaled difference map (displacement field of fluid minus displacement field of glass) of the fields presented in (A) and (B) shows that the glass loses a large part of the forward-flux component present in the fluid.

potentially an observation of literal nearest-neighbor caging behavior, (Fig. 1A-B). Dramatic improvements in computational power and recent advances in spinning-disk confocal technology will make it possible to extend these results to displacement-strain and strain-strain maps that will help us further elucidate this transition.

Now as a post-doc in Prof. Chaikin's lab, my primary project has been to study the structure and dynamics of repulsive colloidal particles adsorbed at the surfaces of liquid droplets[3]. When these particles are adsorbed at flat interfaces, increasing their areal density increases the strength of their repulsive interactions, and forces particles to transition from a disordered fluid to a hexatic liquid-crystal, and then to a triangular crystal. The fundamental structural differences between these phases are determined by the concentration and arrangement of topological lattice defects called disclinations: as described by

[3]This work is still under review at Nature, but copies of the manuscript and SI are available here https://www.dropbox.com/s/7w8pj10uclb0cvd/melting_spheres_20160823.pdf?dl=0, and here https://www.dropbox.com/s/khbczup7d4nlwtm/melting_spheres_SI_20170820.pdf?dl=0.

the theory of Kosterlitz, Thouless, Halperin, Nelson, and Young and verified by extensive experiment and numerical simulation. However, the topology of the sphere constrains the net topological charge of the defects to be +12, and its curvature promotes the proliferation of linear defect aggregates, so-called "scars", that relieve the elastic strain imposed by the curvature. We studied how these defects spontaneously segregate into these scars, and developed a new order parameter to quantify the long-range orientational order that is visually apparent in more densely packed spheres, but which is obscured by the excess topological charge. More relevant to the topic of amorphous solids, we also found that the segregation of lattice defects into isolated "seas" was accompanied by a segregation of particle mobility, which we defined using cage-relative positions and a displacement threshold determined from the histogram of particle displacements.

Unlike displacements measured relative to an absolute lab frame, cage-relative positions are measured from the center-of-mass of a particle's nearest-neighbors; the local character of cage-relative displacements thus makes them ideally suited to detect local cage-breaking events common to glasses and sheared yield-stress fluids. For the particles adsorbed at droplet interfaces, the displacement cut-off that separated in-cage rattling behavior from cage-breaking events was obvious from the histogram of cage-relative particle displacements: I expect a similar threshold will be evident in three-dimensional amorphous solids. Coupling this thresholding process to the microscopic displacement fields produced by our generalized two-point displacement correlations will let us distinguish between displacement fields caused by infinitesimal particle displacements, i.e. the linear-response elastic Green's function of the amorphous solid, from the displacement fields associated with cage-breaking and yielding.

The second major research direction I would like to focus on builds on a new niche that Prof. Chaikin let me carve out within his lab: developing a tool that makes it possible to control the organization and assembly of colloidal particles by modulating the salt concentration of the suspending fluid using supercapacitive electrodes immersed in the medium. Using dealloyed gold electrodes I was able to reduce the concentration of ionic impurities in a solvent that contained highly charged poly(styrenesulfonate) (PSS) microgel particles down to sub-nanomolar concentrations (Fig. 2A-B), while also being able to reduce the ion concentration in a colloid composed of oppositely charged particles slowly enough to allow them to form binary crystals (Fig.2C).

I am now exploring ways to move beyond the simple prototype device — including photolithographically defined electrodes, less expensive capacitive electrode materials, and finer-grained spatial control of voltages and salinity. We recently submitted a proposal

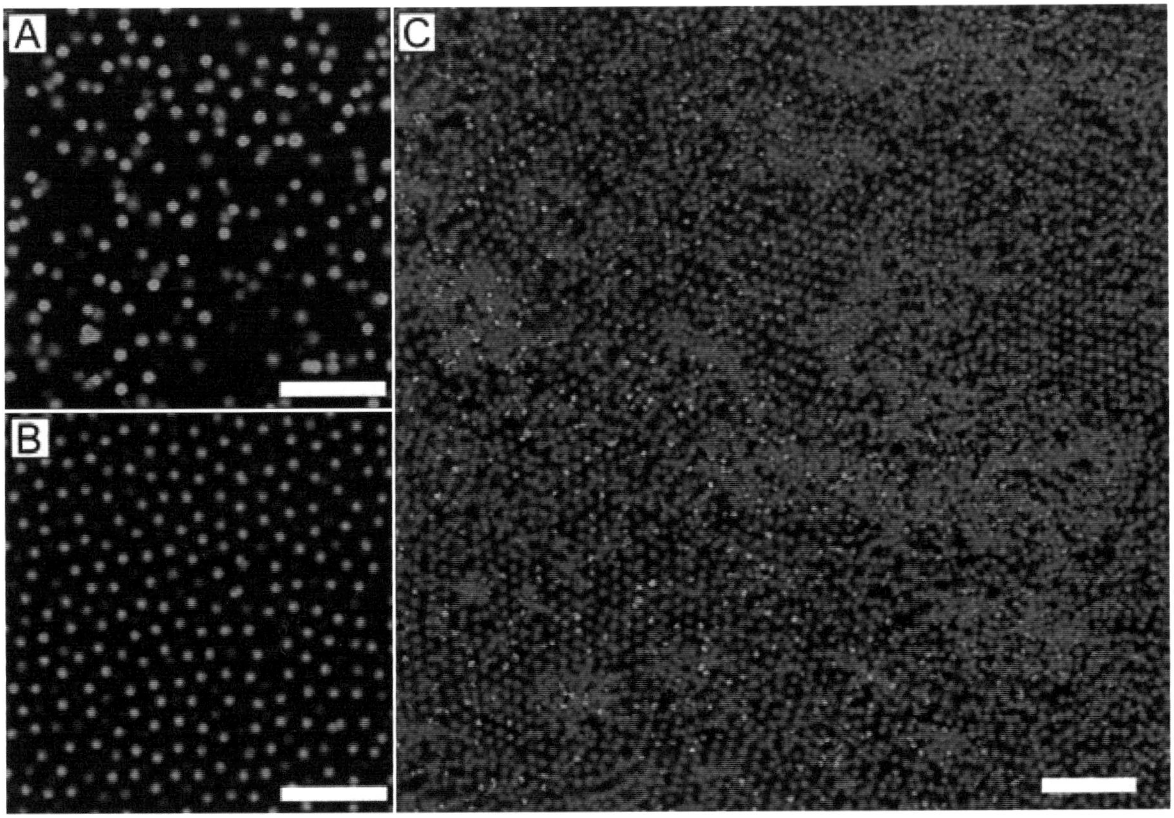

Figure 2. ***in situ* salinity control** Confocal micrographs of PSS microgels dyed by exchanging ordinary counterions for charged fluorescent molecules before (A) and after (B) *in situ* deionization. (C) Dense suspension of oppositely-charged particles in 133 mM LiCl solution, equilibrated at an electrode potential of 25 mV. The slowly increasing attraction between +/- pairs permits reversible bonding and crystallization. Scale bars: 25 μm, 25 μm, and 15 μm respectively

to the NSF's Nanomanufacturing program to support this work[4], where we describe how controlling the long-range electrostatic repulsion between particles can modulate the optical properties of colloidal suspensions: enabling reversible transparency in suspensions of strongly scattering dielectric spheres, and reversible structural coloration in suspensions of silver-coated nanoparticles. We also describe how we will use large-scale interdigitated electrode devices and epitaxial templating to produce gram-scale quantities of binary crystals. We thus expect to reproduce much of the work done with DNA-labeled particles, and extend it to commercially relevant volumes, and to nanoparticles with interesting optical, magnetic, and dielectric properties. Moreover, if this proposal is funded and I secure a position outside of NYU, Prof. Chaikin will arrange a sub-award so that I may take these funds with me, and use them to support a student or post-doc that wants to participate in this project.

[4]A copy of this proposal, titled "Capacitive Control of Colloidal Self-Assembly" and assigned number 1762702, is available at https://www.dropbox.com/s/65ox50ri33nrta0/Deionizer_Proposal_20170922b.pdf?dl=0 .

Teaching statement

Scientists, doctors, or engineers who want to tackle meaningful problems have to be creative and capable of working as part of a team. The classroom and laboratory provide controlled environments were we can help students develop these skills, and I believe that it is our responsibility to make sure that we do so. In my experience as a TA of an advanced undergraduate analytical mechanics class and of a mixed undergraduate-graduate soft matter class, and as a student myself, encouraging or even forcing students to work in groups can help the class cover more advanced material and foster a sense of camaraderie within the working groups: though maybe too often based on a sense of unity against the professorial tyrant.

When properly fostered and supervised, these working groups encourage a form of peer learning that complements, and can even supersede, the instruction provided during the main lecture: for you know a student has really learned something when he or she can explain it to a peer in simple words. However, I believe that it is the responsibility of the teacher to make sure that these groups do not adopt some negative behaviors that can negate their benefits. For example, division of labor economics can push the group to split assignments or cause over-reliance on particularly talented students — at best circumventing the peer learning process. This particular problem can be mitigated by making assignments that cannot be dissected easily and by making them difficult enough that even a very talented student could not complete them alone. However, a more pernicious problem, which I have found to be much rarer, is a kind of negative intra-group dynamic that can border on bullying, which not only hurts the students in the group but is also the death of creative thought.

To be creative is to be fearless. Genuinely good ideas are few and far between, and creative expression requires a sort of safety net, whereby less-than-great ideas are analyzed and discarded without judgment. The instant a member of the team holds back a potentially silly idea for fear of being mocked is the instant creativity dies. It is difficult for a teacher to know when this behavior takes hold, though using TAs — whom the students can sometimes feel more comfortable confiding in — as *de facto* spies can help. Nevertheless, the teacher can set the tone of these interactions by leading by example: taking every idea at face value, analyzing it with the group, and helping the students recognize the strengths and weaknesses of each one.

For smaller classes, the teacher or TA can also monitor the dynamics of the group by turning office hours into a supervised group study session. This approach is still feasible for some larger classes, provided that the number of TAs scales with the number of students. For example, splitting a class of 100 students into 20 groups of 5, which are in turn distributed among 2 TAs that hold two office hour sessions per week, would result in a 5 week rotation for these supervised study sessions: in addition to the weekly lecture-style sections. Including the 2 office hours per week of the teacher into the rotation reduces the length of this cycle to just over 3 weeks, and guarantees that each student has one-on-one time with the head of the class. Moreover, for a class of this size it would also be possible to give students the possibility to opt out of the study group arrangement. For this to work and be fair, this choice would have to be made early on, the problem sets of the individual students would have to be a subset of that given to the groups, and the homework component of the final grade should be greater for those who chose to participate in the groups. Setting the homework component of the final grade at 5 % to 10 % for students who opted out provides little incentive for them to collaborate informally; however, raising this component to 20 % or 25 % for those who do participate may motivate them to work together more effectively, and could help students who do not respond well to high-stress midterm and final exam situations.

In addition to the classroom, this kind of patient, judgment-free environment is especially important for a laboratory setting. First, advisors have an almost unconscionable amount of power over the scientific lives of their students: Any abuse of this power should be identified and dealt with immediately. However, when an advisor analyzes and dissects an idea proposed by a student together with him or her, it helps break down the barriers put up by authority and seniority. Perhaps the most valuable and subversive realization students may arrive at is that the only difference between them and their advisor is experience and age. At this point, no idea is sacred, and everything is up for questioning and revision: which is exactly the kind of attitude every good scientist should have.

Rodrigo Guerra

726 Broadway, Room 864C
New York, NY 10003

October 18, 2017
rodrigo.guerra@gmail.com
(617) 875-1301

Education

- **University of Pennsylvania** — Philadelphia, PA
  BA, Physics and Biochemistry — 2003
- **Harvard University** — Cambridge, MA
  PhD, Physics — 2014
  - Dissertation: Elasticity of Compressed Emulsions

Research Experience

- **University of Pennsylvania** — Philadelphia, PA
  Undergraduate Research Assistant for John Crocker — 2001-2003
- **Schlumberger-Doll Research** — Ridgefield, CT and Cambridge, MA
  Intern for Pabitra Sen — 2004-2006
  - Fluorescence correlation spectroscopy of petroleum and coal asphaltenes
- **Harvard University** — Cambridge, MA
  Graduate Research Assistant for David Weitz — 2003-2014
  - Synthesis of fluorinated surfactants for droplet microfluidics
  - Elasticity of Pickering emulsions
  - Microcellular foaming of pharmaceutical solid solutions
  - Entropic unjamming of surfactant-stabilized emulsions
  - Confocal imaging of jammed emulsions and colloidal glasses
- **New York University** — New York, NY
  Post-doctoral Researcher for Paul Chaikin — 2014-Present
  - Frustrated assembly of colloids on spherical surfaces
  - Spontaneous assembly of hyperuniform structures in periodically shear suspensions
  - Control of colloidal phase behavior by capacitive deionization

Academic Publications and Patents

1. A. B. Andrews, R. E. Guerra, O. C. Mullins, and P. N. Sen, "Diffusivity of Asphaltene Molecules by Fluorescence Correlation Spectroscopy," *J. Phys. Chem. A*, vol. 110, no. 26, pp. 80938097, Jul. 2006.

2. R E. Guerra, K. Ladavac, A. B. Andrews, O. C. Mullins, P. N. Sen, "Diffusivity of coal and petroleum asphaltene monomers by fluorescence correlation spectroscopy," *Fuel*, vol. 86, pp. 20162020, 2007.

3. C. Holtze, R. E. Guerra, J. Agresti, and D. A. Weitz, "Fluorocarbon emulsion stabilizing surfactants," WO2008021123 A121-Feb-2008.

4. K. Ladavac, R. Guerra, D. Kaz, V. N. Manoharan, J. B. Rieger, R. S. Koltzenburg, and D. A. Weitz, "Foams, including microcellular foams, containing colloidal particulates," US20130209520 A115-Aug-2013.

5. K. Ladavac, R. E. Guerra, D. Kaz, V. Manoharan, J. B. Rieger, R. S. Koltzenburg, and D. A. Weitz, "Foams or particles for applications such as drug delivery," US20130202657 A108-Aug-2013.

6. T. E. Kodger, L. H. Cai, R. E. Guerra, A. F. Pegoraro, D. A. Weitz, M. Rubinstein, "Elastomers, methods for their preparation and use," *filed.*

7. L. H. Cai, T. E. Kodger, R. E. Guerra, A. F. Pegoraro, M. Rubinstein, D.A. Weitz, "Soft Poly(dimethylsiloxane) Elastomers from Architecture-Driven Entanglement Free Design," *Advanced Materials*, vol. 27, pp. 5132–5140, 2015.

8. T. E. Kodger, R. E. Guerra, J. Sprakel "Precise colloids with tunable interactions for confocal microscopy," *Scientific Reports* vol. 5:14635, 2015.

9. C. P. Kelleher, A. Wang, G. I. Guerrero-Garca, A. D. Hollingsworth, R. E. Guerra, B. J. Krishnatreya, D. G. Grier, V. N. Manoharan, and P. M. Chaikin, "Charged hydrophobic colloids at an oil–aqueous phase interface", *Physical Review E* 92, 062306, 2015.

10. C. P. Kelleher, R. E. Guerra, A. D. Hollingsworth, and P. M. Chaikin, "Phase Behavior of Charged Colloids at a Fluid Interface", *Physical Review E* 95, 022602, 2017.

11. R. E. Guerra, C. P. Kelleher, A. D. Hollingsworth, and P. M. Chaikin, "Freezing on a Sphere", *Nature, under review.*

12. R. E. Guerra, D. A. Weitz, "Entropic Unjamming of Emulsions", *in preparation*

13. R. E. Guerra, K. Ladavac, D. A. Weitz, "Elasticity of Pickering Emulsions", *in preparation*

14. K. Ladavac, R. E. Guerra, D. M. Kaz, S. Koltzenburg, J. Rieger, U. Hees, D. A. Weitz, V. N. Manoharan, "Particle-packed foams for enhanced dissolution of pharmaceutical actives", *in preparation*

511560 High-Performance Energy Storage and Conversion Devices for Automotive Electrification through A2P Approach

Abstract:

Research Interests:

Automotive electrification provides sustainable pathways to overcome petroleum resource limitations and meet environmental imperatives. Lithium-ion batteries and hydrogen fuel cells are the two leading candidates to power electric-drive vehicles. Battery electric vehicles based on current technology are suitable to small-vehicle low-mileage-per-day applications while the fuel cell is favored for large vehicles with long-range trips. Great efforts have been put to enhance the performance of energy conversion and storage devices. In many cases the device performance is still limited by poor efficiency, short operation life, and high cost due to instability of interface, lack of architecture optimization, and difficulty in large scale fabrication. My research interests focus on the design and synthesis of novel materials from atoms or molecules towards high-performance batteries and fuel cells as final products (A2P approach) based on the understanding of the deviceâ€™s operation mechanisms. The goal is to develop high-performance and low-cost energy storage and conversion devices for practical automotive electrification.

My professional industry experience includes not only research in design and synthesis of novel functional materials for next-generation lithium-ion battery and fuel cells but also fabrication of real-world devices and failure analysis (e.g., fabrication of pouch cells and membrane electrode assembly (MEA), nondestructive and tear-down analysis of high-capacity automotive pouch cells, and post mortem analysis of MEA). My Ph D research mainly focus on synthesis of nanostructured energy materials through various synthesis platforms, such as colloidal chemistry, sol-gel chemistry, chemical vapor deposition, solvothermal synthesis, electrospinning, solid state reaction, aerosol technology and self-assembly approaches.
With the industry demand, synthetic capability and fundamental understanding of energy storage conversion, my future research will focus on materials development for automotive electrification: 1) Design and synthesis of novel materials for high-performance and low-cost energy storage and conversion devices and 2) Deep understanding of synthesisâ€“performance-mechanism relationships through in-situ and ex-situ characterization technologies.

Selected publications:

1. Fang Liu, **Qiangfeng Xiao**, Haobin Wu, Fei Sun, Xiaoyan Liu, Xiaoqiong Bai, Fan Li, Zaiyuan Le, Li Sheng, Ge Wang, Mei Cai, Yunfeng Lu. Regenerative Polysulfide-Scavenging Layers Enabling Lithiumâ€“Sulfur Batteries with High Energy Density and Prolonged Cycling Life. ACS Nano 2017, DOI: 10.1021/acsnano.6b07603.

2. **Qiangfeng Xiao**, Meng Gu, Hui Yang, Bing Li, Cunman Zhang, Yang Liu, Fang Liu, Fang Dai, Li Yang, Zhongyi Liu, Xingcheng Xiao, Gao Liu, Peng Zhao, Sulin Zhang, Chongmin Wang, Yunfeng Lu, Mei Cai. Inward lithium-ion breathing of hierarchically porous silicon anodes. Nature Communications 2015, 6, Article number:8844. doi:10.1038/ncomms9844.

3. **Qiangfeng Xiao**, Mei Cai, Michael P. Balogh, Misle M Tessema, Yunfeng Lu. Symmetric Growth of Pt Ultrathin Nanowires from Dumbbell Nuclei for Use as Oxygen Reduction Catalysts. Nano Research 2012, 5, 145-151.

4. **Qiangfeng Xiao**, Hiesang Sohn, Zheng Chen, Daniel Toso, Matthew Mechlenburg, Z. Hong Zhou, Eric Poirier, Anne Dailly, Haiqiang Wang, Zhongbiao Wu, Mei Cai, Yunfeng Lu. Mesoporous Metal and Metal Alloy Particles Synthesized by Aerosol-Assisted Confined Growth of Nanocrystals. Angew. Chem. Int. Ed. 2012, (hot paper), 51, 10546-10550.

Book Chapter

Qiangfeng Xiao, Bing Li, Fang Dai, Li Yang, Mei Cai. Application of Lithium-Ion Batteries in Vehicle Electrification. Electrochemical Energy: Advanced Materials and Technologies. Publisher: CRC Press, Taylor & Francis Group. Editors: Pei Kang Shen, Chao-Yang Wang, San Ping Jiang, Xueliang Sun, Jiujun Zhang. 2015

Teaching Interests:

Teaching plays a critical role in conveying scientific knowledge to students in the classroom and equipping them with the ability to analyze and solve problems in the real world. I have gained diverse teaching and mentoring experience as Teaching Assistant (TA) during my PhD training at Tulane University and UCLA. Given the teaching experience and the extensive multidisciplinary training including chemical engineering and materials science and engineering, I feel interested in teaching undergraduate and graduate-level core engineering courses, such as Thermodynamics, Transport Phenomena, Reactor Design as well as materials-related courses, such as Polymer Science and Engineering.

505370 Understanding Bacterial Biofilms for Improved Medical and Industrial Processes

Abstract:

Research Interests:

Bacterial biofilms are a community of sessile bacteria which are ubiquitous in the natural environment, industrial settings, and in human hosts. These communities of bacteria form by attaching to a surface and creating a protective layer called the extracellular polymeric substance (EPS). Due to the EPS and other biofilm-specific resistance mechanisms, biofilms are very difficult to eradicate and can cause biofouling of industrial plant equipment, ship hauls, and pipe lines, to name a few. This biofouling can lead to weakened materials, contamination of products, and heat transfer problems. Additionally, these biofilms can form on medically implanted devices, lungs of patients, and hospital equipment. Biofilm infections cannot be readily treated with antibiotics alone since the resistance to antibiotics of the bacteria in a biofilm can increase a hundred-fold, requiring antibiotic concentrations above safe dosages. Although biofilm research has increased our understanding over the last few years, there is still much to learn about biofilms and many applications to be tested and utilized for biofilm mitigation.

The many problems and ubiquitous nature of biofilms create a unique niche for interdisciplinary research. My educational background is in chemical engineering and bioengineering; however, my postdoctoral position is in microbiology providing me a strong base for bridging the gap between the two disciplines. My interest in bacteria and biofilms started in my undergraduate career through the Center for Biofilm Engineering at Montana State University. While there I worked on the isolation and enrichment for magnetotactic bacteria for the development of a natural paramagnetic nanoparticle harvesting process. I also worked on a project to optimize micelle conformation from siderophores for the application of drug vesicles with Dr. Abigail Richards. I completed my doctoral work at the University of Iowa with Dr. Eric Nuxoll and worked with *Pseudomonas aeruginosa* biofilms for the development of an improved means of mitigating biofilms on medically implanted devices without invasive surgery. By using a polymer coating containing iron oxide nanoparticles on a medically implanted device, heat and antibiotics can be released directly to the surface where the biofilm is located using an alternating magnetic field, effectively killing the infection. This technology now has a patent pending. Currently I am a postdoctoral fellow at the University of Colorado Anschutz Medical Campus with Dr. Alexander Horswill. I am working with mutants of methicillin resistant *Staphylococcus aureus*, or MRSA, to determine the genes required for biofilm formation. Between my interdisciplinary research and traditional engineering background I plan to continue research with biofilms to determine natural inhibitor molecules in bacterial interactions and develop novel drug formulas from that information.

Successful Proposals

National Institutes of Health T32 Fellowship for Biocatalysis and Bioprocessing, 2012-2014, 2014-2015

Ballard and Seashore Dissertation Fellowship, 2017

Kammermeyer Graduate Fellowship, 2012-2013

Selected Publications

O'Toole A, **Ricker EB**, Nuxoll E. "Thermal mitigation of *Pseudomonas aeruginosa* biofilms." *Biofouling*, 2015, Vol 31 (8), 665-675.

Ricker EB, Al-Jaafari H, Bader TM, Hundley BS, Nuxoll E. "Thermal Shock Susceptibility and Regrowth of *Pseudomonas aeruginosa* Biofilms." *International Journal of Hyperthermia*, 2017, doi 10.1080/02656736.2017.1347964.

Ricker EB and Nuxoll E. "Synergistic Effects of Antibiotics and Heat on *Pseudomonas aeruginosa* Biofilms." *Biofouling*, in revision.

Ricker EB, Coffel J, and Nuxoll E. "Wireless Heating of *Pseudomonas aeruginosa* Biofilms Using Iron Oxide Nanoparticle Coatings." *Advanced Healthcare Materials*, in preparation.

Teaching Interests:

Teaching Experience

I have taught students from all different backgrounds through classes, tutoring, and mentoring. I have been a teaching assistant at the University of Iowa for chemical reactions engineering (a kinetics course), mass transfer and separation, and an introductory engineering course on engineering problem solving. For each of these courses I was responsible for creating discussion lectures, homework assignments and solutions, grading, and holding office hours in addition to my discussion time to work with the students. I learned to create lectures that introduced problem solving techniques and how to apply them to different situations. I encouraged class participation by asking students what they would suggest as the next step to keep them engaged and thinking critically. I found that the students who engaged more learned the material faster and that an environment with constant positive reinforcement fostered more students of different levels to contribute their opinions. I also achieved greater participations via projects in smaller groups which promoted discussion and demonstrated how the material learned in class could contribute to problems in the engineering world and day to day life. I was recognized for my teaching at the University of Iowa with the Osburn Teaching Award in 2016. I was also a teaching assistant at Montana State University for the introductory course for chemical and biological engineering. In this position, I set up experiments and worked with the students to further their knowledge gained from the experiments. In addition to teaching in a classroom setting I have enjoyed tutoring since 2010. Through my experiences tutoring I learned how to approach a problem from several different angles to help a student grasp the ideas and apply their analytical skills gained from one problem set to another. I also developed a more structured way of introducing ideas that can reach a broad range of students. Working with students outside of the classroom, such as mentoring students in the laboratory is a lot of fun and a great way to bring knowledge to a student in a different setting. I have mentored five undergraduates and one graduate student so far and have greatly enjoyed working with students in the laboratory setting. I was able to get each student to work in the laboratory effectively and encouraged the students to think critically about the experiments and how they could improve them.

Teaching Philosophy

Through my teaching experience I have developed a variety of presentation styles to develop analytical and critical thinking. Analytical approaches to engineering problems are numerous and can be fostered through a combination of interactive lectures and hands-on learning. Exciting and interactive lectures will encourage the students to participate and contribute ideas. Positive reinforcement for participation along with enthusiasm for a topic can make the difference between a forgotten topic and a memorable topic. This can be achieved through demonstrations, encouragement, and questions to the class throughout the lecture. Laboratory experiments and group projects encourage students to think critically about what they are doing while reinforcing the concepts learned in lectures. Projects can also be used to hone the students' communication skills through presentations and writing. These projects can include problems that do not have a defined solution, challenging the students to think critically and apply their knowledge in a new way. In addition to the activities in a classroom, I plan on having students apply their knowledge and problem-solving skills to real world problems through Engineers Without Borders (EWB). I have been a part of several student and professional EWB projects and have grown as a person, intellectually and as a team member. I will either contribute to a current chapter or create an EWB chapter so that students can get field experience and internalize concepts by making them relevant. Through this experience the students will also form professional and personal contacts locally and all over the world. The analytical and critical thinking skills learned through lectures and hands on experiences will create another generation of productive and innovative engineers.

505277 Biomaterial Design for Tissue Engineering, Drug/Gene Delivery and Biomedical Processes

Abstract:

Research Interests:

Biodegradable Polymer based Scaffold Design for Stem Cell Differentiation and Neural Regeneration: Mimicking the complex 3D microenvironment of natural extracellular matrix is crucial to provide an available environment for cellular attachment, growth, proliferation, spreading, migration and differentiation.(1,2) Our research is concentrated on biomaterial design for stem cell-laden implantable scaffolds-based cellular therapies with a particular interest in stem cell differentiation targeting neuroregeneration. We uniquely integrated various physical and chemical cues on porous polymeric films, which possess longitudinal surface micropatterns and surface concentration gradient of neurotropic factors providing directed nerve growth as well as neurotropic factors loaded microparticles throughout the film matrix providing controlled release properties.(3) In addition, we demonstrated that the 3D microstructural and mechanical properties of the scaffolds has profound effect on stem cell differentiation.(4) Our results suggested that the scaffolds with larger pore size and higher porosity and possessing less stiff structure are providing the most favorable environment for differentiation of mesenchymal stem cells (MSCs) into Schwann cell-like phenotypes (SCs) via chemical induction. Recently, we have shown that the electrical stimuli applied through conductive graphene circuits developed on 2D substrates can be used as an alternative strategy for differentiation of MSCs into SCs.(5) This study selected as cover art for Advanced Healthcare Materials Journal, received Star Award from the Society for Biomaterials and resulted in provisional patent filing. Recently, we identified significantly regulated proteins and related cellular pathways upon transdifferentiation of MSCs into SCs via chemical stimuli.(6) This comparative proteomics analysis provided valuable information about specific proteins and pathways that play significant roles in myelination, axonal guidance, cytokine and growth factor secretion. Based on this background, we will focus on designing stem cell-laden, implantable, 3D bioprinted scaffolds to investigate material-cell and cell-cell interactions and *in vitro* and *in vivo* stem cell differentiation and neural regeneration. The future directions will be to serve the ultimate goal of making realistic 3D tissue models that mimic the actual cellular arrangement of a neural tissue using 3D bioprinting.

Nanoscale Drug/Gene Delivery System Design for Cancer Treatment and Theranostics: Currently applied treatments for cancer, involving surgical cytoreduction followed by systemic cytotoxic agent chemotherapy, have not provided desired outcomes and resulted in chemo-resistance and tumor relapse.(7) As an alternative, RNA based therapies (microRNAs (miRNAs) and small interfering RNAs (siRNAs)) that influence major driver pathways in cancer development by directing sequence-specific degradation of target messenger RNAs (mRNA), have been proposed as a potential therapeutic for cancer treatment.(8) However, the potential clinical application of RNA therapy is limited by challenges related to delivery, stability, transport, diffusion, cellular entry, endosomal escape, efficient release and activity in the cytoplasm as well as the co-incorporation of drug molecules and RNAs in a single nano delivery device.(7,8) We have developed stimuli responsive polymer, peptide and inorganic nanoparticle based small interfering RNA (siRNA) delivery systems for cancer treatment. Our works have demonstrated that the pH and temperature responsive, cationic, amphiphilic and biocompatible multi-block copolymer-base systems are able to electrostatically condense and protect siRNA as well as providing enhanced cellular uptake and endosomal escape due to their temperature and pH responsive micellization enhancing the biological siRNA activity.(8) The cell penetrating and fusogenic peptide based peptideplex and conjugate systems are also alternative platforms for effective siRNA delivery.(9) We have also showed an efficient use of gold nanoparticles for theranostics and siRNA delivery applications.(8,10) The gold nanoparticle based multicomponent systems developed through subsequent layer-by-layer addition of siRNA and stimuli responsive polymer or cell penetrating and fusogenic peptide layers resulted in enhanced siRNA delivery and activity.(8,9,10) We have recently concentrated on developing nanoscale delivery platforms for co-delivery of siRNA and/or miRNA with chemotherapeutic drugs to target various intracellular pathways to reduce drug resistance and increase drug activity for ovarian and pancreatic cancer treatment.(11) As future direction, we will focus on the development of functional nanoscale delivery systems for efficient delivery of genome editing tools (Crispr-Cas9-sgRNA), along with chemotherapeutic agents for cancer treatment.

Design of Biomimetic, Polymer based Films for Biomedical Application: The objective of my research is to develop biomimetic synthetic membranes that are responsive to multiple external stimuli and exhibit selective transport capabilities, anti-fouling and self-cleaning features as in biological cell membranes. These bioinspired synthetic membranes are potentially promising for various biomedical applications including cell sorting, cell transfer, protein separation, hemodialysis and so on. We developed various polymer based membranes using dry or dry-wet phase inversion techniques and further carried out surface modification via layer-by-layer approach or surface chemistry. We developed nano/ultrafiltration membranes with enhanced surface properties to prevent chemical or biological fouling for waste water treatment applications.(12,13) We developed active agent incorporated polymeric membranes and developed different controlled release strategies (i.e. changing the membrane structure, incorporating nanoparticles in membrane matrix etc.) to provide sustained release, antibacterial activity and increase the shelf life of foods.(14,15) Recently, we have proposed bioinspired, biomimetic membranes functionalized through surface initiated atom transfer radical polymerization and protein localization to mimic selective separation, anti-fouling and self-cleaning properties of natural cell membranes for biomedical applications. This study received the Baxter Young Investigator Award. We will work on the development of functional polymeric films for biosensor and microfluidics applications.

Teaching Interests:

I am interested in teaching fundamental chemical and biological engineering principles and helping students understand how to use these principles in current technological problems. My teaching philosophy is based on embracing diversity/inclusion and using various learning methods such as, flipped classroom, problem, team and project-based learning, along with technological and online teaching tools in the class. I served as a teaching assistant for various chemical engineering undergraduate core courses and recently taught 3 credit chemical engineering undergraduate course "CHE 160: Chemical Engineering Problems with Computer Applications Laboratory" at Iowa State during 2015 Fall Semester. In addition to teaching core courses of chemical and biological engineering, I am also interested in curriculum development.

References:

1. **Uz, M,** and Mallapragada, S.K.; Smart Materials for Nerve Regeneration and Neural Tissue Engineering. Smart Materials for Tissue Engineering: Applications, Qun Wang (Ed.), Royal Society of Chemistry, 2017, pp.382-408. *(Print ISBN: 978-1-78262-484-4)*
2. E.J. Sandquist, **M. Uz,** A.D. Sharma, B.B. Patel, S.K. Mallapragada, D.S. Sakaguchi, Stem Cells, Bioengineering, and 3-D Scaffolds for Nervous System Repair and Regeneration: G.L. Zhang and L.D. Kaplan (Eds.), Neural Engineering: From Advanced Biomaterials to 3D Fabrication Techniques, Springer International Publishing 2016, pp. 25-81.
3. **Uz, M.,** Sharma, A., Adhikari, P., Sakaguchi, D., and Mallapragada, S.K., "Development of Multifunctional Films for Peripheral Nerve Regeneration Conduits, *Acta Biomat.* (2017). Available online. https://doi.org/10.1016/j.actbio.2016.09.039
4. **Uz, M.,** Buyukoz, M., Sharma, A., Sakaguchi, D., Altinkaya, S., and Mallapragada, S.K., "Gelatin-Based 3D Conduits for Transdifferentiation of Mesenchymal Stem Cells into Schwann-cell Like Phenotypes", *Acta Biomat.*, 53, 293-306 (2017).

5. Das, S.,# **Uz, M.,#** Ding, S., Lentner, M., Hondred, J., Cargill, A., Sakaguchi, D.S., Mallapragada, S., and Claussen, J., "Electrical Differentiation of Mesenchymal Stem Cells into Schwann Cell-Like Phenotypes using Inkjet Printed Graphene Circuits", *Adv. Healthcare Mater.,* 6(7), 1601087 (2017). Journal cover image. doi: 10.1002/adhm.201601087. *(# These authors put equal contribution)*
6. Anup D. Sharma, Jayme Horning, **Metin Uz,** Pawel Ciborowski, Surya K. Mallapragada, Howard E. Gendelman, and Donald S. Sakaguchi; "Proteomics Analysis of Mesenchymal to Schwann Cell Transdifferentiation.", *Journal of Proteomics, June 2017, In Press,* https://doi.org/10.1016/j.jprot.2017.06.011
7. **Uz, M.;** Altinkaya, S. A.; and Mallapragada, S. K.; Stimuli responsive polymer-based strategies for polynucleotide delivery. Journal of Materials Research, April 2017. (DOI: https://doi.org/10.1557/jmr.2017.116)
8. **Uz, M.;** Mallapragada, S. K.; Altinkaya, S. A., Responsive Pentablock Copolymers for siRNA Delivery. RSC Advances 2015, 5 (54), 43515-43527.
9. **Metin Uz,** Volga Bulmus and Sacide Alsoy Altinkaya; Evaluation of Cell Penetrating and Fusogenic TAT-HA2 Peptide Performance in Peptideplex, Multicomponent and Conjugate siRNA Delivery Systems. *(Under review, draft manuscript available upon request)*
10. **Uz, M.;** Bulmus, V.; Alsoy Altinkaya, S., Effect of PEG Grafting Density and Hydrodynamic Volume on Gold Nanoparticle–Cell Interactions: An Investigation on Cell Cycle, Apoptosis, and DNA Damage. Langmuir 2016, 32 (23), 5997-6009.
11. **Metin Uz,** Satyanarayana Rachagani, Surinder K. Batra and Surya K. Mallapragada; Dual Delivery Nanoscale Device for miR-345 and Gemcitabine Co-Delivery to Treat Pancreatic Cancer. *(Under review, draft manuscript available upon request)*
12. **Uz, M.;** Mahlicli, F. Y.; Polat, M.; Altinkaya, S. A., Characterization of Polysulfone Based Hemodialysis Membranes by AFM. *Procedia Engineering* 2012, 44, 1166-1167.
13. **Metin Uz,** Filiz Yasar Mahlicli, Erol Seker, Sacide Alsoy Altinkaya; Silver Release from TiO_2-nAgCl Xerogels Incorporated Mixed Matrix Ultrafiltration PAN Membranes in the Presence of Excess Chloride Ions: Long-term Silver Retaining, Biofouling Resistance and Antibacterial Activity. *(Under review, draft manuscript available upon request)*
14. Ozer P.B.B. ‡; **Uz M. ‡;** Oymaci P.; Altinkaya S.A, Development of a Novel Strategy for Controlled Release of Lysozyme from Whey Protein Isolate Based Active Food Packaging Films. *Food Hydrocolloids* 2016, 61, 877–886. (‡*These authors contributed equally to this work.*)
15. **Uz, M.;** Altınkaya, S. A., Development of Mono and Multilayer Antimicrobial Food Packaging Materials for Controlled Release of Potassium Sorbate. LWT - *Food Science and Technology* 2011, 44 (10), 2302-2309.

505342 Creating Rechargeable Antithrombotic Surfaces for Medical Devices

Abstract:

Research Interests:

My research focuses on developing new synthetic biomaterials and bio-conjugation strategies to induce specific biological responses and enhance the performances of diagnostic or medical devices. My PhD research focused on developing surface modification strategies by utilizing a mussel-mimetic catecholamine polymer as well as glycopeptide-mimetic polymers to create non-fouling surfaces. During postdoctoral training, my research interests are extended to develop a system for creating rechargeable surfaces to reduce thrombosis on the blood contacting medical devices. My goal is to run a multidisciplinary research program that focuses on 1) developing multifunctional biomaterials / bioactive coatings that can deliver therapeutics with a spatial and temporal control, and 2) designing glycopeptide-mimetics for cardiovascular applications. I envision that students within this research program will have strong collaborative interactions with other scientists, engineers, and clinicians, which will further enrich this training environment.

Teaching Interests:

I have established a strong foundation of knowledge in Chemical and Biomolecular Engineering and in the field of biomaterials, molecular and cardiovascular tissue engineering. At this interface between chemistry/ materials, and bioengineering, my research leverages many disciplines including biomaterials design and synthesis, surface characterization techniques, and molecular/ protein engineering. The experience noted above will allow me to teach courses like Biomaterials, Biointerfaces, and Tissue Engineering. I would also be pleased to develop new interdisciplinary courses to contribute to the current department curriculum.

Selected Publications:

1. **Hyun Ok Ham**, Zheng Qu, Carolyn A. Haller, Brent M. Dorr, Erbin Dai, Wookhyun Kim, David R. Liu and Elliot L. Chaikof. *In situ* regeneration of bioactive coatings enabled by directed evolution of Staphylococcus Aureus sortase A, *Nat. Comm.*, 2016; 7: 11140

2. Brent M. Dorr, **Hyun Ok Ham**, Chihui An, Elliot L. Chaikof, David R. Liu. Reprogramming the specificity of sortase enzymes. *Proc. Natl. Acad. Sci. U.S.A.*, 2014; 111 (47): 13343â€"13348.

3. **Hyun Ok Ham**, Sung Hyun Park, Josh W. Kurutz, Igal G. Szleifer, and Phillip B. Messersmith. Antifouling glycocalyx-mimetic peptoids. *J. Am. Chem. Soc.*, 2013; 135 (35): 13015-13022.

4. Hye Sung Kim, **Hyun Ok Ham**, Young Ju Son, Phillip B. Messersmith, and Hyuk Sang Yoo. Electrospun catechol-modified poly(ethyleneglycol) nanofibrous mesh for anti-fouling properties. *J. Mater. Chem. B.*, 2013; 1: 3940â €"3949.

5. **Hyun Ok Ham**, Zhongqiang Liu, K. H. Aaron Lau, Haeshin Lee, and Phillip B. Messersmith. Facile DNA immobilization on surface via catecholamine polymer. *Angew. Chem., Int. Ed.*, 2011; 123(3): 758â€"762.

504519 Creation of Self-Assembled Materials from Recombinant Fusion Proteins for Advanced Biomedical Platforms

Abstract:

Research Interests:

My research interests are to combine characterization of soft matter self-assembly, with recombinant protein materials fabrication, and biofunctional assessment of self-assembled materials for advanced biomedical platforms.

Self-assembly of biomolecules has been harnessed to create novel supramolecular architectures at the nanometer to micrometer scale with high complexity. Among different types of biological building blocks, proteins are potent molecules that can be used for many applications such as therapeutics, sensors, and biocatalysts due to their specific bioactivity based on amino acid sequences and folded structures. Furthermore, the protein self-assembled biomaterials enable us not only to overcome the limitations of soluble proteins such as stability, localization, and recovery but also to enhance their functions with engineered supramolecular structures. The biggest advantage of using proteins as self-assembling building blocks is that the protein domains can be designed to confer specific functions to engineered biomaterials via recombinant technology. Fusion of either a functional protein or an assembly domain to a leucine zipper domain makes the materials design strategy modular, based on the high affinity of the zippers.[1]

To establish a rational strategy to create self-assembled materials from recombinant proteins with tunable structure and size, understanding of their phase behavior as a function of diverse physical parameters including concentration, temperature, buffer pH and ionic strength is critical. Yet, basic assembly principles of recombinant fusion proteins are underdeveloped due to the greater complexity of protein-protein interactions than other molecules. The first goal of my research is to investigate recombinant protein self-assembly behavior in terms of structure and physical property under different environment conditions by using x-ray/neutron scattering, dynamic light scattering, microscopy techniques and quartz crystal microbalance with dissipation energy monitoring (QCM-D). During my Ph.D. program, I have been training in neutron scattering for structural characterization of the macromolecular self-assembled platforms at the nm-scale, along with physical property characterization using QCM-D and AFM.[2-4]

Ultimate goals of my research are to identify fundamental structural knowledge of the self-assembly of recombinant fusion proteins and to use this information to create innovative protein-assembled materials from specifically designed, bioactive globular proteins. For example, functional globular protein-incorporated vesicles can be utilized for drug delivery vehicles, micro-reactors, or bio-sensors, and nanosheets or hydrogel platforms can be developed for wound healing patches or stem cell scaffolds. I believe that self-assembled materials from specifically designed recombinant proteins will create a number of advanced biomedical applications since the biological functions and stimuli-responsiveness can be inserted into the platforms by protein materials design via recombinant technology. Beyond the fabrication of new biomaterials, the fundamental studies on protein self-assembly into higher ordered structures will also give scientific inspiration to other soft matter research fields.

Teaching Interests:

During my career development, I was fortunate enough to have served as a teaching assistant at "Quantum Mechanics" at Seoul National University, and lecturer for an intermediate level course "Interfacial Engineering" at Kyunghee University. I found that my professional experience provided me a broad view that is useful in assisting students with managing projects and assignments. These experiences have built my confidence and an interest in teaching. I can teach the stem courses in Chemical Engineering such as "Physical/Organic Chemistry", "Chemical Process Principles", and "Chemical Engineering Thermodynamics". More specifically, I am confident to teach "Polymer Engineering and Characterization" and "Protein Engineering" for undergraduate students, and current trends in "Interfacial Engineering", "Soft Matter Self-Assembly Principle and Applications", and "Scattering Techniques" for graduate students. I look forward to the opportunity to not only teach existing courses but also work to develop new ones that I can contribute to. Aside from teaching courses, I have a lot of experience in mentoring undergraduate and graduate students during my academic carriers. From these experiences, I have learned the PI's role should be much bigger. I will share my experiences, successes, and struggles with my students to encourage them and create a supportive environment that allows all students to flourish.

Research Experiences:

My academic training during the doctoral program and 3 years of experience working as a postdoctoral researcher prepare me to be an effective researcher and instructor. My doctoral dissertation was conducted in consultation with Prof. Kookheon Char at Seoul National University (School of Chemical and Biological Engineering), under the thesis title of "Functional Multilayer Films with Controlled Nanostructures for *in vitro* Cell Studies". I have investigated stimuli-triggered changes in structure and property of polymer thin film multilayers, which are specially designed for biomedical applications such as drug eluted coatings, cancer cell study platforms.[5-7] During my Ph.D. program, I had been able to strengthen the basics in macromolecular self-assembly and an integrated design of polymer-assembled platforms for biological applications. In addition to my dissertation research, other areas that interest me for future research stem from my goal of developing a new class of biomaterials via recombinant technology. First, I have worked on hybrid microbubbles consisting of recombinant proteins and biocompatible triblock copolymers at the University of Pennsylvania with Profs. Daeyeon Lee and Daniel A. Hammer. I have focused on controlling the size and mechanical properties of the recombinant protein microbubbles generated by microfluidic devices for antivascular ultrasound theranostics.[8] I believe that the experience in using microfluidics, soft lithography to fabricate microbubbles, emulsions, vesicles as well as characterizing the mechanical properties by micropipette aspiration enlarged my research capability in addition to the doctoral background in engineering polymer multilayers. Then, I moved into Georgia Institute of Technology (PI: Julie A. Champion who is an expert in protein engineering) to learn more about the recombinant fusion protein materials. Here, I have designed a new functional fusion protein that can self-assemble into vesicles and engineered the protein vesicles for cancer targeting applications and antibacterial performance. In synergy with my forte on fundamentals of soft matter assembly and postdoctoral experiences in the design of new biomaterials from recombinant proteins, I have been able to develop multipotent self-assembled materials made from the recombinant fusion proteins.

Successful Proposals: The proposal # DMR - 1709428 titled "Protein Vesicles: Understanding Self-Assembly of Fusion Proteins into Vesicles to Engineer Structures and Biofunctional Properties" is being considered for an NSF three-year award at the requested amount of $393,605.

Future Direction:

I am extremely interested in obtaining a faculty position in School of Chemical Engineering, Biomolecular Engineering, and/or Materials Science, where I can contribute to its focus on education and research. I would like to continue my research on creating multipotent self-assembled materials from recombinant proteins and biopolymers and devote to lead campus and professional service activities as a faculty. I am confident that my skills, knowledge and personality are well-aligned with the role.

Selected Publications:

1. **Jang, Y.**; Champion, J. A., Self-Assembled Materials Made from Functional Recombinant Proteins. *Accounts of Chemical Research* **2016,** *49*, 2188-2198.
2. **Jang, Y.**; Seo, J.; Akgun, B.; Satija, S.; Char, K., Molecular Weight Dependence on the Disintegration of Spin-Assisted Weak Polyelectrolyte Multilayer Films. *Macromolecules* **2013,** *46*, 4580-4588.
3. **Jang, Y.**; Akgun, B.; Kim, H.; Satija, S.; Char, K., Controlled Release from Model Blend Multilayer Films Containing Mixtures of Strong and Weak Polyelectrolytes. *Macromolecules* **2012,** *45*, 3542-3549.
4. Son, H.*; **Jang, Y.***; Koo, J.; Lee, J.-S.; Theato, P.; Char, K., Penetration and exchange kinetics of primary alkyl amines applied to reactive poly(pentafluorophenyl acrylate) thin films. *Polym J* **2016,** *48*, 487-495. (*Equal Contribution)
5. **Jang, Y.**; Lee, H.; Char, K.; Nam, J.-M., Transparent, Nanoporous, and Transferable Membrane-Based Cell–Cell Paracrine Signaling Assay. *Advanced Materials* **2015,** *27*, 1893-1899.
6. Lee, H.*; **Jang, Y.***; Seo, J.; Nam, J.-M.; Char, K., Nanoparticle-Functionalized Polymer Platform for Controlling Metastatic Cancer Cell Adhesion, Shape, and Motility. *ACS Nano* **2011,** *5*, 5444-5456. (*Equal Contribution)
7. **Jang, Y.**; Park, S.; Char, K., Functionalization of polymer multilayer thin films for novel biomedical applications. *Korean Journal of Chemical Engineering* **2011,** *28*, 1149-1160.
8. **Jang, Y.**; Jang, W.-S.; Gao, C.; Shim, T. S.; Crocker, J. C.; Hammer, D. A.; Lee, D., Tuning the Mechanical Properties of Recombinant Protein-Stabilized Gas Bubbles Using Triblock Copolymers. *Acs Macro Letters* **2016,** *5*, 371-376.

505475 Decoding the Nature-Designed Codes in Membranes: Applications in Biomedicines and Bioengineering

Abstract:

Research Interests:

The bilayer cell membranes and the monolayer membranes existing at the air-aqueous interfaces in the biological system, such as alveolar interface in lungs, are fascinating examples of nature-designed two-dimensional heterogeneous complex fluids equipped with remarkable properties. How different lipids and proteins specifically and stoichiometrically interact with each other and form static and/or dynamic macromolecular assemblies in membranes have intrigued me over the years, and are the emerging area of membrane research. Cell membrane of each type of cell lines possesses its own unique macromolecular assemblies of lipids/proteins. Moreover, specific macromolecular assemblies in cell membranes of either cell line have their own spatial and temporal regulation based on the activity level, ageing and/or state of health of cells. Similarly, lipids/proteins monolayer (lung surfactant) lying at the air-aqueous interfaces of alveoli in each air-breathing mammals possesses their own specific ratios of constituents. The composition, and the mechanical and physical properties of these lung surfactants evolve with the age, activity and the state of health as well. Therefore, specific and stoichiometric arrangement of lipid and protein molecules in a two-dimensional monolayer and bilayer membranes imparts not only specialized architecture and function to the membranes but also localized nano- to micro-structural hot spots of physical and chemical forces to these soft materials. I believe we can harness such nano- to micro-scale hot spots of unique physical and chemical force maps present on the two-dimensional landscape of membranes for diagnostic and pharmaceutical purposes, as cells utilize membrane-based chemical and physical force maps during migration and specific cell-cell communication.

Future Direction:

My academic career training has been an amalgam of diverse fields of science and engineering, such as materials science, soft-matter physics, biochemistry, biophysics and human physiology and pathology. My strength as a researcher lies in my critical thinking and inherent problem-solving skills. As a researcher I would continue to apply those skills along with multidisciplinary approach for understanding the nature-designed structures, exploring their potentials in different areas of bioengineering and biomedicine, as well as creating biomimetic devices for industrial applications. I would unite and orchestrate the knowledge gained over the period of my academic career in soft-matter physics of membranes for: exploring the hidden potentials in nature-designed two-dimensional monolayer and bilayer membranes.

Although the optical microscopy techniques have improved resolutions, probe-based Atomic force microscopy (AFM) technique is still the highest resolution tool for observing and revealing hierarchical organization of lipids and proteins in cell membranes. Driven by its immense potential and the requirement of scientific community, the AFM technique since its invention in 1980's has been developed for many specialized applications in materials science, such as scanning ion-conductance microscopy, Kelvin probe force microscopy (KPFM), Electrochemical force microscopy, scanning near-field optical microscopy, scanning capacitance microscopy, magnetic force microscopy, just to name some. During my PhD research, I extensively exploited AFM's high-resolution imaging, material characterization and nano-manipulation modes to probe the nanoscale structures in monolayer and multilayered lipids/protein membrane. In addition, I customized the state-of-the-art KPFM technique to unveil the contact potential differences at submicron resolution over the nanoparticle clusters embedded in multilayered structures of lipids. My research lab will customize the probe-based state-of-the-art nanoscopy techniques to characterize the membrane surfaces as well as invest in development of innovative surface characterization techniques and methodology to probe the sub-micron surface properties of native as well as artificial complex membrane models. Revelation of the map of spatial distribution of physical forces on the membrane along with high-resolution structural organization can be breakthrough in translational research and bioengineering. *Detailed information will be presented at poster session.*

Teaching Interests:

The very first task of teaching I took was self-appointed when I had decided to offer free physics and chemistry classes to socio-economically weak students of high school and senior secondary (10+2) level in 2004, which helped me realize the essence of being a teacher and different challenges involved. Thereafter I gained extensive teaching and mentoring experience. At higher levels, my significant teaching experiences are: 1) appointed as a teaching assistant (TA) for *Electrophysiology laboratory course* to participating PhD researchers at Physiology of Nerve Cells Workshop organized by the SERC-DST in 2009 at the Indian Institute of Science, Bangalore. 2) Mentoring and teaching assistant to undergraduate and graduate students during my PhD research at the University of Muenster, Germany, where I taught *lecture courses on Biophysical techniques*, especially Atomic force microscopy and Dark-field electron microscopy, and a *lab course* on Ultramicrotomy technique. 3) Mentored the directed research of two graduate (PhD) and three undergraduate students during my current postdoctoral research.

As a faculty, I will be delighted to teach courses related to biomedical field, bioengineering, soft-matter physics of membrane and my own research projects: *fundamentals of membrane, membranes and their pharmaceutical importance, interfacial rheology and mechanics, nature-inspired bioengineering*, and *biophysical techniques*.

Successful Grant Proposals: Contributed in writing with novel ideas, theories and results in successful NIH R01 (graded 1-2 on Innovation scale by the reviewers) and NSF grant applications submitted by PI.

Postdoctoral Project: "Shear and Dilatational Rheology of 2-D Fluid Membranes at Planar and Curved Interfaces and Their Relation with Micro-structures."

Principal Investigator: Prof. Joseph A. Zasadzinski, Chemical Engineering and Materials Science, University of Minnesota.

PhD Research: "Functional Nanostructures of Pulmonary Surfactant Monolayer Membrane and Its Interaction with Nanoparticles: Atomic Force Microscopy Investigation."

Under supervision of Prof. Hans-Joachim Galla, Institute of biochemistry, and Prof. Rudolf Reichelt, Institute of medical physics and biophysics, University of Muenster, Germany.

Peer-reviewed Publications:

Sachan A. K. and Zasadzinski J. A. (2017). Interfacial Curvature Effects on Morphology and Dynamics of Monolayer Membrane. (Under review in *PNAS*)

Sachan A. K., Choi S. Q., Kim K. H., Tang Q., Hwang L., Lee K. Y. C., Squires T. M., Zasadzinski J. A. (2017). Interfacial Rheology of Coexisting Solid and Fluid Monolayers. *Soft Matter* 13 (7), 1481-1492.

Sachan A. K.* and Galla H. J. (2014). Understanding the Mutual Impact of Interaction between Hydrophobic Nanoparticles and Pulmonary Surfactant Monolayer. *Small* 10 (6), 1069-75.

Cramer S., Tacke S., Bornhorst J., **Sachan A. K.**, Klingauf J., Schwerdtle T., Galla H. J. (2014). The Influence of Silver Nanoparticles on the Blood-Brain and the Blood-Cerebrospinal Fluid Barrier in vitro. *Journal of Nanomedicine & Nanotechnology* 5 (5), 1- 12.

Sachan A. K. and Galla H. J. (2013). Bidirectional Surface Analysis of Monomolecular Membrane Harboring Nanoscale Reversible Collapse Structures. *Nano Letters* 13 (3), 961–966.

Sachan A. K., Harishchandra R. K., Bantz C., Maskos M., Reichelt R., Galla H. J. (2012). High-Resolution Investigation of Nanoparticle Interaction with a Model Pulmonary Surfactant System. *ACS Nano* 6 (2), 1677–1687.

Harishchandra R. K., **Sachan A. K.**, Kerth A., Lentzen G., Neuhaus T., Galla H. J. (2011). Compatible Solutes: Ectoine and Hydroxyectoine Improve Functional Nanostructures in Artificial Lung Surfactants. *BBA – Biomembranes* 1808 (12), 2830–2840.

505599 Engineering Functional Nucleic Acid Nano-Devices

Abstract:

PhD dissertation:

"Single molecule RNA folding studied with optical trapping", UC Berkeley

Advisors:: Ignacio Tinoco, Chemistry; Carlos Bustamante; Physics

Postdoctoral research:

"Shielded covalent probes for imaging and control of gene expression", Caltech

Supervisor: Niles Pierce, Bioengineering

Research Interests:

Control of molecular structure and reactivity at the nanometer scale has been a dream of chemical engineers for years. For far longer, Nature has accomplished this feat with immense variation and scale, relying on bottom-up self-assembly of genetically evolved components to build structures and reaction pathways more complex than anything humans can design. Nucleic acid nanotechnology has emerged as an exciting field in which the programmable hybridization rules of base pairing are used to construct intricate two- and three-dimensional structures, carry out logic operations, and build molecular circuits and motors that operate at the smallest of scales. The primary theme of my proposed research is to leverage the advances of nucleic acid nanotechnology to build *functional* nano-devices capable of carrying out chemical transformations and probing biochemical systems in vitro and in vivo.

My graduate training focused on single-molecule studies of RNA folding, using optical trapping to measure the thermodynamics and kinetics of folding and unfolding reactions as a function of nucleic acid sequence, environmental conditions, and mechanical force. This provided a thorough grounding in the physical interactions that underlie nucleic acid self-assembly, as well as frontier areas such as non-equilibrium statistical mechanics and reaction pathway design.

My postdoctoral research aimed to apply our knowledge of nucleic acid thermodynamics and kinetics to the design and construction of dynamic, conformation-switching molecular nanodevices capable of performing chemical reactions with programmable targeting of biological molecules. I developed 'shielded covalent probes', nucleic acid hairpins that covalently capture target nucleic acid molecules with exquisite sensitivity and near-quantitative yield. These probes are currently being used in in situ imaging and gene expression, and were granted a US Patent.

My current research at the Institute for Molecular Engineering has expanded the applications of nucleic acid self-assembly to multiple length scales and novel chemistries. I have developed new methods for site-selective conjugation of nucleic acids to proteins and peptides, am investigating synthetic base pairing with metal ions for organometallic catalysis (in collaboration with Prof. Hosea Nelson at UCLA), and have led an extensive research program in polyelectrolyte assembly of nucleic acid – peptide nanoparticles with the laboratory of Prof. Matthew Tirrell. The common goal of these projects is applying our knowledge of nucleic acid interactions to produce useful molecular devices that operate on the nano-scale and interact with targets based on their specific genetic sequences, measuring and controlling biological outcomes with the utmost precision.

Future directions:

As a PI, I will apply my skills and training to engineer functional nucleic acid nanodevices that solve several pressing biochemical problems:

Programming chemical reactivity: Using the capabilities to form site-specific covalent linkages between oligonucleotides and proteins that I have developed, I plan to use nucleic acid scaffolds and nanopores to engineer enzymatic cascades in which the product of one reaction immediately becomes the substrate of another, avoiding the need for high intermediate concentrations or costly step-by-step separations. Additionally, I will continue a collaboration developing nucleic acids as conformationally-controllable ligands for organometallic catalysis; the ultimate goal of this project is to produce therapeutic molecules inside the specific cells that need them to treat disease.

Biomolecular phase transitions: In membraneless organelles and pathological aggregates, molecular conformation change drives phase separation in vivo. We recently found that the phase of nucleic acid-peptide complexes is controlled by nucleic acid hybridization. I plan to extend this work to full-length mRNA and proteins in order to determine the physical principles governing phase transitions in normal and pathological states. The same principles will enable the design of nanoparticles for more efficient delivery of therapeutic nucleic acids, a key challenge to realizing the potential of therapies such as siRNA.

Ribonucleoprotein engineering: Nature has evolved protein enzymes that operate with amazing effectiveness, but processing genetic information is primarily the job of nucleic acids, such as those at the core of natural ribonucleoproteins like the ribosome, spliceosome, and sequence specific nucleases like Dicer (RNAi) and CasX (CRISPR). By combining protein enzymes with the sequence recognition capabilities of nucleic acids, I hope to create synthetic ribonucleoproteins capable of therapeutic control of gene expression in vivo.

Teaching Interests:

I believe that teaching and research inform each other, and am excited to help students learn both foundational concepts and cutting-edge developments. As a graduate student, I had the opportunity to TA several courses in physics and chemistry, in both small and large format classes, including one introductory undergraduate class that was in the process of converting from traditional lecture-based to small group active learning. I completed a STEM pedagogy course as a graduate student, and have obtained a certificate from the Caltech Project in Effective Teaching as a postdoctoral researcher. Throughout my career, I have mentored undergraduate and graduate students, and look forward to doing so as a PI. As a faculty member, I am excited to teach students in core chemical engineering courses such as thermodynamics and chemical kinetics, and hope to develop an upper division/graduate course in biomolecular design.

Successful Proposals:

- NIH R01 (assisted, postdoctoral scholar)
- UCBREP Graduate Research and Education in Adaptive Biotechnology Training Grant (primary, graduate student)
- UChicago Institute for Translational Medicine Core Grant (primary, research scientist)

Selected Publications and Patents:

Journal Articles (12 published, 3 in preparation/review; h-index: 9)

- J.R. Vieregg and T-Y D. Tang, Polynucleotides in cellular mimics: coacervates and lipid vesicles.

Curr. Opin. Colloid & Interface Science, 26, 50-57 (2016). DOI: 10.1016/j.cocis.2016.09.004

- J.R. Vieregg, H.M. Nelson, B.M. Stoltz, and N.A. Pierce, Selective nucleic acid capture with shielded covalent probes.

J. Am. Chem Soc., **135**, 9691-9699 (2013). DOI: 10.1021/ja4009216

- P.T.X. Li, J.R. Vieregg, I. Tinoco, Jr., How RNA unfolds and refolds.

Ann. Rev. Biochem. **77**, 77-100 (2008). DOI: 10.1146/annurev.biochem.77.061206.174353

- J.R. Vieregg, W. Cheng, C. Bustamante, I. Tinoco, Jr. Measurement of the effect of monovalent cations on RNA hairpin stability.

J. Am. Chem. Soc. **129**, 14966-73 (2007). DOI: 10.1021/ja074809o

Patents

US Patent # 8,658,780: Triggered Covalent Probes for Imaging and Silencing Gene Expression

495214 Engineering Surfaces to Study Biological Interactions

Abstract:

Hormone disrupting compounds are pervasive; they are found in plastics, pesticides, and medications, among other sources. We have developed a detection strategy based on electrochemical impedance spectroscopy to quantify endocrine disrupting compounds that is both fast and portable. Our method requires no specialized skills to perform, and, rather than responding to individual chemicals, this system reports the total estrogenic activity of a sample using engineered bacteria displaying a native estrogen receptor construct. This approach will be broadly applicable to the detection of chemically dissimilar classes of compounds that bind to a single receptor.

The ability to selectively attach biopolymers, including DNA and peptides, to surfaces is imperative for biological research. Of great interest is their specific placement on gold surfaces. We have developed a reagentless coupling method to electrochemically attach biopolymers to surfaces using the specific oxidation of a catechol to a particular electrode. With our method, commercially-available DNA is rapidly attached to surfaces, yielding complete coupling within minutes. Furthermore, we have full control over the amount of biopolymer on the surface, which can be tailored to the specific application of the device. Using this attachment method, we have formed whole cell-based thin films. Additionally, we have optimized the platform for the specific detection of the endocrine disruptor bisphenol A (BPA), a problematic additive to plastics, using a DNA aptamer on the surface.

Research Interests:

Surface modification for the specific attachment of biomolecules is vital for the effective construction of diagnostics as well as for the study of fundamental biological processes. Developing improved chemistries to attach biopolymers such as DNA and peptides to surfaces will enable the creation of improved devices. These techniques will be utilized to immobilize DNA aptamers, enzymes, metalloproteins, and even whole cells to study clean energy production and intercellular interactions.

Teaching Interests:

Experience:

Protein thermodynamics-In this lecture course, protein binding to ligands was explored, in addition to discussions of chemical concepts behind enzymatic processes, including entropy and enthalpy.

Survey course in biochemistry-In this weekly discussion-based course, a professor discussed their research interests with first-year graduate students. Problem sets related to that weekâ€™s research talk were discussed in biweekly discussion groups.

General chemistry lab-In addition to supervision during the laboratory, problem sets and laboratory reports were evaluated. New experiments were also designed and evaluated for their efficacy in conveying chemical concepts to students.

505334 Induction of Tolerance or Immunity by Targeting Antigens to Specific Antigen Presenting Cells via Synthetic Polymeric Glycosylations

Abstract:

Research Interests:

As a post doc in Dr. Jeff Hubbell's Lab at the University of Chicago, my research has focused on the development of biomaterials that target antigens to specific antigen presenting cells (APCs) for the induction of immunological tolerance or immunity. To this end, I have developed a platform based on synthetic glycopolymers that when tethered to antigens via an intracellular self-immolate linker, target APCs and release antigens in their unmodified form, thus optimizing antigen uptake and presentation by APCs. In the case of Tolerance, autoimmune disorders are driven by aberrant auto-reactive T cell responses. For this reason, therapies that specifically ablate auto-reactive T cells, unlike current therapies that induce global immune suppression, would represent a major step forward in the treatment of autoimmunity. When synthetized from monomers composed of NAc-glucosamine, our glycopolymer-antigen conjugates target antigens to hepatic APCs, which naturally induce tolerogenic T cell responses upon antigen presentation. Antigen-poly(NAc-glucosamine) conjugates are able to knockout antigen-specific T cell responses, and thus prohibit the onset of experimental murine diabetes, as well as induce regulatory T cells that provide long-term antigen-specific tolerance in the same murine diabetes model. Using the same glycopolymer platform with monomers composed of mannose and an APC-activating co-monomer, we are able to efficiently target and activate dendritic cells, which present the delivered antigens to T cells in the presence of cellular signals that induce T cell activation. When tethered to a malaria antigen and delivered via subcutaneous injection, antigen-poly(mannose) conjugates induce robust CD4 and CD8 T cell responses and muster antibodies that prevent malaria infection in human hepatocytes. My lab will focus on the development of immuno-modulatory biomaterials that are capable inducing immunity to infectious disease and malignancy. In addition, we will focus on biomaterials based strategies that use cells from tissue donors to induce tolerance in transplant recipients and eliminate the need for a life-time of suppressive therapies.

Post Doc Funding: Whitaker International Scholars Award

PhD Research: Reactive oxygen sensitive biomaterials for the treatment diseases associated with intestinal inflammation, Georgia Institute of Technology: Bioengineering, School of Chemical Engineering. Advisor: Dr. Niren Murthy.

Master's Research: Molecular thermodynamics of polymeric dendrimers, University of Oklahoma, Chemical Engineering and Material Science. Advisor: Dr. Lloyd L. Lee.

Teaching Interests:

In addition to performing research, I have a very diverse set of teaching experiences. As an undergraduate, I worked for a year as a solo English Teacher to high school and college level students in the town of Clermont-Ferrand, France. Then as a master's student at the University of Oklahoma, I had the chance to TA both lab and lecture courses and was given the award as the best graduate TA during the Fall 2003 semester. After completing my master's degree, I solo taught undergraduate math courses at Florida State University, while I applied to PhD bioengineering programs. My research background in molecular thermodynamics and chemical engineering core course work at Georgia Tech have endowed me with a strong mathematical background that will facilitate my teaching of core chemical engineering courses such as transport phenomenon, engineering math, design, and of course thermodynamics. In addition, my research involving organic synthesis, biomaterials, and protein engineering will enable me to teach and create a variety of electives that will be interesting to undergraduates and graduate students alike.

Selected Publications:

M. Damo, **D.S. Wilson**, E. Simeoni, J.A. Hubbell, "TLR-3 stimulation improves immunity elicited by dendritic cell exosome-based vaccines in a murine model of melanoma", *Sci. Rep.* **5**, 17622, (2015).

C.D. Herman*, **D.S. Wilson***, K.A. Lawrence, X. Ning, R. Olivares-Navarrete, J.K. Williams, R.E. Guldberg, N. Murthy, Z. Swartz, B.D. Boyan, "Rapidly polymerizing injectable click hydrogel therapy to delay bone growth in a murine resynostosis model", *Biomaterials.* **35**, 9698, (2014). * Authors contributed equally to this work

R.E. Whitmire, **D.S. Wilson**, A. Singh, M.E. Levenston, N. Murthy, A.J. Garcia, "Self-assembling nanoparticles for intra-articular delivery of anti-inflammatory proteins", *Biomaterials.* **30**, 7665, (2012).

D.S. Wilson, G. Dalmasso, L. Wang, S. Sitaraman, D. Merlin, N. Murthy., "Orally delivered thioketal nanoparticles loaded with TNF-α-siRNA target inflammation and inhibit gene expression in the intestines", *Nat. Mater.,* **9**, 923, (2010).

D.S. Wilson, L.L Lee, "Chemical Potentials and Phase Equilibria of Lennard-Jones Mixtures: A Self-consistent Integral Equation Approach", *J. Chem. Physics,* **123**, 44512, (2005).

D.S. Wilson, L.L. Lee, "Molecular Recognition and Adsorption Equilibria in Starburst Dendrimers: Gas Structure and Sensing via Molecular Theory", *Fluid Phase Equilibria,* **228-229**, 197, (2005).

505568 Materials Design via Soft-Matter Crystallography

Abstract:

Research Interests:

Designing materials with novel functionality requires detailed understanding and control over structural order. However, engineering structures in self-assembly is a fundamental challenge, in particular in novel materials with complex structures. The focus of my research program is the analysis and description of ordered structures—and the underlying mechanisms and pathways that lead to order—across multiple length scales. Both nano- and mesoscopic materials can exhibit the same highly symmetric and complex structures that form due to a plethora of different interactions. Thus far, the occurrence of these same geometries has been discussed only in the context of the specific physical systems, *e.g.*, electronic interactions in the case of atomistic structures. My research group will investigate the essential criteria underlying ordered structures and structure formation—*i.e.*, nucleation and growth—across length scales, and will exploit this new understanding to design materials with dedicated structures that exhibit desirable properties and behavior.

My work bridges hard and soft matter in novel ways: it leverages techniques developed for the crystallography of hard materials to elucidate the intricate structural order found in soft materials, giving new insight into the complex behaviors that lead to order in soft matter. I develop minimal-model soft-matter systems that exhibit complex structural order previously only found in much more elaborate atomic systems; this suggests that common, simple underlying ordering mechanisms exist, which persist across length scales differing by several orders of magnitude. My complementary research background in hard and soft matter research uniquely enables me to execute this two-fold approach and gain deeper insights into the fundamentals of the structure of materials.

My goal is to establish a research group that works toward reaching a level of understanding in the area of structure prediction—*via* studying structure formation and stability with the tools of chemistry and crystallography, numerical simulations, *ab-initio* calculations, and machine learning—that will finally make engineering materials by design possible.

Teaching Interests:

I am currently mentoring several graduate students in the Chemical Engineering, Materials Science and Engineering, and Physics Departments, as well as the Macromolecular Science and Engineering and the Applied Physics Programs. I also completed a short-course for postdoctoral researchers on "High-Impact Principles and Practices for STEM Education" at the University of Michigan. During my time at ETH, I advised multiple graduate and undergraduate theses and taught weekly exercises in "Crystallography". I also acted as substitute lecturer in this subject, as well as "Methods of Materials' Characterization". I have since taught multiple short courses in Crystallography and Structure Analysis and completed a postdoctoral short-course on teaching at the University of Michigan. I look forward to teaching core engineering and physics courses, as well as developing specialty courses on materials, crystallography, and simulation.

Education & Training:

since 2015: Postdoctoral fellow, Chemical Engineering, University of Michigan, Ann Arbor, USA; *advisor: Sharon C. Glotzer*

2013 - 2015: Postdoctoral fellow, Materials Science, ETH Zurich, Switzerland; *advisor: Walter Steurer*

2009 - 2013: PhD student, Materials Science, ETH Zurich, Switzerland; *thesis advisor: Walter Steurer*

2002 - 2008: Diploma student, Physics, TU Dresden, Germany; *thesis advisor: Dirk C. Meyer*

Selected Honors & Awards:

Outstanding Postdoctoral Fellow Award of the University of Michigan, **2016**.

Max-von-Laue-Preis awarded to a Young Scientist by the German Crystallographic Society (DGK), **2015**.

SNSF Early Postdoc.Mobility fellowship, **02/2015-07/2016**.

Selected Publications *(total: 1 book, 17 papers)*:

W. Steurer, **J. Dshemuchadse**, *"Intermetallics: Structures, Properties, and Statistics"* (**2016**). Oxford University Press, ISBN: 9780198714552. *(608 pages)*

A. T. Cadotte*, **J. Dshemuchadse***, P. F. Damasceno, R. S. Newman, S. C. Glotzer, *"Self-assembly of a space-tessellating structure in the binary system of hard tetrahedra and octahedra"* (**2016**). Soft Matter 12 (34), 7073-7078. *(journal cover)*; *: co-first authors.

E. G. Teich, G. van Anders, D. Klotsa, **J. Dshemuchadse**, S. C. Glotzer, *"Clusters of polyhedra in spherical confinement"* (**2016**). Proc. Natl. Acad. Sci. USA 113 (6), E669–E678. *(highlighted article)*

J. Dshemuchadse, W. Steurer, *"Some Statistics on Intermetallic Compounds"* (**2015**). Inorg. Chem. 54, 1120-1128.

J. Dshemuchadse, W. Steurer, *"More of the 'Fullercages'"* (**2014**). Z. Anorg. Allg. Chem. 640, 693–700.

J. Dshemuchadse, S. Bigler, A. Simonov, T. Weber, W. Steurer, *"A new complex intermetallic phase in the system Al–Cu–Ta with familiar clusters and packing principles"* (**2013**). Acta Crystallogr. B 69, 238–248. *(highlighted article)*

J. Dshemuchadse, D. Y. Jung, W. Steurer, *"Structural building principles of complex face-centered cubic intermetallics"* (**2011**). Acta Crystallogr. B 67, 269–292. *(feature article)*

T. Weber, **J. Dshemuchadse**, M. Kobas, M. Conrad, B. Harbrecht, W. Steurer, *"Large, larger, largest – a family of cluster-based tantalum copper aluminides with giant unit cells. I. Structure solution and refinement"* (**2009**). Acta Crystallogr. B 65, 308–317.

Successful Research Proposals:

I have contributed to successful compute time proposals for XSEDE and INCITE resources. I was awarded an 18-month grant by the Swiss National Science Foundation (SNSF) to work as a postdoctoral fellow in the Glotzer group at the University of Michigan.

Service:

I co-founded the "Young Crystallographers" work group within the Germany Crystallographic Society (DGK) and chaired it for three years, during which I organized and chaired two symposia within the DGK's annual meetings, as well as one standalone meeting of the "Young Crystallographers". I remain a member of the group's executive committee, and am also acting as an elected member and the chair of the prize committee of the "Will-Kleber-Gedenkmünze" award of the DGK.

499016 Molecular Understanding of Physical Phenomena in Soft Materials Design and Process Development

Abstract:

Research Interests:

My group will harvest molecular-level knowledge of physical phenomena such as assembly, association, and interfacial non-bonded interactions, and utilize this knowledge to rationally design soft materials and develop processes. The long-term goal of my group is to create a transformative simulation-experiment framework that enables rapid materials design and process development. This framework will be built starting from three research projects: (1) designing zwitterionic polyelectrolyte systems for energy applications, (2) developing peptide coatings for nanomedical devices, and (3) predicting protein aggregation in biomanufacturing processes.

My PhD and postdoctoral research experiences equip me with the expertise to lead the research of my future group. My PhD research in Prof. Shaoyi Jiangâ€™s group at the University of Washington focused on understanding and designing zwitterionic anti-biofouling materials. Biofouling is a universal problem and zwitterionic materials emerge as promising anti-biofouling candidates. We developed force field parameters for zwitterionic materials, investigated their hydration, self-association, protein interactions, and developed a framework for designing zwitterionic anti-biofouling materials. One of our computationally designs has been proven to resist biofouling in full blood by experiment.

My postdoctoral research in Prof. Carol Hallâ€™s group at North Carolina State University focused on understanding mechanisms governing the formation of a corona of proteins around nanoparticles. The protein corona forms around nanoparticles when they enter the human body; it determines the functions of the nanoparticles. We probed the adsorption of amino acids on nanoparticles, developed a coarse-grained protein-nanoparticle model, explored adsorption isotherms of proteins on nanoparticles, and investigated allosteric effects of nanoparticles on the structure and dynamics of human serum albumin. We are also developing generalized models that correlate nanoparticle-protein interaction energy with protein properties.

Teaching Interests:

Teaching shapes the future of the world: delivering knowledge to and sharing experiences with the next generation are as important as discovering secrets in nature. My teaching philosophy can be summarized in four phrases: (1) active learning, (2) motivating student interest, (3) learning by doing, and (4) balancing the needs of individual students with the needs of the whole class.

I am fully committed to teaching and look forward to developing new courses, mentoring graduate students, and educating undergraduate students. I served as a teaching assistant for three undergraduate chemical engineering courses (Thermodynamics, Process Design, and Introduction to Molecular and Nanoscale Principles), and taught thermodynamics to undergraduate and graduate students. I am confident and happy to teach any chemical engineering course, particularly thermodynamics and modeling-related ones. I also plan to develop new courses that introduce how recent development of computational technologies can help shape chemical engineering at the undergraduate and graduate levels.

505340 Photoautotrophic Synthesis of Designer Polysaccharides

Abstract:

Research Interests:

The synthesis of a variety of complex products can be achieved through microbial bioprocessing. These environmentally-friendly processes can convert renewable resources into the compound of interest without the creation of toxic by-products. Industrial microbiology has relied on organisms, including *E. coli* and *S. cerevisiae*, which require the addition of an organic source of reduced carbon, limiting profitability. While the use of autotrophic microbes could address this issue, the ability to forward engineer these potential microbial cell factories is in its infancy. Synthetic biology tools that provide complex and specific control of gene expression are essential for harnessing these organisms' metabolic potential. I have developed a number of genetic circuits that control transcriptional regulation for the model cyanobacterium, *Synechocystis* sp. PCC 6803. These single-input and multi-input regulators coordinate industrially-relevant signals, including light, oxygen, and the cell's nitrogen status, to control gene expression in the photosynthetic, carbon-fixing host. I have started similar work in the purple nonsulfur bacterium *Rhodopseudomonas palustris* CGA009, a photosynthetic, nitrogen and carbon-fixing bacterium that can also break down aromatic compounds, such as depolymerized lignin, anaerobically.

These synthetic biology tools can now be applied to the production of a number of compounds of interest. One such family of compounds is polysaccharides. Cyanobacteria produce complex polysaccharides, composed of a number of different sugars and non-carbohydrate constituents, which are polymerized and exported to the cell's surface. The cyanobacteria change the composition of these exopolysaccharides (EPS) to help them adapt to a range of environmental conditions. The basics of the EPS enzymatic process are understood, but synthetic control to produce specific functionalities has yet to be achieved. I will utilize my experience developing synthetic biology tools for photoautotrophs to engineer cyanobacteria as novel polysaccharide producing hosts. The potential applications are broad, including water treatment, antivirals, biofilm formation, stress tolerance, phage resistance, and more. The development of a cyanobacterial host for the production of polysaccharides could lead to new and unique compounds that are designed based on the specific need.

Successful Proposals:

Nebraska Corn Board, NSF Graduate Research Fellowship, People, Prosperity and the Planet Student Design Competition for Sustainability

Postdoctoral Project:

"Anaerobic conversion of monolignols into a plastic precursor"

Under the supervision of Dr. Rajib Saha, Department of Chemical and Biomolecular Engineering, University of Nebraska

PhD Dissertation:

"Genetic Circuits for Transcriptional Regulation in *Synechocystis* sp. PCC 6803"

Under the supervision of Dr. Tae Seok Moon, Department of Energy, Environmental & Chemical Engineering, Washington University in St. Louis

Selected Publications:

CM Immethun and TS Moon. in *Synthetic Biology of Cyanobacteria* Genetic Engineering of Cyanobacteria (Springer/Nature Press) Invited chapter (in preparation).

CM Immethun, DM DeLorenzo, CM Focht, D Gupta, CB Johnson, TS Moon. Physical, Chemical and Metabolic State Transcriptional Regulators Expand the Synthetic Biology Toolbox for *Synechocystis* sp. PCC 6803, *Biotechnol Bioeng.* **114**:7 (2017).

N Wan, DM DeLorenzo, L He, L You, **CM Immethun**, G Wang, EEK Baidoo, W Hollinshead, JD Keasling, TS Moon, YJ Tang. Cyanobacterial Carbon Metabolism: Fluxome Plasticity and Oxygen Dependence, *Biotechnol Bioeng.* **114**:7 (2017).

CM Immethun, KM Ng, DM DeLorenzo, YC Lee, B Waldron-Feinstein, TS Moon. Oxygen-Responsive Genetic Circuits Constructed in *Synechocystis* sp. PCC 6803, *Biotechnol Bioeng.* **113**:2 (2016).

BM Berla, R Saha, **CM Immethun**, CD Maranas, TS Moon, HB Pakrasi. Synthetic Biology of Cyanobacteria: Unique Challenges and Opportunities, *Front. Microbio.* **4**:246 (2013).

Teaching Interests:

Knowledge acquired in college covers only a small part of what an engineer will need to know throughout their career. It provides the foundation from which we start solving a problem; yet, the broad, complex and dynamic situations engineers face require the confidence to think creatively and the ability to direct our own learning. Students' development of a firm knowledge base from which they have the confidence to solve the variety of problems they will face throughout their careers is my first educational objective. My students will tackle realistic problems designed to increase their comfort with resolving underdetermined systems, addressing the complex web of needs and constraints, thinking creatively, and communicating effectively. Practicing an organized and dynamically controlled thought process in new learning situations is the second educational objective for my students. By assessing the task at hand, evaluating the available resources, planning the approach, applying and monitoring the planned strategies, reflecting on the success of the planned strategies, and revising by adjusting the approach, securing additional resources and sometimes even re-assessing the task, students become self-directed learners. My courses utilize goal-directed practice with targeted feedback to help them embrace the learning process and realize its utility in solving the problems they will encounter. I returned to academia propelled by the desire to help students construct the foundation from which they can make important contributions to the challenges we face as a society. My instructional strategies and assessment methods are designed to support that goal and my educational objectives.

Selected Teaching Experience:

Teaching Assistant, Chemical Process Dynamics and Control - Received the EECE Graduate Student Teaching Award, based on student evaluations

Advisor, International Genetically Engineered Machine (iGEM) Competition - Washington University in St. Louis and University of Nebraska teams

Mentor, 11 Graduate Rotation and Undergraduate Students, Washington University in St. Louis - 7 are co-authors on my publications

Judge, International Genetically Engineered Machine (iGEM) Competition's Giant Jamboree

Teacher Workshop Instructor, BioBuilder® Educational Foundation

494781 Production of Artificial Cell Membranes Bearing New Characteristics or Behaviors Using "Click" Chemistries

Abstract:

Towards the development of an artificial cell, it is preferable to generate liposomes containing synthetic lipids capable of imparting valuable new characteristics and behaviors to the membrane system. Here, we present a more facile method for the incorporation of distinct moieties onto the phospholipid structure using "click" chemistries. To demonstrate one utility of this system, phospholipids bearing a photocleavable moiety have been synthesized. When these lipids are incorporated into a liposome and irradiated, non-spherical vesicles undergo pinocytosis, enabling uptake of resources from the external environment. This has the potential to greatly expand the functional lifetime of artificial cells leading to greater product generation and getting closer to a "living" cell.

Research Interests:

Building upon my current work on artificial cellular systems, I will be pursuing future research into the dynamics that proteins play in synthetic membrane systems. This work will deal with the dynamics of proteins at a synthetic membrane surface and/or the expression systems that reside inside to produce proteins of interest. Together, this work will allow me to delve deeper into the realm of building an artificial cell, as it pertains both to the understanding of the development of current cellular systems as well as more efficient methods to express proteins for the study of genetic diseases or for the production of protein-based drugs.

Teaching Interests:

My interests lie at the intersection of biology, chemical engineering and materials science. Research at this interface has important implications in a range of fields from medical advancements to valuable, new material design. To foster the continuation of this interplay between fields I believe in the inspiration of young chemical engineering students to pique their interest in biology and the vast knowledge we can borrow from it. For this reason I would like to teach courses such as biology or biochemistry for engineers as well as biomaterials courses. In addition to teaching in the classical classroom setting, I am dedicated to individual mentoring as is evidenced by my work in lab as well as my work as a graduate mentor for the undergraduate Discovery Learning Apprenticeship end of year project presentations and symposium.

505632 Self-Organization in Soft, Active Materials

Abstract:

Research Interests:

Soft and active materials self-organize from macromolecules to form complex structures such as biological cells. Each structure within the cell has distinct properties and function derived from interactions of the constitutive macromolecules. Despite advances in biochemistry that continue to reveal specific interactions, how biomolecular assemblies and soft materials can be regulated and self-organized remain exciting frontiers. In cells, many assemblies, such as cytoskeletal structures, are formed from filaments, which have inherent anisotropy. Others, such as molecular motors, structure the assemblies through enzymatic activity. Building from soft materials, colloid and polymer science, I plan to develop an interdisciplinary and collaborative research program that focuses on experimentally elucidating the physical mechanisms of self-organization in active biological materials. Additionally, I plan to engineer novel colloidal and polymeric materials inspired by the rich set of self-organization strategies gleaned from biological materials.

My postdoctoral research at the University of Chicago focuses on structure formation in active biopolymer materials that derive from the protein filaments and molecular motors which form the dynamic structure that drives cellular deformations. I am particularly interested in intriguing liquid condensed phases of biopolymer filaments that we recently demonstrated with cross-linked actin filaments. I am interested in elucidating the role of anisotropy and activity in structuring these complex filamentous fluids. As a graduate student at the University of California, Santa Barbara, my research focused on deformations in biologically inspired membranes. I investigated supported phospholipid bilayer formation through vesicle adsorption and stress-induced transitions from planar to tubular bilayer. Additionally, I was part of an interdisciplinary team that engineered phospholipid vesicles to target specific viruses with aim of developing novel therapeutic strategies.

Teaching Interests:

I value inquiry-based interactive teaching with a focus on training interdisciplinary, independent scholars. With a Ph.D in Biomolecular Science and Engineering, an interdepartmental program, and undergraduate background in Physics, I have a diverse training from physical and biological perspectives. I look forward to applying my background to teach core courses and am excited to develop courses at intersection of physical sciences, biological sciences, and engineering. I am particularly interested in teaching experimental courses, especially those that emphasize introducing undergraduates to research. As a graduate student, I emphasized inquiry as teaching assistant in Physics and Biology courses at both upper and lower division levels. Throughout my graduate and postdoctoral research, I have enjoyed mentoring several undergraduate and graduate students in interdisciplinary and collaborative research projects.

505224 Tough Gradient Double Network Hydrogels for Artificial Implants

Abstract:

Research Interests:

Gradient material exhibits a continuous variation in any of its physical or chemical properties. For example, tendons, fibrous tissues found in our bodies, illustrate naturally occurring gradient materials; they exhibit a continuous change in modulus, which assists them to achieve efficient connectivity between soft muscle tissues and hard bones. Mimicking such gradients using synthetic hydrogels with tunable stiffness will have a tremendous impact on the development of medical implants. The research program that I envision to develop in my research group will explore new methods for generating tough gradient hydrogels. The strategy will be to synthesize double network hydrogels (DN) by incorporating degradable network strands, wherein the first network can be degraded under acidic conditions while maintaining the second network intact. Such a system will also allow for complete characterization of DN strands after their formation by selective degradation of either one of the two networks to examine the network properties of the other gel. After gaining fundamental insights into the structure of DN, the project will focus on the development of gradient DN (gDN), demonstrating a continuous variation in their network density and modulus. A key advantage of gDN is that a single specimen can be used to investigate the effect of varying a given parameter systematically (*e.g.,* elasticity or molecular concentration). I am also interested in exploring the non-thrombogenic coatings, and the detection method for thrombi formation in blood contacting devices, *e.g.,* left ventricular assist devices (VADs). VADs are used as life support for many patients with heart failure or those waiting for the heart transplantation, which can take months to several years. This project will develop a new technique to target platelets using biocompatible microbubbles (MBs) as ultrasound contrast agents. Such a noninvasive and low-cost molecular imaging technique will have a tremendous impact on the blood contacting device industry, where monitoring thrombus formation (*in-vivo*) is of utmost importance.

Selected Publications

- C. K. Pandiyarajan and Jan Genzer, Effect of Network Density in Surface-Anchored Poly (N-isopropylacrylamide) Hydrogels on Adsorption of Fibrinogen, *Langmuir 2017*, 33, 1974-1983 (Link).
- C. K. Pandiyarajan, Michael Rubinstein, and Jan Genzer, Surface-Anchored Poly (N-isopropyl acrylamide) Orthogonal Gradient Networks, *Macromolecules 2016*, 49 (14), 5076-5083 (Link).
- C. K. Pandiyarajan, Oswald Prucker, and Jürgen Rühe, Humidity Driven Swelling of the Surface Attached Poly (N-alkylacrylamide) Networks, *Macromolecules 2016*, 49 (21), 8254-8264 (Link).
- C. K. Pandiyarajan, Biofunctionalized Hydrogels for Tissue Engineering, *Advanced Science News: Regenerative Medicines, 2016* (Link)
- C. K. Pandiyarajan, Oswald Prucker, Barbara Zieger, and Jürgen Rühe, Influence of the Molecular Structure of Surface-Attached poly (N-alkylacrylamide) Coatings on Blood Platelet Adhesion, *Macromolecular Bioscience, 2013,* 13, 873-884 (Link).
- C. K. Pandiyarajan, Oswald Prucker, Barbara Zieger, and Jürgen Rühe, *Macromolecular Bioscience, 2013,* 13, 7, 817 [Front cover page, Link].

Teaching Interests:

I am qualified and happy to teach undergraduate courses on Organic Chemistry, Inorganic Chemistry, Physical Chemistry, Biochemistry, and Thermodynamics. For graduate students, considering my previous experiences, I can teach advanced courses on Polymer at Interfaces, Proteins at Interfaces, Molecular Spectroscopies and Surface-Analytical Tools (SPR, OWS, ATR and TOF-SIMS) to mention a few.

505353 Transcriptome-Guided Cell and Gene Therapy Strategies to Treat Neurodegeneration

Abstract:

Research Interests:

Neurodegenerative disorders affect large patient populations and cause enormous suffering and socio-economic burden. Every year in the US alone, several million patients suffering from neurodegenerative disorders cost the US economy billions of dollars in healthcare and lost economic productivity. Typically, these disorders lack cures and effective long-term treatments, and as a result patients often succumb to disease 10-20 years after onset. Therefore, new treatment strategies are urgently needed to address the growing burden of neurodegenerative disorders in an increasingly aging population. Stem cell and gene therapy based strategies, when properly designed and implemented, can be powerful treatment options. However, limited understanding of stem cell biology and of target cell populations *in vivo* often hampers such efforts. Fortunately, recent improvements in RNA sequencing can offer increased understanding of cell populations at a single-cell resolution. Leveraging my interdisciplinary training in engineering and biology, I aim to develop a multifaceted approach to treat neurodegenerative disorders using single-cell transcriptomics to guide biomaterial, gene, and cell-based treatment technologies.

Research Experience:

In my doctoral work, I developed non-viral, lipid based vectors for targeted gene delivery to cancer tissue. The optimized vector achieved high target specificity with minimum off-target effects *in vitro* and *in vivo*. In my postdoctoral work, I diversified my skillset with additional tools relevant to disease treatment. First, to help meet the high demand for transplantable cells for cell replacement therapy (CRT), I developed efficient, scalable methods to generate region-specific neurons from human pluripotent stem cells (hPSCs) using a biomaterial scaffold. In parallel, to address some of the major challenges facing successful CRT, I developed cell-instructive biomaterial scaffolds to enhance survival, dispersion, and integration of transplanted neurons. These strategies significantly improved CRT treatment outcomes in rodent models of Parkinson's disease and Huntington's disease. Finally, I used cutting-edge RNA sequencing technology to investigate the effect of culture platforms on stem cell behavior at the single-cell level. Thus, armed with tools to enhance gene and cell therapies, I am excited to develop multifaceted, combinatorial treatment strategies for neurodegenerative disorders.

Education:

B.S., Chemical Engineering, Biology, MIT (2007)

Ph.D., Chemical Engineering and Materials Science, University of Minnesota (2013)

Ph.D. Dissertation:

A Platform for Next-generation Cancer Therapies: Multi-targeted Non-viral Vectors for Site-specific Gene Delivery and Expression.

Supervised by Prof. Efrosini Kokkoli, Department of Chemical Engineering and Materials Science, University of Minnesota

Post-doctoral projects:

a) Developing a hyaluronic acid based transplantation scaffold to enhance post-transplantation survival, dispersion, and integration of hPSC-derived dopaminergic neurons to facilitate cell replacement therapy in a rodent model of Parkinson's Disease.

b) Using hPSC-derived striatal cells generated in a 3D biomaterial to evaluate cell replacement therapy in a progressively degenerating mouse model of Huntington's disease.

c) Using single-cell RNA sequencing to characterize the effect of culture platform on stem cell population heterogeneity.

Supervised by Prof. David V. Schaffer, Department of Chemical and Biomolecular Engineering, University of California, Berkeley

Grants:

California Institute of Regenerative Medicine (CIRM) Postdoctoral Fellowships 2014, 2015 (Both successful)

CIRM RT3 07800 2015 (Helped PI; Successful)

NIH R21 2016 (Helped PI; Unsuccessful)

Selected Publications (5 of 15 total):

1. **M.M. Adil**, Z.S. Erdman, E. Kokkoli, Transfection mechanism of polyplexes, lipoplexes and stealth liposomes in $\alpha_5\beta_1$ bearing DLD-1 colorectal cancer cells, *Langmuir*, 30 (2014) p3802-3810
2. **M.M. Adil**, G.M. Rodrigues, R. Kulkarni, A. Rao, N. Chernavsky, E.W. Miller, D.V. Schaffer, Efficient derivation of hPSC-derived midbrain dopaminergic neurons in a fully defined, scalable 3D biomaterial platform, *Scientific Reports*, 2017 7:40573
3. **M.M. Adil*,** T. Vazin*, B. Ananthanarayanan*, G. M. Rodriques, S. Kumar, D.V. Schaffer, Engineered hydrogels increase the post-transplantation survival of encapsulated hESC-derived midbrain dopaminergic neurons, *Biomaterials,* 136(2017) 1-11
4. **M.M. Adil**, T. Gaj, A. Rao, R Kulkarni, C.M. Fuentes, G.N. Ramadoss, E.W. Miller, DV Schaffer, hESC-derived striatal cells generated using a scalable 3D hydrogel promote recovery in a Huntington's disease mouse model, *In revision*
5. **M.M. Adil**, T. Ashuach, N.E. Chernavsky, G.N. Ramadoss, S. Dudoit, D.V Schaffer Single-cell RNAseq reveals culture platform dependent stem cell heterogeneity, *In preparation*

Future Research

The overarching goal of my lab will be to facilitate treatment of neurodegenerative disorders utilizing biologically-informed engineering strategies. Specifically, I will take three separate, initially independent, routes to address this challenge:

1) **Identifying optimal cells for cell replacement therapy**. CRT is a promising strategy to address neurodegenerative disorders, however clinical efficacy of this approach has only been demonstrated for Parkinson's disease (PD). For many other incurable neurodegenerative diseases, especially ones affecting the striatum, the optimal composition of transplantable cells for successful CRT is currently unknown. One focus in my lab will be to identify the transcriptomic signature of human pluripotent stem cell (hPSC)-derived populations which can facilitate CRT in striatal neurodegenerative disorders. Subsequently, this

transcriptomic information will guide generation of appropriate cell types for effective CRT in striatal neurodegeneration.

2) **Biomaterial cues to guide stem cell fate**. Effective control over the stem cell fate will facilitate a range of biomedical applications including CRT. However, many aspects of stem cell biology are still poorly understood, resulting in heterogeneity in stem cell cultures and low directed differentiation yields with poor reproducibility. Recently, there is increasing evidence that material microenvironments can have profound effects on stem cell fate. In my postdoctoral work, we demonstrated that culture within a defined 3D material can limit unwanted differentiation, increase proliferation, and improve yield of directed differentiation. Additionally, single-cell RNA sequencing revealed platform-specific transcriptomic signatures. Guided by this transcriptomic knowledge, my lab will aim to identify specific material cues that control stem cell fate. Ultimately, this knowledge will inform materials design for effective stem cell differentiation and transplantation in CRT applications.

3) **Targeted gene delivery to the striatum**. Cell replacement therapy, while a promising strategy to treat neurodegenerative disorders, may benefit from combinatorial treatment schemes to maximize treatment benefits. My postdoctoral work showed that while transplanted striatal cells alleviated Huntington's disease symptoms for a period of time, rodent models eventually succumbed to the disease as a result of progressive neurodegeneration. My lab will investigate if therapy may benefit from additional approaches to halt degeneration and enhance graft survival. Leveraging my doctoral training, we will focus on developing effective non-viral vectors to deliver therapeutic genes specifically to supporting cells such as striatal astrocytes, with targeting strategy informed by the transcriptomic profile of the target cells.

In the long-term, I aim to incorporate findings from each of these three tracks into an optimized combinatorial strategy to address striatal neurodegeneration. The multidisciplinary nature of this proposed research provides many opportunities for collaboration, especially with computational and neuroscience labs.

Teaching Interests:

Among the classes in my academic career, I have gained the most from the ones that encouraged critical thinking, interactive learning, and creative problem-solving, while drawing parallels to real-life scenarios with the occasional, relevant anecdote. I strive to implement such features in the classes I will teach.

I taught core Chemical Engineering undergraduate courses as a teaching assistant at the University of Minnesota, where responsibilities included classroom lectures in discussion sections. Additionally, I mentored nine undergraduate students (two during my graduate school and seven during my postdoctoral years) in research; six of my mentees co-authored publications with me, three students went on to medical school, and three students are in graduate school. Several of my mentees completed their undergraduate honors theses under my direct supervision. Furthermore, I have mentored two graduate students on semester-long rotation programs in the lab. Through my undergrad (MIT) and graduate (University of Minnesota) education in leading Chemical Engineering departments, I am well-trained to teach common core curriculum courses, and possess extensive experience in research mentorship. Additionally, I am interested in developing an interactive graduate/senior undergraduate level course to teach and discuss the latest engineering approaches in cell and gene therapy.

505481 Cancer Immunotherapy, Cell Imaging and Drug Delivery from Self-Assembled Structure

Abstract:

Research Interests: Cancer immunotherapy, computational modeling, cell imaging, and drug delivery

Teaching Interests: Colloids, Kinectics, Engineering Mathematics, Bionanotechnology

Cancer immunotherapy, especially monoclonal antibody (mAb)-based solid tumor therapy, is challenging due to various parameters that can impede the tumor delivery and penetration of therapeutic agents. The antigen (Ag)-mediated tumor targeting of mAb may be hampered by the presence of shed antigen in blood circulation because a high shed antigen concentration in the blood could act as a decoy preventing mAbs from binding to antigens expressed on tumor cells. My research is directed toward development of immunotherapy and cancer treatment based on nanomedicine to overcome the challenges in cancer immunotherapy. My research results presented here include: 1. Cancer immunotherapy *via* radiolabeling and image analysis 2. Computer model development for cancer immunotherapy 3. Stem cell imaging. 4. Drug delivery from self-assembled structures. For cancer immunotherapy, the tumor and organ uptake of Cu-64 or Zr-89 amatuximab has been studied and computer models including pharmacokinetics were developed to simulate the biodistribution and tumor uptake of Cu-64 and immunotoxin therapy effects. These studies further can be extended to simulate animal experiments and clinical studies to evaluate therapy effects and tumor progress. Stem cell tracking is still vital to understand cell and disease fate. To understand better the biological activities, stem cell tracking by multimodal superparamagnetic iron oxide nanoparticle (SPION) is also presented. Drug delivery based on self-assembled structure especially with biopolymers, vesicles, and liposomes is presented. Future research will explore personalized therapy based on targeted delivery (e.g., using recombinant DNA, antibody and exosome-like liposome nanoparticles) and drug screening for cancer therapy, computer-aided therapy diagnosis and prediction, and cellular engineering. As a scientist, I joined as reviewers in many journals and performed reviewing activities. Currently I serve as editorial members in some journals. I had submitted K99 grant as a fellow to gain grant process and am preparing for grants.

Navigating the Unpaved Roads and Knowing the Unwritten Rules: Advancement for Teaching-Focused Faculty (Abstract ID# 483241)

Lisa G. Bullard
North Carolina State University

Unlike the tenure track faculty ranks where mentors are routinely assigned and the requirements to obtain tenure are clearly articulated, teaching-focused faculty face a landscape in which there may not be a senior faculty member to serve as a mentor, and the path to promotion may not be clearly defined or defined at all. Departmental voting faculty may not understand how to evaluate the promotion dossier for a teaching-focused faculty member. Departmental resources for professional development may be limited, and the funding landscape is less certain. The good news is that there are travelers who have walked this way before and established some signposts along the way. This paper will provide practical suggestions for teaching-focused faculty who would like to hone their teaching and educational scholarship skills, demonstrate leadership and contribution on- and off-campus, and obtain recognition for their work.

What each institution calls these individuals may vary greatly; some of these might include "teaching professors", "lecturers", "professors of the practice", "non-tenure stream", 'non-tenure track", etc. (For the purposes of this paper, I will refer to the position as a "teaching professor"). Likewise, each university has policies and procedures about promotion and tenure. Make sure that you are aware of those policies and follow them – to the letter. Typically these policies are located on the web site of the Provost or other academic officer. They may also be found on the web site for the Office of Faculty Development or similar organization. These are the "official" documented policies. However, often there are "unwritten" (and sometimes unspoken) policies for promotion in each department. The suggestions below are intended to help non-tenure track faculty best position themselves for contribution, advancement, and promotion. Ready? Let's begin!

Take advantage of opportunities to hone your teaching skills, demonstrate leadership and contribution on- and off-campus, and obtain recognition for your work.

- Find ways to get in front of your faculty and make them aware of your "high impact" contributions – make presentations at faculty meetings or workshops, industrial advisory board meetings, conduct teaching/advising tips workshops within the department.
- Attend on- and off-campus teaching workshops. At some point, you may want to start teaching such workshops yourself!
- Professional societies offer regional and national workshops as well as opportunities for leadership in specific divisions of interest. There may be other relevant professional societies in which you can become involved. Consider running for office or volunteering to chair a committee either at the regional or national level.
- Make conference presentations related to your teaching and/or educational research at regional and/or national conferences. Look closely at what you are already doing that you can share – it doesn't have to be "extra" work.
- Identify relevant educational or trade journals related to your field or area of teaching and consider submitting an article. Some journals accept one-page "Teaching Tips" which can be a good way to get started with a good idea to share that doesn't require a lot of assessment or funding.
- Participate in and take a leadership role in course and curriculum committees, coordinators of advising committees, accreditation teams, etc.
- Apply or ask to be nominated for teaching and advising awards as appropriate. If there is an award that is "expected" for faculty in your role, be intentional about examining the criteria for the award and determining what you need to do to be a strong candidate.
- Find out about TA training for your department or college and offer to present a workshop or participate in an existing training session.
- Consider participation in multidisciplinary or multi-institution research studies in your area of interest or teaching specialty. (Conferences are a great way to meet others who teach in your area or share similar interests).
- Move forward best educational practices across your department and college. Take initiative in promoting teaching effectiveness efforts on campus and within your department.
- Work with your Department Head to become an associate (or full) member of the graduate faculty. This is needed for NSF and other funding applications.

- Identify potential co-advising opportunities for PhD students -- those in discipline areas or those for students are interested in educational sides of fields. This is especially appropriate if being co-PI on a grant will lead to publications and external visibility.
- Negotiate with your department to pay for one conference a year, with the possibility for two if you have a paper published or presented. You might also ask for a laptop in your first year until you get funds to replace it several years in. AIChE and ASEE provide excellent opportunities to present your work and network with others.
- Promote your work (teaching, advising, research, extension, engagement) on Facebook, Twitter, the departmental web site, etc. College publications are always looking for good stories. Don't be too humble!
- Make sure that you are an active and visible part of departmental life. While you may not be part of the DVF, you should still have "a seat at the table" when discussions are held about courses, curricula, faculty hiring, and departmental governance. Make sure that you are included on email communications within the department. Actively participate in faculty meetings and make sure that you are listed in departmental materials (web site, recruiting materials, etc.)

Apply for grants for educational or disciplinary research

- Seek out interactions with potential funding agencies; face-to-face interactions with program directions are invaluable in building relationships and networking.
- Volunteer to be on a review panel by sending a 1-2 page CV to the Director. This is a great way to see both successful and unsuccessful proposals and make contacts/connections in your field.
- Make sure you're submitting your proposal to the right division – call the program director and discuss before spending a lot of time targeting the wrong solicitation.
- Prepare a one-page white paper as a discussion tool before generating a long document.
- Get connected in the "broader impact" area of other proposals. Make sure that the PI lists you as a co-PI or includes funding for your work, funding for your travel to conferences, or some type of recognition/compensation.

Manage Your Statement of Mutual Expectations

- Maintain a *Statement of Mutual Expectations* (SME). Depending on your institution, the name of this document may be different, but it typically documents the mutually-agreed-upon expectations from both you and the department during your appointment. It is a written description of the mix of your realms of responsibility – teaching, research, and extension. It is typically prepared during your first year and is maintained in your personnel folder through your course of service to the University.

- The SME should be reviewed annually with your Department Head, and it should be changed when necessary, e.g. after promotion and when changes to departmental leadership occur.
- You should discuss with your Department Head how your realms of responsibility may need to change if you plan to pursue promotion. For example, it's difficult to be engaged with external collaborators, do educational research, and demonstrate external impact if you are teaching multiple sections of large class each semester. Any changes need to be documented and dated.
- Evaluation for promotion consists of (a) SME (your assignment), (b) departmental rules (the level of quality expected) and (c) your annual review/assessment (are you meeting these expectations?) You should be getting feedback from your Department Head every year. If not – ask.
- Ultimately it is your responsibility to make sure that you are not working on a regular basis outside the realms articulated in your SME. An occasional activity is not a problem, but a regular assignment in a realm outside the SME may raise a red flag. If this is the case, then your SME may need updating.
- Explore your department's or college's strategic plan and put information in the SME that reflects the mission/vision/goals and is aligned with what you want to do now or in the future. The SME can be a starting point for a discussion of what you want your career to look like. It may involve some iteration or negotiation.

Prepare Your Package for Promotion

- Find out what format promotion documents require and begin to document your yearly accomplishments in that format. Number the publications and activities (for those who might just be looking at the numbers).
- Know the policies and regulations that affect your employment at your institution.
- Make sure that you have a formal Peer Evaluation of your teaching on a routine basis, as designated by your department. This is typically a critical component in the review and promotion processes.

- Discuss the issue of promotion with your Department Head at your annual review. Be sure your Department Head knows of your interest to submit your dossier for promotion by the previous spring. Departments typically submit a list of those "going up" in the early fall, when promotion packages are due.
- Request a milestone review by the Departmental Voting Faculty (DVF) in advance of your going up for promotion. This helps the faculty see what you are doing and provides a chance for early feedback.
- Teaching faculty are typically evaluated against the standards of their department and within the context of college and department RPT rules. Talk to your Department Head about his/her understanding of those standards and rules.
- How is "scholarly activity" defined? For teaching-focused faculty, it can be more than what happens in the classroom. This could include service responsibilities, participation/leadership in professional organizations outside the university, attending workshops, sitting on panels, etc. In other words, "talking and thinking about teaching". **"Research"** may be within your discipline or in the area of teaching and learning. The expectation is that you are "scholarly in whatever you do".
- Do successful candidates have to have something to "show" in every realm of responsibility? No. However, to be eligible for promotion, you do need to show that you are doing the items on the SME really well. Remember that "the SME is not aspirational, it's achievable". Other things can be thrown in (administrative assignments, service, etc.) You may have something in the dossier that is not on the SME – you can still include it.
- If you are the first teaching-focused faculty to go up for promotion, or your department is not familiar with the criteria for teaching faculty promotion, benchmark others in your college or at the university. Ask your Department Head if a teaching faculty member from another Engineering department can review your package and provide input and feedback about your contributions.
- Find someone in the DVF to be your advocate/sponsor. They can speak up for you at promotion time. Identify someone who is influential and a thought shaper in the department, and make sure they know what you're doing and what impact you have had.
- When composing your candidate statement, look at the wording of the promotion criteria *and use those same words.* Make it easy for the committee to see how you have met the criteria.

Create Community

Identify other teaching-focused faculty in your department, college, or university, in addition to faculty at other institutions who have similar roles. Besides being a valuable source of new ideas and approaches, it's encouraging to have similar-minded colleagues with whom you can vent – and celebrate. As a group, you might also be able to influence organizational policy around non-tenured promotion procedures or advocate for resources or training. If there's not such a group in your college or university, start one!

Conclusions

The suggestions in this paper are based on the author's personal experience, as well as the experience of other teaching-focused faculty at North Carolina State University and colleagues at other institutions. The title of this paper, which refers to "unpaved roads and unwritten rules", suggests that while your career path as a non-tenured track faculty member may be less predictable, it is also yours to define. Whether your focus is teaching, educational research, advising, administration, or some combination of these, you play a key role in the department and in the lives of your students. If you follow the advice in this paper, there are no guarantees of promotion, but I predict that you'll feel more empowered in your role and understand what is required to seek promotion, if that is your choice.

Oklahoma State University
School of Chemical Engineering

Production and Stabilization of a Fiberless Adenovirus Gene Vector

Grit Kupgan
Anna Condacse*
Josh Ramsey

AIChE Annual Meeting
30 October 2017

Gene Therapy

- Process of treating a disease through the transfer of genetic material, DNA, or RNA, into the patient's cells with a gene delivery vector

- **Applications**
  - Inherited Diseases
  - Acquired Diseases
  - Medical Conditions

- **Challenges**
  - Immunology
  - Targeting
  - Multiple treatments
  - Cost

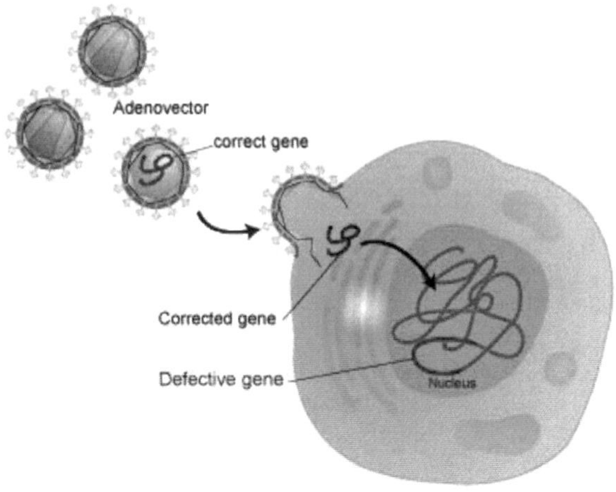

http://www.genome.gov/

2

Ex vivo and *In Vivo* Gene Therapy

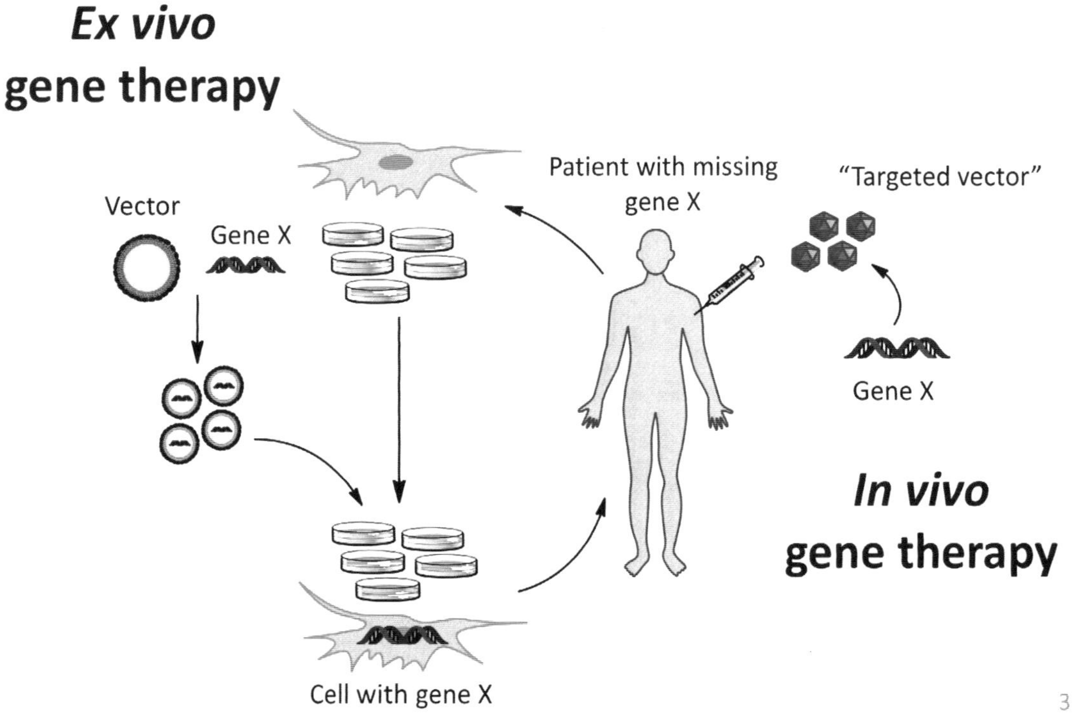

3

Success Stories

- In 1990, Ashanti DeSilva, who had SCID, was treated for the first approved gene therapy trial.

Life Science Foundation

- In 2000, children with SCID-X1 were treated by *ex vivo* gene transfer.

Foundation Fighting Blindness

- In 2009, an 8 year old boy with Leber's congenital amaurosis was successfully treated using AAV vector.

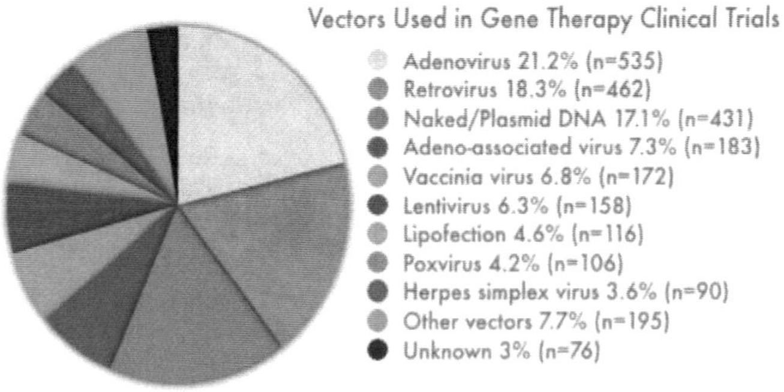

The Journal of Gene Medicine, © 2017 John Wiley and Sons Ltd

Sheridan (2011)
www.wiley.co.uk/genmed/clinical

4

Native Adenovirus Vector

Advantages

- Nonpathogenic / non-oncogenic
- Capable of infecting dividing and non-diving cells
- High level of transgene expression
- Easily purified and large DNA capacity

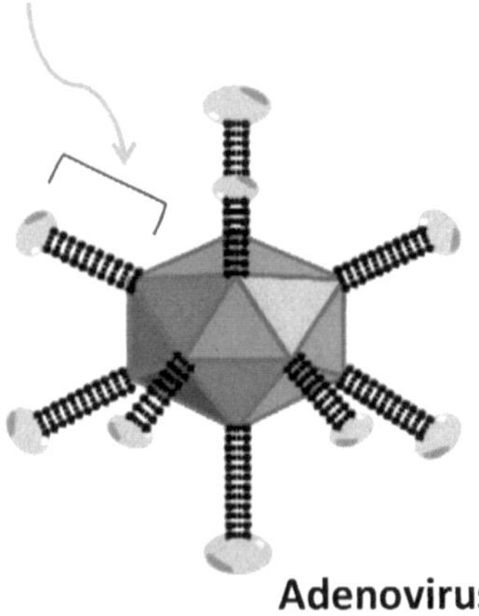

Disadvantages

- Immunogenicity
  - *Fiber protein recognized by neutralizing antibodies*
  - *Fiber initiates innate immunity*
- Promiscuous Tropism
  - *Adenovirus binds to Coxsackie and Adenovirus receptor CAR (+) cells*
  - *Cancer cells are CAR (-)*

4

Fiberless Adenovirus Vector

Advantages

- Nonpathogenic / non-oncogenic
- Capable of infecting dividing and non-diving cells
- High level of transgene expression
- Easily purified and large DNA capacity

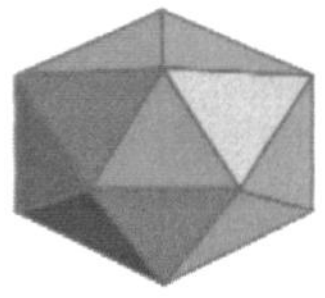

Disadvantages

- ~~Immunogenicity~~
  - ~~*Fiber protein recognized by neutralizing antibodies*~~
  - ~~*Fiber initiates innate immunity*~~
- ~~Promiscuous Tropism~~
  - ~~*Adenovirus binds to Coxsackie and Adenovirus receptor CAR (+) cells*~~
  - *Cancer cells are CAR (-)*

4

The Big Picture

Adenovirus vector without promiscuous tropism

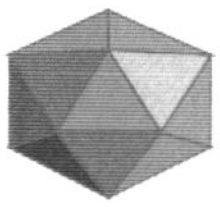

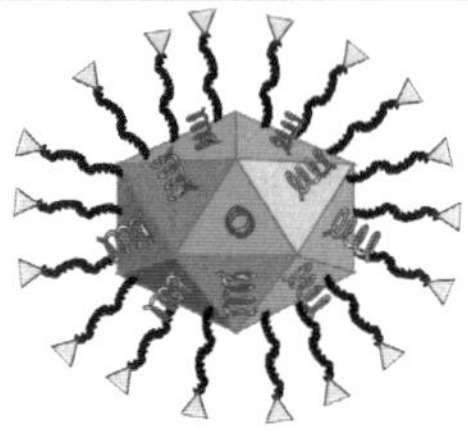

Stable and safe targeted adenoviral vector for *in vivo* gene therapy

Ligands for vector targeting fused with polyethylene glycol (PEG) to reduce immune responses

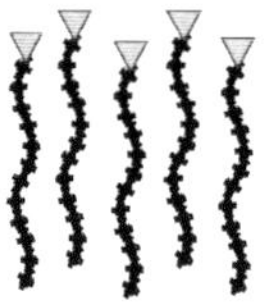

Cell penetrating peptides (CPPs)

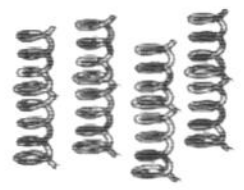

7

Approach

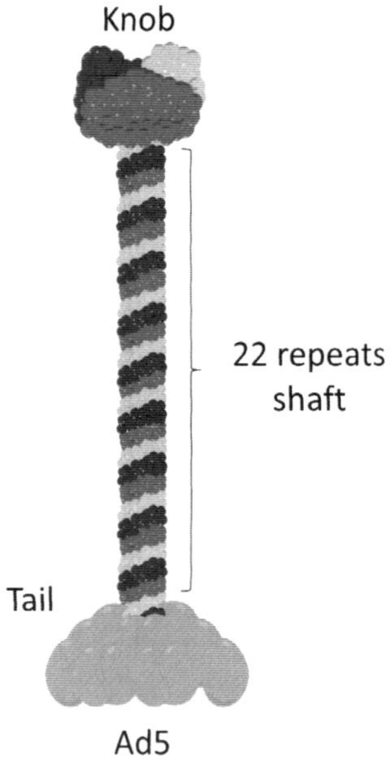

8

Methods

- Perform sequential homologous recombination events in *E. coli* with positive/negative selections

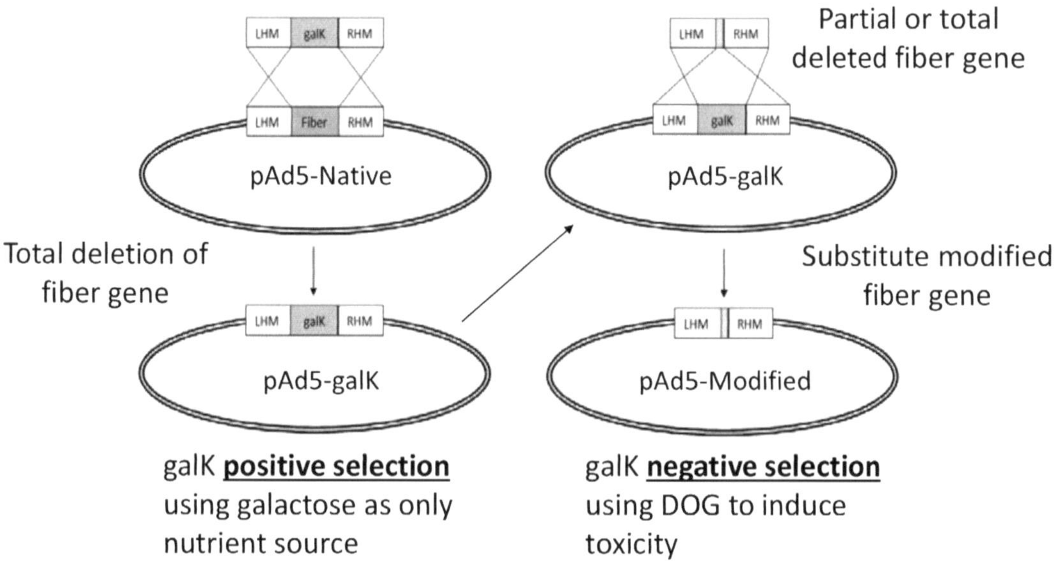

9

Fiberless Virus Stability

- The physical stability of mutants can be ranked as:

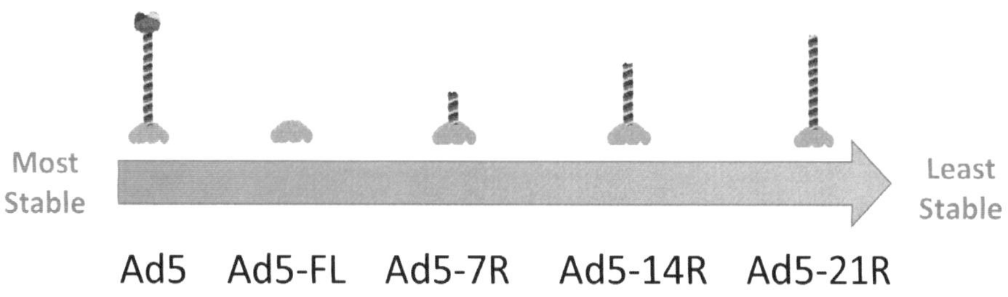

10

Stabilization of Fiberless Adenovirus

Native Adenovirus (Ad5) **Fiberless Adenovirus (Ad5-FL)**

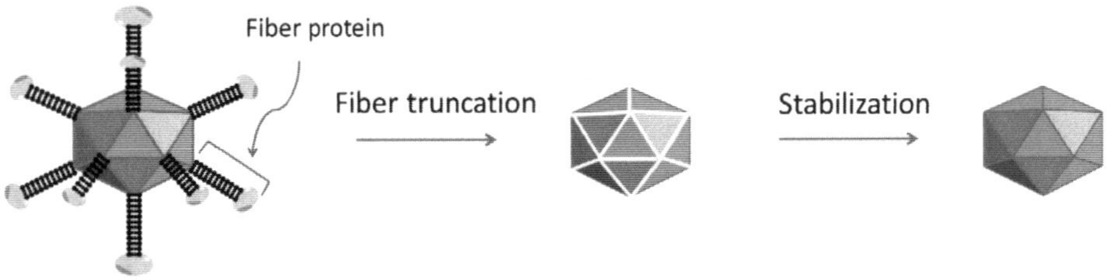

Disadvantages associated with the fiber protein

- Infect many types of cells. Vector prefers to bind CAR receptor (liver, kidney, lung, brain, heart, colon, and many more)
- Fiber initiates innate immunity and is also recognized by neutralizing antibodies

Truncation of the fiber protein

- No negative effects from fiber interactions
- However, the vector has lower stability than the native adenovirus

Methodically stabilize fiberless vector

- The vector has enough stability for gene delivery applications

11

Methods

- Explored effects of pH and temperature of virus stability
- Further improved stability of fiberless adenovirus by adding exipients

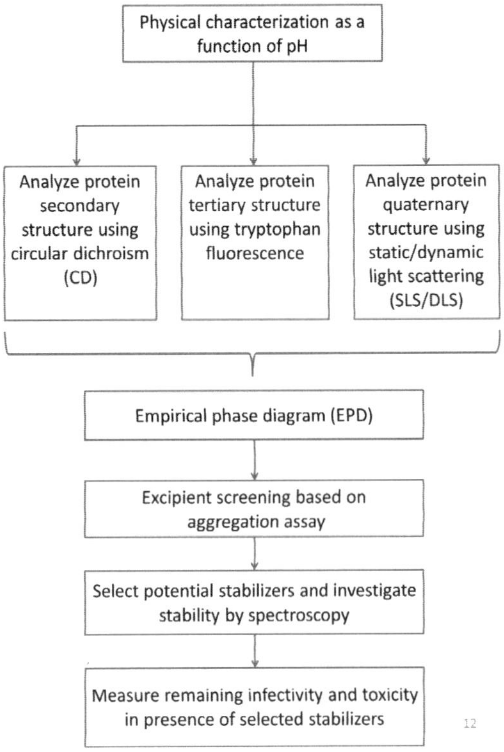

12

Results – The Effect of pH on the Virus

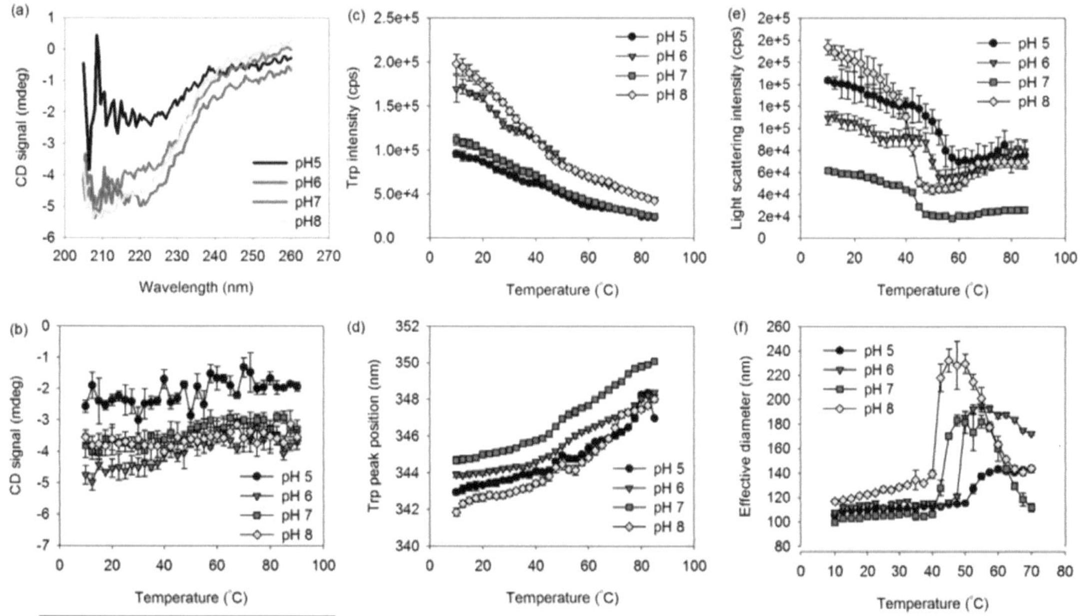

pH	SLS, T_m (°C)	DLS, T_m (°C)
5	51.75 ± 0.02	52.37 ± 0.53
6	49.59 ± 0.03	48.63 ± 0.21
7	43.96 ± 0.04	43.15 ± 0.23
8	41.45 ± 0.27	41.07 ± 0.54

a) CD Spectra
b) Signals from CD at 222 nm
c) Tryptophan Fluorescence
d) Tryptophan Peak Position
e) Static Light Scattering
f) Dynamic Light Scattering

Results- Empirical Phase Diagram

- Empirical phase diagram (EPD) represents overall stability trends from spectroscopy data
- Generated from CD at 222 nm, tryptophan intensity and peak positon, and light scattering data

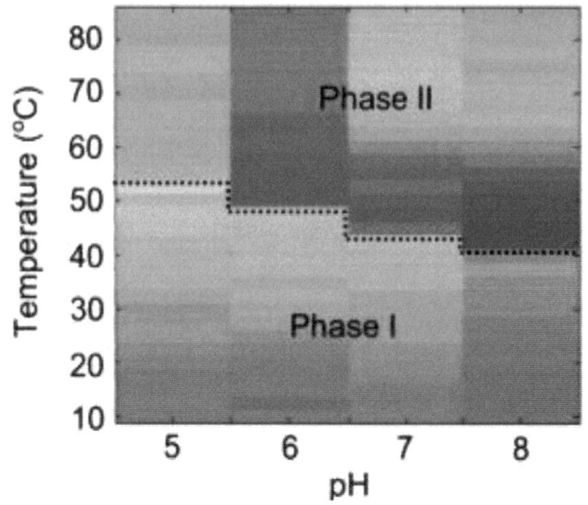

Results – Spectroscopy Analysis

- Ranked by excipient ability to prevent aggregation
- Four best excipients chosen for spectroscopy studies:
  - ➢ 0.1% Pluronic F-68
  - ➢ 0.0075 M Dextran T70
  - ➢ 0.3 M Glycine
  - ➢ 0.3 M Diethanolamine

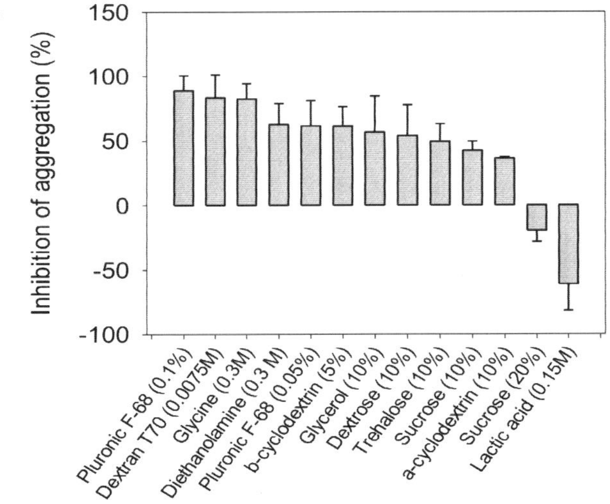

15

Results – Spectroscopy Analysis

- Spectroscopy analysis using selected excipients:
  - a) Tryptophan intensity
  - b) Tryptophan peak position
  - c) Static light scattering
  - d) Dynamic Light Scattering

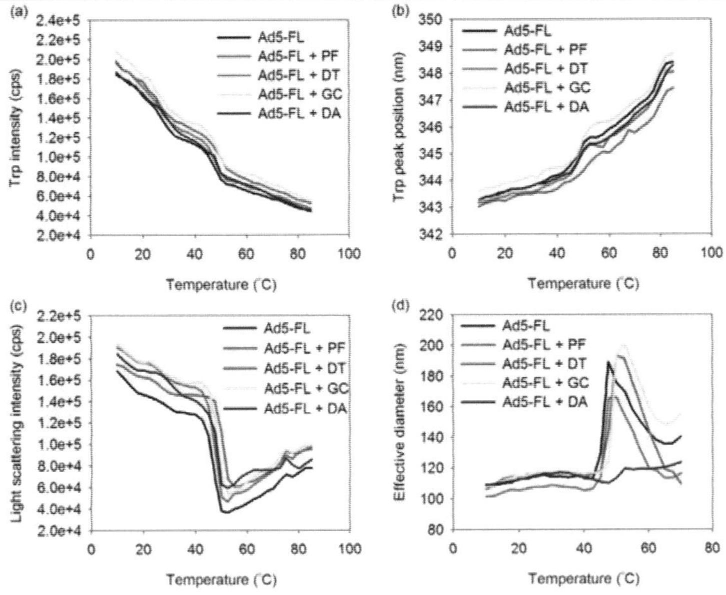

Excipients (M or %, w/v)	SLS, ΔT_m (°C)	DLS, ΔT_m (°C)
Pluronic F-68 (0.1%)	3.82 ± 0.31^a	2.16 ± 0.25^a
Dextran T70 (0.0075M)	0.61 ± 0.70	0.24 ± 0.30
Glycine (0.3M)	1.81 ± 0.70^a	2.96 ± 0.37^a
Diethanolamine (0.3M)	1.37 ± 0.29^a	0.80 ± 0.53^a

[a] The reported ΔT_m were statistically different from Ad5-FL control (p<0.05).

16

Results – Remaining Biological Activity and Toxicity

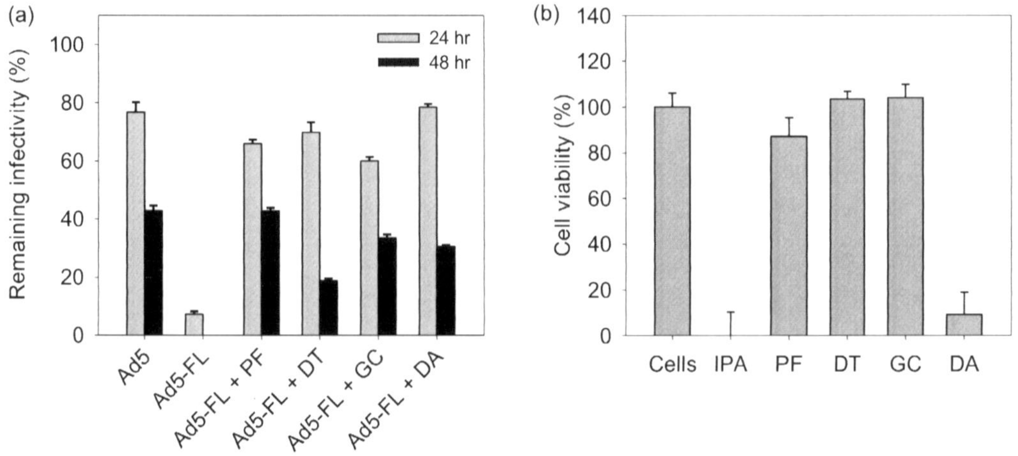

- The remaining biological activities measured by infectivity assay after a 24 h (grey) or 48 h (black) incubation period at 43°C

- The toxicity of different excipients after 48 h

17

Conclusions

- The fiberless virus is more stable than the mutants, but less stable than the native adenovirus

- Stability of fiberless adenovirus is improved in acidic conditions (pH~6)

- Non-ionic surfactants and amino acids improved biological activity

18

References

Kupgan, G. (2014). *The Effect of Fiber Truncations on the Stability of Adenovirus Type 5.* Oklahoma State University Department of Chemical Engineering. Molecular Biotechnology, **56**: 979-991.

Kupgan, G., Choudhari, S., Flynn, N., Nigatu, A., Vupputuri, S., Picking, W., Ramsey, J.(2017). *Identification of Excipients for Stabilizing Fiberless Adenovirus as Pharmaceuticals.* Journal of Pharmaceutical Sciences; Elsevier Inc. 1-8.

19

Acknowledgements

This work was supported in part by:

- Scientist Development Grant through the American Heart Association (AHA)

American Heart Association®
life is why™

- National Science Foundation Career Award

NSF

20

Production and Stabilizaition of a Fiberless Adenovirus

Conclusion – Providing a portion of the shaft protein does not improve stability of the virus when compared to the fiberless adenovirus. Although the fiberless adenovirus is less stable than the native adenovirus, the stability of the fiberless adenovirus can be enhanced through the addition of excipients and regulation of temperature and acidic conditions.

Grit Kupgan

Anna Condacse

Dr. Josh Ramsey

Oklahoma State University

21

499077 Protein Nanocarrier for Targeted Intracellular Delivery of Functional Antibodies

Abstract:

The cell membrane remains a formidable barrier for antibody-based therapies, and efficient intracellular delivery of functional antibodies would enable modulation of intracellular signaling mechanisms and protein-protein interactions involved in various disorders. This study utilized protein engineering techniques to develop a novel protein nanocarrier (HEX) that is capable of delivering functional antibodies to the intracellular environment. Each nanocarrier contains six SPAB antibody-binding domains, and is therefore capable of delivering up to six antibodies. The interaction between the SPAB domain of the nanocarrier and the heavy chain constant region of the antibody is noncovalent, thus allowing the nanocarrier to bind different functional antibodies with the same affinity. We successfully expressed the protein monomers, assembled the functional nanocarrier, and investigated its antibody-delivery properties. Results of cellular studies using HeLa, MCF-7, as well as SK-BR-3 cancer cell lines indicate significant delivery of fluorescent antibodies bound to the nanocarriers compared to soluble antibody control, as measured by fluorescent microscopy and flow cytometry. We have also achieved specific targeting of HER-2 positive breast cancer cells with our nanocarrier design through addition of targeting antibodies. This result demonstrates the significant potential of this design as a therapeutic delivery platform. Ongoing experiments are aimed at utilizing the HEX nanocarrier to deliver intracellular therapeutic antibodies that inhibit cell signaling pathways. In addition to efficient cellular uptake, the highly biocompatible and modular nature of our nanocarrier makes it ideal for further expanding the scope of antibody-based therapeutics to the intracellular environment.

499221 Novel Chemistries and Engineering for the Replacement of Methylenedianiline in Composites

Abstract:

4,4â€™-Methylenedianiline (MDA) is used to produce high temperature epoxy resins as well as other high performance composites; however, MDA is a known carcinogen and liver toxin. Added safety measures to prevent MDA exposure have increased manufacturing costs, driving the search for potential MDA replacements. Alternatives to MDA have not gained significant traction due to reduced performance, reduced confidence in the material, and they offer a small, if any, reduction in toxicity. In this paper, we present our work on developing structure-property-toxicity relationships of various substituted dianalines by examining the effect of molecular architecture on overall polymer performance and toxicity. Novel dianilines derived from both petroleum and biomass with different numbers of methyl and methoxy substituents on the aromatic ring were synthesized to decrease toxicity and carcinogenic aspects of the resulting material while maintaining thermal and mechanical integrity. Structure-toxicity relationships indicate that multiple substituents on the aromatic ring are necessary to severely reduce the toxicity of the dianiline. Structure-property relationships show that methoxy groups reduce the glass transition temperature and thermal degradation temperature while methyl groups only slightly affect these properties. The addition of substituent groups increases the melting point of the dianiline to varying degrees based on substituent placement.

Free Surface Electrospinning of Microemulsions Containing Fenofibrate

Jack Lift
Uyen Phan
Thai Nguyen
Katarina Guzman
Hovhannes Gregorchuk
Advisor: Dr. Keith M. Forward

Chemical and Materials Engineering
California State Polytechnic University, Pomona

Global AIChE Conference – October 2017

CAL POLY POMONA
College of Engineering

1

Pharmaceutical Industry

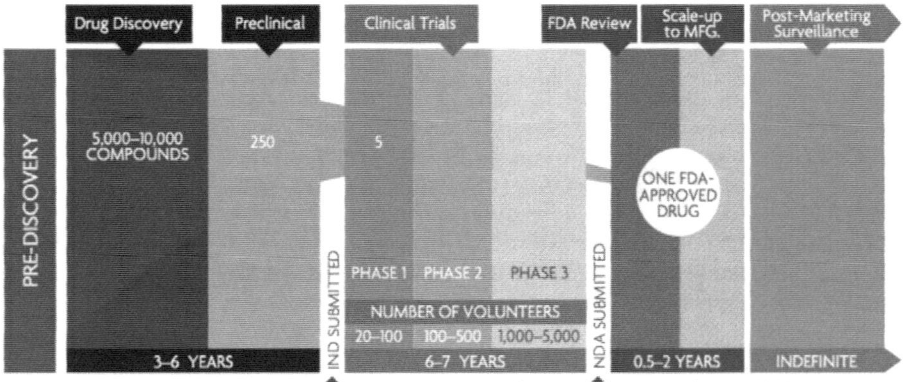

https://www.fda.gov/Drugs/DevelopmentApprovalProcess/SmallBusinessAssistance/ucm053131.htm

CAL POLY POMONA
College of Engineering

2

Motivation

Problems that Plague Pharmaceutical Industry:

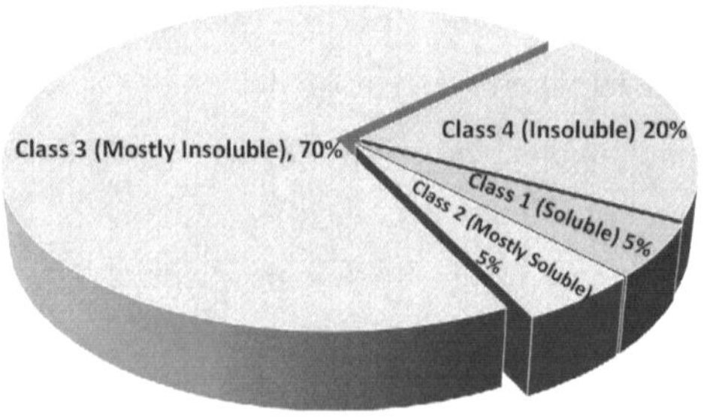

Biopharmaceutics Classification

- ❑ **Bioavailability**
  - ❑ Active Pharmaceutical Ingredients (API's) exhibit poor water solubility
  - ❑ Varied Toxicity in the Body

3

CAL POLY POMONA
College of Engineering

Motivation

Problems that Plague Pharmaceutical Industry:

- ❑ **Productivity**
  - ❑ Batch Process
  - ❑ High Operation Costs
    - ❑ High number of Variables
  - ❑ Waste Production
  - ❑ Work Hazards

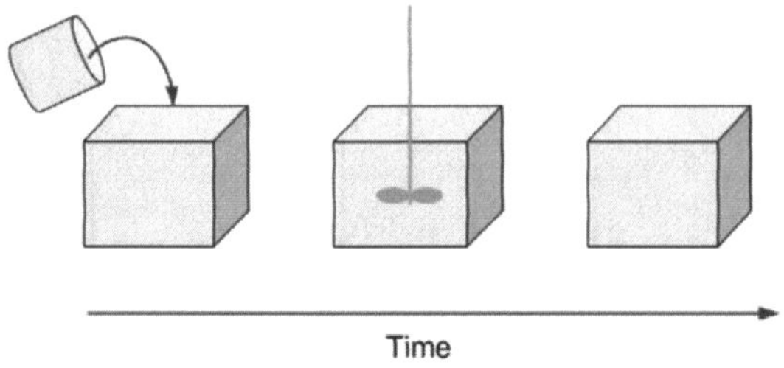

4

CAL POLY POMONA
College of Engineering

Current Methods for Increasing Solubility and Bioavailability of poorly-soluble API

- ❑ **Salt Formation**
  - -Convert API into an ionic salt
- ❑ **Produce in Amorphous-form**
  - -Less structure / order
  - -High microscopic surface area

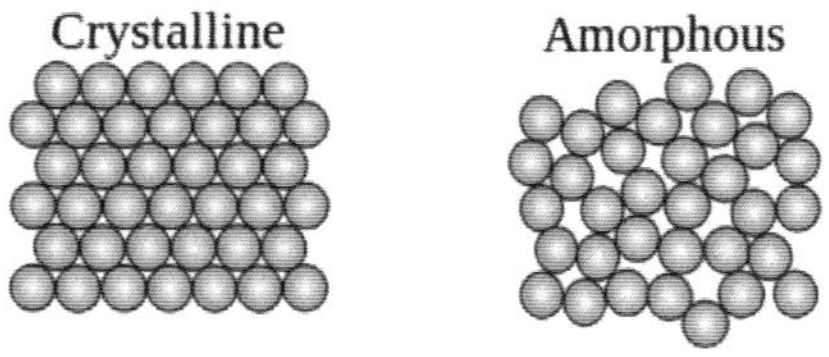

- ❑ **Decrease Domain Size**
  - -Increase effective surface area
  - -Increase dissolution
- ❑ **Addition of Surfactant**
  - -Lower surface tension at hydrophilic and hydrophobic interface.
  - -Promote micelle formation, break crystallinity.

CAL POLY POMONA
College of Engineering

5

Current Methods for Increasing Productivity

- ❑ **Continuous Process**
  - ❑ Increases Time and Cost Efficiency
  - ❑ Accepted by Food and Drug Administration (FDA)
- ❑ **Remove Powder Handling**
  - ❑ Lowers Work Hazards
  - ❑ Triboelectric Charging
- ❑ **Increase Solubility**
  - ❑ Decreases Production Costs

CAL POLY POMONA
College of Engineering

6

Free Surface Electrospinning (FSE) of Microemulsion

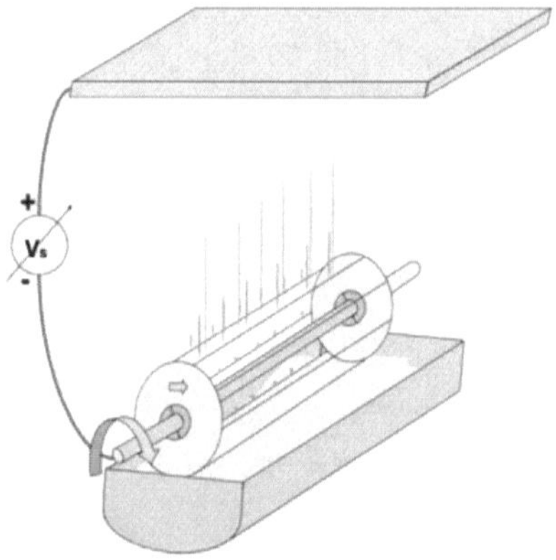

- A cylindrical drum crossed by a wired electrode, swept in a rotary motion, through a bath containing polymeric microemulsion in the presence of a high voltage electric field.
- Liquid droplets are formed on the electrode wire then thinned, stretched, and electrostatic jetted to a grounded plate.
- API encapsulated by nano-fibers are collected on the surface of the grounding plate.

CAL POLY POMONA
College of Engineering

8

Free Surface Electrospinning (FSE) of Microemulsion

- FSE increases Solubility and Bioavailability of API
  - Microemulsion Solution containing surfactant that is also sonicated (creates small domain size)
  - Nano-fibrous mats produced are amorphous.
- FSE is a continuous production process involving liquid phase microemulsions.
- FSE achieves multiple electrostatic jets from a free liquid surface.

http://www.azonano.com/news.aspx?newsID=33048

CAL POLY POMONA
College of Engineering

7

Objective

Increase Bioavailability and Productivity of Powder API's by producing uniform nanofibrous mats encapsulating API through Free Surface Electrospinning of Microemulsion Solutions

9

CAL POLY POMONA
College of Engineering

API Investigation: Fenofibrate ($C_{20}H_{21}ClO_4$)

- **Lowers High Cholesterol and Triglyceride Levels**
- **Inexpensive and Easy to handle**
  - Popular amongst manufacturers and consumers
- **Poor Solubility and Bioavailability**

10

CAL POLY POMONA
College of Engineering

Creating the Microemulsion

Microemulsion Formulation

Organic Phase

- ❑ Fenofibrate
- ❑ Ethyl Butyrate, EB

Surfactant

- ❑ EL-35 (castor oil)

Aqueous Phase

- ❑ Water, Ethanol

Polymers

- ❑ Polyvinylpyrrolidone, PVP
- ❑ Polyethylene Glycol, PEG

11

CAL POLY POMONA
College of Engineering

Addition of Polymer

Polymers added:

- ❑ **PVP - 1.3 Mda**
  - ❑ Commonly used in pharmaceuticals
  - ❑ Bulk material used to **capture microemulsion**
- ❑ **PEG - 35 Kda**
  - ❑ Serves as filler to increase mat thickness

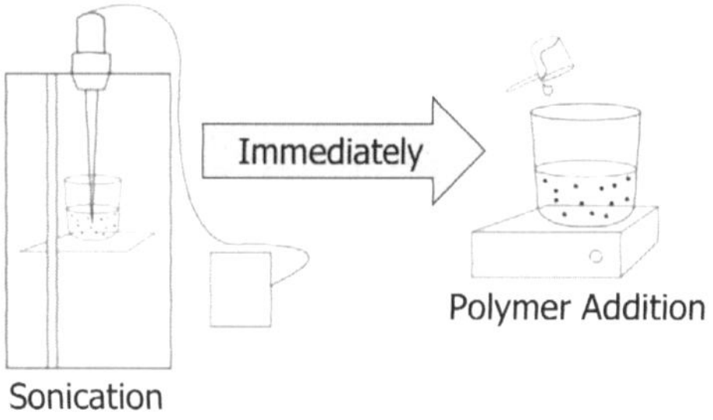

12

CAL POLY POMONA
College of Engineering

Electrospinning Parameters

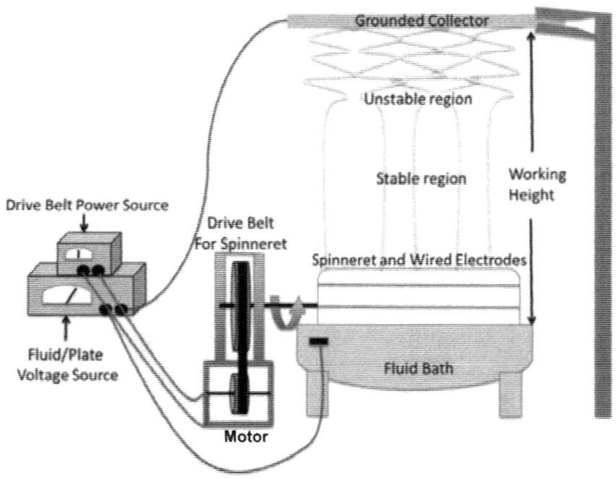

Dependent Variables

- Productivity
- Uniformity and thickness
- Fiber Diameter

Independent Variables

- Applied voltage
- Spinneret speed
- Wire diameter
- Working height
- Humidity
- Temperature
- Polymer Concentration
- Spin Time

CAL POLY POMONA
College of Engineering

13

Results

- ❑ Several solutions of varying microemulsion concentrations were electrospun, and the produced mats were analyzed **macroscopically**
- ❑ Mats quality was determined based on: **Uniformity and Thickness**

CAL POLY POMONA
College of Engineering

14

Mat Quality

4.7 wt% EB 3.8 wt% EL-35	5.7 wt% EB 2.8 wt% EL-35
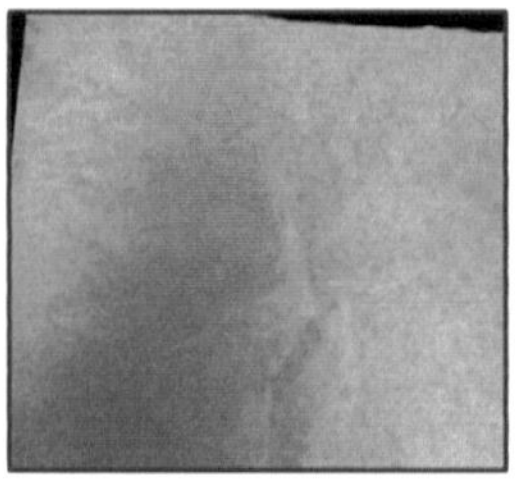	
Undesired Mat	*Improved Mat*

Characteristics

Undesired	Improved
-High viscosity -Thin mat -Splattering	-Low viscosity -Thick mat -Uniform

Electrospin Parameters:

Applied Voltage: *60 kV*
Spinneret Speed: *9 rpm*
Wire Diameter: *0.005 in*
Working Height: *0.42 m*
Relative Humidity: *10%*
Temperature: *~21°C*

15

CAL POLY POMONA
College of Engineering

Conclusion

- ❑ **Summary**
  - ❑ Varying Quality of Fenofibrate mats were produced
  - ❑ The role of **EL-35 surfactant and EB** was studied
- ❑ **Future Research**
  - ❑ Increase Fenofibrate wt % Concentration in Microemulsion solutions
  - ❑ Adjust microemulsion solution concentrations to produce higher quality nanofibrous mats

16

CAL POLY POMONA
College of Engineering

Acknowledgements

We would like to thank our advisor, Dr. Keith Forward
and the work done by previous groups.

17

CAL POLY POMONA
College of Engineering

Questions?

18

CAL POLY POMONA
College of Engineering

499233 Polyvinyl Sulfonic Acid: A Low-Cost RNase Inhibitor for Enhanced RNA Preservation and Retained Function

Abstract:

RNA is a central molecule of life and is being used for many applications including medical diagnostics, protein engineering, and therapeutics against debilitating diseases such as cancer, ALS, and HIV. However, RNA is fundamentally limited by the ubiquitous presence of ribonucleases (RNases) that rapidly degrade RNA. RNA is challenging to work with as it requires special ventilation systems, specific supplies, and continuous treatment to prevent RNase activity. As humans, we constantly excrete RNases, which makes working with RNA particularly difficult.

Here we present work to overcome the challenges of RNA degradation in an economical way with polyvinyl sulfonic acid (PVSA). We demonstrate that PVSA is an RNase inhibitor that inhibits not only the common RNase A, but is also effective in *Escherichia coli* lysate, which is rich in diverse RNases. PVSA was shown to improve the integrity of mRNA transcripts by more than 5-fold as measured by their translational viability. We also demonstrate the ability to remove PVSA rapidly and economically using alcohol precipitation.

Most importantly, PVSA is less than 0.06% the cost of commonly used ribonuclease inhibitors. Using ribonuclease inhibitors is typically essential for working with RNA and is the largest reagent cost (commonly more than 95%) in RNA production and storage. As reported here, PVSA greatly reduces the cost of RNA production. PVSA could also reduce the need for special equipment and supplies making RNA applications more accessible to researchers and more economical for consumers.

499313 A Stimulus-Responsive, *in situ* Forming, Nanoparticle-Laden Hydrogel for Ocular Drug Delivery

Abstract:

Glaucoma and other chronic optic neuropathies are the leading causes of blindness worldwide. These diseases are characterized by damage to the optic nerve at the base of the retina, and most medications targeting optic neuropathies are administered as eye drops. However, since corneal penetration efficiencies are typically <5% for such dosage forms, not only are these formulations ineffective, but the dosage of drug is much higher than required, which incurs higher costs. There is a clear, unmet need for novel transcorneal drug delivery vehicles that achieve higher drug penetration at significantly lower dosages. To this end, we have developed a stimulus-responsive, in situ forming, nanoparticle-laden hydrogel for spatiotemporal and dosage-controlled release of poorly bioavailable drugs into the aqueous humor of the eye. The hydrogel is formulated as a composite of hyaluronic acid (HA) and methylcellulose (MC). Both polymers are biocompatible, highly mucoadhesive and approved by the FDA as GRAS. The amphiphilic nanoparticles are composed of poly(ethylene oxide) (PEO) and poly(lactic acid) (PLA). Experimental design aided the identification of hydrogel composition and nanoparticle content in the formulation for stimulus-responsive switching between thixotropy and temperature-dependent rheopexy. The liquid formulation has a sol-gel transition temperature of 32Â°C, which is the temperature of the ocular surface, and its gelation characteristics permit the formulation to form a thin, uniform coating over the cornea through blinking of the eyelid. These properties facilitate application of the medication as an eye drop immediately prior to the patientâ€™s bedtime, and also permit easy and scalable manufacturing, which greatly simplifies the path towards development of a regulated industrial-scale manufacturing process. We subsequently tested the efficacy of our formulation in whole-eye experiments using Bovine eyeballs by loading the nanoparticles with cannabigerolic acid (CBGA), a molecule that is a close mimic of an approved glaucoma medication. Our formulation exhibits over a 300% increase in transcorneal penetration over control formulations. Our work paves the way for the introduction of novel products targeting ocular diseases to the market.

499285 Single use, in vitro biosensors for the detection of T-tau protein and beta-amyloid 42, biomarkers of neuro-degenerative disorders in PBS & human serum using differential pulse voltammetry (DPV)

Abstract:

A single-use, in vitro biosensor for the detection of T-Tau protein and beta-amyloid 42 in phosphate-buffer saline (PBS) and undiluted human serum was designed, manufactured, and tested. Differential pulse voltammetry (DPV) served as the transduction mechanism. This biosensor consisted of three electrodes: working, counter, and reference electrodes fabricated on a PET sheet. Both working and counter electrodes were thin gold film, 50 nm in thickness. Laser ablation technique was used to define the size and structure of the biosensor. The biosensor was produced using cost-effective roll-to-roll process. Self-assembled monolayers (SAM) of 3-mercaptopropionic acid (MPA) were employed to covalently immobilize the anti-T-Tau (T-Tau antibody) or the beta-amyloid 42 antibody on the gold working electrode. A carbodiimide conjugation approach using N-(3-dimethylaminopropyl)-N'-ethylcarbodiimide hydrochloride (EDC) and N–hydroxysuccinimide (NHS) cross-linked anti-T-Tau to the carboxylic groups on one end of the MPA. A T-Tau protein ladder with six isoforms and beta-amyloid 42 peptides were used in this study. The anti-T-Tau concentration used was 500,000 pg/mL. The T-Tau protein concentration ranged from 1000 pg/mL to 100,000 pg/mL. The antibody concentration of β-amyloid 42 used was 18.75 μg/mL. The concentration range of β-amyloid 42 in this study was from 0.0675 μg/mL to 0.5 μg/mL. DPV measurements showed excellent responses, with a good calibration curve. Thus, a practical tool for simple detection of T-Tau protein and beta-amyloid 42, biomarkers of neuro-degenerative disorders, has been was successfully developed. This tool could also be extended to detect other biomarkers for neuro-degenerative disorders, such as P-Tau protein and S100B protein.

503908 Networking for Nerds: How to Land (or Create) Your Dream Job and Keep Your Career Moving Forward!

Abstract:

Wanna land your dream job? Get ready to network! Most jobs and other game-changing career opportunities are not advertised, and even if they are, there is usually a short-list of candidates already in mind. So how do you find out about and access the 90% of jobs and other opportunities that are "hidden"? In this workshop, we will focus on proven networking strategies and tactics to identify new opportunities, locate decision-makers within organizations, solidify your reputation and brand in the minds of those who hire, and gain access to hidden jobs and game-changing opportunities. Discover how networking and self-promotion can enable you to land or even create your dream job from scratch!

485399 Mixing Experiences

Abstract:

MIXING EXPERIENCES

Abstract

Chemical engineers design and develop processes to convert raw materials into desired products. In order to accomplish this, the raw materials may be mixed, heated, reacted and separated as diagrammed in Figure 1. Each process is designed with a specific order to the steps, and each step requires the design of specific equipment. Chemical engineering students are taught to design both the process and the equipment. Full courses, or significant portions of courses are devoted to the heating, reacting and separating steps. However little course time is devoted to the mixing steps and often hands-on opportunities are not available in laboratories.

Figure SEQ Figure * ARABIC 1. A pictorial view of chemical engineering.

Mix

Heat

React

Separate

Mixing is a central operation in most industries: the food, pharmaceutical, paper, plastics, ceramics and rubber industries all include mixing steps. Despite its widespread use, the process is difficult to analyze. Advances in obtaining experimental measurements and analyzing the data are helping to meet the challenges, but unexpected results due to mixing continue to arise during the scale up of a new process. Inefficiencies due to this lack of understanding have been estimated to cost $1-10 billion per year[1].

Mixing experiments have been published, with the purpose of providing students with practical experience[2, 3]. The

experiments investigate mixing of both Newtonian and non-Newtonian fluids using classic impellers. The experiments provide a general introduction and experience with the dimensionless variables. However, the equipment is not available at all institutions, and only two impellers are used.

To address these issues, students completed a mixing experiment at XX University that was developed with a colleague at Philadelphia Mixing Solutions, Ltd[1]. The necessary equipment is likely available in a standard chemical engineering laboratory. Philadelphia Mixing Solutions generously donated six impellers for the study, and will provide similar impellers to institutions that wish to include mixing in the curriculum. Details of the experiment procedure and analysis will be provided in the full paper.

The experiment was implemented for the first time in the Spring 2017 semester. After a brief introduction to mixing, twelve groups of four students measured the mixing time for an impeller at a variety of stirring rates. They calculated the dimensionless mixing time, Reynolds number and power supplied. A guest speaker from the mixing industry presented mixing theory and its limitations and challenged the students to compare their results to theory. One class period was also devoted to scale up methods. In total, three class periods and two hours in the laboratory were spent on the topic of mixing.

An assessment of student learning showed that after completing the experiment:

á The students had a greater understanding of what is ÒmixedÓ. They were able to apply what they learned in process controls (response to a step change) and determine when the solution was 90% homogeneous.

á The students are conversant in the language of mixing. In conversations with the instructor they correctly used phrases such as radial and axial mixing, impeller Reynolds number and power number. The written reports echoed this.

á The students showed an understanding of the trade off between shear and efficiency among different impellers. In fact, when shown six different impellers, the students can now provide their name and their mixing characteristics.

While these students are certainly not mixing experts, they now have knowledge that will be valuable to future employers. Future offerings will include readings from the literature and small hands on activities during class.

References:

1. Grenville, R.K., ÒMixing in the chemical engineering curriculumÓ, Paper ID #11457, American Society of Engineering Education Annual Conference, Seattle, WA (2105).

2. Ascanio, G., Legros, R., Tanguy, P.A., ÒA Fluid-Mixing Laboratory for ChE UndergraduatesÓ, Chemical Engineering Education 37(3) :296-299 (2003).

3. Pour, S.B., Norca, G.B., Fradette, L., Legros, R., and Tanguy, P.A., ÒSolid-Liquid and Liquid-Liquid Mixing Laboratory for ChE UndergraduatesÓ, Chemical Engineering Education 41(2): 101-106 (2007).

487460 Large Changes in the Okstate Unit Operations Lab

Abstract:

Student reviews of the OKState Unit Operations Laboratory courses were very favorable.Â Regardless, and working alongside members of the Teaching and Learning Center, ChemE instructors made very significant changes to the 2-semester courses. Those changes were spurred, in part, by an industrial perspective of the courses and the students. Among those changes were the following: Outlawing of in-lab cell phones, more experiments for each student, individual testing after every group project, anonymous peer reviews at the end of each course; integrations with lecture courses; purposely covering topics not covered in lecture courses; shelving of experiments with high cost-to-benefit ratios; assigning experiments in advance of lecture class instructions; longer and more diverse oral reports; shorter written reports; focusing on working equations as opposed to differential equationsÂ - and - initiation of a somewhat controversial â€œNever Ever List.â€ Of course, evaluations of these changes led to changes in the changes.

499170 Pharmaceutical Engineering: Curricular Integration and a Liberal Arts Perspective on a Chemical Engineering Elective

Abstract:

Most Chemical Engineering curricula require or allow for one or more upper level electives within the discipline to be taken in the junior and/or senior years. Many of these classes focus on technical depth often directly related to the faculty member's research expertise. This talk presents the success of one such upper level elective that takes a broader approach. In this approach, the course focuses on the aspects of pharmaceuticals that both integrate topics from throughout the chemical engineering core curriculum and highlight the impact of pharmaceuticals from a social and humanistic perspective. This talk will highlight the course structure and approach, pedagogical methods used, specific examples of topics included, as well as data detailing the success and areas for improvement within the class.

Based on instructor availability across the department, the Pharmaceutical Engineering elective course is typically offered every 2-3 years, and has included three offerings since the spring semester of 2011. Each time the class has been offered, it has been fully or nearly fully enrolled. Courses at Bucknell are the equivalent of four semester credit hours, thus the course meets four hours per week throughout the semester. In each instance, at least one two hour block of time is included in order to allow for a broad set of pedagogical approaches to be used. As an elective course, it is not used to specifically assess any ABET criteria; however, the approach used leads to that natural incorporation of almost all aspects of ABET 'a' through 'k'.

The course introduces students to the pharmaceutical industry through the lens of the FDA clinical approval process and pharmaceutical drug development. Group presentations are given both on pharmaceutical companies as well as specific pharmaceutical molecule products. These presentations provide a framework for a discussion of the historical, present and future contributions of the industry. Technical content is delivered in both lecture and problem based formats in areas that include crystallization, extraction, filtration and drying and each area specifically considers the related material and energy balances, materials science, fluid mechanics, heat and mass transfer, thermodynamics, unit operations and process design considerations. Social and humanistic aspects are introduced through both class discussion based on both readings and required documentaries, as well as formal classroom debates where two groups present the technical, economic and social considerations to a topic of current interest. The topical questions that have been considered in these formats include:

- What is the future of antibiotics?
- Should pharmaceutical products be directly advertised to consumers?
- Should pharmaceuticals be prescribed to critically ill patients outside of clinical trials?

Along with the positive outcomes of the class, lessons learned and recommendations both from the student and faculty perspective will be considered.

482691 Updating the Process Controls and Dynamics Course for the 21st Century

Abstract:

The goal of this presentation . . .

In this presentation I will provide a topical outline/course schedule for an updated course in Process Controls and Dynamics and explain the contents including examples. The material is available in a new textbook published by the Taylor and Francis Division of the CRC Press entitled, "Designing Controls for the Process Industries". Supplemental author resources will also be introduced.

Motivation . . .

A few years ago I finally had the opportunity to take over a one semester senior (4th year) course entitled "Process Dynamics and Controls". I was looking forward to this as I had developed substantial applied controls experience during my 16 year industrial career. During that time, process controls underwent a complete transformation from electronic instruments to digital distributed control– based systems. I was excited to see new textbooks that reflected this "revolution" in process controls.

Imagine then my disappointment when I found that all the major textbooks in this field were still following the same format and with essentially the same content as textbook published in the 1960s and 1970s! These books emphasize simplified mathematical descriptions of process dynamics using time-dependent linear ordinary differential equations and their analytical solutions using Laplace transform solution methodologies.

The primary goal of these textbooks appears to be to help the reader understand the dynamics of the proportional-integral-derivative controller mathematically, so that the stability of control loops could be properly evaluated. This was an appropriate approach to the subject matter when process control was performed using a suite of stand-alone electronic controllers but is much less important now that control is most commonly performed using sophisticated, integrated distributed control and automation systems. While stability analysis is still of some interest to process control engineers, modern control system algorithms have reduced the importance from primary to secondary for control engineers.

Another big gap in the current literature is a lack of coverage of batch processes and, more importantly, the integration of batch unit operations within continuous processes (such a step being known as a semi-batch unit operation). Of the 17 most common textbooks that I reviewed, only one gave this topic any coverage at all. Yet almost every continuous commercial pathway includes at least one semi-batch step and batch processing represents a sizeable minority of process pathways employed (particularly in certain industries such as pharmaceutical).

Presentation Contents – an Overview of the Textbook . . .

"Designing Controls for the Process Industries" was conceived to address these deficiencies in the currently available literature. The goal is to completely transform chemical engineering process control and process dynamics education to focus on those aspects that are most important for process engineering in the 21st century.

Instead of starting with the controller, the book starts with the process and then moves on to how basic regulatory control schemes can be designed to achieve the process' objectives while maintaining stable operations. Without a deep understanding of the process itself, the power of the modern plant automation system cannot be fully enabled.

As much as possible, I have tried to follow the International Society for Automation's (ISA) guidelines for process control and instrumentation documentation. Some adjustments to the ISA guidelines were made where these improved the clarity of the concepts presented in the text. Most importantly, all of the process control schemes assume that field signals will be converted into digital form at the field device and that control will be accomplished in a distributed control system or programmable logic controller module(s).

In addition to continuous control concepts, I have embedded process and control system dynamics into the text with each new concept presented. I have also included sections on batch and semi-batch processes within new concept areas where appropriate. Finally, sections on safety automation are also included within concept areas.

The four most common process control loops - feedback, feedforward, ratio, and cascade - are introduced in Chapter 2 and the application of these techniques for process control schemes for the most common types of unit operations are provided in Chapters 3-6. For the practicing engineer, these chapters may prove to be the most useful for designing new control schemes or to help troubleshoot existing process instabilities. By comparing the schemes in these chapters to an existing situation, the engineer may be able to identify poorly designed control schemes. Modification of poorly designed control schemes may be an easy and cost effective way to solve process instability problems. This is often a better approach than to try to "tune" your way out of a problem.

More advanced and less commonly used regulatory control options are presented in Chapter 7 such as override, allocation, and split range controllers. These techniques provide additional ways to increase the overall safety, stability, and efficiency for many process applications.

Chapter 8 introduces the theory behind the most common types of controllers used in the process industries. For those instructors that prefer to start with a "what's inside the box" approach, you might want to go through Chapters 1 and 2, and then jump to Chapter 8 prior to Chapters 3-7. For those instructors who are uncomfortable making a complete transition from the older course formats to that presented in this text, Appendix A provides content on how to solve simple linear ODEs using Laplace transforms while Appendix B provides information on PID controller tuning.

Chapters 9-12 provide various additional plant automation-related subjects. An instructor in a one semester course is unlikely to be able to use all of this material, but has the opportunity to emphasize those aspects that they feel are most important. Personally, I use Chapters 9 and 10. I then use Chapter 11 in a capstone design course. Chapter 12 is probably more appropriate for a graduate level class in process dynamic modeling or as part of an advanced transport phenomena course. However instructors who want to emphasize process modeling in their course may wish to use this material.

490677 Safety Considerations When Designing a New Chemical Engineering Research Laboratory

Abstract:

The University of Kansas recently completed an expansion project for the School of Engineering. The new building added interactive classrooms as well as state-of-the-art research facilities. This presentation provides an example of the safety considerations implemented in the design and construction of our research laboratory in the new building. Safety procedures and equipment which are common in industry have been included in the design and function of the new laboratory. A variety of safety systems will be reviewed including ventilated enclosures, gas handling systems, air monitoring, interlock systems, uninterruptable power and personnel protective equipment. Process safety management procedures will also be highlighted and reviewed.

500520 Chemical Engineering Laboratory at the University of Kansas

Abstract:

A comprehensive chemical engineering (ChE) laboratory course is an important component for adequately preparing engineering students for future careers in the chemical industry. The University of Kansas (KU) Department of Chemical and Petroleum Engineering (CPE) has recognized the importance of such a course and has recently constructed a state-of-the-art unit operations laboratory in their new engineering building. The department has committed significant resources towards the laboratory and recently modified how the course is taught. Chemical Engineering Laboratory I is offered to students in the fall semester of their senior year, and seeks to reinforce ChE fundamentals, technical writing, oral communication, teamwork, leadership skills, and safety practices. Three experiments focused on chemical kinetics, fluid mechanics and thermodynamics provide the connection with theory taught in the core ChE courses. The laboratory provides the first opportunity for undergraduate students to apply the knowledge and skills acquired from coursework to actual chemical engineering experiments. Chemical Engineering Laboratory II is offered to students in the spring semester of their senior year and includes additional elements on mass and heat transfer, controls, design, and cost analysis. The three experiments are focused on the distillation of a methanol-water mixture, production of biodiesel from vegetable oil, and PID control. The lab also exposes students to pilot plant scale equipment and reinforces theories taught in the advanced core ChE courses.

This presentation will provide an overview of the labs taught at KU and a detailed analysis of the recently introduced chemical kinetics and vapor-liquid equilibrium experiments.

482574 Biodiesel Production As a Case Study in Chemical Engineering Senior Laboratory at University of Delaware

Abstract:

Transportation accounts for nearly thirty percent of energy used in the United States, and the majority of that energy is sourced from petroleum-based liquid fuels. Offsetting some of these petroleum products with renewable fuels has a number of environmental benefits. One alternative that has been investigated is "biodiesel," or fatty acid methyl esters (FAMEs) that can be produced from a plant-based feedstock, such as vegetable cooking oils.

Because the production and analysis of biodiesel requires a number of principles from chemical engineering, including mass/energy balances, thermodynamics, and reaction kinetics, it provides an excellent case-study for senior-level chemical engineering students. At the University of Delaware, we have one pilot-scale and one bench-scale biodiesel production system in place for student projects. Here, we present the results from a senior chemical engineering laboratory "special project" aimed at improving the product yield from our pilot scale biodiesel production system.

Biodiesel is produced in bulk by base catalyzed transesterification of triglycerides. In our system, approximately 90 lb of soybean oil is reacted with excess amount of methanol and potassium hydroxide catalyst to produce FAME, glycerol, and water. The aqueous phase (water, glycerol, methanol, and catalyst) is then decanted and the FAME is transferred to a wash tank. The FAME is washed with water to reduce the amount of any remaining impurities. The mixture is decanted again, where there is a clear "water" phase, a cloudy "emulsion" phase, and an oil phase. The oil is then allowed to dry under low vacuum overnight and removed from the system for analysis.

A team of four senior-level undergraduate students was tasked with improving the yield of the baseline process described above, in particular by reducing the amount of product that is lost in the "emulsion" phase. We guided the students through four objectives: (1) establishing a baseline process and detailed standard operating procedure (SOP), (2) developing a hypothesis for why there is significant product loss in the emulsion, (3) designing experiments and collecting data on the bench-scale system, (4) applying the results to the pilot-scale system. At the same time, we worked on the GC analysis with a purpose to validate/improve existing analysis tools and to better quantify main components such as FAME, glycerin and methanol in reaction mixtures.

In this talk we will discuss the learning objectives for the course, the experimental apparatuses we have available at UD for this task, the approach, methods, and results from the project, and suggestions for future educational projects using the bench-scale and pilot-scale biodiesel production systems.

491387 Laboratory and Design Projects in Energy Sustainability Based on Industrial Operations and Data (Power Plants, Sugar Mills, Pilot Plants)

Abstract:

The goal of this educational project is to promote concepts of Energy Sustainability throughout the Chemical, Mechanical and Biological Engineering curricula, in the form of class projects large and small, and remote laboratories. This "Energy Sustainability Remote Laboratory (ESRL)" is housed at http://www.esrl.lsu.edu/, with a variety of learning materials available to any faculty or student wishing to use them. There is both a secure (faculty only) area and a public area. To date there are 15 large-scale modules for the study of energy efficiency and production. Several of these are based on data (some real-time, some archived) from a cogeneration power plant that produces ~20 MW of electricity and 90,000 lb/hr steam. Others are based on operating data from various sugar mills, a nuclear plant, a solar photovoltaic facility, and a crystallization pilot plant.

The modules/projects are appropriate to classes ranging from the early Soph. to late Sr. year. Many can be seamlessly used in the 2-4 week cycle of a typical engineering laboratory course, and even more can be used in lecture-based courses from the Soph. to Senior level, in some cases in as little as 1-3 class periods, as assignments or projects. Several suggested assignments of varying duration are provided, with solutions, in the secure area. Ten faculty at five universities have adapted the modules for use in several classes (Thermo, Intro Mass and Energy Balances, Process Dynamics, Jr. and Sr. Labs, Heat and Mass Transfer), yielding valuable student and faculty feedback. More than 500 students have been affected to date.

503670 Digital Alchemy for Assembly Engineeringc

Abstract:

Back in the day, Stuart Churchill was a pioneer in the use of computers for modeling chemical and thermal processes. Since that time, the role of computing in chemical engineering has become indispensable. In this talk, we present new computational methods we call "digital alchemy" that allow for predictive modeling of thermodynamic assembly, and permit the design of building blocks optimized for assembly into target structures with desired properties. In this approach, standard thermodynamic ensembles are generalized to include alchemical variables and their thermodynamic conjugates, allowing for rapid, real-time computational investigation of large, complex design spaces.

503671 Improved Algebraic, Numerical, and Graphical Representations in Fluid Mechanics

Abstract:

The concept of equivalents is applied to fully developed flows through straight channels, and two applications are introduced: the equivalents that produce equal rates of flow through channels of different cross-section, and the diminishment that occurs between the regimes of laminar and turbulent flow. The validity of the eddy diffusivity and the invalidity of the mixing-length as measures of the shear stress due to turbulence have perhaps never before been demonstrated so clearly. The possible choices of primary variables and the consequences are also examined. The objective of these explorations is an improved understanding of the basic fluid mechanics by teachers, students, and industrial practitioners.

503672 The Scaling of Turbulence Near the Wall and the Churchill Turbulent Flux Correlation: Insights with Lagrangian Simulations

Abstract:

The evaluation of mean velocity and mean temperature in wall-bounded turbulent flows has been based on the "classical" theory of turbulence and the law of the wall. The viscous wall parameters, namely the friction velocity, the friction temperature, and the kinematic viscosity of the fluid have been used to scale not only the mean values of velocity, but also other mean turbulent quantities close to the wall, including the Reynolds stresses and the turbulent velocity and temperature fluctuations. Stuart Churchill, as he questioned the validity of the Reynolds analogy between momentum and heat or mass transfer in his works [1], he realized that extending the scaling of momentum to the scaling of scalar quantities in turbulence was not a sound approach. He proposed a scaling of turbulence that depends on the turbulent momentum flux and the turbulent heat flux. In his AIChE Institute Lecture (in Miami, 1998), Churchill presented a summary of his work on this topic, and closed form solutions for the mean velocity and temperature [2]. The turbulent Prandtl number, the most significant parameter for the modeling of turbulent transport not only in theory but also in modern CFD commercial software packages, was also interpreted physically based on these ideas [3]. We have used direct numerical simulation results and results from Lagrangian simulations [4] for the transport of passive scalars in turbulent channel flows to explore Churchill's ideas on turbulent flow scaling and to develop an updated theory for the scaling of heat transport in turbulent flows [5-7]. His passion to remove empiricisms and approximations from the development of predictive correlations in turbulent transport has been an inspiration. In this talk, we will revisit the Churchill scaling in view of our simulation results, and will discuss the counterintuitive finding that wall turbulence can be used to separate rather than mix substances under special circumstances [8].

1. Churchill, S. W., 1996, "A Critique of Predictive and Correlative Models for Turbulent Flow and Convection," Ind. Eng. Chem. Res., 35, pp. 3122–3140.
2. Churchill, S.W., 2000, "Progress in the Thermal Sciences: AICHE Institute Lecture," AIChE Journal, 46(9), pp. 1704-1722.
3. Churchill, S. W., 2002, "A reinterpretation of the Turbulent Prandtl Number," Ind. Eng. Chem. Res., 41, pp. 6393–6401.
4. Papavassiliou, D.V, 2002, "Turbulent transport from continuous sources at the wall of a channel," Int. J. Heat Mass Transfer, 45(17), pp. 3571-3583.
5. Le, P.M., and Papavassiliou, D.V., 2008, "On the scaling of heat transfer using thermal flux gradients for fully developed turbulent channel and Couette flows," Int. Commun. Heat Mass Transfer, 35(4), pp. 404-412.
6. Srinivasan, C., and Papavassiliou, D. V., 2011, "Prediction of Turbulent Prandtl Number in Wall Flows With Lagrangian Simulations," Ind. Eng. Chem. Res., 50(15), pp. 8881–8891.
7. Srinivasan, C., and Papavassiliou, D.V., 2013, "Heat transfer scaling for wall bounded turbulent flows," Applied Mechanics Review 65(3), Art. 031002.

8. Nguyen, Q., Srinivasan, C., and Papavassiliou, D.V., 2015, "Flow induced separation in wall turbulence", Phys Rev E, 91, Art. 033019.

503674 Flow Boiling Using a Piranha Pin Fin Heat Sink

Abstract:

Flow Boiling using a Piranha Pin Fin Heat Sink

Corey Woodcock, Xiangfei Yu, Yoav Peles[1] and Joel Plawsky

Rensselaer Polytechnic Institute; [1]University of Central Florida

The *Piranha Pin Fin* (*PPF*) microdevice is a heat exchanger engineered to dissipate extreme thermal loads using either dielectric fluids such as HFE7000 or water. Fabrication of the device is CMOS compatible and the first generations of the device have been fashioned from silicon substrates. The PPF operates on the principle that heat transfer would be more efficient if heated fluids could be removed nearly as fast as they are created. This process would keep the driving force as high as possible. Thus the heat exchanger is built as two layers; a primary substrate that dissipates the bulk of the heat and shunts the heated fluids to a booster section that continues to heat the fluid and extends the total surface area of the device. Three generations of devices were fabricated. The first two generations were single-layer systems designed to help understand the operation of the primary layer and optimize the shape and location of the pin fins. The third generation was the first, two-layer system to be fabricated and tested.

Figure 1 Representative PPF array

Correlations for the heat transfer coefficient, h_{tp}, based on the Boiling Number and the Weber Number in the PPF design was developed and validated with Gen II. The discrepancy of heat transfer coefficient between predicted and measured values was within 20%. The ratio of two-phase to single-phase pressure drop was correlated with the Boiling Number (Bo) and the Jakob Number (Ja). The discrepancy of the two-phase multiplier between prediction and experimental measurements were within 12%.

The third generation *PPF* heat sinks were the first series to provide both a primary substrate populated with *PPF*s, as well as a booster substrate populated with solid micro pin fins. The substrates are micro-aligned and fusion-bonded to provide a full three dimensional (3D) architecture with bulk silicon properties. Experimental results from the Gen III *PPF* microdevices provide a proof of concept of the true *PPF* heat sink concept with fluid venting into multiple substrates. At low flowrates (G = 670 [kg/s/m^2]) the Gen III *PPF* heat sink has been shown to dissipate heat loads approaching 1.1 [kW/cm^2] at an exit quality of 95%, while at higher flowrates (G = 3,300 [kg/s/m^2]) the thermal limit of the heat sink remains unknown due to exceeding the available power supply at >1.4 [kW/cm^2].

503673 Transport Problems in the Spirit of Stuart Churchill for Teaching and Research at the University of Michigan

Abstract:

Stuart Churchillâ€™s solutions to important heat transport problems have greatly influence both pedagogy and research. The correlation of Churchill and Chu for the natural convection heat transfer to or from a vertical plate, over an astonishing range of Rayleigh numbers is based partly on a similarity solution of a type common in heat, mass, and momentum transfer for problems in chemical engineering. One such problem is asphaltene or wax transport and deposition onto the wall of an oil pipeline, which can be solved using similar methods, as will be discussed. In addition, such problems, and the wealth of their applications make for excellent motivations and examples in the teaching of graduate or undergraduate transport, as will be discussed in the talk.

489253 Generating Student-Created Exam Solutions: An Activity for Repetition and Reflection

Abstract:

Posted exam solutions are strategically archived by student groups, e.g., a fraternity, to give future group members extra problems to practice or memorize problems that instructors may reuse. In lieu of posting exam solutions, a post-exam homework assignment has been constructed and used over several years in different undergraduate courses. The individual assignment provides an opportunity for all students to review and reinforce difficult course material covered on a recent exam. An exam rubric is provided to explain the distribution of points, so the rubric serves as a guide to completing the assignment. Office hours with the professor and teaching assistants are also available to help students 'catch up' and successfully complete the assignment. If students earned an insufficient score on one or more exam problems, the student writes 1. a complete solution (for their version of the exam in a large course) and 2. at least two complete sentences reflecting on the solution or other points of emphasis suggested by the instructor. If students earn a sufficient score on a problem, they write a very brief statement. For example, "I earned xx points on question 1." where xx is your score, and draw a smiley face. Therefore, all students in the course must complete the assignment to earn homework points. An additional team component provides students an opportunity to check their solutions and discuss the exam outside of class (i.e., a peer instruction activity). For example, spend at least 15 minutes discussing different ways to solve each problem, and write at least two sentences about what you learned from the team's discussion. An example assignment and rubric will be shown and shared freely.

494833 A Student-Created, Open Access, Living Textbook

Abstract:

Most courses in engineering use textbooks to supplement lecture-style learning. Unfortunately, while core Chemical Engineering classes have a well-established repertoire of textbooks, the availability of high quality textbooks for topic-specific elective courses can be limited. A significant challenge related to textbook creation and maintenance for such elective courses is the ever-increasing breadth and depth of knowledge. Many instructors circumvent this lack of a formal textbook through the use of peer-reviewed journal articles. This approach builds a connection between course content and cutting-edge scientific research. However, it can be challenging to provide a broader resource of background knowledge on the field. To augment this approach, we suggest a mechanism to allow for the development of student-led, online textbooks that can provide both general background information on a topic and can be editable in real time to keep pace with scientific developments within a field without requiring the printing of new editions.

We used wikis hosted at <openwetware.org> to allow students to create and update their own textbooks for two classes at the University of Massachusetts Amherst; Tissue Engineering (CHEM-ENG 575), and Microfluidics and Microscale Analysis in Materials and Biology (CHEM- ENG 590E). Each year, the instructor identifies a range of topics that require either the development of a new wiki page, or improvement from previous years. Students then have the flexibility to choose a topic related to their own interests, integrating the idea of a flipped classroom with student self-directed learning. Wiki page creation can then be coupled with other student-led activities such as lectures, discussions, or hands-on design projects. Student response to this approach has been extremely positive, citing reasons such as "wiki pages are better to understand and cheap (free)," that "information was compressed and easier," that they found it "better to learn by contributing," and that the "wiki is well-synthesized, with concise information and good jumping points." Additionally, the integration of wikis into the classroom represents an excellent opportunity to cover many aspects of the current ABET accreditation criteria (a)-(k), many of which are more difficult to integrate into the core chemical engineering curriculum.

References:

1. Course website for Tissue Engineering: http://openwetware.org/wiki/ChemEng_590B
2. Course website for Microfluidics and Microscale Analysis in Materials and Biology: http://openwetware.org/wiki/CHEM-ENG590E

499873 Adapting Best Practices from Middle School Classrooms to Chemical Engineering Courses

Abstract:

Highly effective middle school teachers create lessons that are differentiated, incorporate constant feedback, and maximize classroom time. This talk will review several simple teaching techniques that have been adopted from secondary school classrooms into university-level chemical engineering courses. These include the use of individual white boards, daily drills, and differentiated activities and assessments. These techniques can help structure a class period, address common areas of weakness without sacrificing significant class time, increase student engagement, and prevent â€œteaching to the middle.â€ This talk will demonstrate several techniques and present student feedback on their effectiveness.

498307 Giving a Grade to Teamwork

Abstract:

Throughout the undergraduate engineering curriculum students have the opportunity to work in team settings. These team experiences could be anything from occasionally working together on homework to a full capstone design project experience. Teaming activities are prevalent in industry as well. Because teaming is such an important part of becoming a successful engineer, we should be helping students learn how to work on teams. Indeed even ABET feels so strongly about preparing students for teamwork that it gives teamwork its own Student Outcome.

To emphasize improving team skills and learning how to function on teams in our course, we have been assigning grades to teamwork. While the obvious item to grade is the product of teamwork (for example the project or homework assignment), what is harder to assess is team health. Unless we followed the team around twenty-four seven throughout the term we would not be privy to how teams handle work distribution, decision-making, and conflict. We rely on the teams to report how well their team is functioning.

In our course students work on teams for a full term project. We collect team feedback twice throughout the term through a survey administered by teamwork assessment software. Students are asked a series of questions to assess themselves and their teammates on contribution to the teamâ€™s work and interactions with each other. The instructor and team members review the feedback to get a picture of the current team health. At the midterm feedback review, teams are asked to identify one developmental goal to improve team performance. At the final feedback survey, students are invited to comment on whether they have met their developmental goal. Teamwork grades are based on improvement on the developmental goal, improvement between midterm and final feedback surveys, participation in teamwork assessment process, and the assessment of the product of teamwork (for example final project or written report.)

484928 Laboratory Measurement: Much Ado about Everthying

Abstract:

Today, many Bachelor of Science Chemical Engineers graduate with little laboratory experience. Thus, they have little or no experience with laboratory-scale equipment. They also graduate with little experience at applying statistics to actual data sets.

In this presentation, we discuss an experiment designed to enlighten future Chemical Engineers about laboratory containers and delivery systems and the use of statistics to choose the appropriate container for making analytic-quality dilutions.

We ask students to fill a variety of laboratory containers to an identified mark on the container, then determine the accuracy and precision of the fill. To determine the accuracy and precision of each fill, they must perform the procedure in triplicate, at least. With the resulting data, they can calculate the mean, <x>, and the standard deviation, s, of the procedure. We also ask them to use a set of pipets to deliver a given volume, which is equal to the target volume used when filling each container.

The students are to identify, using statistics, the precision of each container and delivery system. We also ask the students to determine, statistically, which containers are accurate. The first task requires calculating s/<x>; the smaller this ratio, the more precise the result. The second task requires the use of the one-sample t-test, given as

Â t = |<x> -Â Î¼|/(s/N)

where Î¼ represents the target volume and N is the degrees of freedom for the comparison. This t-value is then compared to the t-value from a table of critical t-values. This comparison determines whether the procedure is accurate or not. We also ask the students to determine whether the containers and delivery systems produce similar or dissimilar fills. The students accomplish this task by individually comparing each data set to all other data sets via the t-test, which requiers using the pooled standard deviation for the test data sets.

Finally, we ask the students to make a 50:50 methanol/water dilution in triplicate and compare the variance of the dilution to the summed variance of each step of the procedure; for example $(s^2)_{Dilution}$

$= (s^2)_{Scale} + (s^2)_{Pipet} + (s^2)_{Container}$

From this comparison, they are to identify which step requires refinement. This last task generates significant student/instructor interaction, which we will discuss during our presentation.

497076 Aligning the Unit Operations Laboratory and the National Academy's Grand Challenges

Abstract:

In 2008, The National Academy of Engineers identified several grand challenges which identify significant global needs.[1] These challenges require critical thinking and problem solving skills, open-ended task analysis and teamwork.[2,3]In 2014, we initiated an effort to meet these grand challenges via realignment of our chemical engineering unit operations laboratory. Specifically, we added these grand challenges as inquiry and experientially based learning processes into the typical fluid flow experiments.

Students are now asked to choose a "grand challenge" and link their fundamental experiments to their grand challenge choice so that they can understand how fundamental knowledge leads to profound engineering developments. Students work in teams of 3-5 students to address their chosen challenge. To address some of the challenges associated with teamwork our approach includes a group "team charter"[4] early in the process. This charter addresses leadership, team-function and time management skills. They are required to complete the charter each week which identifies team goals, ground rules, task lists, individual contributions and deadlines. The charter also provides a structure for specific feedback when students compare desired outcomes to weekly accomplishments.[5]

For the final project, students design their own experiment to address their "Grand Challenge." The entire project occurs over the 12 weeks of the laboratory course. This includes fundamental "pre-experiments", safety reviews, statistical analysis, and oral reports that identify the students' progress. Their final design must be supported by evidence from their previous experiments and the final proposal also requires a time-line and budget estimate which must be supported by evidence from their previous experimental experiences. Examples of design projects in the areas of health, sustainability and security have included nanoparticle drug delivery for cancer treatment, carbon sequestration using mineral carbonation, and protecting a water source from biological attack. Our presentation will share how this redesign fills several gaps in engineering education as identified in the NAE and AICHE reports.[2,3]

[1] National Academy of Engineering, http://www.engineeringchallenges.org/

[2] National Academy of Engineering, Educating Engineers, https://www.nap.edu/catalog/18254/educating-engineers-preparing-21st-century-leaders-in-the-context-of

[3] Luo et al., Chemical Engineering Academic/Industry alignment study, //www.aiche.org/sites/default/files/docs/conferences/2015che_academicindustryalignmentstudy.compressed.pdf

[4] Wolfe, J., Team Writing, A Guide to Working in Groups, Bedford/St. Martin's, Boston, 2010

[5] TEAMMATES, http://teammatesv4.appspot.com/

485777 No More Death By Powerpoint! Using the Assertion-Evidence Technical Presentation Model in a ChE Unit Operations Course

Abstract:

No More Death by PowerPoint! Using the Assertion-Evidence Technical Presentation Model in a ChE Unit Operations Course*

It is well-known that effective oral communication skills are critical to the success of chemical engineering (ChE) graduates in the modern workplace, so it is important that ChE instructors provide their students with numerous opportunities to practice oral communication skills through in-class presentations. However, assigning several oral presentations across multiple sections or in large classes can be mentally exhausting for the course instructor, and students may similarly feel that they are simply "turning the crank" and feel bored or unchallenged when faced with the construction (or sit in the audience) of yet another dull technical presentation – these are symptoms of the widely-discussed "Death by Powerpoint" (DBP) syndrome.

Part of the reason DBP syndrome occurs is adherence to the classic bullet/sub-bullet structure espoused in much instruction on technical presentations, which can lead to presentations feeling formulaic and lacking in creativity. In recent years new and exciting presentation paradigms have been developed and adapted for technical presentations. One modern presentation style, the assertion-evidence (AE) model, requires a heavier reliance on visuals rather than text. Recent studies claim that presentations using the AE model lead to significantly better comprehension and recall of advanced concepts by both presenters and audience members compared to common presentation formats.

This work describes the incorporation of oral presentation assignments using the AE model into different sections of a senior ChE Unit Operation laboratory course. In order to determine if assigning the AE model impacted students' improvement in oral presentations, student performance for an experimental cohort using the AE model was compared against a similar control cohort that completed oral presentations using the DBP format. It was found that student improvement in presentations increased when assigning the AE model without negative impact on student grades. Anecdotally, the instructor as well as students enjoyed the change of pace of the AE model compared to the DBP format. Student views on being assigned the AE model were mostly positive (65% of respondents), with student comments indicating they enjoyed the challenge and change of pace using the AE model both as speakers and audience members. Students also commented that they plan to use tenets of the AE model in their professional presentations moving forward. Suggestions for instructors interested in assigning the AE model in their courses are provided.

*NOTICE: This abstract describes an extension of work previously published under the citation M. Cooper, "No More Death by Powerpoint! Using an Alternative Presentation Model in a ChE Unit Operations Laboratory Course", ASEE 2016 Annual Conference, Columbus OH, Paper ID#18898 (2016) and it should be noted portions of the text of this abstract also appear in the ASEE conference proceedings.

509054 Solve this! Fundamental Approach to Problem Solving in Industrial Processes I (Invited Talks)

Abstract:

Solving complex plant problems is an everyday exercise that affects all projects and budgets, and it is very rarely introduced to students at any level. In fact, by the time young engineers are few years into working in industry, most have lost the engineering skills that they developed in school. We hope to encourage future engineers to retain those skills and learn the importance of plant problem-solving.

498677 Road Map for Embedding Ethics into ChE Undergraduate Curricula

Abstract:

The author will outline a number of methodologies and a suggested subject sequencing that can be effectively used to further develop an embedded ethics curricula. This curricula will consist of definitions, examples and suggestions on where to embed ethics into any existing ABET-approved, chemical engineering undergraduate course set. The author has lectured extensively at universities on ethics and a number of ethics related subjects.

504018 Views on Ethics in Undergraduate Education

Abstract:

Engineering programs are expected to graduate students who understand their professional ethical responsibilities. A studentâ€™s experience with ethics often begins with academic integrity. Research on student attitudes and behavior regarding academic integrity will be reviewed with suggestions to help faculty nurture a strong ethical foundation in their students.

495849 Ethical Reasoning in the Engineering Curriculum

Abstract:

Although the provision of ethics in engineering curricula has increased in recent years, the teaching of engineering ethics remains patchy and sporadic in a number of countries. Ethics in research, on the other hand, appears to be at a more mature stage. At the same time, the number of ethical codes, codes of practice and research concordats has flourished, both nationally and internationally. The questions remain: are those codes â€œliveâ€? Have those concordats resulted in an improvement of professional integrity? Is the teaching of ethics different and/or disconnected from ethics in research and in the workplace? Is ethics in chemical engineering different from ethics in other engineering disciplines? The paper analyses those questions in relation to an ever changing landscape, influenced by local and international drivers.

West Point Kicking Mule Brewery

Cadets Jesse Hudgins, Evan Ousley, Denis Glinski

CPT Joshua White, LTC Matthew Armstrong, LTC Corey James, LTC Geoffrey Bull

Department of Chemistry and Life Science

West Point, New York

ABSTRACT

The Kicking Mule Brewing Co. enables members of our chemical engineering club students and beer enthusiasts alike to combine chemical engineering acumen, knowledge and skills by developing a tangible product. Currently, Kicking Mule Brewing Co. aspires to hone in on the scientific aspects of brewing to enter into competitive beer brewing competitions including reaction engineering, heat transfer, and separations. The West Point Chemical Engineering program is process oriented and the Kicking Mule Brewery allows Cadets hands on experience with a batch reaction process. Through a rigorous on-site brewing process, with subject matter experts, our end goal is enhancement of student chemical engineering knowledge with respect to batch reactors and mixtures, heat exchange processes, chemical engineering process control, and bio-chemical reaction rates while simultaneously providing quality products to The United States Corps of Cadets.

PROJECT GOALS

- Produce a West Point beer using raw ingredients, a self-made recipe, and experimental procedures.
- Apply chemical engineering principles such has heat exchange, thermodynamics, batch reactions, mass flow, kinetics and process control.
- Determine reaction rates through ethanol production.
- Combine chemistry, biological processes, process control and chemical engineering principles in a tangible and fun way.
- Process parameters for varied flavors and palette while monitoring cost and profit.

$$(C_6H_{10}O_5)_x \xrightarrow{\text{enzymes}} C_6H_{12}O_6 \xrightarrow{\text{enzymes}} 2CH_3CH_2OH + 2CO_2$$

Starch — Glucose — Ethanol

Figure 1. Fermentation equation

PRECURSORS

- Fermentation requires sugars in order for yeast to produce alcohol (Figure 1).
- In order to make glucose (our reactant), malted barley at different roasts are built into a recipe for desired flavor profile and alcohol content.
- Sugars for fermentation are cooked out of the malted barley after the grain has been milled.

WORT

- Wort is the sugary precursor to beer. This includes all sugar for the reaction as well as hops for bitterness and aroma.
- We create the wort from milled grain. Cooking the grain in an aqueous batch reaction (the mash) activates enzymes that break down long chain sugars into short chain fermentables.
- During the boil, hops are added into the kettle to add bitterness and temper the sweetness.
- Our 3-vessel electric system is a professional pilot system to brew beer from grain. It has a PID controlled heat exchanger to keep the mash at the exact desired temperature.

PICTURE REFERENCES

- Image 1 - http://2012books.lardbucket.org/books/introduction-to-chemistry-general-organic-and-biological/s17-04-reactions-that-form-alcohols.html
- Image 3 - http://www.blichmannengineering.com/products/therminator

EQUIPMENT

- ½ bbl (15.5 gal) three tank, electric heat exchange recirculating mash system (Figure 5).
- Temp controlled Fermenter (25 gal).
- Heat Exchanger and Pump.
- Grain mill.
- Assorted hoses, and instrumentation.

Figure 2: Temp controlled fermenter

PROCESS CONTROL

- PID control is utilized to maintain strict adherence to the set point, which is set by the brewer and recipe dependent.
- Modeling of the control structure and system response allows a comparison of modeled data with real data.

Figure 3: Results of system temperature with PID control.

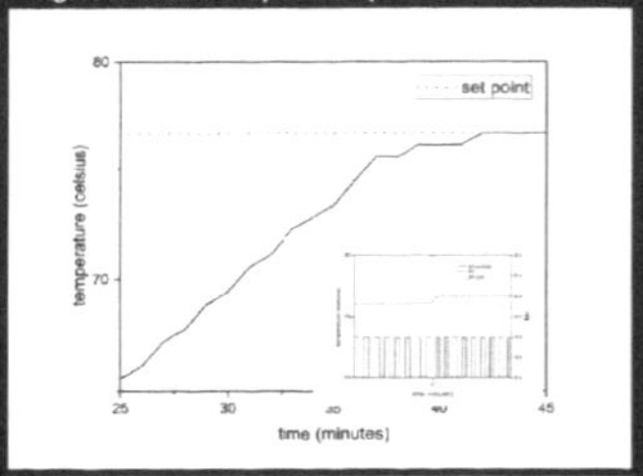

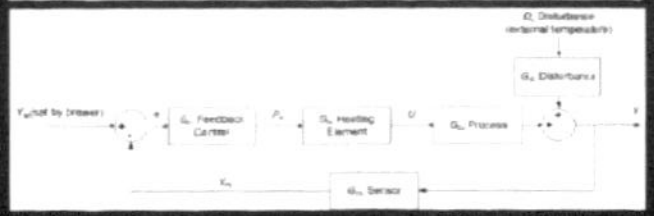

Figure 4: Block diagram of system feedback control.

Figure 5: ½ bbl E-HERMS system

FUTURE RESEARCH

- Optimizing our system and techniques to obtain the highest efficiency of sugar extraction and bring down cost.
- Continue to collect and analyze kinetic data to obtain best-fit parameters.
- Yeast counting to get biomass information to solve the remaining kinetic differential equation.

HEAT EXCHANGE

- Utilizing a countercurrent plate heat exchanger, the wort is chilled from boiling to ≈70ºF (30ºC) utilizing tap water.
- Adjusting flow rates of the wort with a pump, and the countercurrent chilling water with the faucet, we can achieve optimal temperature and flow. This information can be used to calculate heat transfer information of the heat exchanger.
- The heat exchanger was modeled in ChemCAD for energy balance. Correct modeling (Figure 6) of wort properties and sizing of the exchanger will be a future project.

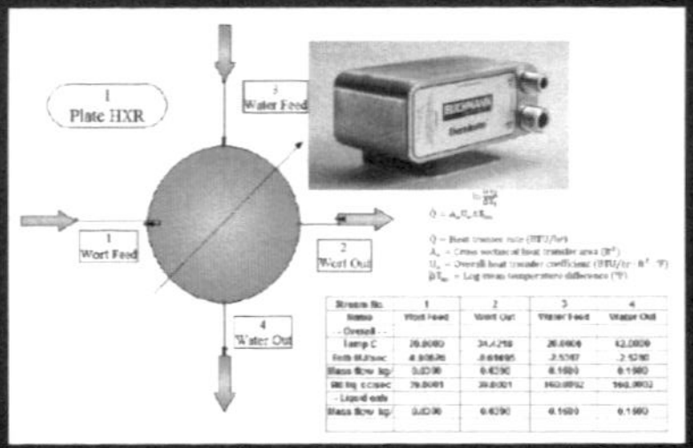

Figure 6: ChemCAD model of plate-type heat exchanger.

REACTION RATES AND FERMENTATION

- Different fermentation temperatures will alter reaction rates.
- The higher the temperature of the fermenter, the higher the reaction rate and the more vigorous the fermentation.
- We collected data on several fermentations and the results to start modeling kinetic data.
- Data was collected using a refractometer, hydrometer and a GC mass spec.

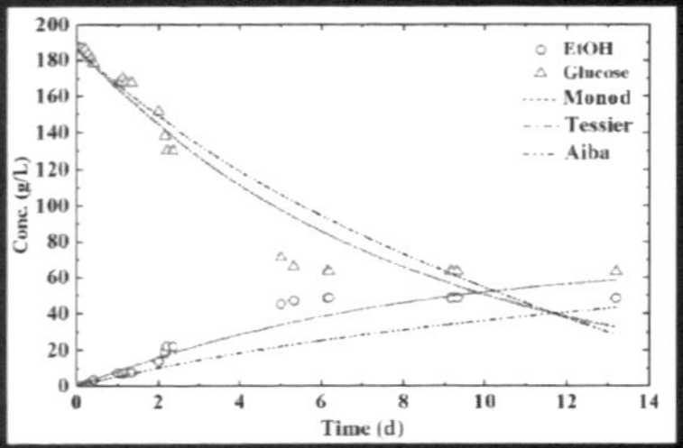

Figure 7: Kinetic model of fermentation

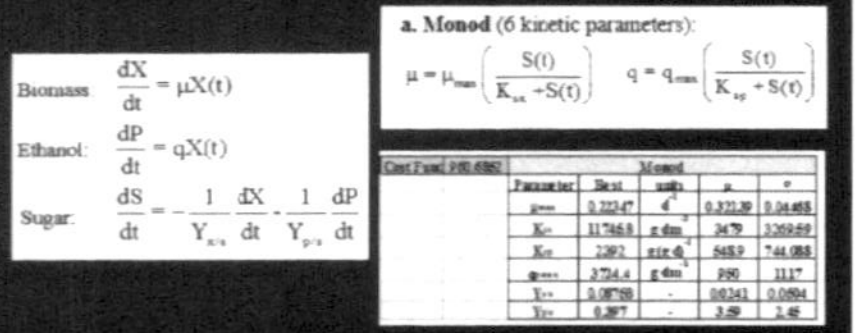

Figure 8: Governing equations/parameter values for kinetics

PACKAGING

- After fermentation is complete, the uncarbonated beer is filtered and transported to a keg.
- Once carbonated and refrigerated, the beer is now ready to serve.
- The beer is sold by DCA* at the on-post First Class Club which gives the cadets a real-world economics model to determine profitability and viability of the process.

ACKNOWLEDGMENTS

- Directorate of Cadet Activities for funding and a venue to sell Kicking Mule Beer.

499372 Beer Experiments - a Hopping Success

Abstract:

In the Biochemical Engineering Laboratory (CHE 194) course, students learn bioprocess engineering experimental techniques for eight weeks including microbial fermentation, protein purification, ultrafiltration, chromatography and enzyme kinetics.

Following the modular labs, a true inquiry lab experience is had by the students. Student teams submit a proposal for an experiment of their choice, using a previously published rubric for design of experiments. Five weeks is allotted for the inquiry experiment so they have sufficient time to repeat parts of the experiment that didnâ€™t work the first time. Students rate the experience highly on the end-of-semester evaluations.

Beer is a popular subject for many chemical engineering students. An experiment has been added into the chromatography module giving the students the opportunity to calibrate the high pressure liquid chromatograph (HPLC) instrument using a standard for alpha acids and then they can bring their favorite beers and test them for alpha acid content. Beer hops confer alpha acids into the beverage that have some health benefits, such as antioxidant activity, as well as anti-microbial activity for preserving the beer early during the fermentation when the alcohol level is low. Hops are added during the boil because the solubility of the alpha acids is higher in hot water than cold. Popular now is also the dry-hopping process whereby hops are added after the boil and the ethanol produced during the fermentation enables the solubilization of the alpha acids. Students have done several experiments related to beer over the years. The alpha acid HPLC module will be presented here as well as the inquiry experiment design and two of the most interesting beer experiments designed and executed by students over the years.

496488 Why Is the Mustard in the Fridge?Â a Fun Assignment on the Thermodynamics of Food Safety

Abstract:

Activity is often introduced in chemical engineering thermodynamics courses in the context of modeling non-ideal solutions in vapor-liquid equilibrium, but this is only the first of many useful applications. One of these applications is estimation of osmotic pressure, often in the context of drinking water purification or creation of isotonic medical saline solution. In both of these examples, the activity of water in solution (a_w) is the key value in the analysis. For both my thermodynamics and applied food science courses, I created an assignment that focuses on yet another important application of a_w: food safety. Prior to refrigeration, extremes of osmotic pressure were used to prevent or reduce microbial spoilage of foods through addition of salt, sugar, acid, or a combination thereof. This assignment, which works both in-class and as homework, starts with published food labels and a few assumed molecular masses and results in fairly accurate determinations of whether or not a given food should be â€œshelf stableâ€ without refrigeration. The assignment can also be extended to ask students to invent a recipe for preserving fruits and vegetables that result in answers remarkably similar to traditional versions of jam, raisins, and pickles.

499784 Improving Conceptual Knowledge and Retention in Introduction to Engineering Thermodynamics

Abstract:

By the end of our sophomore level Introduction to Engineering Thermodynamics course (CBEN 210), students should be able to demonstrate both procedural and conceptual knowledge. Procedural knowledge is the ability to solve domain problems and usually yields a final numerical answer.[1] Conceptual knowledge is understanding of the principles governing a domain.[2] Data collected over the course of two years in CBEN 210 showed that students consistently performed better on procedural problems than conceptual problems. Conceptual knowledge can be more challenging because it requires organizing key principles, like content experts, and it is frequently coupled with misconceptions, which are hard to undo.[3] In this study, we identified three conceptual topics that students performed poorly on and initiated three in-class interventions to address the deficiency. The three topic areas were isenthalpic flow through a valve, isentropic work, and entropy generation. The interventions were all interactive activities that tied the thermodynamic concept to a real-word application. For example, isenthalpic flow through a valve was examined by the students as a technique to measure steam quality in a process line. The impact of these interventions will be measured with a concept inventory exam administered 8 -12 months following enrollment in CBEN 210. Since our chemical engineering students take a second course in thermodynamics, their ability to retain knowledge beyond the end of the sophomore level class is important. The test group contains approximately 70 students and will be compared to a control group from the previous year that did not receive the interventions. Correlations between thermodynamic knowledge retention and CBEN 210 performance, class timing (8 or 12 months), and other factors such as GPA will be examined.

[1] R. Taraban *et al*, *J. Eng. Ed.*, **96**, 57, 2007.

[2] R. A. Streveler *et al*, *J. Eng. Ed.*, **97**, 279, 2008

[3] P. N. Van Meter *et al*, *J Eng. Ed.*, **105**, 245, 2016.

494839 Tax Revenue Laffer Curve from Thermodynamics Perspective

Abstract:

It has been a long tradition of economists and scientists making parallel analogy between economics and thermodynamics, establishing phenomenological theory of economic systems. In this contribution, the proposed economic temperature and economic entropy identification and subsequently developed economic production and growth models were used to obtain the quantitative relationship between tax revenue, payroll tax as well as corporate/personal income tax, and the wealth creation efficiency of an economic entity. The derived expression for corporate/personal income tax exhibits a quadratic dependence of wealth creation efficiency and tax rate, which agrees well with the famous tax revenue Laffer curveâ€”There are always two tax rates that yield the same revenues. The expression also indicates that the wealth creation efficiency improvement plays a pivotal role in the pay off of the revenue lost by prior tax rate cut. The derived mathematical revenue expression together with the proposed monetary rule may be used for future fiscal policy design and optimization.

492365 The Art of the Game: Infusing Thermodynamics Learning with Classic Household Games

Abstract:

We all understand the drab that comes with learning Chemical Engineering Thermodynamics in general. Now try teaching it at 8am and one can immediately see the conundrum that presents itself. Altering the classroom learning environment in unusual ways can spark new life into students who do not necessarily appreciate thermodynamics for what it is worth in their later courses. In this presentation, the use of classic household games to change the dynamics of the classroom will be discussed. From MONOPOLY to Blackjack, creating an incentive to learn Thermodynamics (and other concepts) leads to a transformation of the student that promotes growth and turns the classroom into a fun learning environment, even at 8am in the morning.

Industrial Safety Curriculum for Chemical Engineering Education

Juanita Miller, PE, CFEI

Dr. David Rockstraw, PE, PhD

Student Outcomes – CHME 448

- Core Course
- Fulfills the requirements of ABET
  - *The curriculum must include the engineering application of these basic sciences to the design, analysis, and control of chemical, physical, and/or biological processes, including the* <u>*hazards associated with these processes.*</u>

Student Outcomes – CHME 448

- Provides a broad overview
  - ***catalog description:*** *An introduction to the fundamentals of chemical process safety, including toxicology, industrial hygiene, source models, fires and explosions, relief systems, hazard identification, risk assessment, environmental fate and transport, hazardous waste generation, pollution prevention, and regulatory requirements. Chemical engineering majors must earn C or better in this course.*

NM STATE College of Engineering

Engineering Is Discov

Course Development History

- PSM course in a "box"
  - Single disaster focus – Texas City
  - Not sufficient materials
- Broad focus
  - All types of hazards
  - Multiple disaster examples
- Instructor with industry experience

NM STATE College of Engineering

Engineering Is Discov

Course Materials

- Textbook –
  - "Chemical Process Safety" by Daniel Crowl & Joseph Louvar
- Other text materials
  - "What Went Wrong" by Trevor Kletz
- CSB Videos on disasters
- Real world projects

Course Syllabi

- Introduction & hazard identification
- Process safety management (PSM)
- Occupational safety and health (OSHA) overview
- Environmental regulations
- Toxicology
- Industrial hygiene
- Process hazard assessments

Course Syllabi

- Hazard sources
- Toxic releases and dispersion models
- Fires and explosions
- Concepts to prevent fires and explosions
- Valves and piping related to safety and reliability
- Types of reliefs and relief sizing
- Safety procedures and designs
- Ethics

Personal Safety as well as Process Safety

- Regulations
  - OSHA General Industry
  - EPA – RCRA, CWA, CAA, CERCLA etc.
- Industrial Hygiene
  - Hazard communications
  - Toxicology
  - Exposure controls
    - PPE
    - Administrative
    - Engineering

Hazard Identification and Risk Analysis

- Concepts
  - What can go wrong ?
  - How likely is it ?
  - What are the consequences ?
- Methods
  - What If
  - FMEA
  - HAZOP
  - Fault Tree

College of Engineering
NM STATE
Engineering Is Discov

Valves and Piping Model – Individual effort

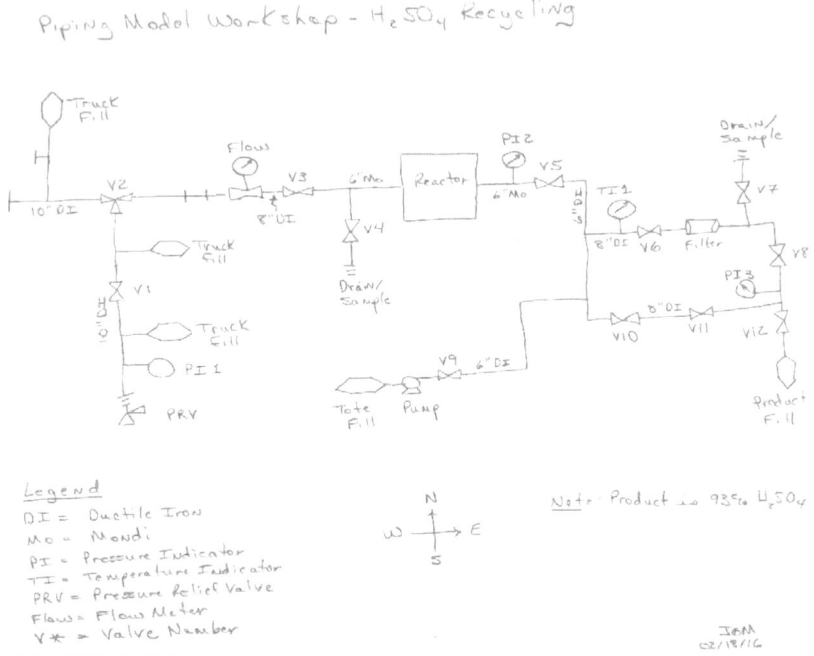

College of Engineering
NM STATE
Engineering Is Discov

Scenario for Valves & Piping Model

- Plant built many years prior and expanded/modified through the years
  - Piping material and size differences
  - Reactor replacement
- Potential contaminates in feed material
- Feed delivered in different size containers
- Filter maintenance procedures
- Future expansion plans
- Safety information and regulations
- Personnel staffing, work shifts and overtime

NM STATE College of Engineering

Engineering Is Discov

Valves and Piping Model - Questions

- Regulations
  - Hazard communications
  - PSM, RMP or CFATS
- Valve and piping design
  - What protects piping from exposure
  - Compare seat and seal materials
- Maintenance
  - Exposure controls
  - Design for ease of use and hence safety
- Release scenarios
  - Valve leak
  - Piping release
  - Containment sizing

NM STATE College of Engineering

Engineering Is Discov

Projects – Team Effort

- Real World Scenario
  - Evaluate a system
  - Consider strengths and weaknesses from P&ID drawings
  - Much more detail than block diagrams
  - Divide plant, or areas of a plant, into nodes
  - Allow teams to analyze hazards in each node

Candy Factory – Block Diagram

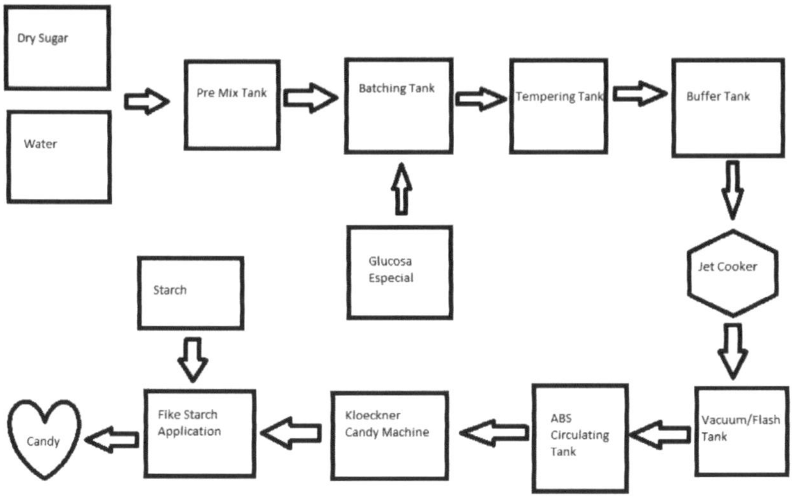

Industrial Safety – CHME 448

Questions?

NM STATE College of Engineering

Engineering Is Discov

Using AIChE's Concept Warehouse to Help Teach Process Safety- Related Engineering Science

Bruce K. Vaughen
AIChE Center for Chemical Process Safety (CCPS)

Keywords: Education, Undergraduate, Concept Warehouse, Curriculum, Process Safety

Abstract

The AIChE Concept Warehouse is a web-based instructional tool that enables chemical engineering instructors to better provide their students with concept-based instruction. Whereas some students may focus more on rote memorization and learning rather than on conceptual understanding, it is known that concept-based instruction helps students use their knowledge better when given new situations, especially after graduation. The Concept Warehouse can be used in addition to other tools, such as homework assignments, to help reinforce the concepts fundamental to understanding the engineering sciences being addressed.

The focus of this presentation is to describe a potential use for the Concept Warehouse when discussing the engineering design of equipment specifically being used to help reduce process safety risks. By integrating the key concepts for these engineering controls within the context of effective process safety systems, sets of warehouse "pallets of information" can be created and used. The background for developing these pallets includes our AIChE Code of Ethics, which states first that we "hold paramount the safety, health and welfare of the public and protect the environment in performance of [our] professional duties." Thus, it is essential that everyone handling hazardous materials manage them safely to protect people from the material's harmful properties, such as their toxicity, flammability, explosivity and/or reactivity.

In addition, recent developments for effectively managing process safety programs have focused on three essential foundations: safety culture and leadership, process safety systems, and operational discipline. This presentation will describe the key concepts for each of these foundations, the key concepts of the eight process safety systems required to effectively manage the process safety risks, and then provide some specific engineering design concept-related examples that are discussed in a process safety course or the senior design class. The engineering designs include preventive and mitigative controls, from the equipment's inherently safer design to the engineering controls, such as pressure relief systems, which are used to help reduce the risks. The concepts behind the design and management of these engineering controls are essential for students, helping them understand and successfully manage their future process safety risks once they graduate.

1 Introduction

This paper provides background on AIChE's Education Division's Concept Warehouse and how process safety-related engineering science concepts are being introduced to into the Concept Warehouse. The Concept Warehouse provides instructors with a real time interactive teaching tool which is used during lectures. The instructor can assess how the class is doing and, based on the results, adjust their discussion to clarify ideas which are not being understood by the students. Several process safety concepts are described before showing how they can be used in the Concept Warehouse. By integrating the key concepts for these engineering controls within the context of effective process safety systems, sets of warehouse "pallets of information" can be created and used.

This paper then provides some examples of questions that have been developed to date and briefly discusses the future directions envisioned by the author for the Concept Warehouse. These examples include questions on the fundamental concepts for effectively managing process safety program and a concept behind the engineering science used when designing a pressure relieve valve PRV. The PRV is but one example of safeguarding equipment specifically designed used to help reduce process safety risks.

The future for adding new Concept Warehouse process safety-related questions is discussed in context of the AIChE Foundation (through its Undergraduate Process Safety Initiative), through the Safety and Chemical Engineering Education (SAChE) program (its AIChE Academy Undergraduate Process Safety Courses) and the AIChE Center for Chemical Process Safety (CCPS) books and workshops.

2 The AIChE "Concept Warehouse"

The AIChE Education Division's Concept Warehouse is a web-based instructional tool that enables chemical engineering instructors to better provide their students with concept-based instruction [1]. Whereas some students may focus more on rote memorization and learning rather than on conceptual understanding, it is known that concept-based instruction helps students use their knowledge better when given new situations, especially after graduation. The Concept Warehouse can be used in addition to other tools, such as homework assignments, to help reinforce the concepts fundamental to understanding the engineering sciences being addressed.

The Concept Warehouse contains questions and interactive problems from fundamental chemical engineering course subjects (each designated as a "pallet"), such as material balances, energy balances, and fluid dynamics, etc. (**Figure 1**). While many subjects have had time for questions to be developed and entered, the number of process safety-related questions being populated has only just begun. The effort to enter more questions and interactive tools is ongoing at the time this paper was being written.

When the Concept Warehouse is used for interactive reviews during lectures, instructors can easily enter and use their own questions, access and use questions already entered, obtain real-time feedback from clickers, smart phones, or laptops, immediately assess how the students are doing, and adjust their lecture content based on results of the assessment.

Figure 1
The Image of the Concept Warehouse with an Instructor as the Fork Truck Driver
[1]

3 The Context of and Definitions for Process Safety-related Terms

This section briefly covers some historical context for having essential process safety-related content in a chemical engineering curriculum and defines some common process safety-related terms used in industry.

3.1 AIChE Code of Ethics

Our AIChE Code of Ethics, states first that we "hold paramount the safety, health and welfare of the public and protect the environment in performance of [our] professional duties [2]." This is the premise by which sets the basis for process safety: Our efforts are dedicated to those who lost their lives because we lost control of our hazardous materials and energies.

3.2 The Incident at T2 Laboratories, Inc.

In 2007 a runaway reaction and explosion that occurred at the T2 Laboratories, Inc., caused four fatalities and injured 32 people in the surrounding community [3]. As a result of the US Chemical Safety Board's investigation, these two recommendations were issued:

- [AIChE to] Work with the Accreditation Board for Engineering and Technology, Inc. (ABET) to add reactive hazard awareness to baccalaureate chemical engineering curricula requirements.
- [AIChE to] Inform all student members about the Process Safety Certificate Program and encourage program participation.

In 2010, the Safety and Chemical Engineering Education (SAChE) Committee made recommendations with regard to student understanding in process safety to meet the ABET curriculum requirements [4]. As a result, process safety first went into effect as a criterion for

Chemical Engineering programs in the 2012-13 ABET accreditation cycle. The ABET criterion regarding safety in the chemical engineering curriculum [5]:

> "The curriculum must provide a thorough grounding in the basic sciences including chemistry, physics, and/or biology, with some content at an advanced level, as appropriate to the objectives of the program. The curriculum must include the engineering application of these basic sciences to the design, analysis, and control of chemical, physical, and/or biological processes, including the hazards associated with these processes. This criterion can be addressed by (i) a dedicated course and/or (ii) incorporating safety into other courses."

3.3 Definitions

For the purposes of this paper, and in context of the literature and process safety glossaries currently available, some common process safety definitions have been slightly modified and are listed as follows:

Process Safety Hazard: Intrinsic properties of materials and processes that can either directly or through release of energy lead to harmful effects, such as toxicity, flammability, explosivity, and reactivity ("instability").

Process Safety Incident: A combination of conditions and events that result in toxic releases, fires, explosions, or runaway reactions that harm people, damage the environment, and cause property and business losses.

Process Safety Risk: A measure of potential harm associated with process activities and events, based on frequency (e.g., years) and potential consequences (e.g., injury, property damage, environmental). The risk (harm) is a function of the frequency times the consequence.

Barriers: The engineering and administrative controls designed to reduce the frequency of or reduce the consequences (the harm) from an incident - the "loss of containment" of hazardous materials and energies. Engineering control includes computers and pressure relief valves; administrative controls include procedures (**Figure 2**).

Process Safety and Risk Management ("PSM"): The principles and systems applied to identify, understand and control process hazards to reduce process safety risks and protect the safety and health of people, neighbors and the environment.

Figure 2
The Barrier Model
Adapted from [6]

4 Safely Managing Hazardous Materials and Energies

This section briefly describes the concepts behind the process safety program foundations, the eight process safety systems and some process safety engineering science and design. Although there are many books on how to apply process safety systems (the practice), this paper focuses on the concepts described in the book published by Klein and Vaughen [6]. In addition, there are many engineering texts that describe the math (the theory) and equations used to model dispersions and determine the pressure relief valve design or, this paper focuses solely on the math depicted in Crowl and Louvar [7].

4.1 The Process Safety Program Foundations

Klein and Vaughen have published a book describing the three foundations which are required for a group to have an effective process safety program [6]. The program ensures that process safety systems integrate the process safety elements to ensure that process hazards are identified, evaluated, and understood Such that appropriate risk management controls are provided to achieve and maintain safe and reliable operations. The groups include everyone who works processes and equipment handling hazardous materials or energies, including university laboratories, petrochemical or pharmaceutical industries, and refineries. The three foundations are depicted in **Figure 3**: safety culture and leadership, process safety systems, and operational discipline. The key concept for each of these foundations is listed in **Table 1**.

Figure 3
The Three Foundations for Effective Safety Programs
[6]

Table 1
The Concepts for Each of the Process Safety Program Foundations
[6]

Foundation	Key Concept – Objectives
Safety Culture and Leadership	The normal way things are done at a facility, company, or organization, reflecting expected organizational values, beliefs, and behaviors That set the priority, commitment, and resource levels for safety programs and performance.
Process Safety Systems	Systems integrating the process safety elements to ensure that process hazards are identified, evaluated, and understood such that appropriate risk management controls are provided to achieve and maintain safe and reliable operations.
Operational Discipline	Following system and procedure requirements correctly every time. "What you do when no one is looking"

4.2 The Process Safety Systems

There are eight Process Safety Systems required to effectively manage the process safety risks. The purpose – the concept behind them - for each of these systems is shown in **Table 2.**

Table 2
The Concepts for Each of the Process Safety Systems
[6]

Process Safety System		Purpose - Concept
1	Design Safe Processes	Design safe processes to ensure that appropriate technology and safeguards are incorporated to control process risks and achieve safe and reliable operations.
2	Identify and Assess Process Hazards	Identify and assess process hazards to ensure that process risks can be properly evaluated and managed.
3	Evaluate and Manage Process Risks	Evaluate and manage process risks to ensure safe process design and that process safety systems are properly designed, implemented, and maintained.
4	Operate Safe Processes	Operate safe processes by ensuring that personnel are provided appropriate procedures and training to develop capability in operations and process safety.
5	Maintain Process Integrity and Reliability	Maintain process integrity and reliability by ensuring that appropriate tests and inspections, preventive maintenance, and quality assurance programs are conducted.
6	Change Processes Safely	Change processes safely by ensuring that process, equipment, system, and organizational changes are evaluated and authorized and that operational readiness reviews are conducted.
7	Manage Incident Response and Investigation	Manage incident response and investigation by ensuring that personnel are trained how to respond during emergencies and that incidents are thoroughly investigated.
8	Monitor Process Safety Program Effectiveness	Monitor process safety program effectiveness by ensuring that appropriate metrics are selected, evaluated, and used to measure and improve performance.

4.3 Process Safety-related Engineering Science and Design

This section briefly describes some of the theoretical concepts, including dispersion models, used when designing process safety safeguarding equipment. The safeguards (the barriers) include preventive and mitigative safeguard design and pressure relief design.

Dispersion calculations are well-established and can be used to predict how far a dispersion cloud can travel [7]. A typical dispersion represented by a vapor cloud's plume is depicted in **Figure 4.** One parameter used in the calculations is the dispersion coefficient (σ), as is shown in **Equation 1.**

$$\sigma_x^2 = \frac{1}{2}(C)^2(ut)^{2-n}$$

Equation 1 [7]

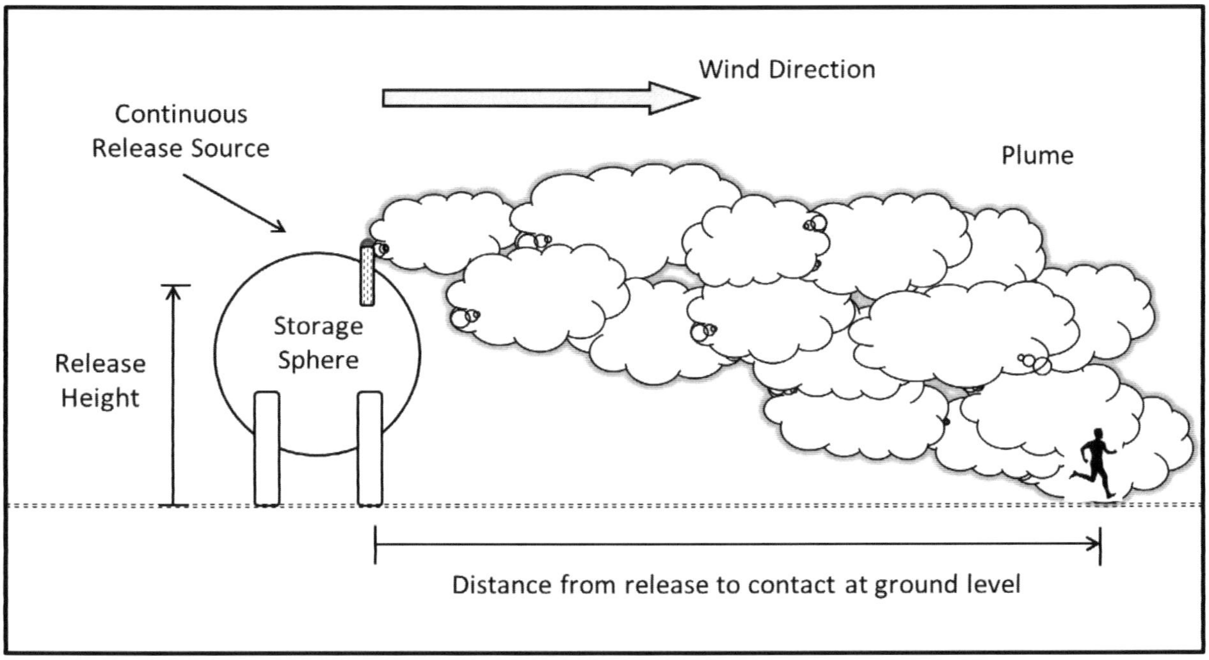

Figure 4
A Depiction of a Gaseous Plume from an Elevated Release Point

The design of pressure protection systems is well-established to ensure that the relief device will protect the processing equipment and piping. A cross sectional view of a Pressure Relief Valve (PRV) used for protection from potential overpressures is shown in **Figure 5.** One parameter used in the PRV design calculations is the vent area (A), as is shown in **Equation 2.**

$$A = \frac{Q_v}{C_o\sqrt{2g_c}}\sqrt{\frac{\rho}{\Delta P}}.$$

Equation 2 [7]

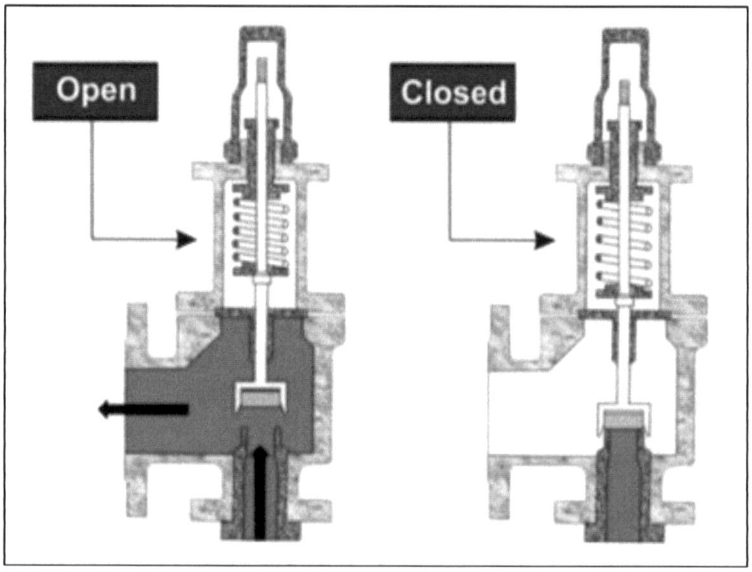

Figure 5
A Cross Sectional View of a Pressure Relief Valve (PRV)
[8]

5 Example Process Safety-related Questions

Although the pallets of information for Process Safety-related are being developed and are being added to the Concept Warehouse, this program is still in its infancy with a limited number of questions available for review and use. However, several questions do exist at the time of this paper, of which three are described in more detail in this section. These represent but a sample of the potential questions which are going to be entered in the future.

5.1 An Example of a Process Safety Program Concept Question

"Name the essential foundations for an effective process safety program:

- O *Process Safety Systems*
- O *Safety Culture and Leadership*
- O *Management of Change*
- O *Operational Discipline"*

The students must select the three correct answers from the four options. As shown in **Figure 1**, the three foundations are: Process Safety Systems, Safety Culture and Leadership, and Operational Discipline [6].

5.2 An Example of an Engineering Science Concept Question

"The atmospheric dispersion coefficient (σ) is affected by which of the following:

- ○ *Downwind direction*
- ○ *Vertical direction*
- ○ *Crosswind direction*
- ○ *Atmospheric stability*

The students must select all four answers (see **Equation 1** [7]).

5.3 An Example of an Engineering Design Concept Question

"Both Reactors R-101 and R-102 have Pressure Relief Valve (PRV) designed to relieve each reactor's potential overpressure of 25 psi (1.7 atm.) to a common pressure relief piping system. The emergency relief piping systems – a "header" – leads to the common flare system.

If the operating pressure inside R-101 is greater than the operating pressure in R-102 (all other conditions being identical), then the diameter for the PRV on R-101 will most likely be ________ the diameter for the PRV on R-102.

- ○ *greater than*
- ○ *less than*
- ○ *is equal to*

The students must select "greater than" for the correct answer, as the vent area increases with a greater differential pressure (see **Equation 2** [7]). Since the area of the circular seat equals πr^2, the diameter (2r) must be greater for the PRV located on Reactor R-101.

6 Summary

This paper presents a few examples of questions currently in the "process safety pallet" in the AIChE Concept Warehouse. The concepts behind the design and management of these engineering controls are essential for students, helping them understand and successfully manage their process safety risks in their engineering roles once they graduate.

7 Continuing Our Journey

The efforts for populating the Concept Warehouse are being performed in conjunction with the following AIChE initiatives and programs:

1) The AIChE Foundation - through its Undergraduate Process Safety Initiative [9])
2) The Safety and Chemical Engineering Education (SAChE) Certificate program – through support from the AIChE Foundation by development and implementation of online AIChE Academy [10])

3) The AIChE Center for Chemical Process Safety (CCPS) publications [11], CCPS Faculty Process Safety Workshops [12], and CCPS support for the Undergraduate Student Process Safety Boot Camps [13],
4) The AIChE Chem-E-Car Competition® – with process safety specific reviews from volunteers in the AIChE Safety & Health Division [14], including the Annual meeting's mandatory Chem-E-Car Competition® Safety Inspection and Chem-E-Car Poster Competition, and
5) The annual AIChE Student Design Competition – with process safety-specific reviews including inherently safer design concepts from volunteers in the AIChE Safety & Health Division [16].

8 Acknowledgements

The author would like to thank Jim Klein and Dan Crowl for their many discussions over these past two decades. In addition, I am extending special appreciation to the many colleagues who have been a part of my journey: those who showed me how to manage process safety risks and those who showed me how not to manage them.

9 References

1 AIChE Education Division Concept Warehouse, http://jimi.cbee.oregonstate.edu/concept_warehouse/
2 AIChE Code of Ethics, https://www.aiche.org/about/code-ethics
3 US Chemical Safety Board, *T2 Laboratories, Inc., Runaway Reaction,* Report No. 2008-3-I-FL, issued in September 2009.
4 AIChE Safety and Chemical Engineering Education (SAChE), *SAChE Recommendations for ABET Safety Content in Chemical Engineering*, February 18, 2010, http://www.sache.org/SACHEGuidelinesForABET.pdf
5 Accreditation Board for Engineering and Technology, Inc. (ABET), *Criteria for Accrediting Engineering Programs*, http://www.abet.org/
6 Klein, James A., and Bruce K. Vaughen, 2017, *Process Safety: Key Concepts and Practical Applications*, CRC Press, Boca Raton, FL.
7 Crowl, Daniel A. and Joseph F. Louvar, *Chemical Process Safety, Fundamentals with Applications, 3rd Edition*, 2011, Prentice Hall, Upper Saddle River, NJ.
8 Industrial Valve Store, http://www.industrialvalvestore.com/pressure-relief-valve
9 AIChE Undergraduate Process Safety Initiative, https://www.aiche.org/ccps/community/undergraduate-process-safety-learning-initiative
10 Safety and Chemical Engineering Education (SAChE) Certificate Program https://www.aiche.org/ccps/community/technological-communities/safety-and-chemical-engineering-education-sache/certificate-program
11 Center for Chemical Process Safety (CCPS) Publications, https://www.aiche.org/ccps/resources/publications
12 Center for Chemical Process Safety (CCPS) Faculty Workshops, https://www.aiche.org/ccps/resources/conferences/events/faculty-workshops
13 Undergraduate Student Process Safety Workshops, https://www.aiche.org/ccps/resources/conferences/events/student-bootcamps

14 AIChE Chem-E-Car Competition®, https://www.aiche.org/community/students/chem-e-car
15 AIChE Safety and Health Division Student Design Competition Award for Safety, https://www.aiche.org/community/awards/safety-and-health-division-student-design-competition-award-safety

Simulation Solutions, Inc.

Minds-On/Hands-On Training for Operators™

"Process Safety Education Using Simulators in a Chemical Engineering Operations Center Experience"

Mr. Donald C. Glaser
President

Mr. Matthew Garvey
Project Engineer

Simulation Solutions, Inc.
179 Avenue at the Common, Suite 1
Shrewsbury, NJ 07702-4804
dgalser@simulation-solutions.com

Robert G. Bozic, Ph.D.
Lieutenant Colonel U.S. Army(Retired)
Lecturer in Chemical Engineering Design and
Chemical Engineering Process Safety
Department of Chemical Engineering
Columbia University in the City of New York
robert.bozic@columbia.edu

COLUMBIA ENGINEERING
The Fu Foundation School of Engineering and Applied Science

Outline

- Background/Motivation
- Safety Recommendations for ABET Feb 2010
- CHEN E4501 Chemical Engineering Process Safety
- Trade Show Experience
- Learning from History (Case Study & Design)
- Using Simulations to Teach and **Evaluate Process Safety Education**
- **Operations Center Experience**
- Acknowledgements and References
- Conclusion and Future Work

2

Case Study: Marsh and McLennan Reports 14th (1992) and 18th (1998) Editions

 3

https://www.marsh.com/us/insights/research/the-100-largest-losses-in-the-hyrdocarbon-industry-1974-2015.html

Cause of Losses by Percent

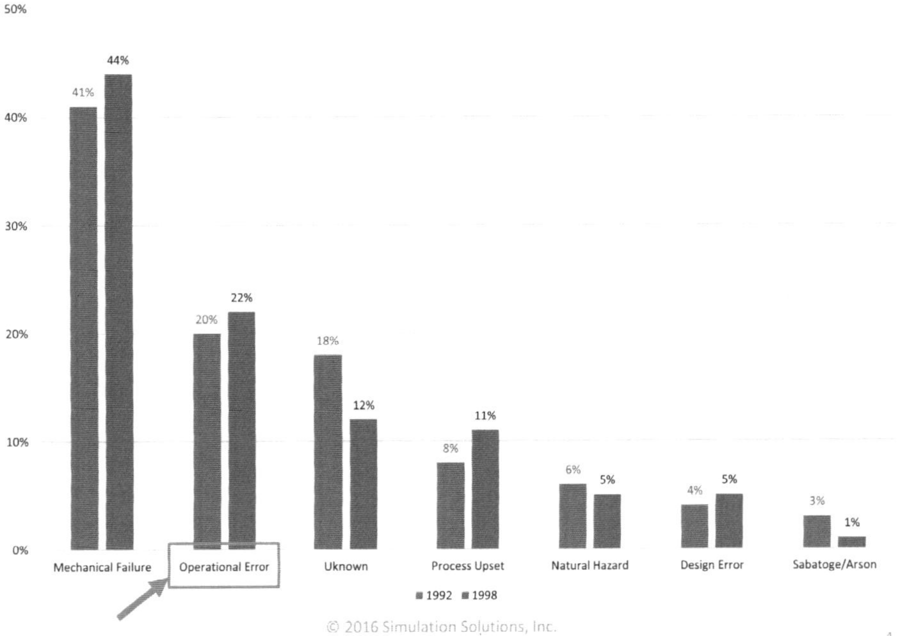

 4

Case Study: Marsh and McLennan Reports 14th (1992) and 18th (1998) Editions

Background and Motivation

"SAChE Recommendations for ABET Safety Content in Chemical Engineering February 18, 2010"
"The Safety and Chemical Engineering Education (SAChE) Committee makes the following recommendations with regard to student understanding to meet ABET curriculum requirements.

1. The graduate must understand the importance of process safety and the resources and commitment required. This should include the important incidents that define process safety, and how these incidents affected the practice of chemical engineering.

2. The graduate must be able to **characterize the hazards** associated with chemicals and other agents. This must include toxic, flammable, and reactive hazards.

3. The graduate must understand and be able to apply concepts of **inherently safer design**.

4. The graduate must understand how to **control and mitigate hazards to prevent accidents**. This should include generally accepted management systems, plant procedures and designs to prevent accidents.

5. The graduate should be **familiar with the major regulations** that impact the safety of chemical plants.

6. The graduate should **understand the consequences of chemical plant incidents** due to acute and chronic chemical releases and exposures.

7. The graduate should be reasonably proficient with **at least one hazard identification procedure**.

8. The graduate should have an introduction to the process of **hazard evaluation and risk assessment**.
Resource materials to assist in meeting these requirements can be found on the SAChE web."

5

(www.sache.org accessed 13 Apr 2015)

Background and Motivation

"ABET Curriculum Statement approved in October 2011"

"1. Curriculum
The curriculum must provide a thorough grounding in the basic sciences including chemistry, physics, and biology, with some content at an advanced level, as appropriate to the objectives of the program. The curriculum must include the engineering application of these basic sciences to the design, analysis, and control of chemical, physical, and/or biological processes, including the hazards associated with these processes."

6

(www.abet.org accessed 13 Apr 2015)

CHEN E4501 Chemical Engineering Process Safety Course Description

"Aimed at seniors and graduate students. Provides classroom experience on chemical engineering process safety as well as Safety in Chemical Engineering certification. Process safety and process control emphasized. Application of basic chemical engineering concepts to chemical reactivity hazards, industrial hygiene, risk assessment, inherently safer design, hazard operability analysis, and engineering ethics. Application of safety to full spectrum of chemical engineering operations."

((http://www.columbia.edu/cu/bulletin/uwb accessed 27 Sep 2015)

CHEN E4501 Chemical Engineering Process Safety Course Significance

"Chemical Engineering Process Safety will be used anywhere in the world across the complete range of chemical engineering operations. Process Safety is current recommended ABET curriculum content by the Safety in Chemical Engineering Safety Committee. Academia and industry contributed to the content of applying chemical engineering concepts to chemical reactivity hazards, industrial hygiene, risk assessment, inherently safer design, and hazard operability analysis. With the ultimate aim of safer operations in the chemical engineering profession, the applications of the Chemical Engineering Process Safety will make the world safer."

7

(Fall 2017 Course Proposal)

CHEN E4501 Chemical Engineering Process Safety Course Objectives

1. Learn about the importance of process safety and the resources and the commitment required. This should include the important incidents that define process safety, and how these incidents affected the practice of chemical engineering.

2. Learn how to characterize the hazards associated with chemicals and other agents to include toxic, flammable, and reactive hazards.

3. Learn how to apply concepts of inherently safer design.

4. Learn how to control and mitigate hazards to prevent accidents.

5. Become familiar with the major regulations that impact the safety of chemical plants.

6. Learn about the consequences of chemical plant incidents due to acute and chronic chemical releases and exposures.

7. Become familiar with the process of hazard identification and evaluation, and risk assessment

8. Become familiar with engineering ethics.

8

(Spring 2017 CHEN 4501 Syllabus)

Course Design and SAChE Integration

Week 1: Chemical Engineering Process Safety Course Overview
Intro to Process Safety (Chp 1) / Intro to SAChE (www.sache.org)
Lab Safety / Chemical Reactivity (Chp 8) / Hazard Identification (Chp 11)
SAChE Chemical Reactivity Hazards (Robert Johnson)
Week 2: Chemical Reactivity (Chp 8) and Hazard Identification (Chp 11)
SAChE Chemical Reactivity Hazards
Industrial Chemical Hygiene (Chp 3) and Toxicology (Chp 2)
Risk Assessment (Chp 12)
Week 3. Design/Safety Project Start Up
SAChE Chemical Process Safety in the Chemical Process Industries (Dan Crowl)
Week 4: SAChE Risk Assessment (Ralph Pike)
Prevention of Fires and Explosions (Chp 6 and Chp 7)
SAChE Dust Explosion Control (Louvar)
Week 5: Chemical Reactor safety
Safety Procedure and Designs (Chp 13)
SAChE Runaway Reactions (Amy Theis)
Week 6: Inherently Safer Design and Engineering Ethics
Intro to Reliefs (Chp 9), Relief Sizing (Chp 10), Order of the Engineer.
SAChE Inherently Safer Design certificate
Week 7: Design/Safety Project Progress Review 1
Week 8: Design/Safety Project Progress Review 1
HAZOPS

Situational Awareness (Weeks 1–6)

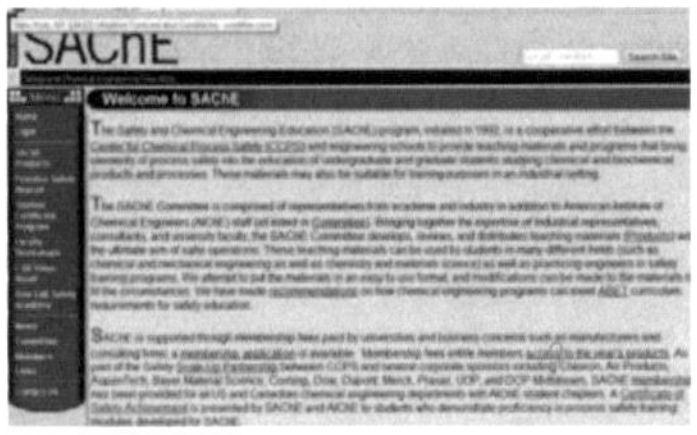

(www.sache.org 27 Sep 2015)

Course Design and SAChE Integration

Week 9: INTERPHEX Exhibition assignment at Javits Center, New York, NY
Guest Lecture Simulation Solutions Inc
Week 10: Guest lecture: Simulation Solutions Inc
Operations Center Exercise
Week 11: Operations Center Exercise- After Action Review
HAZOPs Review
SAChE Process Safety Lessons Taught from Experience (Wiley)
Week 12: Design/Safety Project Progress Review 2
Week 13: Design/Safety Project Progress Review 2
Presentation of Accident Investigation Case Studies
Week 14: Presentation of Accident Investigation Case Studies
Week 15: Study Days
Week 16: Final Design/Safety Project Due

Situational Understanding (Weeks 9–15)

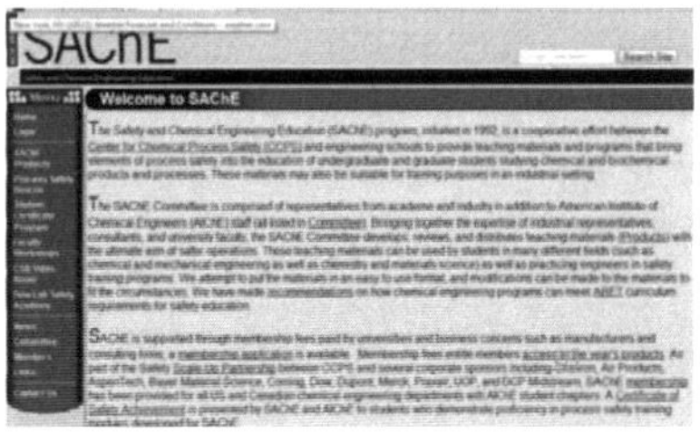

(www.sache.org 27 Sep 2015)

Trade Show Assignment INTERPHEX 2017
Javits Center 655 W 34th St, New York, NY 10001

(Photo by R.G. Bozic)

1. Emphasis inherently safer design

-A one page report on **one** vender you visit at the INTERPHEX

-The structure of your INTERPHEX report should be as follows: title page, an executive summary, results, discussion of what you learned addressing the areas of Inherently Safer Design: Moderate, Substitute, Minimize, and Simplify, and your conclusions.

Learning from History
Case Study and Design

Case study presentations Spring 2017

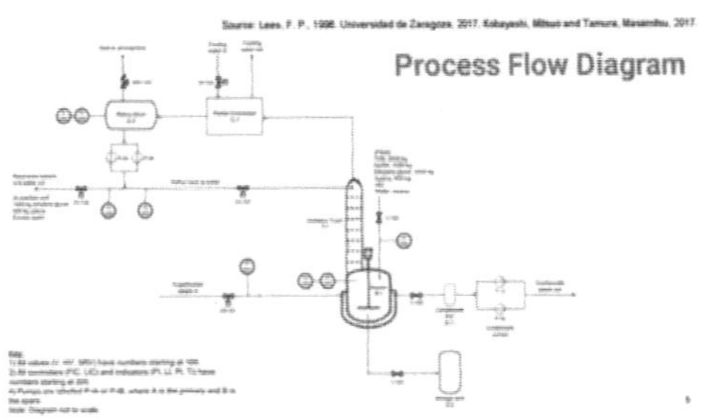

1. Emphasis on Process Flow Diagram to develop a common operating picture

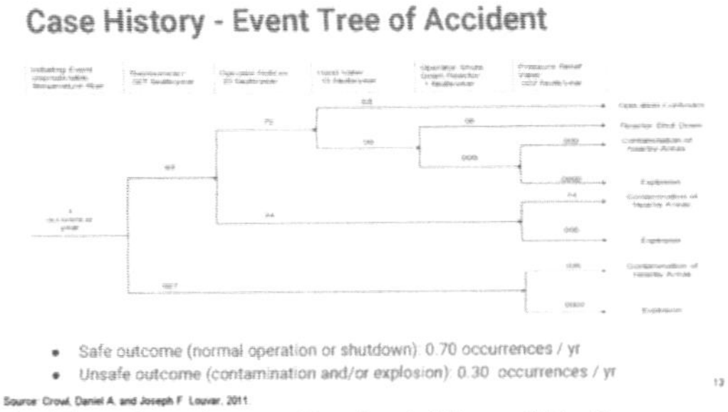

2. Emphasis on Probability of Failure

Learning from History
Case Study Format

Oral Communication Skills- (Briefing During Class)

Written Communication Skills- (Slides turned in to the instructor)

Process Flow Diagrams

Case History (the accident, the event that produces injury/death/damage, and the hazard, the chemical or physical condition that can cause damage. Identify and evaluate the hazards involved. Determine what controls were used, and what should or could have been used. Define the scenario, a description of the events that result in an accident; the indecent, the loss of cantonment of material/energy, the consequence, result of accident, and the likelihood, expected probability of the event. Remember to address, risk- as defined, Risk = f(frequency, consequence) of the accident and determine what aspects of inherently safer design, moderate, substitute, minimize, and simplify, could have been used in order avoid such an accident)

Lessons Learned

Documentation/References

(Crowl, Daniel A. and Louvar, Joseph, 2011, Chp1)

13

Learning from History
Design

Oral Communication Skills - (In Class Progress Review)

Written Communication Skills- (Report turned in to the instructor)

Process Flow and Process Control Block Diagram Diagrams

Design Results (Outline what you recommend as a replacement and what your major conclusions are from your design and safety analysis. Design Results include summary of ASPEN simulation results, your safety analysis results from the Process Hazard Analysis Risk Assessment, Event Tree Analysis, Layer Of Protection Analysis, HAZOPS Analysis. Make sure you address aspects of Inherently Safer Design: Moderate, Substitute, Minimize, and Simplify. In this project, use Aspen Plus 8.6 for simulating your material and energy balances. As appropriate, make use of good engineering assumptions in order to complete this design project. Remember to assume, check, and validate assumptions. Keep a record of your assumptions and include them in your presentation.)

Discussion (Discussion of major decisions for your design results and your safety analysis)

Documentation/References

(Crowl, Daniel A. and Louvar, Joseph, 2011, Chp1)

14

Learning from History
8 Significant Disasters (Case Studies)

Morganton, NC Jan 2006 (explosion)

Anacortes, Washington Apr 2010 (explosion)

Flixborough, England, June 1974 (pipe rupture)

Jacksonville, Florida, Dec 2007 (explosion)

Texas City, Texas May, 2005 (explosion)

Patterson, NJ Apr 1998 (explosion)

Bhopal, India, Dec 1984 (toxic release)

Seveso, Italy, Jun 1976 (runaway reactor, toxic release)

15

(www.csb.gov accessed 31 Jan 2016)

(Crowl, Daniel A. and Louvar, Joseph, 2011, Chp1)

Evaluating Chemical Engineering Safety Training Results

COLUMBIA UNIVERSITY

Donald C. Glaser & Matthew Garvey
Simulation Solutions, Inc.

23 Mar 2017 and 28 Mar 2017

16

Objectives

- Learn more about the importance of process safety and the resources & commitment required.
- Learn how to control and mitigate hazards to prevent accidents.
- Learn about operator training through the use of simulators.
- Review process control and chemical engineering design concepts.

17

Simulation Solutions, Inc. Outline

- Overview of Operator Training
- Review of Operator Safety & INSTO™ Methodology
- Introduction to P&ID / DCS / Outside Operator
- SSI "What-If" Exercises
- SSI Troubleshooting Exercises
- Q&A for Students

18

Simulators provide life-like Training for New and Experienced Operators

Laboratory Experience

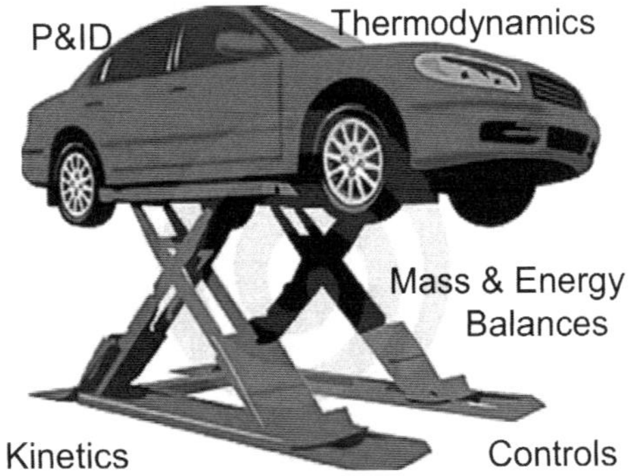

Real World Experience

19

Minds-On/Hands-On Troubleshooting Process

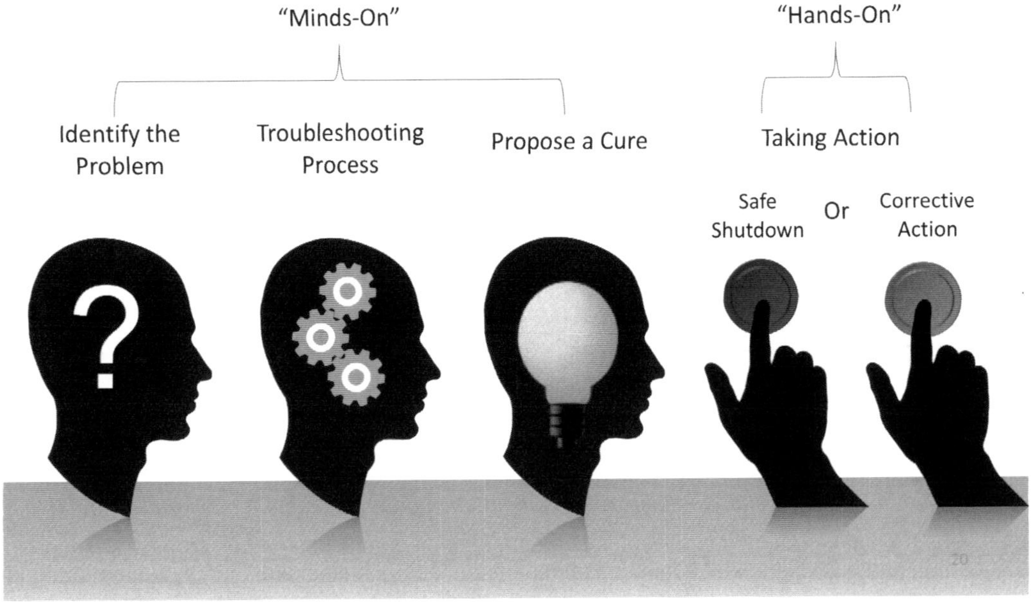

20

Grafting an *Operations Mindset*™ to New Operators

Identification:
Learn how to come onto a new plant and figure out what's there and how it is connected using P&IDs, the plant itself, and the controls.

Normal Operations:
Understand how the plant runs under normal conditions and learn/develop a sense of "expected results" when making operational moves on a plant.

Start-Up and Shut-Down:
Follow detailed, and time critical procedures for start-up and shutdown, starting with a sense of the proper order and important safety issues.

Troubleshooting and Upsets:
Learn to approach plant problems/upsets and troubleshoot them in a systematic way. Learn how to implement safe solutions and/or call for an emergency shut down.

Optimization and Operating Strategies:
Learn the safe limits of operation given the current set of equipment, controls, feedstock, etc. in order to maximize productivity/profits.

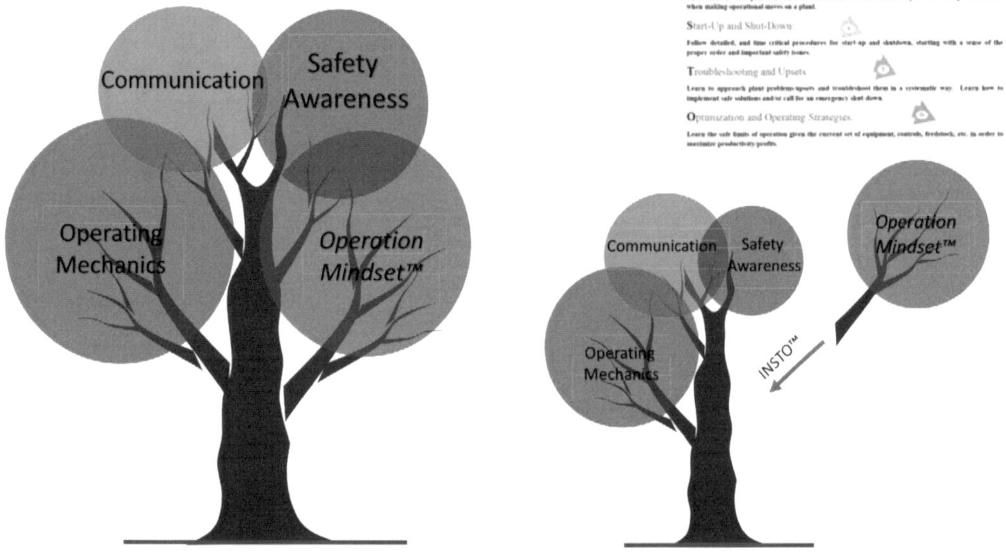

"Old School" Operator
Numerous Opportunities for Experiential Learning

"New School" Operator
Vastly Different Circumstances

21

Coursework Flow

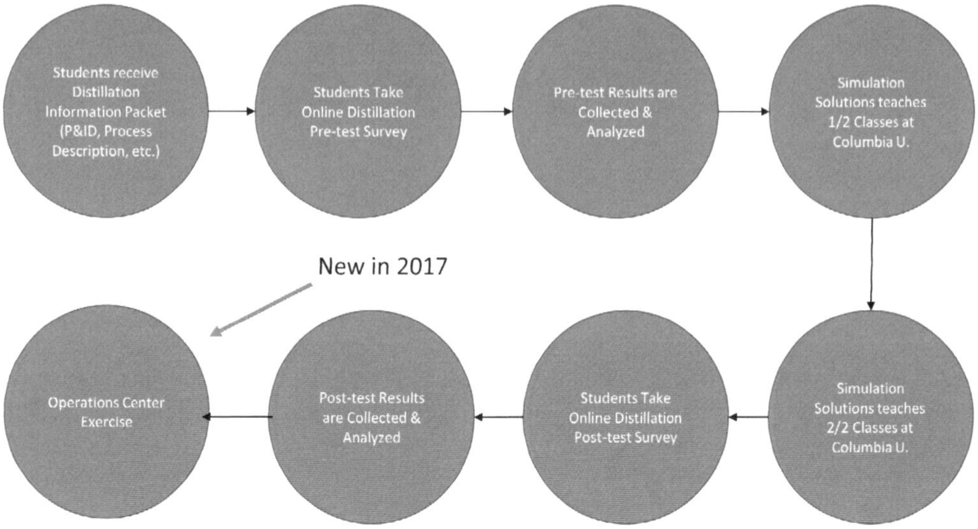

22

Coursework Flow

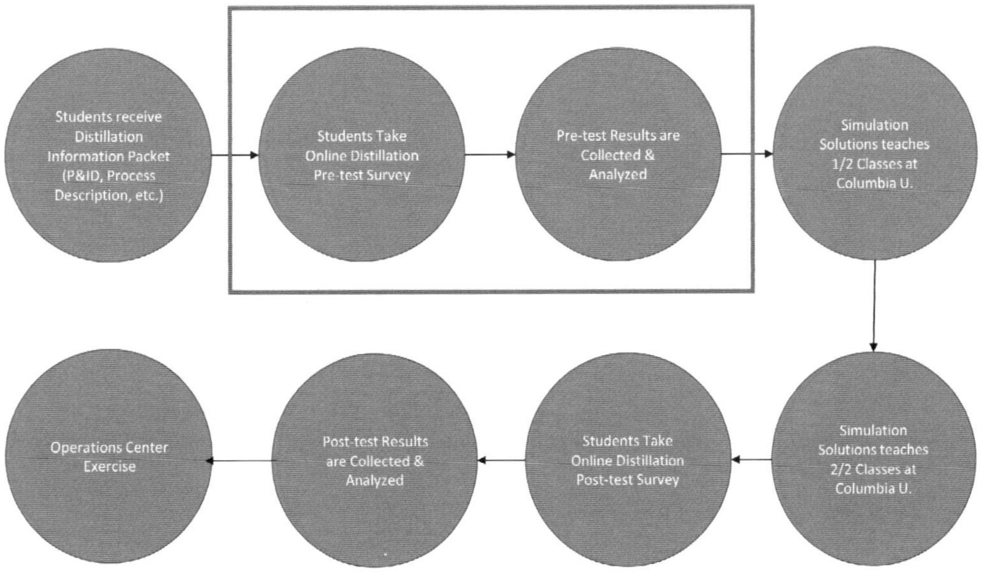

23

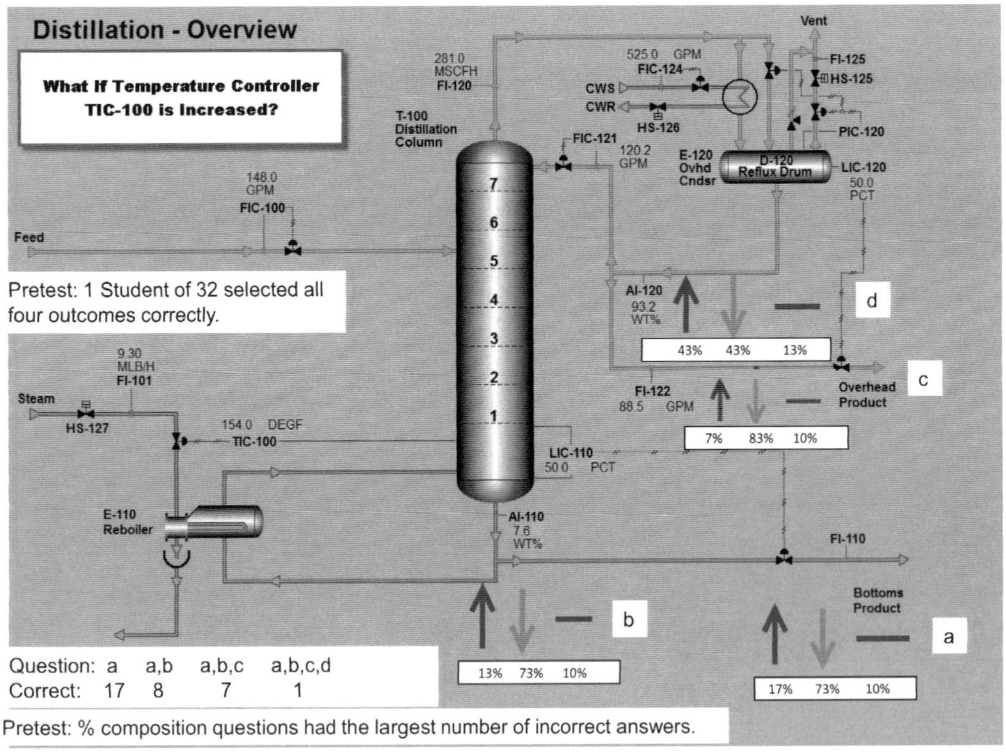

SAChE/ Course Objective/ Lesson Objective "Learn how to control and mitigate hazards to prevent accidents."

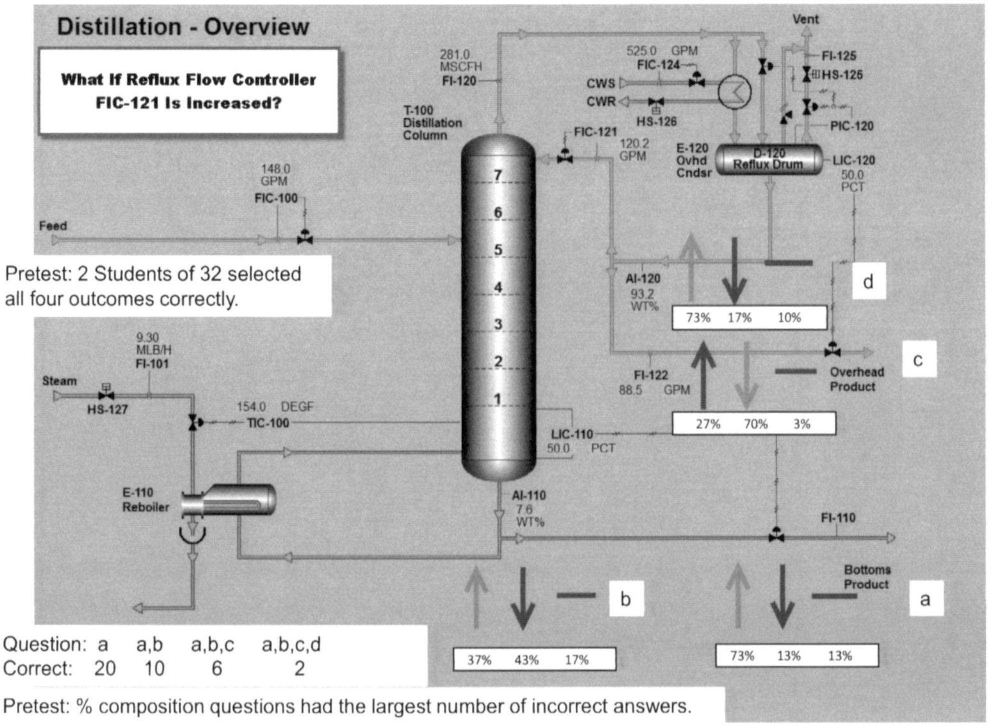

SAChE/ Course Objective/ Lesson Objective "Learn how to control and mitigate hazards to prevent accidents."

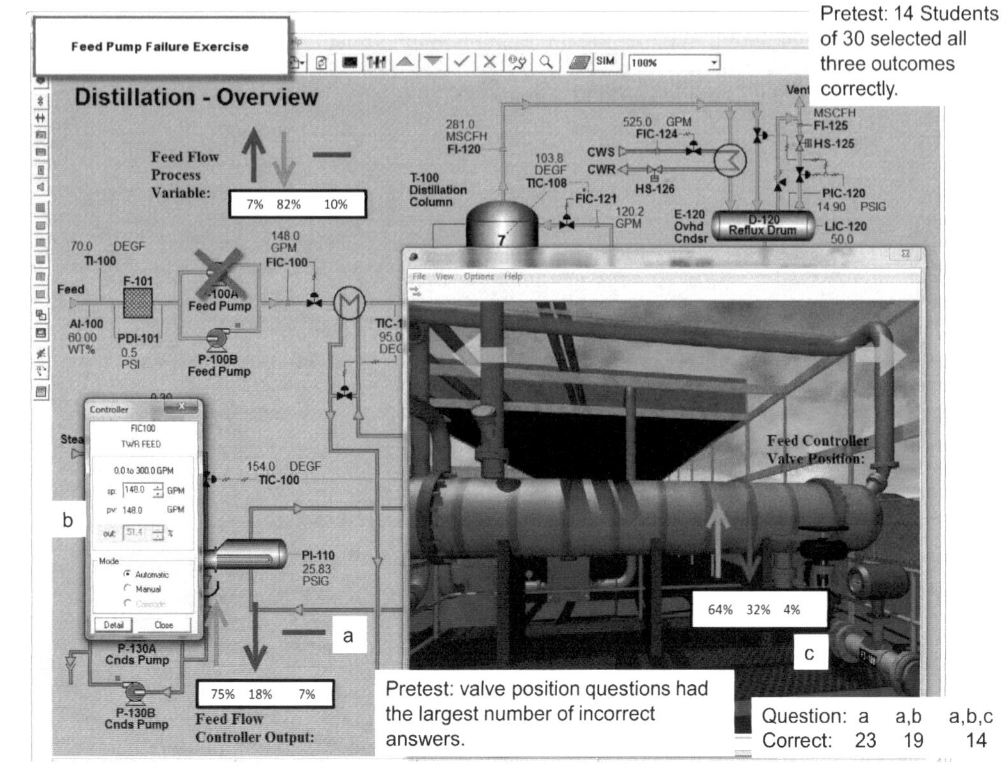

SAChE/ Course Objective/ Lesson Objective "Learn how to control and mitigate hazards to prevent accidents."

Feed Pump P-100A Failure Recovery Steps

Step No.	Operator Action	Number of Students Remaining with Proper Sequence (28 Total)
1.	Place Feed Flow Controller in Manual	14
2.	Lower Feed Controller Output to 10%	9
3.	Start Spare Feed Pump, P-100B	8
4.	Increase Feed Flow Output until Process Variable is at the Design Flow Rate	7
5.	Place Feed Flow Controller into Automatic Mode	7

27

SAChE/ Course Objective/ Lesson Objective "Learn how to control and mitigate hazards to prevent accidents."

Coursework Flow

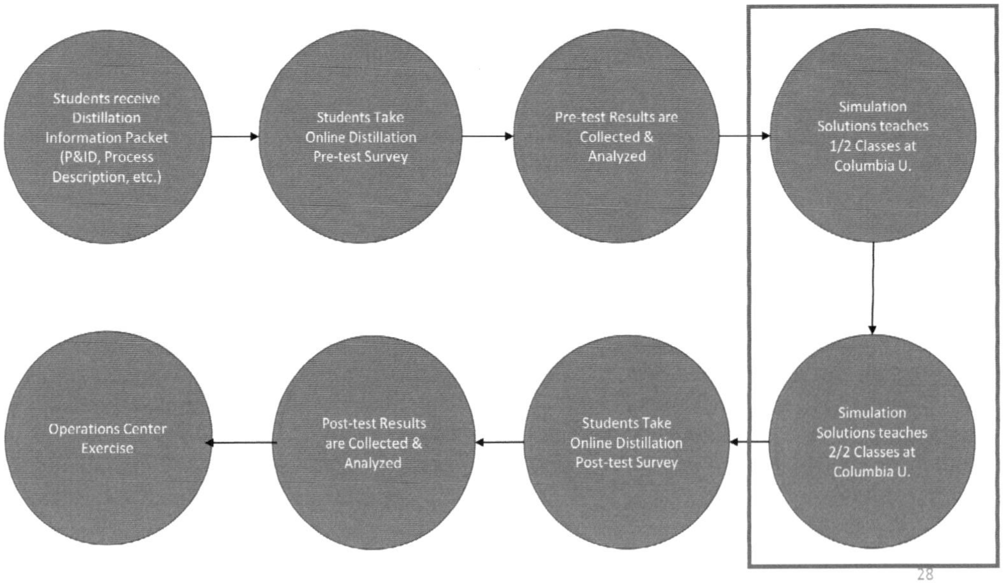

28

Simulator Demonstrations

Distillation

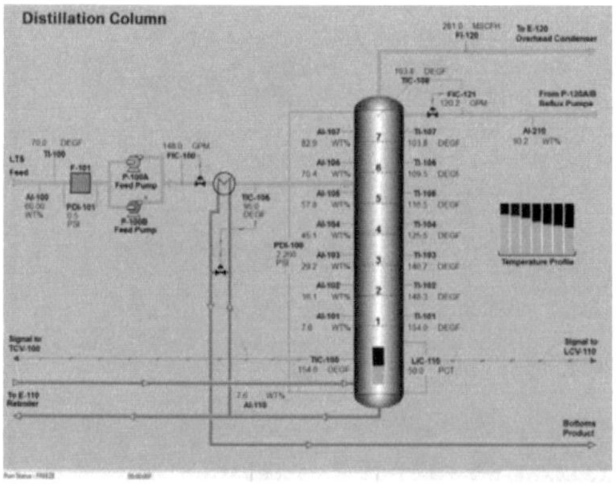

"Minds-On/Hands-On" Training™

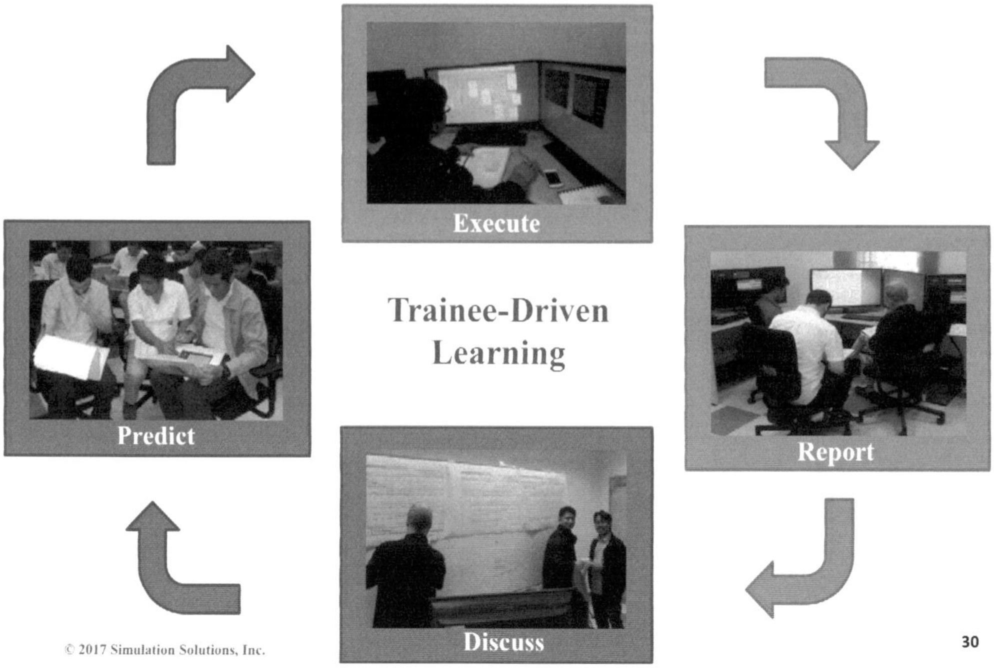

Don Glaser and Matt Garvey of Simulations Solutions Inc. instruct the CHEN E4501 Chemical Engineering Process Safety Class on aspects of operator training

Yi Ling Yang and Anqi Sun run a distillation simulation during a "What if?" exercise in class during the Simulation Solutions Inc. Guest Lecture in the CHEN E4501 Chemical Engineering Process Safety course.

Desha Dike and Aria Perkins run a distillation simulation during a "What if?" exercise in class during the Simulation Solutions Inc. Guest Lecture in the CHEN E4501 Chemical Engineering Process Safety course.

33

Jaren Shapiro and Farah Taufiq predict actions and then observe simulation during the Simulation Solutions Inc. Guest Lecture in the CHEN E4501 Chemical Engineering Process Safety course.

34

Distillation
"What-If" Flip Charts

How will the Bottom of the Tower react if Feed Flow is increased?

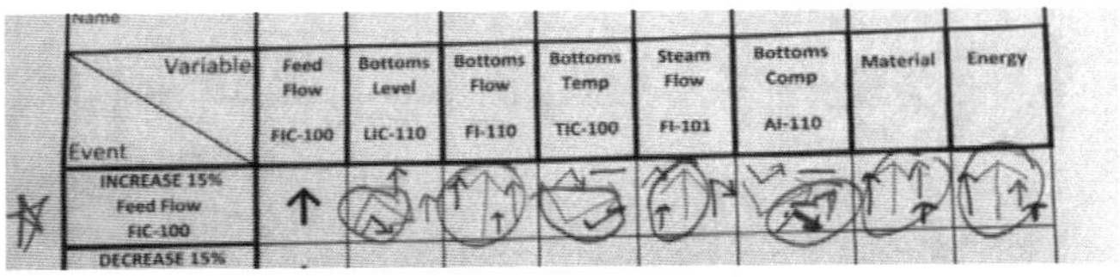

How will the Top of the Tower react if Reboiler Steam is Increased?

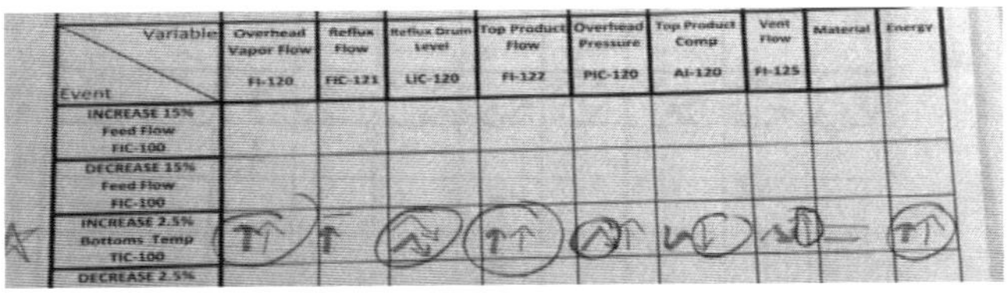

How will Product Flows & Compositions react if Overhead Pressure Increases?

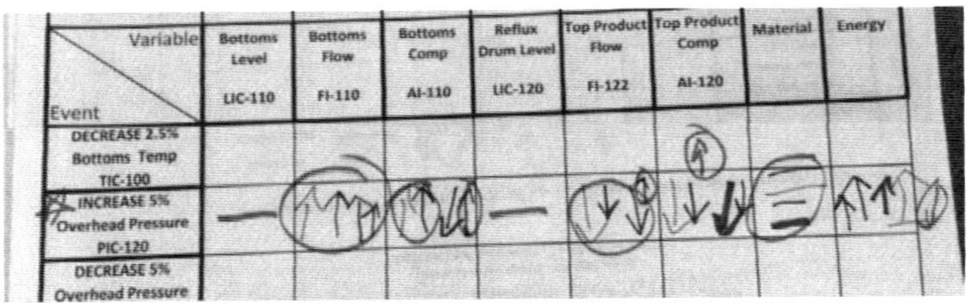

35

Distillation
"What-If" Discussion

- Students were asked to predict how Controllers and Indicators would respond to a Setpoint change in a Distillation Simulator. (i.e. Increasing Feed Flow Rate, Tower Temperature, Reflux Flow Rate, Overhead Pressure).
- Students did not always correctly predict controller responses. For example: When more Feed is put down a Distillation Column the bottoms level will rise, but eventually come back to set point given the controller in this system. This is achieved by the Bottoms Flow increasing. Some students only predicted that this level would rise, and not return to setpoint.
- Another area of trouble was predicting Product Stream purities. For example when the Overhead Pressure was increased in the system, many students predicted that the Top Product would have less "lights" in the stream. An increase in Pressure will actually sweeten the Top Product Stream.

36

Troubleshooting Exercise Results – Loss of Cooling Water to the Tower

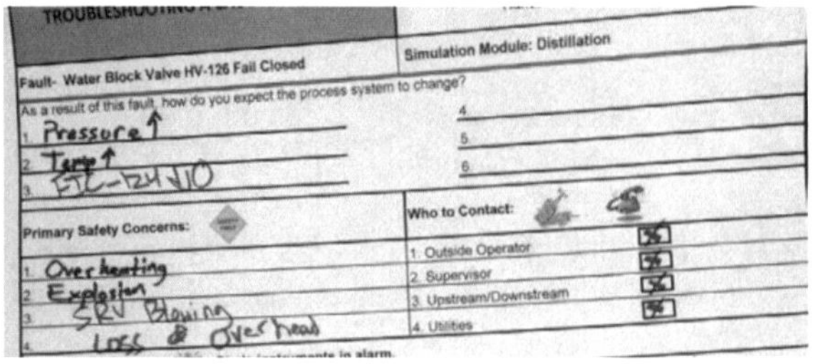
TROUBLESHOOTING

Simulation Module: Distillation

Fault- Water Block Valve HV-126 Fail Closed

As a result of this fault, how do you expect the process system to change?

1. Pressure ↑
2. Temp ↑
3. FIC-124 ↓10
4.
5.
6.

Primary Safety Concerns:

1. Over heating
2. Explosion
3. SRV Blowing
4. Loss & Overhead

Who to Contact:

1. Outside Operator
2. Supervisor
3. Upstream/Downstream
4. Utilities

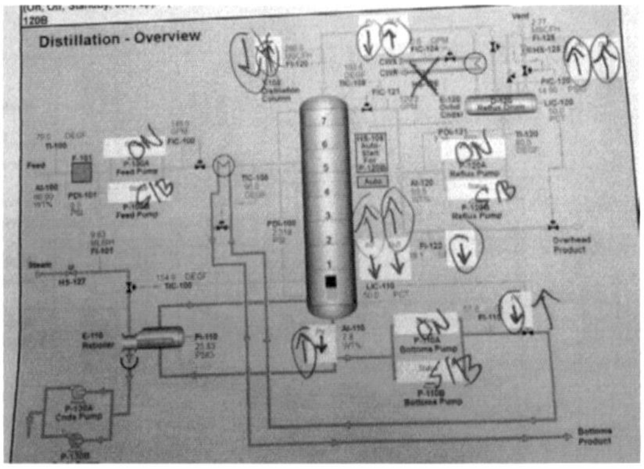

Troubleshooting Exercise Discussion

- Students were asked to predict Controller and Indicator Responses if Cooling Water to the Tower went to zero, due to a closed block valve.
- A focus was put on the "big picture" safety concerns of this fault, a few safety concerns students listed were:
  - Over Pressurization of Tower
  - Safety Relief Valve Blowing
  - Loss of Overhead Product (Distillate) Flow
- An emphasis was placed on considering downstream and upstream effects of a fault within your unit.

Student Survey Results 2017 (Graded on Scale 1-5)

	For each statement below, rank from 1-5, where 5 indicates strong agreement and 1 indicates strong disagreement.	CHEN E4501 Lessons 18 and 19	Standard Deviation
1	As a result of lessons 18 and 19 my ability to understand the importance of process safety and the resources and commitment required has improved. (course objective 1)	4.59	0.57
2	As a result of lessons 18 and 19 my ability to understand how to control and mitigate hazards to prevent accidents has improved. (course objective 4)	4.44	0.64
3	As a result of lessons 18 and 19 my understanding of operator training through the use of simulators has improved..	4.63	0.56
4	As a result of lessons 18 and 19 my understanding of process control and chemical engineering design concepts has improved.	4.48	0.58

31of 31 students responded.

Student Survey Results 2017 (1 of 2)

General Comments about Simulation Solutions Safety Course Spring at Columbia University

CHEN E4501 Chemical Engineering Lessons 19 and 20

- Worthwhile program. Hopefully, in the future it'll be great if we can have the class in a computer lab since Mac's can't run the software.
- Really helpful because it was interactive. Feel more confident in understanding of safe procedures and hazard mitigation.
- It is interesting to use the simulators to understand preventing hazards in engineering design.
- Fun class and learned a lot.
- Fun, but I felt a bit rushed.
- Software was very helpful for visualizing real time effects of process variable manipulator. Exercises were well developed.
- Great presentation. The workbook is an effective tool.
- The lesson was useful in illuminating the hazards of process control but perhaps more links to the concepts from class would be useful.
- I appreciate the operator training since we usually don't get hands-on experience with engineering concepts. The guest lecturers were nice and thorough.
- You help me learn a lot.
- Very Helpful.
- Great lessons.

Student Survey Results 2017 (2 of 2)

General Comments about Simulation Solutions Safety Course Spring at Columbia University
CHEN E4501 Chemical Engineering Lessons 19 and 20

- Class time is short. What we need to learn is much. I truly did not understand some content.
- Slightly rushed towards the end of lecture, but otherwise informative and helpful.
- Great experience to use software on simulation.
- I really like the addition of simulator training.
- Simulation Soln representatives were very helpful.
- Useful course
- These two lessons are really meaningful.
- It's fun. I feel closer to the real world.:-)
- Great lesson! Works great seeing simulation!
- Having a visual and industry view of controls has helped me understand where the controls course we took was mostly theoretical.

41

Coursework Flow

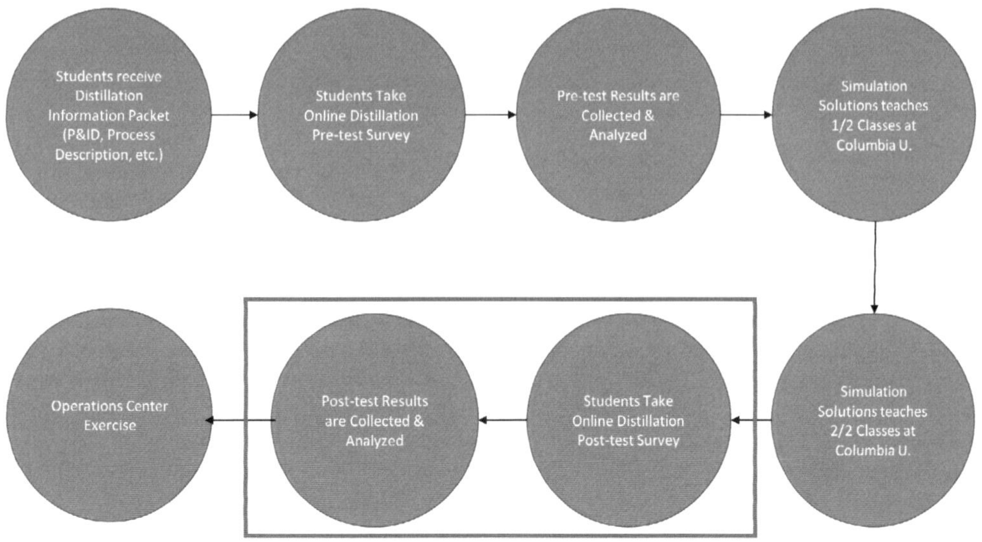

42

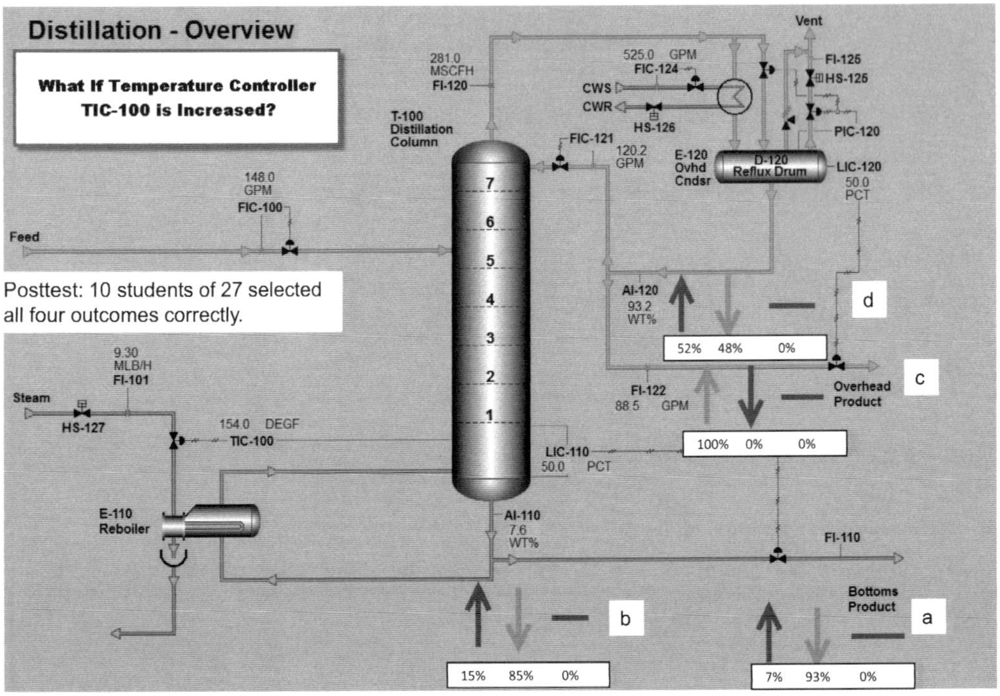

Question: a a,b a,b,c a,b,c,d
Correct: 25 22 22 10

Post-test: % composition questions had the largest number of incorrect answers.

SAChE/ Course Objective/ Lesson Objective "Learn how to control and mitigate hazards to prevent accidents."

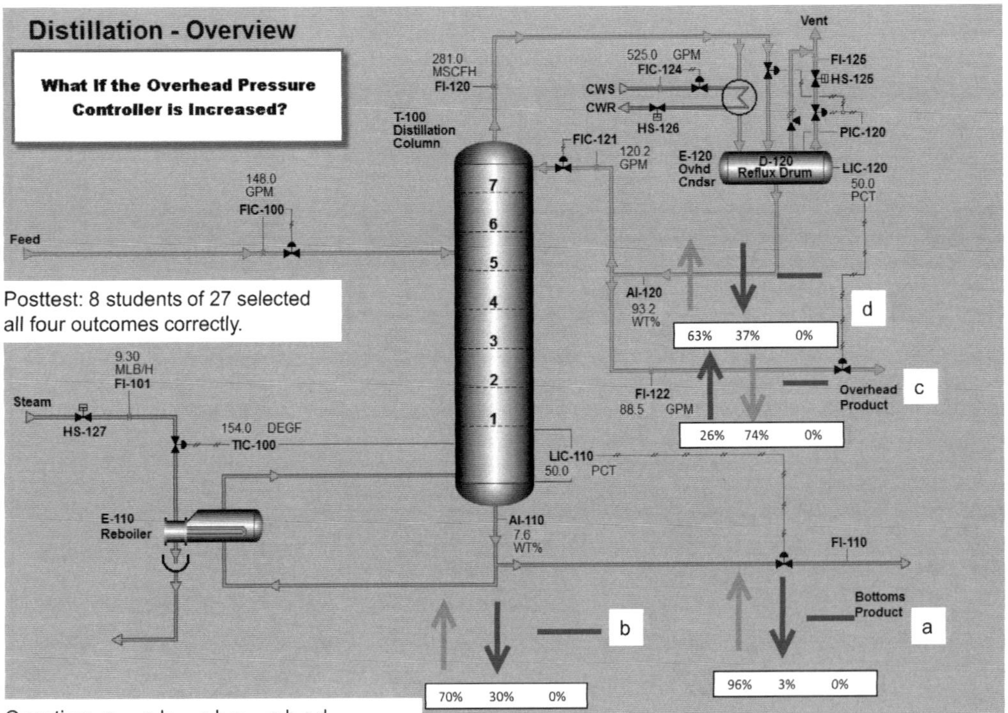

Question: a a,b a,b,c a,b,c,d
Correct: 26 18 15 8

SAChE/ Course Objective/ Lesson Objective "Learn how to control and mitigate hazards to prevent accidents."

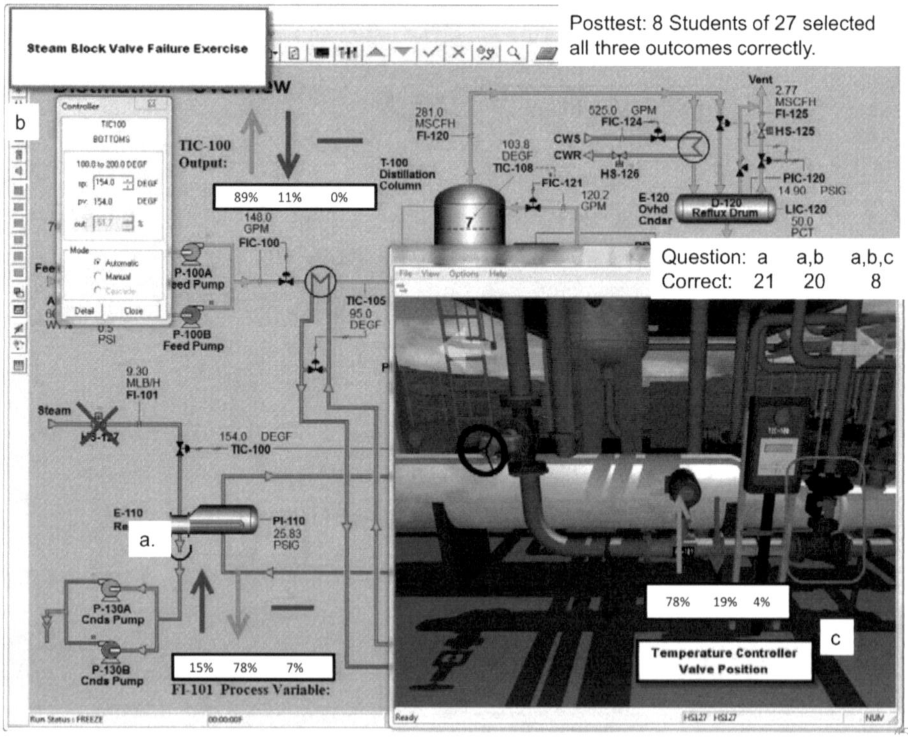

SAChE/ Course Objective/ Lesson Objective "Learn how to control and mitigate hazards to prevent accidents."

Condensate Pump Failure Recovery Steps

Step No.	Operator Action	Number of Students Remaining with Proper Sequence (27 Total)
1.	Place Temperature Controller in Manual	22
2.	Lower Temperature Controller Output from 100% to the Design Rate of 50%	17
3.	Start Spare Condensate Pump, P-130B	13
4.	Adjust Temperature Controller Output until the Design Temperature of 154 degF is achieved.	13
5.	Place Temperature Controller in Automatic Mode	13

46

SAChE/ Course Objective/ Lesson Objective "Learn how to control and mitigate hazards to prevent accidents."

Comparing Pre-test and Post-test Data

- Students showed substantial improvement in recovering from a plant upset in a safe manner in order to mitigate and control a hazard.
- Students showed improvement in predicting how both product flow rates and product purities would respond to disturbances in a distillation column.

47

Lessons Learned Simulator Exercise

<u>Lessons learned</u>

By scheduling 2 Project Progress Reviews Prior to the case study and design assignment, students were encouraged to go in a direction that required them to do the case study work prior to the Progress Review. This seemed to help decrease procrastination.

Students were inspired and enjoyed opportunities to use computer software that simulated process equipment.

Student performance on the pretest and post test showed a marked improvement. With only two 75 min classes, and a small amount of individual “stick time” on the simulator, student ability to predict outcomes improved considerably.

Having a pre-test and post-test seemed to increase interest in the event as well as provide the major means by which performance was evaluated.

Ability to understand overall design goal including product rates and separation needs improvement.

48

Operation Center Exercise

826 Mudd Building, Omar A. Davidson Multi-Media Room (L to R) Don Glaser, Matt Garvey, and Ariel Sanchez, test connection with screens set up for the Operations Center Exercise

Operation Center Exercise

Group 1, Process Engineers, talk to the operator about what is appearing on the display screen while some students observe the action in 826 Mudd Building, Omar A. Davidson Multi-Media Room, Thursday 30 March 2017

Operation Center Experience

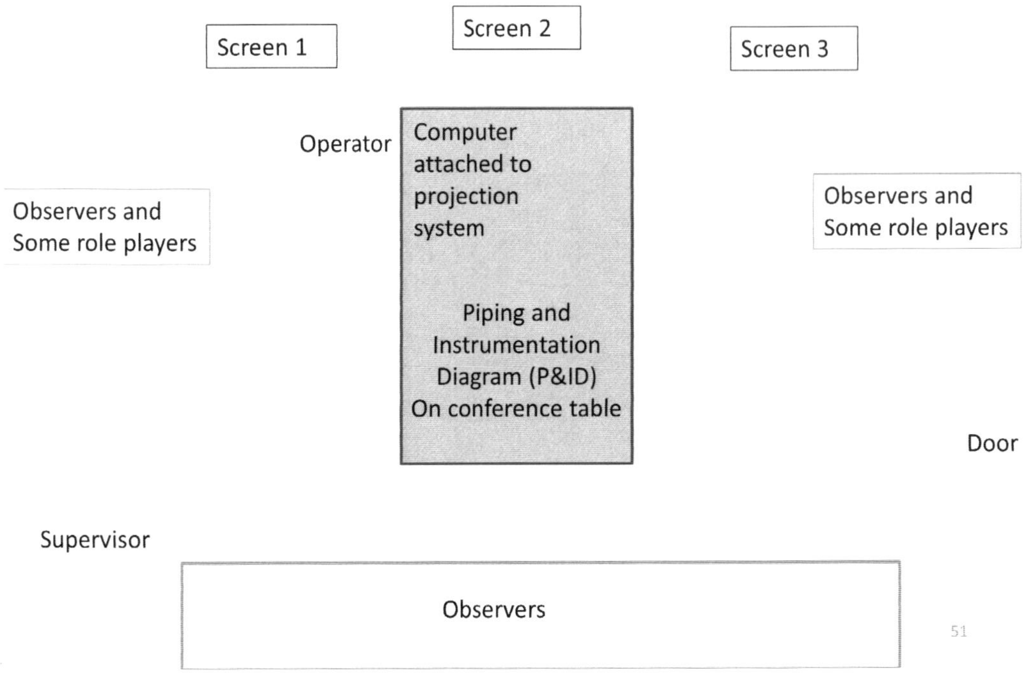

51

COLUMBIA | ENGINEERING

2. Review Time Line and sequence of events

	1:10 PM	1:15 PM	1:20 PM	1:25 PM	1:30 PM	1:35 PM	1:40 PM	1:45 PM	1:50 PM	1:55 PM	2:00 PM	2:05 PM	2:10 PM	2:15 PM	2:20 PM	2:25 P
Group 1time (min)	0	5	10	15	20	25										
Group 2 time (min)						0	5	10	15	20	25					
Group 3 time (min)											0	5	10	15	20	25
Distillation Column Critcal Event		Utilities - CW Block Valve HS-126 Fail close Outside operator /utilites		Equipment - Bottoms Pump, P-110A Failure. Motor Failure, Light Goes Red			People-Vent Block Valve, HV-125 Fails Closed		Equipment - Condensate Pump P-130A Motor Failure, Light Goes Red			People-Block valve fail close-HS127		Equipment - TCV-100 Steam Leak		
Process Engineer Unit	Shift change				Shift change	Shift change					Shift change	Shift change			Shift change	
Utilites		Cooling Water Fail														
1st Responder			1st Responders													
Contractor/Repairman													Contractor/Repairman			
Press Release								Press Release								

TRANSCENDING DISCIPLINES, TRANSFORMING LIVES   COLUMBIA | ENGINEERING

Spring 2017 CHEN E4501 Chemical Engineering Process Safety
Operation Center Exercise
After Action Review

Agenda

1. Define After Action Review

2. Review Time Line and sequence of events

3. Each group briefly presents their Issue, Discussion, and Recommendation with class follow up on each topic.

TRANSCENDING DISCIPLINES, TRANSFORMING LIVES   COLUMBIA | ENGINEERING

Spring 2017 CHEN E4501 Chemical Engineering Process Safety
Operation Center Exercise
After Action Review

1. Define After Action Review-

•"An after-action review (AAR) is a professional discussion of an event, focused on performance standards, that enables soldiers to discover for themselves what happened, why it happened, and how to sustain strengths and improve on weaknesses. It is a tool leaders and units can use to get maximum benefit from every mission or task. It provides--

•Candid insights into specific soldier, leader, and unit strengths and weaknesses from various perspectives.

•Feedback and insight critical to battle-focused training.

•Details often lacking in evaluation reports alone."
(http://www.au.af.mil/au/awc/awcgate/army/tc_25-20/chap1.htm accessed 31 Mar 2017)

Other References:
Forbes: (https://www.forbes.com/sites/jeffboss/2016/12/01/dont-skimp-on-the-after-action-review-6-reasons-why/#17902202ba3d

Harvard Business Review: https://hbr.org/product/hbr-tools-after-action-reviews/TLAAR1-ZIP-ENG)

"What do I think is the problem?"

"What are the indicators?

"How certain am I of the indicators?"

"What is available to me to solve the problem?"

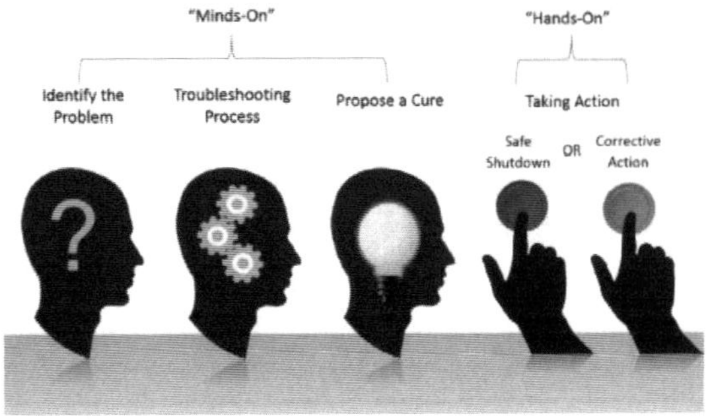

Minds-On/Hands-On Training for Operators™
www.simulation-solutions.com

Do "METT-T" Analysis

Course of Action Development (restrictions/constraints)

"What if? Analysis"

Take Action

How do I know the solution is working? (Process variables as indicators of successful/unsuccessful action)

Review ways to prevent problems in the future.

Update Standard Operating Procedures

2. Review Time Line and sequence of events

Day 3 – Thursday, 30 March 2017 Operations Center Exercise Day (826 Mudd:)

Three process engineer teams will rotate through plant operations operating the distillation column with a role player and simulator scenario activities. Three groups will be selected at the beginning of the semester as process engineer teams. Other groups will be selected as observer/recorder groups. Some groups will be selected as role player groups. Each group will be required to research the roles of the group assignment in the chemical process industry regarding regulations/best practices/and actions taken to ensure process safety. The observer/recorder groups will research all player roles and be ready to observe and record the actions of the Operations Center Exercise Day. On the second day of this exercise, the observer team will report observations of the actions taken by the process engineer teams and role players over the designated time segments.

Process Engineer Teams: Group 1, Group 4, Group 8

Role Player Teams:

- 1st Responder(Fire and Paramedic): Group 2
- Hot Work Contract Repair: Group 5
- Press Agents: Group 7

Observer/Controller Teams: Group 3, Group 6

The operations center exercise will start with a shift change from the chemical engineers from Simulation solutions to the first student group. From that point forward the student groups will do shift changes according to the schedule below. The teams need to be prepared for events during the exercise such as but not limited to: Shift Change, 1st Responder Coordination, Mass Media/Press, Hot Work Contractors, Pump Failure, Valve Failure, Instrument Failure, and Inclement Weather.

Process Engineer Team Shifts will cycle as follows:

1:10 PM Shift Change from Simulations Solutions to Group 1

1:35 PM Shift Change from Group 1 to Group 4

2:00 PM Shift Change from Group 4 to Group 8.

TRANSCENDING DISCIPLINES, TRANSFORMING LIVES — COLUMBIA | ENGINEERING

Spring 2017 CHEN E4501 Chemical Engineering Process Safety
Operation Center Exercise
After Action Review

Lessons Learned

- **Recognizing Root Cause vs Symptoms**
- **Alarm Rationalization**
- **Inter-Agency Coordination**
- **Time Rate of Change of Deviations or Corrections**

Acknowledgements

Thanks to the Columbia Students of CHEN E4501!
Big Thanks to Don Glaser and Matthew Garvey and Ariel Sanchez!
Go Lions!

Simulation Solutions, Inc.
Minds-On/Hands-On Training for Operators™

58

Conclusion and Future Work

Continued success with initial integration of SaChE Recommendations into curriculum.

Continue a "Back to Basics" approach with Process Flow and Process Control Diagrams

Continue the approach related to operator training and evaluation of process safety education and examining ways to incorporate more predictive style exercises with the use of simulators.

Refinement of process simulation in the classroom with CHEN E4500 Principles of Process Design Course

Future work: on standards for process simulation and Operations Center Role Playing Scenarios in the classroom with CHEN E4501 Chemical Engineering Process Safety Class

Add a design component using ASPEN to the distillation problem to renew emphasis on key design variables.

59

References

ASPENTECH V8.8 Software (used for design work)

Beer, David and David McMurrey, *A Guide to Writing as an Engineer 2nd Edition*. Wiley, New York, 2005

Crowl, Daniel A. and Joseph F. Louvar, *Chemical Process Safety" Fundamentals with Applications 3rd Edition*, Pearson, New York, 2011

Felder, Richard M. and Ronald W. Rousseau, *Elementary Principles of Chemical Processes 3rd Edition*, 2000, John Wiley and Sons, New York, New York. Section 4.3a Flowcharts, p90-92

Graham, Nasby, "Using Process Flowsheets as Communication Tools" *Chemical Engineering Progress*, Volume 108, No. 10.Oct 2012. p36-44, Available in Chemical Engineering Progress on line

i.ytimg.com/vi/8A1xSCUtB-M/hqdefault.jpg accessed 22 Jan 2015

Perry, Robert H. and Don W Green, *Perry's Chemical Engineering Handbook*, 7th Edition, 1997, McGraw Hill, New York, New York

Seborg, Dale E., Thomas F. Edgar, and Duncan A. Mellichamp, *Process Dynamics and Control*. Wiley Inc, New York, New York (1989) p7-9

Walker, Vornel, "Designing a Process Flowsheet" *Chemical Engineering Progress*, Volume 105, No.5. May 2009. p15-21, Available in Chemical Engineering Progress on line

www.aiche.org/academy/courses/sache/sache-certificate-program accessed 19 Jan 2015

www.aiche.org/ccps/community/students-teachers-and-communities/certificate-safety-achievement accessed 19 Jan 2015

www.bing.com/images/search?q=jacksonville%2C%20florida%20chemical%20plant%20dec%202007&qs=n&form=QBIRMH&pq=jacksonville%2C%20florida%20chemical%20plant%20dec%202007&sc=0-0&sp=-1&sk=#view=detail&id=757F70F7B398E519B65D6CACB55A59E57E50CB72&selectedIndex=5 accessed 22 Jan 2015

www.bing.com/images/search?q=pasadena%20texas%20oct%201989&qs=n&form=QBIR&pq=pasadena%20texas%20oct%201989&sc=0-19&sp=-1&sk=#view=detail&id=1A52930A92CCEC5F28D374DBCC19EA634FE9A02F&selectedIndex=1 22 Jan 2015

www.csb.gov accessed 3 Feb 2015 used for case study videos

www.csb.gov/assets/1/7/Tesoro_Anacortes_2014-May-01.pdf accessed 28 Jan 2015

www.csb.gov/assets/1/19/Final_Report_small.pdf accessed 28 Jan 2015

https://www.marsh.com/us/insights/research/the-100-largest-losses-in-the-hyrdocarbon-industry-1974-2015.html accessed 2 May 2017

www.sache.org accessed 2 Jan 2015, used for SaChE

www.srai.org/wp-content/uploads/2010/06/pic2.jpg accessed 22 Jan 2015

www.toxipedia.org/download/attachments/9353/Seveso.chem.jpg?version=1&modificationDate=1238498605000&api=v2 andhttp://www.bing.com/images/search?q=Seveso+Disaster+Affected+Area&Form=IQFRDR#view=detail&id=A9CEBE175583891AADF628A0E49D369903208B4E&selectedIndex=0 accessed 22 Jan 2015

www.bing.com/images/search?q=port%20wentworth%20georgia%20feb%202008&qs=n&form=QBIR&pq=port%20wentworth%20georgia%20feb%202008&sc=0-24&sp=-1&sk=#a accessed 22 Jan 2015

www.bing.com/images/search?q=texas%20city%20mar%202005&qs=n&form=QBIR&pq=texas%20city%20mar%202005&sc=8-11&sp=-1&sk=#view=detail&id=F2A258E4AD124686AFE250ED468FBB03FA6C7066&selectedIndex=6

60

Back Up Slides Start Here

61

Student Survey Results 2016 (1 of 2)

General Comments about Simulation Solutions Safety Course Spring 2016 at Columbia University
CHEN E4501 Chemical Engineering Lessons 19 and 20

- good as expected
- I'm still confused about the some basic concepts regarding controllers. The speech may have went too quickly
- I hope that we have more reading material on this since the lecture is a bit fast and overwhelming
- Good
- I thought it was engaging and fun! I don't feel like an expert now but it was a good introduction to operator controls and applied concepts. I feel with further training this would really prepare someone to work as an operator
- very interesting and informative
- This experience has greatly improved from last year. Through expanding the lesson to two classes it was much easier to absorb the material and it does not feel as rushed. Additionally, it is very interesting to experience a practical application of chemical engineering rather than theoretical. (most likely the grader who took the course in 2015)
- Personally, the first lesson felt hard to follow but the second lesson I caught on and felt I had gained and improved my knowledge of process control
- These 75 minutes felt like 15 mins. Very interesting. But I think if this can be 2 weeks, it would help understand controls even better. Thanks Dr. Bozic
- Zero complaints. Great presenters, practical applicability was very helpful, only suggestion would be to have the exercises be a little more guided to hear more of the proper mindset in finding solutions and allowing us to do more exercises.
- Really likes working with simulator. Was a lot of fun, wish we were able to do more, but it covered a good amount given the time. Definitely Recommend!
- Really enjoyed and learned much from them!
- It was better than last time. Do it again.
- It was really fun to have an interactive class. These lessons were a great way to see the inner though process of a future engineer.

62

Student Survey Results 2016 (2 of 2)

General Comments about Simulation Solutions Safety Course Spring 2016 at Columbia University
CHEN E4501 Chemical Engineering Lessons 19 and 20

- Helpful for basic understanding of chemical engineering simulating. I can't thoroughly understand details but it's overall good.
- The contents on What-if problem was great. Hope to learn more on hazard control topics.
- It was great to learn the operating process by using simulation software, which helped a lot to improve my understanding in terms of various situations.
- was really informative, interactive and fun.
- it helped me have a global view of control system and the entire chemical reactor process.
- It's really interesting to learn about the engineering design and simulators. Definitely should keep the lessons in the future!
- useful information would prefer all answers to worksheet so I can review on my own & a copy of the simulation program
- These lessons were useful but it would be better if we can run the software on macbook too for everyone to get the experience for using the simulator
- I wasn't sure about the presentations/simulations at first but ended up having a great time learning about process control/design concepts. My understanding has definitely improved, even though I don't quite understand everything yet. Before the lessons I had no idea how to read the provided diagram but now I feel like I have a better understanding.
- very helpful to learn the process control
- Process was very cool. It may be worthwhile to mention that the software for the demo isn't very demanding- i.e. I was initially concerned that the software being loaded would be a huge file size, something like Chem CAD or ASPEN, that it wouldn't be able to run on my (deleted word) computer. It worked fine though.

2015 to 2017 Comparison

For each statement below, rank from 1-5, where 5 indicates strong agreement and 1 indicates strong disagreement.	2017 CHEN E4501 Lesson 18 and 19	2017 Standard Deviation	2016 CHEN E4501 Lessons 19 and 20	2016 Standard Deviation	2015 CHEN E4501 Lesson 21 and 22	2015 Standard Deviation
As a result of CHEN E4501 lessons 18 and 19 my ability to understand the importance of process safety and the resources and commitment required has improved. (course objective 1)	4.58	0.56	4.67	0.62	4.80	0.50
As a result of CHEN E4501 lessons 18 and 19 my ability to understand how to control and mitigate hazards to prevent accidents has improved. (course objective 4)	4.45	0.62	4.33	0.73	4.40	0.58
As a result of lessons 18 and 19 my understanding of operator training through the use of simulators has improved.	4.68	0.54	4.63	0.63	4.56	0.65
As a result of CHEN E4501 lessons 18 and 19 my understanding of process control and chemical engineering design concepts has improved.	4.48	0.57	4.48	0.89	4.64	0.57

490266 Training Our Upcoming Chemical Engineers By Simulating an Industrial Setting: A Classroom Case-Study on Waste Cellulose Valorization

Abstract:

Chemical engineers are trained in a broad, multidisciplinary environment. This is necessary since they are typically employed in an industrial environment in which not only (chemical) engineering is key, but also knowledge of informatics, mechanics, economics and even human psychology. At Ghent University, in Belgium, our chemical engineering students are trained in every aspect of their future professional career. One of the more prominent, truly multidisciplinary courses in this respect is the 'cross-course project' in the third bachelor year. This is a mandatory, project-based course in which a group of students tackle a real-life industrially relevant engineering problem. Each year, new project topics are defined, based either on the actual need of an industrial partner or on a more fundamental problem encountered by a research institute. By simulating such an industrial setting, the students get a clear picture of their future professional career.

Figure 1: a schematic representation of a typical wastewater treatment plant [1]

One of the project topics was put forward by an industrial partner specialized in wastewater treatment. Apart from organic matter, domestic wastewater contains small, solid particles, mostly composed of cellulose fibers originating from toilet paper. Without specific actions, these fibrous particles would reach the aeration basin resulting in an extra load for the sludge in the basin. Aiming at creating added value, an extra fine-meshed sieve was installed upstream of the aeration basin, allowing the recovery of these solid particles. As a result, an additional solid stream was obtained, almost entirely composed of wet cellulose fibers. The challenge for the students was to propose and design a solution to valorize this cellulosic waste stream. As a result, not only the scientific feasibility of their solution needed to be evaluated, but also the economic and industrial feasibility.

Figure 2: cellulose conversion measured at specific time intervals in a batch reactor. Solid lines represent the Michaelis-Menten kinetic model

After an initial brainstorm session, a two-stage conversion process was developed. In a first step, the cellulose fibers are hydrolyzed to glucose after which the glucose is fermented to bioethanol. The cellulose hydrolysis was performed using a mixture of commercially available cellulase enzymes in a lab-scale batch reactor. The reactant cellulose was provided by dissolving fresh toilet paper in water by stirring. During reaction, samples were drawn from the reactor mixture at well-defined time intervals and the unreacted cellulose was filtered from the mixture. For analytical

purposes, sulfuric acid was added to the samples, which converted the glucose into hydroxymethylfurfural of which the concentration can be measured by UV-VIS absorbance [2]. By varying the reaction conditions, such as temperature and enzyme concentration, the kinetics of the enzymatic transformation could be mapped experimentally. These data were subsequently used to construct a Michaelis-Menten kinetic model. In the second stage of the process, glucose was converted into ethanol by the addition of baker's yeast. To determine the enzymatic kinetics of the glucose, a similar approach was followed as in the first step. A second Michaelis-Menten kinetic model was constructed which could simulate the experimentally measured ethanol production rates quite well. In order to measure reliable kinetic data with a minimal experimental error, a critical mindset of the students in the lab was proven to be essential. Through trial-and-error, the students got aware of a number of potential sources of unreliability, e.g., cellulose fibers still present in the sample, poor cellulose solubility in water, undesired oxygen during fermentation.

Figure 3: Aspen Plus® model of an industrial plant for the conversion of cellulose into glucose and further fermentation to ethanol

Using both kinetic models and industrial wastewater capacities, several preliminary commercial configurations were designed and evaluated via the process modelling software Aspen Plus®. Although this valorization step was feasible on the laboratory scale, some key challenges remain for its industrial application, e.g., the limited solubility of the cellulose and the loss of the homogeneous enzymes after reaction. Overall, the project resulted in a critical evaluation by the students of biomass as raw material in the chemical industry, expressing both opportunities and challenges of a bio-based society.

From a management perspective, the students worked as a synergetic team by appointing specific roles to every team member. They learned to be critical but also appreciated each other's contributions. Apart from field work and reporting, the students had weekly meetings with the responsible professors and, sometimes, the industrial partner from which they received valuable feedback. Finally, their presentation skills were trained by having to present their progress twice, orally, for a broad audience of chemical engineers and students. The students also participated in an international Student Research Conference where they gave a presentation of their work, which earned them the first prize from both the jury as well as the audience.

[1] W. Noorderzijlvest. Available: www.noorderzijlvest.nl

[2] A. A. Albalasmeh, A. A. Berhe, and T. A. Ghezzehei, "A new method for rapid determination of carbohydrate and total carbon concentrations using UV spectrophotometry," *Carbohydrate polymers*, vol. 97, no. 2, pp. 253-261, 2013.

Acknowledgement

The authors would like to thank Bert Biesemans, Christophe Naessens, Laura De Saedeleer and Laura Truyens for their commitment and hard work during the project.

University College Dublin
School of Chemical & Bioprocess Engineering

Irish Engineering Graduates Advancing Global Manufacturing Competitiveness: Design Simulation for the Process Industries

An inter-institutional initiative for Chemical Engineering Education in Ireland

Federico Orefice, Darragh Coakley, Philip Donnellan, David Dorran, Noel Duffy, Brian Freeland, Carmel Hensey, Witold Kwapinski, Damian Mooney, Gearoid Ó Súilleabháin, Jorge Oliveira, Michael O'Mahony, Steve Cropper, Brian Glennon, Joe Hannon, John Milne, Kevin Smyth & **Patricia Kieran**

CIT DCU DIT TL JACOBS IChemE

Irish Pharma/Biopharma Industry

"The Biopharmaceutical industry has made a capital investment of approx. $8 billion in new facilities in Ireland, most.... in the last 10 years" **IDA Ireland**

- 9 of the world's Top 10 pharma companies
- Annual exports: $51.7bn (2014)
- World's 6th largest exporter of pharma products
- 25,000+ employed; 60% of ChemE grads (2012-16)

... & Academia

"Ireland has one of the youngest and most highly educated populations in Europe" **IDA Ireland**

- Strong collaboration between Pharma/Biopharma Industry & Academia; developing curricula reflect industry feedback

Context & Motivation
Overview
Partners & Stakeholders
Structure
Sample Output
Distillation & Simulation
Evaluation
Future work & Conclusions

Irish Engineering Graduates Advancing Global Manufacturing Competitiveness: Design Simulation for the Process Industries

Motivation

Continuous growth and improvement in Chemical/Process Engineering curricula, partially driven by industry feedback & needs.

- Pharma/Biopharma applications of simulation tools, e.g. in developing optimization strategies
- Companies value graduates with strong simulation & modelling skills.

... in addition [Mott, Leay & Silverstein, 2013]

- Learning to use simulation software develops students critical thinking skills
- Appropriate use of simulation tools supports inductive learning (knowledge by observation), with reduced physical resource and time demands

... However,

- Limited relevant research to support pedagogical basis

Context & Motivation
Overview
Partners & Stakeholders
Structure
Sample Output
Distillation & Simulation
Evaluation
Future work & Conclusions

Irish Engineering Graduates Advancing Global Manufacturing Competitiveness: Design Simulation for the Process Industries

Project Overview

Strategy & supporting resources for sustainably embedding effective use of **discipline-specific simulation tools** in Chemical Engineering and cognate programmes, for the **achievement of professionally-relevant graduate skills**

Deliverables

Media-rich, reusable, shared learning resources on open, online platform for **Chromatography, Fermentation, Process Control, Distillation**

Resources Formats

- Screencasts
- Videos
- Animations
- E-Learning presentations
- Simulations
- Assessments

Context & Motivation
Overview
Partners & Stakeholders
Structure
Sample Output
Distillation & Simulation
Evaluation
Future work & Conclusions

Irish Engineering Graduates Advancing Global Manufacturing Competitiveness: Design Simulation for the Process Industries

Academic Partners

All Irish Chemical/Process Engineering & Biopharma Schools

University College Dublin	Cork Institute of Technology	University College Cork	University of Limerick	Dublin City University	Dublin Institute of Technology

Accreditation Body

IChemE

"The global professional institution, operating worldwide, building and supporting community and network of professionals involved in Chemical Engineering disciplines."

Stakeholders

NATIONAL FORUM FOR THE ENHANCEMENT OF TEACHING AND LEARNING IN HIGHER EDUCATION

National Forum for the Enhancement of T&L in Higher Education.
"Putting teaching and learning at the centre of sectoral enhancement and innovation."

- Context & Motivation
- Overview
- **Partners & Stakeholders**
- Structure
- Sample Output
- Distillation & Simulation
- Evaluation
- Future work & Conclusions

Irish Engineering Graduates Advancing Global Manufacturing Competitiveness: Design Simulation for the Process Industries

External Partners

nibrt

apc

JACOBS

DynoChem

- Context & Motivation
- Overview
- **Partners & Stakeholders**
- Structure
- Sample Output
- Distillation & Simulation
- Evaluation
- Future work & Conclusions

External Partners

nibrt

apc

JACOBS

DynoChem

The National Institute for Bioprocessing Research & Training

"...a unique learning experience for trainees in an environment that replicates the most modern industrial bioprocessing facility."
"... research in key areas of bioprocessing in collaboration with industry." **www.nibrt.ie**

Irish Engineering Graduates Advancing Global Manufacturing Competitiveness:
Design Simulation for the Process Industries

External Partners

nibrt

apc

JACOBS

DynoChem

APC supports top pharma & biopharma companies by providing solutions to accelerate the delivery of quality, life-changing medicines. **www.approcess.com**

Irish Engineering Graduates Advancing Global Manufacturing Competitiveness: Design Simulation for the Process Industries

External Partners

nibrt

apc

JACOBS

DynoChem

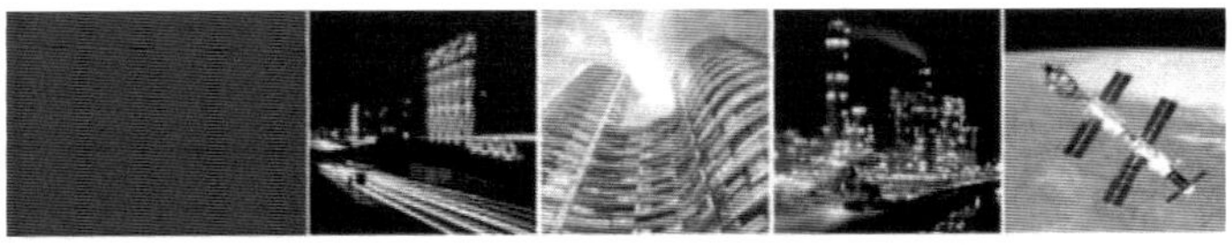

JACOBS

Jacobs globally manages engineering projects. Jacobs' Irish offices have developed particular expertise in pharma/biopharma.
www.jacobs.com

- Context & Motivation
- Overview
- **Partners & Stakeholders**
- Structure
- Sample Output
- Distillation & Simulation
- Evaluation
- Future work & Conclusions

Irish Engineering Graduates Advancing Global Manufacturing Competitiveness: Design Simulation for the Process Industries

External Partners

nibrt

apc

JACOBS

DynoChem

Scale-up Systems

DynoChem, industry standard software for pharmaceutical process development and simulation **www.scale-up.com**

- Context & Motivation
- Overview
- **Partners & Stakeholders**
- Structure
- Sample Output
- Distillation & Simulation
- Evaluation
- Future work & Conclusions

Irish Engineering Graduates Advancing Global Manufacturing Competitiveness: Design Simulation for the Process Industries

Target Areas

Chromatography | Fermentation | Process Control | Distillation | Simulation Tools

- Context & Motivation
- Overview
- Partners & Stakeholders
- **Structure**
- Sample Output
- Distillation & Simulation
- Evaluation
- Future work & Conclusions

Irish Engineering Graduates Advancing Global Manufacturing Competitiveness: Design Simulation for the Process Industries

Target Areas

Chromatography | Fermentation | Process Control | Distillation | Simulation Tools

Approach

Underpinning Theory

Core Chemical Engineering Principles

Experimental Systems

Building from shared expertise/facilities

Simulation

Training in structured, experimentally-based modelling & simulation

- Context & Motivation
- Overview
- Partners & Stakeholders
- **Structure**
- Sample Output
- Distillation & Simulation
- Evaluation
- Future work & Conclusions

Resource Suite: *Chromatography*

UCD DUBLIN nibrt

Experimental Systems

Lab-Scale: Äkta Start
- Simple protein separation

Pilot-Plant
- Fully-instrumented chromatography skid
- 8-50L columns

Resources

- **Lab-Scale:** experiment, video, lab manual, data, exercises
- **Pilot-Plant**: Column packing & operation videos, exercises
- **Simulation:** ASPEN exercise

Context & Motivation
Overview
Partners & Stakeholders
Structure
Sample Output
Distillation & Simulation
Evaluation
Future work & Conclusions

UCD DUBLIN
Irish Engineering Graduates Advancing Global Manufacturing Competitiveness: Design Simulation for the Process Industries
ChemE ie

Resource Suite: *Fermentation*

UCD DUBLIN DCU nibrt

Experimental Systems

Lab-Scale:
- 5L bioreactor

Pilot-Scale:
- 15L bioreactor

Resources

- **Gowning:** training package
- **Bioreactor:** Assembly videos
- **Database**: *S. cerevisiae*, batch, fed-batch
- **Simulation**: Excel & DynoChem

Context & Motivation
Overview
Partners & Stakeholders
Structure
Sample Output
Distillation & Simulation
Evaluation
Future work & Conclusions

Irish Engineering Graduates Advancing Global Manufacturing Competitiveness: Design Simulation for the Process Industries

Resource Suite: *Distillation & Simulation*

Introduced in Year 3 of UCD Chemical & Bioprocess Engineering programmes

TOPIC	CONTENT
ASPEN Plus Intro	Overview - ThermoData evaluation - FLASH model – Sensitivity Tool
DynoChem Intro	Overview - VLLE model
Lab-Scale Batch Distillation	Context - Lab-Scale Experimental System - Experimental Data - Exploring the Data
Pilot-Scale Batch Distillation with Reflux	Pilot-Scale experiment (varying reflux ratio) - Experimental Data ASPEN Batch Modeller simulation – Exploring beyond Experimental Conditions
Lab-Scale Solvent Swap	Context - Lab-Scale Experimental System – DynoChem for Experiment Planning Experiment – Exploring the Data – DynoChem Scale-up Exercise
Continuous Distillation	Context - Aspen Plus (DSTWU, RADFRAC)
Analytical Methods	EtOH-Water Analysis using RI – Calibration Curve – Use of PTR46 Refractometer
ASPEN Plus	2+ unti model – REACTOR - Model Analysis Tools

- Context & Motivation
- Overview
- Partners & Stakeholders
- Structure
- Sample Output
- **Distillation & Simulation**
- Evaluation
- Future work & Conclusions

Irish Engineering Graduates Advancing Global Manufacturing Competitiveness: Design Simulation for the Process Industries

Resource Suite: *Solvent Swap Distillation*

DynoChem

Experimental Systems

Chemical System: Ethanol-Toluene
Lab-Scale: 500mL

Simulation Tools

DynoChem

Resources

- **Context:** Solvent Swap Distillation & Operation Modes
- **Lab-Scale**: experiment, video, lab manual, data, exercises
- **Simulation**: DynoChem Scale-up exercise

- Context & Motivation
- Overview
- Partners & Stakeholders
- Structure
- Sample Output
- **Distillation & Simulation**
- Evaluation
- Future work & Conclusions

Resource Suite: *Pilot-Scale Batch Distillation (with Reflux)*

Experimental Systems

Chemical System: Ethanol-Water
Pilot-Scale: 12L, 15 trays

Simulation Tools

Aspen Batch Modeler

Resources

- **Pilot-Plant**: System description & operation (with sample collection)
- **Sample Analysis:** PTR 46 Refractometer video
- **Simulation:** Aspen exercise

Irish Engineering Graduates Advancing Global Manufacturing Competitiveness: Design Simulation for the Process Industries

Preliminary Evaluation:

Based on your experience, what software tools do you believe to be most useful in your company?

- Recent (2012-2016) UCD Graduates: n = 51
- Current (2017-18) Year 4 Students (with industrial placement experience, 2-15 months): n = 27

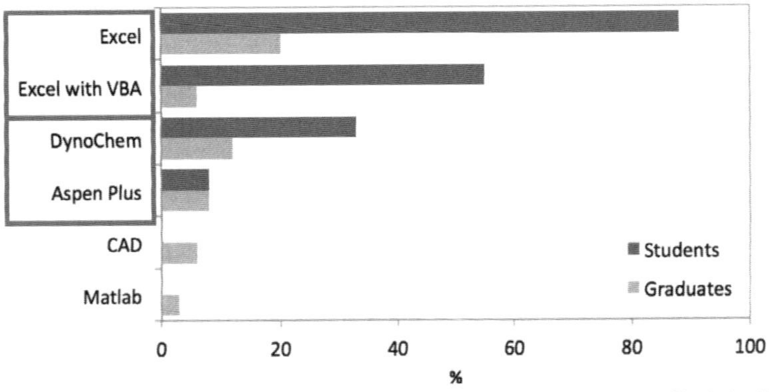

* Other software tools identified: Minitab, C++, gPROMS, SchedulePro, BlueBeam, Simulation Solutions

Survey – Student Evaluation of Resources

Usefulness of Introduced Resources in improving…

Resource	… modelling skills	… understanding Chem Eng Principles & Critical Thinking
DynoChem Intro	> 50%: useful / very useful	> 40%: useful / very useful
Aspen Plus Intro	> 50%: useful / very useful	> 40%: useful / very useful
Pilot-Scale Expt, with Aspen Batch Modeler	> 70%: useful / very useful	> 75%: useful / very useful

Student Comments

- "…simulation tools are a much better way to learn than the more theoretical methods."
- "DynoChem is widely used in industry… very beneficial… before starting my placement."
- "…packages were complex … may have needed more time to introduce them properly."

Irish Engineering Graduates Advancing Global Manufacturing Competitiveness: Design Simulation for the Process Industries

What Next? Continuous process of Investigation, Updating and Improvement

Questions

… Basing on **Students feedback**:
- Are the project's outputs (and teaching strategy) useful?
- Could they be applied earlier in the programme?

… Basing on **Industry feedback**:
- Are project's outputs producing forecasted effects on new graduates?
- Are there more topics to investigate?

… Basing on **Recently Graduates feedback**:
- Is there any area you would have been happy to have investigated better?

Context & Motivation
Overview
Partners & Stakeholders
Structure
Sample Output
Distillation & Simulation
Evaluation

University College Dublin
School of Chemical & Bioprocess Engineering

Irish Engineering Graduates Advancing Global
Manufacturing Competitiveness:
Design Simulation for the Process Industries

Thank you for your attention!

501697 Tutorial: SMART-CN Education Modules for Senior Undergraduate or Graduate Engineering Curriculum

Abstract:

Tutorial: SMART-CN Education Modules for Senior Undergraduate or Graduate Engineering Curriculum

Debalina Sengupta, Yinlun Huang, Thomas F. Edgar, Cliff I. Davidson, Mario R. Eden, Mahmoud M. El-Halwagi*

* corresponding author, el-halwagi@tamu.edu

Session: (Education Division) 04A00 Free Forum on Engineering Education: Junior and Senior Years

The Sustainable Manufacturing Advances in Research and Technology (SMART) project funded by National Science Foundation aims to bridge the gap between academic knowledge discovery and industrial technology innovation for sustainable manufacturing. The SMART project involves a multidisciplinary team which has created an interdisciplinary and international Research Coordination Network (RCN) through the joint effort among a number of leading academic laboratories, centers, non-government organizations, and major manufacturing industries.

Meeting the educational goals of the SMART CN has been achieved through the creation of coordination among member universities and developing online modules specific to sustainable manufacturing. These modules are highly structured classroom-classroom-ready materials for use by instructors at the undergraduate or graduate level. The modules can also be used by industrial entities aiming to further their knowledge in sustainable manufacturing through self-study.

The presentation will provide an overview of this educational component of the project with emphasis on the dissemination of modules supported by computer tools for incorporation into the engineering curriculum. These modules include topics like life cycle assessment, green design and engineering, green chemistry, process intensification, process integration, optimization, and process safety among the core module elements. Test-use data, classroom use experience and general student feedback will be shared. These can be adopted by chemical, mechanical, civil, environmental and other engineering disciplines. Extensive case studies demonstrating these concepts are also provided for classroom teaching, homework assignments and term projects. The modules already available through the CACHE website have been reviewed by corporate sustainability teams, as well as academic instructors.

Reference: http://cache.org/super-store; http://www.research.che.utexas.edu/susman/edu.html

502062 Dual Learning in (Chemical) Engineering: From Theory to Practice

Abstract:

Dual Learning in (chemical) Engineering: From Theory to Practice

Laureano Jiménez Esteller[1], Dieter Thomas Boer[2]

[1]Department of Chemical Engineering, University Rovira i Virgili, Tarragona, Spain

[2]Department of Mechanical Engineering, University Rovira i Virgili, Tarragona, Spain

Laureano.Jimenez@urv.cat

The first aspect to clarify about dual education is that we can find a plethora of (almost) synonyms that refer to very similar concepts: apprenticeship, alternance training, work-based learning, on-the-job training…

In any case, we can define dual learning as a system that combines apprenticeships in a company and class-based learning in the same course. This system is regulated in many countries (At least one pathway offering a combination of work-based and school based learning in the vast majority of EU countries), but its practical application is very marginal, except in notably Germany, Austria, and Switzerland (and recently in South Korea). In fact, there is no single model of alternance programmes in the EU, but a continuum of types of programs that integrate work-based learning. When developing/improving alternance schemes, significant time required before the benefits/improvements are obvious or measureable.

There is little analysis of the costs and benefits of schemes were learners spend a limited amount of time in the workplace. But it is clear that perform an internship in a company or develop the final year project in a company is not dual learning.

The advantages of this educational system are clearly far beyond any reasonable doubt, but of course it is not perfect.

Main advantages:

- There is some sort of labor contract (paid).
- The training is developed under real conditions.
- Trainees also benefit from coworkers experience.

· Hard and soft skills are simultaneously developed.

· Facilitates school-to-work transition.

· Companies obtained an improved retention for employers.

Main disadvantages:

· Training is expensive. School-based is financed by government budgets (national or regional). Alternance schemes is primarily financed by employers: governments financially incentivize (grants, subsidies, tax reliefs...).

· Companies have to follow many regulations.

· Quality of placements and their learning potential: SME/micro companies. Some companies are often so specialized that are unable to train apprentices in a range of knowledge, skills and competences.

· Strict quality controls to avoid situations where apprenticeships are used as cheap labour or substitute unskilled employees

· More student effort required.

· Universities should also follow organizational changes.

· Is an elitist, valid just a few top-students.

Some basic questions before implementing any dual system are the following:

1. Need of new curricula in selected engineering areas at bachelor/master level?
    a. Inclusion of entrepreneurship aspects.
2. Implement dual education?
    a. Bridging the gap between academia & industry
    b. Strengthening relations with external economic and social stakeholders, creating/developing cooperation and promoting employment of graduates.
3. Resistance to change to dual education:
    a. Develop and implement management of educational processes.
    b. Establishment of a unified procedure of evaluation.
4. Regulations involved:
    a. Dead times.
    b. Deadlines.
5. Attitude of companies:
    a. To what extent they will be involved.
    b. Personal interest of professionals.

c. Access to external stakeholders: public bodies dealing with labour market

At the TEEDE project (Towards Excellence in Engineering Curricula for Dual Education), the objective is to develop dual programs, and the first steps that have been already done are the following: Level: master or bachelor?; existing or new program? (Offered as a parallel track to an existing school-based scheme); Support/attitude of companies.

Disclaimer:

This project has been funded with support from the European Commission. This communication reflects the views only of the author, and the Commission cannot be held responsible for any use which may be made of the information contained therein.

Acknowledgements:

Co-funded by the Erasmus+ Programme of the European Union

eu_flag_co_funded_pos_%5Brgb%5D_left

A Graduate Course in Research Data Management

UNIVERSITY OF WYOMING

Lawrence O. Schmidt
Brinkerhoff Geology Library
University of Wyoming
Laramie, WY 82071

Joseph H. Holles
Department of Chemical Engineering
University of Wyoming
Laramie, WY 82071

Introduction

- Effective practices in research data management (RDM) are integral to research:
  - It leads to thoughtful and thorough use of the data for sharing and publication.
  - It has become a required part of federal funding for research.
- Students with advanced degrees are, by definition, leaders in their future companies.
- RDM knowledge is essential:
  - To effectively and economically obtain and maintain research data.
  - To ensure that research data is preserved and accessible.
  - These students will develop and modify RDM policies at their companies.

Motivation

- The National Academy of Sciences has advocated for:
  - Stewardship of research data to ensure integrity and data accessibility.
  - Development of standards and policies regarding the dissemination and management of data.
  - Data management training for all researchers.
- Federal agencies (NSF, NIH, USGS) are requiring the submission of a data management plan (DMP) when submitting funding proposals
- Literature studies have shown that faculty:
  - Understand the need for RDM education.
  - Acknowledge they cannot provide adequate instruction.
- Graduate students are usually expected to carry out their own data management for their research.

There is a need for thorough and systematic education of graduate students in the area of Research Data Management.

Previous Approaches to Teaching RDM

- Two general approaches have been tried:
  - Standalone courses – usually in information science programs. Taught by data experts.
    - Advantages: In depth material coverage.
    - Disadvantages: Attracting students to courses outside their discipline.
  - Seminars/workshops – commonly offered through libraries. Taught by librarians or IT professionals.
    - Advantages: Require less time.
    - Disadvantages: Not for credit and have retention issues.

Current Approach

- Graduate course in Research Data Management.
  - Taught in a specific discipline.
  - Student focused and not discipline focused.
- Designed to provide in-depth RDM as suggested by NAS and NSF.
- Course was co-taught by:
  - Librarian
    - Broad knowledge of RDM tools.
    - Experience with standards and data repositories.
  - Active faculty researcher
    - Research focused examples and experience.

Leverage expertise from both sides

Course Details

- 3 credit, semester long graduate elective course.
- 42 class periods of 50 minutes
- Initial offering: 10 students – 9 ChE and 1 CE
- Required textbook:
  - "Data Management for Researchers: Organize, Maintain, and Share your Data for Research Success" by Kristin Briney.
- Other resources:
  - Managing and Sharing Research Data by Corti et al.
  - RDM course at Oregon State:
    - https://jlsc-pub.org/articles/abstract/10.7710/2162-3309.1246/
  - DMPtool: https://dmptool.org/
  - Data Curation Profile: http://datacurationprofiles.org/

Course Goals

- Expose the students to broad concepts and best practices of RDM.
- Introduce students to federal data mandates and granting agency requirements and how to address them.
- Bring in outside experts to demonstrate specific areas of RDM.
- Provide a focused application of RDM to active research projects.

Schedule

Week	Class	Topic	Week	Class	Topic
1	1	Introduction/Syllabus	9	24	DCP Draft/Revision
	2	What is Research Data?		25	DCP Draft/Revision
	3	RDM and Sharing Mandates		26	Guest – PI on Multi-University Data Intensive Project
2		Holiday	10	27	Guest –RDM in the Humanities
	4	Overview of Data & Lifecycles (1 & 2)		28	DCP Draft/Revision
	5	Planning Your Research Project (3)		29	DCP Draft/Revision
3	6	Organization, File Naming and Structure (5)	11	30	Long Term Storage & Preservation (9)
	7	Lab notebooks & Readme files (4)		31	Guest – RDM for Human Subjects (7)
	8	DMP Tool		32	Data Sharing (10)
4	9	Resources at Univ., National, & International	12	33	Student Projects
	10	Tools Support		34	Student Projects
	11	Reference Managers		35	Student Projects
5	12	Citation Management	13	36	Student Projects/Help Session
	13	Data Curation Profile (DCP)			Thanksgiving
	14	Setting up for Interview			Thanksgiving
6	15	Class Canceled	14	37	Data Sharing and Governance (10 & 11)
	16	Guest - Metadata		38	Guest –Data for Re-use
	17	Guest – Orchid Profile		39	Improving Data Analysis (6)
7	18	DCP Preview	15	40	Student Presentation
	19	DCP Exercise		41	Student Presentation
	20	Guest – Data Management on Interdisciplinary Project		42	Post-Assessment & Student Feedback
8	21	Guest - Canceled	16		Final Exam Week
	22	Guest – Univ. Data Archive and Management			
	23	DCP Draft/Revision			

Numbers in parenthesis refer to chapters of the textbook.

- Broad concepts in RDM:
    - Approximately 10 lectures from the text.
    - 5 additional lectures on mandates, DMPtool and reference managers.
- Guest experts on specific applications of RDM:
    - 8 lectures
- Application of RDM to active research projects:
    - Approximately 11 lectures.
    - 4 days for project work.
    - 2 days for student presentations.

Assignments

Topic		Objectives
Perceptions of Data		Holistic examination of data; define student knowledge base
Data Lifecycle		Overview of all aspects of data management
DMP Tool		Experience with DMPs and creating DMPs
DCP Module Refinement		Critical examination RDM details
Guest Speaker Reflections		Potential application of speaker's experience to student RDM
Final Project		Application of the DCP developed by the students to campus research faculty
A	Planning Document	Establish roles and tasks; examination of DCP as applied to researcher; practice session
B	Interview Session	Interview of researcher to gain knowledge for development of DCP
C	Combined Document	Synthesis of individual material from interview into one document for refinement into DCP
D	Post Interview Reflection	Examination of positives/negatives of interview and DCP template.
E	Data Curation Profile	Suggested RDM best practices for the researcher
F	Presentation Outline	Distillation of DCP experience into presentable format
G	Presentation	Sharing of experience/knowledge with the broader class; Presentation skills
H	Student Presentation Reflections	Potential Application of results/observation from other groups to student's RDM
Student Data Reflection		Self-examination of the student's RDM and potential changes/additions to data management from taking the course

- Individual Assignments – reinforce topics from class.
- Student reflections on guest speakers – applications to the students RDM.
- Generation of a DCP.
- Student reflection on their current RDM practices.

Individual Assignments

- Small assignments on Perceptions of Data and Data Lifecycle.
- DMPtool: This is an on-line resource designed to help investigators develop a DMP to be submitted with a grant proposal.
  - Access requires university participation.
  - Pick your funding agency and tool will help you complete a DMP with the required components.
  - Students completed DMP based on their funding agency and their current research project.
- Student Data Reflection:
  - Holistic examination of their current RDM practices.
  - Contemplate and suggest potential changes based on all aspects of the course.

DMP Tool

Guest Speakers and Student Reflections

- 8 guest speakers presented lectures – goal: provide specific examples of RDM outside the instructor's areas of expertise.
- 2 speakers covered/reinforced areas of the text:
  - Managing sensitive data.
  - Data reuse.
- 3 speakers discussed different aspects of the same project:
  - Day-to-day data management by IT expert.
  - University's data curation archive and repository management.
  - PI on multi-university data intensive project.
- Each student completed a student reflection for each speaker to consider:
  - RDM issues associated with the lecture.
  - Possible applications and changes to the students RDM.

ARCC

Advanced Research Computing Center

WyCEHG

WYOMING CENTER FOR ENVIRONMENTAL HYDROLOGY AND GEOPHYSICS

Data Curation Profile

- Final Project for the course: Goal – Focused application of RDM to an active research project.
- DCP provides a means for researchers to:
  - Thoughtfully and systematically consider their needs for the data.
  - Guide the management and curation of the data through its lifecycle.
- Used the DCP Toolkit from Purdue as a starting point.
- Several classes reviewing and revising the Purdue DCP
  - Goal – Focus the DCP on the researcher.

Data Curation Profiles Toolkit

Final Project - DCP

- Used the student modified DCP as the basis.
- Divided the class into four groups of two/three students.
- Each group worked with a research active faculty across campus.
- Project:
  - Prepare for the interview.
  - Interview the researcher about their RDM needs/practices.
  - Developed a DCP with RDM best practices for that faculty member/research project.
- Presentation – method for each group to share experiences and observations.
- Student reflection – individual assignment to consider potential applications of the DCP to students own RDM.

Assessment

- Pre- and post-course assessment was determined in 8 areas of RDM.
- Student rated knowledge for each area increased by an average 1.15 points.
- Hypothesis testing demonstrated that pre- and post- means were different.

- Students who felt capable of writing a DMP for their research increased from 2 to 9.
- Students with protocols for managing their research data increased from 4 to 8.

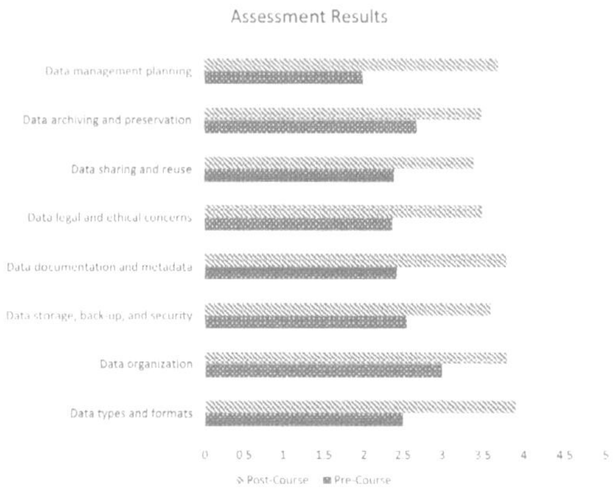

Lessons Learned

- Combined instruction from a librarian and research active faculty is an effective way to teach the course.
- This approach provides both:
  - Theory – from the librarian – knowledge of archives and tools.
  - Practice – from the faculty – experience on how data is collected, stored, and shared.
- There is usually a broad need across campus for this type of course.
  - Consider offering as Engineering Science (ES), through the Graduate School (GRAD) or Library Science (LIB).
  - Parts can be taught through seminars.
  - Incorporate aspects into existing courses:
    - Graduate seminar series.
    - Introductory graduate mentoring classes.

Feedback

- DCP discussions really focused the students on data from the researchers perspective.
- DCP project in teams allowed a balance between learning and work required.
- Discipline diversity in the interviewed faculty brought in multiple perspectives.
- Use and variety of guest speakers was well received.
  - Several literature reports have also noted this positive experience.
- Student reflections were also well received.
  - Allowed the student to consider RDM issues multiple times but also when ideas were fresh.

RDM in the Undergraduate Curriculum

- Entirety of graduate course is probably not appropriate for undergraduates.
- Select appropriates places in the undergraduate curriculum to incorporate specific aspects:
  - Target students doing research:
    - Undergraduate research course.
    - Honors students for honors project.
  - Start small – DCP for the students part of the research project.
  - Encourage them to place their data in university repositories.
    - Invite campus expert to discuss the university resources and show them how to do this.
  - Show them how to find and use data from open repositories:
    - For analysis itself.
    - For comparing to the students work.
- Some concepts might be appropriate for senior design class or other pre-professional course.

Conclusions

- Combined instruction from a librarian and research active faculty provides an effective way to combine expertise for teaching this course.
- Students were prepared to develop DMPs as required by funding agencies.
- Students were prepared to develop and use DCPs for their research projects.
- Student knowledge growth was demonstrated in 8 areas of RDM.
- Students are prepared to become leaders in the area of RDM for their future employers.
- Website: https://doi.org/10.5072/FK2MG7P42S

NCEES

advancing licensure for engineers and surveyors

EFFECTIVE AND EFFICIENT USE OF THE FUNDAMENTALS OF ENGINEERING EXAM FOR OUTCOMES ASSESSMENT

AICHE 2017 ANNUAL MEETING

David Whitman, PhD, PE
NCEES Past-President
University of Wyoming, Retired

NCEES
advancing licensure for engineers and surveyors

Overview

- Applicability of the FE exam for assessment
- ABET student outcomes that can be measured
- Computer-based FE exam format
- Various methods to use FE exam results for outcomes assessment
- Self-study examples and closing the loop
- Questions and answers

NCEES

Applicability of the FE Exam for Assessment

- It is a direct method of objective assessment with comparisons of institutional results against national results.
- Assessment does not use pass rates but how students perform on individual exam areas.
- Because there are over 55,000 FE examinees per year, it provides high reliability.
- The FE can be used as an assessment tool with a pool of all graduates or with a self-selecting pool.

NCEES

ABET Outcomes Assessment Possible with FE Exam

a) An ability to apply **knowledge of mathematics, science, and engineering**

b) An ability to design and conduct experiments, as well as to **analyze and interpret data**

c) An ability to **design a system, component, or process** to meet desired needs

NCEES

ABET Outcomes Assessment Possible with FE Exam (cont.)

(e) An ability to **identify, formulate, and solve engineering problems**

(f) An understanding of **professional and ethical responsibility**

(k) An ability to use the techniques, skills, and modern **engineering tools** necessary for engineering practice

NCEES

Computer-Based Testing (CBT) Format

- The FE became a CBT exam in January 2014
  - Shorter in time and fewer questions
  - Presented in a different format

NCEES

CBT Format (cont.)

- Length
  - The appointment time at test centers is 6 hours.
    - Tutorial–8 minutes
    - Nondisclosure agreement–2 minutes
    - Exam time–5 hours, 20 minutes with a 25-minute scheduled break after approximately 55 questions
    - Brief post-exam survey
  - Total of 110 questions

NCEES

CBT Format (cont.)

- Testing opportunities
  - Testing windows
    - January–March
    - April–June
    - July–September
    - October–December
- Test center locations
  - Pearson VUE test centers

NCEES

CBT Format (cont.)

- *FE Reference Handbook*
  - Provided electronically with the exam as a searchable PDF
  - Available for free download and for purchase as a hard copy at http://ncees.org/engineering/fe/
- *NCEES Examinee Guide*
- Computer-based practice exams

NCEES

CBT Format (cont.)

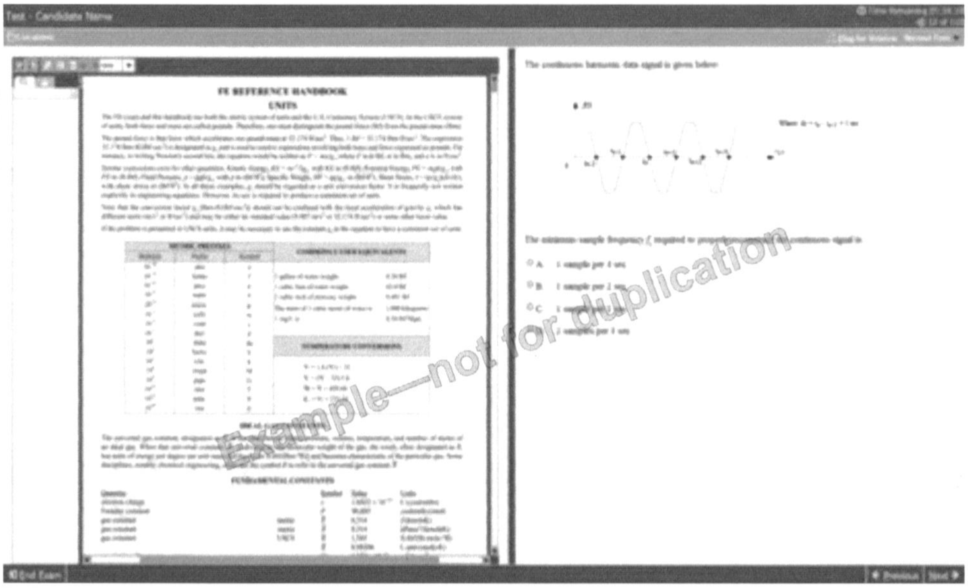

NCEES

CBT Format (cont.)

- Content of the exam
  - 7 free-standing discipline-specific exams
    - Chemical, Civil, Electrical and Computer, Environmental, Industrial and Systems, Mechanical, Other Disciplines
  - FE no longer has a common morning portion (breadth module)
    - Selected exam topics that were previously on the breadth module are now tested within each discipline-specific exam.

NCEES

FE Chemical Engineering Exam Specification

- Mathematics
- Probability and Statistics
- Engineering Sciences
- Computational Tools
- Materials Science
- Chemistry
- Fluid Mechanics/Dynamics
- Thermodynamics
- Material/Energy Balances
- Heat Transfer
- Mass Transfer and Separation
- Chemical Reaction Engineering
- Process Design and Economics
- Process Control
- Safety, Health, and Environment
- Ethics and Professional Practice

Additional details can be found at http://ncees.org/engineering/fe/

NCEES

CHE Example of Engineering Design

13. Process Design and Economics

A. Process flow diagrams and piping and instrumentation diagrams
B. Equipment selection (e.g., sizing and scale-up)
C. Cost estimation
D. Comparison of economic alternatives (e.g., net present value, discounted cash flow, rate of return, expected value and risk)
E. Process design and optimization (e.g., sustainability, efficiency, green engineering, inherently safer design, evaluation of specifications)

NCEES

So, what actual data are available, and what can you do with the data?

NCEES

Subject Matter Report

Subject Matter Report (cont.)

- Reports are generated twice a year.
  - In July for the January–June testing period (Spring)
  - In January for the July–December testing period (Fall)
- The report is specific to
  - An institution,
  - Students within an engineering degree program at that institution, and
  - The discipline-specific exam that those students completed.

NCEES

Subject Matter Report (cont.)

- Data are provided for all examinees testing within 12 months of graduation (either before or after graduating).
- Only first-time takers are included.
- Random guessers are removed from the report.
- National performance data, with standard deviation information, are also provided for the same degree program and same discipline-specific exam.

NCEES

Subject Matter Report (cont.)

- For each topic, the students' performance is given as a Performance Index on a scale of 0–15.
- The Performance Index is indirectly related to the average number of questions answered correctly.
- This is necessary because each examinee receives a different set of questions within each topic area.
  - Any one examinee's set of questions may be harder or easier than another examinee's set.

NCEES

Subject Matter Report (cont.)

- Getting the data
  - NCEES sends links to reports directly to an institution via email.
  - If you don't know, NCEES can tell you who receives your institution's reports.
  - Reports also include information on the specific institution's examinees who took the FE or PE exam more than 12 months after graduation.

NCEES

The specifics of using the FE exam for outcomes assessment

NCEES

Getting Started

- **Faculty** should be involved.
- Identify areas of strength.
- Acknowledge areas that are not emphasized.
- Set program-specific goals for each area.

NCEES

Suggested Analysis Techniques

- Choose your longitudinal time basis.
  - Performance from multiple examination windows
  - Academic year performance
- Choose your presentation method.
  - Ratio method
  - Scaled score method

NCEES

Ratio Method

- The ratio score is simply the ratio between the program's performance index (P.I.) in any topic area and the P.I. of the comparator performance.
  - Ratio score = Program P.I./Comparator P.I.

NCEES

Comparison of Ratios by Subject Area

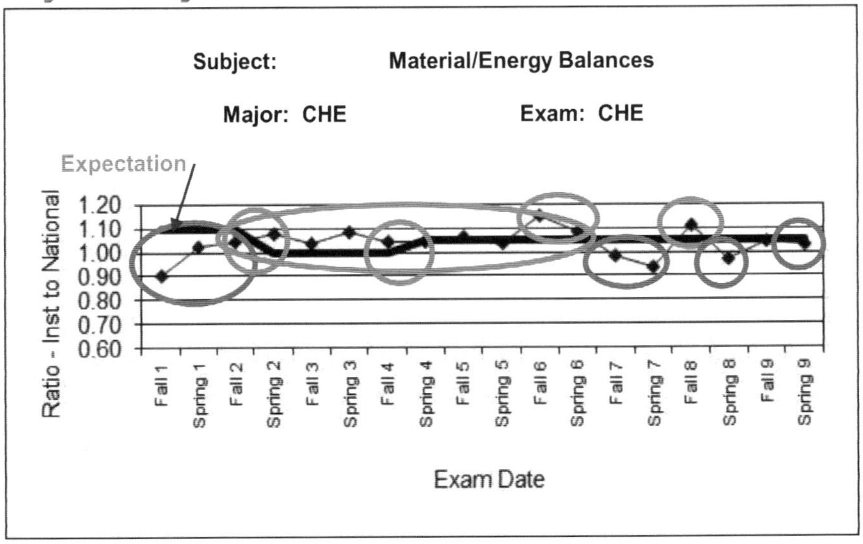

NCEES

Comparison of Ratios by Subject Area

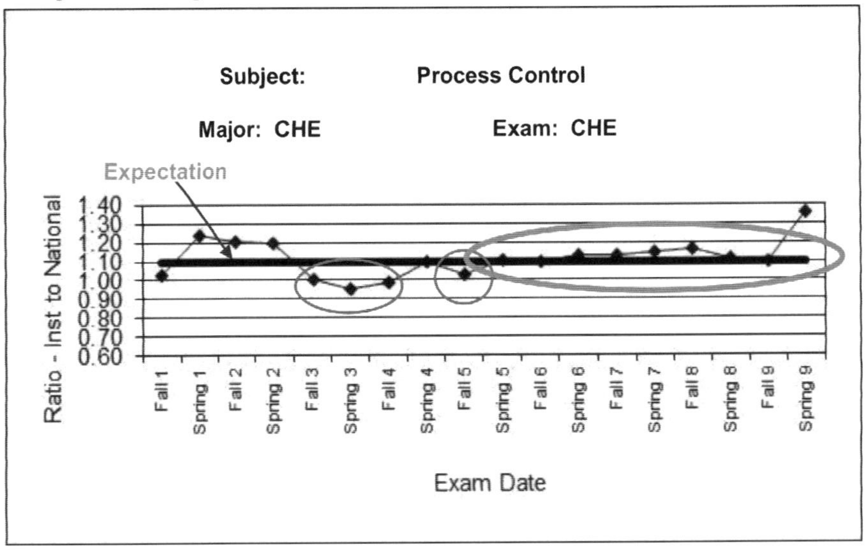

NCEES

Comparison of Ratios by Subject Area

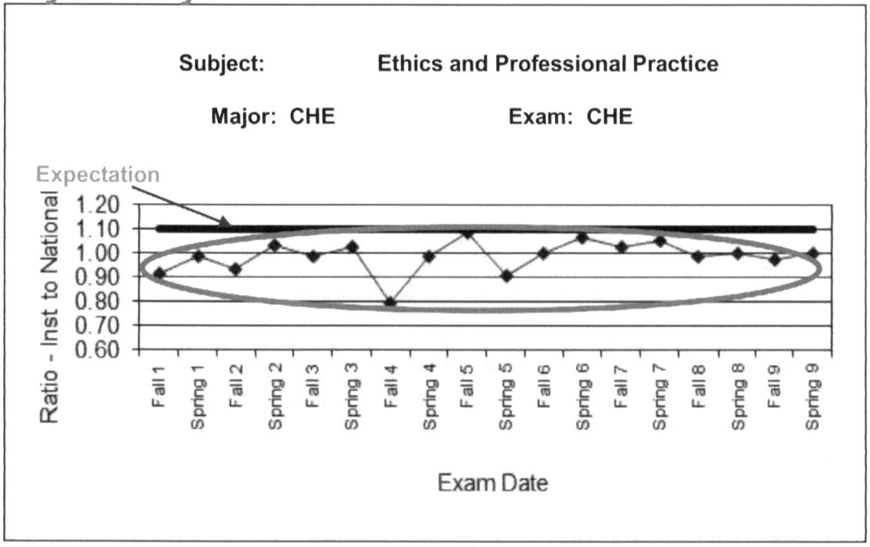

NCEES

Effect of Academic Year Averaging

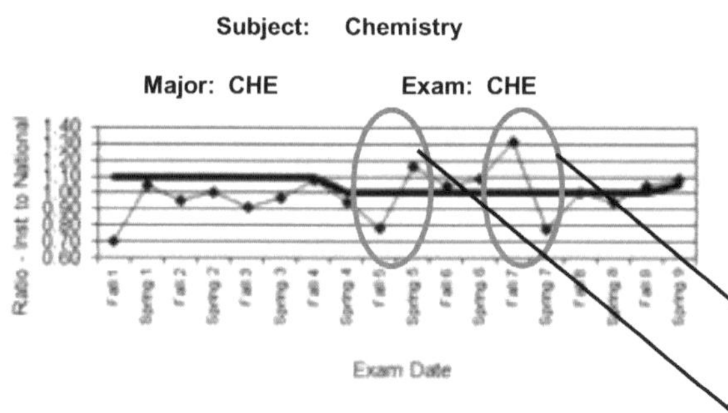

Causes a smoothing of the data—note the Fall/Spring 5 and Fall/Spring 7 results compared to the AY 5 and AY 7 academic year results.

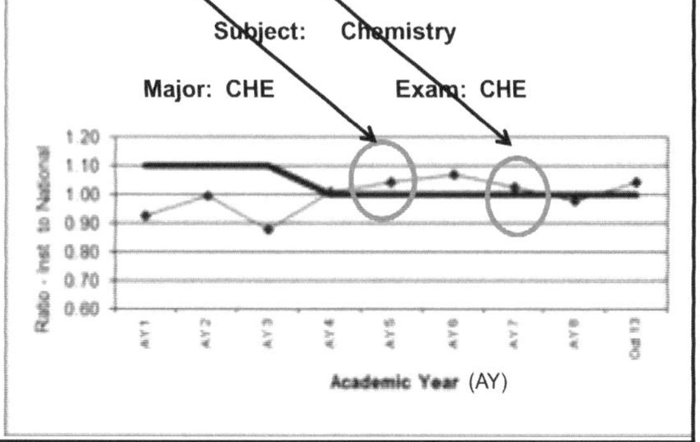

NCEES

Scaled Score Method

Define Scaled Score (S.S.)

$$S.S. = \frac{\text{PI for Univ X} - \text{PI comparator}}{\text{PI comparator standard deviation}} \pm \frac{1}{\sqrt{\#\ \text{of takers at Univ X}}}$$

NCEES

Comparison Using Scaled Score by Subject Area

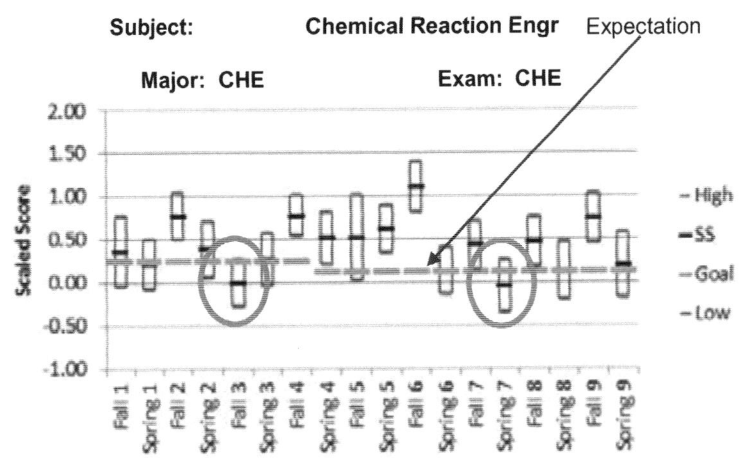

NCEES

Comparison Using Scaled Score by Subject Area

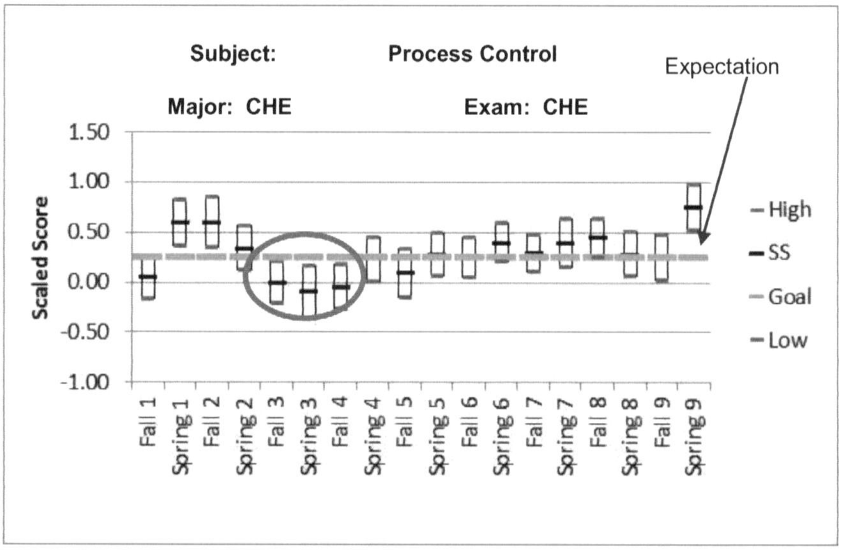

NCEES

Academic Year Averaging

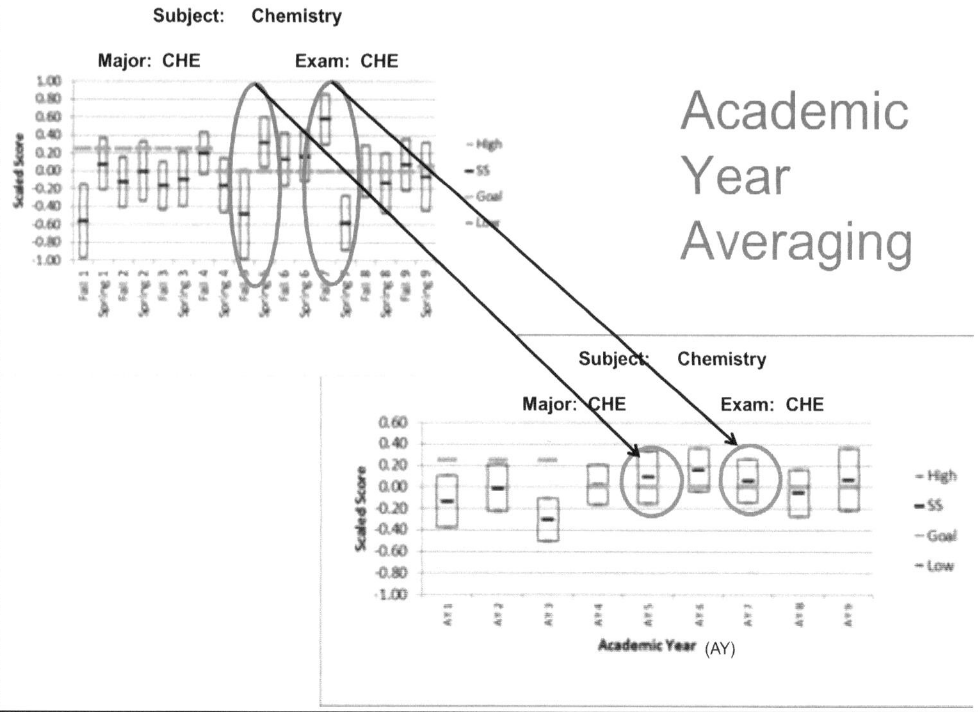

NCEES

Examples of Assessing Some of the ABET (a)–(k) Outcomes

NCEES

Ability to Measure ABET (a) Math and Science

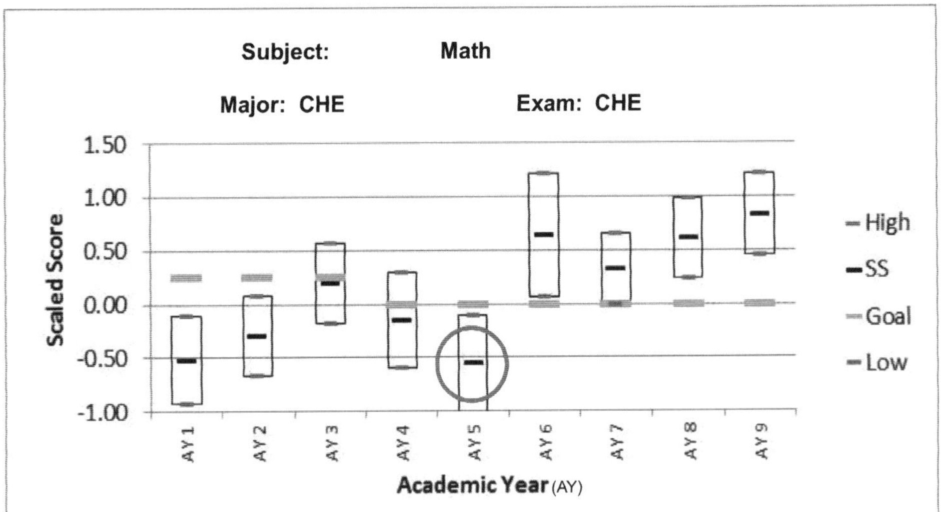

NCEES

Ability to Measure ABET (c) Design

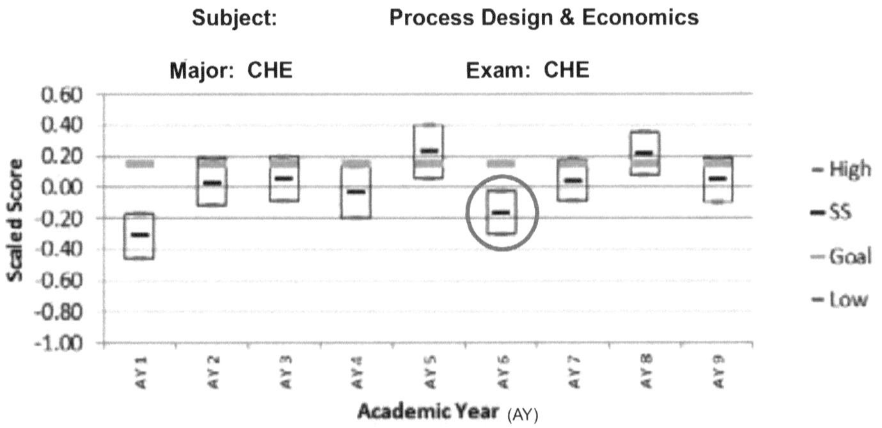

NCEES

Ability to Measure ABET (e) Solve Engineering Problems

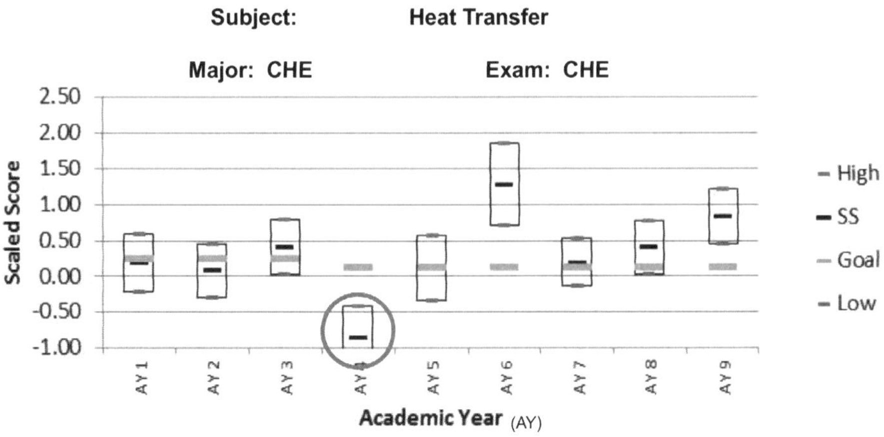

NCEES

Ability to Measure ABET (k) Engineering Tools

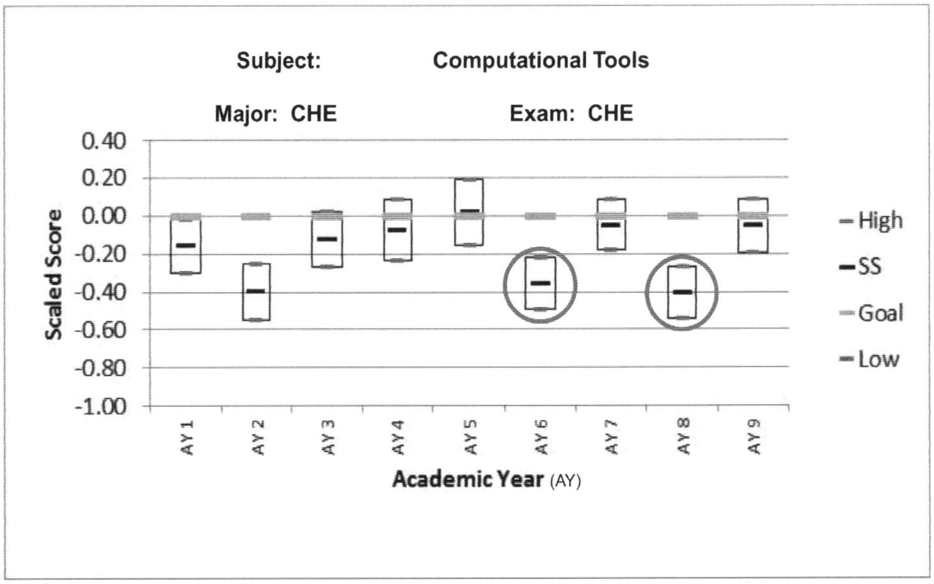

NCEES

Conclusions

- Discipline-specific assessment information can be gleaned from this exam.
- FE exam provides a direct, quantitative assessment technique.
- The Subject Matter Report provides comparative data.
- NCEES sends a link to the Subject Matter Report directly to your institution via email.

NCEES

Conclusions (cont.)

- The FE exam is one effective assessment tool to be used as part of your institution's full assessment package.

NCEES

Additional Resources

For more information on reports, email fereports@ncees.org

NCEES

Additional Resources

For further information, contact:

Cheryl Warren, Ph.D., P.E.
NCEES Exam Development Engineer
800-250-3196, ext. 5472
email: cwarren@ncees.org

NCEES

Questions?

NCEES

Dr. John Wagner P.E.

McKetta Department of Chemical &
Bioprocess Engineering
Trine University

My Background

- Passed the FE as a senior in ChE at University of Colorado – 1987
- Worked at Shell Oil– Deer Park TX
- Rice University for Ph. D. – 1994
- Accepted a Faculty Position at Trine University
- Passed PE in 1997
- Have been on the NCEES FE Committee since 1998

How we use the FE Exam fits into to the Assessment in ChE at Trine

Assessment Tool	Student Outcomes											
	a	b	c	d	e	f	g	h	i	j	k	l
FE Exam	X		X		X						X	X
CAE (Internal Comprehensive Exam)	X	X	X		X							X
Exit Interview /Survey	X	X	X	X	X	X	X	X	X	X	X	X
PAB - Professional Advisory Board						X	X			X		
Math Exam - Internal Comprehensive	X											
SAChE Student Program												X
Key Assessment Courses												
Process Measurements Lab												X
Numerical Methods											X	
Unit Operations Lab I		X										X
Unit Operations Lab II		X		X								
Kinetics	X											
Design I	X				X						X	X
Design II			X			X	X	X	X	X		

Color Code

Program Educational Objective 1	-	Technical Competency
Program Educational Objective 2	-	Problem Solving, Design, Communication and Team Skills
Program Educational Objective 3	-	Professionally Obligated

FE Exam is our only External, Objective Quantitative Assessment Tool

	Objective	Subjective
Internal	CAE Math Exam	Exit Interview Key Assessment Courses
External	FE Exam SAChE Certificate	PAB

Assessment Type	Student Outcomes											
	a	b	c	d	e	f	g	h	i	j	k	l
Objective, External	X		X		X						X	X
Objective, Internal	X	X	X		X							X
Subjective, External						X	X			X		
Subjective, Internal	X	X	X	X	X	X	X	X	X	X	X	X

FE Exam Results, 2010-2013

		2010		n=	9	2011		n=	12	2012		n=	18	2013		n=	18	4 yr Average	
AM Subject	#Q	Avg	Nat. Avg	Dev	Scaled	Avg	Nat. Avg	Dev	Scaled	Avg	Nat. Avg	Dev	Scaled	Avg	Nat. Avg	Dev	Scaled	Scaled	AM Subject
Mathematics	19	58	66	-8	-0.46	63	74	-11	-0.70	65	79	-14	-0.99	67	76	-9	-0.59	-0.717	Mathematics
Engineering Prob & Stats	8	61	65	-4	-0.20	69	69	0	0.00	47	54	-7	-0.33	62	61	1	0.05	-0.12	Engineering Prob & Stats
Chemistry	11	85	89	-4	-0.34	89	85	4	0.31	80	85	-5	-0.39	77	79	-2	-0.15	-0.158	Chemistry
Computers	8	71	76	-5	-0.29	69	76	-7	-0.40	68	77	-9	-0.60	81	82	-1	-0.06	-0.338	Computers
Ethics & Business Prac	8	77	76	1	0.05	80	73	7	0.43	80	85	-5	-0.33	82	80	2	0.13	0.036	Ethics & Business Prac
Engineering Econ	10	82	66	16	0.76	66	65	1	0.06	81	71	10	0.53	78	65	13	0.59	0.485	Engineering Econ
Eng Mechanics (Statics)	8	64	58	6	0.27	40	41	-1	-0.04	60	58	2	0.10	51	57	-6	-0.27	-0.02	Eng Mechanics (Statics)
Eng Mechanics (Dynamics)	5	78	68	10	0.42	48	62	-14	-0.54	36	42	-6	-0.23	40	42	-2	-0.08	-0.147	Eng Mechanics (Dynamics)
Strength of Materials	8	36	39	-3	-0.17	39	40	-1	-0.05	31	38	-7	-0.35	44	40	4	0.21	-0.081	Strength of Materials
Material Properties	8	67	65	2	0.11	61	60	1	0.05	51	56	-5	-0.25	69	62	7	0.35	0.061	Material Properties
Fluid Mechanics	8	69	71	-2	-0.09	69	62	7	0.28	63	66	-3	-0.16	69	72	-3	-0.15	-0.054	Fluid Mechanics
Electricity and Magnetism	11	56	55	1	0.05	50	55	-5	-0.23	45	61	-16	-0.77	44	50	-6	-0.31	-0.381	Electricity and Magnetism
Thermodynamics	8	51	54	-3	-0.13	58	64	-6	-0.30	76	76	0	0.00	82	73	9	0.45	0.058	Thermodynamics
Score	120	65	66	-0.3	-0.049	63	65	-2.3	-0.12	62	68	-5.8	-0.36	66	66	-0.3	-0.04	-0.159	Score
PM Subject																			PM Subject
Chemistry	6	57	58	-1	-0.05	85	77	8	0.40	56	70	-14	-0.65	78	74	4	0.24	-0.052	Chemistry
Mat/Energy Balances	9	75	72	3	0.16	72	73	-1	-0.05	70	68	2	0.09	82	73	9	0.45	0.186	Mat/Energy Balances
Chem Thermodynamics	6	63	62	1	0.05	64	58	6	0.30	65	65	0	0.00	64	59	5	0.21	0.139	Chem Thermodynamics
Fluid Dynamics	6	83	79	4	0.22	43	48	-5	-0.27	53	61	-8	-0.32	65	57	8	0.32	-0.023	Fluid Dynamics
Heat Transfer	6	74	74	0	0.00	71	73	-2	-0.10	68	80	-12	-0.60	73	71	2	0.09	-0.181	Heat Transfer
Mass Transfer	6	46	47	-1	-0.05	69	60	9	0.42	61	69	-8	-0.37	50	50	0	0.00	-0.036	Mass Transfer
Chem Reaction Engr	6	65	68	-3	-0.13	81	77	4	0.17	62	69	-7	-0.30	68	71	-3	-0.14	-0.123	Chem Reaction Engr
Proc Dsgn Econ Opt	6	72	70	2	0.10	85	75	10	0.50	73	72	1	0.05	89	76	13	0.60	0.326	Proc Dsgn Econ Opt
Computer Usage and Chem	3	78	67	11	0.41	61	67	-6	-0.23	69	75	-6	-0.23	89	68	21	0.70	0.168	Computer Usage and Chem Eng
Process Control	3	41	35	6	0.23	92	68	24	0.90	80	68	12	0.40	89	75	14	0.53	0.517	Process Control
Safety Health and Environ	3	85	81	4	0.17	50	53	-3	-0.13	54	61	-7	-0.21	76	75	1	0.04	-0.053	Safety Health and Environ
		67	66	1.7	0.079	71	67	3.6	0.161	64	69	-4.6	-0.206	74	67.65	6.1	0.264	0.064	
Score	60																		

Trine was not a testing site starting 2014. We stopped requiring the FE and our participation dropped to only 3 or 4 students/year

Students graduating in 2019 will once again be required to take the FE

Dr. Lisa G. Bullard, PE

Chemical and Biomolecular Engineering
North Carolina State University

NC STATE UNIVERSITY

My Background

- Passed the FE as a senior at NC State
- PhD → Eastman Chemical Company (Engineering & Construction Division)
- Took/passed the PE after two years of industry experience (plus 2 from PhD)
  - "Unwritten rules" about promotion
  - Changes in state regulations
  - Company support
- Faculty member at NC State

NC STATE UNIVERSITY

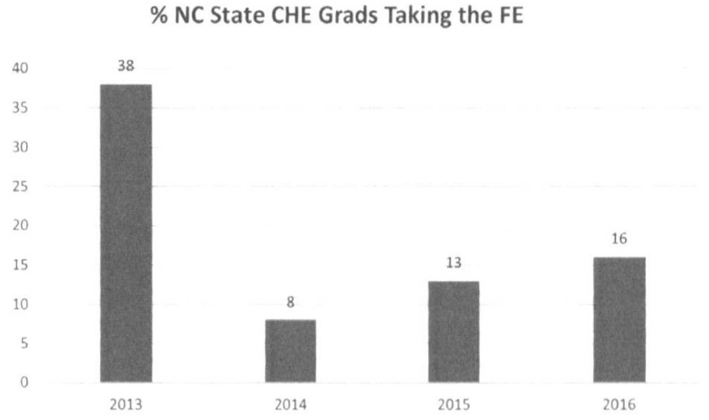

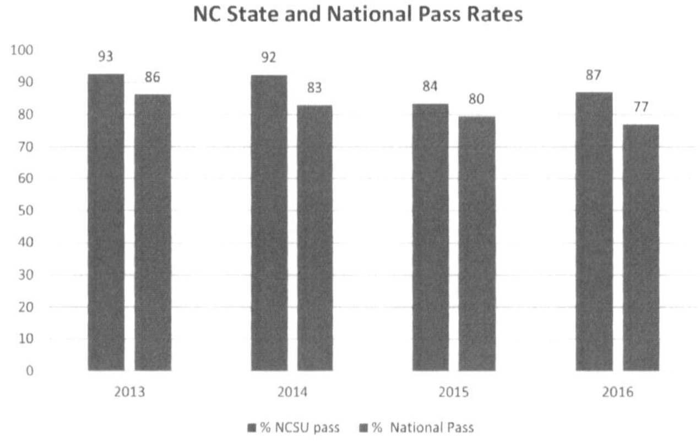

NC STATE UNIVERSITY

Issues/Questions

- Do companies hiring our students expect/promote the PE?
  - Industry
  - Functional role
- When is the best time for students to take the exam – in school or after graduation?
- Increased cost to students
- Resources for test preparation

500727 Design Challenge Parleys As a Conduit for Growing Student Expert Thinking in the Classroom

Abstract:

Increasingly, millennials enter classrooms that make a shift away from the traditional stand-and-deliver style of lectures to one of collaborative learning experiences. There is a rising need for instructors to adapt course content to support students as active learners capable of using the knowledge and skills they bring into the learner-centered educational setting. Instructors often approach student learning from a blank slate instead of building on students' assets to grow expert thinking, allowing them to adapt in different settings. The Department of Chemical and Biological Engineering (CBE) at the University of New Mexico (UNM), which serves a very diverse population, has begun introducing design challenges as an asset-based approach into core courses. Asset-based approaches nurture the knowledge and prior experience of students to support learning and engagement. The design challenges are aligned to course content, are introduced through a design brief and a challenge video, and completed collaboratively. The design challenges are realistic and client-driven, engaging students in applying course content to solve industry, research, and community problems. The community-inspired challenges in particular provide an opportunity for students to build on their past experiences and knowledge.

In this paper, we present a pedagogical strategy that we developed as a way to support students to learn from one another and come to consensus about key decisions: parley sessions. In our previous work, we showed how they fit within a suite of active learning strategies to support students to make progress on a design challenge. Our purpose in this paper is to illustrate how the structure of parley sessions fostered learning, and particularly how they supported students to build on their past experiences and knowledge. Students completed surveys, providing demographic information and data about their past experiences. We video recorded parley sessions and collected student work on the design challenges. We found that students were able to locate peer-reviewed research and information from reputable sources, and that they fostered this practice in each other: we also observed students making requests from peers to legitimize their work with citations and argue from evidence to support claims. We also found that students from rural communities and lower-socio economic backgrounds were able to leverage assets such as regional knowledge about local industries, site amenities, economics and environmental impact in their design decisions. Overall, students grew in their ability to reason from various pieces of knowledge, including diverse and contradictory points of view. These findings support parley sessions as a pedagogical strategy providing abundant opportunities for students to develop expert thinking in the classroom.

497221 Course Design Vs. Student Experience: To What Extent Do We Agree on What Happens in Class?

Abstract:

Student-centered instructional practices like active, collaborative, and problem-based learning have shown great promise in improving student cognitive and affective outcomes in engineering courses. We hypothesized, based on self-determination theory, that seven course attributes that support active learning would also cultivate student's motivation in class and curiosity about the course subject: 1) Relatively small class size for 2) elective courses with 3) an interdisciplinary student composition, where 4) students work on "real" problems through 5) a flexible design process with many paths and solutions and 6) are encouraged to apply broad perspectives (ethics, sustainability, economics...) to ultimately create a 7) physical artefact or prototype. Our work demonstrated that yes, these attributes are associated more strongly with courses with high student intrinsic motivation and curiosity than with courses lacking these elements.

We also noticed an intriguing discrepancy in the data; when students were asked to describe their courses, students in the same course didn't agree on the presence of these elements. Some students described the course as including real problems, while others in the same course did not, even though the instructor's intent was to include real problems in the course. We have redesigned our post-course survey to more closely examine this discrepancy and explore its roots and implications. Students and faculty in nine courses in the 2016-17 academic year were asked about their course experiences relating to the seven factors listed above. This presentation describes work-in-progress, as the survey is anticipated to continue into the 2017-18 academic year.

490548 A Longitudinal Study Regarding the Impact of a Teamwork Skill Building Activity

Abstract:

Many intervention strategies regarding development of teamwork skills are either time intensive, costly, or ineffective. In the Fall of 2013, we developed an activity using the commercial board game, Pandemic, to show students the value of certain teamwork skills and to develop those skills. With those students now on the verge of graduating, we collected both quantitative and qualitative data to assess the long-term impacts of the activity. In this presentation, we will compare the quantitative teamwork skill gains between "Pandemic-trained" students and traditional students. In addition, we will summarize the students' self-perceived gains to validate and triangulate the quantitative results. Throughout the talk, we will also briefly discuss what constitutes this work as a Scholarship of Teaching and Learning project and the challenges of ensuring research integrity.

500735 Identifying Factors That Aid Students in Developing "Engineering Intuition"

Abstract:

Research in psychology has indicated that "intuition" is the ability to subconsciously and quickly draw from previous experience to make a (better) decision. In engineering, intuition is a highly desirable characteristic of the practicing engineer, but an intangible learning objective in the classroom.

In an effort to better understand how students develop engineering intuition, we have compared common factors among students who naturally display high engineering intuition with respect to technology-aided problem solutions. We administered open-ended simulation or other technology-aided problems to chemical and aerospace engineering students. In each case, the problem was solvable but not necessarily realistic. Students were asked to provide a recommendation or prompted to otherwise comment on the feasibility of the solution. Evaluating the solution feasibility required students to integrate knowledge not explicitly stated in the problem, but within the body of their expected knowledge. We refer to these as "stretch" problems, as they "stretch" beyond the calculations directly asked for in the problem.

Students' success with assessing "stretch" problems' feasibility was studied with respect to a number of factors: internship experience, program of study, ideal job choice, overall GPA, engineering GPA, STEM GPA, and learning preferences. Internship experience and GPA each showed statistically significant correlations with students' engineering intuition, and further inspection revealed distinct trends with respect to not only correctly answering the problem, but also with students' likelihood to attempt the problem. While it may not be surprising that internship experience and GPA rose to the forefront of factors studied, it is interesting to note that students with high overall GPA were more likely to both attempt "stretch" problems and answer them correctly. This GPA trend did not hold true when engineering or STEM-specific GPA was isolated.

This multi-disciplinary study is ongoing at two U.S. non-PhD granting institutions. Short-term future work includes expanding the number of study sites and disciplines to more strongly support identified factors. We will then work to better understand what aspects of these factors or experience contribute to development of "engineering intuition", and develop course interventions that strive to incorporate similar aspects to the classroom experience in an attempt to better enable students to develop "engineering intuition."

496971 Using Survey Data to Assess Chemical Engineering Student Acumen

Abstract:

American Institute of Chemical Engineering

Matthew J. Armstrong, Geoffrey Bull, Andrew Biaglow

1) Department of Chemistry and Life Science, United States Military Academy,

West Point, NY 10996

Using Survey Data to Assess Chemical Engineering Student Acumen

Engineers graduating from ABET-accredited programs are expected to be able to identify engineering problems. The skill is identified in Criterion 3 of the *Criteria for Programs*, which lists the required student outcomes. [1] According to student outcome (e), at the time of graduation, students will have "an ability to identify, formulate, and solve engineering problems." A considerable amount of information is available on theory and practice relating to this outcome. [2-5]. However, while most assessment protocols do a good job on formulation and solving, it is difficult to find information on the direct assessment of students' ability to *identify* engineering problems. Identification is almost never directly assessed. [6]

The ability to identify an engineering problem is an important skill that has been identified by ABET as part of its accreditation criteria. Specifically, student outcome (e) in Criterion 3 requires the ability to identify, formulate, and solve engineering problems. A great deal has been written about how engineers formulate and solve problems, but very little information is available on how engineers identify problems, even less on how this skill is assessed. For the most part, engineering programs focus on assessing formulation and solution, and assume that identification is a natural part of that process. In this presentation, we discuss a test that we have developed to directly assess our student's ability to identify engineering problems. The results show that our students possess the ability to a degree, and that this ability develops with the amount of student experience. Furthermore, by comparing the test results from those of the students to those obtained form more experienced engineers, we are able to directly assess the degree to which our students have developed the ability. [6]

We expand the study longitudinally (30% more data) with the addition of new graduating class data from our most recent graduating chemical engineers, as well as present a novel new visual presentation strategy to show how the classes compare to previous classes, as well as how each class compares to department faculty who have taken the survey. In addition we compare our results to other departments at West Point.

References

[1] Engineering Accreditation Commission, "Criteria for Accrediting Engineering Programs, 2016-2017," 2016. [Online]. Available: http://www.abet.org/accreditation/accreditation-criteria/criteria-for-accrediting-engineering-programs-2016-2017/.

[2] Foundation Coalition, "Outcome e an ability to identify, formulate, and solve engineering problems," 25 February 2005. [Online]. Available: http://www.foundationcoalition.org/home/keycomponents/assessment_eval/outcome_e.html.

[3] D. Woods, A. Hrymak, R. Marshall, P. Wood, C. Crowe, T. Hoffman, J. Wright, P. Taylor, K. Woodhouse and C. Bouchard, "Beveloping Problem Solving Skills: The McMaster Problem Solving Program," *Journal of Engineering Education,* pp. 75-91, 1997.

[4] M. Shaw, Engineering Problem Solving: A Classical Perspective, Norwich, New York: Noyes Publications / William Andrew Publishing, 2001.

[5] R. Felder and R. Brent, "Designing an dTeaching Courses to Satisfy the ABET Engineering Criteria," *Journal of Engineering Education,* vol. 92, no. 1, pp. 7-25, 2003.

[6] M. Armstrong, G. Bull, and A. Biaglow. Assessment of Student Ability to Identify Engineering Problems. (2016) ASEE Mid-Atlantic Section Conference. Hofstra University.

http://www.hofstra.edu/academics/colleges/seas/seas-fall16-asee-conference-papers.html

496995 Using E-Portfolio's to Assess ABET Outcomes and Student Learning

Abstract:

Student outcomes describe what students are expected to know and be able to do by the time of graduation.[1] Evidence of meeting these outcomes can take many forms. In this presentation, we discuss the use of E-Portfolios – Portfolios that contain work completed by students throughout their academic careers placed on a digital platform. We have been working with a group of about 62 students (Class of 2019) to form their own portfolios. The students tie their work to the ABET a through k criteria by placing examples of their work into pre-defined folders labelled as such. We'll share our "learning curve" experiences with the optimism that this is one sustainable way to consider demonstrating student outcomes during an ABET visit. Another benefit is the involving of students directly into understanding how their learning experiences tie together to complete their preparation as a chemical engineer.

[1] ABET, http://www.abet.org/

491325 A 'Cards Against Humanity'-Style Card Game for Increasing Engineering Students Awareness of Ethical Issues in the Profession

Abstract:

Across all engineering disciplines, engineering design solutions frequently have major, long-lasting impacts on societal development. Since the process of technical innovation occurs in increasingly complex social exchanges, engineers are frequently confronted with social and ethical dilemmas in their professional lives. Yet, many engineering curricula do not incorporate these issues in formal ways into engineering education. Many widely publicized failures of complex engineering systems can be traced back to lapses in judgment on either ethical or societal impact axes, including the Volkswagen Diesel Engine scandal, the BP Gulf Oil Spill, the Challenger and Columbia space shuttle disasters, and recently, the Flint, Michigan Water Crisis.

In this work, we explore the application of game-based and game-inspired learning in an engineering education setting. Giving students the opportunity during their education to recognize the wider social and ethical impacts of the profession - through multimedia simulation, role-playing games, case-based learning, and review of other, fictionalized cases - can give them opportunities to reflect on the need to identify complex situations in future settings, as well as a safe environment in which to explore, make mistakes, and discuss the ramifications of various decisions in authentic contexts. Ultimately the goal is to better prepare young engineers to tackle ethically the current and future challenges that have tended to be underemphasized in traditional engineering curriculums.

In this particular application, we have developed a beta version of a 'Cards Against Humanity' (CAH)-type card game focused on ethical, societal, and social questions within an engineering context. For those not familiar with the 'Cards Against Humanity' game model, a group of players each draws a number of 'response cards' (White Cards). One player then draws a Black Card, which contains a question, situation, or fill-in-the-blank. The other players look at all of the potential responses they have in their hand (the White Cards), and attempt to make the best response out of the combination of the Black Card and their White Card. The player who read the Black Card judges all of the other player responses, makes a selection of the winning condition (in Cards against Humanity, this is usually the 'funniest' response combination), awards points, and then play continues with a different player drawing another Black Card and reading another scenario.

The CAH model has been adapted previously in other educational settings. [refs] As discussion of ethical, societal, or social scenarios could prove to be controversial, using a game-based approach may allow participants to engage in discussion of the topics and concepts in a more playful, abstract manner, allowing freer responses and more meaningful exploration of the topics, as they are safely confined within a game construct.

The white card and black card prompts are responses are currently being crowd-sourced via a diverse group of engineering faculty, education faculty, and engineering students. In the Spring 2017 semester, the game will be beta-tested with select groups of engineering students. Results of this play testing and lessons learned will be discussed.

500773 Incorporating Diversity and Bias Awareness in a Technical and Professional Communication Course

Abstract:

The ability to work on diverse teams is commonly accepted as a desirable trait for the professional engineer. While many instructors attempt to create diverse team experiences by balancing teams for performance, gender, or other factors, the diversity that can be achieved on teams is limited by the class composition. This varies greatly by institution type (large public institution vs. small private institution), location (city vs. rural), and degree program (highly structured vs. flexible).

Recognizing that there are limitations to our ability to expose students to authentic experiences on diverse teams, it is important to engage them in thinking about diversity as they may encounter it in the "real world." To this end, a diversity module was added to a Technical and Professional Communication course. The pilot version of this module included: 1) defining and exploring implicit biases; 2) defining and reflecting on stereotype threat "traps"; and 3) a culture game that highlights how we can observe, analyze, and adapt to perceived differences in order to achieve common goals.

While the course has always required some consideration of diversity to achieve the understanding audience learning objective, these concepts had not explicitly been considered previously. Student feedback on the module was highly positive, and indicated a deeper level of thinking on diversity issues (anecdotally) when compared to previous classes.

Future work includes adopting the diversity module as a regular component of the Technical and Professional Communication course, and incorporating socially-relevant engineering topics in technical courses as well.

501337 Providing Access to Clean Water: Activities for Freshman Engineering Courses and Outreach

Abstract:

This talk will describe an adaptable module on water quality that has been used for middle school students through college engineering freshman. It will describe the motivation for the activity, modifications for different time constraints and audiences, and evaluation data from engineering summer camps and freshman "Introduction to Engineering" courses. The National Academy of Engineering's Engineering Messages and Grand Challenges for Engineering highlight the important role of the engineering profession in addressing societal challenges. Surveys of high school and engineering freshman have shown that students have particular interest in the grand challenge "Provide Access to Clean Water." In its simplest form, the module provides a basic introduction to filtration, nanotechnology, and the importance of water quality and scarcity issues. This version of the activity takes approximately 1.5 hours. More detailed versions can incorporate a broad range of science and engineering concepts including filtration, nanotechnology, concentration measurement and unit conversion, learning Excel, citizen science, and remediating biological contamination. The full version used in "Introduction to Engineering" classes, takes approximately one week of class time (three sessions plus homework). It includes a reading assignment on the water scarcity and quality issues, an industry speaker from a related field, an in class lab activity, an out of class activity, and a report. The in class activity involves filtering water from a local natural water source through different media, including filter paper impregnated nanosilver via a simple microwave oven technique. The students use a combination of test strips, iDip handheld water meters that send data to their cell phones, and traditional plate streaking techniques to make a variety of measurements on the water both pre and post filtration through the various media. For homework, the students checkout one of the meters to gather data on additional water sources of their choosing. The resulting data is compiled and used to teach Excel skills such as plotting data with error bars, and basic statistic functions. At the end of the activity, the students write a report, which consolidates the whole activity, incorporates specific calculations in concentration measurement and unity conversion, and provides students experience in written technical communication.

Supplement Homework with Problems from a Freely Available Repository

Mcketta Department of
Chemical & Bioprocess Engineering

Trine University

John Wagner, Allen Hersel, & Amanda Malefyt

Learning objectives of the presentation
by the end of this presentation you should be able to:

1) Query problems from the repository

2) Describe the two basic kinds of files on the repository
    text and computer based

3) Retrieve the default files of those two forms

4) Describe some of the options for Text and Computer based problems

5) Create an alternate text version of the file

6) Describe the file type you will need to modify the computer based version

Conceptual Structure of the Repository

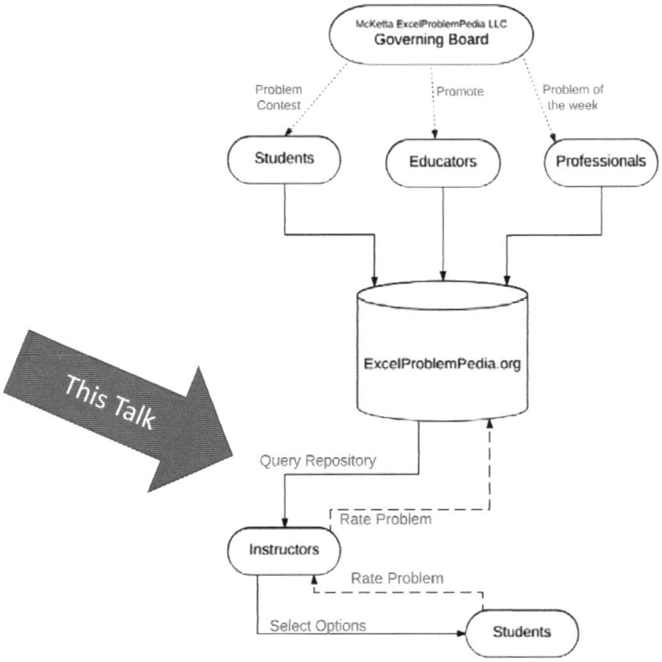

Goals of the Problem Repository System

Fast & Easy

Flexible

Good Problems

Follow one example from the repository – The original problem is on the LearnChemE website

A gas mixture containing 85 mole % N_2 and the balance *n*-hexane flows through a pipe at a rate of 100.0 m^3/h. The pressure is 2.00 atm absolute and the temperature is 100° C.

a) What is the molar flow rate of the gas in kmol/h?

b) Is the gas saturated? If not, at what temperature would it have to be cooled to in order for hexane to start condensing?

c) To what temperature would the gas have to be cooled to at constant pressure in order to condense 80% of the hexane?

LearnChemE
Chemical Engineering Screencasts

Each of the highlighted regions below are parameters that could be altered to create a variable parameter problem

A gas mixture containing 85 mole % N_2 and the balance *n*-hexane flows through a pipe at a rate of 100.0 m^3/h. The pressure is 2.00 atm absolute and the temperature is 100° C.

a) What is the molar flow rate of the gas in kmol/h?

b) Is the gas saturated? If not, at what temperature would it have to be cooled to in order for hexane to start condensing?

c) To what temperature would the gas have to be cooled to at constant pressure in order to condense 80% of the hexane?

LearnChemE
Chemical Engineering Screencasts

To find these problems-go to Website

Put in Search Criteria

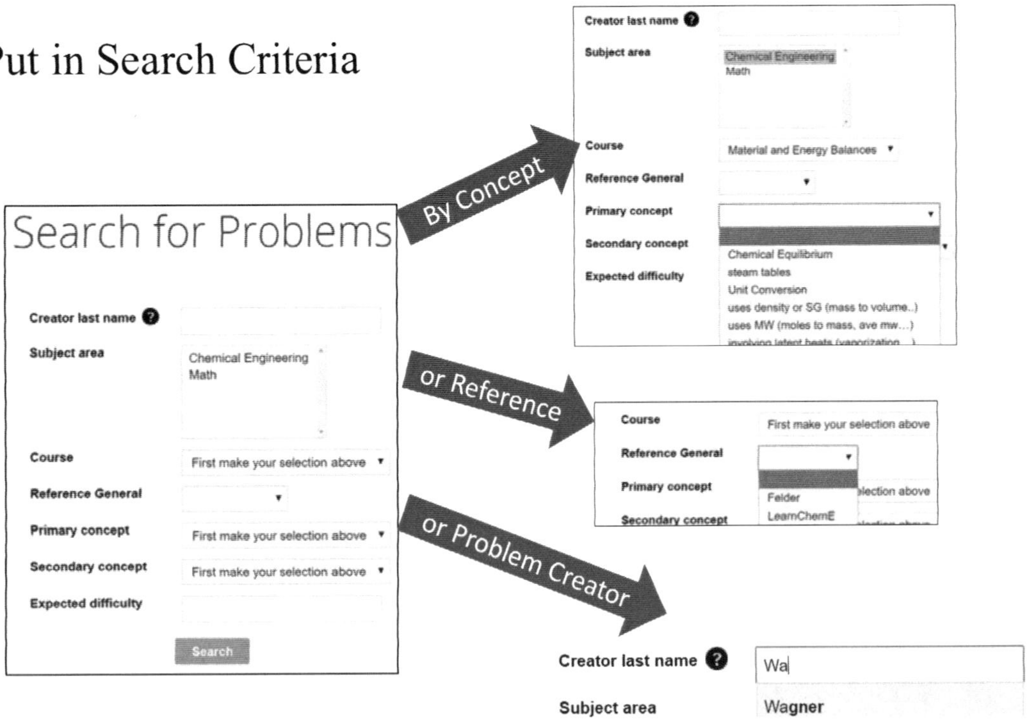

Scroll through problems that meet your criteria

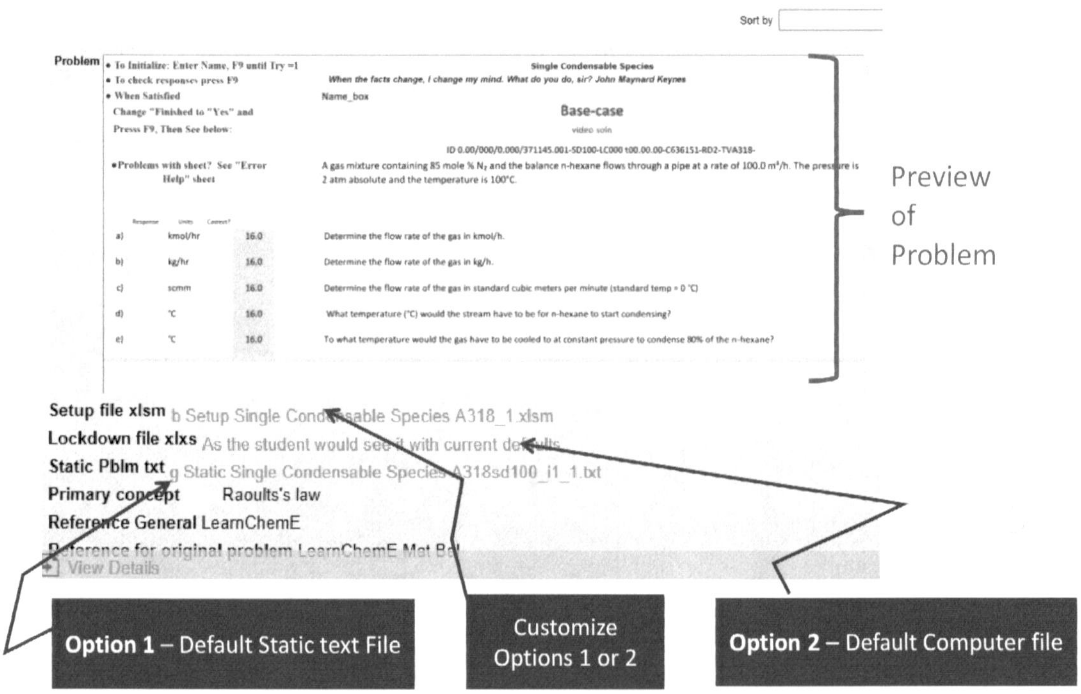

Option 1 - Text file

A gas mixture containing 85 mole % N2 and the balance n-hexane flows through a pipe at a rate of 100.0 m³/h. The pressure is 2 atm absolute and the temperature is 100°C.

a) Determine the flow rate of the gas in kmol/h.
b) Determine the flow rate of the gas in kg/h.
c) Determine the flow rate of the gas in standard cubic meters per minute (standard temp = 0 °C)
d) What temperature (°C) would the stream have to be for n-hexane to start condensing?
e) To what temperature would the gas have to be cooled to at constant pressure to condense 80% of the n-hexane?

Problem Statement

Define and Prepare:
i) Draw a large, neat, and simple picture of the problem (e.g. boxes and lines) including labeled knowns and identifying the value(s) the problem is asking for. re-read the problem and make sure that your diagram has all the relevant information from the problem statement.
ii) For the values that the problem is asking for, are there inherent or practical limits or is any real number reasonable? If so what are they for each value.
iii) What equations or principle (e.g. ideal gas law, conservation of mass...) will you use to solve this problem and what limitations are inherent in those equations and what limitations will you impose? (for example steady state)
iv) Is the problem self contained or will you need any supplemental information such as chemical or physical property data - if so what information do you think you will need and where will you get it?
v) What level of math will you need for this problem? arithmetic, algebra, quadratic equation, trial and error, regression, trig, calculus
vi) Specifically, what tool will you use for this problem? for example - paper and calculator, spreadsheet, math program (e.g. Mathcad, MATLAB, polymath), simulator or specialized software. Why did you pick that tool?
vii) If you do not get the answer - what will you check first, what will you check second.

Preliminary Problem Definition Activities

Option 1 a text file (cont.)

Reflect:
I) Did you get the problem correct on the first try? If not give a brief description of the mistake you made.
II) Lookup the hydrocarbon in the problem above on Wikipedia. How is it made and what is it used for? Check the SDS sheet. What is its flashpoint? What is a flashpoint?
III) If the absolute pressure of the stream were doubled, and all other conditioned remained the same as the original problem, how would the standard flowrate change?

Reflection Questions

Answers:
a) 6.531 kmol/hr

b) 240 kg/hr

c) 2.44 scmm

d) 34.83 °C

e) 2.52 °C

Answers to the Numerical Part

Solution Outline:

```
MW_sub=MW([@subst])
mw_n2=MW("nitrogen")
Mwbar=([@[y_N2]]*[@[mw_n2]]+(100-[@[y_N2]])*[@[MW_sub]])/100
T1_K=[@T1]+273.15
ndot(kmol/h)=100/22.415*273.15/[@[T1_K]]*[@[P_atm]]
mdot(kg/hr)=[@[ndot(kmol/h)]]*[@Mwbar]
SCMM=[@[ndot(kmol/h)]]*22.415/60
P_subst(mm)=(100-[@[y_N2]])/100*[@[P_atm]]*760
TdP=AntoineT([@subst],[@[P_subst(mm)]])
n_vap=(100-[@[y_N2]])*0.2
y_80=[@[n_vap]]/([@[y_N2]]+[@[n_vap]])
Pvap_80_mm=[@[y_80]]*[@[P_atm]]*760
T_80=AntoineT([@subst],[@[Pvap_80_mm]])
```

Solution Outline

Customize Option 1
get a Text file with different parameters

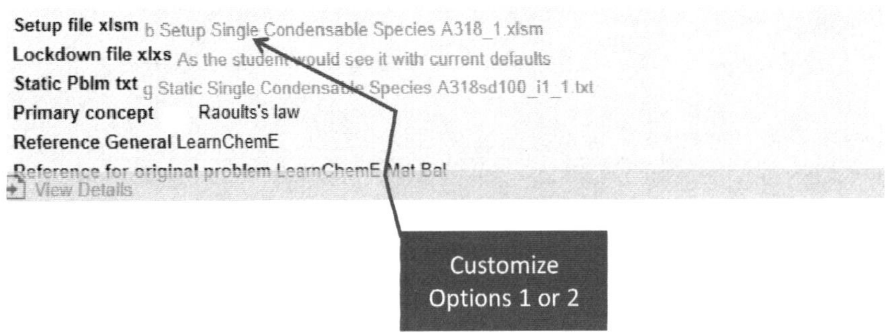

Option 1' – get a different text file

Download xlsm file by clicking on it, Open it up – it should look like

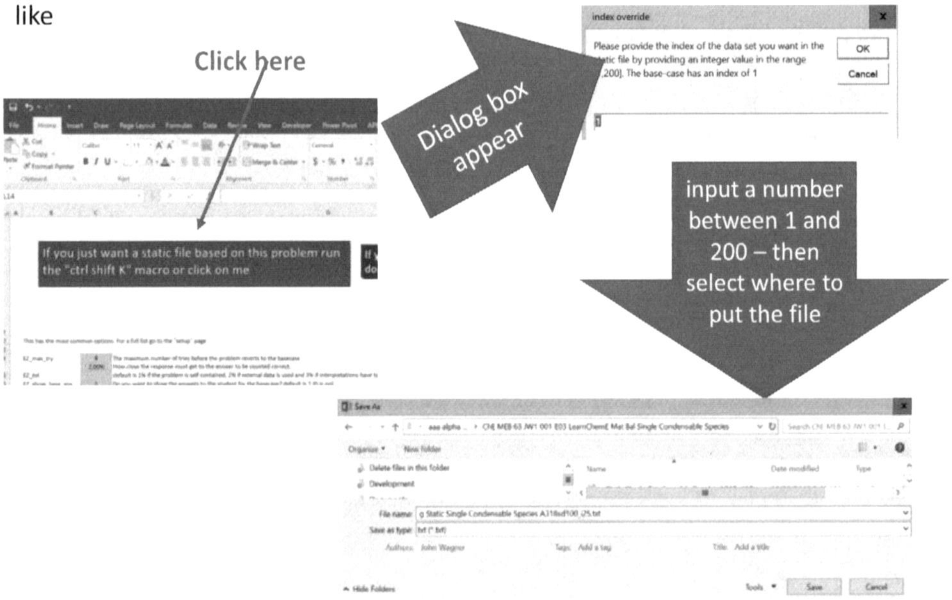

Same example with some of the parameters changed

A gas mixture containing 77.2 mole % N2 and the balance n-heptane flows through a pipe at a rate of 100.0 m³/h. The pressure is 2.4 atm absolute and the temperature is 280°C.

a) Determine the flow rate of the gas in kmol/h.
b) Determine the flow rate of the gas in kg/h.
c) Determine the flow rate of the gas in standard cubic meters per minute (standard temp = 0 °C)
d) What temperature (°C) would the stream have to be for n-heptane to start condensing?
e) To what temperature would the gas have to be cooled to at constant pressure to condense 80% of the n-heptane?

For the Text file options

- Copy text from file into a word processer – keep / edit the parts you want

- Good for paper based review, quiz or test questions

Option 2
Computer file with Creators Defaults

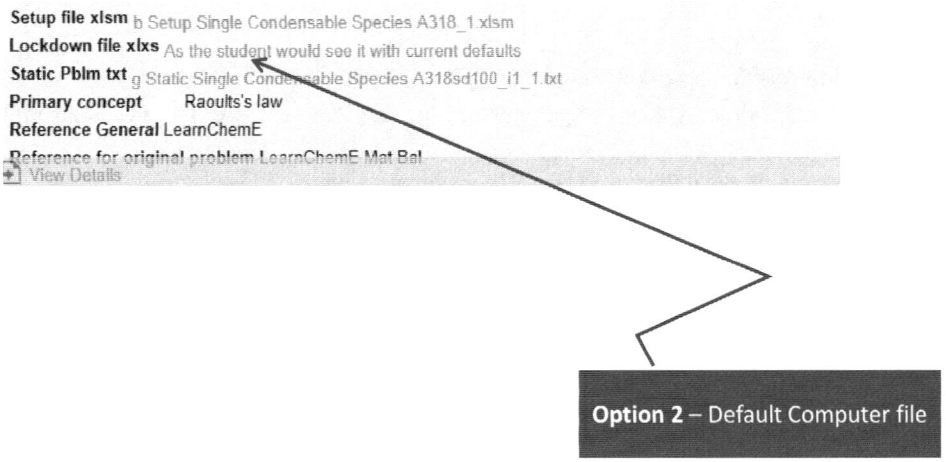

Option 2 – computer based problems - download directly the file that will be distributed to the students but with the creators defaults

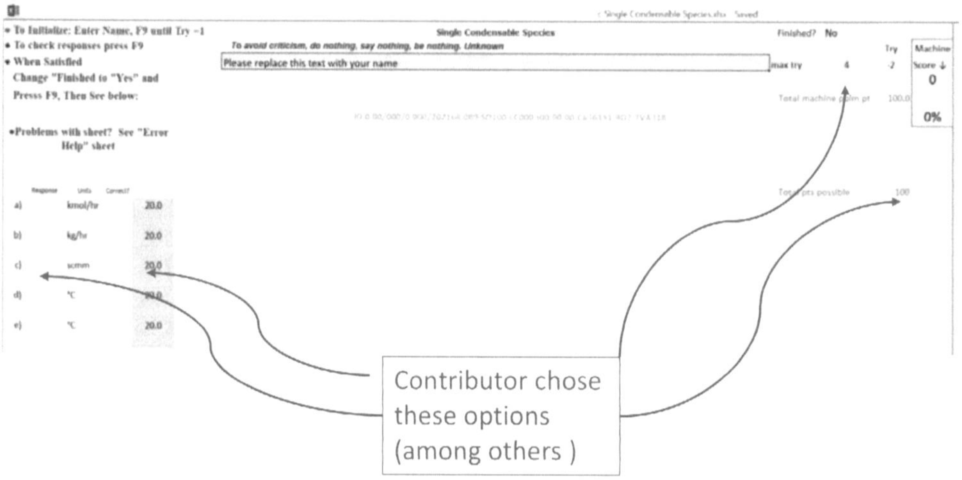

Option 2 (cont.) showing a student attempt

- To Initialize: Enter Name, F9 until Try =1
- To check responses press F9
- When Satisfied
  Change "Finished to "Yes" and
  Press F9, Then See below:

- Problems with sheet? See "Error Help" sheet

Single Condensable Species

Improve by 1% a day, and in just 70 days, you're twice as good. Alan Weiss

Ima Student

Finished? No

		Try	Machine Score ↓
max try	4	4	80
Total machine pblm pts	80.0		100%
Total Refect pts	20		
Total pts possible	100		

A gas mixture containing 77.2 mole % N_2 and the balance n-heptane flows through a pipe at a rate of 100.0 m^3/h. The pressure is 2.4 atm absolute and the temperature is 280°C.

	Response	Units	Correct?		
a)	5.29	kmol/hr	YES	16.0	Determine the flow rate of the gas in kmol/h.
b)	235.20	kg/hr	YES	16.0	Determine the flow rate of the gas in kg/h.
c)	1.98	scmm	YES	16.0	Determine the flow rate of the gas in standard cubic meters per minute (standard temp = 0 °C)
d)	79.15	°C	YES	16.0	What temperature (°C) would the stream have to be for n-heptane to start condensing?
e)	42.18	°C	YES	16.0	To what temperature would the gas have to be cooled to at constant pressure to condense 80% of the n-heptane?

Single Condensable Species

Please select the statement that best describes the following

A. difficulty of the problem

2. Fairly Easy - Solution path was straight forward but had several opportunities for mistakes

B. effectivness of the problem

4. Effective - Problem gave me effective practice or helped me understand or apply important concepts

C. your performance on the problem

3. OK - I definitely have some room for improvement in solving these types of problems

Press F9 then Respond to all of the following queries labeled with upper case Roman numbers then ctrl-P to Print

I. Did you get the problem correct on the first try? If not give a brief description of the mistake you made.

II. Lookup the hydrocarbon in the problem above on Wikipedia. How is it made and what is it used for? Check the SDS sheet. What is its flashpoint? What is a flashpoint?

III. If the absolute pressure of the stream were doubled, and all other conditioned remained the same as the original problem, how would the standard flowrate change?

If the student runs out of tries

- To Initialize: Enter Name, F9 until Try =1
- To check responses press F9
- When Satisfied
  Change "Finished to "Yes" and
  Press F9, Then See below:

- Problems with sheet? See "Error Help" sheet

Single Condensable Species

Improve by 1% a day, and in just 70 days, you're twice as good. Alan Weiss

Ima Student

Base-case

video soln

Finished? No

		Try	Machine Score ↓
max try	4	1	0
Total machine pblm pts	80.0		0%
Total Refect pts	20		
Total pts possible	100		

A gas mixture containing 85 mole % N_2 and the balance n-hexane flows through a pipe at a rate of 100.0 m^3/h. The pressure is 2 atm absolute and the temperature is 100°C.

	Response	Units	Correct?		Ans	
a)		kmol/hr		16.0	6.53E+00	Determine the flow rate of the gas in kmol/h.
b)		kg/hr		16.0	2.40E+02	Determine the flow rate of the gas in kg/h.
c)		scmm		16.0	2.44E+00	Determine the flow rate of the gas in standard cubic meters per minute (standard temp = 0 °C)
d)		°C		16.0	1.48E+01	What temperature (°C) would the stream have to be for n-hexane to start condensing?
e)		°C		16.0	2.52E+00	To what temperature would the gas have to be cooled to at constant pressure to condense 80% of the n-hexane?

Contributor chose these options but all students get the same base-case

If you would like to select these options

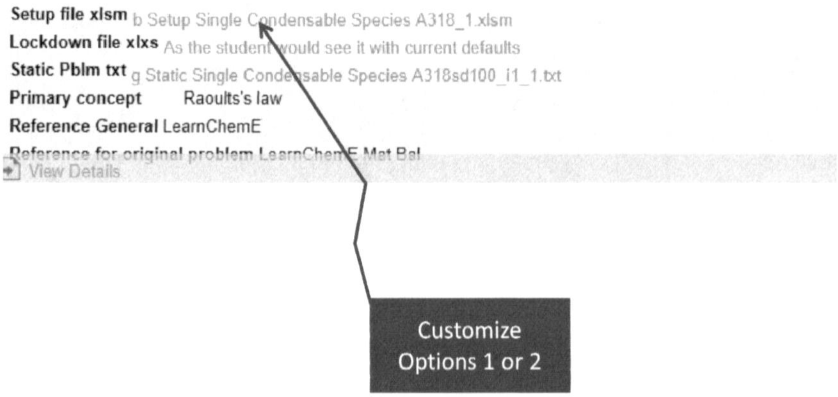

Common Option changes - use Setup-EZ Sheet

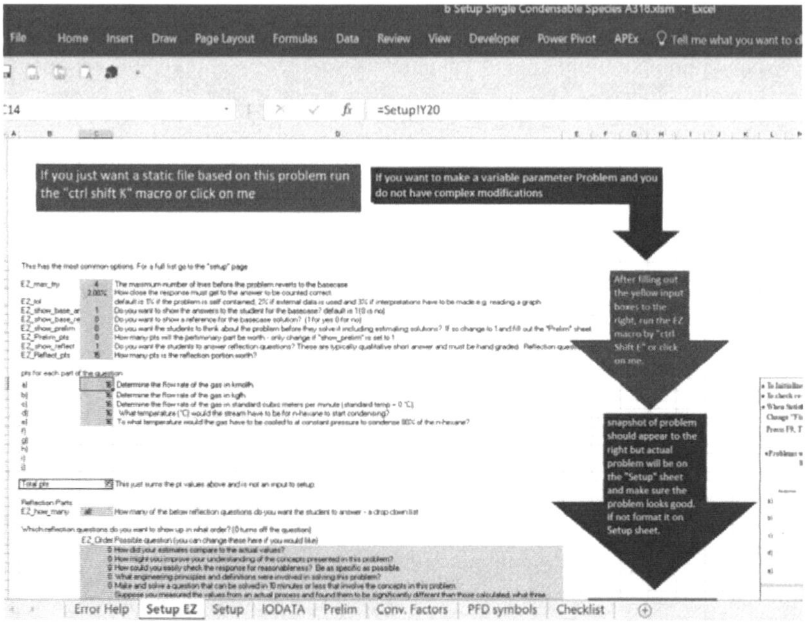

For **all** problem Options - use Setup Sheet – dark yellow are input cells

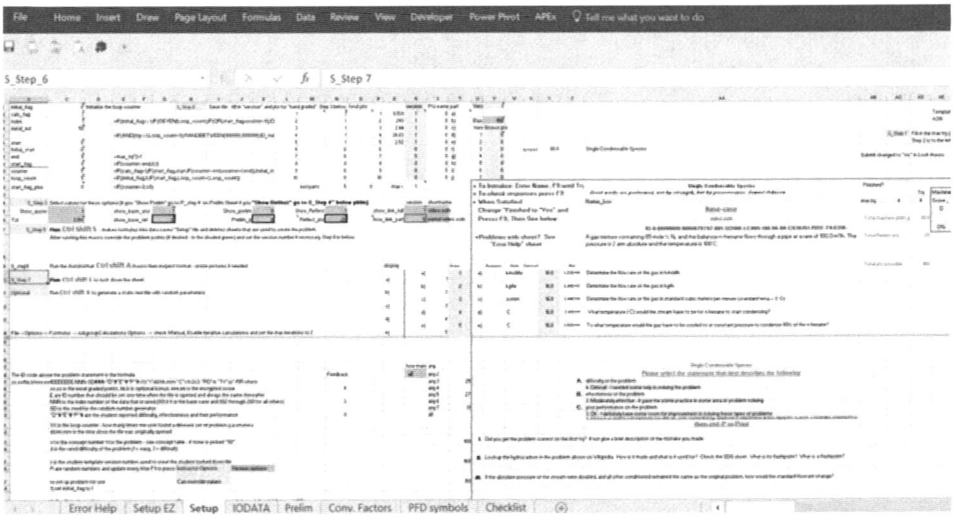

For Example – the same problem modified

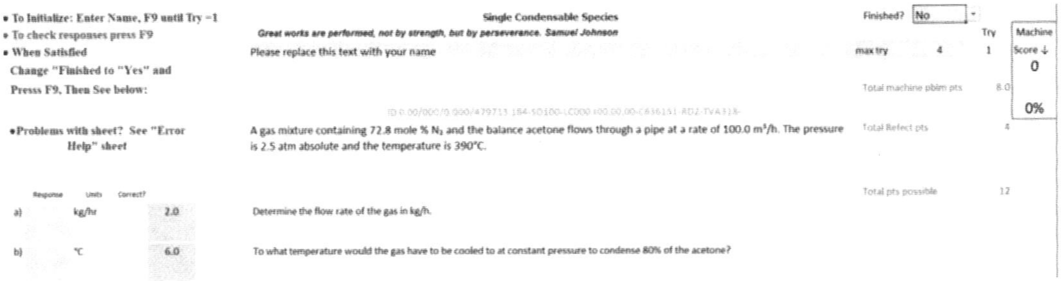

Summary

A problem repository located at Excelproblempedia.org has been started that is:

- Fast and easy to find problems
- Flexible – useful for both text based and computer based problems – both can be modified
- Problems incorporate planning and reflection parts to bring out higher level thinking

Learning objectives of the presentation

by the end of the presentation you should be able to:

1) Query problems from the repository

2) Describe the two basic kinds of files on the repository

3) Retrieve the default files of those two forms

4) Describe some of the options for the two forms

5) Create an alternate text version

6) Describe the file type you will need to modify the computer based version

Future Work

1) More problem in more subjects
2) Instructional videos / tutorials
3) Develop the Grader / Feedback module
4) Initialize problem contests and incentives for problem creation
5) Web based student interface

Questions?

484933 Liquid-Liquid Extraction: Last but Not Least

Abstract:

When we ask chemical engineering students to list equilibrium stage-wise processes, distillation is the first such process on the list, followed by stripping, then, maybe, absorption. They generally donâ€™t list liquid-liquid extraction. What is the reason for this over-sight? First, it comes late in a Unit Operations semester; thus, if the course falls behind the course calendar, it gets dropped. Second, liquid-liquid extraction involves the use of phase diagrams to calculate the number of equilibrium stages for a process; hence, considerable effort must be expended teaching their use and the arcane mathematics required for the calculation. Third, there are no simple liquid-liquid extraction experiments from which students can gain hands-on experience. Taken *in toto*, most graduating B.S. Chemical Engineers have little knowledge of liquid-liquid extraction.

This presentation investigates liquid-liquid extraction heuristically; i.e., using dimensional analysis. It also discusses a simple, inexpensive laboratory experiment that highlights the pertinent concepts of a dimensional analysis of liquid-liquid extraction.

485553 Portable Wet Process Control Laboratory for Every Student's Desk and Home

Abstract:

Overview

With the dramatic decrease in the price of microcontrollers such as the Arduino and of compatible instrumentation it is now possible to have each student build an inexpensive individual portable wet experimental lab that can be transported between the classroom and home. The only classroom requirement is an adequate supply of power outlets. The software/hardware system described here can be used for an entire series of experiments that form the basis of a Process Control laboratory class or in a class that combines Process Control theory and experiments. The setup also enables a wet lab experience for remote online students.

Using the Arduino UNO microcontroller attached via a USB port to a laptop, a software/hardware system has been designed and tested that controls the temperature of a water-filled cylindrical container with thin walls of high thermal conductivity (e.g., a soda can) by low frequency pulse width modulation (PWM) of a 125W beverage heater (immersion coil). This is done using an enclosed and finger-protected solid state relay. In addition, two 5V fans powered from the Arduino (peak draw current 180 mA each) are used to increase the maximum cooling rate, the fraction of time the heater is on at safe temperatures, and to provide disturbances. The fan velocity can be changed by using Arduino-provided low frequency PWM of a logic-level MOSFET which controls both fans. The temperature is measured using a DS18B20 waterproof sensor. Running an Excel VBA program (PLX-DAQ version 2), the experimental results are continually tabulated and plotted in Excel. Although a version of the class software uses an Excel sheet for the user interface (UI), the preferred version uses the Arduino Serial Monitor. To make this possible, a software serial port is created and connected to a second laptop USB port via a TTL serial to USB cable. The entire cost of the experimental setup (excluding the laptop) is about $80.

The experimental system is shown disassembled for transportation in Fig. 1 and assembled in the classroom in Fig.2. Assembly involves connecting nine DuPont wires and requires less than 5 minutes.

Main Components

1. Temperature Probe

A DS18B20 probe [1, 2] in waterproof housing was selected due to the combination of good performance in the target temperature range and low cost. It is digital and outputs to a single UNO digital input/output pin. It has resolution 9 to 12 bits and requires about 750 ms to complete a reading at the highest resolution. It is supported by the OneWire library originally developed by Jim Studt [3]. The DS18B20 can operate in normal mode or in parasitic mode where it powers up from the communication wire. The normal mode is used and the probe is powered from the UNO 5V pin (the current drawn is only up to 1.5 mA).

2. Immersion Heater and Relay

Safety considerations put an upper limit to the temperature at which the experiments should be run and evaporative losses become significant at higher temperatures. Therefore the target range of heater operation was selected to be 25-60 ËšC (and the software implements a safety shutdown if the temperature exceeds 75 ËšC). Given these temperatures, a minimal power heater is desired. The lowest-power widely-available heater that could be located is the 125W Lewis N' Clark beverage immersion heater. To manipulate its heating output, a 0-5V input relay is controlled by the UNO using low frequency (the program currently uses 4 Hz) pulse width modulation (PWM). As the setup presented here requires at least 15 minutes inactive time for many experiments, it is intended for usage in a classroom where the instructor lectures while the experiment progresses. This prohibits the selection of noisy electromagnetic relays. Solid state relays (SSRs) are silent, but there is no commercial unit that integrates a suitable SSR with an AC outlet. Several SSRs were investigated and an option is the OMRON G3NA-210B (load 24-240VAC, input 5-24VDC), which is totally enclosed and has finger protection over the screw terminals to prevent user shocks. Its cost is relatively low, it has an LED that indicates when it is closed, and it does not require a heat sink for this application. This relay must be coupled with a power cord (where the heater plugs in), and a polarized two-prong cord is recommended. The hot wire must be separated, cut and stripped, and then connected to the relay screw terminal. Although this is a simple operation, to avoid any danger to the students, a professional or an experienced instructor should connect the power cords to the SSRs.

3. Cooling Fans and Their Control

To increase the maximum rate of cooling and the fraction of time the heater is on in the target temperature range of 25-60 ËšC, the water-filled container is sandwiched between two 5V cooling fans (see Fig. 1). The fan speeds can be manipulated via PWM allowing them to be used as disturbance inputs. As the current drawn by each fan is much higher than the 20 mA UNO PWM pins can safely provide, the speed of both fans is manipulated via PWM of a single logic level N-channel MOSFET. A 10 kohm pull-down resistor is used to ensure proper MOSFET output at all times, and two rectifier diodes have been added to protect the MOSFET from reverse voltages as explained in [4, 5]. In addition, a 220 ohm current-limiting resistor has been added to protect the UNO as suggested in [6, 7]. The MOSFET gate is controlled by PWM using an UNO digital pin. To obtain optimal fan performance and reduce fan noise, the PWM frequency had to be lowered from the default 488 Hz to the lowest available, 31 Hz, using the code provided in [8]. To reduce cost and avoid the need of an additional outlet, the UNO is powered through a laptop USB port. In this case the maximum available current is 370 mA. A fan that perfectly satisfies this need is the Evercool EC8025M05CA, which draws a peak current of 180 mA and has a steady draw of 140 mA.

4. Real-Time Output to Excel and User Interface

The Parallax data acquisition tool, PLX-DAQ [9], is a software add-on for Microsoft Excel that can be used for real-time output (and therefore plotting) from the UNO to Excel. Unfortunately, Parallax no longer supports it, and it can only be used with earlier 32-bit versions of Excel. Fortunately, Jonathan Arndt developed a new version of PLX-DAQ with full Office support (32 bit and 64 bit and versions up to Office 2016) [10]. Two versions of the main class software have been developed. One uses an Excel sheet for the UI (Fig. 3). But this interface requires a minimum sampling period of 15 s to properly process inputs from Excel. The second version (Fig. 4) enables sampling periods as low as 3 s by using as UI the Arduino Serial Monitor. To enable two serial communications (one for Excel and one for the Serial Monitor) a second serial port is created via software [11]. A TTL serial to USB cable is required for the software port and the SiLabs CP2102 chip works well with all versions of Windows if the driver instructions posted in [12] are followed.

5. Control Algorithm

The default algorithm used is the velocity form of the PID control law modified so as to remove derivative kick and proportional kick [13]. Low-pass filtering can be used for the derivative term (and is recommended). For instructional purposes, a flag replaces the velocity form with the position form of the PID law. This incorporates anti-reset windup, bumpless tuning changes, and bumpless transition from manual to automatic following methods discussed by Brett Beauregard [14].

Initial Class Offering

The first classroom use of this system will take place this summer in a class of 40-50 students. The introductory lecture will feature a student contest in which they manipulate the heater output manually and must respond to pre-programmed set-point and disturbance changes (the winner is the one who finishes the exercise with the lowest average squared error). Other planned class experiments include experimental determination of the parameters of an energy balance model, obtaining transfer functions from step changes, obtaining transfer functions from pulse changes, feedback performance of different tuning methods, the effect of filtering, and the effect of time delay. In addition, the experimental set-up will be used for homework assignments. The class will be assessed by comparing student test performance to that of the previous simulations-based class offering, and by extensive student surveys.

References

1. http://www.best-microcontroller-projects.com/ds18b20.html
2. http://datasheets.maximintegrated.com/en/ds/DS18B20.pdf
3. http://bildr.org/2011/07/ds18b20-arduino/
4. http://enricosimonetti.com/control-a-motor-speed-with-arduino/
5. http://bildr.org/2012/03/rfp30n06le-arduino/
6. http://www.oddwires.com/using-a-mosfet-to-control-a-dc-motor/
7. http://fritzing.org/media/fritzing-repo/projects/m/mosfet-motor-example/images/MOSFET%20Motor.png
8. http://playground.arduino.cc/Code/PwmFrequency?action=sourceblock&num=2
9. https://www.parallax.com/downloads/plx-daq
10. https://forum.arduino.cc/index.php?topic=437398.0
11. https://www.arduino.cc/en/Reference/softwareSerial
12. https://learn.adafruit.com/adafruits-raspberry-pi-lesson-5-using-a-console-cable/software-installation-windows
13. http://www2.widener.edu/~crn0001/Engr314/Digital%20PID%20Controllers-2.pdf
14. http://brettbeauregard.com/blog/2011/04/improving-the-beginners-pid-introduction

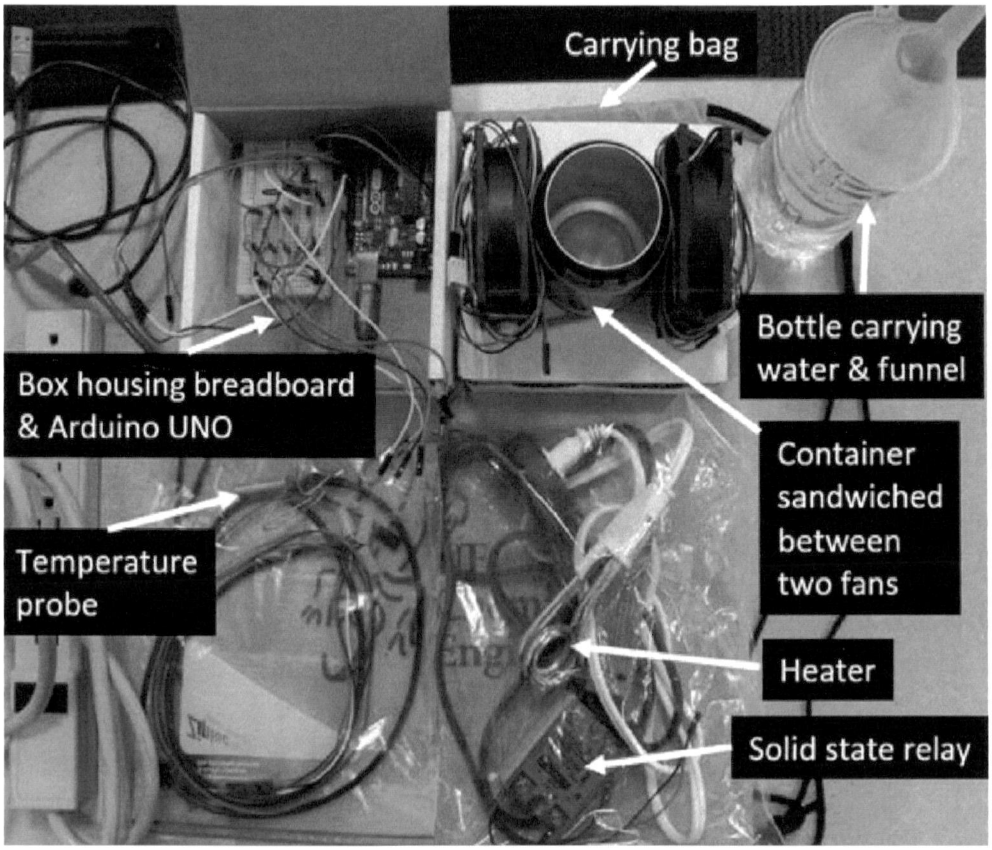

Fig. 1: The experimental system disassembled for transportation.

Fig. 2: Assembled experimental system in the classroom.

Fig. 3: Excel tabular and plot output with Excel sheet as the UI.

Fig. 4: Excel tabular and plot output with Serial Monitor as the UI.

505332 Vertically Integrated IoT, Off Grid Control-Sensors, Advancing with Arduino, Android App Shells, Programming, Rapid Big Data: The Kitchen Sink

Abstract:

There are a number of "internet-of-things", *i.e.* IoT, topics that have emerged since the turn of the century. Chemical engineers have the ability to contribute to and use IoT as stepping stone means to learn new skills, apply classic principles, and extend them to current topics such as the fusion of data-controls-safety. This workshop segment will focus on a vertically integrated project that brings together societal stakeholders in the context of technology, education, and arts. The overall project makes use of culturally relevant arts theme in Richmond, Virginia "the river city", resulting in what could be called a "boundary object". Stakeholders include primary & higher education students-teachers, community, and professional societies with extensions to industry. The focus of this segment involves technology and the repurposing of an urban space into an "urban commons", resulting in a multifunctional learning center. The project allows community and higher education to participate in integrated science technology education and mathematics, *i.e.*STEM, while incorporating art. This segment demonstrates advanced topics of developing integrated affordable electronic-controls IoT projects. Specifically, environmental sensors, a heater, a motor and Arduino shield development are integrated in an off of the grid electrical system. The segment demonstrates electronic development principles, programming, trouble shooting, and data collection. Extensions are made from chemical process control to data collection and wireless networking. A particular focus of discussed aspects will include purposeful and flexible interfaces for future modifications. Flexibility is introduced because it is expected that higher education students would interface with community to introduce upgrades or modifications in addition to teaching-learning infusions at the community level. The experience is akin to working with chemical processes in an industrial setting.

508515 Re-situating the Professional Formation of Engineers at Oregon State University

Abstract:

Re-situating the Professional Formation of Engineers at Oregon State University

Jim Sweeney, Milo Koretsky, Susan Nolen, Michelle Bothwell, Susannah Davis, Devlin Montfort

While industry has long called for changes to how engineers and computer scientists are prepared, much of the core coursework remains virtually the same as it was 50 years ago. To address this mismatch between industry and education, the National Science Foundation has funded a set of projects that aim to *REvolutionize Engineering and Computer ScienceDepartments* (RED), serving as leaders in institutional change. We describe early results and ongoing work in a RED project in the School of Chemical, Biological, and Environmental Engineering (CBEE) at Oregon State University. We seek to re-situate the faculty/staff/student community and curricular/co-curricular student learning experience. Our work is centered on the creation of a holistic, inclusive, and professionally-based learning environment for our students. We are working to promote change through establishing a culture of inclusion and shifting student learning environments from highly sequestered activities to more realistic and consequential work that more closely resembles actual engineering workplace. These efforts require an important and concurrent shift in unit culture (e.g. values, norms, policies and procedures).

504781 ExxonMobil's Diversity and Inclusion Efforts – Creating a Premier Global Workforce that Works to Help Power the World's Progress

Abstract:

At ExxonMobil, the diversity and inclusion of thought, skill, knowledge and culture make ExxonMobil more competitive, more resilient and better able to navigate the complex and constantly changing global energy business. Our diversity and inclusion efforts are comprehensive. They range from attracting and developing a premier global workforce within a work environment that values unique perspectives, encourages individual growth and treats all employees with respect, to educational partnerships and community development initiatives designed to improve the quality of life where we live and work.

This presentation is a view of ExxonMobil's technology focus areas and the Diversity and Inclusion efforts being used to create a global workforce ready to address them.

489242 Quantifying Reading and Online Homework Completion Using an Interactive Material and Energy Balances Textbook

Abstract:

Interactive textbooks provide instant feedback to students as well as to the professor. Last year, features of an interactive textbook from zyBooks for a material and energy balance course were detailed, including scaffolded question sets and animations. This year, new features include over 170 automatically graded homework problems and interactive material on using spreadsheets for solving course problems. The machine-graded, personalized homework, which are called challenge activities, include rolling numbers, randomized content, and scaffolded questions across 3 to 6 levels per activity. The increasing difficulty was verified by student success rates in 2017. Quantifying student usage will be central to this contribution. In 2016, student reading/participation averaged 87% over the entire interactive web book over the entire semester with 100 students, which far exceeds the less than 30% reading statistics in the literature. Additionally, statistically significant higher textbook reading scores were observed for students earning A and B final course grades compared to C, D, and F final grades as well as female students compared to male students. Data being generated in early 2017 will be compared to the earlier findings related to course grades and reading.

493310 An Open-Access Gate-to-Gate Life Cycle Assessment for Graduate Researchers

Abstract:

As graduate students become masters of niche areas in narrow sub-disciplines, it is easy for them to lose sight of the bigger picture. Whether itâ€™s the feasibility of the project, or potential impacts, it can be difficult for them to place their work in a larger context. Here we discuss a project-based research assessment for a graduate-level chemical engineering class using resources that may be freely used and adapted. Based off of a simplified Life Cycle Assessment, a materials and energy inventory, this excel based evaluation seeks to quantitatively assess the social, economic and ecological boons and burdens of their own research compared to industrial and academic processes in the literature. Performing cost, toxicity, and environmental analyses allows students to weigh the pros and cons of specific processes, and students learn that few decisions are simple to make. Next students evaluate their work and importantly, suggest possible improvements to the assessment itself. By conducting manageable and practical investigation of their own research, students learn how to contextualize their innovations and understand the relationship between how their choices in the laboratory can result in safer, cheaper, and more benign products and practices. Just as this evaluation is built on open access content and strengthened through iterative student and educator generated advances, simply put, this process is really an opportunity for students to dig into the data, as they mine the literature for a variety of parameters. Using technology to spur collaboration, and whole systems thinking, this project balances what young research would like to do, against what information is available, highlighting next steps. Understanding the nexus of energy, materials and people may seem to be a complex art, but by utilizing available materials to spark inquiry, it can be engineered into a craft.

490487 Chemical Engineering Beyond Politics; A Futuristic World Educational System

Abstract:

Chemical engineering beyond politics; a futuristic world educational system

S. Rohani[1,2], S. Agah[1,3], S. Dini[1,4], E. Hemmat[1], B. Mottahed[1], A. Parvizi-Majidi[1], F. Samadi[1], F. Sefidvash[1]

[1]Baha'i Institute for Higher Education (BIHE)

[2]Department of Chemical and Biochemical Engineering, Western University, Canada, Email: srohani@uwo.ca

[3]Chemical and Biomolecular Engineering, Rice University, TX, USA

[4]Mechanical Engineering Department, Western New England University, Springfield, Massachusetts, USA

We offer an internet-based chemical engineering program beyond borders with no tuition fees to the students and minimum overall costs to the institutions offering the program.

We and several other professionals voluntarily established a chemical engineering department within an online university in Iran called BIHE (Bahá'í Institute for Higher Education).

In early 1980's, following the cultural revolution in Iran, the professors and students in all disciplines who confessed their belief in the Baha'i Faith, were expelled from the universities across Iran. In the following years up to the present time, the Baha'i youth have been denied the right to higher education. International appeals initiated by the worldwide Baha'i community to right this obvious human rights violation did not achieve a meaningful conclusion. Consequently, the Baha'i community in Iran with the help of the then expelled professors and under the guidance of the international administrative body of the Faith, the Universal House of Justice, formed a community-based university, the Baha'i Institute for Higher Education (BIHE) in 1987. As an initiative, *the New York Times* called it "an elaborate act of communal self-preservation, to educate those who were unjustly denied access to higher education"[1]. The origin of BIHE was, thus, rooted in a spirit of purposefulness, and dedication. The belief in the power of BIHE has continued to conduct its courses using accredited professors and in accordance with the highest academic standards.

The unique circumstances surrounding BIHE, whereby its professors teach without compensation and its staff and students participate in university activities at a great personal risk, have forced it since 2005 to combine traditional classroom instruction with online learning to unwittingly become a leader in hybrid education.

Although BIHE issues degrees to all of its graduates, Iranian authorities cannot certify these degrees since the BIHE is not recognized as a degree granting institution. However, even in the absence of such certification, BIHE graduates have been admitted into post-graduate programs in universities across North America, Europe, Asia and Australia. So far, more than 90 universities worldwide have recognized the quality of BIHE courses, and accepted them for credit, allowing BIHE graduates to continue their education abroad[2].

In September 2011, two Nobel Peace Prize laureates, East Timor past President José Ramos-Horta and Archbishop Desmond Tutu, wrote an open letter to the international academic community calling on it to condemn the Iranian government's policies towards the Baha'is, to evaluate BIHE curricula, and consider accepting its credits and to offer online curricula to students in Iran who would otherwise be deprived of the right to higher education[3].

Despite these pressures, the BIHE has been functioning with ever-increasing efficiency. The classrooms and the laboratories of this fledgling educational system are the homes and basements of the Baha'i families. The professors offer their voluntary services with dedication and the students fueled by their love for learning attend regular classes, via internet. There are many graduates of the BIHE who have completed their PhD and higher degree programs in high ranking institutions and currently working as professionals, abroad. A few who upon the completion of their post-graduate degrees chose to return to Iran and serve as faculty members in BIHE, faced imprisonment and harsh treatment by the Iranian government. The BIHE students in various cities in Iran attend classes offered via internet in their home towns and every month travel to Tehran for in-person classes and possible laboratory work. While in Tehran, they are hosted by families who open their homes for this purpose. The classes are offered by the professors via internet with the state-of-the-art technology. The classes are often held in the evenings allowing the students to work during the day.

The BIHE faculty and students in Iran have consistently faced oppressive conditions. However through the efforts of dedicated faculties and staffs in Iran, and an ever-increasing Affiliated Global Faculty from universities around the world, BIHE now offers 38 university-level programs including Chemical Engineering, Chemistry and Industrial Chemistry, Civil Engineering, Architecture, Computer Science, Social Sciences and Humanities, Psychology, and Dentistry.

In 2009, the authors with the invitation and help of a core group residing in Iran launched a new major in chemical engineering. To fulfill the teaching assistant requirement, applied chemistry students who were interested studied some relevant courses intensively in two years along with their own major to serve as teaching assistants. The organizing group reached out to several universities and communities outside and inside Iran to find volunteers to teach chemical engineering courses. Since 2012, a group of 14 individuals from different professional and industrial backgrounds got together to facilitate this process. Since then we have met online each month and consulted about how to best offer the curriculum. The purpose of the chemical engineering program is to offer an outstanding undergraduate degree program and train chemical engineers with excellent technical and leadership skills, integrity, and social responsibility to serve the citizens of Iran and the world.

First year courses consist of General Chemistry 1 and 2 and Chemistry Laboratories, General Mathematics 1 and 2, Physics 1 and 2 and Labs, electricity, Computer Programming, and Humanities.

In Year 2 students take more common basic courses which include Organic Chemistry 1 and 2 and Laboratory, Basic Calculations in Chemical Engineering, Chemical Engineering Thermodynamics 1 and 2, Differential Equations, Fluid Mechanics, and Humanities.

In Year 3 more chemical engineering core courses are taught. These include Mass Transfer, Heat Transfer, Statics, Strength of Materials, Numerical Analysis, Unit Operations 1, Kinetics and Reactor Design, Applied

Math, Introduction to Electrical Engineering, a Technical Elective, and Humanities.

In the final year Unit Operations 2, Chemical Process Control, Modelling and Simulation using software packages such as HYSYS, Introduction to Material Science, Capstone Plant Design, two more Technical Electives and a course in Professionalism and Business are taken.

The technical electives are offered in Waste Water Treatment, Oil and Gas and Petrochemistry, Introduction to Environmental Engineering, Introduction to Biomedical Engineering and a non-technical elective Entrepreneurship for Chemical Engineers at the fourth year level. This provides a unique synergy and integration of courses offered in our program, so that all of our graduates have a holistic appreciation of process and plant design, which includes environmental aspects.

In all courses, emphasis is placed on creativity, innovation, problem solving through teamwork, communications, and reflection – the skills necessary to practice engineering in any discipline.

The design experience in the Chemical Engineering program is realized through the Capstone Design course, as well as the design components existing in the majority of Program Compulsory and Technical Elective courses. The aim is to expose and train students in handling open-ended engineering problems. Solutions must consider safety requirements, codes, standards, specifications, health, environment, social issues, and other pertinent constraints. Students will receive training with regard to economics, business, legal matters, and interpersonal and communication skills. Design courses also provide training in spreadsheets, process simulation, equipment selection, sizing and costing, and cash flow analysis.

We also proposed and managed several research projects under Scientific Advisory Committee since January 2016. This Committee is formed to help facilitate and accelerate the research activities of BIHE that will train our students in conducting research projects and publishing the outcomes in well-known journals.

We praise the students of the BIHE for their ultimate sacrifice and their dedication in pursuit of education. We also offer the operation of the BIHE as a world-wide educational model that can operate at low-cost and high efficiency in many less fortunate countries and remote areas that do not have access to higher education.

[1] http://www.nytimes.com/1998/10/29/world/iran-closes-university-run-covertly-by-the-bahais.html

2 http://bihe.org/index.php?option=com_content&task=view&id=146&Itemid=236

3 http://www.huffingtonpost.com/desmond-tutu/iran-bahai-_b_978090.html

494531 Enhance Learning Experience By Augmented Reality Tools

Abstract:

The increase in the popularity of modern technologies among students is increasingly challenging the status quo of the educational system. Pairing teaching and learning strategies with appropriate technology trends can greatly enhance the motivation to learn. Ergonomics and human factors can play a critical role in analyzing the abilities and behaviors of students and employing the gained knowledge in designing tools, products, and systems that are more effective for accomplishing a given task. Wearable technologies equipped with augmented reality capabilities offer a lot of promise for education and training especially for the applied science disciplines. In this work, we utilize wearable technologies, namely the Google Glass and Microsoft HoloLens, to immerse students in an augmented reality environment that guides actions at every step of the experiment via intelligent audio-visual feedback. In addition, we share our experiences over the tools development and first pilots.

499625 Chemical Engineering Mobile Apps

Abstract:

Although we have seen a nearly complete shift among chemical engineering students from desktop to laptop computers, it has merely created an incremental change in their education. However, the benefits of mobile devices in engineering education still remain largely untapped. There is a great need and opportunity to put chemical engineering data and problem-solving tools in studentsâ€™ (and faculty membersâ€™) pockets. This presentation will discuss 5+ years of effort and motivation behind the development of several chemical engineering-oriented mobile apps, including *Chemical Engineering AppSuite*. Through the use of embedded analytics, measuring user demographics and interaction with the apps reveals interesting details about what kinds of calculations and what types of data chemical engineers are most interested in and use most frequently. While using to these mobile apps has a number of benefits for expediting calculations and facilitating access to data, the associated analytics can help shape teaching chemical engineering in the 21st Century.

495506 Process Design: Learning in Chemical Engineering

Abstract:

As chemical engineers, we are trained to identify and model the fundamental mechanisms of chemical and biological processes towards innovation and reliable production. This disposition is the cornerstone of a discipline that continues to make key contributions in traditional (e.g., petrochemical, pharmaceutical) and emerging (e.g., microelectronics, biotechnology, alternative energy) fields. This *chemical engineering approach* can also be applied to the complex processes of learning. Thus, it is not surprising that chemical engineers, like Phil Wankat, have been pioneers in the emerging field of engineering education research, where scholars seek systematic and generalizable understanding of the learning process and learning systems. This talk will focus on examples from the authorâ€™s research of learning processes and ways that he has explored foundational mechanisms and connected them to innovations and reliable delivery of course and unit programs.

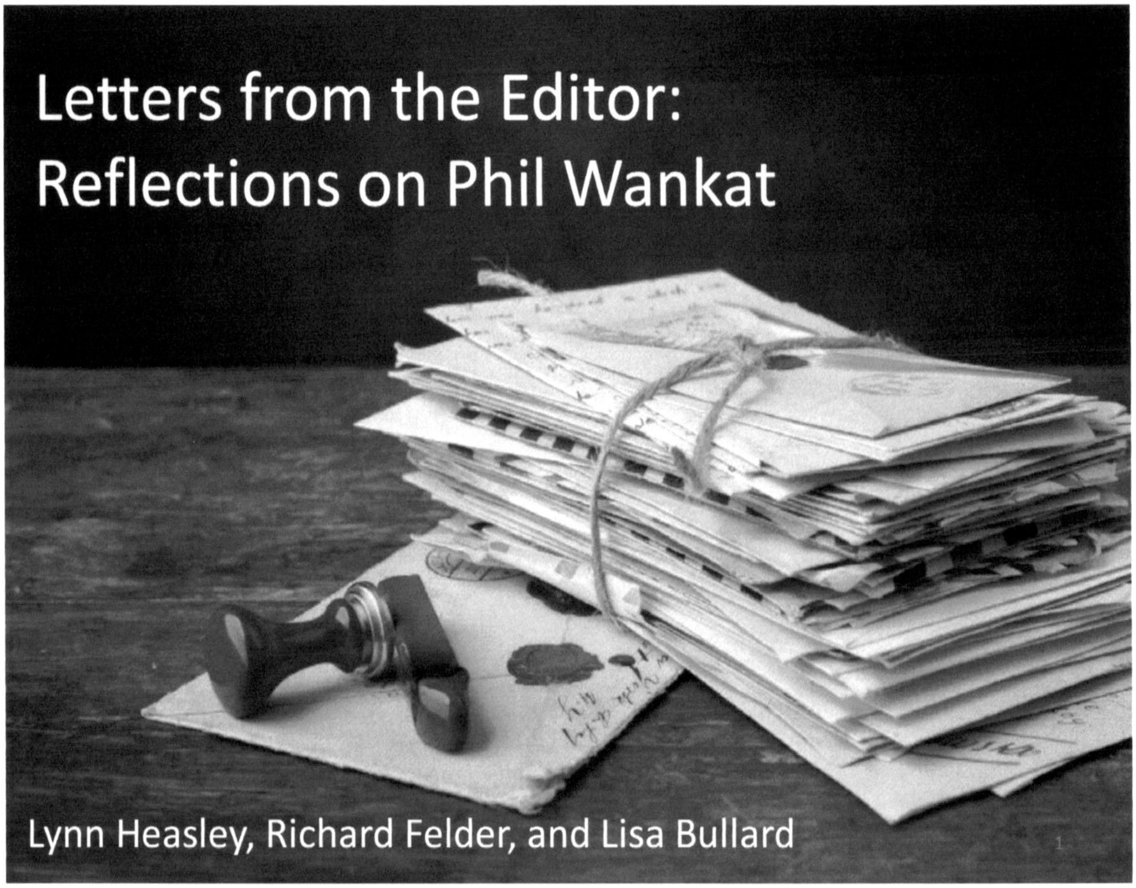

"I am interested in your comments on my e-mail that describes the situation."

2

"Make any suggestions on diplomacy."

3

"Sooner or later we will have to deal with this. Since you are more diplomatic than I am, I suggest you be the lead and respond."

4

"Prof. [Blank]: Unfortunately, you appear to have misinterpreted our letter. It was meant to be a polite way of rejecting your paper. We wish you luck directing it or future revisions of it elsewhere."

5

"The author needs to look at the journal to see what types of papers are published."

6

"Please send the authors the following. They will be unhappy."

7

"In a nutshell, the current paper is not tightly focused, is too long, has unnecessary detail, does not follow CEE style guidelines, and needs to be rewritten."

8

"I must have sinned, and Dr. [Blank] is the penalty for my sins."

9

"Lynn, You have no idea how delighted I am to receive e-mails from you that start, *'This one has a lot of hallmarks of past problem papers.'* "

10

“There were a couple of places where your silence was golden.”

11

“I do not think we need to review this book. It does not appear to be teaching oriented (do not say this to the author, as he will argue it is).”

12

"Dr. [Blank] is a problem. He ignores the advice on the most important point and changes minor points. I am not willing to inflict this revision on the reviewers. I am saying no, but with the carrot that if he follows the original advice, we will send it out for re-review."

13

"This is not the first time we have had rather far-out materials. I am willing to give him one trial to develop an acceptable paper, but I am not willing to play extended games trying to save the world."

14

$5.00
1983
INDIANA
MIGRATORY
WATER FOWL
STAMP
Void after Feb. 29, 1984

"Enjoy!"

15

"Sorry for the late response, but I was in Scotland for vacation and refused to be sucked in by my e-mail."
INDIANA TROUT STAMP
1951 $1
DEPARTMENT OF CONSERVATION
DIVISION OF FISH AND GAME

16

"They have been good sports about fixing things."

17

"Unfortunately, Dr. [Blank]'s Teaching Tip did not receive as much attention from me as it deserved. Upon rereading it, I think it will benefit from more work. Going over items too quickly when I have too much to do is a recurring problem. I will have to try and slow down."

Followed shortly after by:

"On second thought, let's go with Dr. [Blank]'s current version. I do not want start requesting late changes of authors because I do not have my act together. Better to get my act together."

18

"Some people are just more fun to work with than others."

19

Image Citations

1. www.istockphoto.com
2. https://about.usps.com/news/national-releases/2016/pr16_048.htm
3. http://www.stockunlimited.com/vector/white-house-stamp_1525799.html
4. https://www.pinterest.com/bluememaw/usa-state-birds-postage-stamps/
5. https://arago.si.edu/category_2028905.html
6. http://www.theus50.com/indiana/stamp.php
7. http://www.indianastampclub.org/
8. http://www.brookmanstamps.com/Netcat/DUCK/in.htm
9. https://postalmuseum.si.edu/stampstakeflight/firstusairmail.html
10. http://www.brookmanstamps.com/Netcat/DUCK/in.htm
11. https://www.mysticstamp.com/Products/United-States/4290/USA/
12. https://www.123rf.com/photo_34876810_usa-and-indiana-state-flag--old-postage-stamp.html
13. https://www.flickr.com/photos/shookphotos/5794984932
14. http://www.brookmanstamps.com/Netcat/DUCK/in.htm
15. https://about.usps.com/news/national-releases/2016/pr16_048.htm
16. https://postalmuseum.si.edu/stampstakeflight/firstusairmail.html
17. https://waterfowlstampsandmore.com/gallery-3/indiana-trout-stamps/
18. http://www.indianastampclub.org/
19. http://www.brookmanstamps.com/Netcat/DUCK/in.htm

20

487550 Blending Teaching, Research and Writing Textbooks

Abstract:

In the early years of teaching engineering as an academic discipline instead of an apprenticeship, teaching was the main function of professors. Research was one of several methods to stay abreast of changes in engineering and to stay intellectually active. This view remained common even after the Second World War had illustrated the power of research. For example, the opening of NSF brought forth the comment, "it would in fact be surprising if more than 500 or 1000 engineering professors would be interested in a federal research program" [1, p. 60]. A combination of war experiences, the availability of research money, the ASEE Grinter report, and the shock of Sputnik in 1957 led to the engineering science revolution in engineering education [2]. This change took time but was in place when I attended Purdue. Currently, about 2/3 of professors at research institutions are more interested in research than teaching [3, p. 421]. Regardless of one's interest, how do professors blend teaching, research and service, and in the process balance their lives? I will present my story of teaching, research, and writing textbooks in both separations and education. I am well aware that my story may be "a tale told by an idiot, full of sound and fury, signifying nothing." [4]

1. Stewart, *IRE Trans. Educ.*, **1** (2) 54–61, 1958.

2. Froyd, Wankat and Smith, *Proceedings of the IEEE, 100,* 1344-1360 (2012)

3. Wankat and Oreovicz, *Teaching Engineering,* 2nd ed., Purdue University Press: West Lafayette, IN (2015).

4. Shakespeare, *The Tragedy of Macbeth,* speech by Macbeth, probably first performed in 1606.

493227 The Women Event: Engaging High School Girls and Their Parents in STEM

Abstract:

The Women's Outreach in Materials, Energy, and Nanobiotechnology (WOMEN) Event has been organized annually by Cornell University's Chemical and Biomolecular Engineering Graduate Women's Group (CBE Women) since 2010. The mission of the WOMEN Event is: (1) to provide important information about STEM fields, careers paths in engineering, and the college application process to female high school students, and (2) to develop leadership skills of female CBE graduate students who design and direct the WOMEN Event. Key features of the WOMEN Event that set it apart from other similar outreach programs aimed at women include targeting the event to high school students from rural areas and devoting a significant portion of the program to STEM careers and the college experience.

Every year, 20 to 40 10th grade girls and their parents are invited to Cornell for the WOMEN Event. Over the past 8 years, more than 200 girls have participated. These girls are from rural upstate New York, where 66% of women do not attend or complete college (statusofwomendata.org). In these areas, girls encounter fewer female role models in STEM and their parents may be unable to offer guidance on STEM career options or how to prepare for them. The WOMEN event program includes hands-on lab modules, college admission information sessions, talks highlighting STEM career opportunities, and a panel discussion with current Cornell students, all free of charge. Engaging lab modules created specifically for the WOMEN event illustrate topics in chemical engineering, including materials design, bioengineering, and process development. One unique aspect of our programming is engagement with parents. While students participate in lab activities, the parents participate in an information session on college financial aid and admissions. The parents also listen to talks from women faculty to gain insight into the many paths STEM careers can take. In the final laboratory module, parents and their daughters conduct experiments together. By reaching out to both girls and their parents, we hope to not only interest the girls in STEM but also enable their parents to support them in growing that interest into a career. After the 2016 event, 90% of the girls indicated they intend to enter a STEM field and the number of attendees interested in an engineering career doubled. While 4 past participants of the event have gone on to study engineering at Cornell, this is not intended to be a recruitment event. We instead aim to provide an experience that encourages girls to pursue their STEM passions wherever that leads them.

In addition to inspiring the girls who attend, the WOMEN Event provides a valuable learning experience for the graduate student volunteers. Five graduate students have given conference presentations about the WOMEN event. In 2017, half of the volunteers and leaders of the WOMEN event were women; a notable achievement in a department that has 30% women MS and PhD students. Volunteers have received recognition through the departmental Hooey Award, Cornell's Cook Award, and the College of Engineering Alumni Association Award for Student Group Leadership.

500867 Comparison of Web-Based and Lecture-Based Training Approaches to Educate High-School Students with Simulink Modeling Skills

Abstract:

The math performance of US high-school students was ranked 38th out of out of 71 countries in the most Programme for International Student Assessment (PISA) in 2015 [1]. In addition, a report from President's Council of Advisors on Science and Technology shows that US would have one million deficits in technical workers in STEM fields if the STEM education is not improved in the next decade [2]. Therefore, it is urgent for US educators to create new approaches to attract more high-school students in the STEM fields, especially in math, which deals with challenging equation and symbolic operations. One way to address this issue is providing interesting modeling projects along with a powerful software that students can reply on for math operation. We did an "experiment" in Summer 2014 by introducing MATLAB Simulink to a local high-school junior student and providing him a project in which he developed Simulink Model for a microbial fuel cell (MFC). That project was quite successful, as the student presented his work in ASEE 2015 Mid-Atlantic Spring Conference [3]. The success of that project is based upon the following two factors: 1) MATLAB Simulink provides friendly user-interface so that students can build ordinary differential equation (ODE) models in Simulink like playing Lego; 2) the microbial fuel cell project deals with converting organic compounds in waste water to bioenergy.

On the basis of the feedback from the student, we further extended our previous projects to web-based training modules to attract and train more high-school students. The web-based training idea comes from another ASEE project of ours in which we prepared biology review and virtual experiment/training videos to enhance learning in biochemical engineering courses [4]. In the web-based training modules, two videos were developed to educate students step by step on how to build two ODE models, one for water tank and the other for the continuous flow stirred-tank reactor (CSTR) [5], in MATLAB Simulink. A video was developed to introduce students the background information and application examples of MFC. Students then constructed the ODE model of a microbial fuel cell, which consists of four differential equations, 9 constrained equations and 25 parameters. The instructor offered Skype meetings to help students in their projects. The developed training modules were also delivered to another group of students using the traditional lecture-based teaching approaches. Anonymous surveys were given to both web-based and lecture-based groups of students. Survey results showed that the web-based approach was effective in educating high-school students with Simulink modeling skills. On the other hand, the lecture-based group of students had slightly better performance than the web-based group.

References:

[1] webpage: http://www.pewresearch.org/fact-tank/2017/02/15/u-s-students-internationally-math-science/

[2] President's Council of Advisors on Science and Technology, Engage to excel: producing one million additional college graduates with degrees in science, technology, engineering, and mathematics (Executive Office of the President of the United States, 2012).

[3] C. Ekaputra, Z. Huang. "Educating High School Students in Process Simulation and Control with a Simulink-Based Controller Design for Microbial Fuel Cells. ASEE Mid-Atlantic Spring 2015, Villanova, PA, 2015.

[4] J.J. Elmer, N.K. Comolli, W.J. Kelly, Z. Huang. "Preparation of Biology Review and Virtual Experiment/Training Videos to Enhance Learning in Biochemical Engineering Courses," 2015 ASEE Annual Conference and Exposition, Seattle, Washington, June 2015.

[5] Dale E. Seborg, Thomas F. Edgar, Duncan A. Mellichamp, Francis J. Doyle, III. Process Dynamics and Control, Third Edition, Wiley, 2010.

501376 Nanotechnology & Engineering Grand Challenges

Abstract:

This talk highlights outreach activities on potential nanotechnology solutions to the NAEâ€™s Engineering Grand Challenges developed as part of an NSF Nanotechnology Undergraduate Education grant and a Math Science Partnership. Most of the activities were designed for school classrooms or summer camps, but they have also been adapted for teacher training and public outreach events such as NanoDays and university open houses. This talk will provide information about these activities and evaluations of their impact on participants' science and engineering knowledge and attitudes. The value of creating adaptable activities which are easily implemented by undergraduate and graduate engineering students to reach a range of outreach needs and maximize impact while minimizing stress will be highlighted. The incorporation of consumer grade technology tools into activities will also be highlighted. Specific examples that will be described include activities on reverse engineering the brain and renewing and improving urban infrastructure.

501605 Going Beyond Demonstrations to "Chose Your Own Adventure" Engineering Experiences for Service-Learning K-12 Outreach Opportunities for 3rd Year Engineering Students and Enhanced Student Engagement for 1st Year Engineering Students

Abstract:

The enthusiasm of STEM teachers and early exposure to STEM fields has been shown to strongly influence a student's selection of engineering as a major/career. Due to the somewhat abstract and often hazardous nature of chemical engineering compared to other engineering fields, it is often difficult to create engaging chemical engineering experiences for K-12 students. To help address this issue, my students and I have created over 70 engineering adventure experiences which have been presented to over 6,000 K-12 students. We have focused on hands-on interactive demonstrations that use as many senses as possible; however, feedback from our experiences has inspired us to modify our approach to involve a higher level of active learning. Here we present our efforts to create open-ended, hypothesis-driven, "chose your own adventure" chemical engineering experiences that are safe, fun, inexpensive, and easily tailored to elementary, middle, or high school students. Survey response data from 3rd year engineering students who developed and presented these experiences is presented demonstrating their desire to be involved in service-learning opportunities and the impact it had on their understanding of basic engineering concepts.

Inspired by the above experience, we have also created a "Chose Your Own Adventure" Engineering Experience for 1st Year Engineering Students. The experience is constrained to address an Engineering Grand Challenge as Described by the National Academy of Engineering and students must complete the Engineering Design Cycle. However, beyond this and approval from risk management, the experience is completely open ended. To maximize intrinsic motivation and engagement, student teams are assigned based on interest in a given Engineering Grand Challenge. Here we also report the impact of such an open-ended experience during the 1st Year on student engagement and excitement for continuing in an engineering major.

494840 Building Block Air Quality Sensors

Abstract:

Communicating chemical engineering research to the general public can be difficult, but hands-on activities can help to both engage the layperson and bridge the gaps in understanding. We have developed an effective teaching module to introduce students to the science of air quality and the ways in which we, as engineers, may quantify it.

This module has students design build and test light scattering air quality sensors, using an arduino board and several low-cost electronic components. The sensor housings are assembled out of Legos or Lego-like building blocks, allowing students to attempt many design types. Once assembled the sensors are tested with mist or smoke, and assessed for their ability to detect changes in air quality and their susceptibility to noise. They may also be tested for other design considerations, such as material cost. The module needs approximately a one-hour class period to execute and can be used as a hands-on project for teams of up to 4 students. The difficulty of the module can be easily scaled from the middle school to the undergraduate level, and this activity could be appropriate for an introductory undergraduate lab.

This work is part of a NSF project (#1648756) to place air quality sensors in K-12 schools and create a real-time air quality map. K-12 students in this project are trained to become citizen scientists, and this module has been used to introduce them to the relevant sensor concepts; the commodity sensors integrated into a custom printed circuit board, which these classrooms will be hosting use the same sensing concepts as the student-built Lego sensors. We have performed this module for over 100 students and have received very encouraging feedback and significant attention from local and national press, leading to increased interest in participation in this research.

489245 Developing Spreadsheet Skills Using an Interactive Material and Energy Balances Textbook

Abstract:

Spreadsheets are a central problem-solving tool for chemical engineers. Interactive textbooks provide an opportunity to introduce students to spreadsheets. An interactive textbook from zyBooks for a material and energy balance course incorporates over 80 animations, 170 automatically graded homework problems, 200 homework/example problems, and 900 clicks to read the entire book. While screenshots of the different features will be included, interactive material on using spreadsheets will be highlighted. Introductory content on spreadsheets, formulas, functions, and charting is applied in later sections solving material and energy balance problems. For example, completing repetitive heat capacity and enthalpy calculations in a spreadsheet will be introduced using animations and immediately applied through interactive questions and machine-graded problems.

497636 Using Student Developed Comics to Promote Learning of Transport Phenomena Concepts

Abstract:

Anecdotal evidence indicates that current undergraduate engineering students spend numerous hours per week reading popular science-based Internet comics, such as xkcd.com and phdcomics.com, for example. Thus, it makes sense that comics can be leveraged as a learning modality for engineering students. The use of comics in STEM education is not new, however, they have yet to be utilized extensively in higher education to teach engineering. Having taught a senior level mass transport course numerous times, a common challenge for students that persists is applying mathematical concepts to describe binary diffusion. Thus, here we will present the implementation of a project in which teams of students developed comic books with a goal of deriving and explaining the most general form of Fickâ€™s first law of diffusion to an average freshman chemical engineering student. This was piloted in a senior level chemical engineering course, CHE 4131: Transfer Science III: Diffusion and Diffusive-Convective Mass Transport at Tennessee Technological University and will also be conducted during the Spring 2017 semester of an elective transport phenomena course offered at the University of Connecticut. Initial feedback from students was positive, revealing a desire to be creative and explore complex engineering topics from a unique perspective.

491761 Nature Inspired Chemical Engineering: Development of a New Course on an Emerging Topic

Abstract:

In 2015, a senior undergraduate and Master's-level elective course on Nature Inspired Chemical Engineering (NICE) was introduced to complement the research activities of the multi-disciplinary EPSRC "Frontier Engineering" Centre for Nature Inspired Engineering (http://cnie.org.uk)at University College London (UCL). This course has now been taught twice. We will share our methods in developing the course and the student experience.

The overarching vision of the Centre is to use nature as a guiding platform to seek transformative solutions to engineering grand challenges. Rather than *ad hoc*biomimicry, the NICE methodology seeks to first understand the fundamental mechanisms underpinning desired properties, like efficiency, scalability or robustness in natural systems. These mechanisms are then applied to achieve similar properties in engineered systems, while being cognizant that chemical engineering applications typically have different constraints from those found in biological systems. The NICE approach is systematic and of broad applicability, making it useful as a methodology for engineers to learn and practice in seeking innovative solutions. This prompted the development of a NICE course for chemical engineering students.

This course features traditional lectures and coursework, and is supplemented by active learning and new pedagogical approaches. These approaches include annotated journal article readings (with online question and discussion forums), laboratory demonstrations, and a term-long group project. The group project involves multiple written and oral presentations, which serve as milestones, where the students are given constructive feedback. Each project is different, driven by student creativity and interest in a particular chemical engineering application. The teams seek inspiration from nature to discover an innovative pathway to resolve challenges faced by more conventional approaches of chemical engineering.

At last year's AIChE Annual Meeting, we presented our first experiences teaching the NICE course. Our poster at the meeting this year will give an overview of the course structure, and principally focus on the student-driven projects over the past two years. These projects illustrate how the NICE approach leads budding engineers to think beyond often-presented examples of biomimicry, such as gecko-inspired adhesive tapes and lotus-leaf inspired hydrophobic coatings. Projects vary from bioreactor mixing inspired by the hydrodynamic changes induced by tubercles on humpback whale fins, to protein-based drug crystallization inhibitors. Other examples include projects concerning the optimal placement and orientation of photovoltaic panels, novel gas sensors inspired by spider webs, and the intensification of polymerization reactors by seeking inspiration from pattern formation induced by ocean waves.

497987 The "Cilindro Rotador" As a Pedagogical Tool for Complex Engineering Systems

Abstract:

The engineering description of multiphase systems (such as organs within the human body or packed-bed reactors in the petrochemical industry) represents a challenging pedagogical task for the instructor and a steep learning curve for the students. Both students and instructors face a multitude of domains with different scales and with interconnecting interfaces that need an intimate understanding before physical concepts can be put into mathematical models. Such systems usually involve various transport processes with either homogeneous or heterogeneous reactions; they reflect different types of geometries that anchor irregular pore shapes, and they are comprised of different types of materials with various properties In reaching a successful description, students must bring together "disconnected" concepts from a variety of courses in a coherent manner; they need to assess the assumptions in each phase that match the relevant scale and, therefore, connect them with microscopic or macroscopic models. Their skill set should allow them to identify suitable boundary conditions that capture the interfacial physics. In all, the analysis is a daunting learning task for the students. Within the framework of the Renaissance Foundry Model, this study is exploring the use of well-characterized physical or laboratory devices as pedagogical tools to help the students acquire the skill set to describe complex systems. In particular, we will present elements and pedagogical functions of the "Cilindro Rotador" to integrate concepts learned in different courses, train the students in the art of simplifying assumptions, identifying suitable scales, and assessing proper conditions for the interfacial boundaries.

496187 Chem Quest: An Adaptive Pre-Freshman Online Chemistry Course

Abstract:

The Chemical Engineering and Food Technology (CEFT) cluster in Singapore Institute of Technology (SIT) offers different undergraduate programs â€“ (1) Chemical Engineering, (2) Pharmaceutical Engineering, and (3) Food Technology. A pre-freshman Chemistry course is necessary because 95% of the matriculated students are polytechnic diploma holders from diverse programs, including marginally related programs (e.g. Logistics).

Chem Quest is an SIT initiative to bridge the knowledge gap between pre-university and freshman **Chemistry**. In order to reach as many **pre-freshmen** as possible, Chem Quest is offered as an **online** course. Recognising that the Chemistry competencies of pre-freshmen is at differing levels, Chem Quest is also made **adaptive**, where learners with lower Chemistry competencies undergo a more vigorous course while learners with higher Chemistry competencies experience Chem Quest as a refresher course. Chem Quest will be initially rolled out to pre-freshmen in the CEFT cluster.

501765 Peer Mentoring in Graduate School - Fostering Diversity to Achieve Scholarly Excellence

Abstract:

A peer mentoring program was implemented in the Chemical Engineering Department at the University of Michigan with the aim of building and sustaining a more diverse graduate program. The "peer mentor" program consists of selected senior PhD student "mentors" paired with groups of 6 first-year PhD student "mentees". Peer mentor groups are selected to maximize diversity with respect to gender, ethnicity, and research interests. Groups meet regularly in both academic and social settings to facilitate social bonds between students. Mentees indicate increased collaboration and social cohesion in the department versus PhD students who did not participate in the peer mentor program. Importantly, these gains were maintained in second-year PhD students after the program. Mentors also benefited from the opportunity to lead a group of first-year graduate students, serving as invaluable training for mentors considering career in academia. Overall, a majority of program participants would recommend the program to other Chemical Engineering departments.

496847 Developing a Graduate Student Professional Development Course

Abstract:

Graduates from PhD and Masters programs in science and engineering not only go on to work in academia, but approximately 55% of them join the industrial workforce.[1] It is generally acknowledged that there is a gap between the knowledge, skills and attributes (KSAâ€™s) that industry and academics require and those that are acquired by students in the academic environment.[2,3]

The primary skills developed in graduate programs include research, publishing and presenting. However, there are a number of transferrable skills that can be further addressed in graduate programs; the most frequently identified skills include communication, mentorship and leadership.[4] Other necessary skills include: research ethics; research development; technology commercialization; entrepreneurship; data science and big data skills; science policy; governance, risk and compliance; time management and project management skills; and developing cultural competency and intercultural teamwork skills.[5]

Based on graduate student and faculty surveys, there were several recommendations for improvement to the graduate program which include, active learning exercises for more student engagement as well as further development of the necessary knowledge, skills and attributes to bridge the professional development skills gap. A course was developed to provide these necessary skills and in addition an E-portfolio system was developed to measure whether or not the future course changes lead to effective changes in studentsâ€™ KSAâ€™s.

1. National Science Foundation, Science and Engineering Indicators 2014, https://www.nsf.gov/statistics/seind14/index.cfm/overview, (2014).

2. Council of Graduate Schools and Educational Testing Service , Pathways Through Graduate School and Into Careers, Report from the Commission on Pathways Through Graduate School and Into Careers., Princeton, NJ: Educational Testing Service (2012).

3. Grafton, A., Grossman, J., No More Plan B: A Very Modest Proposal for Graduate Programs in History. https://www.historians.org/publications-and-directories/perspectives-on-history/october-2011/no-more-plan-b (2011).

4. McClurkin, J, et. al., Development of Industry Modules for Engineers Pursuing Advanced Degrees, 121st ASEE Annual Conference & Exposition, ASEE, 2014

5. Denecke, D., Feaster, K., Stone, K., Professional Development: Shaping Effective Programs for STEM Graduate Students, Washington, DC: Council of Graduate Schools, (2017).

497025 Re-Situating Learning and Shifting Culture in ChE at OSU

Abstract:

This poster presents our efforts towards *revolutionizing* the undergraduate learning environment in the School of Chemical, Biological, and Environmental Engineering at Oregon State University (CBEE). While we currently provide students many innovative learning opportunities, we are concerned that the extent to which these efforts are marginalized and isolated may deter from their influence. To address this issue, we are in the process of creating a more holistic, inclusive, professionally-based learning environment woven through both curricular requirements and co-curricular opportunities for students. We specifically target social inequality by seeking to create engineering educational systems and interpersonal interactions that are professionally and personally life-affirming for all people across their differences. We seek to catalyze change through construction of a culture of inclusion and a shift in our learning environments from sequestered activities to realistic, consequential work. This requires a fundamental change in the nature of department culture (values, norms and structure).

Project activity is structured around four "pillars" that include: (1) curricular redesign of nine core sophomore- and junior-level studio classes to include more *realistic, consequential work* leveraging research-based pedagogies like *problem-based learning* and *model-eliciting activities*; (2) growing faculty and students' *capacity to engage issues of inclusivity* by shifting their cognitive and affective knowledge of power and privilege; (3) planning and implementing *student professional development pods*, longitudinally mixed student teams where students help one another understand the university experience and how it relates to professional practice; and (4) implementing *formal changes in governing policies and procedures* within CBEE.

This work is supported by the National Science Foundation pilot program *Revolutionizing Engineering Departments* (RED) that is aligned with the NSF Engineering (ENG) Directorate's multi-year initiative, the *Professional Formation of Engineers*, to create and support an innovative and inclusive engineering profession for the 21st Century.

501983 A Cross-Discipline Heat Exchanger Project

Abstract:

Due to the cross-disciplinary nature of modern day engineering, it is necessary for students to gain strong communication, interpersonal, and teamwork skills in order to be prepared for successful careers in industry. Through project based learning, students in different courses and majors interacted with each other to work on different aspects of a common problem. After an initial offering of a cross-disciplinary heat exchanger project that required chemical engineering students and mechanical engineering students to work together in a simulated consulting environment, modifications to the project were made to improve the learning experience for the students in the second iteration of the project.

The project was assigned to junior level chemical engineering students in CME 425: Heat and Mass Transfer. They were asked to design a basic shell and tube heat exchanger and to solicit designs for a compact heat exchanger from junior level mechanical engineering students in ME 325: Elements of Heat Transfer. The mechanical engineering groups acted as consulting firms, with the burden of communication between the groups being placed on the students.

The reporting requirements for the mechanical engineering students were reduced between the first and second iterations to require a more appropriate level of effort from the students. In both iterations, an initial meeting was held between the client groups and consultant groups to establish the parameters of the project. However, all follow-up questions were communicated through technical memos. Preliminary design proposals were submitted by the consulting groups half-way through the project, at which point the client groups had the opportunity to address any misconceptions. At the end of the project, both the client and consultant groups presented their work to the class and submitted a technical memo to the instructor detailing their calculations.

In the first iteration of the project, the consultant groups were allowed to choose any flow configuration for their compact heat exchanger. While this led to many interesting designs, it also resulted in the many of the students attempting to analyze systems that were beyond the scope of an undergraduate heat transfer course. In the second iteration, A set of design constraints were placed on the students, requiring them to design a counter flow heat exchanger with rectangular channels. They were allowed to modify geometric parameters: including wall thickness, channel size, and number of channels. This limited the scope of the project, which ensured that the students would be attempting a problem that could be solved using the tools provided in the textbook and covered in class.

Students completed surveys to determine their perspectives on the project, any skills gained, and their attitudes towards the other engineering discipline. Here we present the details of the project, and the outcomes of the project after the second iteration changes.

491321 Measuring Impacts of Course Changes in Introduction to Chemical Engineering

Abstract:

In an effort to improve the retention of a more diverse student population, we have made changes to our freshman-level Introduction to Chemical Engineering course at the University of Delaware. Some of these changes include the scheduling of the course, the adoption of active learning techniques (think-pair-share, worksheets with gaps) in both the main lecture and the recitation sections, as well as the restructuring of recitation sections to improve the teacher-student ratio in these sections.

Simultaneous to making these changes, we have planned to collect and analyze particular sets of data: (1) enrollment and grade distribution data for previous years of the course, (2) responses to specific questions on the end-of-term course evaluations, and (3) concept maps generated in the final week of the introductory course. We seek to demonstrate statistically significant differences in the "DFW" rate (rate at which students earn a grade of D or F or withdraw from the course) particularly for underrepresented groups (female students and non-white, non-Asian students), as well as assess whether there are notable changes to student attitudes regarding the course. Further, we will use established measures to evaluate the "strength" of student concept maps to determine whether there are significant changes in the ways students communicate and organize their understanding of the course objectives.

In our presentation, we will further discuss the deliberate changes made to the course, as well as report the preliminary findings from our data collection and analysis. We hope to also elicit feedback from those in attendance for improving the rigor of this analysis.

493771 Seeing Is Learning: Showing First-Year Undergraduate Students a Preview of Senior-Level Chemical Engineering Experiments

Abstract:

To get their child interested in the science of baseball, a parent would likely take them to the ballpark to watch a game first or play catch with them on the playground before teaching them the dynamics of the motion of spherical objects through the air. Once the child has taken to the game, they would be a lot more motivated to learn the theory behind various aspects of the game. It is this rather intuitive sequence that we are attempting to institute with incoming chemical engineering undergraduate students at Kansas State University.

As part of the study, students in the first or second semester of their chemical engineering program will be invited to participate in a demonstration of selected experiments from the unit operations laboratory that they would be doing down the line in their senior year. Emphasis will be on observation rather than the theory behind the experiments, with students getting the opportunity to see start-up, steady-state operation, and shutdown of different pieces of equipment. The exercise is meant to be informal, where students can ask questions, take notes and record videos for posterity. Participation in the exercise will be made optional and left to the individual students. The demonstrations will be restricted to groups of no more than 6 students, with the total time commitment for each student being 2 hours at most. There will be no accompanying test or assignment; students will simply be asked to give feedback about the experience on an anonymous survey. Finally, each participating student will be given one credit for the assembly course (ChE 015). For context, every chemical engineering undergraduate student must have at least 24 assembly credits before graduation.

Response rate of students to the exercise along with feedback from the attending students will be analyzed and presented at the conference. Data collected in the study, in conjunction with the comments and suggestions from members of the chemical engineering education community will be used in the implementation of such laboratory demonstrations in a more structured and mandatory manner as part of the freshman course on current topics in chemical engineering (ChE 110) at Kansas State University.

493936 Visualization of Newtonian Pipe Flow Around Static Mixers: 3-D Printing Applications in the ChE Curriculum

Abstract:

A laboratory demonstration to study Newtonian pipe flow around a static mixer was designed and constructed. The equipment constructed is simple, cost-effective and can be used to study and visualize pipe flow and the effect of static mixers. The experimental system consists of ½ in. diameter clear Tygon® tubing, a flow meter and pressure gauges. Water and a water-based dye were used as fluids for all experiments. The demonstration module is portable, self-contained and can be connected to laboratory water. A standard helical static mixer was designed using SolidWorks® and constructed from PLA using a 3-D printer. The designed experiment can be used to introduce students to chemical engineering applications of 3-D printing and prototyping. A 1/16 in. diameter stainless steel tube was used to introduce blue dye into the center of the flow tube. This allowed for the visualization of fluid mixing in the system. Successful validation of the equipment included experiments with and without the static mixer. All visual observations were recorded and image analysis software was used to provide a semi-quantitative measure of mixing quality. The equipment allows for pressure drop measurements to investigate the effect of the static mixer on pressure drop and on the Reynolds Number v. friction factor relationship for pipe flow. The module described here can be used in fluid mechanics courses for standard pipe flow studies, visualization of flow characteristics, and for mixing studies with low viscosity Newtonian fluids. The experiments can be highly visual and excellent for use in K-12 outreach efforts.

496285 An Introduction to Design for Chemical Engineering Undergraduates

Abstract:

An Introduction to Design for Chemical Engineering Undergraduates

Daryl R. Williams

Department of Chemical Engineering, Imperial College London, South Kensington Campus, London SW7 2AZ, UK

(d.r.williams@imperial.ac.uk +44 (0) 0207 594 5611)

Imperial College runs a 4 year MEng degree course in chemical engineering, All students start the course as chemical engineers, and the first year curriculum contains a number of core CE courses including fluid mechanics, heat and mass transport and engineering design.

In the first-year Design course, we introduce students in their first semester of study to problem solving and group work, both of which figure in the activities of a professional engineer. This is achieved by introducing students to the different type of problems they are likely to encounter. Our aims are to make students better able to: § define problems and responsibilities; § distinguish between different sorts of problems; § deal with design problems for which there are many solutions by identifying the data needed to tackle the problem, and generating, evaluating and presenting solutions; § work as a member of a group by contributing effectively, sharing work and involving all members of a team.

The students should appreciate that for many problems there are no unique solutions, there may be wrong ones, but also many correct ones, which is best for a certain situation depend on many factors, some of which may be beyond the control of the design team. Students should learn that every member of the team is important, not just the loud ones, and should begin to develop skills to extract the best from all. The students should learn that there is a huge amount of information available and that it is important to ask the correct questions

The course is run as a series of problems classes, the problem is posed, the students consider the problem, ask questions, either individually to the teaching team or in a plenary session and then attempt the design, asking further questions when necessary

For assessment the students will undertake a design and be assessed on the quality of the design brief they write, an

oral presentation they give and a poster they construct concerning their designs.

This lecture will review the course structure and outputs as well as the teaching strategy of this early stage course.

498079 The Engineering and Computing Residential Living and Learning Community at the University of South Carolina

Abstract:

To help students with the transition to university living, the University of South Carolina has created the Engineering and Computing Community (ECC), a residential living and learning community. The ECC helps students in the College of Engineering and Computing with a strong focus on entering freshmen. Many activities throughout the year help the ECC students develop personally, professionally, academically, and socially. A one-hour professional development course has been offered for these students, focusing on various enrichment activities. In the 2016-2017 academic year, multiple factors apparently helped improve the success of the community. These include:

1. Linked courses were implemented for the first time in the ECC. Multiple sections of Calculus and Pre-Calculus were reserved initially for ECC members. This allowed students to take key courses with other students that were living in the same building.

2. University tutoring services relocated to the first floor of the ECC residential living and learning community. This allowed ECC students to have easy access to help with coursework.

3. Multiple Resident Mentors (RM) hired from the USC College of Engineering and Computing. These RMs help serve as role models for the ECC students, guiding them toward improved academic success.

4. For the first time, USC College of Engineering and Computing students became a majority in the ECC residential living and learning community building. Having a high concentration of engineering and computing students may help with ECC academic success.

While facing many ongoing obstacles, ECC activities appear to be helping students with their initial progression in the College of Engineering and Computing at USC.

500908 Progressive Development of Students' Capacities to Engage in Functional Teaming

Abstract:

Engineering work relies on effective collaboration and communication among diverse groups of people in many roles, including: engineers, scientists, managers, technicians, end-users, among others. While all engineering educational programs require student teamwork, instruction on effective and inclusive practices are often missing. When present they may be *ad hoc* and only addressed in the senior year. Moreover, outcomes are rarely assessed. As part of our unit's efforts to revolutionize the undergraduate learning environment, we are strategically designing and implementing a scaffolded and progressive approach to growing students' capacities to engage in inclusive teaming, where diverse perspectives are encouraged and valued. We call this approach *functional teaming*. We utilize a studio model where students regularly work in teams in ten courses in their first three years. By coordinating student team experiences in these courses, we aim to develop productive interaction practices. The studio teaming experiences prepare them to engage in more comprehensive team work in the senior year laboratory and design courses.

In this paper, we will report on preliminary implementation of functional teaming, identify key aspects that frame functional teaming, with particular attention given to identifying: (i) the role status plays in group interaction, i.e., to what degree does an idea depend on who is saying it rather than the idea itself; and (ii) work that is "team worthy," i.e., is the problem / issue presented to the student team challenging enough to benefit from multiple perspectives and various slices of understanding? This work is supported by the National Science Foundation pilot program *Revolutionizing Engineering Departments* (RED) that is aligned with the NSF Engineering (ENG) Directorate's multi-year initiative, the *Professional Formation of Engineers*, to create and support an innovative and inclusive engineering profession for the 21st Century.

501939 The Use of Numerical Worksheets for Material and Energy Balances Course

Abstract:

Experience in using a user-friendly software, Mathcad, in the sophomore material and energy balance course is discussed. Example problems considered for illustration deal with solution of linear algebraic equations (parameter estimation, mass balances) and nonlinear algebraic equations (equilibrium calculations for multiple reactions, vapor-liquid equilibrium calculations for mixtures, solutions of equations of state). The capabilities of Mathcad are of significant benefit in accelerating the learning and strengthening the fundamental knowledge base.

503356 Overview of Chemical, Bioengineering, Environmental, and Transport Systems Division (CBET)

Abstract:

The National Science Foundation (NSF) is an independent federal agency created by Congress in 1950 "to promote the progress of science; to advance the national health, prosperity, and welfare; to secure the national defenseâ€¦" With an annual budget of $7.5 billion (FY 2016), we are the funding source for approximately 24 percent of all federally supported basic research conducted by America's colleges and universities. In many fields such as mathematics, computer science and the social sciences, NSF is the major source of federal backing.

503358 Highlights of CBET Cluster on Chemical and Biochemical Systems

Abstract:

The National Science Foundation (NSF) is an independent federal agency created by Congress in 1950 "to promote the progress of science; to advance the national health, prosperity, and welfare; to secure the national defenseâ€¦" With an annual budget of $7.5 billion (FY 2016), we are the funding source for approximately 24 percent of all federally supported basic research conducted by America's colleges and universities. In many fields such as mathematics, computer science and the social sciences, NSF is the major source of federal backing.

503360 Highlights of CBET Cluster on Engineering Biology and Health

Abstract:

The Division of Chemical, Bioengineering, Environmental, and Transport Systems (CBET) has reorganized and refocused several of the programs within its Engineering Biology and Health cluster. This effort was undertaken to clarify the scope of each of the programs and to minimize programmatic overlap. There are five programs in the EBH cluster: Biophotonics, Cellular and Biochemical Engineering (CBE), Disability and Rehabilitation Engineering (DARE), Engineering of Biomedical Systems (EBMS), and Nano-Biosensing. Four programs have been refocused (all but Biophotonics), and two of those (CBE, Nano-Biosensing) have welcomed new program directors during the past 12 months. These changes have driven and been driven by shifts in research emphasis and scope. These changes will be presented and discussed, as will areas of overlap between programs within the cluster, and with associated programs in other divisions and Directorates at NSF. Recently developed and developing funding initiatives involving programs within the cluster will be described, as will the funding portfolios for each program in the cluster.

496409 Laboratories in the Chemical Engineering Curriculum: Current State and Best Practices

Abstract:

Hands-on experiences with laboratory and pilot-scale equipment are one of the hallmarks of the undergraduate chemical engineering experience. Lab experiences require significantly more space, infrastructure, and instructional time than do classroom sections or computational studios, and this combined with recent growth in enrollment means labs are getting a hard look at many institutions.

This session focuses on the undergraduate laboratory experience and will present results from the most recent AIChE Survey on Unit Operations lab as well as invited speakers on standard and innovative approaches to this important experience.

AUTHOR INDEX

AUTHOR INDEX

AUTHOR INDEX

AUTHOR INDEX

AUTHOR INDEX

AUTHOR INDEX

AIChE
120 Wall Street, FL 23
New York, NY 10005-4020

ISBN 978-1-5108-5793-3
9 781510 857933

CURRAN ASSOCIATES INC.
proceedings.com

American Institute of Chemical Engineers

7th Topical Conference on Natural Gas Utilization 2007

Held at the 2007 AIChE Spring National Meeting

AIChE

April 22-27, 2007
Houston, Texas, USA